VOYAGES

THROUGH THE UNIVERSE

This recent image from the Hubble Space Telescope shows "cometary knots" in the Helix Nebula, a shell of glowing gas expelled by a dying star about 450 light years away in the constellation of Aquarius. A wind of hot gas from the star collides with the shell of denser gas the star emitted about 10,000 years earlier. The collision fragments the inner part of the shell into denser, finger-like droplets that superficially resemble comets we see in our solar system. However, the heads of the "comets" in the Helix Nebula are typically twice the size of our entire planetary system (twice the diameter of the orbit of Pluto). Each tail stretches away from the star for about 100 billion miles. (R. O'Dell, K. Handron & NASA) *(Inset)* A view of the full Helix Nebula taken with the Anglo-Australian telescope. (© 1979 Anglo-Australian Telescope Board)

VOYAGES
THROUGH THE UNIVERSE

Andrew Fraknoi
Chair, Astronomy Department
Foothill College

David Morrison
Chief, Space Science Division
NASA Ames Research Center

Sidney Wolff
Director
National Optical Astronomy Observatories

®

SAUNDERS COLLEGE PUBLISHING
Harcourt Brace College Publishers

FORT WORTH   PHILADELPHIA   SAN DIEGO   NEW YORK   ORLANDO   AUSTIN

SAN ANTONIO   TORONTO   MONTREAL   LONDON   SYDNEY   TOKYO

Text Typeface: New Caledonia
Composition: Progressive Information Technologies
Publisher: John Vondeling
Acquisitions Editor: Jennifer Bortel
Developmental Editor: Jennifer Bortel
Picture Development Editor: George Semple
Managing Editor: Carol Field
Senior Project Editor: Anne Gibby
Copy Editor: Judy Patton
Proofreader: Michele Gitlin
Manager of Art and Design: Carol Bleistine
Art Director: Carol Bleistine
Illustration Supervisor: Susan Kinney
Art and Design Coordinator: Kathleen Flanagan
Text and Cover Designer: Ruth A. Hoover
Text Artwork: George V. Kelvin/Science Graphics
          Rolin Graphics
Director of EDP: Tim Frelick
Production Manager: Alicia Jackson
Marketing Manager: Marjorie Waldron
Editorial Assistant: Tara Pauliny
Marketing Coordinator: Karen Milstein

Cover Credit: Hubble Space Telescope image. Gas pillars in M16, *Eagle Nebula*, a nearby star-forming region 7,000 light years away, in the constellation Serpens. (Jeff Hester and Paul Scowen, Arizona State University, and NASA)
Frontispiece: *Helix Nebula*. (© 1979 Anglo-Australian Telescope Board/David Malin)

Printed in The United States of America

Voyages Through the Universe

ISBN: 0-03-020053-9

Library of Congress Catalog Card Number: 96-68673

9 0 1 2 3 4 5   032   10 9 8 7

About the Authors

Andrew Fraknoi is the Chair of the Astronomy Department at Foothill College near San Francisco and an Educational Consultant for the Astronomical Society of the Pacific (where he directs Project ASTRO, a program to bring astronomers into elementary and junior high school classrooms). From 1978 to 1992 he was Executive Director of the Society, as well as Editor of *Mercury* Magazine and the *Universe in the Classroom* Newsletter. He has taught astronomy and physics at San Francisco State University, Cañada College, and the University of California Extension Division. He is author of *The Universe in the Classroom*, co-author of *Effective Astronomy Teaching and Student Reasoning Ability*, and scientific editor of *The Planets and The Universe*, two collections of science and science-fiction literature. In the past 22 years he has presented over 400 public lectures on astronomical topics. For five years he was the lead author of a nationally syndicated newspaper column on astronomy, and he appears regularly on radio and television explaining astronomical developments. He has received the Annenberg Foundation Prize of the American Astronomical Society and the Klumpke-Roberts Prize of the Astronomical Society of the Pacific for his contributions to the public understanding of astronomy. Asteroid 4859 was named Asteroid Fraknoi in 1992 in recognition of his work in astronomy education.

David Morrison received his Ph.D. from Harvard University. He was at the University of Hawaii from 1969 to 1988, where his positions included Professor of Astronomy, Chair of the Astronomy Graduate Program, Director of the Infrared Telescope Facility at Mauna Kea Observatory, and University Vice-Chancellor for Research and Graduate Education. Dr. Morrison currently heads the space science program at the NASA Ames Research Center. His primary research interests are in planetary science. Dr. Morrison is the author of more than 120 professional articles and of several books, including *The Planetary System, Cosmic Catastrophes, Exploring Plane-*

tary Worlds, and three other astronomy texts from Saunders. He has served as President of the Astronomical Society of the Pacific, Chair of the Astronomy Section of the American Association for the Advancement of Science, and President of the Planetary Commission of the International Astronomical Union. Dr. Morrison has received the Klumpke-Roberts Prize of the Astronomical Society of the Pacific for contributions to public understanding of science, and two medals for Outstanding Leadership from NASA for his contributions to the Galileo mission and to protecting the Earth from asteroid impacts. A celestial object, Asteroid 2410 Morrison, is named for him.

Sidney C. Wolff received her Ph.D. from the University of California at Berkeley, and then joined the Institute for Astronomy at the University of Hawaii. During the seventeen years Dr. Wolff spent in Hawaii, the Institute for Astronomy developed Mauna Kea into the world's premier international observatory. Dr. Wolff became Associate Director of the Institute for Astronomy in 1976 and Acting Director in 1983. She earned international recognition for her research, particularly on stellar atmospheres and what they can tell us about the evolution, formation, and composition of stars. In 1984, she was named Director of the Kitt Peak National Observatory, and in 1987 became Director of the National Optical Astronomy Observatories. She is the first woman to head a major observatory in the United States. As Director of NOAO, Dr. Wolff and her staff of 460 oversee facilities used annually by nearly 1000 visiting scientists. Recently, Dr. Wolff has also been acting as Director of the Gemini Project, which is an international program to build two state-of-the-art 8-m telescopes. Dr. Wolff has served as President of the Astronomical Society of the Pacific and is the second woman to be elected President of the American Astronomical Society. She is also a member of the Board of Trustees of Carleton College, a liberal arts school that excels in science education. Dr. Wolff is the author of more than 70 professional articles and a book, *The A-Type Stars: Problems and Perspectives.*

Preface for the Student

In college textbooks, there is a long tradition that the preface of the book is read by the instructor and the rest of the book by the student. Still, many students begin reading the preface (it does come first) and then wonder why it doesn't say much to them.

So, we begin our book with a preface for student readers. It's not a preface about the subject matter of astronomy, which is introduced in the Prologue, but a preface that tells you a little about the book and gives you some hints for the effective study of astronomy. (Your professor will probably have other, more specific suggestions for doing well in your class.)

Astronomy, the study of the universe beyond the confines of our planet, is one of the most exciting and rapidly changing branches of science. Even scientists from other fields often confess to having had a lifelong interest in astronomy, though they may now be doing something more practical, such as biology, chemistry, or engineering. There are fewer than 10,000 *professional* astronomers in the world; but astronomy has a large group of *amateur astronomers* who spend many an evening with a telescope under the stars observing the sky, and who occasionally make a discovery, such as a new comet or exploding star.

Many people are fascinated just to read about bizarre objects that astronomers are uncovering, such as black holes and quasars. Others are intrigued by the scientific search for planets or life in other star systems. And many people like to follow the challenges of space exploration, such as the repair of the Hubble Space Telescope by the Shuttle astronauts, or the Galileo mission to probe the giant planet Jupiter. Hearing about an astronomical event in the news media may be what first sparked your interest in taking an astronomy course.

But some of the things that make astronomy so interesting also make it a challenge for the beginning student. The universe is a big place full of objects and processes that do not necessarily have familiar counterparts here on Earth. Like a visitor to a new country, it will take you a while to feel familiar with the territory or the local customs. Astronomy, like other sciences, has its own special vocabulary, and keeping up with the pace of discovery in astronomy is a monumental challenge.

To assist students taking their first college-level course in astronomy, we have built a number of special features into this book, and we invite you to make use of them:

- All technical terms are printed in **boldface** type the first time they are used and clearly defined in the text; their definitions are listed alphabetically in Appendix 3 (the glossary), so you can refer to them at any time. The summaries at the end of each chapter also include these boldface terms as a review.

- The book begins with a historical summary of astronomy and then surveys the universe, starting at home and finishing with the properties of the entire cosmos. But don't worry if your instructor doesn't assign all the chapters or doesn't assign the chapters in order. Throughout the book "directional signs" lead you to earlier material you need to know before tackling the current section.

- We use tables to bring together numerical data for your convenience. For example, some tables summarize the important properties of each planet in the solar system (including the Earth). Students who want to see more of the data that astronomers use can investigate the appendices at the back of the book, which give the latest information on many aspects of astronomy.

- Figure captions clearly describe what phenomena or objects students are looking at. In many textbooks, captions are afterthoughts, with only a few words of description, but in this book, we have scrutinized each figure and asked what would help clarify the diagram or image.

- Each chapter ends with a summary of the essential points in the chapter, plus review questions, thought questions, and problems to help you "process" what you have learned.

- Suggestions for further reading are included for students who want or need to learn more about a particular topic. These books and articles are written at the same introductory level as this text.

- Appendix 1 is a guide to some of the more interesting astronomy sites on the World Wide Web.

Here are a few suggestions for studying astronomy that come from good teachers and good students from around the country:

- First, the best advice we can give you is to be sure to leave enough time in your schedule to study the material in this class *regularly*. It sounds obvious, but it is not very easy to catch up with a subject like astronomy by trying to do everything just before an exam. Try to put aside some part of each day, or every other day, when you can have uninterrupted time for reading and studying astronomy.

- Try to read each assignment in the book twice, once before it is discussed in class, and once afterwards. Take notes or use a highlighter to outline ideas that you may want to review later. Also, take some time to coordinate the notes from your reading with the notes you take in class. Many students start college without good note-taking habits. If you are not a good note-taker, get some help. Many colleges and universities have student learning centers that offer short courses, workbooks, or videos on developing good study habits. Take a little time and find out what your school has to offer.

- Form a small astronomy study group with people in your class; get together as often as you can and discuss the topics that may be giving group members trouble. Make up sample exam questions and make sure everyone in the group can answer them confidently. If you have always studied alone, you may at first resist this idea, but don't be too hasty to say no. Study groups are very effective ways of discussing new information, or learning a foreign language, or studying law or astronomy.

- Before each exam, do a concise outline of the main ideas discussed in class and presented in your text. Compare your outline with those of other students as a check on your own study habits.

- If you find a topic in the text or in class especially difficult or interesting, don't hesitate to make use of the resources in your library for additional study.

- *Don't be too hard on yourself!* If astronomy is new to you, many of the ideas and terms in this book will be unfamiliar. And astronomy is like any new language; it may take a while to become a good conversationalist. Practice as much as you can, but also realize that it is natural to be overwhelmed by the vastness of the universe and the variety of things that are going on in it.

We hope you enjoy reading this text as much as we enjoyed writing it. We are always glad to hear from students who have used the text and invite you to send us your reactions to the book and suggestions for how we can improve future editions. We promise you we will read and consider every serious letter we receive. You can send your comments to Andrew Fraknoi, Astronomy Department, Foothill College, 12345 El Monte Rd., Los Altos Hills, CA 94022, USA. (Please note that we will not send you the answers to the chapter problems or do your homework for you, but all other thoughts are welcome.)

Andrew Fraknoi, David Morrison, and Sidney Wolff
July 1996

Preface for the Instructor

Voyages **Through the Universe** is a new astronomy text produced with today's students in mind—it is designed for non-science majors who may even be a little intimidated by science, and who approach astronomy with more interest than experience. With features that make it appropriate for everyone from university business majors to first-time community college students, **Voyages** is written to draw in and engage all readers, while preserving the accuracy and timeliness that our colleagues expect of us.

This new text is based in part on **Realm of the Universe,** by the late George Abell, David Morrison, and Sidney Wolff, but has been completely rethought and rewritten to make it an even more useful tool for teaching and learning astronomy. We have made the language friendly and inviting and have used examples drawn from everyday experience. Vignettes from the lives of astronomers and occasional touches of humor make this a book that students will actually *enjoy* reading.

Organization and Special Features

The book is not too long for a one-semester course, yet not so brief that important topics have been omitted. It does not overwhelm the student with detail, but instead focuses on the major threads and overarching ideas that illuminate the relationships among the various branches of astronomy. We probably have a bit less jargon than most textbooks, but we have not sacrificed any of the key concepts that you would want to see students learn in a basic course.

We have worked hard to include many of the latest ideas and discoveries in astronomy, not merely for their novelty, but for their value in advancing the quest for a coherent understanding of the universe. In each case, we have tried to fit the latest research results into a wider context and to explain clearly what they mean. Among the recent topics included in the text are the discovery of planets around other stars, the results from both the Galileo probe and Comet Shoemaker-Levy 9 impacts on Jupiter, the discovery of a number of the icy members of the Kuiper belt, new candidates for black holes in our galaxy, recent measurements of the age of the universe from several different observing groups, clearer evidence for supermassive black holes at the centers of galaxies, and much more.

We portray astronomy as a human endeavor and have tried to include descriptions and images of some of the key men and women who have created our science over the years. Illustrations also include many of the latest images from the Hubble Space Telescope and other space instruments, as well as an up-to-date collection of color images from ground-based observatories around the world. Many of the planetary images are second- or third-generation corrected views, not merely the first releases rushed out for the news media. Full-color diagrams are used as teaching tools, not as cosmetic devices. Figure captions contain full explanations of what the student should be seeing and understanding.

- The book is written as a coherent story. However, because the authors know that many instructors follow an order of topics that is different from theirs, the sections are modular.

- We do not expect students to remember every concept introduced in previous chapters. Unobtrusive verbal "sign posts" are inserted throughout to help students find where a concept was defined or explained in detail or to briefly review a key idea that may have been introduced many chapters ago. We want every student to use the book as an easy navigational tool through the world of astronomy. A complete glossary is supplied in Appendix 3.

- A carefully written Prologue introduces the basic ideas and vocabulary of astronomy and makes sure all students start their study of the universe at the same point. The Epilogue summarizes key ideas about cosmic evolution and then applies them to the quest for life elsewhere.

- Appendix 1, written with David Bruning, lists a wide range of useful World Wide Web sites in astronomy that are accessible to students.

Special Sections and Boxes

The chapters in **Voyages Through the Universe** feature a number of highlighted sections designed to help non-science students appreciate the breadth of astronomy without distracting them from the main narrative.

- *Making Connections.* These special boxes show how astronomy connects to students' experiences with other fields of human endeavor and thought, from poetry to engineering, from popular culture to natural disasters.

- *Thinking Ahead.* Each chapter begins with a stimulating question about the material that follows.

- *Voyagers in Astronomy.* These profiles of noted astronomers focus not only on their work, but on their lives and human dimensions.

- *Astronomy Basics.* Fundamental science ideas and terms that other texts just assume students know are explained carefully.

- *Seeing for Yourself.* Students get familiar with the sky and everyday astronomical phenomena through observations using simple equipment.

- *Chapter Summary.* A concise overview that enumerates all important ideas and lists important new terms in boldface.

- *Review Questions, Thought Questions,* and numerical *Problems.* Questions that allow you a wide latitude for testing student understanding. Many can be used directly for discussion sections or essay exams.

RedShift CD-ROM

The award-winning *RedShift* software (Version 1, 2), published by Maris Multimedia, expands **Voyages Through the Universe** from a static presentation to a dynamic simulation of many aspects of astronomy. The dual-platform CD-ROM allows students to

- view realistic models of the planets and main satellites in the solar system

- identify over 300,000 stars, nebulae, and galaxies
- simulate astronomical events over the course of 15,000 years

- view more than 700 full-screen photographs (including a number by David Malin)

- access the *Penguin Dictionary of Astronomy* (with over 2,000 entries)

- navigate through surface maps of Earth, Moon, and Mars.

We have included some end-of-chapter exercises using *RedShift* (written by David Bruning of *Astronomy* magazine.)

RedShift is a valuable learning tool while your course is in progress. It is also an enjoyable piece of recreational software which students can use to explore the universe long after their academic experience is completed. The CD-ROM may be packaged with the text for a very low price.

Ancillaries

In addition to *RedShift,* qualified adopters of **Voyages Through the Universe** can receive

- *The Cosmos in the Classroom: A Resource Guide for Teaching Astronomy* by Andrew Fraknoi. This manual is a rich compilation of teaching ideas and resources for both novice and veteran instructors. It includes annotated listings of the best non-technical books and articles in astronomy organized by subject; listings of outstanding slides, videos, and software for each chapter, with addresses and phone numbers of suppliers; topics for discussion and for writing papers; resource guides for exploring the lives of astronomers; ideas for historical and interdisciplinary topics and on-site observation; and helpful appendices to make a teacher's job smoother.

- *The Saunders Internet Guide for Astronomy* by David Bruning, Randy Reddick, and Elliot King. This wonderful new handbook for both instructors and students is a thorough, up-to-date introduction to the internet and the World Wide Web. The first part reviews the history and current state of the internet and explains everything from simple e-mail to Multi-User Dungeons, with special attention to those applications useful in higher education. The second part features a thorough annotated listing of World Wide Web sites related to astronomy and astronomy education.

- *The Voyages Instructor's Manual/Test Bank.* The instructor's manual test bank contains answers to thought questions and problems in the textbook, as well as a host of multiple-choice test questions for use in a variety of classroom settings.

- *ExaMaster Computerized Test Bank* for Windows and Macintosh. ExaMaster features all the questions from the printed test bank in a format that allows instructors to edit them, add questions, and print assorted versions of the same test.

- The *Saunders Astronomy Transparency Collection.* This collection contains 205 overhead transparencies of conceptually based artwork. The enlarged reproductions contain figures from Saunders' astronomy textbooks, as well as supplemental illustrations that complement the text figures. A detailed guide arranged by topic accompanies the collection. These images are also available as slides.

- The *Saunders Astronomy Collection Supplement, 1994,* features 25 additional images from the Hubble Space Telescope and major observatories.

- *Voyages Through the Universe Transparency Collection.* This collection includes 25 of the most current Hubble and ground-based telescope photographs as well as striking and informative illustrations from the textbook.

- *Saunders MediaActive CD-ROM to accompany Voyages Through the Universe.* This exclusive CD-ROM contains all the diagrams and tables from the textbook and is a superb presentation tool to be used with such software packages as Powerpoint™, Persuasion™, and Saunders' LectureActive™.

- *Saunders Astronomy Videodisc: The Solar System.* This two-sided CAV disc contains 1000 still images and 45 minutes of video showing some of the best spacecraft images of the solar system. A bar code manual and LectureActive™ presentation software accompany every disc.

The Registered Adopters Program

Instructors who fill out a brief survey can become **Registered Voyagers Adopters.** There is no cost involved, and registered adopters will receive

- invitations to special events at astronomy meetings (where they can talk with the authors and other adopters)

- first access to new ancillary materials

- updates on new developments in astronomy, as well as new teaching techniques and tools

- an opportunity to have direct input into the planning of future editions

To become a registered adopter, talk with your Saunders representative, call us toll-free at 1-800-939-7377 (ask for Karen Milstein), e-mail us at voyages@saunderscollege.com, or write to: Karen Milstein, Marketing Coordinator, Saunders College Publishing, Public Ledger Building, Suite 1250, 150 S. Independence Mall West, Philadelphia, PA 19106.

Saunders College Publishing may provide complimentary instructional aids and supplements or supplement packages to those adopters qualified under our adoption policy. Please contact your sales representative for more information. If as an adopter or potential user you receive supplements you do not need, please return them to your sales representative or send them to:

Attn: Returns Department
Troy Warehouse
465 South Lincoln Drive
Troy, MO 63379

Let Us Hear from You

Unlike stars and planets, textbooks in astronomy do not exist in a vacuum. All three authors have benefited tremendously over the years from the advice of colleagues and students who teach astronomy. We want to be sure that we continue to make changes and updates in Saunders' astronomy texts that will be the most useful to you.

Therefore, we welcome comments and suggestions about the text and the ancillaries and ideas for how future materials can be made more effective. Address your cards and letters to: Andrew Fraknoi, Astronomy Dept., Foothill College, 12345 El Monte Rd., Los Altos Hills, CA 94022 or e-mail: FRAKNOI@ADMIN.FHDA.EDU.

Acknowledgments

We would like to thank the many colleagues and friends who have provided information, images, and encouragement, including: Charles Avis, Charles Bailyn, Bruce Balick, Judy Barrett, Roger Bell, Michael Bennett, Roy Bishop, Chip Clark, Richard Dreiser, George Djorgovski, Alex Filippenko, Lola Fraknoi, Michael Friedlander, Alan Friedman, Ian Gatley, Paul Geissler, Margaret Geller, Cheryl Gundy, Heidi Hammel, Bob Havlen, Todd Henry, Scott Hildreth, George Jacoby, William Keel, Geoff Marcy, Jeff McClintock, Edward McNevin III, Michael Merrill, Jacqueline Mitton, Janet Morrison, Donald Osterbrock, Michael Perryman, Carle Pieters, Edward Purcell, Axel Quetz, Jessica Richter, Dennis Schatz, Rudy Schild, Maarten Schmidt, Joe Tenn, Ray Villard, Althea Washington, Adrienne Wasserman, and Richard Wolff.

David Bruning played an essential role in the development of the text, by writing the "Using RedShift" sections, as well as the first draft of Appendix 1. George Kelvin rendered some of the fine color diagrams that grace the book. We are grateful to Bill Hartmann, John Spencer, Don Davis, and Don Dixon for permission to reproduce their astronomical paintings, and to David Malin for his assistance and his superb astronomical photographs.

We benefited very much from the suggestions of the following reviewers of preliminary drafts of the text:

Grady Blount
Texas A&M University, Corpus Christi

Michael Briley
University of Wisconsin, Oshkosh

David Buckley
East Stroudsburg University

John Burns
Mt. San Antonio College

Paul Campbell
Western Kentucky University

Eugene R. Capriotti
Michigan State University

George L. Cassiday
University of Utah

John Cunningham
Miami-Dade Community College

Grace Deming
University of Maryland

Miriam Dittman
DeKalb College

Gary J. Ferland
University of Kentucky

George Hamilton
Community College of Philadelphia

Adrian Herzog
California State University, Northridge

Ronald Kaitchuck
Ball State University

William C. Keel
University of Alabama

Steven L. Kipp
Mankato State University

Jim Lattimer
SUNY, Stonybrook

Robert Leacock
University of Florida

Terry Lemley
Heidelberg College

Bennett Link
Montana State University

Charles H. McGruder III
Western Kentucky University

Stephen A. Naftilan
Claremont Colleges

Anthony Pabon
DeAnza College

Cynthia W. Peterson
University of Connecticut

Andrew Pica
Salisbury State University

Terry Richardson
College of Charleston

Margaret Riedinger
University of Tennessee, Knoxville

Jim Rostirolla
Bellevue Community College

Michael L. Sitko
University of Cincinnati

John Stolar
West Chester University

Charles R. Tolbert
University of Virginia

Steve Velasquez
Heidelberg College

David Weinrich
Moorhead State University

David Weintraub
Vanderbilt University

Mary Lou West
Montclair State University

Dan Wilkins
University of Nebraska, Omaha

J. Wayne Wooten
Pensacola Junior College

No book project of this complexity could succeed without the diligent efforts of many people at the publisher's. We very much appreciate the assistance of John Vondeling, Vice-President and Publisher; Jennifer Bortel, Acquisitions Editor; Anne Gibby, Senior Project Editor; Carol Bleistine, Manager of Art and Design; Alicia Jackson, Production Manager; George Semple, Picture Development Editor; Margie Waldron, Director of Marketing; Tara Pauliny, Editorial Assistant; Karen Milstein, Marketing Coordinator; and other members of the staff at Saunders College Publishing, for graciously accommodating the needs of three authors with busy schedules and strong opinions.

We dedicate this book to Alex Fraknoi, who came into the universe during the same period this book was planned and born. May a future edition of the text still be helping students with their exploration of the universe by the time he gets to college.

Andrew Fraknoi, David Morrison, and Sidney Wolff
July 1996

Contents Overview

Contents

Galileo Orbiter Mission

In June 1996, the Galileo Orbiter began its two-year mission to survey the four large satellites of Jupiter. Although problems with the spacecraft antenna will limit the quantity of data radioed back to Earth, it is apparent that the quality of the images and spectra is superb, promising spectacular advances beyond the photos returned by the two Voyager spacecraft during their flybys 17 years earlier. Reproduced on these pages are examples of the first Galileo images of the moons Ganymede, Europa, and Io.

See Chapter 11 for more on these intriguing worlds.

Figure 1
A very high resolution view of the older, heavily cratered terrain on Ganymede, showing a region only a few miles across. The complex, hilly terrain and indications of the presence of both light- and dark-colored materials on the surface were among the new information revealed as the spacecraft passed much closer to Ganymede than had been possible with Voyager. (NASA/JPL)

Figure 2
Two of the new better-detail Galileo photos of Ganymede are superposed on the image of the same area transmitted by Voyager in 1979. The mountainous areas of the satellite are now seen to be much more rugged than had been suggested previously, indicating a more violent geological history for this moon and suggesting the action of plate tectonics of Ganymede at some time in the distant past. (NASA/JPL)

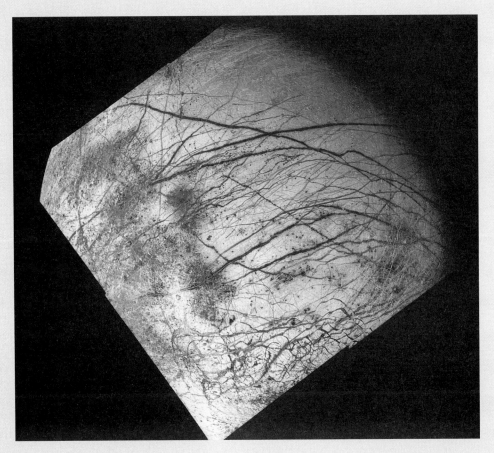

Figure 3
The first Galileo views of ice-covered Europa were better than the Voyager images but still far inferior to the two photos of Ganymede shown above. Still, the complex cracks on the surface hint of the possibility of a global ocean beneath the icy crust. If Europa really has large bodies of liquid water today, it is the only other location in the solar system beside our own planet Earth that possesses liquid water. (NASA/JPL)

Figure 4
Three early views of Io, imaged by the Galileo spacecraft from a great distance, show the multihued surface of this volcanically active satellite and demonstrate that its color has indeed changed in the 17 years since Voyager. These changes are the result of several large-scale volcanic eruptions that have taken place since the moon was last mapped. (NASA/JPL)

A mosaic of images of distant galaxies of stars, taken with four cameras aboard the repaired Hubble Space Telescope in March 1994. The area of the sky shown in this image is extremely small—about the size of President Roosevelt's eye on a dime held at arm's length. The brilliant galaxy that fills the small image (bottom right) lies on the outskirts of the Coma Cluster, a rich group of galaxies so far away that light from it takes 300 million years to reach us. Everything on this image that is *not* a round dot is a galaxy of billions of stars (the round dots are stars in our own Milky Way Galaxy). Except for the spiral-shaped galaxy at left, all the other galaxies in this view are significantly further away than the Coma Cluster, and are shown with great clarity from the Hubble's position above the Earth's atmosphere. They present a marvelous array of shapes and colors, challenging astronomers to understand their birth and evolution. (Courtesy of William Baum and NASA)

Prologue and Brief Tour of the Universe

We invite you to come with us on a series of voyages to explore the universe as astronomers understand it today. Out there we will find vast, awesome, and magnificent realms—full of objects that have no counterparts on Earth. Nevertheless, we will learn to call the universe our home, and we will see that its evolution has been directly responsible for our presence on planet Earth today.

Everybody, at last, is getting nosy. I predict that in our time astronomy will become the gossip of the marketplace.

Galileo speaking in Bertolt Brecht's play *Galileo*.

1

Figure P.1
This image of Mars was constructed at the U.S. Geological Survey in Flagstaff, Arizona, by combining many individual images taken by the Viking spacecraft starting in 1976. The view is centered on the Valles Marineris (Mariner Valley) complex of canyons, which is as long as the United States is wide. (USGS)

Along our journey we will encounter:

- a "grand canyon" system so large that on Earth it would stretch from Los Angeles to Washington, D.C. (Figure P.1).
- a tiny moon whose gravity is so weak that one good throw from its surface can put a baseball into orbit.
- a giant planet made mostly of gas and liquid that offers no surface for future explorers to stand on.
- a collapsed star so dense that to duplicate it we would have to squeeze every human being on Earth into a single raindrop.
- exploding stars whose violent end could wipe a planet orbiting a neighboring star clean of all its life forms (Figure P.2).
- a cannibal galaxy with billions of stars that has already consumed a number of its smaller galaxy neighbors, and is not yet finished finding new victims.
- the radiation echo that is the faint but unmistakable signal of the creation event itself.

It is these kinds of discoveries that make astronomy today such an exciting and fascinating field for both scientists and people in other walks of life. But we will explore more than just the objects in our universe and the latest discoveries about them. We will pay equal attention to the process by which we have come to understand the realms beyond our Earth and will examine the tools we need to increase that understanding.

Astronomy today must still be carried out mostly from Earth. We gather information about the rest of the cosmos from the messages that the universe is kind enough to send our way. And since the *stars* are the fundamental building blocks of the universe, decoding the message of starlight has been a central challenge and triumph of modern astronomy. By the time you are done reading this text, you will know a bit about how to read that message and how to understand what it is telling us.

The Nature of Astronomy

Astronomy is defined as the study of the objects that lie beyond the atmosphere of our planet Earth, and of the processes by which these objects interact with one another. But, as we will see, it is much more than that. It is also humanity's attempt to organize what we learn into a clear history of the universe, from the instant of its birth in the explosion we call the Big Bang to the present moment in which you are finishing this sentence.

Putting this history together is an audacious undertaking—especially by a creature of modest stature, living on a rocky ball circling a nondescript star in the suburbs of the Milky Way—and we are surely far from finished with it. Throughout the book, we emphasize how much science is a "progress report" that constantly changes as

new techniques and instruments allow us to probe the universe more deeply.

In considering the history of the universe, we will see again and again that the cosmos *evolves*—it changes in profound ways over long periods of time. Although you may sometimes read that the concept of evolution is controversial in science, this is simply not true. Few ideas are better established than the notion that the universe is not the same today as it was long ago. You will see, for example, that the universe could not have produced the readers of this book during the first generation of stars after the Big Bang. The ingredients in the recipe for an astronomy student simply did not exist then in sufficient numbers. The universe still had to make the carbon, the calcium, the oxygen, the iron needed to construct something as interesting and complicated as you. Today, many billions of years later, the universe has evolved into a more hospitable place for life. Tracing the evolutionary processes that continue to shape the universe is one of the most important (and satisfying) parts of modern astronomy.

The Nature of Science

Science—unlike religion or philosophy—accepts nothing on faith. The ultimate judge in science is always the experiment or observation: what nature itself reveals. Science, then, is not merely a body of knowledge, but a *method* by which we attempt to understand nature and how it behaves. This method begins with many observations over a period of time. From the trends in the observations, scientists may come up with a *model* of the particular phenomenon we want to understand. Such models are always approximations of nature itself, subject to further testing.

To take a concrete astronomical example, ancient astronomers constructed a model (partly from observations, partly from philosophical beliefs) that the Earth was the center of the universe, and that everything moved around it. At first, the available observations of the Sun, Moon, and planets could be fit to this model, but eventually, better observations required the model to add circle after circle to the movements of the planets to keep the Earth at the center. As the centuries went by, and improved instruments were developed for keeping track of celestial objects more precisely, the old model (even with a huge number of circles) could no longer explain all the observed facts. As we will see in Chapter 1, a new model, with the Sun at the center, fit the experimental evidence better. After a period of philosophical struggle, it became accepted as our view of the universe.

When first proposed, new models or ideas are sometimes called *hypotheses*. Many students today think that there can be no new hypotheses in a science such as astronomy—that everything important has already been learned. Nothing could be further from the truth. Throughout this book you will find discussions of recent, and occasionally still controversial, hypotheses in astronomy—concerning, for instance, the significance for life on our planet of huge chunks of rock and ice that hit the Earth; the existence of vast quantities of invisible "dark matter" which could make up the bulk of the universe; and the presence of a strange "black hole" at the center of the Milky Way. All such hypotheses are based on difficult observations done at the forefront of our technology, and all require further testing before we fully incorporate them into our standard astronomical models.

This last point is crucial: if there is no possible way of testing a hypothesis, it does not belong in the realm of science. In some cases, theories (especially those about the very largest structures in the universe, or the very smallest structures inside the atom) may not lend themselves to testing with our *current* instruments. Many years may pass before the appropriate experiments or observations can be performed. But ultimately, all scientific ideas must be testable, and, moreover, must stand up to being tested again and again in as many ways as we can muster. If the experiments or observations do not support a model or hypothesis, its proponents must be ready to modify it or let it go, no matter how fond of it they may have grown.

It is this *self-correcting* aspect of science that sets it off from most human activities. Scientists spend a great deal of time questioning and criticizing one another. No project is funded and no report is published without extensive *peer review*—that is, without careful examination by other scientists in the same field. While in other areas, young people are often taught to accept the authority of their elders without question, in science (after proper training) everyone is encouraged to try new and better experiments, and to challenge any and all hypotheses.

This is one of the reasons science has made, and is making, such dramatic progress. An undergraduate science major today knows more science and math than did Isaac Newton, one of the most brilliant scientists who ever lived. Even in this *introductory* astronomy course you will learn about objects and processes that a few generations ago no one even dreamed existed. While the domain of science is limited, within that domain its achievements have been glorious.

The Laws of Nature

Over the centuries, scientists have extracted from countless observations certain fundamental principles, called *scientific laws*. These are, in a sense, the rules of the game nature plays. One remarkable discovery about nature that underlies everything you will read in this book is that the same laws apply everywhere in the universe. The rules that govern the behavior of gravity on Earth, for example, are the same rules that determine the motion of two stars in a system so far away that your unaided eye cannot find them in the sky.

Note that without the existence of such universal laws, we could not do much astronomy. If each pocket of the universe had not only different objects, but completely different rules, we would have little chance of interpreting what happened in other "neighborhoods." But the consistency of the laws of nature gives us enormous power to understand distant objects without traveling to them and learning the local laws. In the same way, if every U.S. state or Canadian province had completely different laws, it would be very difficult to carry out commerce or even to understand the behavior of people in different regions. But a consistent set of laws allows us to apply what we learn or practice in one state to any other state.

You might ask whether the laws of the natural world could (under the right circumstances) be suspended. This is an enticing fantasy, but, despite many attempts, not a shred of scientific evidence has been found to support such an idea. While it would be nice if we could suspend the law of gravity and float away from our porch through effort of will alone, in real life such experiments generally result in broken bones.

This is not to say that our models or rules cannot change. New experiments and observations can lead to new or more sophisticated models—models that can even include new phenomena and new laws about their behavior. The theory of relativity proposed by Albert Einstein is a perfect example of such a transformation, one that took place not long ago. It led us to predict and, more recently to catch, indirect glimpses of a strange new class of objects astronomers call black holes. But wishing isn't going to bring such new models into existence; only the patient process of observing nature ever more finely can reap such rewards.

One important problem about describing scientific models has to do with the limitations of language. When we try to describe complex phenomena in everyday terms, the words themselves may not be adequate to the job. For example, you may have heard the structure of the atom likened to a miniature solar system. While some aspects of our modern model of the atom do remind us of planetary orbits, many other aspects are fundamentally different.

This is why scientists often prefer to describe their theories using equations rather than words. In this book, designed to introduce the field of astronomy, we have used mainly words to discuss what scientists have learned, and have avoided math beyond basic algebra. But if this course piques your interest and you go on in science, more and more of your studies will involve the language of mathematics.

Numbers in Astronomy

You may have heard a television reporter refer to a large number (such as the national debt) as "astronomical." In astronomy, we do have to deal with distances on a scale you may never have thought about before, and with numbers larger than any you may have encountered. Most students take a while to learn to navigate among the millions and billions that astronomers tend to throw about in their everyday discussions. With some practice, you will become just as good at it!

By the way, if you sometimes have trouble sorting out millions and billions, take heart. Even so distinguished a group as the Presidential Advisory Committee on the Future of the U.S. Space Program, in a 1990 report, listed the distance to the planet Uranus as 1.7 million miles, when in fact it is 1.7 *billion* miles away. A million (1,000,000) is a thousand times less than a billion (1,000,000,000). If Uranus were that close, we would all have noticed it—since it would be bigger and brighter than the Moon in our skies.

In this text we adopt two approaches that make dealing with astronomical numbers a little bit easier. First, we use a system for writing large and small numbers called *power-of-ten notation* (or sometimes *scientific notation*). This system is very appealing because it does away with the huge number of zeros that can seem overwhelming to the beginner.

In scientific notation, if you want to write a figure like $490,000 (which is definitely NOT the starting salary for an astronomer, but might be for a national television star), you write 4.9×10^5; the little number after the ten, called an *exponent*, keeps track of the number of times you have to multiply ten together to get the number you want. In our example, ten is multiplied by itself five times, and $10 \times 10 \times 10 \times 10 \times 10$ gives 100,000. Multiply 100,000 by 4.9 and you get our astronomical starting salary. Another way to remember the basics of this notation is to note that five is the number of places you have to move the decimal point to the right to convert 4.9 to 490,000. If you are encountering this system for the first time, we

suggest you look at Appendix 4 for more information and examples.

Small numbers are written with negative exponents. Three millionths (0.000003) is expressed as 3.0×10^{-6}. One reason this notation is so popular among scientists—trust us, it is, even if you at first don't like it—is that it makes arithmetic a lot easier. To multiply two numbers in scientific notation, you need only add their exponents: thus $10^3 \times 10^9 = 10^{12}$. To divide numbers, just subtract exponents.

The second way we try to keep numbers simple is to use a consistent set of units—the *international metric system*. Unlike the British system, in which a completely arbitrary number such as 5280 feet equals a mile, metric units are related by powers of ten: a kilometer, for example, equals a thousand meters. The metric system, which has been adopted by every major country in the world except the United States, is summarized in Appendix 5, and should become part of your vocabulary if you want to be ready for the future (when everyone will be using metric units).

Light Years

To give you a chance to practice scientific notation, and to set the scene for the tour of the universe in the next section, let's define a common unit astronomers use to describe distances in the universe. A *light year* (LY) is the distance that light travels in one year. Since light always travels at the same speed, and since its speed turns out to be the fastest possible speed in the universe, it makes a good standard for keeping track of distances. Some students complain about this name for a unit of distance—light *years* seem to imply that we are measuring time. But this mix-up of time and distance is common in everyday life as well—for example, when we tell a friend to meet us at a movie theater that's 20 minutes away.

So how many kilometers are there in a light year? If you are new to scientific notation, we suggest you work through all the steps of the following example for yourself on a separate sheet of paper. First, in case you are not yet a metric system fan, we should tell you that a kilometer is about 0.6 miles. Light travels at the amazing pace of 3×10^5 km per second (km/s). Think about that—light covers 300,000 km every second. In one second, it can travel seven times around the circumference of the Earth; a commercial airplane, in contrast, would take about two days to go around once, not counting time to refuel.

Now that we know how far light goes in a second, we can calculate how far it goes in a year. There are 60 (6×10^1) seconds in each minute, and 6×10^1 minutes in every hour. Thus light covers 3×10^5 km/s $\times 3.6 \times 10^3$ s/h $= 1.08 \times 10^9$ km/h. There are 24 or 2.4×10^1 hours in a day, and 365.24 (3.65×10^2) days in a year. The product of those two numbers is 8.77×10^3 h/year. Multiplying that by 1.08×10^9 km/h gives 9.46×10^{12} km in a light year. That's almost 10 trillion kilometers that light

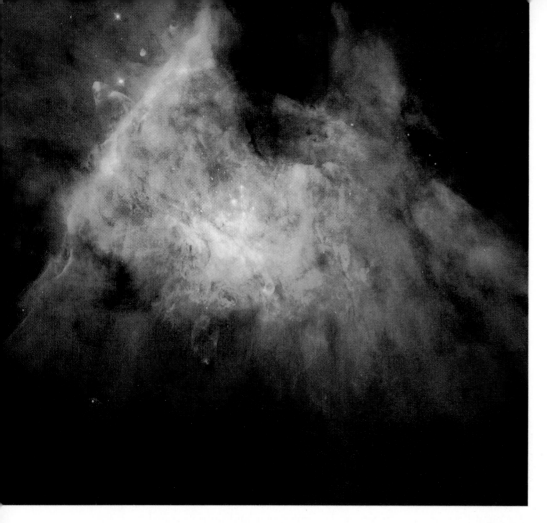

Figure P.3
This beautiful cloud of cosmic raw material (gas and dust from which new stars and planets are being made) called the Orion Nebula is about 1500 LY away. That's a distance of roughly 1.4×10^{16} km—a pretty big number! The picture we see is actually a seamless mosaic of 15 smaller images taken with the Hubble Space Telescope. The field of view is about 2.5 LY wide and shows only a small part of a vast reservoir of mostly dark material. The gas and dust in this region is illuminated by the intense light from a few extremely energetic adolescent stars in the neighborhood. (C. R. O'Dell and NASA)

covers in a year. A string 1 LY long could fit around the circumference of the Earth 236 million times!

You might think that such a long unit would more than reach to the nearest star. But the stars are far more remote than our imaginations (or episodes of *Star Trek*) might lead us to believe. Even the nearest star is 4.3 LY away—more than 40 trillion km. Other stars visible to the unaided eye are hundreds or even thousands of light years away (Figure P.3). This is why astronomers are skeptical that UFOs are extraterrestrial spacecraft coming here across vast distances, briefly picking up two rural fishermen or loggers, and then going straight home. It seems like such a small reward for such a large investment.

Consequences of Light Travel Time

There is another reason the speed of light is such a natural unit of distance for astronomers. Information about the universe comes to us almost exclusively via radiation (of which light is one example), and all such radiation travels at the speed of light—that is, 1 LY every year. This sets a limit on how quickly we can learn about events in the universe. If a star is 100 LY away, the light we see from it tonight left that star 100 years ago and is just now arriving in our neighborhood. The soonest we can learn about any changes in that star—its blowing up, for example—is 100 years after the fact. For a star 500 LY away, the radiation we detect tonight left 500 years ago, and is carrying 500-year-old news.

Some students, accustomed to CNN and other news media known for "instant world coverage," at first find this frustrating. You mean when I see that star up there, they ask, I won't know what's actually happening there *now* for another 500 years? But that's not really the right way to think about the situation. For astronomers, *now* is when the light reaches us here on Earth. There is no way for us to know anything about that star (or other object) until its radiation reaches us; despite the fondest dreams of science-fiction writers, instant communication through the universe is not possible.

But what at first may seem a great frustration is actually a tremendous benefit in disguise. If astronomers really want to piece together what has happened in the universe since its beginnings, they need to find evidence about each epoch of the past. Where can we find evidence today about cosmic events that occurred billions of years ago? Unfortunately, the universe doesn't leave written records (or even decent videotapes) of its main activities!

The delay in the arrival of light provides such evidence automatically. The farther out in space we look, the longer the light has taken to get here, and the longer ago it left its place of origin. By looking billions of light years out into space, astronomers are actually seeing billions of years into the past. In this way, we can reconstruct the history of the cosmos and get a sense of how it has evolved over time.

This is one reason astronomers strive to build telescopes that can collect more and more of the faint light (and other radiation) the universe sends us. The more light we collect, the fainter the objects we can make out. On average, fainter objects are farther away, and can thus tell us about periods of time even deeper in the past. New instruments, such as the Hubble Space Telescope (Figure P.4) and Hawaii's Keck Telescope (see Chapter 5), are giving astronomers views of deep space and deep time better than any we have had before.

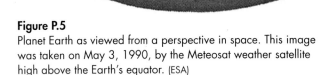

Figure P.4
The Hubble Space Telescope, shown here being repaired aboard the Space Shuttle *Endeavour* in December 1993, is an example of the new generation of astronomical instruments in space. (NASA)

Figure P.5
Planet Earth as viewed from a perspective in space. This image was taken on May 3, 1990, by the Meteosat weather satellite high above the Earth's equator. (ESA)

A Tour of the Universe

Let us now take a brief introductory tour of the universe as astronomers understand it today, just to get acquainted with the sorts of objects and distances we will encounter throughout the text. We begin at home, with the Earth, a nearly spherical planet about 13,000 km in diameter (Figure P.5). A space traveler entering our planetary system would easily distinguish the Earth by the large amount of liquid water that covers some two-thirds of its crust. If the traveler had equipment to receive radio or television signals, or came close enough to see the lights of our cities at night, she would soon find signs that this water planet has intelligent life (depending on what television channel the traveler tuned to, that conclusion might need to be changed to "semi-intelligent" life).

Our nearest astronomical neighbor is the Earth's satellite, commonly called the Moon. Figure P.6 shows the Earth and the Moon to scale on the same diagram. Notice how small we have to make these bodies to fit them on the page with the right scale. The Moon's distance from Earth is about 30 times the Earth's diameter, or approximately 384,000 km, and it takes about a month for the

Moon to revolve around the Earth. The Moon's diameter is 3476 km, one fourth the size of the Earth.

Light (or radio waves) takes 1.3 seconds to travel between the Earth and the Moon. If you've seen videos of the Apollo flights to the Moon, you may recall that there was a delay of about 3 seconds between the time Mission Control asked a question and the time the astronauts replied. The reason is not that the astronauts were thinking slowly, but that it took the radio waves almost 3 seconds to make the round trip.

The Earth revolves around our star, the Sun, which is about 150 million km away—approximately 400 times as far as the Moon. We call the average Earth–Sun distance an *astronomical unit* (AU), because in the early days of astronomy it was the most important measuring standard. Light takes a little more than 8 minutes to travel one AU, which means our latest news from the Sun is always 8 minutes old.

It takes the Earth one year (3×10^7 s) to go around the Sun at our distance; to make it around, we must travel at approximately 110,000 km/h. (If you, like many students in the United States, still prefer miles to kilometers, you might find the following trick helpful. To convert kilometers to miles, you can multiply kilometers by 0.6. Thus 110,000 km/h becomes 66,000 mi/h—a fast clip no matter what units you use.) Since gravity holds us firmly to the Earth and there is no resistance to the Earth's motion in the vacuum of space, we participate in this breakneck journey without being aware of it day by day.

The diameter of the Sun is about 1.5 million km; our Earth could fit comfortably inside one of the minor eruptions that occur on the surface of our star. If the Sun were reduced to the size of a basketball, the Earth would be a small apple seed some 30 m from the ball.

The Earth is only one of nine planets that we have discovered revolving around the Sun. These planets, along with their satellites and swarms of smaller bodies, make up the *solar system,* what we might call the family of the Sun. A planet is defined as a body of some significant size that orbits a star and does not produce its own light. (If a large body consistently produces its own light, it gets to be called a star.) We are able to see the nearby planets in our skies only because they reflect the light of our local star, the Sun. If the planets were much further away, the tiny amount of light that they manage to reflect would not be visible to us. The few planets we have so far been able to

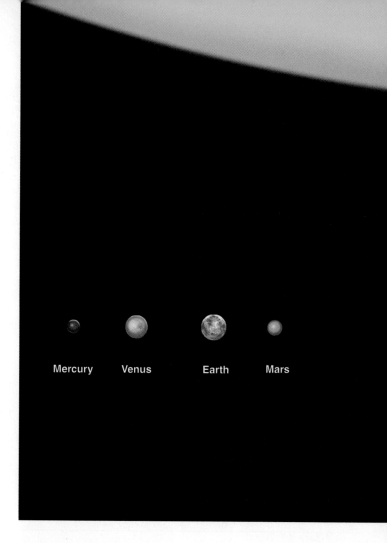

Mercury   Venus   Earth   Mars

discover orbiting other stars were found from the pull their gravity exerts on their parent stars, not from their light.

Jupiter, the largest planet in the solar system, is about 143,000 km in diameter, 11 times the size of the Earth (Figure P.8). Its distance from the Sun is five times the Earth's, or 5 AU. On the scale where the Sun is a basketball, Jupiter is about the size of a grape, and is located about 150 m from the basketball. The orbits of the planets are shown schematically in Figure P.9.

Usually, the most distant planet is little Pluto, but currently (until the year 1999), Pluto's strange orbit actually carries it inside the orbit of Neptune. Pluto's average

Figure P.6
The Earth and Moon, drawn to scale.

Figure P.7
The Sun and the planets shown to scale. Notice the size of the Earth compared to the giant planets.

orbital distance, however, is about 40 AU or 5.9 billion km from the Sun. On our basketball scale, Pluto is a grain of sand about 1 km from the ball.

The Sun is our local star and all the other stars are also suns—enormous balls of glowing gas that generate vast amounts of energy by nuclear reactions deep within. We will discuss the fascinating processes that make the stars shine in more detail later in the book. The other stars look faint only because they are so very far away. Continuing our basketball analogy, Proxima Centauri, the nearest star beyond the Sun, which is 4.3 LY away, would be almost 7000 km from the basketball.

Figure P.8
The largest planet in our solar system is Jupiter. We could fit almost eleven Earths side by side into its equator; and it contains as much mass as all the other planets combined. This image, which also shows one of its large satellites, was taken in February 1979 when the Voyager 1 spacecraft flew by. (JPL/NASA)

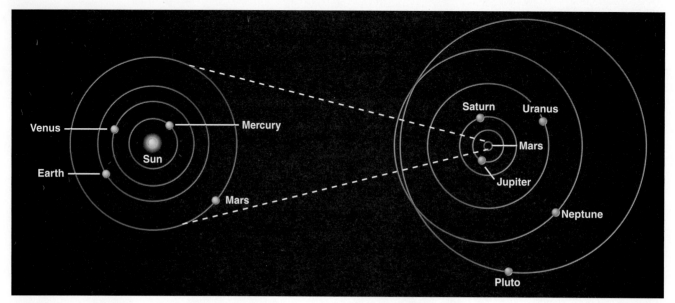

Figure P.9
The orbits of the planets in our solar system. (left) The inner planets. (right) The outer planets (note the change of scale).

When we look up at the star-studded country sky on a clear night, all the stars visible to the unaided eye turn out to be part of a single collection of stars we call the Milky Way Galaxy, or just simply our Galaxy. (When referring to the Milky Way, we capitalize Galaxy; when talking about *other* galaxies of stars, we use lowercase.) The Sun is one of hundreds of billions of stars that compose the Galaxy; its extent, as we will see, staggers the human imagination.

Let's make a rough scale drawing showing the stars within 10 LY of the Sun (Figure P.10). The small circle labeled (a) represents a sphere 10 LY in radius centered on the Sun. We find roughly ten stars in this sphere. Now we

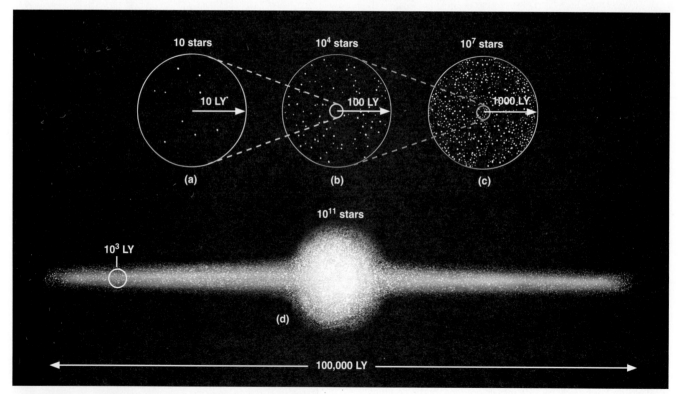

Figure P.10
The distribution of stars around the Sun within (a) 10 LY, (b) 100 LY, (c) 1000 LY, and (d) the Galaxy (with its disk seen edge-on).

Figure P.11
This galaxy of billions of stars, called by its catalog number M83, is about 10 million LY away and is thought to be similar to our own Milky Way Galaxy. Here we see the giant wheel-shaped system face-on, as if we were looking down on the disk of stars. (Cerro Tololo Interamerican Observatory/NOAO)

change scale: the circle labeled (b) represents a sphere 100 LY in radius. Note that all of (a) is now just a small circle in the center. Sphere (b) contains about 10,000 (10^4) stars, far too many to make the job of counting them pleasant, or the job of naming them reasonable. And yet in going out to a distance of 100 LY, we have traversed only a tiny part of the Milky Way Galaxy.

Now let's draw a circle of radius 1000 LY (c), in which, once again, our previous circle is just a small center. Within the 1000-LY sphere, we find some 10 million (10^7) stars. In the bottom half of Figure P.10, we change scale and examine the entire Galaxy, a wheel-shaped system whose visible diameter is roughly 100,000 LY (seen edge-on in our figure). Our Galaxy looks like a giant frisbee with a small ball in the middle. If we could move outside our Galaxy and look down on the disk of the Milky Way from above, it would probably resemble the galaxy in Figure P.11, its spiral structure outlined by the blue light of hot adolescent stars.

The Sun is somewhat less than 30,000 LY from the center of the Galaxy, in a location with nothing much to distinguish it. From our position inside the Milky Way Galaxy, we cannot see through to its far rim (at least not with ordinary light) because the space between the stars is not completely empty. It contains a sparse distribution of gas (mostly the simplest element, hydrogen) intermixed with tiny solid particles that we call interstellar dust. This gas and dust collects into enormous clouds in many places in the Galaxy, becoming the raw material for future generations of stars. Figure P.12 shows a beautiful image of the disk of the Galaxy as seen from our vantage point.

Typically, the interstellar material is so extremely sparse that the space between stars is a far, far better vacuum than anything we can produce in terrestrial laboratories. Yet, the dust in space, building up over thousands of light years, can block the light of more distant stars. Like the distant buildings that disappear from our view on a smoggy day in Los Angeles, the more distant regions of the Milky Way cannot be seen behind the layers of interstellar smog. Luckily, astronomers have found that stars and raw material shine with various forms of invisible radiation that do penetrate the smog, and so we have been able to build up a pretty good map of the Galaxy.

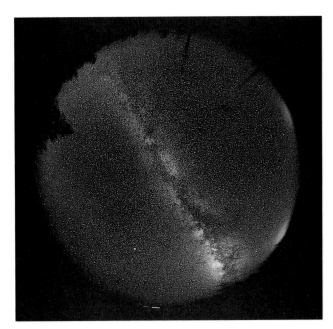

Figure P.12
Because we are inside the Galaxy, we see its disk in cross-section flung across the sky like a great white avenue of stars. This full-sky view, taken with a special lens from the summit of Mount Graham in southern Arizona, shows the Milky Way, with its myriad stars and dark "rifts" of dust. The outer circle is the horizon, and you can see, in addition to fir trees and test towers, the lights of several Arizona cities around it. (Roger Angel, Stewart Observatory/University of Arizona)

Figure P.13

In a cloud of cosmic raw material called the Rosette Nebula, a cluster of bright, hot stars can be seen forming in the upper left.

(Image by David Malin; © Anglo-Australian Telescope Board)

Recent observations, however, have also revealed a rather surprising and disturbing fact. There appears to be more—much more—to the Galaxy than meets the eye (or the telescope). From various investigations, we have evidence that much of our Galaxy is made of material that we cannot at present observe directly with our instruments. We therefore call this component of the Galaxy *dark matter*. We know the dark matter is there by the pull its gravity exerts on the stars and raw material we *can* observe, but what this dark matter is made of and how much of it exists remains a mystery. (And, as you will see, this dark matter is not confined to our Galaxy; it appears to be an important part of other star groupings as well.)

By the way, not all stars live by themselves, as the Sun does. Many are born in double or triple systems, with two or three stars revolving about each other (and even larger numbers of partners possible). Because the stars influence each other in such close systems, multiple stars allow us to measure characteristics that we cannot discern from observing single stars. In a number of places, enough stars have formed together that we recognize them as *star clusters* (Figure P.13). Some of the largest of the more than a thousand star clusters that astronomers have cataloged contain hundreds of thousands of stars and take up volumes of space hundreds of light years across.

Because stars live a long time (compared to the people who like to watch them), you often hear them referred to as eternal. But in fact, no star can last forever. Since the

"business" of stars is making energy, and energy production requires some sort of fuel to be used up, eventually all stars will run out of fuel and go out of business. (This news should not make you rush out to stock up on thermal underwear—our Sun still has at least 7 or 8 billion years to go.) But ultimately, the Sun and all stars *will* die, and it is in their death throes that some of the most intriguing and important processes of the universe stand revealed. For example, we will see that many of the atoms that now make up our bodies were once inside stars. These stars exploded at the ends of their lives, recycling their material back into the reservoir of the Galaxy. In this sense, all of us are literally made of "star dust."

The Universe on the Large Scale

In a very rough sense, you could think of the solar system as your house or apartment, and the Galaxy as your town, made up of many houses and buildings. In the 20th century, astronomers were able to show that just as our planet is made up of many, many towns, so the universe is made up of enormous numbers of galaxies. (We define the universe to be everything that exists that is accessible to our observations.) Galaxies stretch as far in space as our telescopes can see, many billions of them within the reach of modern instruments. When they were first discovered, some astronomers called galaxies "island universes" and the term is aptly descriptive: galaxies do look like islands of stars in the vast, dark seas of intergalactic space (see the opening figure for this chapter).

The nearest galaxy, just discovered in 1993, is a small one that lies 75,000 LY from the Sun in the direction of the constellation Sagittarius, where the smog in our own Galaxy makes it especially difficult to discern. (A *constellation* is one of the 88 sections into which astronomers divide the sky, each named after a prominent star pattern within it.) The existence of this Sagittarius dwarf galaxy is, in fact, still controversial—and will require other observations before all astronomers are convinced. Beyond it lie two other small galaxies, about 160,000 LY away. First recorded by Ferdinand Magellan's crew as he sailed around the world, these are called the Magellanic Clouds (Figure P.14). All three of these small galaxies are satellites of the Milky Way, interacting with it through the force of gravity. Ultimately, all three may even be swallowed by our much larger Galaxy.

The nearest large galaxy is a spiral quite similar to our own, located in the constellation of Andromeda and thus often called the Andromeda Galaxy; it is also known by one of its catalog numbers, M31 (Figure P.15). Given the number of galaxies, no reasonable person would suggest giving all of them proper names. M31 is about 2 million LY away and, along with the Milky Way, it is part of a small cluster of over 30 galaxies that we call the *Local Group*.

At distances of 10 to 15 million LY we find other small galaxy groups, and then at about 50 million LY there

Figure P.14
The Large Magellanic Cloud, at a distance of 160,000 LY, is one of the Milky Way's closest neighbor galaxies, and contains billions of stars. The reddish regions are *nebulae,* places where gas and dust are being illuminated by the energetic light of young stars. The brightest nebula, at left center, is nicknamed the Tarantula. (Photo taken with the U.K. Schmidt Telescope; © Anglo-Australian Telescope Board)

Figure P.15
The Andromeda Galaxy (M31) is a spiral-shaped collection of stars similar to our own Milky Way. Its flat disk of stars looks blue because its light is dominated by the vigorous light output of hot blue stars. The bulge in its center, on the other hand, contains older stars, which are predominantly yellow and red. Two smaller galaxies can be seen on either side of M31. (Tony Hallas)

is a more impressive system, with thousands of member galaxies, called the Virgo Cluster. We have discovered that galaxies occur mostly in clusters, both large and small (Figure P.16). Some of the clusters themselves form into larger groups called *superclusters.* Our Local Group, as well as the Virgo Cluster, are part of such a supercluster, which stretches over a diameter of at least 60 million LY. We are just beginning to explore the structure of the universe at these enormous scales, and are already finding some unexpected results.

At even greater distances, where many ordinary galaxies are too dim to see, we find the *quasars.* These are brilliant centers of galaxies, glowing with the light of some extraordinarily energetic process. One theory suggests that perhaps a giant *black hole* is swallowing whole neighborhoods of raw material. (We will describe the bizarre objects called black holes once we have discussed the lives of the stars. If you are not a black hole fan now, we predict you will be by the time you've finished this book.) Whatever the quasars are, their brilliance makes them the most distant beacons we can see in the dark oceans of space. They allow us to probe the universe 10 billion or more LY away, and thus 10 billion or more years in the past.

With the quasars we see a substantial way back to the Big Bang explosion that marks the beginning of time. Beyond the quasars, we can detect only the feeble glow of the explosion itself, filling the universe and thus coming to

Figure P.16
The central portion of a remote cluster of galaxies seen with the Hubble Space Telescope. The light the Hubble sees from these galaxies left them over 4 billion years ago, when the universe was roughly two-thirds its present age. (Alan Dressler and NASA)

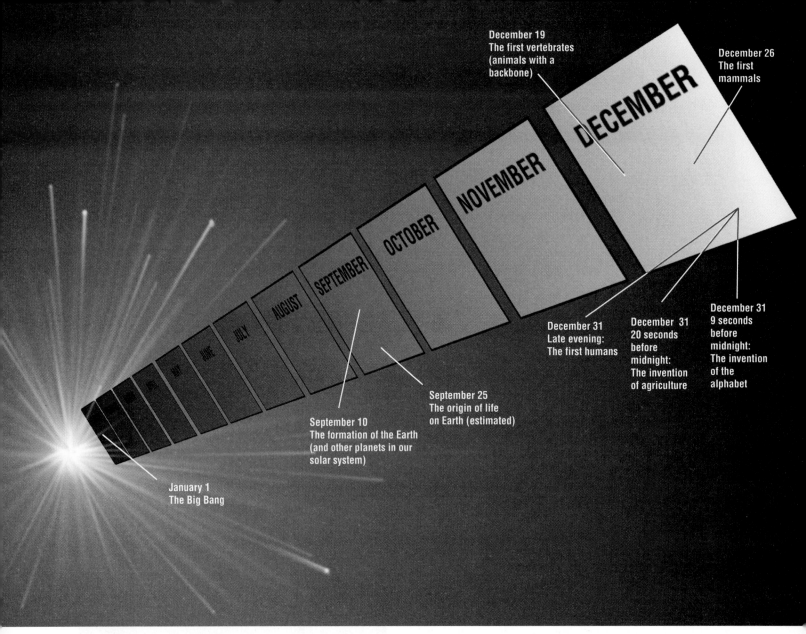

Figure P.17
On a cosmic calendar, where the time since the Big Bang is compressed into one year, creatures we would call human do not emerge on the scene until the evening of December 31.

us from all directions in space. The discovery of this "afterglow of creation" was one of the most exciting events in 20th-century science, and we are still exploring the many things it has to tell us about the earliest times of the universe.

For now, we might just note that such ideas are far easier to state than to discover. Measurements of the properties of galaxies and quasars in remote locations require large telescopes, sophisticated light-amplifying devices, and painstaking labor. Every clear night, at observatories around the world, astronomers and students are at work on such mysteries as the large-scale structure of the universe, mostly observing one star or one galaxy at a time, and fitting their results into the tapestry of our understanding.

The Universe of the Very Small

The foregoing discussion should impress on you that the universe is extraordinarily large and extraordinarily empty. The universe on average is 10,000 times emptier than our Galaxy. Yet, as we have seen, even the Galaxy is mostly empty space. The air we breathe has about 10^{19} atoms in each cubic centimeter—and we think of air as pretty empty stuff. In the interstellar gas of the Galaxy, there is about *one* atom in every cubic centimeter. Intergalactic space is so sparsely filled that to find one atom, on average, we must search through a cubic *meter* of space! Most of the universe is fantastically empty; places that are dense, such as the bodies of our readers, are tremendously rare.

Yet even the familiar solids, such as this textbook, are mostly space. If we could take such a solid apart, piece by piece, we would eventually reach the molecules of which it is formed. Molecules are the smallest particles into which matter can be divided while still retaining its chemical properties. A molecule of water (H_2O), for example, consists of two hydrogen atoms and one oxygen atom, bonded together.

Molecules, in turn, are built up of atoms, which are the smallest particles of an element that can still be identified as that element. For example, an atom of gold is the smallest possible piece of gold (although one atom of gold won't impress your sweetheart very much). Nearly 100 different kinds of atoms (elements) exist in nature, but most of them are rare, and only a handful account for more than 99 percent of everything with which we come in contact. The most abundant elements in the cosmos today are listed in Table P.1; think of this table as the "greatest hits" of the universe when it comes to elements.

All atoms consist of a central, positively charged nucleus, surrounded by negatively charged electrons. The bulk of the matter in each atom is found in the nucleus, which consists of positive protons and electrically neutral neutrons all tightly bound together in a very small space. Each element is defined by the number of protons in its atoms: thus any atom with 6 protons in its nucleus is called carbon, any with 50 protons is called tin, and any with 69 protons is called ytterbium. (Ytterbium, as you can probably guess, is not big on the cosmic hit parade of elements, but we like its name. For a list of the elements, see Appendix 13.)

The distance from an atomic nucleus to its electrons is typically 100,000 times the size of the nucleus itself. This is why we say that even solid matter is mostly space. The typical atom is far emptier than the solar system out to Pluto. (The distance from the Earth to the Sun, for example, is only 100 times the size of the Sun.) This is another reason atoms are not like miniature solar systems.

Remarkably, physicists have discovered that everything that happens in the universe, from the smallest atoms to the largest superclusters of galaxies, can be explained through the action of only four forces: gravity, electro-magnetism (which combines the actions of electricity and magnetism), and two forces that act at the nuclear level. We will get to know these forces in much more detail throughout the book, but the fact that there are four forces (and not a million, or just one) has puzzled physicists and astronomers for many years, and has led to a quest for a unified picture of nature. We will return to this quest in chapter 28.

A Conclusion and a Beginning

If you are typical of students new to astronomy, you have probably reached the end of our tour in this Prologue with mixed emotions. On the one hand, you may be fascinated by some of the new ideas you've read about, and eager to learn more. On the other, you may be feeling over-

TABLE P.1
The Cosmically Abundant Elements

Element*	Symbol	Number of Atoms per Million Hydrogen Atoms
Hydrogen	H	1,000,000
Helium	He	80,000
Carbon	C	450
Nitrogen	N	92
Oxygen	O	740
Neon	Ne	130
Magnesium	Mg	40
Silicon	Si	37
Sulfur	S	19
Iron	Fe	32

*Elements are arranged in order of the atomic number, which is the number of protons in each nucleus.

whelmed by the number of topics we have covered, and the number of new words and ideas we have introduced. Learning astronomy is a little like learning a new language—at first it seems there are so many new expressions that you'll never learn them all, but with practice, you soon develop facility with them.

At this point you may also feel a bit small and insignificant, dwarfed by the cosmic scales of distance and time. Such a feeling is not a bad thing from time to time—sometimes we wish more of our politicians and film stars felt this way. And just before a difficult exam, or when you've ended a treasured relationship, it can certainly help to see your problems in a cosmic perspective. But there is another way to look at what we've learned from our first glimpses of the cosmos.

Let us consider the history of the universe from the Big Bang to today, and compress it, for easy reference, into a single year. (We have borrowed this idea from Carl Sagan's Pulitzer-Prize-winning book, *The Dragons of Eden,* published in 1977 by Random House.) On this scale, the Big Bang happened at the first moment of January 1, the solar system formed around September 9, and the oldest rocks we can date on Earth go back to the beginning of October (Figure P.17).

Where in this "cosmic year" does the origin of human beings fall? The answer turns out to be the evening of December 31! The invention of the alphabet doesn't occur until the 50th second of 11:59 P.M. on December 31. And the beginnings of modern astronomy are a mere fraction of a second before the New Year. Seen in a cosmic context, the amount of time we have had to study the stars is minute, and our success at piecing together as much of the story as we have is remarkable.

Certainly, our attempts to understand the universe are not complete. Most likely, your grandchildren's grandchildren will find some of the contents of this book a bit primitive. But as you read our current progress report on the exploration of the universe, take a few minutes every once in a while just to savor how much we have already learned.

As the Earth rotates, the stars appear to move across the sky, as shown in this 10-hour exposure. The dome houses the Anglo-Australian Telescope. (© Anglo-Australian Telescope Board)

CHAPTER 1

Observing the Sky: The Birth of Astronomy

Thinking Ahead

Much to your surprise, a member of the Flat Earth Society moves in next door. He believes that the Earth is flat and that all the NASA images of a round Earth are either faked, or simply show the round (but flat) disk of the Earth from above. How could you prove to this neighbor that the Earth is really round? (You can find some of our answers in the chapter and in a box at the end of the chapter, but try to figure out some ways by yourself before peeking at our suggestions! You might want to discuss ideas with other students in the class.)

If the Lord Almighty had consulted me before embarking upon Creation, I should have recommended something simpler.

Statement attributed to Alfonso X, King of Castile, after having the Ptolemaic system explained to him.

Today, few people really spend much time looking at the night sky. When we see patterns in the dark, we're much more likely to be looking at a television or movie screen than at the heavens over our heads. But in ancient days, before electric lights and television robbed so many people of the splendor of the sky, the stars and planets were an important aspect of everyone's daily life. All the records that we have—on paper and in stone—show that ancient civilizations around the world noticed, worshipped, and tried to understand the celestial lights and fit them into their own view of the world. These ancient observers found both majestic regularity and never-ending surprise in the motions of the heavens.

The Babylonians and Greeks, for example, believing the planets to be gods, studied their motions in hopes of understanding the presumed influences of those planet-gods on human affairs;

thus developed the religion of astrology. Through their careful study of the planets, however, the ancient Greeks and later the Romans were also laying the foundations of the science of astronomy. As we shall see, their crowning achievement was Claudius Ptolemy's marvelously complicated model in the 2nd century A.D. Ptolemy's system predicted the positions of planets with good precision for hundreds of years, and it was not substantially improved upon until the European Renaissance in the 16th and 17th centuries.

In this chapter we shall take a look at the night sky as seen with the unaided eye, and also examine some of the interesting history of how we came to understand the realm above our heads.

1.1

The Sky Above

Our senses suggest to us that the Earth is the center of the universe—the hub around which the heavens turn. This **geocentric** (Earth-centered) view was what almost everyone believed until the Renaissance. After all, it is simple, logical, and seemingly self-evident. Furthermore, the geocentric perspective reinforced those philosophical and religious systems that taught the unique role of human beings as the central focus of the cosmos.

However, the geocentric view happens to be wrong. One of the great themes of intellectual history is the overthrow of the geocentric perspective. Let us therefore take a look at the steps by which we re-evaluated the place of our world in the cosmic order.

The Celestial Sphere

If you go on a camping trip or live far from city lights, your view of the sky on a clear night is pretty much identical to that seen by people all over the world before the invention of the telescope. Gazing up, you get the impression that the sky is a great hollow dome, with you at the center (Figure 1.1). The top of that dome, the point directly above your head, is called the **zenith,** and where the dome meets the Earth is called the **horizon.** It's easy to see the horizon as a circle around you from the sea or a flat prairie, but from most places where people live today, the horizon is hidden by mountains, trees, buildings, or smog.

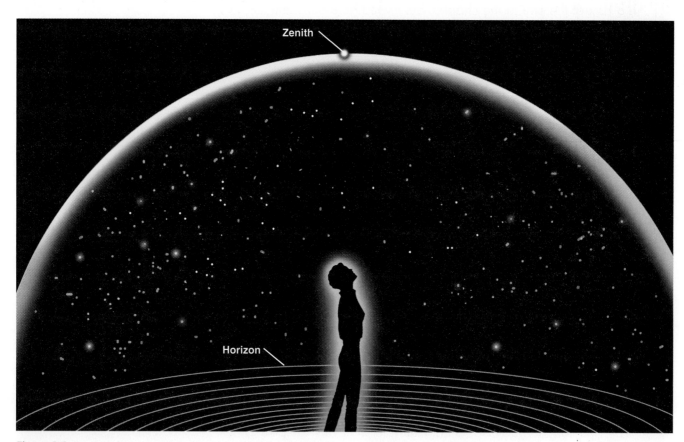

Zenith

Horizon

Figure 1.1
The dome of the sky, as it appears to a naive observer. The horizon is where the sky meets the ground, and the observer's zenith is the point directly above his head.

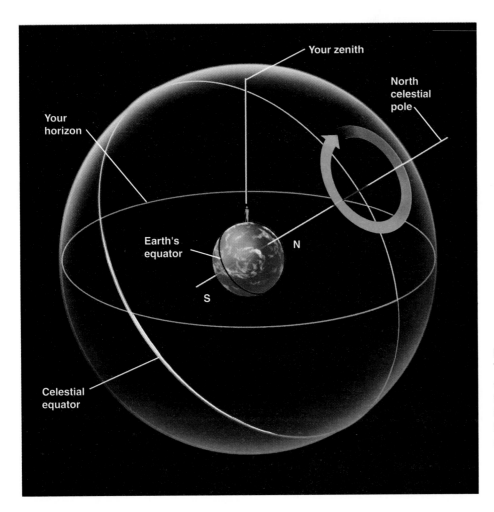

Figure 1.2
The (imaginary) celestial sphere around the Earth, on which objects in the sky can turn. Note the axis through the Earth. In reality, it is the Earth that turns around this axis, creating the illusion that the sky revolves around us.

If you lie back in an open field and observe the night sky for hours, as ancient shepherds and travelers regularly did, you will see stars rising on the eastern horizon (just as the Sun and Moon do), moving across the dome of heaven in the course of the night, and setting on the western horizon. Watching the sky turn like this night after night, you might eventually get the idea that the dome of the sky is really part of a great sphere that is turning around you, bringing different stars into view as it turns. The early Greeks regarded the sky as just such a **celestial sphere** (Figure 1.2). Some apparently thought of it as an actual sphere of transparent crystalline material, with the stars embedded in it like tiny jewels.

Today, of course, we know that it's not the celestial sphere that turns as the night and day proceed, but rather the planet on which we live. We can put an imaginary stick through the Earth's north and south poles; we call this stick our planet's **axis.** It is because the Earth turns on this axis every 24 hours that we see the Sun, Moon, and stars rise and set with clockwork regularity. Nevertheless, it is sometimes still convenient to talk about the celestial dome or sphere to help us keep track of objects in the sky. There is even a special theater, called a *planetarium*, in which we project a simulation of the stars and planets onto a white dome.

As the celestial sphere rotates, the objects on it maintain their positions with respect to one another. A grouping of stars like the Big Dipper has the same shape during the course of the night, although it turns with the sky. During a single night, even objects that we know to have significant motions of their own, such as the nearby planets, seem fixed relative to the stars. Only the *meteors*—brief "shooting stars" that flash into view for just a few seconds—move appreciably with respect to the celestial sphere. This is because they are not stars at all. Rather, they are small pieces of cosmic dust, burning up as they hit the Earth's atmosphere.

We can use the fact that the entire celestial sphere seems to turn together to help us set up systems for keeping track of what things are visible in the sky and where they happen to be at a given time.

Celestial Poles and Celestial Equator

To help orient us in the turning sky, astronomers use a system that extends our Earth's special points into the sky. We extend the line of the Earth's axis upward: the points where the axis meets the celestial sphere are defined as the **north celestial pole** and the **south celestial pole.** As the Earth rotates about its axis, the sky appears to turn

Figure 1.3
Time exposure showing trails left by stars as a consequence of the apparent rotation of the celestial sphere. The bright trail at top center was made by Polaris (the North Star), which is very close to the north celestial pole. (National Optical Astronomy Observatories)

in the opposite direction about those celestial poles (Figure 1.3). We also (in our imaginations) throw the Earth's equator onto the sky and call this the **celestial equator.** It lies halfway between the celestial poles, just as the Earth's equator lies halfway between our planet's poles.

Now let's imagine how riding on different parts of the spinning Earth affects our view of the sky. The apparent motion of the celestial sphere depends on your *latitude* (position north or south of the equator). Bear in mind that the Earth's axis is pointing at the celestial poles, so those two points on the sky do not appear to turn.

If you stood at the North Pole of the Earth, for example, you would see the north celestial pole overhead, at your zenith. The celestial equator, 90° from the celestial poles, lies along your horizon. As you watched them during the course of the night, the stars would all circle around the celestial pole, with none rising or setting (Figure 1.4). Only that half of the sky that is north of the celestial equator is ever visible to an observer at the North Pole. Similarly, an observer at the South Pole would see only the southern half of the sky.

If you were at the Earth's equator, on the other hand, you would see the celestial equator (which after all is just an "extension" of the Earth's equator) pass overhead through your zenith. The celestial poles, being 90° from the celestial equator, must then be at the north and south points on your horizon. As the sky turns, all stars rise and set; they move straight up from the east side of the horizon and set straight down on the west side. During a 24-hour period, all stars are above the horizon exactly half the time. (Of course during some of those hours, the Sun is too bright for us to see them.)

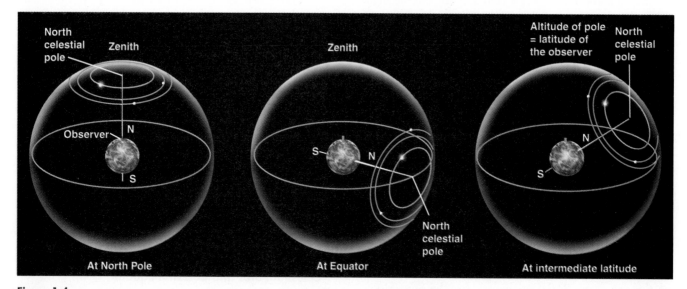

Figure 1.4
The turning of the sky looks different depending on your latitude on Earth. (a) At the North Pole, the stars circle the zenith and do not rise and set. (b) At the Equator, the celestial poles are on the horizon, and the stars rise straight up and set straight down. (c) At intermediate latitudes, the north celestial pole is at some position *between* overhead and the horizon. Its angle turns out to be equal to the observer's latitude. Stars rise and set at an angle to the horizon.

And what would an observer in the intermediate latitudes of the United States or Europe see? In our case, the north celestial pole is neither overhead nor on the horizon, but in between. It appears above the northern horizon at an angular height, or *altitude*, equal to the observer's latitude (Figure 1.4). In San Francisco, for example, where the latitude is 38° N, the north celestial pole is 38° above the horizon.

For an observer at 38° N latitude, the south celestial pole is 38° below the southern horizon. As the Earth turns, the whole sky seems to pivot about the north celestial pole. For this observer, stars within 38° of the North Pole can never set. They are always above the horizon, day and night. This part of the sky is called the north **circumpolar zone.** To observers in the continental United States, the Big and Little Dippers and Cassiopeia are examples of star groups in the north circumpolar zone. On the other hand, stars within 38° of the south celestial pole never rise. That part of the sky is the south circumpolar zone. To most U.S. observers, the Southern Cross is in that zone. (Don't worry if you are not familiar with those star groupings. We will introduce them more formally later on.)

At this particular time, there happens to be a star very close to the north celestial pole. It is called Polaris, the pole star, and has the distinction of being the star that moves the least amount as the northern sky turns each day. (Because it moved so little while the other stars moved much more, it played a special role in the mythology of several Native American tribes, for example.)

ASTRONOMY BASICS

What's Your Angle?

Astronomers measure how far apart objects appear in the sky by using angles. By definition, there are 360° in a circle, and therefore the entire celestial sphere contains 360°. The half-sphere or dome of the sky then contains 180° from horizon to opposite horizon. Thus if two stars are 18° apart, their separation spans about 1/10 of the dome of the sky. To give you a sense of how big a degree is, the full Moon is about half a degree across.

Rising and Setting of the Sun

We have described the movement of stars in the night sky. The stars continue to circle during the day, but the brilliance of the Sun makes them difficult to see. (The Moon is still easily seen in the daylight, however.) On any given day, we can think of the Sun as being located at some position on the hypothetical celestial sphere. When the Sun

rises—that is, when the rotation of the Earth carries the Sun above the horizon—sunlight is scattered by the molecules of our atmosphere, filling our sky with light and hiding the stars that are above the horizon.

For thousands of years, astronomers have been aware that the Sun does more than just rise and set. It gradually changes position on the celestial sphere, moving each day about 1° to the east relative to the stars. Very reasonably, the ancients thought this meant that the Sun was slowly moving around the Earth, taking a period of time we call one **year** to make a full circle. Today, of course, we know that it is the Earth which is going around the Sun, but the effect is the same: the Sun's position in our sky changes day to day. We have a similar experience if we walk around a campfire at night; we see the flames appear successively in front of each person seated about the fire.

The Sun's apparent path around the celestial sphere each year is called the **ecliptic.** Each day the Sun rises about 4 min later with respect to the stars; the Earth must make just a bit more than one complete rotation (with respect to the stars) to bring the Sun up again.

As we look at the Sun from different places in our orbit, we see it projected against different stars in the background, or we would, at least, if we could see the stars in the daytime. In practice, we must deduce what stars lie behind and beyond the Sun by observing the stars visible in the opposite direction at night. After a year, when the Earth has completed one trip around the Sun, the Sun will appear to have completed one circuit of the sky along the ecliptic.

The ecliptic does not lie along the celestial equator but is inclined to it at an angle of about 23°. In other words, the Sun's annual path in the sky is not lined up with the Earth's equator. This is because our planet's axis of rotation is tilted by about 23° from the plane of the ecliptic (Figure 1.5). The inclination of the ecliptic is the reason the Sun moves north and south in the sky as the seasons change. In Chapter 3 we will discuss the progression of the seasons in more detail.

Fixed and Wandering Stars

The Sun is not the only object that moves among the fixed stars. The Moon and each of the five planets visible to the unaided eye—Mercury, Venus, Mars, Jupiter, and Saturn—also slowly change their positions from day to day. During a single day, the Moon and planets all rise and set as the Earth turns, just as the Sun and stars do. But like the Sun, they have independent motions among the stars, superimposed on the daily rotation of the celestial sphere. Noticing these motions, the Greeks of 2000 years ago distinguished between what they called the fixed stars, those that maintain fixed patterns among themselves through many generations, and the wandering stars, or **planets.** The word *planet*, in fact, means "wanderer" in Greek.

Today we do not regard the Sun and Moon as planets, but the ancients applied the term to all seven of the mov-

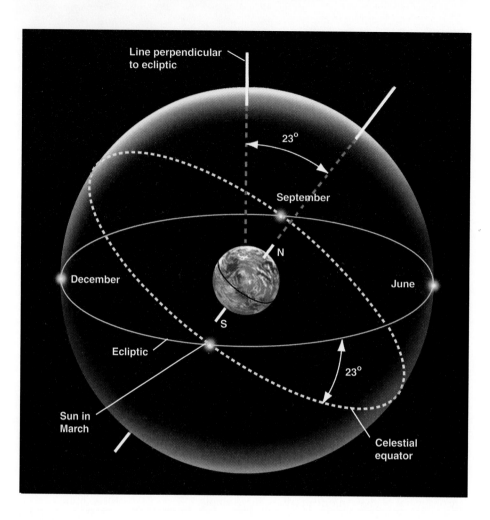

Figure 1.5
The celestial equator is tilted by 23° to the ecliptic. As a result, North Americans and Europeans see the Sun north of the celestial equator and high in our sky in June, and south of the celestial equator and low in the sky in December.

ing objects in the sky. Much of ancient astronomy was devoted to observing and predicting the motions of these celestial wanderers. The Moon, being the Earth's nearest celestial neighbor, has the fastest apparent motion; it completes a trip around the sky in about one month. (The Moon moves about 12°, or 24 times its own width, each day.)

The individual paths of the Moon and planets in the sky all lie close to the ecliptic, although not exactly on it. This is because the paths of the planets about the Sun, and of the Moon about the Earth, are all in nearly the same plane, as if they were circles on a huge sheet of paper. The planets and Moon are thus always found in the sky within a narrow 18°-wide belt called the **zodiac,** that is centered on the ecliptic. (The root of this term is the same as that of the word *zoo* and means a collection of animals: many of the patterns of stars within the zodiac belt reminded the ancients of animals, such as a fish or a goat.)

How the planets appear to move in the sky as the months pass is a combination of their actual motions plus the motion of the Earth about the Sun; consequently their paths are somewhat complex. As we shall see, this complexity has fascinated and challenged astronomers for centuries.

Constellations

The backdrop for the motions of the "wanderers" in the sky is the canopy of stars. If there were no clouds in the sky and we were on a flat plain with nothing to obstruct our view, we could see about 3000 stars with the unaided eye. To find their way around such a multitude, the ancients found groupings of stars that made some familiar geometric pattern or (more rarely) resembled something they knew. Each civilization found its own patterns in the stars, much like a modern Rorschach test in which you are asked to find images in a set of inkblots. The ancient Chinese, Egyptians, and Greeks, among others, found their own groupings or **constellations** of stars helpful in navigating among them and in passing their star-lore on to their children.

You may be familiar with some of the star patterns we still use today, such as the Big and Little Dippers, or Orion the hunter, with his distinctive belt of three stars (Figure 1.6). However, many of the stars we see are not part of a distinctive star pattern at all, and a telescope can reveal millions of stars too faint for the eye to see. Therefore, in the early decades of the 20th century, astronomers from many countries got together and decided to establish a worldwide system for organizing the sky.

Figure 1.6
The winter constellation of Orion, the hunter, as illustrated in the 17th-century atlas by Hevelius. (J. M. Pasachoff and the Chapin Library)

Today, we use the term *constellation* to mean one of 88 arbitrarily drawn sectors into which we divide the sky, much as the United States has been divided into 50 states. The modern boundaries between the constellations are imaginary lines in the sky running north-south and east-west, so that each point in the sky falls in a specific constellation. The constellations are listed in Appendix 14. Whenever possible, we have named each modern constellation after the Latin translation of one of the ancient Greek star patterns that lies within it. Thus the constellation of Orion is a kind of box on the sky, which includes, among many other objects, the stars that made up the ancient picture of the hunter. Some people use the term **asterism** to denote an especially noticeable star pattern within a constellation (or, sometimes, spanning parts of several constellations). For example, the Big Dipper is an asterism within the constellation of Ursa Major, the Big Bear.

Students are sometimes puzzled because the constellations seldom resemble the people or animals for which they were named. In all likelihood, the Greeks themselves did not name groupings of stars because they looked like actual people or objects (any more than the outline of Washington State resembles George Washington). Rather, they named sections of the sky in honor of the characters in their mythology and then fitted the star configurations to the animals and people as best they could.

1.2

Ancient Astronomy

Let us now look briefly back into history. Much of modern Western civilization is derived in one way or another from the ideas of the ancient Greeks and Romans, and this is true in astronomy as well. However, many other ancient cultures also developed sophisticated systems for observing and interpreting the sky.

Astronomy Around the World

The Babylonian, Assyrian, and ancient Egyptian astronomers knew the approximate length of the year. The Egyptians of 3000 years ago, for example, adopted a calendar based on a 365-day year. They kept careful track of the rising time of the bright star Sirius in the predawn sky, whose cycle happened to correspond with the flooding of the river Nile. The Chinese also had a working calendar and determined the length of the year at about the same time. They recorded comets, bright meteors, and dark spots on the Sun. (Many types of astronomical objects were introduced in the Prologue of this book. If you are not familiar with some of their names, you may want to look at the Prologue.) Later, Chinese astronomers kept careful records of "guest stars"—those that are normally too faint to see, but suddenly flare up to become visible to the naked eye for a few weeks or months. We still use some of these records in studying stars that exploded a long time ago.

The Mayan culture in Central America developed a sophisticated calendar based on the planet Venus, and they made astronomical observations from sites dedicated to this purpose a thousand years ago. The Polynesians learned to navigate by the stars over hundreds of kilometers of open ocean, thus allowing them to colonize new islands far away from where they began.

In the British Isles, before the widespread use of writing, ancient people used stones to keep track of the motions of the Sun and Moon. We still find some of the great stone circles they built for this purpose, dating from as far back as 2800 B.C. The best known of these is Stonehenge, which is discussed in Chapter 3.

Early Greek and Roman Cosmology

Our concept of the cosmos—its basic structure and origin—is called **cosmology,** a Greek word. Before the invention of telescopes, humans had to depend on the simple evidence of their senses for a picture of the universe. The ancients developed cosmologies that combined their direct view of the heavens with a rich variety of philosophical and religious symbolism.

At least 2000 years before Columbus, educated people in the eastern Mediterranean region knew the Earth was round. Belief in a spherical Earth may have stemmed from the time of Pythagoras, a philosopher and mathematician who lived 2500 years ago. He believed circles and spheres to be "perfect forms" and suggested that the Earth should be a sphere. As evidence that the gods liked spheres, the Greeks cited the fact that the Moon is a sphere, using evidence we will describe below.

Figure 1.7

A lunar eclipse, with the Moon moving into the Earth's shadow. Note the curved shape of the shadow—evidence for a spherical Earth that has been recognized since antiquity. (Kent Wood)

The writings of Aristotle (384–322 B.C.), the tutor of Alexander the Great, summarize many of the ideas of his day. They describe how the progression of the Moon's phases—its changing shape—results from our seeing different portions of the Moon's illuminated hemisphere during the month (see Chapter 3). Aristotle also knew that the Sun has to be further away from the Earth than is the Moon, because occasionally the Moon passes exactly between the Earth and Sun and temporarily hides the Sun from view. We call this a *solar eclipse* (see Chapter 3).

Aristotle cited two convincing arguments that the Earth must be round. First is the fact that as the Moon enters or emerges from the Earth's shadow during an eclipse of the Moon, the shape of the shadow seen on the Moon is always round (Figure 1.7). Only a spherical object always produces a round shadow. If the Earth were a disk, for example, there would be some occasions when the sunlight would be striking it edge-on, and its shadow on the Moon would be a line.

As a second argument, Aristotle explained that travelers who go south a significant distance are able to observe stars that are not visible further north. And the height of the North Star—the star nearest the north celestial pole—decreases as a traveler moves south. On a flat Earth, everyone would see the same stars overhead. The only possible explanation is that the traveler must have moved over a curved surface on the Earth, showing stars from a different angle. (See the box at the end of the chapter for more ideas on proving the Earth is round.)

One brave school of Greek thought, whose best-known exponent was Aristarchus of Samos (310–230 B.C.), even suggested, long before Copernicus, that the Earth was moving around the Sun. But Aristotle and most of the ancient Greek scholars rejected this idea. One of the reasons for their conclusion was their understanding that if the Earth moved about the Sun, they would be observing the stars from different places along Earth's orbit. This would mean that the apparent directions of nearby stars in the sky would change during the year relative to more distant stars. (In a similar way, we see foreground objects appear to move against a more distant background whenever we are in motion. When we ride on a train, the trees in the foreground appear to shift their position relative to distant hills as the train rolls by. Unconsciously, we use this phenomenon all of the time to estimate distances around us.)

The apparent shift in the direction of an object as a result of the motion of the observer is called **parallax.** We call the shift in the apparent direction of a star due to the Earth's orbital motion *stellar parallax.* The Greeks made dedicated efforts to observe stellar parallax, even enlisting the aid of Greek soldiers with the clearest vision, but to no avail. The brighter (and presumably nearer) stars just did not seem to shift as they observed them in the spring and then again in the fall.

This either meant that the Earth was not moving, or that the stars had to be so tremendously far away that the parallax shift was immeasurably small. A cosmos of such enormous extent required a leap of imagination that most ancient thinkers were not prepared to make, and so they retreated to the safety of the Earth-centered view, which would dominate western thinking for nearly two millenia.

Measurement of the Earth by Eratosthenes

The Greeks not only knew the Earth was round, they were also able to measure its size. The first fairly accurate determination of the Earth's diameter was made about 200 B.C. by Eratosthenes, a Greek living in Alexandria, Egypt. His method was a geometrical one, based on observations of the Sun.

The Sun is so distant from the Earth compared with its size that the Sun's light rays intercepted by all parts of the Earth approach it along essentially parallel lines. To see why this is, look at Figure 1.8. Take a source of light near the Earth, say at position A. Its rays strike different parts of the Earth along diverging paths. From a light source at B, or at C, still farther away, the angle between rays that strike opposite parts of the Earth is smaller. The more distant the source, the smaller is the angle between the rays. For a source infinitely distant, the rays travel along parallel lines.

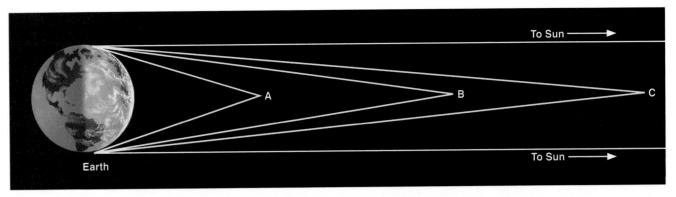

Figure 1.8
The more distant an object, the more nearly parallel are the rays of light coming from it.

The Sun is not, of course, infinitely far away, but light rays striking the Earth from a point on the Sun diverge from one another by an angle far too small to be observed with the unaided eye. As a consequence, if people all over the Earth who could see the Sun were to point at it, their fingers would all be essentially parallel to one another. (The same is also true for the planets and stars, an idea we will use in our discussion of how telescopes work.)

Eratosthenes noticed that on the first day of summer at Syene, Egypt (near modern Aswan), sunlight struck the bottom of a vertical well at noon. This indicated that Syene was on a direct line from the center of the Earth to the Sun. At the corresponding time and date in Alexandria, he observed that the Sun was not directly overhead but slightly south of the zenith, so that its rays made an angle with the vertical equal to about 1/50 of a circle (7°). Since the Sun's rays striking the two cities are parallel to one another, why would the two rays not make the same angle with the Earth's surface? Eratosthenes reasoned that the curvature of the Earth is responsible and that this could be used to measure how big the Earth is.

Alexandria, he realized, must be 1/50 of the Earth's circumference north of Syene (Figure 1.9). Alexandria had been measured to be 5000 stadia north of Syene (the *stadium* was a Greek unit of length, derived from the length of the racetrack in a stadium). Eratosthenes thus found that the Earth's circumference must be 50 × 5000, or 250,000 stadia.

It is not possible to evaluate precisely the accuracy of Eratosthenes' solution because there is doubt about which of the various kinds of Greek stadia he used as his unit of distance. If it was the common Olympic stadium, his result was about 20 percent too large. According to another interpretation, he used a stadium equal to about 1/6 km, in which case his figure was within one percent of the correct value of 40,000 km. Even if his measurement was not exact, his success at measuring the size of our planet by using only shadows, sunlight, and the power of human thought was one of the greatest intellectual achievements in history.

Hipparchus

Perhaps the greatest astronomer of antiquity was Hipparchus, born in Nicaea in what is present-day Turkey. He erected an observatory on the island of Rhodes in the period around 150 B.C., when the Roman Republic was increasing its influence throughout the Mediterranean region. There he measured as accurately as possible the directions of objects in the sky, compiling a pioneering star catalog with about 850 entries. He designated celestial coordinates for each star, specifying its position in the sky, just as we can specify the position of a point on the

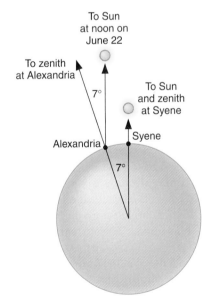

Figure 1.9
Eratosthenes' method of determining the size of the Earth. The Sun's rays come in parallel, but because the Earth's surface curves, a ray at Syene comes straight down, while a ray at Alexandria makes an angle of 7° with the vertical. That means, in effect, that at Alexandria the Earth's surface has curved away from Syene by 7° out of 360°, or 1/50 of a full circle. Thus the distance between the two cities must be 1/50 the circumference of the Earth.

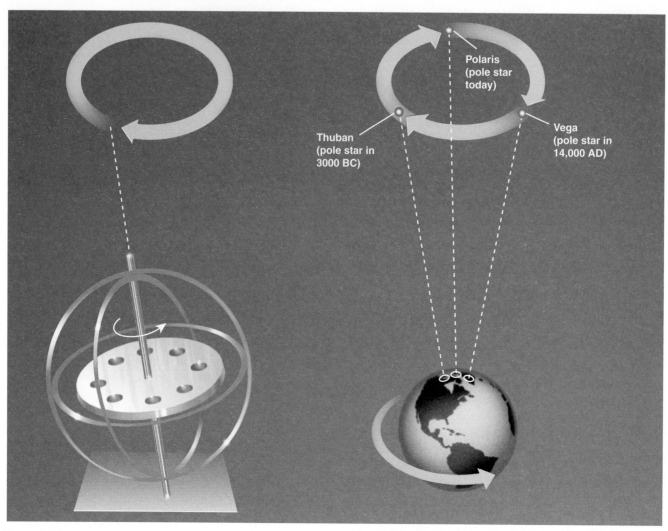

Figure 1.10

Just as the axis of a rapidly spinning top slowly wobbles in a circle, so the axis of the Earth wobbles in a 26,000-year cycle. Today the north celestial pole is near the star Polaris, but about 5000 years ago it was close to a star called Thuban, and in 14,000 A.D. it will be closest to the star Vega.

Earth by giving its latitude and longitude. He also divided the stars into six **magnitudes,** according to their apparent brightness. He called the brightest ones "stars of the first magnitude," the next brightest group "stars of the second magnitude," and so forth. This system, in modified form, still remains in use today (see Chapter 16).

By observing the stars and comparing his data with older observations, Hipparchus made one of his most remarkable discoveries: the position in the sky of the north celestial pole had altered over the previous century and a half. Hipparchus correctly deduced that this had happened not only during the period covered by his observations, but was in fact happening all the time: the direction around which the sky appears to rotate changes continuously.

You will recall that the north celestial pole is just the projection of the Earth's North Pole into the sky. If the

north celestial pole is wobbling around, then the Earth itself must be doing the wobbling. Today we understand that the direction in which the Earth's axis points does indeed change slowly but regularly, a motion we call **precession.** If you've ever watched a child's spinning top wobble, you have observed a similar kind of motion. The top's axis describes a path in the shape of a cone, as the Earth's gravity tries to topple it over (Figure 1.10).

Because our planet is not an exact sphere, but bulges a bit at the equator, the pull of the Sun and Moon cause it to wobble like a top. It takes about 26,000 years for the Earth's axis to complete one circle of precession. But as a result of this motion, where our axis points in the sky changes as time goes on. While Polaris is the star closest to the pole today (it will reach its very closest point around 2100 A.D.), the star Vega in the constellation of Lyra will be the north star in the year 14,000 A.D.

Ptolemy

The last great astronomer of antiquity was Claudius Ptolemy (or Ptolemaeus), who flourished in Roman Alexandria about 140 A.D. He compiled a 13-volume series on astronomy known today by its Arabic name, the *Almagest* (meaning "The Greatest"). The *Almagest* does not deal exclusively with Ptolemy's own work, for it includes a compilation of the astronomical achievements of the past, principally those of Hipparchus. In fact, it is our main source of information about Greek astronomy.

Ptolemy's most important contribution was a geometrical representation of the solar system that predicted the motions of the planets with considerable accuracy. Hipparchus, not having enough data on hand to solve the problem himself, instead amassed observational material for posterity to use. Ptolemy supplemented this material with new observations of his own and produced a cosmological model that endured more than a thousand years, until the time of Copernicus.

The complicating factor in explaining the motions of the planets is that their apparent wanderings in the sky result from the combination of their own motions with the Earth's orbital revolution. As we watch the planets move from our vantage point on Earth, it's a little like watching a car race from inside one of the cars. Sometimes other cars pull ahead, while at other times we might pass opponents' cars, making them seem to move backwards for a while with respect to our car.

Figure 1.11a shows the motion of the Earth and a planet farther from the Sun, such as Mars or Jupiter. The Earth travels around the Sun in the same direction as the other planet and in nearly the same plane, but its orbital speed is faster. As a result, it periodically overtakes the planet, like a faster race car on the inside track. The figure shows the directions where we see the planet in the sky at different times. In Figure 1.11b, we see the resulting apparent path of the planet among the stars.

Normally, planets move eastward in the sky over the weeks and months as they orbit the Sun. But from positions *B* to *D* in our figure, as the Earth passes the planet it appears to drift backward, moving west in the sky. Even though it is actually moving to the east, the faster Earth has overtaken it and seems, from our perspective, to be leaving it behind.

As the Earth rounds its orbit toward position *E*, the planet again takes up its apparent eastward motion in the sky. The temporary apparent westward motion of a planet as the Earth swings between it and the Sun is called **retrograde motion.** Such backward motion is much easier for us to understand today, now that we know the Earth is one of the moving planets and not the unmoving center of all creation. But Ptolemy was faced with the far more complex problem of explaining such motion while assuming a stationary Earth.

Furthermore, because the Greeks believed that celestial motions had to be circles, Ptolemy had to construct a model using circles alone. To do it, he needed dozens of

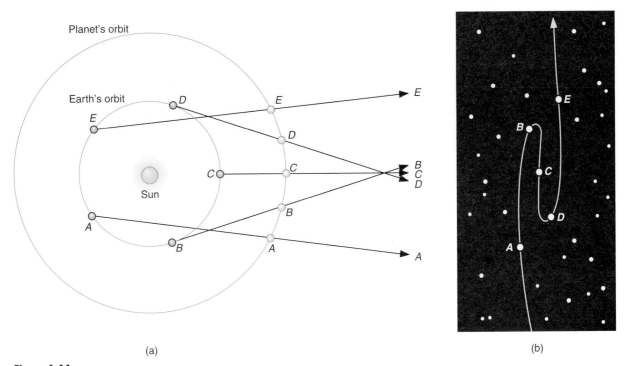

(a)   (b)

Figure 1.11
Retrograde motion of a planet beyond the Earth's orbit. (a) Actual positions of the planet and the Earth. (b) The apparent path of the planet as seen from the moving Earth, against the background of stars.

circles, some moving around other circles, in a complex edifice that makes a modern viewer dizzy. But we must not let our modern judgment cloud our admiration for Ptolemy's achievement. In his day, a complex universe centered on the Earth was perfectly reasonable and, in its own way, quite beautiful.

Ptolemy solved the problem of explaining the observed motions of planets by having each planet revolve in a small orbit called an **epicycle.** The center of the epicycle then revolved about the Earth on a circle called a deferent (Figure 1.12). When the planet is at position x in Figure 1.12 on the epicycle orbit, it is moving in the same direction as the center of the epicycle; from Earth the planet appears to be moving eastward. When the planet is at y, however, its motion is in the opposite direction to the motion of the epicycle's center around the Earth. By choosing the right combination of speeds and distances, Ptolemy succeeded in having the planet moving westward at y at the correct speed and for the correct interval of time—thus replicating retrograde motion with his model.

However, we shall see in the next chapter that the planets, like the Earth, travel about the Sun in orbits that are ellipses, not circles. Their actual behavior cannot be represented accurately by a scheme of uniform circular motions. In order to match the observed motions of the planets, Ptolemy had to introduce additional circles in his model, and center the deferent circles not on the Earth, but at a point some distance from the Earth called the equant point. All of this considerably complicated his scheme.

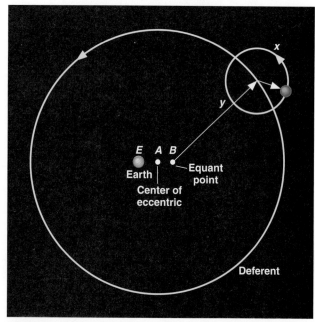

Figure 1.12
Ptolemy's cosmological system. Each planet orbits around a small circle called an epicycle. The system is not exactly centered on the Earth but on a point called the equant. The Greeks needed all this complexity to explain the motions in the sky, because they believed that the Earth was stationary.

It is a tribute to the genius of Ptolemy as a mathematician that he was able to develop such a complex system to account successfully for the observations of planets. It may be that Ptolemy did not intend his cosmological model to describe reality, but merely to serve as a mathematical representation that allowed him to predict the positions of the planets at any time. Whatever his thinking, his model, with some modifications, was accepted as absolute authority in the Christian and (later) Muslim world.

1.3

Astrology and Astronomy

Many ancient cultures regarded the planets and stars as representatives or symbols of the gods or other supernatural forces that controlled their lives. For them, the study of the heavens was not an abstract subject, but was directly connected to the life-and-death necessity of understanding the actions of the gods and currying favor with them. Before the time of our scientific perspective, everything that happened in nature, from the weather, to diseases and accidents, to celestial surprises like eclipses or new comets, was thought to be an expression of the whims or displeasures of the gods. Any signs that helped people understand what these gods had in mind were extremely important to them.

The movements of the seven "wanderer" objects through the realm of the sky—the Sun, Moon, and five planets visible to the unaided eye—clearly must have special significance in such a system of thinking. Most ancient cultures associated these seven objects with various supernatural rulers in their pantheon, and kept track of them for religious reasons. Even in the comparatively sophisticated Greece of antiquity, the planets had the names of gods and were credited with having the same powers and influences as the gods whose names they bore. From such ideas was born the ancient system called **astrology,** still practiced today, in which the positions of these bodies among the stars of the zodiac are thought to hold the key to understanding what we can expect from life.

The Beginnings of Astrology

Astrology began in Babylonia about three millennia ago. The Babylonians, believing that the planets and their motions influenced the fortunes of kings and nations, used their knowledge of astronomy to guide their rulers. When the Babylonian culture was absorbed by the Greeks, astrology gradually came to influence the entire Western world and eventually spread to the Orient as well.

By the 2nd century B.C., the Greeks democratized astrology by developing the idea that the planets influence every individual. In particular, they believed that the configuration of the Sun, Moon, and planets at the moment of birth affected a person's personality and fortune, the doctrine called *natal astrology*. Natal astrology reached its zenith with Ptolemy 400 years later. As famous for his astrology as for his astronomy, Ptolemy compiled the *Tetrabiblos*, a four-volume treatise on astrology that remains the "bible" of the subject. It is essentially this ancient religion, older than Christianity or Islam, that is still practiced by today's astrologers.

The Horoscope

The key to natal astrology is the **horoscope,** a chart showing the positions of the planets in the sky at the moment of an individual's birth. In charting a horoscope, the planets (including the Sun and Moon, classed as planets by the ancients) must first be located in the zodiac. For the purposes of astrology, the zodiac is divided into 12 sectors called signs, each 30° long. Each sign is named after a constellation in the sky through which the Sun, Moon, and planets are seen to pass—the sign of Virgo after the constellation of Virgo, for example.

When someone today casually asks you your "sign," what they are asking for is your "sun-sign"—which zodiac sign the Sun was in at the moment you were born. But more than 2000 years have passed since the signs received their names from the constellations. Because of precession, the constellations of the zodiac slide westward along the ecliptic, going once around the sky in about 26,000 years. Thus today the real stars have wobbled over by about 1/12 of the zodiac—about the width of one sign. In most forms of astrology, however, the signs have remained assigned to the dates they had when astrology was first set up. This means that the astrological signs and the real constellations are out of step; hence the sign of Aries, for example, now occupies the constellation of Pisces. When you look up your sun-sign in a newspaper astrology column, because of precession the name of the sign associated with your birthday is no longer the name of the constellation in which the Sun was actually located when you were born. To know that constellation, you must look for the sign *before* the one that includes your birthday.

A complete horoscope shows the location of each planet in the sky by indicating its position in the appropriate sign of the zodiac. However, as the celestial sphere turns (owing to the rotation of the Earth), the entire zodiac moves across the sky to the west, completing a circuit of the heavens each day. Thus the position in the sky (or "house" in astrology) must also be calculated. There are more-or-less standardized rules for the interpretation of the horoscope, most of which (at least in Western schools of astrology) are derived from the *Tetrabiblos* of Ptolemy. Each sign, each house, and each planet, the last supposedly acting as a center of force, is associated with particular matters.

The detailed interpretation of a horoscope is therefore a very complicated business, and there are many schools of astrological thought on how it should be done.

Although some of the rules may be standardized, how each rule is to be weighed and applied is a matter of judgment—and "art." It also means that it is very difficult to tie astrology down to specific predictions or to get the same predictions from different astrologers.

Astrology Today

Astrologers today use the same basic principles laid down by Ptolemy nearly 2000 years ago. They cast horoscopes (a process much simplified by the development of appropriate computer programs) and suggest interpretations. Sun-sign astrology (which you read in the newspapers and many magazines) is a recent simplified variant of natal astrology. Although even professional astrologers do not place much trust in such a limited scheme, which tries to fit everyone into just 12 groups, sun-sign astrology is taken seriously by many people (perhaps because it is so commonly discussed in the media). In a recent poll of teenagers in the United States, more than half said they "believed in astrology."

Today, with our knowledge of the nature of the planets as physical bodies, as well as our understanding of human genetics, it is hard to imagine that the positions of these planets in the sky at the moment of our birth could have anything to do with our personality or future. There are no known forces, not gravity or anything else, that could cause such effects. (For example, a simple calculation shows that the pull of the obstetrician delivering a newborn baby is greater than that of Mars.) Astrologers thus have to argue that there are unknown forces exerted by the planets that depend on their configurations with respect to one another and that do not vary according to the distance of the planet—forces for which there is not a whit of solid evidence.

Another curious aspect of astrology is its emphasis on planet configurations at *birth*. What about the forces that might influence us at conception? Isn't our genetic makeup more important for determining our personality than the circumstances of our birth? Would we really be a different person if we had been born a few hours earlier or later, as astrology claims? (Back when astrology was first conceived, birth was thought of as a moment of magic significance, but today we understand a lot more about the long process that precedes it.)

Actually, very few thinking people today buy the claim that our entire lives are predetermined by astrological influences at birth. But many people apparently believe that astrology has validity as an indicator of affinities and personality. A surprising number of Americans make judgments about people—whom they will hire, associate with, and even marry—on the basis of astrological information. To be sure, these are difficult decisions, and you might argue that we should use any relevant information that might help us to make the right choices. But does astrology actually provide any useful information on human personality? This is the kind of question that can be tested

Testing Astrology

In response to modern public interest in astrology, scientists have carried out a wide range of statistical tests to assess its predictive power. The simplest of these examine sun-sign astrology to determine whether—as astrologers assert—some signs are more likely than others to be associated with such objective measures of success as winning Olympic medals, earning high corporate salaries, or achieving elective office or high military rank. (You can make such a test yourself by looking up the birthdates of all members of Congress, for example, or all members of the U.S. Olympic Team.) Are our political leaders somehow selected at birth by their horoscopes and thus more likely to be Leos, say, than Scorpios?

You don't even need to be specific about your prediction in such tests. After all, many schools of astrology disagree about which signs go with which personality characteristics. To demonstrate the validity of the astrological hypothesis, it would be sufficient if the birthdays of all our leaders clustered in one or two signs in some statistically significant way. Dozens of such tests have been performed, and all have come up completely negative: the birthdates of leaders in all fields tested have been found to be randomly distributed among the signs. Sun-sign astrology does not predict anything about a person's future occupation or strong personality traits.

In a fine example of such a test, two statisticians examined the re-enlistment records of the Marine Corps. We suspect our readers will agree that it takes a certain kind of personality to not only enlist, but *re-enlist* in the Marines. If sun-signs can predict strong personality traits, as astrologers claim, then the re-enlistees (with similar personalities) should have been distributed preferentially in those one or few signs that matched the personality of someone who loves being a Marine. In fact, the re-enlistees were distributed completely randomly among all the signs.

More sophisticated studies have also been done, involving full horoscopes calculated for thousands of individuals. The results of all of these studies are also negative: none of the systems of astrology has been shown to be at all effective in connecting astrological aspects to personality, success, or finding the right person to love.

Other tests show that it hardly seems to matter what a horoscope interpretation says, as long as it is vague enough, and as long as each subject feels it was personally prepared just for him or her. The French statistician Michel Gaugelin, for example, sent the horoscope interpretation for one of the worst mass-murderers in history to 150 people, but told each recipient that it was a "reading" prepared exclusively for him or her. Ninety-four percent of the readers said they recognized themselves in the interpretation of the mass-murderer's horoscope.

Geoffrey Dean, an Australian researcher, reversed the astrological readings of 22 subjects, substituting phrases that were the opposite of what the horoscope actually said. Yet his subjects said that the resulting readings applied to them just as often (95%) as the people to whom the original phrases were given. Apparently, those who seek out astrologers just want guidance, and almost any guidance will do.

using the scientific method (see the "Making Connections" box above).

The results of hundreds of tests are all the same: there is no evidence that natal astrology has any predictive power, even in a statistical sense. Why then do people often seem to have anecdotes about how well their own astrologer advised them? Effective astrologers today use the language of the zodiac and the horoscope only as the outward trappings of their craft. Mostly they work as amateur therapists, offering simple truths that clients like or need to hear. (Recent studies have shown that just about any sort of short-term therapy makes people feel a little better. This is because the very act of talking about our problems is in itself beneficial.)

Astrology itself has no basis in scientific fact, however. It is an interesting historical system, left over from pre-scientific days and best remembered for the impetus it gave people to learn the cycles and patterns of the sky. From it grew the science of astronomy, which is our main subject for discussion.

1.4

The Birth of Modern Astronomy

Astronomy made no major advances in medieval Europe, where the prevailing philosophy was acceptance of dogmatic authority. Astrology was widely practiced, however, and an interest in the motions of the planets was thus kept alive. The birth and expansion of Islam after the 7th century A.D. led to a flowering of Arabic and Jewish cultures that preserved, translated, and added to many of the astronomical ideas of the Greeks. (Many of the names of the brightest stars, for example, are today taken from the Arabic, as are such astronomical terms as zenith.)

As European culture began to emerge from its long Dark Age, trading with Arab countries led to a rediscovery of ancient texts such as the *Almagest,* and to a reawakening of interest in astronomical questions. This time of rebirth (in French, *Renaissance*) in astronomy was clearly embodied in the work of Copernicus.

Copernicus

One of the most important events of the Renaissance was the displacement of Earth from the center of the universe, an intellectual revolution initiated by a Polish cleric in the 16th century. Nicolaus Copernicus (1473–1543) was born in the ancient city of Torun. His training was in law and medicine, but his main interests were astronomy and mathematics. His great contribution to science was a critical reappraisal of the existing theories of planetary motion and the development of a new Sun-centered, or **heliocentric,** model of the solar system. Copernicus concluded that the Earth is a planet and that all the planets circle the Sun. Only the Moon orbits about the Earth (Figure 1.14).

He described his ideas in detail in his book *De Revolutionibus Orbium Coelestium (On the Revolution of Celestial Orbs)*, published in 1543, the year of his death. By this time, the old Ptolemaic system (like a cranky old machine) needed significant adjustments to predict the positions of the planets correctly. Copernicus wanted to develop an improved theory from which to calculate planetary positions, but in doing so, he was himself not free of all traditional prejudices.

He began with several assumptions that were common in his time, such as the idea that the motions of the heavenly bodies must be made up of combinations of uniform circular motions. But he did not assume (as most people did) that the Earth had to be in the center of the universe, and he presented a defense of the heliocentric system that was elegant and persuasive. His ideas, although not widely accepted until more than a century after his death, were much discussed among scholars and ultimately had a profound influence on the course of world history.

One of the objections raised to the heliocentric theory was that if the Earth were moving, we would all sense or feel this motion. Solid objects would be ripped from the surface, a ball dropped from a great height would not strike the ground directly below, and so forth. But a moving person is not necessarily aware of that motion. We have all experienced seeing an adjacent train, car, or ship appear to move, only to discover that it is we who are moving.

Copernicus argued that the apparent annual motion of the Sun about the Earth could be represented equally well by a motion of the Earth about the Sun. He also reasoned that the apparent rotation of the celestial sphere

Figure 1.13
Nicolaus Copernicus (1473–1543), cleric and scientist, played a leading role in the emergence of modern science. While he could not prove that the Earth revolves about the Sun, he presented such compelling arguments for this idea that he turned the tide of cosmological thought and laid the foundations upon which Galileo and Kepler so effectively built in the following century. (Yerkes Observatory)

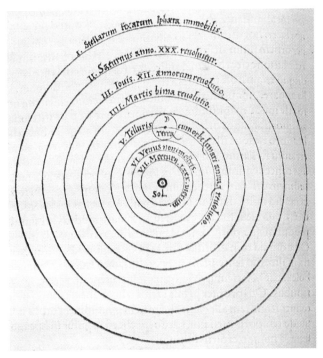

Figure 1.14
Heliocentric plan of the solar system in the first edition of Copernicus' *De Revolutionibus.* Notice the word "Sol" for Sun in the middle. (Crawford Collection, Royal Observatory, Edinburgh)

Figure 1.15
The phases of Venus according to the heliocentric theory. As Venus moves around the Sun we see changing illumination of its surface, just as we see the face of the Moon illuminated differently in the course of a month.

could be explained by assuming that the Earth rotates while the celestial sphere is stationary. To the objection that if the Earth rotated about an axis it would fly into pieces, Copernicus answered that if such motion would tear the Earth apart, the still faster motion of the much larger celestial sphere required by the geocentric hypothesis would be even more devastating.

The Heliocentric Model

The most important idea in Copernicus' *De Revolutionibus* is that the Earth is one of six (then known) planets that revolve about the Sun. Using this concept, he was able to work out the correct general picture of the solar system. He placed the planets, starting nearest the Sun, in the correct order: Mercury, Venus, Earth, Mars, Jupiter, and Saturn. Further, he deduced that the nearer a planet is to the Sun, the greater its orbital speed. With his theory, he was able to explain the complex retrograde motions of the planets without the necessity for epicycles, and to work out a roughly correct scale for the solar system.

Copernicus could not *prove* that the Earth revolves about the Sun. In fact, with some adjustments the old Ptolemaic system could have accounted as well for the motions of the planets in the sky. But Copernicus pointed out that the Ptolemaic cosmology was clumsy and lacking in the beauty and symmetry of its successor.

In Copernicus' time, in fact, few people thought there were ways to prove whether the heliocentric or the older geocentric system was correct. A long philosophical tradition, going back to the Greeks and defended by the Catholic Church, held that pure human thought combined with divine revelation represented the path to truth. Nature, as revealed by our senses, was suspect. For example, Aristotle had reasoned that heavier objects (having more of the quality that made them heavy) must fall to Earth faster than lighter ones. This is absolutely incorrect, as any simple experiment dropping two balls of different weights will show. But in Copernicus' day, experiments did not carry much weight (if you will pardon the expression); Aristotle's brilliant reasoning was more convincing.

In this environment, there was little motivation to carry out observations or experiments to distinguish between competing cosmological theories (or anything else). It should not surprise us, therefore, that the heliocentric idea was debated for more than half a century without any tests being applied to determine its validity. (In fact, in the American colonies, the older geocentric system was still taught at Harvard University in the first years after it was founded in 1636.)

Contrast this with the situation today, when scientists rush to test each new theory and do not accept any ideas until the results are in. For example, when two researchers at the University of Utah announced in 1989 that they had discovered a way to achieve nuclear fusion (the process that powers the stars) at room temperature, other scientists at more than 25 laboratories around the United States attempted to duplicate this feat within a few weeks—without success, as it turned out.

How would we look at Copernicus' model today? When a new hypothesis or theory is proposed in science, it must first be checked for consistency with what is already known. Copernicus' heliocentric idea passes this test, for it allows planetary positions to be calculated at least as well as does the geocentric theory. The next step is to see what predictions the new theory makes that differ from those of competing ideas. In the case of Copernicus, one example is the prediction that, if Venus circles the Sun, it should go through the full range of phases just as the Moon does, whereas if it circles the Earth, it should not (Figure 1.15). But in those days, before the telescope, no one imagined testing this prediction.

Galileo and the Beginning of Modern Science

Many of the modern scientific concepts of observation, experimentation, and the testing of hypotheses through careful quantitative measurements were pioneered by a man who lived nearly a century after Copernicus. Galileo Galilei (1564–1642), a contemporary of Shakespeare, was

Figure 1.17
The leaning bell tower of the Cathedral at Pisa. While living in Pisa, Galileo carried out experiments to show that objects of different mass fall at the same rate. According to legend, one of his experiments consisted of dropping cannonballs of different weights from this tower. (David Morrison)

Figure 1.16
Galileo Galilei (1564–1642) advocated that we perform experiments or make observations to ask nature its ways. When Galileo turned the telescope to the sky, he found that things were not the way philosophers had supposed. (Yerkes Observatory)

born in Pisa (Figure 1.17). Like Copernicus, he began training for a medical career, but he had little interest in the subject and later switched to mathematics. He held faculty positions at the Universities of Pisa and Padua, and eventually he became mathematician to the Grand Duke of Tuscany in Florence.

Galileo's greatest contributions were in the field of mechanics, the study of motion and the actions of forces on bodies. It was familiar to all persons then, as it is to us now, that if a body is at rest, it tends to remain at rest and requires some outside influence to start it in motion. Rest was thus generally regarded as the natural state of matter. Galileo showed, however, that rest is no more natural than motion.

If an object is slid along a rough horizontal floor, it soon comes to rest because friction between it and the floor acts as a retarding force. However, if the floor and object are both highly polished, the body, given the same initial speed, will slide farther before coming to rest. On a smooth layer of ice, it will slide farther still. Galileo reasoned that if all resisting effects could be removed, the object would continue in a steady state of motion indefinitely. In fact, he argued, a force is required not only to start an object moving from rest but also to slow down, stop, speed up, or change the direction of a moving object. You will appreciate this if you have ever tried to stop a rolling car by leaning against it, or a moving boat by tugging on a line.

Galileo also studied the way bodies **accelerate,** or change their speed, as they fall freely or roll down a ramp. He found that such bodies accelerate uniformly; that is, in equal intervals of time they gain equal increments in speed. Galileo formulated these newly found laws in precise mathematical terms that enabled future experimenters to predict how far and how fast bodies would move in various lengths of time.

Sometime in the 1590s Galileo adopted the Copernican hypothesis of the solar system. In Roman Catholic Italy, this was not a popular philosophy, for the Church authorities still upheld the ideas of Aristotle and Ptolemy, and they had powerful political and economic reasons for insisting that the Earth was the center of creation. It was primarily because of Galileo that in 1616 the Church issued a prohibition decree stating that the Copernican doctrine was "false and absurd" and not to be held or defended.

Galileo's Astronomical Observations

It is not certain who first conceived of the idea of combining two or more pieces of glass to produce an instrument that enlarged images of distant objects, making them appear nearer. The first such *telescopes* that attracted much notice were made by the Dutch spectacle maker Hans Lippershey in 1608. Galileo heard of the discovery and, without ever having seen an assembled telescope, constructed one of his own with a three-power magnification, which made distant objects appear three times nearer and larger (Figure 1.18).

On August 25, 1609, Galileo demonstrated a telescope with a magnification of 9 to government officials of the city-state of Venice. By a magnification of 9, we mean that the linear dimensions of the object being viewed appeared nine times larger or, alternatively, that the objects appeared nine times closer than they really were. There were were obvious military advantages associated with a device for seeing distant objects. For his invention, Galileo's salary was nearly doubled, and he was granted lifetime tenure as a professor. (His university colleagues were outraged, particularly since the invention was not even original.)

Others had used the telescope before Galileo to observe things on Earth. But, in a flash of insight that changed the history of astronomy, Galileo realized that he could turn the power of the telescope toward the heavens. Before using his telescope for astronomical observations, Galileo had to devise a stable mount, and he improved the optics to provide a magnification of 30. Galileo also needed to acquire confidence in the telescope.

At that time, the eyes were believed to be the final arbiter of truth about sizes, shapes, and colors. Lenses, mirrors, and prisms were known to distort distant images by enlarging, reducing, or even inverting them. Galileo undertook repeated experiments to convince himself that what he saw through the telescope was identical to what he saw up close. Only then could he begin to believe that the miraculous phenomena revealed in the heavens were real.

Beginning his astronomical work late in 1609, Galileo found that many stars too faint to be seen with the naked eye became visible with his telescope. In particular, he found that some nebulous blurs resolved into many stars, and that the Milky Way—the strip of whiteness across the night sky—was also made up of a multitude of individual stars.

Examining the planets, he found four satellites, or moons, revolving about Jupiter, with periods ranging from

Figure 1.18
Some telescopes used by Galileo. The longer one has a wooden tube covered with paper and a lens 26 mm across. (Instituto e Museo di Storia della Scienza di Florenza)

just under 2 days to about 17 days. This discovery was particularly important because it showed that not everything has to revolve around the Earth, and that there could be centers of motion that are themselves in motion. Defenders of the geocentric view had argued that if the Earth were in motion the Moon would be left behind, because it could hardly keep up with a rapidly moving planet. Yet here were Jupiter's satellites doing exactly that! (To recognize this discovery and honor his work, NASA has named the spacecraft now exploring the Jupiter system *Galileo*.)

With his telescope, Galileo was also able to carry out the test of the Copernican theory mentioned above, based on the phases of Venus. Within a few months he had found that Venus goes through phases like the Moon, showing that it must revolve about the Sun, so that we see different parts of its daylight side at different times (see Figure 1.15). These observations could not be reconciled with any model in which Venus circled about the Earth.

Galileo observed the Moon and saw craters, mountain ranges, valleys, and flat dark areas that he thought might be water. These discoveries showed that the Moon might be not so dissimilar to the Earth—suggesting that the Earth, too, could belong to the realm of celestial bodies.

After Galileo's work, it became increasingly difficult to deny the Copernican view, and slowly the Earth was dethroned from its central position in the universe and given its rightful place as one of the planets attending the Sun. Galileo himself had to appear before the Inquisition to answer charges that his work was heretical, and he was condemned to house arrest. His books were on the Church's forbidden list until 1836, although in countries where the Roman Catholic Church held less sway, they were widely read and discussed. Not until 1992 did the Catholic Church publicly admit that it had erred in the matter of censoring Galileo's ideas.

The new ideas of Copernicus and Galileo began a revolution in our conception of the cosmos. It eventually became evident that the universe is a vast place, and that the Earth's role in it is relatively unimportant. The idea that the Earth moves around the Sun like the other planets raised the possibility that they might be worlds themselves, perhaps even supporting life. And as the Earth was demoted from its position at the center of the universe, so too was humanity. The universe, despite what we may wish, does not revolve around us, and we must find a more modest place for ourselves in the scheme of the cosmos.

Most of us take these things for granted today, but four centuries ago such concepts were frightening and heretical for some, immensely stimulating for others. The pioneers of the Renaissance started the European world along the path toward science and technology that we still tread today. For them, nature was rational and ultimately knowable, and experiments and observations provided the means to reveal its secrets.

How Do We Know the Earth Is Round?

SEEING FOR YOURSELF

In addition to the two ways (from Aristotle's writings) discussed in this chapter, you might also reason as follows:

1. Let's watch a ship leave its port and sail into the distance on a clear day. On a flat Earth, you would just see the ship get smaller and smaller as it sailed away. But that's not what we actually observe. Instead, ships sink below the horizon, with the hull disappearing first and the mast remaining visible for a while longer. Eventually only the top of the mast can be seen, as the ship sails around the curvature of the Earth; then finally the ship disappears.

2. The Space Shuttle circles the Earth once every 90 minutes or so. Photographs taken from the Shuttle and other satellites show that the Earth is round from every perspective.

3. Suppose you made a friend in each time zone of the Earth. Then you could call all of them in the same hour and ask, "Where is the Sun?" On a flat Earth, each caller would give you roughly the same answer. But on a round Earth you would find that for some, the Sun would be high in the sky, while for others it would be rising, or setting, or completely out of sight (and this last group of friends would be upset with you for waking them up).

Observing the Planets

At almost any time of the night, and at any season, you can spot one or more bright planets in the sky. All five of the planets known to the ancients—Mercury, Venus, Mars, Jupiter, and Saturn—are more prominent than any but the brightest stars, and they can be seen even from urban locations if you know where and when to look. One way to tell planets from bright stars is that planets twinkle less.

Venus, which stays close to the Sun from our perspective, appears either as an "evening star" in the west after sunset, or as a "morning star" in the east before sunrise. It is the brightest object in the sky after the Sun and Moon. It far outshines any real star, and under the most favorable circumstances it can even cast a visible shadow. Some young military recruits have tried to shoot Venus down as an approaching enemy craft or UFO!

Mars, with its distinctive red color, can be nearly as bright as Venus when close to the Earth, but normally it remains much less conspicuous. Jupiter is most often the second-brightest planet, approximately equaling in brilliance the brightest stars. Saturn is dimmer, and it varies considerably in brightness, depending on whether its rings are seen nearly edge-on (faint) or more widely opened (bright).

Mercury is quite bright, but few people ever notice it because it never moves very far from the Sun (it's never more than 28° away in the sky) and is always seen against bright twilight skies. There is a story (probably apocryphal) that even the great Copernicus never saw the planet Mercury.

True to their name, the planets "wander" against the background of the "fixed" stars. Although their apparent motions are complex, they reflect an underlying order, upon which the heliocentric model of the solar system, as described in this chapter, was based. The positions of the planets are often listed in newspapers (sometimes on the weather page), and clear maps and guides to their locations can be found each month in such magazines as *Sky & Telescope* and *Astronomy* (available at most libraries). There are also a number of computer programs (including shareware and commercial programs) that allow you to calculate and display where the planets are on any night.

Activities:

1. If either Venus or Mercury is visible, plot its position with respect to the background stars and note the date. Use a map of the relevant portion of the sky, or sketch the location of the brighter stars. Wait a few days and repeat the observation. Estimate how many degrees the planet moves per day. (The width of your thumb, held straight out at arm's length, covers about 1° of sky; your fist at arm's length covers about 10°.) If you make enough observations, you can estimate each planet's maximum distance in degrees from the Sun.

2. Use binoculars to see if you can determine the shape of Venus. This experiment will be easier when Venus is on the same side of the Sun as the Earth.

3. Determine which of the outer planets—Mars, Jupiter, and Saturn—are visible. For each visible planet, draw its position with respect to the background stars and note the date. Repeat this experiment every few days for two months or as long as each planet is visible. Is the planet moving east or west? Did it appear to change direction? Did you observe retrograde motion for any of the planets?

Summary

1.1 The direct evidence of our senses supports a **geocentric** perspective, with the **celestial sphere** pivoting on the **celestial poles** and rotating about a stationary Earth. We see only half of this sphere at one time, limited by the **horizon;** the point directly overhead is our **zenith.** The Sun's annual path on the celestial sphere is the **ecliptic,** a line that runs through the center of the **zodiac,** the 18°-wide strip of sky within which we always find the Moon and planets. The celestial sphere is organized into 88 **constellations,** or sectors.

1.2 Ancient Greeks such as Aristotle recognized that the Earth and Moon are spheres, and understood the phases of the Moon, but because of their inability to detect stellar **parallax,** they rejected the idea that the Earth moves.

Eratosthenes measured the size of the Earth with surprising precision. Hipparchus carried out many astronomical observations, making a star catalog, defining the system of stellar **magnitudes,** and discovering **precession** from the apparent shift in the position of the north celestial pole. Ptolemy of Alexandria summarized classical astronomy in his *Almagest;* he explained planetary motions, including **retrograde motion,** with high accuracy using a model centered on the Earth. This geocentric model, based on combinations of uniform circular motion using the **epicycles,** was accepted as authority for more than a thousand years.

1.3 The ancient religion of **astrology,** whose main contribution was a heightened interest in the heavens, began in Mesopotamia. It reached its peak in the Greco-Roman world, especially as recorded in the *Tetrabiblos* of Ptolemy. Natal astrology is based on the assumption that the positions of the planets at the time of our birth, as described by a **horoscope,** determines our future. However, modern tests clearly show that there is no evidence for this, even in a broad statistical sense.

1.4 Nicolaus Copernicus introduced the **heliocentric cosmology** to Renaissance Europe in his book *De Revolutionibus*. Although he retained the Aristotelian idea of uniform circular motion, Copernicus suggested that the Earth is a planet and that the planets all circle about the Sun, dethroning the Earth from its position at the center of the universe. Galileo Galilei was the father of both modern experimental physics and telescopic astronomy. He studied the **acceleration** of moving objects, and in 1610 began telescopic observations, discovering the nature of the Milky Way, the large-scale features of the Moon, the phases of Venus, and four satellites of Jupiter. Although accused of heresy for his support of the heliocentric cosmology, Galileo's observations and brilliant writings convinced most of his scientific contemporaries of the reality of the Copernican theory.

Review Questions

1. From where on Earth could you observe all of the stars during the course of a year? What fraction of the sky can be seen from the North Pole?

2. Describe a practical way to determine in which constellation the Sun is found at any time of the year.

3. What is a constellation as astronomers define it today? What does it mean when an astronomer says, "I saw a comet in Orion last night" ?

4. Give four ways to demonstrate that the Earth is round.

5. Explain why we see retrograde motion of the planets, according to both geocentric and heliocentric cosmologies.

6. Draw a picture that explains why Venus goes through phases the way the Moon does, according to the heliocentric cosmology. Does Jupiter also go through phases as seen from the Earth? Why?

7. In what ways did the work of Copernicus and Galileo differ from the traditional views of the ancient Greeks and of the Catholic Church?

Thought Questions

8. Show with a simple diagram how the lower parts of a ship disappear first as it sails away from you on a spherical Earth. Use the same diagram to show why lookouts on old sailing ships could see farther from the masthead than from the deck. Would there be any advantage to posting lookouts on the mast if the Earth were flat? (Note that these nautical arguments for a spherical Earth were quite familiar to Columbus and other mariners of his time.)

9. Parallaxes of stars were not observed by ancient astronomers. How can this fact be reconciled with the heliocentric hypothesis?

10. From our modern perspective, what are the strongest arguments you could use to convince a skeptical neighbor that the Earth is not stationary but goes around the Sun?

11. Design an experiment to test whether or not the planets and their motions influence human behavior.

12. Why do you think so many people believe in astrology? What psychological needs does such a belief system satisfy?

13. Consider three cosmological perspectives: (1) the geocentric perspective; (2) the heliocentric perspective; and (3) the modern perspective in which the Sun is a minor star on the outskirts of one galaxy among billions. Discuss some of the cultural and philosophical implications of each point of view.

Problems

14. The Moon moves relative to the background stars. Go outside at night and note the position of the Moon relative to nearby stars. Repeat the observation a few hours later. How far has the Moon moved? (For reference, the diameter of the Moon is about 1/2°.) Based on your estimate of its motion, how long will it take for the Moon to return to the position relative to the stars in which you first observed it?

15. The north celestial pole appears at an altitude above the horizon that is equal to the observer's latitude. Identify Polaris, the North Star, which lies very close to the north celestial pole. Measure its altitude. (This can be done with a protractor. Alternatively, your fist, extended at arm's length, spans a distance approximately equal to 10°.) Compare this estimate with your latitude. (Note that this experiment cannot be easily performed in the Southern Hemisphere because Polaris itself, of course, is not visible, and no bright star is located near the south celestial pole.)

16. Suppose Eratosthenes had found that at Alexandria at noon on the first day of summer, the line to the Sun makes an angle of 30° with the vertical. What then would he have found for the Earth's circumference?

17. Suppose Eratosthenes' results for the Earth's circumference were quite accurate. If the diameter of the Earth is 12,740 km, evaluate the length of his stadium in kilometers.

18. Suppose you are on a strange planet and observe, at night, that the stars do not rise and set but circle parallel to the horizon. Now you walk in a constant direction for 8000 miles, and at your new location on the planet you find that all stars rise straight up in the east and set straight down in the west, perpendicular to the horizon.

    a. How could you determine the circumference of the planet without any further observations?
    b. What evidence is there that the Greeks could have done what you suggest?
    c. What is the circumference, in miles, of that planet?

Suggestions for Further Reading

Culver, B. and Ianna, P. *Astrology: True or False.* 1988, Prometheus Books. The best skeptical book about astrology.

Ferris, T. *Coming of Age in the Milky Way.* 1988, Morrow. A general history of our understanding of the organization of the universe.

Fraknoi, A. "Your Astrology Defense Kit" in *Sky & Telescope,* Aug. 1989, p. 146. A review of the tenets and tests of astrology.

Gingerich, O. "How Galileo Changed the Rules of Science" in *Sky & Telescope,* Mar. 1993, p. 32.

Gingerich, O. "From Aristarchus to Copernicus" in *Sky & Telescope,* Nov. 1983, p. 410.

Krupp, E. *Beyond the Blue Horizon: Myths and Legends of the Sun, Moon, Stars, and Planets.* 1991, HarperCollins.

Superb introductory book on the sky stories of ancient cultures.

Krupp, E. *Echoes of the Ancient Skies.* 1983, Harper & Row. An excellent beginners' guide to ancient monuments and astronomical systems.

Kuhn T. *The Copernican Revolution.* 1957, Harvard U. Press. Classic study of the changes wrought by Copernicus' work.

Ravetz, J. "The Origin of the Copernican Revolution" in *Scientific American,* Oct. 1966.

Reston, J. *Galileo: A Life.* 1994, HarperCollins. A well-written, popular-level biography by a journalist.

Sagan, C. *Cosmos.* 1980, Random House. The chapter entitled "Backbone of Night" focuses on the ancient Greek astronomers.

Using **REDSHIFT**™

1. To see how many stars are visible in the night sky, choose the *Objects Filter* item in the *Display* menu and click on *Stars.* Set the bottom slider to magnitude 6 (magnitude is a measure of a star's brightness), which corresponds to what you might see at a dark, country observing site.

Next, set the faint *Magnitude* to 3.5, which simulates what it is like observing from the light-polluted skies of urban areas. Check which *Sky Map* best matches your view of the sky.

2. Choose the *Markers* item in the *Display* menu. Find the zenith, the celestial equator, the ecliptic, and the celestial poles.

What point does the vernal equinox mark?

3. Find Polaris in Ursa Minor using the *Find Objects* command in the *Controls* menu. Use Stars for the Object Type and Proper Names for Catalogue.

Where is the north celestial pole with respect to Polaris?

How high is Polaris above the horizon at your latitude? (Use the *Status Panel* to display the altitude above the horizon of an object marked by the cursor.) Change the latitude and watch how the altitude of Polaris changes.

4. To follow the precession of Earth's rotational pole, turn *Markers* on and find the north celestial pole. Choose *Constellations* under the *Display* menu. Turn the *Constellation Figures* on and *Boundaries* off and set the faintest star magnitude to 4. Step ahead and back in 1000-year incre-

ments to watch how the stars' positions change with time. Toward the end of *RedShift's* time limit of 10,000 A.D. you can watch Vega in Lyra approach the north celestial pole.

Has any bright star besides Polaris marked the pole in the last 5000 years?

5. To watch the retrograde motion of a planet, select *Equatorial View* under *Telescope Views* and turn off the Sun, moons, and all planets except for Jupiter in the *Planets* filter under the *Objects* filter. Set the *Date* to October 1, 1993. Center on Jupiter, set the *Time Step* to 10 days, select *Display Paths,* and step ahead a year.

Do the inner planets, Mercury, and Venus show retrograde motion?

6. Follow the phases of Venus by centering and locking on the planet, setting its magnification to 1000, and turning the other planets off (leave the Sun on). Set the *Time Step* to 15 days and watch the planet for a year.

At what phase does Venus appear closest to Earth? Can you account for Venus' phases with a geocentric cosmology?

In February 1984, astronaut Bruce McCandless, having let go of the tether to the Space Shuttle, became the first human satellite of planet Earth. Here he is shown riding the Manned Maneuvering Unit, whose nitrogen gas jets enabled him to move around in orbit and ultimately return to the Shuttle. As he circled our planet, he was subject to the same basic laws of motion that govern all other satellites. (NASA)

CHAPTER 2

Orbits and Gravity

Thinking Ahead

To reach a scientific understanding of the natural world, scientists had to figure out the basic rules or laws of nature. Thinking back over the different laws you may have studied in your earlier classes, which do you think are the most fundamental for describing how the world works?

If we could look down on the solar system from somewhere out in space, far above the plane of the planets' orbits, interpreting planetary motions would be much simpler. But the fact is, we must observe the positions of all the other planets from our own moving planet, and scientists of the Renaissance did not know the nature of the Earth's orbit any better than the orbits of the other planets. Their problem, as we saw in Chapter 1, was that they had to deduce the nature of planetary motion using only observations of the other planets' positions in the sky. To solve this complex problem more fully, better observations and better conceptual models were needed.

One had to be a Newton to notice that the moon is falling, when everyone sees that it doesn't fall.

Paul Valery in *Analects* (Collected Works, Vol. 14, 1966).

The Laws of Planetary Motion

At about the time that Galileo was beginning his experiments with falling bodies, the efforts of two other scientists dramatically advanced our understanding of the motions of the planets. These two astronomers were the observer Tycho Brahe and the mathematician Johannes Kepler. Together, they placed the speculations of Copernicus on a sound mathematical basis and paved the way for the work of Isaac Newton in the next century.

Tycho Brahe

Three years after the publication of Copernicus' *De Revolutionibus*, Tycho Brahe (1546–1601) was born to a family of Danish nobility. He developed an early interest in astronomy and as a young man made significant astronomical observations. Among these was a careful study of what we now know was an exploding star that flared up to great brilliance in the night sky. His growing reputation gained him the patronage of the Danish King Frederick II, and at the age of 30 Tycho was able to establish a fine astronomical observatory on the North Sea island of Hveen (Figure 2.1). Tycho was the last and greatest of the pre-telescopic observers in Europe.

FIGURE 2.1

A stylized engraving showing Tycho Brahe using his instruments to measure the altitude of a celestial object above the horizon. Note that the scene includes hints of the grandeur of Tycho's observatory. (Granger Collection, New York)

FIGURE 2.2

Johannes Kepler (1571–1630), German mathematician and astronomer. His discovery of the basic quantitative laws that describe planetary motion placed the heliocentric cosmology of Copernicus on a firm mathematical basis. (AIP/Niels Bohr Library)

At Hveen Tycho made a continuous record of the positions of the Sun, Moon, and planets for almost 20 years. His extensive and precise observations enabled him to note that the positions of the planets varied from those given in published tables, which were based on the work of Ptolemy. But Tycho was an extravagant and cantankerous fellow, and he accumulated enemies among government officials. When his patron, Frederick II, died in 1597, Tycho was forced to leave Denmark. He took up residence near Prague, where he became Court Astronomer to the Emperor Rudolf of Bohemia. There, in the year before his death, Tycho found a most able young mathematician, Johannes Kepler, to assist him in analyzing his extensive planetary data.

Kepler

Kepler (1571–1630) was born into a poor family in the German city of Württemberg and lived much of his life amid the turmoil of the Thirty Years' War. He attended college at Tübingen and studied for a theological career. There he learned the principles of the Copernican system and became converted to the heliocentric hypothesis. A Protestant refugee from his Catholic homeland, Kepler went to Prague to serve as an assistant to Tycho, who set him to work trying to find a satisfactory theory of planetary motion—one that was compatible with the long series of observations made at Hveen.

But Tycho, jealous of the young Kepler's brilliance, was reluctant to provide him with much material at any one time, for fear that Kepler would discover the secrets of the universal motions by himself, thereby robbing Tycho of some of the glory. It was not until after Tycho's death in 1601 that Kepler obtained possession of the majority of the priceless records. Their study occupied most of Kepler's time for more than 20 years.

Kepler's most detailed study was of Mars, for which the observational data were the most extensive. In 1609 he published the first results of his work in *The New Astronomy;* it is there that we find his first two laws of planetary motion. Their discovery was a profound step in the development of modern science.

The Orbit of Mars

Kepler began his research under the assumption that the orbits of planets were circles, but the observations contradicted this idea. Working with data for Mars, he eventually discovered that the orbit of that planet had the shape of a flattened circle, or **ellipse.** Next to the circle, the ellipse is the simplest kind of closed curve, belonging to a family of curves known as conic sections (Figure 2.3).

An ellipse is a curve for which the sum of the distances from any point on the ellipse to two points inside the ellipse is always the same. These two points inside the ellipse are called its **foci** (singular: **focus**). This property suggests a simple way to draw an ellipse. The ends of a length of string are tied to two tacks pushed through a sheet of paper into a drawing board, so that the string is slack. If a pencil is pushed against the string, making the string taut, and then slid against the string around the tacks (Figure 2.4), the curve that results is an ellipse. At any point where the pencil may be, the sum of the dis-

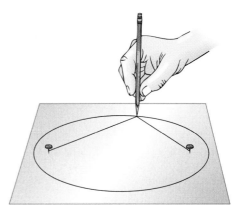

FIGURE 2.4
Drawing an ellipse with two tacks and a string. The tacks are the foci of the ellipse.

tances from the pencil to the two tacks is a constant length—the length of the string. The tacks are at the two foci of the ellipse.

The widest diameter of the ellipse is called its **major axis.** Half this distance—that is, the distance from the center of the ellipse to one end—is the **semimajor axis,** which is usually used to specify the size of the ellipse. For example, the semimajor axis of the orbit of Mars, which also turns out to be the planet's average distance from the Sun, is 228 million km.

The shape (roundness) of an ellipse depends on how close together the two foci are, compared with the major axis. The ratio of the distance between the foci to the length of the major axis is called the **eccentricity** of the ellipse. If an ellipse is drawn using strings and tacks, the length of the major axis is the length of the string, and the eccentricity is the distance between the tacks divided by the length of the string.

If the foci (or tacks) are in the same place, the eccentricity is zero and the ellipse is just a circle; thus a circle is an ellipse of zero eccentricity. We can make ellipses of various shapes by varying the spacing of the tacks (as long as they are not farther apart than the length of the string). The greater the eccentricity, the more elongated is the ellipse, up to a maximum eccentricity of 1.0.

The size and shape of an ellipse are completely specified by its semimajor axis and its eccentricity. Kepler found that Mars has an elliptical orbit, with the Sun at one focus (the other focus is empty). The eccentricity of the orbit of Mars is only about 0.1; its orbit, drawn to scale, would be practically indistinguishable from a circle. Yet the difference is critical for understanding planetary motions.

Kepler generalized his result in his first law and said that the orbits of all the planets are ellipses. Here was a decisive moment in the history of human thought: it was not necessary to have only circles in order to have an acceptable cosmos. The universe could be a bit more complex than the Greek philosophers had wanted it to be.

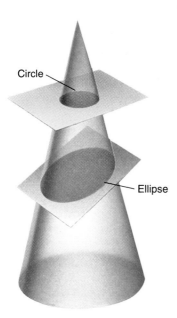

Circle

Ellipse

FIGURE 2.3
The circle and the ellipse are both formed by the intersection of a plane with a cone. This is why both curves are called conic sections.

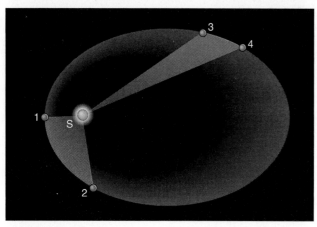

FIGURE 2.5

Kepler's second law: the law of equal areas. A planet moves most rapidly on its elliptical orbit when at position 1, nearest the Sun, which is at one focus of the ellipse. The orbital speed of the planet varies in such a way that in equal intervals of time a line between the Sun and the planet sweeps out equal areas. Thus the area swept out from 1 to 2 is the same as that from 3 to 4. Note that the eccentricities of the planets' orbits are substantially less than shown here.

Kepler's second law deals with the speed with which each planet moves along the ellipse. Working with Tycho's observations of Mars, Kepler discovered that the planet speeds up as it comes closer to the Sun and slows down as it pulls away from the Sun. He expressed the precise form of this relation by imagining that the Sun and Mars are connected by a straight, elastic line (Figure 2.5).

When Mars is closer to the Sun (positions 1 and 2 in the figure), the elastic line is not stretched as much, and the planet moves rapidly. Further from the Sun, as in positions 3 and 4, the line is stretched a lot; hence the planet does not move as fast. As Mars travels in its elliptical orbit around the Sun, the elastic line sweeps out areas of the ellipse (the colored regions in our figure). Kepler found that in equal intervals of time, the areas swept out in space by this imaginary line are always equal; that is, the area of the region from 1 to 2 is the same as that from 3 to 4.

This is also a general property of the orbits of all planets. A planet in a circular orbit always moves at the same speed, but in an elliptical orbit the planet's speed varies considerably, following the rule Kepler found.

Laws of Planetary Motion

Kepler's first two laws of planetary motion describe the shape of a planet's orbit and allow us to calculate the speed of its motion at any point in the orbit. Kepler was pleased to have discovered such fundamental rules, but they did not satisfy his quest to understand planetary motions. He wanted to know why the orbits of the planets were spaced as they are, and to find a mathematical pattern in their orbital periods—a "harmony of the spheres,"

as he called it. For many years he worked to discover mathematical relationships governing planetary spacings and the time each planet took to go around the Sun.

In 1619, Kepler succeeded in finding a simple algebraic relation that links the semimajor axes of the planets' orbits and their periods of revolution. The relation is now known as Kepler's third law. It applies to all of the planets, including the Earth, and it provides a means for calculating their relative distances from the Sun.

Kepler's third law takes its simplest form when the period is expressed in years (the revolution period of the Earth), and the semimajor axis of the orbit is expressed in terms of the Earth's average distance from the Sun, called the **astronomical unit (AU).** One astronomical unit is equal to 1.5×10^8 km. Kepler's third law is, then,

$$(\text{distance})^3 = (\text{period})^2$$

For example, this relationship tells us how to calculate Mars' average distance from the Sun (the semimajor axis of its orbit) from its period of 1.88 years. The period squared is $1.88 \times 1.88 = 3.53$, and the cube root of 3.53 is 1.52, which is equal to the semimajor axis in astronomical units. In other words, to go around the Sun in a little less than 2 years, Mars must be 50% farther from the Sun than Earth is.

Kepler's three laws of planetary motion can be summarized as follows:

Kepler's First Law Each planet moves about the Sun in an orbit that is an ellipse, with the Sun at one focus of the ellipse.

Kepler's Second Law The straight line joining a planet and the Sun sweeps out equal areas in space in equal intervals of time.

Kepler's Third Law The squares of the planets' periods of revolution are in direct proportion to the cubes of the semimajor axes of their orbits.

These three laws provided a precise geometric description of planetary motion within the framework of the Copernican system. With these tools, it was possible to calculate planetary positions with undreamed-of precision. But Kepler's laws are purely descriptive; they do not help us understand what forces of nature constrain the planets to follow this particular set of rules. That step was left to Newton.

Newton's Great Synthesis

It was the genius of Isaac Newton (1643–1727) that found a conceptual framework that completely explained the observations and rules assembled by Galileo, Brahe, Kepler, and others. Newton was born in Lincolnshire, England, in the year after Galileo's death (Figure 2.6). Against the advice of his mother, who wanted him to stay

FIGURE 2.6
Isaac Newton (1643–1727), whose work on the laws of motion, gravity, optics, and mathematics laid the foundations of much of physical science. (AIP/Niels Bohr Library)

home and help with the family farm, he entered Trinity College at Cambridge in 1661 and eight years later was appointed Professor of Mathematics there. Among Newton's contemporaries in England were architect Christopher Wren, authors Samuel Pepys and Daniel Defoe, and composer G. F. Handel.

Newton's Laws of Motion

As a young man in college, Newton became interested in natural philosophy, as science was then called. He worked out some of his first ideas on mechanics and optics during the plague years of 1665 and 1666, when students were sent home from college. Newton, a moody and often difficult man, continued to work on his ideas in private, even inventing new mathematical tools to help him deal with the complexities involved. Eventually, his friend Edmund Halley prevailed on him to collect and publish the results of his remarkable investigations on motion and gravity. The result was a volume that set out the underlying system of the physical world, *Philosophiae Naturalis Principia Mathematica.* The *Principia,* as the book is generally known, was published at Halley's expense in 1687.

At the very beginning of the *Principia,* Newton states three laws that he presumes to govern the motions of all objects:

Newton's First Law Every body continues doing what it is already doing—being in a state of rest, or moving uniformly in a straight line—unless it is compelled to change by an outside force.

Newton's Second Law The change of motion of a body is proportional to the force acting on it, and is made in the direction in which that force is acting.

Newton's Third Law To every action there is an equal and opposite reaction (or, the mutual actions of two bodies upon each other are always equal and act in opposite directions).

In the original Latin, the three laws contain only 59 words, but those few words set the stage for modern science. Let us examine them more carefully.

Interpretation of Newton's Laws

Newton's first law is a restatement of one of Galileo's discoveries, which Newton called the conservation of **momentum,** a measure of a body's motion. The law states that in the absence of any outside influence, a body's momentum remains unchanged. We use this word in everyday expressions as well, as in "This bill in Congress has a lot of momentum; it seems unstoppable."

In other words, a stationary object stays put, and a moving object keeps moving. Momentum depends on three factors. The first is *speed*—how fast a body moves (zero if it is stationary). The second is the *direction* in which the body is moving. Scientists use the term *velocity* to describe both speed and direction. For example, 20 km/h due south is velocity, while 20 km/h is speed. The third factor in momentum is what Newton called *mass.* Mass is a measure of the amount of matter in a body, as we will discuss further below.

As mentioned in the last chapter, it is not so easy to see this rule in action in the everyday world. Objects in motion do not remain in motion because outside forces are always acting. One important force is friction, which tends to slow things down. If you roll a ball down the sidewalk, it eventually comes to a stop because the rubbing of the ball against the sidewalk exerts a force. But out in space between the stars, where friction is negligible, objects could in fact continue to move (coast) indefinitely. Newton's first law is sometimes called the *law of inertia,* inertia being the tendency of objects (and legislatures) to keep doing what they are already doing.

The momentum of a body can change only under the action of an outside influence. Newton's second law defines *force* in terms of its ability to *change momentum.* A force (which you can think of as a push or a pull) has both size and direction. When force is applied to a body, the momentum changes in the direction of the applied force. This means that a force is required to change either the speed or the direction of a body, or both—that is, to start it moving, to speed it up, to slow it down, to stop it, or to change its direction.

Any such change in an object's state of motion is called **acceleration.** Newton showed that the acceleration of a body was proportional to the force. The harder you kick your astronomy textbook after a difficult exam, the faster it will fly across the room. How much a force will accelerate an object is also determined by the object's mass. If you kick a tennis ball with the same force that you should remain constant. Therefore, any change of momentum within the system must be balanced by another change that is equal and opposite to it, so that the momentum of the entire system is not changed.

This means that forces in nature do not occur alone: we find that in each situation there is always a pair of forces that are equal to and opposite each other. If a force is exerted on an object, it must be exerted by something else, and the object will exert an equal and opposite force on that something. Suppose that during that same difficult midterm, another student runs screaming from the room, jumps out a (not very high) window, and lands on the ground. The force pulling him down (as we will see in the next section) is the gravitational force between him and the Earth. Both he and the Earth must suffer the same total change of momentum because of the influence of this mutual force. So both the student and the Earth are accelerated by each other's pull. However, the student does much more of the moving. Because the Earth has enormously greater mass, it can experience the same change of momentum by accelerating only a very small amount. Many things fall toward the Earth all the time, but the acceleration of our planet as a result is far too small to be noticed or measured.

A more obvious example of the mutual nature of forces between objects is familiar to all who have batted a baseball. The recoil of the bat shows that the ball exerts a force on it during the impact, just as the bat does on the ball. Similarly, when a rifle is discharged, the force pushing the bullet out of the muzzle is equal to that pushing backward upon the gun and the person shooting it.

This, in fact, is the principle behind jet engines and rockets: the force that discharges the exhaust gases from the rear of the rocket is accompanied by force that shoves the rocket forward. The exhaust gases need not push against air or the Earth; a rocket actually operates best in a vacuum (Figure 2.7).

FIGURE 2.7
The U.S. Space Shuttle at launch, powered by three liquid fuel engines burning liquid oxygen and liquid hydrogen, with two solid fuel boosters. (NASA)

kick the textbook, the tennis ball—having less mass—can be accelerated to a greater speed.

Newton's third law is the most profound. Basically, it is a generalization of the first law, but it also gives us a way to define mass. If we consider a system of two or more objects isolated from outside influences, Newton's first law says that the total momentum of the system of objects

Mass, Volume, and Density

Before we go on to discuss Newton's other work, we want to take a brief look at some terms that will be important to sort out clearly. We begin with *mass*, which is a measure of the amount of material in an object. Today, we think of it as the number of atoms or molecules that make up an object.

Volume is a measure of the physical space occupied by a body, say in cubic centimeters or liters. In short, the volume is the "size" of an object; it has nothing to do with its mass. A penny and an inflated balloon may both have the same mass, but they have very different volumes.

The penny and balloon are also very different in **density,** which is a measure of how much mass we have

TABLE 2.1
Densities of Materials

Material	Density (g/cm³)
Gold	19.3
Lead	11.4
Iron	7.9
Earth (bulk)	5.6
Rock (typical)	2.5
Water	1.0
Wood (typical)	0.8
Insulating foam	0.1
Silica gel	0.02

FIGURE 2.8
The conservation of angular momentum is demonstrated by a spinning figure skater. When she brings her arms in, their distance from the spin center is small, and so her speed increases. When her arms are out, their distance from the spin center is greater, so she slows down.

per unit volume. Specifically, density is the ratio of mass to volume. Note that often in everyday language we use "heavy" and "light" as indications of density (rather than weight), as, for instance, when we say that iron is heavy or that puff pastry is light.

The units of density that will be used in this book are grams per cubic centimeter (g/cm³) or, alternatively, metric tons per cubic meter.[1] If a block of some material has a mass of 300 g and a volume of 100 cm³, its density is 3 g/cm³. Familiar materials span a considerable range in density, from artificial materials such as plastic insulating foam (less than 0.1 g/cm³) to gold (19 g/cm³) (Table 2.1). In the astronomical universe, much more remarkable densities can be found, all the way from a comet's tail (10^{-16} g/cm³) to a neutron star (10^{15} g/cm³).

To sum up, then, mass is "how much," volume is "how big," and density is "how tightly packed."

Angular Momentum

The concept of **angular momentum** is a bit more complex, but it is important for understanding many astronomical objects. Angular momentum is a measure of the momentum of an object as it rotates or revolves about some fixed point. Whenever we deal with the revolution of spinning objects, from planets to galaxies, we have to consider angular momentum. The angular momentum of an object is defined as the product of three quantities: its mass, its velocity, and its distance from the fixed point around which it turns.

If these three quantities remain constant—that is, if the motion takes place at a constant speed and at a fixed distance from the spin center—then the angular momentum is also a constant. So, ignoring air resistance, if you tie a string to your astronomy textbook and twirl it around your head at constant speed, you will have a system with constant angular momentum.

More generally, angular momentum is constant, or conserved, in any rotating system in which no external forces act, or in which the only forces are directed toward or away from the spin center. An example of such a system is a planet orbiting the Sun. Kepler's second law is an example of the conservation of angular momentum. When a planet approaches the Sun on its elliptical orbit, the distance to the spin center decreases; the planet speeds up to keep the angular momentum the same. Similarly, when the planet is further from the Sun, it revolves more slowly.

Just as a planet speeds up when it approaches the Sun, a shrinking cloud of dust or a star collapsing on itself (both situations you will encounter as you read on) increases its spin rate as it contracts. There is less distance to spin center, so the speed goes up to keep angular momentum the same. The concept is also illustrated by figure skaters, who bring their arms and legs in to spin more rapidly, and extend their arms and legs to slow down (Figure 2.8). (You can duplicate this yourself on a well-oiled piano stool, by starting yourself spinning slowly with your arms extended, and then pulling your arms in.)

2.3

Universal Gravity

The Law of Gravity

Newton's laws of motion showed that, left to themselves, objects at rest stay at rest, while those in motion continue moving uniformly in a straight line. Thus it is the *straight line*, not the circle, that defines the most natural state of motion. In the case of the planets, then, some *force* must be bending their paths from straight lines into ellipses. That force, Newton proposed, is gravity.

---

[1] Generally we use the standard metric (or SI) units in this book. The proper metric unit of density is kg/m³. But to most people g/cm³ provides a more meaningful unit because the density of water is exactly 1 g/cm³. Density expressed in g/cm³ is sometimes called specific density or specific weight.

Although that may sound obvious today, in Newton's time gravity was something associated with the Earth alone. Everyday experience shows us that the Earth exerts a gravitational force upon objects at its surface. If you drop something off a leaning tower in Pisa, it falls toward the Earth, accelerating as it falls. Newton's insight was that the Earth's gravity might extend as far as the Moon and produce the acceleration required to curve the Moon's path from a straight line and keep it in its orbit. He further speculated that gravity is not limited to the Earth, but that there is a general force of attraction between all material bodies. If so, the attractive force between the Sun and each of the planets could keep each in its orbit.

Once Newton boldly hypothesized that there is a universal attraction among all bodies everywhere in space, he had to determine the exact nature of the attraction. The precise mathematical description of that gravitational force had to dictate that the planets move exactly as Kepler had observed them to (as expressed in Kepler's three laws). Also, the law of gravity had to predict the correct behavior of falling bodies on the Earth, as observed by Galileo. How must gravitational force depend on distance in order for these conditions to be met?

The answer to this question required mathematical tools that had not yet been developed. But this did not deter Isaac Newton, who invented what we today call calculus to deal with this problem. Eventually he was able to conclude that the force of gravity must drop off with increasing distance between the Sun and a planet (or between any two objects) in proportion to the inverse square of their separation. In other words, if a planet were twice as far from the Sun, the force would be $1/2^2$ or $1/4$ as large. Put the planet three times farther away, and the force is $1/3^2$ or $1/9$ as strong.

Newton also concluded that the gravitational attraction between two bodies must be proportional to their masses. The more mass an object has, the stronger the pull of its gravity. Expressed as a formula, the gravitational attraction between any two objects is given by one of the most famous formulas in all of science:

$$\text{force} = GM_1M_2/R^2$$

where M_1 and M_2 are the masses of the two objects, and R is their separation. The number represented by G is called the constant of gravitation. With such a force and the laws of motion, Newton was able to show mathematically that the only orbits permitted were exactly those described by Kepler's laws.

Newton's law of gravity works for the planets, but is it really universal? The gravitational theory should also predict the observed acceleration of the Moon toward the Earth, falling around the Earth at a distance of about 60 times the radius of the Earth, as well as of an object (say an apple) dropped near the Earth's surface. The falling of an apple is something we can measure quite easily; can we use it to predict the motions of the Moon?

Newton's theory says that the force on (and therefore the acceleration of) an object toward the Earth should be inversely proportional to the square of its distance from the center of the Earth. Objects like apples at the surface of the Earth ($R = 1$ Earth radius, the distance from its center) are observed to accelerate downward at 9.8 meters per second per second (9.8 m/s^2).

Therefore, if the law holds, the Moon, 60 Earth radii from its center, should experience an acceleration toward the Earth that is $1/60^2$, or 3600 times less—that is, about 0.00272 m/s^2. This is precisely the observed acceleration of the Moon in its orbit. What a triumph! Imagine the thrill Newton must have felt to realize he had discovered and verified a law that holds for the Earth, apples, the Moon, and, as far as we know, everything in the universe!

Putting the math aside, what Newton's law suggests is that gravity is a "built-in" property of mass. Wherever masses occur, they will interact via the force of gravity. The more mass there is, the greater the force it can exert. Here on Earth, the largest concentration of mass is, of course, the planet we stand on, and its pull dominates the gravitational interactions we experience. But *everything* with mass attracts everything else with mass anywhere in the universe. For example, you and this textbook (each having mass) attract each other via gravity. If you let the book fall, however, it is pulled much more strongly by the Earth, so it falls to the Earth, not toward you. But if you and the textbook somehow found yourselves out in space, far from any stars or planets, you would discover that gravitational attraction would start to pull you together.

Newton's law also suggests that gravity never becomes zero. It quickly gets weaker with distance, but it continues to act to some degree no matter how far away you get. The pull of the Sun is stronger at Mercury than at Pluto, but it can be felt far beyond Pluto, where (as we shall see) we have good evidence that it continues to make enormous numbers of smaller bodies move around huge orbits. And its pull joins with the pull of billions of other stars to create the gravitational pull of the Milky Way Galaxy in which we live. That force, in turn, can make other smaller galaxies orbit around the Milky Way, and so forth.

Why is it then, you may ask, that the astronauts aboard the Shuttle appear "weightless," and we see images on television of people and objects floating in the spacecraft? The astronauts in the Shuttle are only a few hundred kilometers above the surface of the Earth, not really a significant distance compared to the size of the Earth. Gravity is certainly not a great deal weaker that much farther away. The astronauts feel weightless for the same reason that passengers in an elevator whose cable has broken or in an airplane whose engines no longer work feel weightless. They are *falling*, and in free fall, they accelerate at the same rate as everything around them, including their spacecraft and a sandwich that got away from their grip. Thus, *relative* to the Shuttle, the astronauts experience no additional force and so feel "weightless." Unlike the falling elevator passengers, however, the astronauts are falling *around* the Earth, not *to* the Earth.

Orbital Motion and Mass

Kepler's laws are descriptions of the orbits of objects moving according to Newton's laws of motion and the law of gravity. Knowing that gravity is the force that attracts planets toward the Sun, however, allowed Newton to rethink Kepler's third law. Recall that Kepler had found a relationship between the period of a planet's revolution and its distance from the Sun. Newton's law of gravity can be used to show mathematically that this relationship is actually

$$D^3 = (M_1 + M_2) \times P^2$$

As explained in Section 2.1, we express distances in astronomical units and periods in years. But Newton's formulation introduces the additional factor of the masses of the Sun (M_1) and planet (M_2), both expressed in units of the Sun's mass.

How did Kepler miss this factor? In units of the Sun's mass, the mass of the Sun is 1; in units of the Sun's mass, the mass of a typical planet is a negligibly small fraction, and so ($M_1 + M_2$) is very, very close to 1. This makes Newton's formula look almost the same as Kepler's. The tiny mass of the planets is the reason that Kepler did not realize that both masses had to be included in the calculation. There are many cases in astronomy, however, in which we *do* need to include the two mass terms—for example, when two stars or two galaxies orbit one another.

But including the mass term allows us to use this formula in a new way. If we can measure the motions (distances and periods) of objects acting under their mutual gravity, the formula will permit us to deduce their masses. For example, we can calculate the mass of the Sun by using the distances and periods of the planets, or the mass of Jupiter by noting the motions of its moons. Indeed, Newton's reformulation of Kepler's third law is one of the most powerful concepts in astronomy. Our ability to deduce the masses of objects from their motions is key to understanding the nature and evolution of many astronomical bodies. We will use this law repeatedly throughout this text in calculations that range from the orbits of comets to the interactions of galaxies.

2.4

Orbits in the Solar System

Description of an Orbit

Celestial mechanics is the study of the motions of astronomical objects, using gravitational theory. The path of an object through space is called its orbit, whether that object is a spacecraft, planet, star, or galaxy. An orbit, once

TABLE 2.2
Orbital Data for the Planets

Planet	Semimajor Axis	Period	Eccentricity
	(AU)	(yr)	
Mercury	0.39	0.24	0.21
Venus	0.72	0.62	0.01
Earth	1.00	1.00	0.02
Mars	1.52	1.88	0.09
(Ceres)	2.77	4.60	0.08
Jupiter	5.20	11.86	0.05
Saturn	9.54	29.46	0.06
Uranus	19.19	84.07	0.05
Neptune	30.06	164.80	0.01
Pluto	39.60	248.60	0.25

determined, allows the future positions of the object to be calculated. When only two objects are involved (such as a planet in orbit about the Sun), three quantities are required to describe the orbit. These are the *size* (the semimajor axis), the *shape* (the eccentricity), and the *period* of revolution.

Two points in any orbit have been given special names. The place where the planet is closest to the Sun (*Helios* in Greek) is called the **perihelion** of its orbit, and the place where it is farthest away and moves the most slowly is the **aphelion**. For a satellite orbiting the Earth (*Geos* in Greek), the corresponding terms are **perigee** and **apogee.**

Orbits of the Planets

Today, Newton's work enables us to calculate and predict the orbits of the planets with marvelous precision. We know nine planets, beginning with Mercury closest to the Sun and extending outward to Pluto. The average orbital data for the planets are summarized in Table 2.2.

One small note about Pluto: it has the largest semimajor axis of any planet (almost 40 AU, or 5 billion km). However, it is not at present the most distant planet. Because Pluto is currently near the perihelion of its peculiar orbit, it is actually closer to the Sun than is Neptune. During the final 20 years of the 20th century, Neptune has the distinction of being the most distant planet, at 30 AU from the Sun.

According to Kepler's laws, Mercury must have the shortest period of revolution (88 Earth days); thus it has the highest orbital speed (averaging 48 km/s). At the opposite extreme, Pluto has a period of 249 years and an average orbital speed of just 5 km/s.

All of the planets have orbits of rather low eccentricity. The most eccentric orbits are those of Mercury (0.21) and Pluto (0.25); all of the rest have eccentricities of less than 0.1. It is fortunate for the development of science that Mars has an eccentricity greater than many of the other planets, for otherwise the pretelescopic observa-

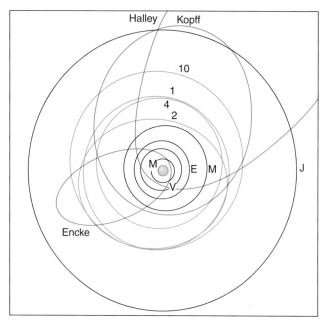

FIGURE 2.9
Orbits of typical comets and asteroids compared with those of the planets Mercury, Venus, Earth, Mars, and Jupiter (black circles). Shown in red are three comets: Halley, Kopff, and Encke. In blue are the four largest asteroids: 1 Ceres, 2 Pallas, 4 Vesta, and 10 Hygeia.

tions of Tycho would not have been sufficient for Kepler to deduce that its orbit had the shape of an ellipse rather than a circle.

The planetary orbits are also confined close to a common plane, which is near the plane of the Earth's orbit (the ecliptic). The strange orbit of Pluto is inclined about 17° to the average, but all of the other planets lie within 10° of the common plane of the solar system.

Orbits of Asteroids and Comets

In addition to the nine planets, there are many smaller objects in the solar system. Some of these are natural satellites that orbit all of the planets except Mercury and Venus. In addition, there are two classes of smaller objects in heliocentric orbits: the asteroids and the comets. Both asteroids and comets are believed to be small chunks of material left over from the formation process of the solar system; their properties will be discussed in some detail in Chapter 12.

Asteroids and comets differ from each other in composition and in the natures of their orbits. In general, the asteroids have orbits with smaller semimajor axes than do the comets (Figure 2.9). The great majority of them lie between 2.2 and 3.3 AU, in the region known as the **asteroid belt.** As you can see in Table 2.2, the asteroid belt is in the middle of a gap between the orbits of Mars and Jupiter. It is because these two planets are so far apart that stable orbits of small bodies can exist in the region

between them. Table 2.2 shows the orbit for the largest asteroid, Ceres.

Comets generally have orbits of larger size and greater eccentricity than those of the asteroids. Typically, the eccentricities of their orbits are 0.8 or higher. According to Kepler's second law, therefore, they spend most of their time far from the Sun, moving very slowly. As they approach perihelion, the comets speed up and whip through the inner parts of their orbits rapidly.

2.5
Motions of Satellites and Spacecraft

Space Flight and Satellite Orbits

The law of gravity and Kepler's laws describe the motions of Earth satellites and interplanetary spacecraft as well as the planets. Sputnik, the first artificial Earth satellite, was launched by what was then called the Soviet Union on October 4, 1957. Since that time, thousands of satellites have been placed into orbit around the Earth, and spacecraft have also orbited the Moon, Venus, Mars, and Jupiter.

Once an artificial satellite is in orbit, its behavior is no different from that of a natural satellite, such as our Moon. If the satellite is high enough to be free of atmospheric friction, it will remain in orbit forever, following Kepler's laws in a perfectly respectable way. However, although there is no difficulty in maintaining a satellite once it is in orbit, a great deal of energy is required to lift the spacecraft off the Earth and accelerate it to orbital speed.

To illustrate how a satellite is launched, imagine a gun firing a bullet horizontally from the top of a high mountain, (Figure 2.10a—adapted from a similar diagram by Newton, shown in Figure 2.10b). Imagine, further, that the friction of the air could be removed and that nothing can get in the bullet's way; other mountains, buildings, and so on, are all absent. Then the only force that acts on the bullet after it leaves the muzzle is the gravitational force between the bullet and Earth.

If the bullet is fired with some velocity we can call v_a, it continues to have that forward speed, but meanwhile the gravitational force acting upon it pulls it downward toward the Earth, where it strikes the ground at point a. However, if it is given a higher muzzle velocity, v_b, its higher forward speed carries it farther before it hits the ground. This is because, regardless of its forward speed, the downward gravitational force is the same. Thus, this faster-moving bullet strikes the ground at point b.

If our bullet is given a high enough muzzle velocity, v_c, the curved surface of the Earth causes the ground to tip out from under it so that it remains the same distance above the ground and falls around the Earth in a complete circle. The speed needed to do this—called the **circular satellite velocity**—is about 8 km/s or about 17,500 miles per hour (mph) in more familiar units.

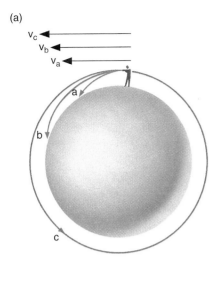

(a)

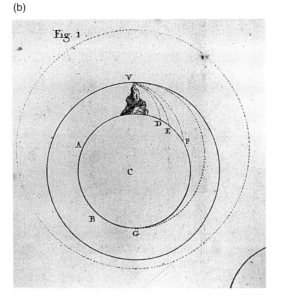

(b)

FIGURE 2.10
(a) Firing a bullet into a satellite orbit. In cases *a* and *b*, the velocity is not enough to prevent gravity from pulling the bullet back to Earth. (b) A diagram by Newton in his *De Mundi Systematic,* 1731 edition, illustrating the same concept shown in panel (a). (Crawford Collection, Royal Observatory, Edinburgh)

Each year more than 50 new satellites are launched into orbit by such nations as Russia, the United States, China, Japan, India, and Israel, as well as by the European Space Agency (ESA), a consortium of European nations (Figure 2.11). Most satellites are launched into low Earth orbit, since this requires the minimum launch energy. At the orbital speed of 8 km/s, they circle the planet in about 90 min.

Low-Earth orbits are not stable indefinitely, since the drag generated by friction with the thin upper atmo-

FIGURE 2.11
A plot of all known unclassified satellites and satellite debris in Earth orbit larger than about the size of softball as of 1995. (U.S. Space Command, NORAD)

sphere eventually leads to a loss of energy and "decay" of the orbit. (In the terms of Newton's work, we can say that friction is another outside force acting on the satellite.) Upon re-entering the denser parts of the atmosphere, most satellites are burned up by atmospheric friction, although some solid parts may reach the surface. Of course, such piloted rockets as the U.S. Shuttle and other recoverable payloads are designed to survive re-entry intact.

Interplanetary Spacecraft

The exploration of the solar system has been carried out largely by robot spacecraft sent to the other planets. To escape Earth, these craft must achieve **escape velocity,** the speed needed to move away from the Earth forever, which is about 11 km/s (about 25,000 mph). After this they coast to their targets, subject only to minor trajectory adjustments provided by small thruster rockets on board. In interplanetary flight, these spacecraft follow Keplerian orbits around the Sun, modified only when they pass near one of the planets.

While close to its target, a spacecraft is deflected by the planet's gravitational force into a modified orbit, either gaining or losing energy in the process. By carefully choosing the aim point in a planetary encounter, controllers have actually been able to use a planet's gravity to redirect a flyby spacecraft to a second target. Voyager 2 (Figure 2.12) used a series of gravity-assisted encounters to yield successive flybys of Jupiter (1979), Saturn (1980), Uranus (1986), and Neptune (1989). The Galileo spacecraft, launched in 1989, flew past Venus once and Earth twice to gain the energy required to reach its ultimate goal at Jupiter.

If we wish to orbit a planet, we must slow the spacecraft with a rocket when the spacecraft is near its destination, allowing it to be captured into an elliptical orbit. Additional rocket thrust is required to bring a vehicle down from orbit for a landing on the surface. Finally, if a return trip to Earth is planned, the landed payload must include enough propulsive power to repeat the entire process in reverse.

2.6

Gravity with More Than Two Bodies

Until now we have considered the Sun and a planet (or a planet and one of its satellites) as nothing more than a pair of bodies revolving around each other. Actually, all the planets exert gravitational forces upon one another as well. These interplanetary attractions cause slight variations from the orbits that would be expected if the gravitational forces between planets were neglected. Unfortunately, the problem of treating the motion of a body that is under the gravitational influence of two or more other bodies is very complicated and can be handled properly only with large computers.

The Interactions of Many Bodies

Here's an example of what we mean: suppose you have a cluster of a thousand stars all orbiting a common center

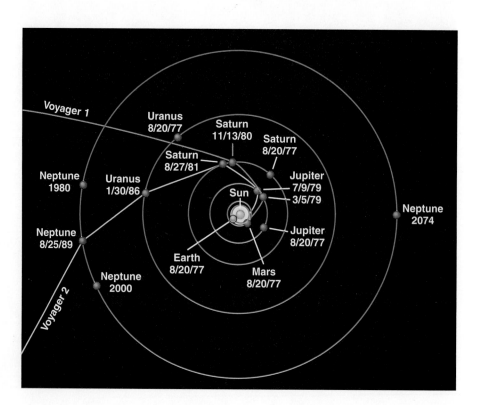

FIGURE 2.12
The flight paths of Voyager 1 (red) and Voyager 2 (yellow) through the outer solar system, taking advantage of the gravitation of each planet to adjust the trajectory toward the next target. (NASA/JPL)

FIGURE 2.13
Supercomputers at NASA Ames Research Center are capable of tracking the motions of more than a million objects under their mutual gravitation. (NASA/ARC)

(such clusters are quite common, as we shall see). If we know the exact position of each star at any given instant, we can calculate the combined gravitational force of the entire ensemble on any one member of the cluster. Knowing the force on the star in question, we can therefore find how it will accelerate. If we know how it was moving to begin with, we can then calculate how it will move in the next instant of time, thus tracking its motion.

However, the problem is complicated by the fact that the other stars are also moving and thus changing the effect they will have on our star. Therefore, we must simultaneously calculate the acceleration of each body produced by the combination of the gravitational attractions of all the others in order to track the motions of all of them, and hence of any one. Such complex calculations have been carried out with modern computers to track the evolution of hypothetical clusters of stars with up to a million members (Figure 2.13).

Within the solar system, the problem of computing the orbits of planets and spacecraft is somewhat simpler. We have seen that Kepler's laws, which do not take into account the gravitational effects of the other planets on an orbit, really work quite well. This is because these additional influences are very small in comparison to the dominant gravitational attraction of the Sun. Under such circumstances, it is possible to treat the effects of other bodies as small *perturbations* (or disturbances) on the

force exerted by the Sun. During the 18th and 19th centuries, mathematicians developed many elegant techniques for calculating perturbations, permitting them to predict very precisely the positions of the planets. Such calculations eventually led to the discovery of a new planet in 1846.

Discovery of Neptune

The discovery of the planet Neptune was one of the high points in the development of gravitational theory. In 1781, William Herschel, a musician and unpaid astronomer, accidently discovered the seventh planet: Uranus. It happens that Uranus had been observed a century before, but in none of those earlier sightings was it recognized as a planet; rather it was simply recorded as a star.

By 1790, an orbit had been calculated for Uranus using observations of its motion in the decade following its discovery. Even after allowance was made for the perturbing effects of Jupiter and Saturn, however, it was found that Uranus did not move on an orbit that exactly fit the earlier observations of it made since 1690. By 1840, the discrepancy between the positions observed for Uranus and those predicted from its computed orbit amounted to about 0.03°—an angle barely discernible to the unaided eye but still larger than the probable errors in the orbital

FIGURE 2.14
John Couch Adams (1819–1892) and Urbain J. J. Leverrier (1811–1877) share the credit for discovering the planet Neptune. (Yerkes Observatory)

calculations. In other words, Uranus just did not seem to move on the orbit predicted from Newtonian theory.

In 1843, John Couch Adams (Figure 2.14), a young Englishman who had just completed his studies at Cambridge, began a detailed mathematical analysis of the irregularities in the motion of Uranus to see if they might be produced by an unknown planet. His calculations indicated the existence of a planet more distant than Uranus from the Sun. In October 1845, Adams delivered his results to George Airy, the British Astronomer Royal, informing him where in the sky to find the new planet. We now know that Adams' predicted position for the new body was correct to within 2°, but for a variety of reasons, Airy did not follow up right away.

Meanwhile, French mathematician Urbain Jean Joseph Leverrier, unaware of Adams or his work, attacked the same problem and published its solution in June 1846. Airy, noting that Leverrier's predicted position for the unknown planet agreed to within 1° with that of Adams, suggested to James Challis, director of the Cambridge Observatory, that he begin a search for the new object. The Cambridge astronomer, having no up-to-date star charts of the Aquarius region of the sky where the planet was predicted to be, proceeded by recording the positions of all the faint stars he could observe with his telescope in that location. It was Challis' plan to repeat such plots at

intervals of several days, in the hope that the planet would distinguish itself from a star by its motion. Unfortunately, he was negligent in examining his observations; although he had actually seen the planet, he did not recognize it.

About a month later, Leverrier suggested to Johann Galle, an astronomer at the Berlin Observatory, that he look for the planet. Galle received Leverrier's letter on September 23, 1846, and, possessing new charts of the Aquarius region, found and identified the planet that very night. It was less than a degree from the position Leverrier predicted. The discovery of the eighth planet, now known as Neptune (the Latin name for the god of the sea), was a major triumph for gravitational theory, for it dramatically confirmed the generality of Newton's laws. The honor for the discovery is properly shared by the two mathematicians, Adams and Leverrier.

The discovery of Neptune was not a complete surprise to astronomers, who had long suspected the existence of the planet based on the "disobedient" motion of Uranus. On September 10, 1846, two weeks before Neptune was actually found, John Herschel, son of the discoverer of Uranus, remarked in a speech before the British Association, "We see [the new planet] as Columbus saw America from the shores of Spain. Its movements have been felt trembling along the far-reaching line of our analysis with a certainty hardly inferior to ocular demonstration."

Astronomy and the Poets

When Copernicus, Kepler, Galileo, and Newton formulated the fundamental rules that underlie everything in the physical world, they changed much more than the face of science. For some, they gave humanity the courage to let go of old superstitions and see the world as rational and manageable; for others, they upset comforting, ordered ways that had served humanity for centuries, leaving only a dry, mechanical clockwork universe in their wake.

Poets of the time reacted to such changes in their own work and debated whether the new world picture was an appealing or frightening one. John Donne (1573–1631), in a poem called *Anatomy of the World,* laments the passing of the old certainties:

> The new philosophy [science] calls all in doubt,
> The element of fire is quite put out;
> The Sun is lost, and th' earth, and no man's wit
> Can well direct him where to look for it.

(Here the "element of fire" refers also to the sphere of fire, which medieval thought placed between the Earth and the Moon.)

By the next century, poets like Alexander Pope were celebrating Newton and the Newtonian world view. Pope's famous couplet, written upon Newton's death, goes:

> Nature, and nature's laws lay hid in night.
> God said, Let Newton be! and all was light.

In his 1733 poem, *An Essay on Man,* Pope revels in the complexity of the new views of the world, incomplete though they are:

> Of man, what see we, but his station here,
> From which to reason, to which refer? . . .

> He, who thro' vast immensity can
>   pierce,
> See worlds on worlds compose one
>   universe,
> Observe how system into system
>   runs,
> What other planets circle other suns,
> What vary'd being peoples every star,
> May tell why Heav'n has made us as we are. . . .
> All nature is but art, unknown to thee;
> All chance, direction, which thou canst not see;
> All discord, harmony not understood;
> All partial evil, universal good:
> And, in spite of pride, in erring reason's spite,
> One truth is clear, whatever is, is right.

Poets and philosophers continue to debate whether humanity was exalted or debased by the new views of science. The 19th-century poet Arthur Hugh Clough (1819–1861) cries out in his poem *The New Sinai:*

> And as of old from Sinai's top God said that God is one,
> By science strict so speaks He now to tell us, there is None!
> Earth goes by chemic forces; Heaven's a Mecanique Celeste!
> And heart and mind of humankind a watchwork as the rest!

(A "mechanique celeste" is a clockwork model to demonstrate celestial motions.)

The 20th-century poet Robinson Jeffers (whose brother was an astronomer) saw it differently in a poem called *Star Swirls:*

> There is nothing like astronomy to pull the stuff out of
>   man.
> His stupid dreams and red-rooster importance:
> Let him count the star-swirls.

Summary

2.1 Tycho Brahe was the most skillful of the pretelescopic astronomical observers. His accurate observations of planetary positions provided the data used by Johannes Kepler to derive the three fundamental laws of planetary motion that bear his name: (1) planetary orbits are **ellipses** (a figure described by its **semimajor axis** and **eccentricity**) with the Sun at one **focus;** (2) in equal intervals, a planet's orbit sweeps out equal areas; and (3) if times are expressed in years, and distances in **astronomical units,** the relationship between period (P) and semimajor axis (D) of an orbit is given by ($P^2 = D^3$).

2.2 In his *Principia,* Isaac Newton established the three laws that govern the motion of objects: (1) bodies continue at rest or in uniform motion unless acted upon by an outside force; (2) an outside force causes an acceleration (and changes the **momentum**) of an object; and (3) for each action there is an equal and opposite reaction. Momentum is a measure of the motion of an object and depends on both its mass and its velocity. **Angular momentum** is a measure of motion of a spinning or revolving object. The **density** of an object is its mass divided by its volume.

2.3 Gravity, the attraction of all mass for all other mass, is the force that keeps the planets in orbit. Newton's law of gravity relates gravitational force to mass and distance $(F = GM_1M_2/R^2)$. Newton was able to show the equivalence of gravitational force (weight) on Earth to the gravitational force between objects in space. When Kepler's laws are re-examined in the light of gravitational theory, it becomes clear that the masses of both Sun and planet are important for the third law, which becomes $(M_1 + M_2) \times P^2 = D^3$. Mutual gravitational effects permit us to calculate the masses of astronomical objects, from comets to galaxies.

2.4 The lowest point in a satellite orbit is its **perigee,** and the highest point is its **apogee** (corresponding to **perihelion** and **aphelion** for an orbit about the Sun).

The planets all follow orbits about the Sun that are nearly circular and in the same plane. Most asteroids are found between Mars and Jupiter in the **asteroid belt,** while comets generally follow orbits of high eccentricity.

2.5 The orbit of an artificial satellite depends on the circumstances of its launch. The **circular satellite velocity** at the Earth's surface is 8 km/s, and the **escape velocity** is 11 km/s. There are many possible interplanetary trajectories, including those that use gravity-assisted flybys of one object to redirect the spacecraft toward its next target.

2.6 Gravitational problems that involve more than two interacting bodies are much more difficult to deal with than two-body problems. They require large computers for accurate solutions. If one object dominates gravitationally, it is possible to calculate the effects of a second object in terms of small perturbations. This approach was used by Adams and Leverrier to predict the position of Neptune from its perturbations of the orbit of Uranus and thus discover a new planet mathematically.

Review Questions

1. State Kepler's three laws in your own words.

2. Why did Kepler need Tycho Brahe's data to formulate his laws?

3. Which has more mass: an armful of feathers or an armful of lead? Which has more volume: a kilogram of feathers or a kilogram of lead? Which has more density: a kilogram of feathers or a kilogram of lead?

4. Explain how Kepler was able to find a relationship (his third law) between the periods and distances of the planets that did not depend on the masses of the planets or the Sun.

5. Write out Newton's three laws of motion in terms of what happens with the momentum of objects.

6. What planet has the largest:
  a. semimajor axis
  b. speed of revolution around the Sun
  c. period of revolution around the Sun
  d. eccentricity

7. Why do we say that Neptune was the first planet to be discovered through the use of mathematics?

Thought Questions

8. Is it possible to escape the force of gravity by going into orbit about the Earth? How does the force of gravity in the Russian space station Mir (orbiting 500 km above the Earth's surface) compare with that on the ground? (*Hint:* The Earth's gravity acts as if all the mass were concentrated at the center of the Earth. Is Mir significantly farther from the Earth's center than the Earth's surface is?)

9. What is the momentum of a body whose velocity is zero? How does Newton's first law of motion include the case of a body at rest?

10. Evil space aliens drop you and your astronomy instructor

1 km apart out in space, far from any star. Discuss the effects of gravity on each of you.

11. A body moves in a perfectly circular path at constant speed. Are there forces acting in such a system? How do you know?

12. As air friction causes a satellite to spiral inward closer to the Earth, its orbital speed increases. Why?

13. Use a history book or an encyclopedia to find out what else was happening in England during Newton's lifetime, and discuss what trends of the time might have contributed to his accomplishments and the rapid acceptance of his work.

14. What is the semimajor axis of a circle of diameter 24 cm? What is its eccentricity?

15. If 24 g of material fills a cube 2 cm on a side, what is the density of the material?

16. Draw an ellipse by the procedure described in the text, using a string and two tacks. Arrange the tacks so that they are separated by one-tenth the length of the string. Comment on the appearance of your ellipse. This (if you have been careful in your construction) is approximately the shape of the orbit of Mars.

17. The Earth's distance from the Sun varies from 147.2 million to 152.1 million km. What is the eccentricity of its orbit? (*Hint:* The distance between the foci of the ellipse is twice the distance between the Sun and the center of the ellipse. To find this, you need to first find the center of the ellipse by dividing the major axis in two. It helps to draw a diagram for yourself.)

18. Look up the revolution periods and distances from the Sun for Venus, Earth, Mars, and Jupiter. Calculate D^3 and P^2 (in the units specified in the text) and verify that they obey Kepler's third law.

19. What would be the period of a planet whose orbit has a semimajor axis of 4 AU? Of an asteroid with a semimajor axis of 10 AU?

20. What is the distance from the Sun (in astronomical units) of an asteroid with a period of revolution of eight years? What is the distance of a planet whose period is 45.66 days?

21. Newton showed that the periods and distances in Kepler's third law depend on the masses of the objects. What would be the period of revolution of the Earth (at 1 AU from the Sun) if the Sun had twice its present mass?

22. By what factor would a person's weight at the surface of the Earth be reduced if the Earth had its present mass but eight times its present volume? Its present size but only one-third its present mass?

Suggestions for Further Reading

Casper, M. *Kepler.* 1959, Collier. Biography by a respected historian.

Christianson, G. "The Celestial Palace of Tycho Brahe" in *Scientific American,* Feb. 1961, p. 118.

Christianson, G. "Newton's *Principia:* A Retrospective" in *Sky & Telescope,* July 1987, p. 18.

Cohen, I. "Newton's Discovery of Gravity" in *Scientific American,* Mar. 1981, p. 166.

Gingerich, O. *The Eye of Heaven: Ptolemy, Copernicus and Kepler.* 1993, American Institute of Physics Press.

King-Hele, D. and Eberst, R. "Observing Artificial Satellites" in *Sky & Telescope,* May 1986, p. 457.

Koestler, A. *The Sleepwalkers: A History of Man's Changing Vision of the Universe.* 1959, Macmillan. A journalist's recreation of Renaissance astronomical developments.

Thoren, V. *The Lord of Uraniborg.* 1990, Cambridge U. Press. Definitive modern study of Brahe's life and work.

Wilson, C. "How Did Kepler Discover His First Two Laws" in *Scientific American,* Mar. 1972.

Using **REDSHIFT**™

1. To track the orbits of the planets, click on *set location* and set the location to heliocentric, latitude to 90°, and distance to 10 AU, turn off the stars and deep-sky objects, and center and lock on the Sun. To get a Ptolemaic view, center and lock on Earth; use a waterbased pen to mark on the screen the position of Mars every 20 days.

During which part of the orbits does retrograde motion occur?

2. To demonstrate Kepler's Second Law, watch the motion of the asteroid Ganymede. Set the view to heliocentric with a latitude of 90° and a distance of 20 AU. Center and lock on the Sun, and turn off all the planets and stars. In the *Asteroids* panel, set the diameter range to 10 to 1000 km and the eccentricity range between 0.5 to 1.0. One asteroid, Ganymede, should appear on the screen. Set the *Time Step* to 50 days and mark Ganymede's motion on the screen with a waterbased pen.

Compare the skinny triangles when Ganymede is far from the Sun to the fat triangles swept out when it is near the Sun.

3. Find the dates when Earth is closest and farthest from the Sun by setting the *Location* to Earth and centering and locking on the Sun. Use *Find Object* to find the Sun. Click on *Reports* in the *Information* window, then choose *Planet Reports* for the current year. Use the *Planet Report* for the entire year to estimate these dates. Then use trial and error to find the exact dates by clicking on the Sun and reading its distance from the information window.

Why do the dates of perihelion and aphelion differ from the dates of the solstices?

As captured with a fish-eye lens aboard the Space Shuttle on December 9, 1993, the Earth hangs above the Hubble Space Telescope as it is repaired. The reddish continent is Australia, its size and shape distorted by the special lens. Since the seasons in the Southern Hemisphere are opposite ours, it is summer in Australia on this December day. (NASA)

CHAPTER 3

Earth, Moon, and Sky

Thinking Ahead

If the Earth's orbit is nearly a perfect circle (as we saw in the previous chapters), why is it hotter in summer and colder in winter? And why are the seasons in Australia or Peru the opposite of those in the United States or Europe?

The story is told that Galileo, as he left the Hall of the Inquisition following his retraction of the doctrine that the Earth rotates and revolves about the Sun, said under his breath, "But nevertheless it moves." The story is perhaps apocryphal, but certainly Galileo knew that the Earth was in motion, whatever the Catholic Church said.

It is the motions of the Earth that produce the seasons and give us our measures of time and date. The Moon's motions around us provide the concept of the month and the cycle of lunar phases. In this chapter we will examine some of the basic phenomena of our everyday world in their astronomical context.

I did not expect, from any accounts of preceding eclipses that I had read, to witness so magnificent an exhibition as that which took place.

Francis Baily in "Some Remarks on the Total Eclipse of the Sun on July 8th, 1842" (*Memoirs of the Royal Astronomical Society,* Vol. 15, 1846).

Earth and Sky

Locating Places on the Earth

Let's begin by fixing our position on the surface of the Earth. As we discussed in Chapter 1, the Earth's axis of rotation defines the locations of its North and South Poles, and of its equator halfway between. Two other directions are also defined by the Earth's motions: east is the direction toward which the Earth rotates, and west is its opposite. At almost any point on the Earth, the four directions—north, south, east, and west—are well defined, despite the fact that our planet is round rather than flat. The only exceptions are exactly at the North and South Poles, where the directions east and west are ambiguous (since the poles do not turn).

We can use these ideas to define a system of *coordinates* attached to our planet. Such a system, like the layout of streets and avenues in Manhattan or Salt Lake City, helps us find where we are or want to go. Coordinates on a sphere are a little more complicated than on a flat surface. We must define circles on the sphere that are equivalent to the rectangular grid that specifies position on a plane.

A **great circle** is any circle on the surface of a sphere whose center is at the center of the sphere. For example, the Earth's equator is a great circle on the Earth's surface, halfway between the North and South Poles. We can also imagine a series of great circles that pass through the North and South Poles. These circles are called **meridians;** they each cross the equator at right angles.

Any point on the surface of the Earth will have a meridian passing through it (Figure 3.1). This meridian specifies the east-west location, or *longitude,* of that place. By international agreement (and it took many meetings

Figure 3.2
The Royal Greenwich Observatory in England—the internationally agreed-upon zero point of longitude on the Earth. Here tourists can stand and straddle the exact line where longitude "begins." (© 1993 Hal Berol/Visuals Unlimited)

for the world's countries to agree), your longitude is defined as the number of degrees of arc along the equator between your meridian and the one passing through Greenwich, England.

Why Greenwich, you might ask? Every country wanted 0° longitude to pass through its own capital. Greenwich, the site of the old Royal Observatory (Figure 3.2), was selected because it was between continental Europe and the United States, and because it was the site for much of the development of a way to measure longitude at sea. Longitudes are measured either to the east or the west of the Greenwich meridian from 0° to 180°. As an example, the longitude of the clock-house benchmark of the U.S. Naval Observatory in Washington, D.C., is 77.066° W.

Your *latitude* (or north-south location) is the number of degrees of arc you are away from the equator along your meridian. Latitudes are measured either north or south of the equator from 0° to 90°. As an example, the latitude of the previously mentioned Naval Observatory benchmark is 38.921° N. The latitude of the South Pole is 90° S.

Locating Places in the Sky

Positions in the sky are measured in a manner very similar to that used on the surface of the Earth, except that instead of latitude and longitude, astronomers use coordinates called declination and right ascension. In denoting positions of objects in the sky, it is often convenient to make use of the fictitious celestial sphere, a concept, we recall, that many early peoples accepted literally. We saw in Chapter 1 that the sky appears to rotate about points above the North and South Poles of the Earth—the north celestial pole and the south celestial pole. Halfway between the celestial poles, and thus 90° from each, is the

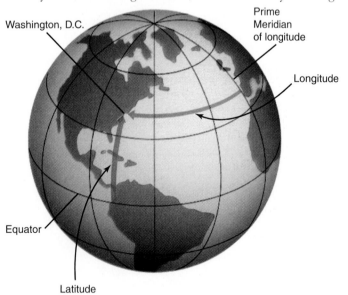

Washington, D.C.

Prime Meridian of longitude

Longitude

Equator

Latitude

Figure 3.1
The latitude and longitude of Washington, D.C.

Figure 3.3
The Foucault Pendulum. (© Bob Emott Photographer)

celestial equator, a great circle on the celestial sphere that is in the same plane as the Earth's equator.

Declination on the celestial sphere is measured the same way that latitude is measured on the sphere of the Earth: from the celestial equator toward the north (positive) or south (negative). So Polaris, the star near the north celestial pole, has a declination of almost $+90°$.

Right ascension (RA) is like longitude, except that instead of Greenwich, its arbitrarily chosen point of origin is the *vernal equinox*, a point in the sky where the ecliptic (the Sun's path) crosses the celestial equator. RA can be expressed in units of angle (degrees) or in units of time. This is because the celestial sphere seems to turn around the Earth once a day, and so the $360°$ of RA that it takes to go once around the celestial sphere can just as well be set equal to 24 hours. Then each $15°$ of arc is equal to 1 hour of time. The hours can be further subdivided into minutes. For example, the celestial coordinates of the bright star Vega (Appendix 14) are RA 18 h 36.2 m ($= 279.05°$) and declination $+38.77°$.

One way to visualize these circles in the sky is to imagine the Earth as a transparent sphere with the terrestrial coordinates (latitude and longitude) painted on it with dark paint. Imagine the celestial sphere around us as a giant ball, painted white on the inside. Then imagine yourself at the center of the Earth, with a bright lightbulb in the middle, looking out through its transparent surface to the sky. The terrestrial poles, equator, and meridians will be projected as dark shadows on the celestial sphere, giving us the system of coordinates in the sky.

The Foucault Pendulum

We have seen that the apparent rotation of the celestial sphere could be accounted for either by a daily rotation of the sky around a stationary Earth, or by the rotation of the Earth itself. Since the 17th century, it has been generally accepted that it is the Earth that turns, but not until the 19th century did the French physicist Jean Foucault provide a direct and unambiguous demonstration of this rotation. In 1851, he suspended a 60-m pendulum weighing about 25 kg from the domed ceiling of the Pantheon in Paris, and then started the pendulum swinging evenly. If the Earth had not been turning, there would have been no force to alter the pendulum's plane of oscillation and so it would have continued tracing the same path. Yet after a few minutes it was apparent that the pendulum's plane of motion was turning. Foucault explained that it was not the pendulum that was shifting, but rather the Earth that was turning underneath it (Figure 3.3).

3.2
The Seasons

One of the fundamental facts of life at the mid-latitudes, where most of this book's readers live, is that there are significant variations in the heat we receive from the Sun over the course of the year. We thus divide the year into *seasons*, each with its different amount of sunlight. The difference between seasons gets more pronounced the farther north or south from the equator we travel. And the seasons in the Southern Hemisphere are the reverse of what we find in the northern half of the Earth. With these observed facts in mind, let us ask what causes the seasons.

According to recent surveys, most people believe that the seasons are the result of the changing distance between the Earth and the Sun. It sounds reasonable at first: it *should* be colder when the Earth is farther from the Sun. But the facts don't bear out this hypothesis. While the Earth's orbit around the Sun is an ellipse, its distance from the Sun varies by only about three percent. That's not enough to cause significant variations in the Sun's heating. To make matters worse for people in North America who hold this hypothesis, the Earth is actually closest to the Sun in January, when the Northern Hemisphere is in the middle of winter. And if distance were the governing factor, why would the two hemispheres have opposite seasons? As we shall show, the seasons are in fact caused by the $23°$ tilt of the Earth's axis relative to the plane in which we circle the Sun.

Figure 3.4

The Earth at different seasons as it circles the Sun. During our winter, the Southern Hemisphere "leans into" the Sun and is illuminated more directly. In summer, it is the Northern Hemisphere that is leaning into the Sun and has longer days. In spring and autumn, the two hemispheres receive more equal shares of sunlight.

The Seasons and Sunshine

As we saw in Chapter 1, an equivalent way to look at our path around the Sun is to pretend that the Sun moves around the Earth (on a circle called the ecliptic) during the course of the year. Figure 3.4 shows the Earth's path around the Sun. Note that the Earth's axis is shown tilted by 23°. As the Earth travels around the Sun, in June the Sun is north of the celestial equator and thus favors the Northern Hemisphere of the Earth. In the figure, we see the Northern Hemisphere "leaning into" the Sun and being more directly illuminated in June. In December, the situation is reversed: the Sun is now favoring regions south of the equator. In our figure, it is the Southern Hemisphere that leans into the Sun, and the Northern Hemisphere that leans away. In September and March, the Sun crosses the celestial equator, and so the two hemispheres are equally favored.

How does the Sun's favoring one hemisphere translate into making it warmer for us down on the surface of the Earth? There are two effects we need to consider. When we lean into the Sun, sunlight hits us at a more direct angle and is more effective at heating the Earth's surface (Figure 3.5). You can get a similar effect by shining a flashlight onto a wall. If you shine the flashlight straight on (if the wall "leans into" the flashlight, so to speak), you get an intense spot of light on the wall. But if you hold the flashlight at an angle (if the wall "leans out" of the beam), then the spot of light will be more spread out. In the same way, the sunlight in June is more direct and intense on the Northern Hemisphere, and hence more effective at heating.

The second effect has to do with the length of time the Sun spends above the horizon (Figure 3.6). Even if you've never thought about astronomy before, we're sure you have observed that the days get longer in summer and

Figure 3.5

The Sun's rays in summer and winter. In summer the Sun appears high in the sky, and its rays hit the Earth more directly, spreading out less. In winter the Sun is low in the sky, and its rays spread out over a much wider area, becoming less effective at heating the ground.

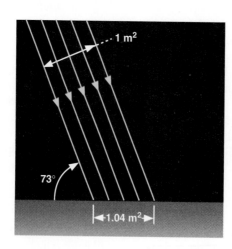
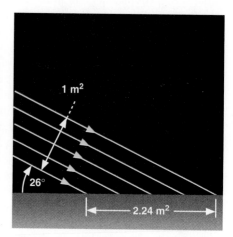

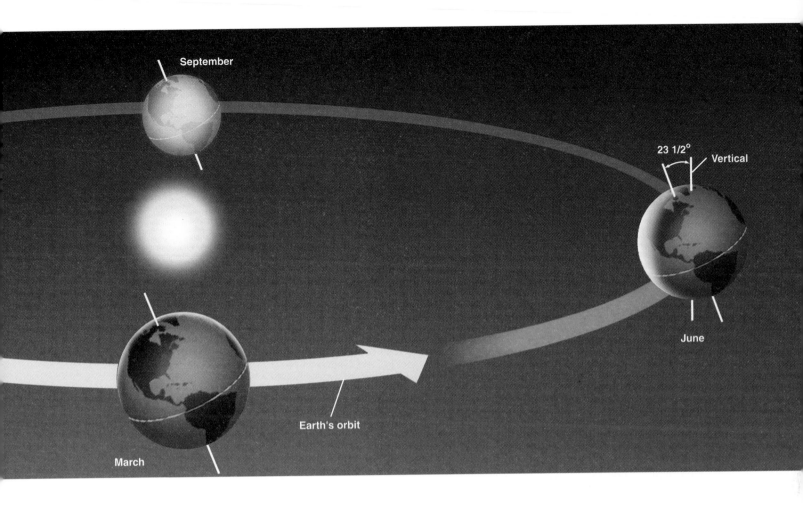

September

23 1/2° Vertical

June

Earth's orbit

March

shorter in winter. In June, the Sun is north of the celestial equator and spends more time with those who live in the Northern Hemisphere. It rises high in the sky, and is above the horizon in the United States for as much as 15 hours. Thus the Sun not only heats us with more direct rays, but has more time to do it each day. (Notice that our gain is the Southern Hemisphere's loss. There, the June Sun is low in the sky, meaning fewer daylight hours. In Chile, for example, June is a colder, darker time of year.) Let's look at what the Earth looks like at some specific dates of the year, when these effects are at their maximum.

On about June 22 (the date we who live in the Northern Hemisphere call the *summer solstice* or sometimes the first day of summer), the Sun shines down most directly upon the Northern Hemisphere of the Earth. It appears 23° north of the equator and thus on that date passes through the zenith of places on the Earth that are at 23° N latitude. The situation is shown in detail in Figure 3.7. To a person at latitude 23° N (near Hawaii, for example), the Sun is directly overhead at noon. (This latitude, where the Sun can appear at the zenith at noon on the first day of summer, is called the *Tropic of Cancer.*)

We also see in Figure 3.7 that the Sun's rays shine down past the North Pole. As the Earth turns on its axis, the North Pole will always be illuminated by the Sun; all places within 23° of the pole have sunshine for 24 hours on the first day of summer. The Sun is as far north on this date as it can get; thus, 90° − 23° or 67° N is the southernmost latitude where the Sun can ever be seen for a full 24-hour period (the "land of the midnight Sun"). That circle of latitude is called the *Arctic Circle.*

Many early cultures scheduled special events around the summer solstice to celebrate the longest days and thank their gods for making the weather warm. This required them to keep track of the lengths of the days and the northward trek of the Sun in order to know the right day for the "party." (You can do the same thing by watching for several weeks, from the same observation point, where the Sun rises or sets relative to a fixed landmark. In summer, the Sun will rise farther and farther north of east, and set farther and farther north of west, reaching the maximum around the summer solstice.)

Now look at the South Pole in Figure 3.7. On June 22, all places within 23° of the South Pole—that is, south of what we call the *Antarctic Circle*—do not see the Sun at all for 24 hours. Day and night are both dark.

The situation is reversed six months later, about December 22 (the date of the *winter solstice* or the so-called first day of winter in the Northern Hemisphere), as shown

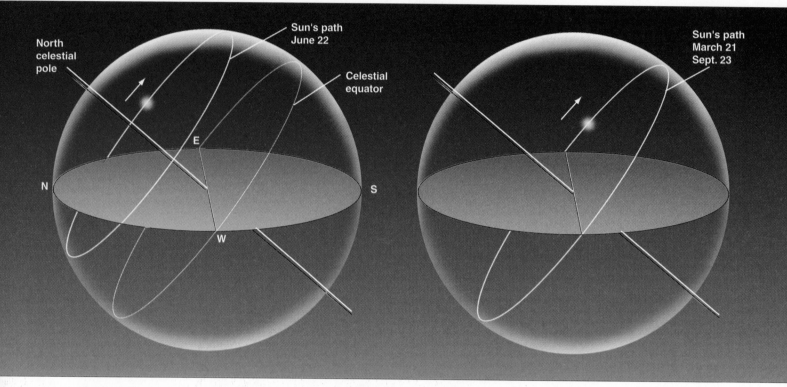

Figure 3.6
The path of the Sun in the sky and the resulting length of the day for different seasons. On June 22, the Sun rises north of east and sets north of west; for observers in the Northern Hemisphere of the Earth, it spends about 15 hours above the horizon. On December 22, the Sun rises south of east and sets south of west; it spends only 9 hours above the horizon, which means short days and long nights (and a strong need for people to hold celebrations to cheer themselves up). On March 21 and September 21, the Sun spends an equal amount of time above and below the horizon.

in Figure 3.8. Now it is the Arctic Circle that has the 24-hour night, and the Antarctic Circle that has the midnight Sun. At latitude 23° S, called the *Tropic of Capricorn*, the Sun passes through the zenith at noon. Days are longer in the Southern Hemisphere and shorter in the north. In the United States and southern Europe, we might get only 9 or 10 hours of sunshine during the day. It is winter in the Northern Hemisphere and summer in the Southern Hemisphere.

Every culture that developed north of the equator has a celebration around December 22 to help people deal

Figure 3.7
The Earth on June 22, the summer solstice in the Northern Hemisphere.

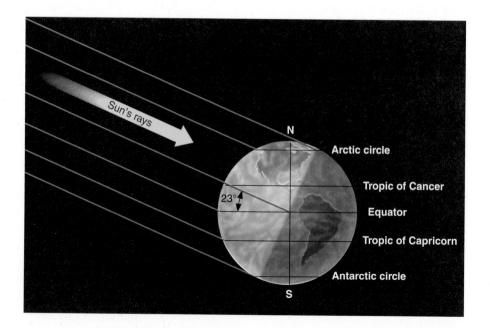

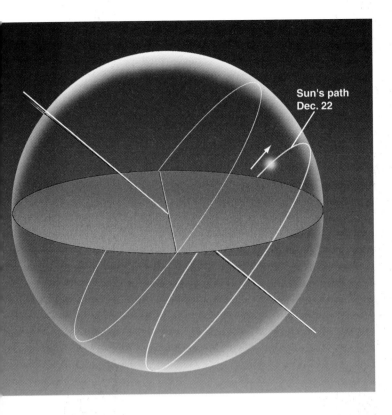
Sun's path
Dec. 22

though few of us now feel the need to pause while eating the sacrificial turkey to ask the sun god to make the days longer).

Halfway between the solstices, on about March 21 and September 23, the Sun is on the celestial equator. From Earth, it appears above our planet's equator, and favors neither hemisphere. Every place on the Earth then receives exactly 12 hours of sunshine and 12 of night. The points where the Sun crosses the celestial equator are called the *vernal* (spring) and *autumnal* (fall) *equinoxes.*

The Seasons at Different Latitudes

The seasonal effects are different at different latitudes on Earth. At the equator, for instance, all seasons are much the same. Every day of the year, the Sun is up half the time, so there are always 12 hours of sunshine and 12 hours of night. Local residents define the seasons by the amount of rain rather than by the amount of sunlight. As we travel north or south, the seasons become more pronounced, until we reach extreme cases in the Arctic and Antarctic.

At the North Pole, all celestial objects that are north of the celestial equator are always above the horizon and, as the Earth turns, circle around parallel to it. The Sun is north of the celestial equator from about March 21 to September 23, and so at the North Pole, the Sun rises when it reaches the vernal equinox and sets when it reaches the autumnal equinox. Each year there are six months of sunshine at each pole, followed by six months of darkness.

Clarifications About the Real World

In our discussions so far, we have been describing the rising and setting of the Sun and stars as they would appear

with the depressing lack of sunlight and the often dangerously cold temperatures. Originally, this was often a time for huddling with family and friends, for sharing the reserves of food and drink, for rituals asking the gods to return the light and heat and turn the cycle of the seasons around. Many cultures constructed elaborate devices for anticipating when the shortest day of the year was coming. Stonehenge in England (see Section 3.4) is probably one such device, built long before the invention of writing. In our own time, we continue the winter solstice tradition with Christmas and Chanukah celebrations (al-

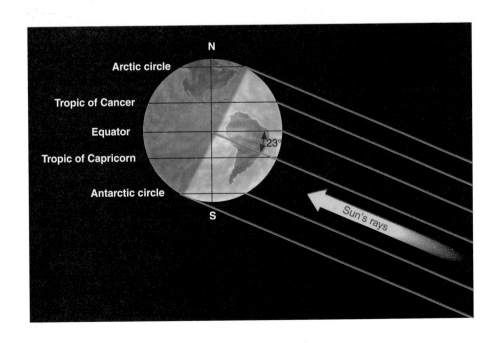

Figure 3.8
The Earth on December 22, the winter solstice in the Northern Hemisphere.

if the Earth had little or no atmosphere. In reality, however, the atmosphere has the curious effect of allowing us to see a little way "over the horizon." This effect is a result of refraction, the bending of light in a medium such as air or water, a phenomenon we will discuss in Chapter 5. Because of this atmospheric refraction, the Sun appears to rise earlier, and to set later, than it would if no atmosphere were present.

In addition, the atmosphere scatters light and provides some twilight illumination even when the Sun is below the horizon. Astronomers define morning twilight as beginning when the Sun is 18° below the horizon, while evening twilight extends until the Sun sinks more than 18° below the horizon.

These atmospheric effects require small corrections in many of our statements about the seasons. At the equinoxes, for example, the Sun appears to be above the horizon for a few minutes longer than 12 hours, and below the horizon for less than 12 hours. These effects are most dramatic at the Earth's poles, where the Sun actually rises more than a week before it reaches the celestial equator. And, as a consequence of twilight, the period of real darkness at each pole lasts for only about three months of each year, rather than six.

Also, you probably know that the summer solstice (June 22) is not the warmest day of the year, even if it is the longest. The hottest months in the Northern Hemisphere are July and August. This is because our weather involves the air and water covering the Earth's surface, and these large reservoirs do not heat up instantaneously. Just as a swimming pool does not get warm the moment the Sun rises, but is warmest late in the afternoon after it has had time to absorb the Sun's heat, so the Earth gets warmer after it has had a chance to absorb the extra sunlight that is the Sun's summer gift to us. In the same way, the coldest times of winter are a month or more after the winter solstice.

3.3

Keeping Time

The measurement of time is based on the rotation of the Earth. Throughout most of human history, time has been reckoned by the positions of the Sun and stars in the sky. Only recently have mechanical and electronic clocks taken over this function in regulating our lives.

The Length of the Day

The most fundamental astronomical unit of time is the day, measured in terms of the rotation of the Earth. There is, however, more than one way to define the day. Usually, it is the rotation period of the Earth with respect to the Sun, called the **solar day.** After all, for most people sunrise is more important than the rising time of Arcturus or

some other star, so we set our clocks to some version of sun-time. However, astronomers also use a **sidereal day,** which is defined in terms of the rotation period of the Earth with respect to the stars.

A solar day is slightly longer than a sidereal day, because (as you can see from Figure 3.9) the Earth moves a significant distance along its path around the Sun in a day. Suppose we start the day when the Earth's orbital position is at A, with both the Sun and some distant star (located in direction C) being above an observer at point O on the Earth. When the Earth has completed one rotation with respect to the distant star, C is again above O. However, notice that the Sun has not yet reached a position above O, because of the movement of the Earth along its orbit from A to B. To complete a solar day, the Earth must rotate an additional amount, equal to 1/365 of a full turn. The time required for this extra rotation is 1/365 of a day, or about 4 min, so the solar day is about 4 min longer than the sidereal day.

Since our ordinary clocks are set to solar time, stars rise 4 min earlier each day. Astronomers prefer sidereal time for planning their observations, because in that system, a star rises at the same time every day.

Apparent Solar Time

Apparent solar time is reckoned by the actual position of the Sun in the sky (or, during the night, its position below the horizon). This is the kind of time indicated by sundials, and it probably represents the earliest measure of time used by ancient civilizations. Today we adopt the middle of the night as the starting point of the day, and measure time in hours elapsed since midnight.

During the first half of the day, the Sun has not yet reached the meridian (the great circle in the sky that passes through our zenith). We designate those hours as before midday (*ante meridiem*, or A.M.). We customarily start numbering the hours after noon over again, and designate them by P.M. (*post meridiem*) to distinguish them from the morning hours.

Although apparent solar time seems simple, it is not really very convenient to use. The exact length of an apparent solar day varies slightly during the year. The eastward progress of the Sun in its annual journey around the sky is not uniform because the speed of the Earth varies slightly in its elliptical orbit. Another reason is that the Earth's axis of rotation is not perpendicular to the plane of its revolution. Thus apparent solar time does not advance at a uniform rate. After the invention of clocks that ran at a uniform rate, it became necessary to abandon the apparent solar day as the fundamental unit of time.

Mean Solar Time and Standard Time

Mean solar time is based on the *average* value of the solar day over the course of the year. A mean solar day contains exactly 24 hours, and it is what we use in our everyday timekeeping. Although mean solar time has the advantage of progressing at a uniform rate, it is still inconvenient for practical use because it is determined by the position of the Sun. For example, noon occurs when the Sun is overhead. But because we live on a round Earth, the exact time of noon is different as you change your longitude by moving east or west.

If mean solar time were strictly observed, people traveling east or west would have to reset their watches continually as the longitude changed, just to read the local mean time correctly. For instance, a commuter traveling from Oyster Bay on Long Island to New York City would have to adjust the time on the trip through the East River tunnel, because Oyster Bay time is actually about 1.6 min more advanced than that of Manhattan. (And just imagine an airplane trip in which the flight attendant gets on the intercom every minute, saying please reset your watch for local mean time.)

Until near the end of the last century, every city and town in the United States did keep its own local mean time. With the development of railroads and the telegraph, however, the need for some kind of standardization became evident. In 1883, the nation was divided into four standard time zones (now five, including Hawaii and Alaska). Within each zone, all places keep the same standard time, the local mean solar time of a standard line of longitude running more or less through the middle of each zone. Now travelers reset their watches only when the time change has amounted to a full hour. Pacific standard time is 3 hours earlier than eastern standard time, a fact that becomes painfully obvious in California when someone on the East Coast forgets and calls you at 5 A.M.

For local convenience, the boundaries between the U.S. time zones are chosen to correspond to divisions between states. Since 1884, standard time has been in use around the world by international agreement (with 24 time zones circling the globe). Almost all countries have adopted one or more standard time zones, although one of the largest nations, India, has settled on a half-zone, being 5 1/2 hours from Greenwich standard.

Daylight saving time is simply the local standard time of the place plus 1 hour. It has been adopted for spring and summer use in most states in the United States, as well as in many other countries, to prolong the sunlight into evening hours, on the apparent theory that it is easier to change the time by government action than it would be for individuals or businesses to adjust their own schedules to produce the same effect. It does not, of course, "save" any daylight at all.

The International Date Line

The fact that time is always more advanced to the east presents a problem. Suppose you travel eastward around the world. You pass into a new time zone, on the average, about every 15° of longitude you travel, and each time you dutifully set your watch ahead an hour. By the time you

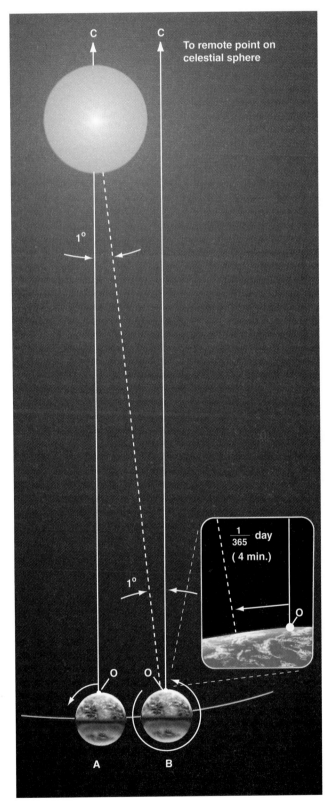

Figure 3.9
The difference between a sidereal and a solar day. This is a top view, looking down as the Earth orbits the Sun. Because the Earth moves around the Sun (roughly 1° per day), after one complete rotation of the Earth relative to the stars, we do not see the Sun in the same position.

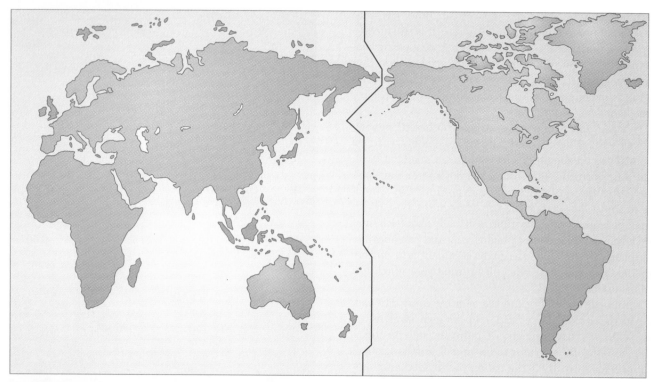

Figure 3.10
The international date line is an arbitrarily drawn line on the Earth where the date changes. So that neighbors do not have different days, the line is located where the Earth's surface is mostly water.

have completed your trip, you have set your watch ahead through a full 24 hours and thus gained a day over those who stayed at home.

The solution to this dilemma is the **international date line,** set by international agreement to run approximately along the 180° meridian of longitude. The date line runs about down the middle of the Pacific Ocean, although it jogs a bit in a few places to avoid cutting through groups of islands and through Alaska (Figure 3.10). By convention, at the date line the date of the calendar is changed by one day. Crossing the date line from west to east, thus advancing your time, you compensate by decreasing the date; crossing from east to west, you increase the date by one day. We simply must accept that the date will differ in different cities at the same time. A well-known example is the date when the Imperial Japanese Navy bombed Pearl Harbor in Hawaii, known in the United States as Sunday, December 7, 1941, but taught to Japanese students as Monday, December 8.

3.4

The Calendar

The Challenge of the Calendar

"What's today's date?" is one of the most common questions you can ask (usually when writing a check, or wor-

ried about the next exam). Long before the era of digital watches that tell the date, people used calendars to help measure the passage of time.

There are two traditional functions of any calendar. First, it must keep track of time over the course of longer spans, allowing people to anticipate the cycle of the seasons and to honor special religious or personal anniversaries. Second, in order to be useful to a large number of people, a calendar must use natural time intervals that everyone can agree on—those defined by the motions of the Earth, Moon, and sometimes even the planets. The natural units of our calendar are the *day,* based on the period of rotation of the Earth; the *month,* based on the period of revolution of the Moon about the Earth; and the *year,* based on the period of revolution of the Earth about the Sun. Difficulties have resulted from the fact that these three periods are not commensurable—that is, one does not divide evenly into any of the others.

The rotation period of the Earth is, by definition, 1.0000 day. The period required by the Moon to complete its cycle of phases, called the lunar month, is 29.5306 days. The basic period of revolution of the Earth, called the tropical year, is 365.2422 days. The ratios of these numbers are not very convenient for calculations. In other words, it turns out that our celestial clocks are pretty incompatible. This is the historic challenge of the calendar, dealt with in various ways by different cultures.

Early Calendars

Even the earliest cultures were concerned with the keeping of time and the calendar. Particularly interesting are monuments left by Bronze Age people in northwestern Europe, especially the British Isles. The best preserved of the monuments is Stonehenge (Figure 3.11), about 13 km from Salisbury in southwest England. It is a complex array of stones, ditches, and holes arranged in concentric circles. Carbon dating and other studies show that Stonehenge was built during three periods ranging from about 2800 B.C. to 1500 B.C. Some of the stones are aligned with the directions of the Sun and Moon during their risings and settings at critical times of the year (such as the summer and winter solstices), and it is generally believed that at least one function of the monument was connected with the keeping of a calendar.

The Maya in Central America, who thrived between the 2nd and 10th centuries A.D., were also concerned with the keeping of time. The Mayan calendar was as sophisticated as, and perhaps more complex than, contemporary calendars in Europe. The Maya did not attempt to correlate their calendar accurately with the length of the year or lunar month. Rather, their calendar was a system for keeping track of the passage of days and for counting time far into the past or future. Among other purposes, it was useful for predicting astronomical events—for example, the positions of Venus in the sky (Figure 3.12).

The ancient Chinese developed an especially complex calendar, largely limited to a few privileged hereditary court astronomer–astrologers. In addition to the motions

Figure 3.12
Ruins of the Caracol, a Mayan observatory at Chichin Itza in the Yucatan, Mexico, dating from around 1000 A.D. (David Morrison)

of the Earth and Moon, they were able to fit in the approximately 12-year cycle of Jupiter, which was central to their system of astrology. The Chinese still preserve some aspects of this system in their cycle of 12 "years"—the Year of the Dragon, the Year of the Pig, and so on—that are defined by the position of Jupiter in the zodiac.

Our Western calendar derives from Greek calendars dating from at least the 8th century B.C. They led, eventually, to the *Julian calendar* introduced by Julius Caesar, which approximated the year at 365.25 days, fairly close to the actual value of 365.2422. The Romans achieved this approximation by declaring years to have 365 days each, with the exception of every fourth year. The *leap year* was to have one extra day, bringing its length to 366 days and thus making the average length of the year in the Julian calendar 365.25 days.

The Romans had dropped the almost impossible task of trying to base their calendar on the Moon as well as the Sun, although a vestige of older lunar systems can be seen in the fact that our months have an average length of about 30 days.

Figure 3.11
Part of Stonehenge, an ancient monument built between 2800 and 1500 B.C., and used to keep track of the motions of the Sun and Moon. Today, heedless tourists and vandals have disturbed and chipped away at the stones to such a degree that the site is now fenced in, and entry is restricted. (David Morrison)

Astronomy and the Days of the Week

The week seems independent of celestial motions, although its length may have been based on the time between quarter phases of the Moon. In Western culture, the seven days of the week are named after the seven "wanderers" that the ancients saw in the sky: the Sun, Moon, and five planets visible to the unaided eye (Mercury, Venus, Mars, Jupiter, and Saturn).

In English, we can easily recognize Sun-day, Moon-day, and Saturn-day, but the other days are named after the Norse equivalents of the Roman gods that gave their names to the planets. In languages more directly related to Latin, the correspondences are clearer. Wednesday, Mercury's day, for example, is Mercoledi in Italian, Mercredi in French, and Miercoles in Spanish. Mars gives its name to Tuesday (Martes in Spanish), Jupiter or Jove to Thursday (Giovedi in Italian), and Venus to Friday (Vendredi in French).

But there is no reason that the week *has* to have seven days rather than five or eight. It is interesting to speculate that if we had lived in a planetary system where fewer planets were visible, we might well have had a shorter week.

However, lunar calendars remained in use in other cultures, and Islamic calendars are still primarily lunar rather than solar.

The Gregorian Calendar

Although the Julian calendar (which was adopted by the early Christian Church) represented a great advance, its average year still differed from the true year by about 11 min, an amount that accumulates over the centuries to an appreciable error. By 1582, that 11 min per year had added up to the point where the first day of spring was occurring on March 11 instead of March 21. If the trend were allowed to continue, eventually the Christian celebration of Easter would be occurring in early winter. Pope Gregory XIII, a contemporary of Galileo, felt it necessary to institute further calendar reform.

The Gregorian calendar reform consisted of two steps. First, ten days had to be dropped out of the calendar to bring the vernal equinox back to March 21; by proclamation, the day following October 4, 1582, became October 15. The second feature of the new Gregorian calendar was a change in the rule for leap year, making the average length of the year more closely approximate the tropical year. Gregory decreed that three of every four century years, all leap years under the Julian calendar, would be common years henceforth. The rule was that only century years divisible by 400 would be leap years. Thus, 1700, 1800, and 1900—all divisible by 4 but not by 400—were not leap years in the Gregorian calendar. On the other hand, the years 1600 and 2000, both divisible by 400, were leap years. The average length of this Gregorian year, 365.2425 mean solar days, is correct to about one day in 3300 years.

The Catholic countries immediately put the Gregorian reform into effect, but countries under control of the Eastern Church and most Protestant countries did not adopt it until much later. It was 1752 when England and the American colonies finally made the change. By parliamentary decree, September 2, 1752, was followed by September 14. Although special laws were passed to prevent such abuses as landlords collecting a full month's rent for September, there were still riots, and people demanded their 12 days back. Russia did not abandon the Julian calendar until the time of the Bolshevik revolution. The Russians then had to omit 13 days to come into step with the rest of the world.

3.5

Phases and Motions of the Moon

After the Sun, the Moon is the brightest and most obvious object in the sky. Unlike the Sun, it does not shine under its own power, but merely glows with reflected sunlight. If you were to follow its progress in the sky for a month, you would observe a cycle of phases, with the Moon starting out dark and getting more and more illuminated by sunlight over the course of about two weeks. After the Moon's disk becomes fully visible, it then begins to fade, returning to dark about two weeks later. These changes fascinated and mystified many early cultures, who came up with marvelous stories and legends to explain the cycle of the Moon. Even in the modern world, many people don't understand what causes the phases, thinking that they are somehow related to the shadow of the Earth. Let us see how the phases can be explained by the motion of the Moon relative to the bright light-source in the solar system, the Sun.

Lunar Phases

Although we know the Sun moves 1/12 of its path around the sky each month, for purposes of explaining the phases, we can assume that the Sun's light comes from roughly the same direction during the course of a four-week lunar cycle. The Moon, on the other hand, moves completely around the Earth in that time. As we watch the Moon from our vantage point on Earth, how much of its face we see illuminated by sunlight depends on the angle the Sun makes with the Moon.

Here is a simple experiment to show you what we mean: Stand about 6 ft in front of a bright electric light in a completely dark room (or outdoors at night) and hold in your hand a small round object such as a tennis ball or an orange. Your head can then represent the Earth, the light the Sun, and the ball the Moon. Move the ball around your head (making sure you don't cause an eclipse by blocking the light with your head). You will see phases just like those of the Moon on the ball. (Another good way to get acquainted with the phases and motions of the Moon is to follow our satellite in the sky for a month or two, recording its shape, its direction from the Sun, and when it rises and sets.)

Let's examine the Moon's cycle of phases using Figure 3.13. The Moon is said to be *new* when it is in the same general direction in the sky as the Sun (position *A*). Here its illuminated (bright) side is turned away from us, while its dark side is turned toward us. In this phase it is invisible; its dark, rocky surface does not give off any light of its own. Since the new moon is in the same part of the sky as the Sun, it rises at sunrise and sets at sunset.

But the Moon does not remain in this phase long, because it moves 12° eastward each day. A day or two after the new phase, the thin *crescent* first appears as we see a small part of the Moon's illuminated hemisphere. It has moved into a position where it now reflects a little sunlight toward us on one side. The bright crescent increases in size on successive days as the Moon moves farther and farther around the sky away from the direction of the Sun (position *B*). Because the Moon is moving eastward away from the Sun, it rises later and later each day, like a student during summer vacation.

After about one week, the Moon is one-quarter of the way around its orbit (position *C*) and so we say it is at the *first quarter* phase. Now half of the Moon's illuminated side is visible to Earth observers. Because of its eastward motion, the Moon now lags about one-quarter of the day

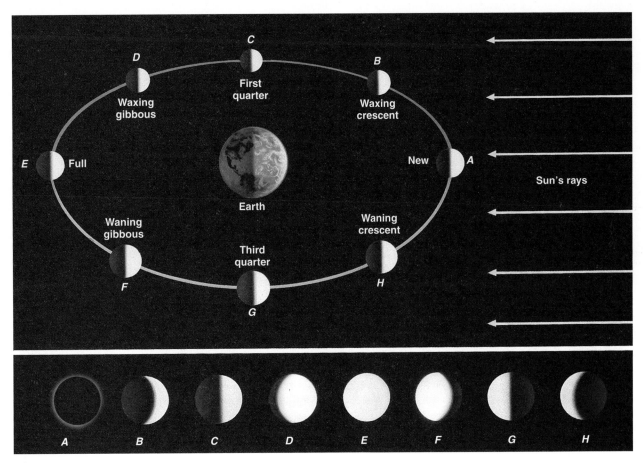

Figure 3.13

Phases of the Moon during the course of a complete cycle. The upper part shows a perspective from space, with the Sun off to the right in a fixed position. Now imagine yourself standing on the Earth, facing the Moon in each part of its orbit around the Earth. From Earth, you would see the phases shown in the strip below. (Please note that the distance of the Moon from the Earth is not to scale in this diagram.)

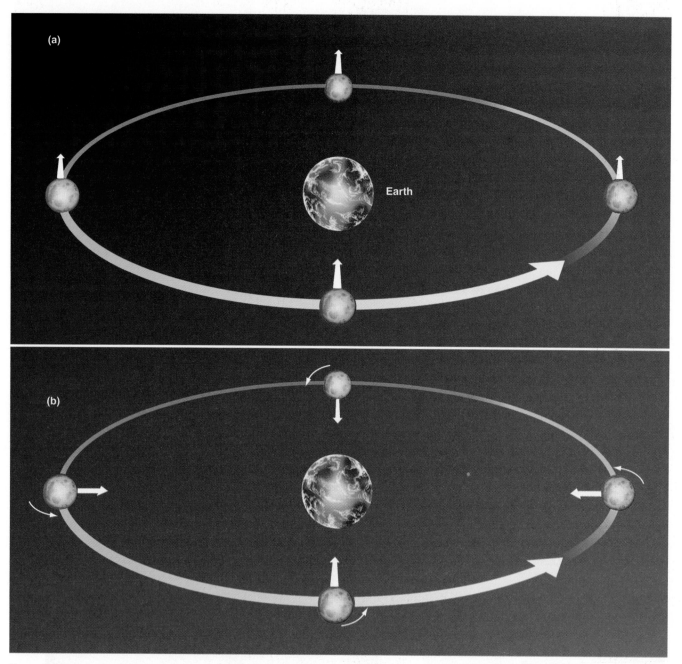

Figure 3.14

In this figure we stuck a white arrow into a point on the Moon to keep track of its sides. (a) If the Moon did not rotate as it orbited the Earth, it would present all of its sides to our view; hence the white arrow would only point directly toward the Earth in the bottom position on the diagram. (b) Actually, the Moon rotates in the same period that it revolves, so we always see the same side (the white arrow keeps pointing to the Earth).

behind the Sun, rising roughly around noon and setting roughly around midnight.

During the week after the first quarter phase, we see more and more of the Moon's illuminated hemisphere (position *D*), a phase that is called waxing (or growing) *gibbous*. Eventually, the Moon arrives at position *E* in our figure, where it and the Sun are opposite each other in the sky. The side of the Moon turned toward the Sun is turned toward the Earth, and we have *full* phase.

When the Moon is full, it lags half an orbit and half a day behind the Sun (and so they are opposite each other in the sky). This means the Moon does the opposite of

what the Sun does, rising at sunset and setting at sunrise. Note what that means in practice. The completely illuminated (and thus very noticeable) Moon rises just as it gets dark, remains in the sky all night long, and sets as the Sun's first rays are seen at dawn. And when is it highest in the sky and most noticeable? At midnight, a time made famous in generations of horror novels and films.

Folk wisdom has it that more crazy behavior is seen during the time of the full moon (the Moon even gives a name to crazy behavior—"lunacy"). But, in fact, statistical tests of this "hypothesis," involving thousands of records from hospital emergency rooms and police files, do not reveal any correlation with the phases of the Moon. As many homicides occur during the new moon or the crescent moon as during the full moon. Most investigators believe that the real story is not that more crazy behavior *happens* on nights with a full moon, but rather that we are more likely to *notice* such behavior with the aid of a bright celestial light that is up all night long.

During the two weeks following the full moon, the Moon goes through the same phases again in reverse order (points *F*, *G*, and *H* in Figure 3.13), returning to new phase after about 29.5 days. About a week after full moon, for example, the Moon is now at *third quarter*—meaning that it is three-quarters of the way around, not that it is three-fourths illuminated. In fact, half of the visible side of the Moon is again dark. At this phase, the Moon is now rising around midnight and setting around noon.

Note that there is one thing quite misleading about Figure 3.13. If you look at the Moon in position *E*, it appears as if, while full in theory, its illumination would in fact be blocked by a big fat Earth, and hence we would not see anything on the Moon except the Earth's shadow. In reality, the Moon is nowhere near as close to the Earth (nor is its path so identical with the Sun's in the sky) as this diagram might lead you to believe. The Moon is actually 30 Earth diameters away from us, and in the Prologue you will find a diagram that shows the two bodies to scale. In reality, the Earth's shadow misses the Moon most months, and we regularly get treated to a full moon. The times when the Earth's shadow does fall on the Moon are called lunar eclipses and will be discussed in Section 3.7.

You can see from this tour of the Moon's phases and times of rising and setting that the writer of the old song that went "I've got the Sun in the morning, and the Moon at night" did not quite remember his college astronomy course. It is just during the full phase that the Moon is up only at night. At other times of the month, it may be visible all morning (third quarter) or all afternoon (first quarter) in the daytime sky.

The Moon's Revolution and Rotation

The Moon's sidereal period—that is, the period of its revolution about the Earth measured with respect to the stars—is a little over 27 days: 27.3217 days to be exact. The time interval in which the phases repeat—say from full to full—is 29.5306 days. The difference is again the fault of the Earth's motion around the Sun. The Moon must make more than a complete turn around the moving Earth to get back to the same phase with respect to the Sun. As we saw, the Moon changes its position on the celestial sphere rather rapidly: even during a single evening, the Moon creeps visibly eastward among the stars, traveling its own width in a little less than 1 hour. The delay in moonrise from one day to the next, caused by this eastward motion, averages about 50 min.

The Moon *rotates* on its axis in exactly the same time that it takes to *revolve* about the Earth. As a consequence (as shown in Figure 3.14), the Moon always keeps the same face turned toward the Earth. The differences in its appearance from one night to the next are due to changing illumination by the Sun, not to its own rotation. You sometimes hear the back side of the Moon (the side we never see) called the "dark side." This is a misunderstanding of the real situation: which side is light and which is dark changes as the Moon moves around the Earth. The back side is dark no more frequently than the front side. Since the Moon rotates, the Sun rises and sets on all sides of the Moon.

3.6

Ocean Tides and the Moon

Anyone living near the sea is familiar with the twice-daily rising and falling of the tides. It was clear early in history that tides must be related to the Moon, because the daily delay in high tide is the same as the daily delay in the Moon's rising. A satisfactory explanation of the tides, however, awaited the theory of gravity, supplied by Newton.

The Pull of the Moon on the Earth

The gravitational forces exerted by the Moon at several points on the Earth are illustrated in Figure 3.15. These forces differ slightly from one another because the Earth is not a point, but has a certain size; all parts are not

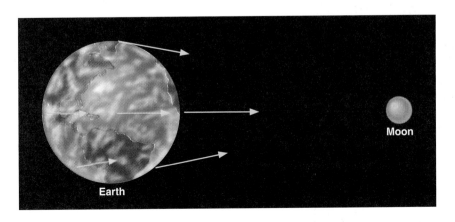

Figure 3.15
The Moon's differential attraction of different parts of the Earth. (Note that the differences have been exaggerated for educational purposes.)

equally distant from the Moon, nor are they all in exactly the same direction from the Moon.

The Earth, moreover, is not perfectly rigid. As a result, the differences among the forces of the Moon's attraction on different parts of the Earth (called differential forces) cause the Earth to distort slightly. The side of the Earth nearest the Moon is attracted toward the Moon more strongly than is the center of the Earth, which in turn is attracted more strongly than is the side opposite the Moon. Thus, the differential forces tend to stretch the Earth slightly into a *prolate spheroid* (a football shape), with its long diameter pointed toward the Moon.

If the Earth were made of water, it would distort until the Moon's differential forces over different parts of its surface came into balance with the Earth's own gravitational forces pulling it together. Calculations show that in this case the Earth would distort from a sphere by amounts ranging up to nearly 1 m. Measurements of the actual deformation of the Earth show that the solid Earth *does* distort, but only about one-third as much as water would because of the great rigidity of the Earth's interior.

Since the tidal distortion of the solid Earth amounts at its greatest to only about 20 cm, the Earth does not distort enough to balance the Moon's differential forces with its own gravity. Hence objects at the Earth's surface experience tiny horizontal tugs, tending to make them slide about. These so-called *tide-raising* forces are too insignificant to affect solid objects like astronomy students or rocks in the Earth's crust, but they do affect the waters in the oceans.

The Formation of Tides

The tide-raising forces, acting over a number of hours, produce motions of the water that result in measurable tidal bulges in the oceans. Water on the side of the Earth facing the Moon is drawn toward it, piling up to greater depths there, with the greatest depths at the point just below the Moon. On the side of the Earth opposite the Moon, water flows to produce a tidal bulge there as well (Figure 3.16).

Figure 3.17
Minas Basin seen at low and high tide. (Courtesy Nova Scotia Tourism)

Note that the tidal bulges in the oceans do not result from the Moon's compressing or expanding the water, nor from the Moon's lifting the water "away from the Earth." Rather, they result from an actual flow of water over the Earth's surface toward the regions below and opposite the Moon, causing the water to pile up to greater depths at those places (Figure 3.17).

In the idealized (and, as we shall see, oversimplified) model just described, the height of the tides would be only a few feet. The rotation of the Earth would carry an observer at any given place alternately into regions of deeper and shallower water. An observer being carried toward the regions under or opposite the Moon, where the water was deepest, would say, "The tide is coming in"; when carried away from those regions, the observer would say, "The tide is going out." During a day, the observer would be carried through two tidal bulges (one on each side of the Earth) and so would experience two high tides and two low tides.

The Sun also produces tides on the Earth, although it is less than half as effective a tide-raising agent as is the Moon. The actual tides we experience are a combination of the larger effect of the Moon and the smaller effect of the Sun. When the Sun and Moon are lined up (at new moon or full moon), the tides produced by the Sun and Moon reinforce each other and so are greater than normal

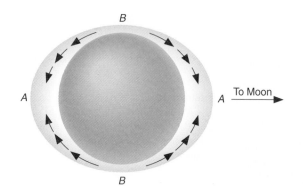

Figure 3.16
Tidal bulges in an "ideal" ocean.

George Darwin
and the Slowing of the Earth

The rubbing of the water over the face of the Earth involves an enormous amount of energy. Over long periods of time, the friction of the tides is actually slowing down the rotation of the Earth. Our day gets longer by about 0.002 seconds each century. While that seems very small, such tiny changes can add up over millions and billions of years.

Although the Earth's spin is slowing down, the angular momentum (see Chapter 2) in a system such as the Earth–Moon system cannot change. So some other spin motion must speed up to take the extra angular momentum. The details of what happens were worked out over a century ago by George Darwin (1845–1912), the son of naturalist Charles Darwin. George Darwin had a strong interest in science, but studied law for six years and was admitted to the bar. However, he never practiced law, returning to science instead and eventually becoming a professor at Cambridge University. He was a protégé of Lord Kelvin, one of the great physicists of the 19th century, and became interested in the long-term evolution of the solar

Darwin. (© Royal Society)

system. He specialized in making detailed (and difficult) mathematical calculations of how orbits and motions will change over geologic time.

What Darwin calculated for the Earth–Moon system was that the Moon will slowly spiral outward, away from the Earth. As it moves farther away, it will orbit less quickly (just as planets farther from the Sun move more slowly in their orbits). Thus the month will get longer. Also, since the Moon will be more distant, total eclipses of the Sun (see Section 3.7) will no longer be visible from Earth.

The day and the month will both continue to get longer, although bear in mind that the effects are very gradual. The calculations show that ultimately—billions of years in the future—the day and the month will be the same length (about 47 of our present days), and the Moon will be stationary in the sky over the same spot on the Earth.

(Figure 3.18). These are called *spring tides* (the name is connected not to the season, but to the idea that higher tides "spring up"). Spring tides are approximately the same, whether the Sun and Moon are on the same or opposite sides of the Earth, because tidal bulges occur on both sides. When the Moon is at first quarter or last quarter (at right angles to the Sun's direction), the tides produced by the Sun partially cancel the tides of the Moon, making them lower than usual. These are called *neap tides*.

The "simple" theory of tides, described in the preceding paragraphs, would be sufficient if the Earth rotated very slowly and were completely surrounded by very deep oceans. However, the presence of land masses stopping the flow of water, the friction in the oceans and between oceans and the ocean floors, the rotation of the Earth, the wind, the variable depth of the ocean, and so on, all complicate the picture. This is why in the real world, some places have very small tides while in other places huge tides become tourist attractions.

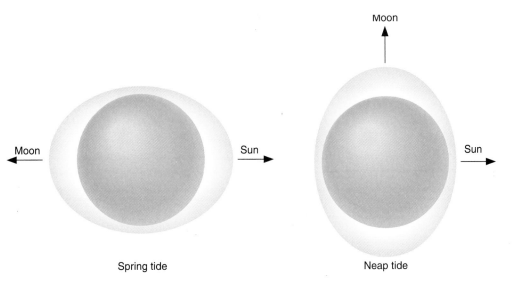

Spring tide

Neap tide

Figure 3.18
Tides caused by different alignments of the Sun and Moon. (a) Spring tides. (b) Neap tides.

Eclipses of the Sun and Moon

One of the fortunate coincidences of living on Earth at the present time is that the two most prominent astronomical objects, the Sun and the Moon, have nearly the same apparent size in the sky. Although the Sun is about 400 times larger in diameter than is the Moon, it is also about 400 times farther away, so both the Sun and the Moon have the same angular size—about 1/2°. As a result, the Moon, as seen from the Earth, can appear to cover the Sun, producing one of the most impressive events in nature.

Any solid object in the solar system casts a shadow by blocking the light of the Sun from a region behind it; this shadow becomes apparent whenever another object moves into it. In general, an **eclipse** occurs whenever any part of either the Earth or the Moon enters the shadow of the other. When the Moon's shadow strikes the Earth, people within that shadow see the Sun at least partially covered by the Moon; that is, they witness a **solar eclipse.** When the Moon passes into the shadow of the Earth, people on the night side of the Earth see the Moon darken in what is called a **lunar eclipse.**

The shadows of the Earth and the Moon consist of two parts: a dark cone where the shadow is darkest, called the *umbra,* and a lighter, more diffuse region of darkness called the *penumbra.* As you can imagine, the most spectacular eclipses occur when a body enters the umbra. Figure 3.19 illustrates the appearance of the Moon's shadow and what the Sun and Moon would look like from different points within the shadow.

If the path of the Moon in the sky were identical to the path of the Sun (the ecliptic), we might expect to see an eclipse of the Sun and the Moon each month, whenever the Moon got in front of the Sun or into the shadow of the Earth. However, the Moon's orbit is tilted relative to the plane of the Sun's orbit by about 5° (imagine two hula hoops with a common center, but tilted a bit). As a result, most months the Moon is sufficiently above or below the Sun to avoid an eclipse. But when the two paths cross (twice a year), it is then "eclipse season" and eclipses are possible.

Eclipses of the Sun

The apparent or angular sizes of both Sun and Moon vary slightly from time to time, as their distances from the Earth vary. (Figure 3.19 shows the distance of the *observer* varying at points A through D, but the idea is the same.) Much of the time, the Moon looks slightly smaller than the Sun and cannot cover it completely, even if the two are perfectly aligned. However, if an eclipse of the Sun occurs when the Moon is somewhat nearer than its average distance, the Moon can completely hide the Sun, producing a total solar eclipse. A total eclipse of the Sun occurs at those times when the umbra of the Moon's shadow reaches the surface of the Earth.

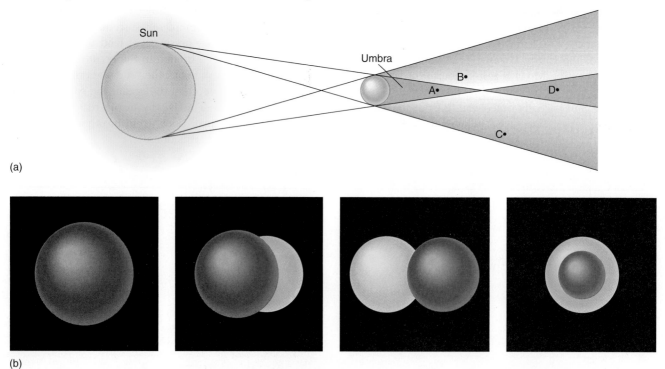

Figure 3.19

(a) The shadow cast by a spherical body (the Moon, for example). Notice the dark umbra and the lighter penumbra. Four points in the shadow are labeled with letters. (b) What the Sun and Moon would look like in the sky at the four labeled points.

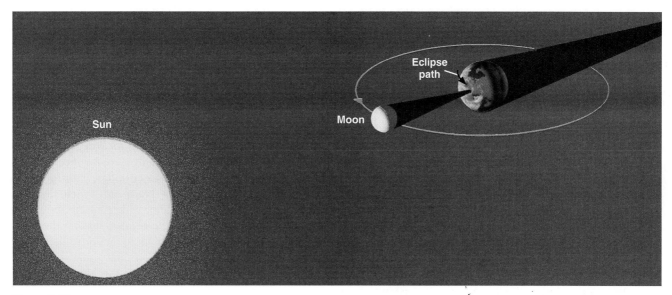

Figure 3.20
Geometry of a total solar eclipse (not to scale). Notice the shadow of the Moon on the Earth.

The geometry of a total solar eclipse is illustrated in Figure 3.20. If Sun and Moon are properly aligned, then the Moon's darkest shadow intersects the ground at a small point on the Earth's surface. Anyone on the Earth within the small area covered by the tip of the Moon's shadow will, for a few minutes, be unable to see the Sun and will witness a total eclipse. At the same time, observers on a larger area of the Earth's surface who are in the penumbra will see only a part of the Sun eclipsed by the Moon: we call this a partial solar eclipse.

As the Moon moves eastward in its orbit, the tip of its shadow sweeps eastward at about 1500 km/h along a thin band across the surface of the Earth; the total solar eclipse progresses along this band. The thin zone across the Earth within which a total solar eclipse is visible (weather permitting) is called the eclipse path. Within a region about 3000 km on either side of the eclipse path, a partial solar eclipse is visible. It does not take long for the Moon's shadow to sweep past a given point on Earth. The duration of totality may be only a brief instant; it can never exceed about 7 min.

Because a total eclipse of the Sun is so spectacular, it is well worth trying to see one if you can. There are people whose hobby is "eclipse chasing" and who brag about how many they have seen in their lifetimes. Since much of the Earth's surface is water, eclipse chasing can involve lengthy boat trips (and often requires air travel as well). As a result, eclipse chasing is rarely within the budget of a typical college student. Nevertheless, a list of future eclipses is given for your reference in Appendix 9.

Appearance of a Total Eclipse

What can you see if you are lucky enough to catch a total eclipse? A solar eclipse begins when the Moon just begins to silhouette itself against the edge of the Sun's disk. A partial phase follows, during which more and more of the Sun is covered by the Moon. About an hour after the eclipse begins, the Sun becomes completely hidden behind the Moon. In the few minutes immediately before this period of totality begins, the sky noticeably darkens; some flowers close up, and chickens may go to roost. As an eerie twilight suddenly descends during the day, other animals (and people) may get disoriented.

In the last instant before totality, the only parts of the Sun that are visible are those that shine through the lower valleys in the Moon's irregular profile and line up along the periphery of the advancing edge of the Moon—a phenomenon called Baily's beads, after the British astronomer who first described them. A final flash of sunlight through a lunar valley produces a brilliant flare on the disappearing crescent of the Sun: eclipse watchers call this "the diamond ring." During totality the sky is dark enough that planets become visible in the sky, and usually the brighter stars do as well.

As Baily's beads disappear and the bright disk of the Sun becomes entirely hidden behind the Moon, the Sun's remarkable corona flashes into view (Figure 3.21). The **corona** is the Sun's outer atmosphere, consisting of sparse gases that extend for millions of miles in all directions from the apparent surface of the Sun. It is ordinarily not visible because the light of the corona is feeble compared with that from the underlying layers of the Sun. Only when the brilliant glare from the Sun's visible disk is blotted out by the Moon during a total eclipse is the pearly white corona visible.

The total phase of the eclipse ends, as abruptly as it began, when the Moon begins to uncover the Sun. Gradually the partial phases of the eclipse repeat themselves, in reverse order, until the Moon has completely uncovered the Sun.

Figure 3.21
The corona visible during the July 11, 1991, total solar eclipse photographed from near La Paz, Mexico, by a dedicated eclipse chaser. (©1991 Stephen J. Edberg)

Eclipses of the Moon

A lunar eclipse occurs when the Moon enters the shadow of the Earth. The geometry of a lunar eclipse is shown in Figure 3.22. The Earth's dark shadow is about 1.4 million km long, so at the Moon's distance (an average of 384,000 km) it could cover about four full moons. Unlike a solar eclipse, which is visible only in certain local areas on the Earth, a lunar eclipse is visible to everyone who can see the Moon. Since a lunar eclipse can be seen (weather per-

mitting) from the entire night side of the Earth, lunar eclipses are observed far more frequently from a given place on Earth than are solar eclipses.

An eclipse of the Moon is total only if the Moon's path carries it through the Earth's umbra. If the Moon does not enter the umbra completely, we have a partial eclipse of the Moon. Sometimes the Moon can miss the umbra altogether, and just enter the larger penumbra of the Earth: in that case we have a *penumbral* eclipse of the Moon. The Moon's light does not decrease very much in the penumbra, so these eclipses are often hard to notice.

A lunar eclipse can take place only when the Sun, Earth, and Moon are in a line, which means the Moon will be in full phase before the eclipse, making the darkening even more dramatic. About 20 min or so before the Moon reaches the dark shadow, it dims somewhat as the Earth partly blocks the sunlight. As the Moon begins to dip into the shadow, the curved shape of the Earth's shadow upon it soon becomes apparent.

Even when totally eclipsed, the Moon is still faintly visible, usually appearing a dull coppery red. The illumination on the eclipsed Moon is sunlight that has been bent into the Earth's shadow by passing through the Earth's atmosphere.

After totality the Moon moves out of the shadow, and the sequence of events is reversed. The total duration of the eclipse depends on how closely the Moon's path approaches the axis of the shadow. For an eclipse where the Moon goes through the center of the Earth's shadow, each partial phase consumes at least 1 hour, while totality can last as long as 1 hour and 40 min. Total eclipses of the Moon occur, on the average, about once every two or three years.

Figure 3.22
Geometry of a lunar eclipse (not to scale). The Moon is full as seen from Earth at A; at position B, the Moon has begun to move into the Earth's shadow; at position C, the eclipse is over and the Moon is moving out of the Earth's shadow. (Note that the distance the Moon moves in its orbit during the eclipse has been exaggerated here for educational purposes.)

How to Observe Solar Eclipses

It is extremely dangerous to look at the Sun; even a brief exposure can damage your eyes. Normally, few sane people are tempted to do this, because it is painful (and something your mother told you never to do!). But during the course of an eclipse, the temptation to take a look is strong. Don't give in. The fact that the Moon is covering part of the Sun doesn't make the uncovered part any less dangerous to look at. Still, there are perfectly safe ways to follow the course of a solar eclipse.

The easiest technique is to make a pinhole projector. Take a piece of cardboard with a small (1-mm) hole punched in it and hold it several feet above a light surface, such as a concrete sidewalk or a white sheet of paper, so that the hole is "aimed" at the Sun. The hole produces a fuzzy but adequate image of the eclipsed Sun (see diagram).

You can also cover up all but about a dime's worth of a hand mirror, and use the small mirror surface that remains to project an image of the eclipsed Sun onto a wall. This takes a little more practice.

Although there are safe filters for looking at the Sun directly, people have suffered eye damage by looking through improper filters (or no filter at all). For example, neutral-density photographic filters are not safe, for they transmit infrared radiation that can cause severe damage to the retina. Also unsafe are smoked glass, completely exposed color film, and many other homemade filters.

It is safe to look at the Sun directly when it is totally eclipsed, even through binoculars or telescopes. Unfortunately the total phase, as we discussed, is all too brief. But if you know when it is coming (and going), be sure you look, for it's an unforgettably beautiful sight. And, despite the ancient folklore that presents them as dangerous times to be outdoors, the partial phases of eclipses—as long as you are not looking directly at the Sun— are certainly not any more dangerous than being out in sunlight.

During past eclipses, unnecessary panic has been created by uninformed public officials acting with the best intentions. There were two marvelous total eclipses in Australia in the 20th century, during which townspeople held newspapers over their heads for protection, and schoolchildren cowered indoors with their heads under their desks. What a pity that all those people missed what would have been one of the most memorable experiences of their lifetimes!

How to watch a solar eclipse safely during its partial phases using a pinhole projector.

Summary

3.1 The terrestrial system of latitude and longitude makes use of the **great circles** called **meridians.** An analogous celestial coordinate system is called **right ascension (RA)** and **declination,** with the vernal equinox serving as reference point (like the prime meridian at Greenwich on the Earth). These coordinate systems help us locate any object on the celestial sphere. The Foucault pendulum is a way to demonstrate that the Earth rather than the sky is turning.

3.2 The familiar cycle of the seasons results from the 23° tilt of the Earth's axis of rotation. At the *summer solstice,* the Sun is higher in the sky and its rays strike the Earth more directly. The Sun is in the sky for more than half the day and can heat the Earth longer. At the *winter solstice,* the Sun is low in the sky and up for fewer than 12 hours. At the *vernal* and *autumnal equinoxes,* the Sun is on the celestial equator, and we get 12 hours of day and night. The seasons are different at different latitudes.

3.3 The basic unit of astronomical time is the day (either the **solar day** or the **sidereal day**). **Apparent solar time** is based on the position of the Sun in the sky, while **mean solar time** is based on the average value of a solar day during the year. By international agreement, we define 24 time zones around the world, each with its own standard time. The convention of the **international date line** is necessary to reconcile times in different parts of the Earth.

3.4 The fundamental problem of the calendar is to reconcile the incommensurable lengths of the day, the month, and the year. Most modern calendars, beginning with the Roman (Julian) calendar of the 1st century B.C., neglect the problem of the month and concentrate on achieving the correct number of days in a year by using such conventions as the leap year. Today, most of the world has adopted the Gregorian calendar established in 1582.

3.5 The Moon's monthly cycle of phases results from the changing angle of its illumination by the Sun. The full moon is visible in the sky only during the night; other phases are visible during the day. Since its period of revolution is the same as its period of rotation, the Moon always keeps the same face toward the Earth.

3.6 The twice-daily ocean *tides* are primarily the result of the Moon's differential gravitational force on the material of the Earth's crust and ocean. These tidal forces cause ocean water to flow into two tidal bulges on opposite sides of the Earth; each day the Earth rotates through these bulges. Actual ocean tides are complicated by the additional effects of the Sun, and by the shape of the coasts and ocean basins.

3.7 The Sun and Moon have nearly the same angular size (about 0.5°). A **solar eclipse** occurs when the Moon moves between the Sun and the Earth, casting its shadow on a part of the Earth's surface. If the eclipse is total, the observer is in the Moon's *umbra*, the light from the bright disk of the Sun is completely blocked, and the solar atmosphere (the **corona**) comes into view. Solar eclipses take place rarely in any one location, but they are among the most spectacular sights in nature. A **lunar eclipse** takes place when the Moon moves into the Earth's shadow; it is visible (weather permitting) from the entire night hemisphere of the Earth.

Review Questions

1. Discuss how latitude and longitude on Earth are similar to declination and right ascension in the sky.

2. What is the latitude of the North Pole? The South Pole? Why has longitude no meaning at the North and South Poles?

3. Make a table showing each main phase of the Moon and roughly when the Moon rises and sets for each phase. During which phase can you see the Moon in the middle of the morning? In the middle of the afternoon?

4. What are the advantages and disadvantages of apparent solar time? How is the situation improved by introducing mean solar time and standard time?

5. What are the two ways that the tilt of the Earth's axis causes the summers in the United States to be warmer than the winters?

6. Why is it difficult to construct a practical calendar based on the Moon's cycle of phases?

7. Explain why there are two high tides and two low tides every day. Strictly speaking, should it be a 24-hour period during which there are two high tides? If not, what should the interval be?

8. What is the phase of the Moon during a total solar eclipse? During a total lunar eclipse?

Thought Questions

9. Where are you on the Earth according to the following descriptions (refer back to Chapter 1 as well as this chapter)?
   a. The stars rise and set perpendicular to the horizon.
   b. The stars circle the sky parallel to the horizon.
   c. The celestial equator passes through the zenith.
   d. In the course of a year, all stars are visible.
   e. The Sun rises on September 23 and does not set until March 21 (ideally).

10. In countries at far northern latitudes, the winter months tend to be so cloudy that astronomical observations are nearly impossible. Why can't good observations of the stars be made at those places during the summer months?

11. What is the phase of the Moon if it
   a. rises at 3:00 P.M.?
   b. is highest in the sky at 7:00 A.M.?
   c. sets at 10:00 A.M.?

12. A car accident occurs around midnight on the night of a full moon. The driver at fault claims he was blinded momentarily by the Moon rising on the eastern horizon. Should the police believe him?

13. The secret recipe to the ever-popular veggie burgers in the college cafeteria is hidden in a drawer in the director's office. Two students decide to break in and get their hands on it, but they want to do it a few hours before dawn on a night when there is no Moon, so they are less likely to be caught. What phases of the Moon would suit their plans?

14. Your granduncle, who often exaggerates events in his own life, tells you about a terrific adventure he had on February 29, 1900. Why would this story make you suspicious?

15. One year, when money is no object, you enjoy your birthday so much that you want to have another one right away. You get into your supersonic jet. Where should you and the people celebrating with you travel? From what direction should you approach? Explain.

16. Suppose you lived in the crater Copernicus on the side of the Moon facing the Earth.
   a. How often would the Sun rise?
   b. How often would the Earth set?
   c. During what fraction of the time would you be able to see the stars?

17. In a lunar eclipse, does the Moon enter the shadow of the Earth from the east or west side? Explain why.

18. Describe what an observer at the crater Copernicus would see while the Moon is eclipsed. What would the same observer see during what would be a total solar eclipse as viewed from the Earth?

19. How would you prove that it is the Earth, rather than the sky around us, that is rotating every 24 hours? (Of course, you can just repeat the explanation in this chapter, but we'd like you to find another proof. *Hint:* How does the spinning of the Earth affect the oceans and the atmosphere?)

Problems

20. a. If a star rises at 8:30 P.M. tonight, approximately what time will it rise two months from now?
  b. What is the altitude of the Sun at noon on December 22 as seen from a place on the Tropic of Cancer?

21. Suppose the tilt of the Earth's axis were only 16°. What, then, would be the difference in latitude between the Arctic Circle and the Tropic of Cancer? What would be the effect on the seasons compared with that produced by the actual tilt of 23°?

22. Consider a calendar based entirely on the day and the month (the Moon's period from full phase to full phase). How many days are there in a month? Can you figure out a scheme analogous to leap year to make this calendar work? Can you also incorporate the idea of a week into your lunar calendar?

23. Show that the Gregorian calendar will be in error by one day in about 3300 years.

Suggestions for Further Reading

Aveni, A. *Empires of Time: Calendars, Clocks, and Cultures.* 1989, Basic Books.

Bartky, I. and Harrison, E. "Standard and Daylight Savings Time" in *Scientific American,* May 1979.

Brown, H. *Man and the Stars.* 1978, Oxford U. Press. Contains good sections on the history of astronomical time-keeping.

Coco, M. "Not Just Another Pretty Phase" in *Astronomy,* July 1994, p. 76. Moon phases explained.

DeVorkin, D. *Practical Astronomy.* 1986, Smithsonian Institution Press.

Gingerich, O. "Notes on the Gregorian Calendar Reform" in *Sky & Telescope,* Dec. 1982, p. 530.

Harris, J. and Talcott, R. *Chasing the Shadow: An Observer's Guide to Eclipses.* 1994, Kalmbach.

Kluepfel, C. "How Accurate Is the Gregorian Calendar?" in *Sky & Telescope,* Nov. 1982, p. 417.

Littmann, M. and Willcox, K. *Totality.* 1991, U. of Hawaii Press. Fine introduction to science and lore of eclipses.

Olson, D. and Doescher, R. "Lincoln and the Almanac Trial" in *Sky & Telescope,* Aug. 1990, p. 184. How he used the Moon.

Pasachoff, J. and Ressmeyer, R. "The Great Eclipse" in *National Geographic,* May 1992, p. 30. About the July 1991 solar eclipse, with spectacular photographs.

Rey, H. *The Stars: A New Way to See Them.* 1976, Houghton Mifflin. Good introduction to time, the seasons, and celestial coordinates.

Using **REDSHIFT** ™

1. To follow the path of the Sun through the seasons, select *Horizon* view, viewing due south. Set the TIME to 12:00 P.M. and the LOCATION to 90° longitude and 45° latitude. Use a waterbased marker to mark on the screen the Sun's position every 10 days through the year. Mark some dates corresponding to significant points of the path:

When is the Sun highest in the sky? When does it appear lowest?

When do the equinoxes occur? Why do the loops have different heights?

2. For another view of Earth's seasons, turn off the stars and deep-sky objects and select only the Sun and Earth to display in an equatorial view. Set the LOCATION to heliocentric with a latitude of 90°, longitude of 270°, and distance of 3 AU. Center the Sun and Lock on it. Note the Julian and calendar dates when Earth appears exactly to the left of the Sun, below it, due right, and exactly above it.

What do these dates correspond to?

How many days elapse between these points in Earth's orbit?

Why is there a difference between the intervals?

3. Follow lunar phases by setting the *Follow Planet* panel to Earth. To see the view from the Sun, choose the inward position at the maximum distance. How does the Moon appear during a month-long orbit? Now set the panel to an above view.

How does the Moon appear during a month-long orbit?

Translate this view to that which an earthbound observer would have.

4. To see why eclipses do not occur at every New and Full Moon, view the Earth for the current year from the *maximum distance* at the leading position in the *Follow Planet* panel.

How many solar eclipses occur during the year? Confirm your result using the eclipse finder.

The Very Large Array of radio telescopes in New Mexico seen with a rainbow. Just as water droplets in our atmosphere can disperse sunlight into a rainbow of colors, so instruments called spectrometers allow astronomers to break the light from planets, stars, or galaxies into their component colors. The details of these colors can help us decipher what conditions are like in the objects from which they originate. (Photo by Martha Haynes and Riccardo Giovanelli, Cornell University)

CHAPTER 4

Radiation and Spectra

Thinking Ahead

The nearest star is so far away that the fastest spacecraft humans have built so far would take almost 100,000 years to get there. Yet we very much want to know what material this neighbor star is composed of, and how it differs from our own Sun. How can we learn about the chemical makeup of stars that we cannot hope to visit or sample?

In astronomy, most of the objects that we study are completely beyond our reach. The temperature of the Sun is so high that a spacecraft would be destroyed by its heat long before reaching the solar surface, and the stars are much too far away to visit in our lifetimes with the technology now available. Even light, which travels at a speed of 300,000 km/s, takes more than four years to reach us from the *nearest* star. If we want to learn about the Sun and stars, or even about most members of our solar system, we must rely on techniques that allow us to analyze them from a distance.

And here we are in luck. Coded into the light and other kinds of radiation that reach us from objects in the universe is a wide range of information about what those objects are like and how they work. If we can decipher this "cosmic code" and read the messages it contains, we can learn an enormous amount about the cosmos without ever having to leave the Earth or its immediate environment.

In the last hundred years the power of interpreting the messages brought by light has increased greatly. We are in touch with the stars.

Sir William Bragg in *The Universe of Light* (1933, G. Bell and Sons).

The light and other radiation we receive from the stars and planets is generated by processes at the atomic level—by changes in the way the parts of the atom interact and move. Thus, to appreciate how light is generated we must first explore how atoms work. There is a bit of irony in the fact that in order to understand some of the largest structures in the universe, we must become acquainted with some of the smallest.

Notice that we have twice used the term "light and other radiation." One of the key ideas explored in this chapter is that light is not unique—it is merely the most familiar example of a much larger family of radiation that can carry information to us.

4.1

The Nature of Light

As we saw in earlier chapters, Newton's theory of gravity accounts for the motions of the planets as well as the objects on the Earth. Application of this theory to a variety of problems dominated the work of scientists for nearly two centuries. In the 19th century, many physicists turned to the study of electricity and magnetism, which—as we shall see in a moment—are intimately connected with the production of light.

The scientist who played a role in this field analogous to Newton's role in the study of gravity was physicist James Clerk Maxwell (1831–1879), born and educated in Scotland (Figure 4.1). Inspired by a number of ingenious experiments that showed an intimate relationship between electricity and magnetism, Maxwell developed a single theory that describes both with just a small number of elegant equations. It is this theory that allows us to understand the nature and behavior of light.

Figure 4.1
James Clerk Maxwell (1831–1879) unified electricity and magnetism into a coherent theory. (American Institute of Physics, Niels Bohr Library)

Maxwell's Theory of Electromagnetism

We will look at the structure of the atom in more detail in Section 4.4, but we begin by noting that the typical atom consists of several types of particles, a number of which have not only mass, but an additional property called electric charge. In the nucleus (central part) of every atom there are *protons,* which are positively charged, while outside the nucleus there are *electrons,* which have a negative charge.

Maxwell's theory deals with these electric charges and their effects, especially when they are moving. In the vicinity of an electric charge, another charge feels a force of attraction or repulsion: opposite charges attract; like charges repel. When charges are not in motion, we observe only this electric attraction or repulsion. But if charges are in motion (as they are inside every atom, and in a wire carrying a current), then we measure another force, called *magnetism.*

Magnetism was well known for much of recorded human history, but its cause was not understood until the 19th century. Experiments with electric charges clearly demonstrated that magnetism was the result of moving charged particles. Sometimes the motion is clear, as in the coils of heavy wire that make an industrial electromagnet. Other times it is more subtle, as in the kind of magnet you buy in a hardware store, in which many of the electrons inside the atoms are spinning in roughly the same direction—it is their aligned motion that causes the material to become magnetic.

We use the word *field* in physics to describe the action of forces that one object exerts on other distant objects. For example, we say the Sun produces a *gravitational field* that controls the Earth's orbit, even though the Sun and the Earth do not come directly into contact. Using this terminology, we can say that stationary electric charges produce *electric fields,* and moving electric charges produce *magnetic fields.*

But the relationship between electric and magnetic phenomena is even more profound. Experiments showed that changing magnetic fields could produce electric currents (and thus changing electric fields). And changing electric currents could (in turn) produce changing magnetic fields. So once begun, it appeared that electric and magnetic field changes could continue to trigger each other.

Maxwell analyzed what would happen if electric charges were *oscillating* (moving constantly back and forth) and found that the resulting pattern of electric and magnetic fields would spread out and travel rapidly through space. Something similar happens when your finger or a nervous frog moves up and down in a pool of water. The disturbance moves outward and creates a pattern we call a *wave* in the water (Figure 4.2). You might first think that there must be very few situations in nature where electric charges would oscillate, but this is not at all the case. As we shall see, because they are warm, all

Figure 4.2
An oscillation in a pool of water creates an expanding disturbance called a wave. (© 1993, Comstock, Inc.)

atoms and molecules (which consist of charged particles) oscillate back and forth; the resulting electromagnetic disturbances are among the most common phenomena in the universe.

Maxwell was able to calculate the speed at which an electromagnetic disturbance moves through space; he found that it is equal to the speed of light, which had been measured experimentally. On that basis he speculated that light was one form of a family of possible electric and magnetic disturbances we call **electromagnetic radiation,** a conclusion that was again confirmed in laboratory experiments. When light (reflected from the pages of an astronomy textbook, for example) enters a human eye, its changing electric and magnetic fields stimulate nerve endings, which then transmit the information contained in these changing fields to the brain.

Since the word *radiation* will be used frequently in this book, it is important to understand what it means. In everyday language, radiation is often used to describe certain kinds of energetic subatomic particles released by radioactive materials in our environment. (An example is the kind of radiation used to treat cancer.) But this is *not* what we mean when we speak of radiation in an astronomy text. Radiation, as used in this book, is a general term for light, x rays, and other forms of electromagnetic waves. This radiation provides almost our only link with the universe beyond our own solar system.

The Wave-like Characteristics of Light

The changing electric and magnetic fields in radiation are quite similar (as hinted above) to the waves that can be set up on a quiet pool of water. In both cases the disturbance travels rapidly outward from the point of origin and can use its energy to disturb other things farther away. (For example, in the case of water, the expanding ripples could disturb the peace of a frog sleeping on a leaf in the same pool.) In the case of electromagnetic waves, the radiation generated by a transmitting antenna full of charged particles at your local radio station can, sometime later, disturb a group of electrons in your radio antenna and bring you the news and weather while you are getting ready for class or work in the morning.

But the waves generated by charged particles differ from water waves in some profound ways. Water waves require water as a *medium* to travel in; sound waves, to give another example, are pressure disturbances that require air to travel through. But electromagnetic waves do *not* require any medium at all; the fields generate each other, and so can move through a vacuum (such as outer space). This was such a disturbing idea to 19th-century scientists that they actually made up a medium to fill all of space, one for which there was not a single shred of evidence, just so light waves could have something to travel through: they called it the *aether.* Today we know there is no aether, and that electromagnetic waves have no trouble at all moving through empty space (as all the starlight visible on a clear night must surely be doing).

The other difference is that *all* electromagnetic waves move at the same speed (the speed of light), which turns out to be the fastest possible speed in the universe. No matter where electromagnetic waves are generated, and no matter what other properties they have, when they are moving (and not interacting with matter) they move at the speed of light. Yet you know from everyday experience that waves like light are not all the same; for example, we perceive that light waves differ from one another in a property we call color. Let's see how we can denote the differences among the whole broad family of electromagnetic waves.

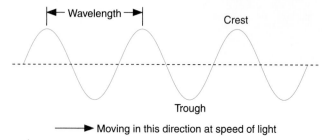

Figure 4.3
Electromagnetic radiation has wave-like characteristics. The wavelength (λ) is the distance between crests, the frequency (*f*) is the number of cycles per second, and the speed (*c*) is the distance the wave covers over time.

The nice thing about a wave is that it is a repeating phenomenon. Whether it is the up-and-down motion of a water wave or the changing electric and magnetic fields in a wave of light, the pattern of disturbance repeats in a cyclical way. Thus any wave motion can be characterized by drawing a series of crests and troughs (Figure 4.3); moving up a full crest and down a full trough completes one cycle. The horizontal length covered by one cycle—the distance between one crest and the next, for example—is called the **wavelength.** Ocean waves provide an analogy: the wavelength is the distance separating successive wave crests.

Various forms of electromagnetic radiation differ from one another in their wavelengths. Those with the longest waves, ranging up to many kilometers in length, are called *radio waves.* Forms of electromagnetic energy of successively shorter wavelengths are called *infrared, light, ultraviolet, x rays,* and *gamma rays.* All these forms (described in greater detail in Section 4.2) are the same basic kind of energy and could be thought of as different kinds of light. For visible light, our eyes (and minds) perceive different wavelengths as different colors; red, for example, is the longest visible wave, and violet is the shortest. The main colors of visible light from longest to shortest wavelength can be remembered using the mnemonic ROY G. BIV—for red, orange, yellow, green, blue, indigo, and violet.

We can also characterize different waves by their **frequency,** the number of wave cycles that pass per second. If you count ten crests moving by each second, for example, then the frequency is 10 cycles per second (cps). In honor of Heinrich Hertz, the physicist who—inspired by Maxwell's work—discovered radio waves, a cps is also called a *Hertz (Hz).* Take a look at the dial of your radio, for example, and you will see the channel assigned to each radio station on the dial characterized by its frequency, usually in units of KHz (kiloHertz or thousands of Hertz) or MHz (mega or millions of Hertz).

Wavelength and frequency are related to each other because all electromagnetic waves travel at the same speed. To see how this works, imagine a parade in which everyone is forced by prevailing traffic conditions to move at exactly the same speed. You stand on a corner and watch the waves of marchers come by. First you see row after row of very skinny fashion models. Because they are not very wide, a good number of the models can fit by you each minute; we can say they have a high frequency. Next, however, come several rows of circus elephants. Since the elephants are quite large, and marching at the same speed as the models, far fewer of them can march past you per minute: they must be content with a lower frequency. Mathematically, we can express this relationship as

$$c = \lambda f$$

where the Greek letter for "l"—lambda or λ—is used to denote wavelength, and *c* is the scientific symbol for the speed of light. In words, the formula can be expressed as follows: For any wave motion, the speed at which a wave moves equals the frequency times the wavelength. Waves with longer wavelengths have lower frequencies.

Light as a Photon

The electromagnetic wave model of light (as formulated by James Clerk Maxwell) was one of the great triumphs of 19th-century science. When, in 1887, Heinrich Hertz actually made invisible electromagnetic waves on one side of a room and detected them on the other, it ushered in a new era that led to the modern era of telecommunications. However, by the turn of the century, new, more sophisticated experiments had revealed that light behaves in certain ways that simply cannot be explained by the wave model. Reluctantly, physicists had to accept that sometimes light (and all other electromagnetic radiation) behaves more like a "particle"—or at least a self-contained packet of energy—than a wave. We call such packets of electromagnetic energy **photons.**

The fact that light behaves like a wave in certain experiments, and like a particle in others, was a very surprising and unlikely idea that was, at first, as confusing to physicists as it seems to students. After all, our common sense says that waves and particles are opposite concepts. A wave is a repeating disturbance that by its very nature is not just in one place, but spreads out. A particle, on the other hand, is something that can only be in one place at any given time. Strange as it sounds, though, countless experiments now confirm that electromagnetic radiation can sometimes behave like a wave, and at other times like a particle. But then again, perhaps we shouldn't be surprised that something that always travels at the "speed limit" of the universe, and doesn't need a medium to travel through, might not obey our everyday "common-sense" ideas. The confusion that this wave–particle duality of light caused in physics was eventually resolved by the introduction of a more complicated theory of waves and particles, now called *quantum mechanics.* (This is one of the most interesting fields of modern science, but mostly beyond the scope of our book.)

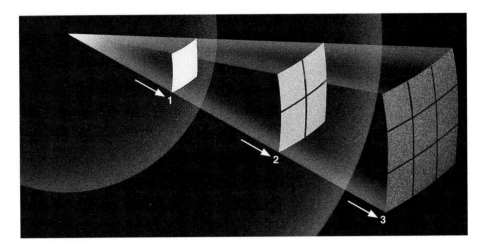

Figure 4.4
The inverse-square law for light. As light energy radiates away from its source, it spreads out in such a way that the energy passing through a unit area decreases as the square of the distance from its source.

In any case, you should now be prepared when you see scientists (or your authors) sometimes discuss electromagnetic radiation as if it consisted of waves, and at other times refer to it as a stream of photons. A photon carries a specific amount of energy (it is a packet of energy after all). That energy is proportional to the frequency of the same radiation when considered as waves. High-frequency waves are the same as photons with high energy, while low-frequency waves are the same as photons with low energy. Among the colors of visible light, violet-light photons have the highest energy, red the lowest.

Propagation of Light

Let's think for a moment about how light from a light bulb moves through space. The expansion of light waves is extremely democratic: they travel away from the bulb not just toward your eyes, but in all directions; they must therefore cover an ever-widening space. Yet the total amount of light available can't change once the light has left the bulb. This means that, as the same expanding shell of light covers a larger and larger area, there must be less and less of it in any given place. Light (and all other electromagnetic radiation) gets weaker and weaker as it gets farther and farther from its source.

The increase in area is proportional to the square of the distance that the radiation has traveled (Figure 4.4). If we stand twice as far from the source, our eyes will intercept two squared, or four, times less light. If we stand ten times farther from the source, we get ten squared, or 100 times less light. You can see how this weakening means trouble for sources of light at astronomical distances! A typical star emits about the same total energy as does the Sun, but even the nearest star is about 270,000 times farther away, and so it appears about 73 billion times fainter. No wonder that the stars, which close-up would look more or less like the Sun, look like mere pinpoints of light from far away.

This idea, that the **apparent brightness** of a source (how bright it looks to us) gets weaker with distance in the way we have described, is known as the **inverse-square law** of light propagation. In this respect, the propagation of radiation is similar to the effects of gravity. Remember that the force of gravity between two attracting masses is also inversely proportional to the square of their separation.

The Electromagnetic Spectrum

Types of Electromagnetic Radiation

Objects in the universe send us an enormous range of electromagnetic waves. To make discussing them manageable, scientists have divided the full **electromagnetic spectrum** (or range) of such waves into some arbitrary categories. These are shown in Figure 4.5, with some information about the waves in each part or band of the spectrum.

Electromagnetic radiation with the shortest wavelengths, not larger than 0.01 nanometer (nm), is called **gamma rays.** (One nm is 10^{-9} m; see Appendix 5.) The name comes from the third letter of the Greek alphabet; gamma rays were the third kind of radiation discovered coming from radioactive atoms when physicists were first investigating their behavior. Because gamma rays carry a lot of energy, they can be dangerous for living tissues. We shall see that gamma radiation is generated deep in the interior of stars by reactions among atomic nuclei. Gamma rays coming from space interact with atoms high in the Earth's atmosphere and are absorbed before they reach the ground; thus they can only be studied using instruments in space.

Electromagnetic radiation with wavelengths between 0.01 nm and 20 nm is referred to as **x rays.** These are now familiar to all of us from visits to the doctor and dentist; being more energetic than light, x rays are able to penetrate soft tissues and so allow us to make images of the shadows of the bones within them. While x rays can penetrate a short length of human flesh, they are stopped by the large numbers of atoms in the Earth's atmosphere

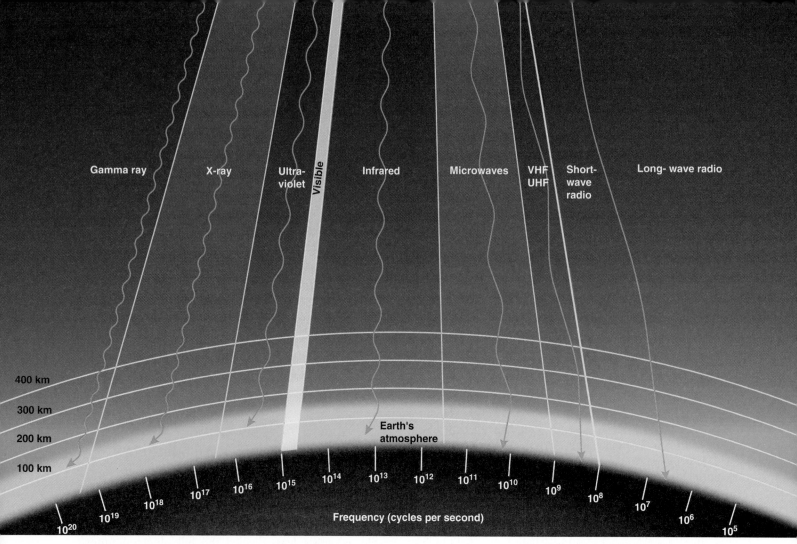

Gamma ray   X-ray   Ultra-violet   Visible   Infrared   Microwaves   VHF UHF   Short-wave radio   Long-wave radio

400 km
300 km
200 km
100 km

Earth's atmosphere

10^{20} 10^{19} 10^{18} 10^{17} 10^{16} 10^{15} 10^{14} 10^{13} 10^{12} 10^{11} 10^{10} 10^{9} 10^{8} 10^{7} 10^{6} 10^{5}

Frequency (cycles per second)

Figure 4.5
The bands of the electromagnetic spectrum and how well the Earth's atmosphere transmits them.

with which they interact. Thus, x-ray astronomy (like gamma-ray astronomy) could not develop until humanity had ways of sending instruments above our atmosphere (Figure 4.6).

Radiation intermediate between x rays and visible light is **ultraviolet** (meaning higher energy than violet). Outside the world of science, ultraviolet light is sometimes called "black light" because our eyes cannot see it. Ultraviolet radiation is mostly blocked by a section of the Earth's atmosphere called the *ozone* layer, but a small fraction of the ultraviolet rays from our Sun do penetrate to cause sunburn or, in extreme cases, skin cancer in human beings. Ultraviolet astronomy is also best done from space.

Electromagnetic radiation with wavelengths between roughly 400 and 700 nm is called **visible light,** because these are the waves that human vision is sensitive to. As we shall see, this is also the band of the electromagnetic spectrum where the Sun gives off the greatest amount of radiation. These two observations are not coincidental: human eyes evolved to see the kinds of waves that the Sun produced most effectively. The more light there was to

see by, the more efficiently we could evade predators and make babies before we got stepped on or eaten (which, in a crude way, is what life is all about from the point of view of your genetic material). Visible light penetrates the Earth's atmosphere effectively, except when it is temporarily blocked by passing clouds.

In 1672, in the first paper that he submitted to the Royal Society, Newton described an experiment in which he permitted sunlight to pass through a small hole and then through a prism. Newton found that sunlight, which gives the impression of being white, is actually made up of a mixture of all the colors of the rainbow (Figure 4.7). While all the colors of light may impinge on objects in our world, not all of them are reflected. When white light from the Sun or from a light fixture hits a pair of blue jeans, the dye in the jeans absorbs all the colors except blue. Some of the blue colors are reflected back into our eyes, and our brain interprets that particular set of wavelengths as blue.

Between visible light and radio waves are the wavelengths of **infrared** or heat radiation. Astronomer William Herschel first discovered infared in 1800, while

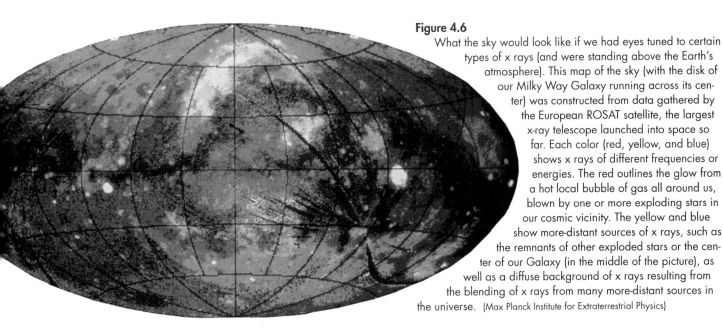

Figure 4.6
What the sky would look like if we had eyes tuned to certain types of x rays (and were standing above the Earth's atmosphere). This map of the sky (with the disk of our Milky Way Galaxy running across its center) was constructed from data gathered by the European ROSAT satellite, the largest x-ray telescope launched into space so far. Each color (red, yellow, and blue) shows x rays of different frequencies or energies. The red outlines the glow from a hot local bubble of gas all around us, blown by one or more exploding stars in our cosmic vicinity. The yellow and blue show more-distant sources of x rays, such as the remnants of other exploded stars or the center of our Galaxy (in the middle of the picture), as well as a diffuse background of x rays resulting from the blending of x rays from many more-distant sources in the universe. (Max Planck Institute for Extraterrestrial Physics)

trying to measure the temperatures of different colors of sunlight spread out into a spectrum. He noticed that when he accidentally pushed his thermometer beyond the reddest color, it still registered heating due to some invisible energy coming from the Sun. This was our first hint about the existence of the other (invisible) bands of the electromagnetic spectrum, although it would take many decades for our full understanding to develop.

A "heat lamp" radiates mostly infrared radiation, and the nerve endings in our skin are sensitive to this band of the electromagnetic spectrum. Infrared waves are absorbed by water and carbon dioxide molecules in the Earth's atmosphere, and for this reason, infrared astronomy is best done from high mountaintops, high-flying airplanes, and spacecraft.

All the electromagnetic waves longer than infrared are called **radio** waves, but this is so broad a category that we generally divide it into several subsections. Among the most familiar of these are **microwaves,** used in shortwave communication and microwave ovens; radar waves, used at airports and by the military; FM and television waves, which carry the news and entertainment without which it is now impossible to imagine modern culture; and AM radio waves, which were the first to be developed for broadcasting. These different categories range in wavelength from a few millimeters to hundreds of meters. Other radio radiation can have wavelengths as long as several kilometers.

With such a wide range of wavelengths, you will not be surprised to learn that not all radio waves interact with the Earth's atmosphere the same way. Microwaves are absorbed by water vapor (which is what makes them effective in heating foods with high water contents in microwave ovens). FM and TV waves are not absorbed and can travel easily through our atmosphere.

We hope this brief survey has left you with one strong impression. Although visible light is what most people associate with astronomy, the light that our eyes can see is only a tiny fraction of the broad range of waves that nature permits and that the universe sends us. Today we un-

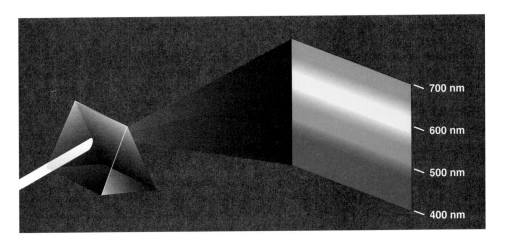

Figure 4.7
When we pass a beam of white sunlight through a prism, we see a rainbow-colored band of light that we call a continuous spectrum.

700 nm

600 nm

500 nm

400 nm

TABLE 4.1

Electromagnetic Radiation

Type of Radiation	Wavelength Range (nm)	Radiated by Objects at This Temperature	Typical Sources
Gamma rays	Less than 0.01	More than 10^8 K	Produced in nuclear reactions. Requires very high energy processes.
X rays	0.01–20	10^6–10^8 K	Gas in clusters of galaxies; supernova remnants; solar corona.
Ultraviolet	20–400	10^4–10^6 K	Supernova remnants; very hot stars.
Visible	400–700	10^3–10^4 K	Stars.
Infrared	10^3–10^6	10–10^3 K	Cool clouds of dust and gas; planets; satellites.
Radio	More than 10^6	Less than 10 K	No astronomical objects this cold; radio emission produced by electrons moving in magnetic fields (synchrotron radiation).

derstand that judging some astronomical phenomenon by using only the light we can see is like hiding under the table at a big dinner party and judging all the guests by nothing but their shoes. There's a lot more to each person than meets our eye under the table. As we discussed in the Prologue, it is very important for those who study astronomy today to avoid being *light chauvinists*—those who respect only the information seen by their eyes while ignoring the information gathered by instruments sensitive to other bands of the electromagnetic spectrum.

Table 4.1 summarizes the bands of the electromagnetic spectrum and indicates the temperatures and types of typical astronomical objects that emit each kind of electromagnetic radiation. Many of the objects mentioned in the table will not be familiar to you this early in the course, but you can return to this table as you learn more about the types of objects astronomers study.

Radiation and Temperature

Some astronomical objects emit mostly infrared radiation, others mostly visible light, and still others mostly ultraviolet radiation. What determines the type of electromagnetic radiation emitted by the Sun, stars, and other astronomical objects? The answer turns out to be their *temperature.*

At the microscopic level, everything in nature is in motion. A *solid* is composed of molecules and atoms that are in continuous vibration; they move back and forth in place, but their motion is much too small for our eyes to make out. A *gas* consists of molecules that are flying about freely at high speed, continually bumping into one another and bombarding the surrounding matter. The energy of these random molecular and atomic motions is called *heat.* The hotter the solid or gas, the more rapid the motion of the molecules or atoms; the temperature of

something is thus just a measure of the average energy of the particles that make it up.

This motion at the microscopic level is responsible for much of the electromagnetic radiation on Earth and in the universe. As atoms and molecules move about and collide, or vibrate in place, their charged particles generate electromagnetic radiation (as we saw in Section 4.1). The characteristics of this radiation are determined by the temperature of the atoms and molecules that give rise to them. In a hot solid or gas, for example, the individual particles vibrate in place or move rapidly from collision to collision, and so the emitted waves are, on average, more energetic. In very cool material, the particles have low-energy atomic and molecular motions, and thus generate lower energy waves.

Radiation Laws

To understand in more quantitative detail the relationship between temperature and electromagnetic radiation, it is useful to consider an idealized object that (unlike your sweater or your astronomy instructor's head) does not reflect or scatter any radiation, but absorbs all the electromagnetic energy that falls on it. Such an object is called a **blackbody.** The energy that is absorbed heats the blackbody, causing the atoms and molecules in it to begin vibrating or moving around at increasing speeds. Thus the blackbody will radiate electromagnetic waves. By absorbing radiation, the blackbody heats up until it is emitting energy at the same rate that energy is being absorbed. We want to discuss such an idealized object, because, as you will see, stars are not bad approximations of blackbodies.

The radiation from a blackbody has several characteristics, as illustrated in Figure 4.8. This figure shows the number of waves (or total energy) emitted by blackbodies of different temperatures at each wavelength. First of all, notice that a blackbody with a temperature higher than absolute zero emits some energy at *all* wavelengths within a given range. We say that it gives off a *continuous spectrum* of waves. This is because in any solid or gas, some molecules or atoms vibrate or move between collisions slower than average, and some move faster than average. So when we look at the electromagnetic waves emitted, we find a broad range or spectrum of energies and wavelengths. There are more waves emitted at the *average* vibration or motion rate (the tall part of each curve), but if we have a large number of atoms or molecules, some waves will be detected at each wavelength.

Second, note that a blackbody at a higher temperature emits *more* energy at all wavelengths than does a cooler one. In a hot gas, for example, the atoms have more collisions and give off more waves at each possible energy or wavelength. In the real world of stars, this means that hotter stars give off more energy at every wavelength than do cooler stars.

Third, the graph shows us that the higher the temperature, the shorter the wavelength at which the maximum energy is emitted. As you might expect, hot objects give off their average waves at shorter wavelengths (higher frequencies, higher energies per photon) than do cool objects. You may have observed examples of this rule in everyday life. When a burner on an electric stove is turned on low, it emits only heat, which is infrared radiation, but does not yet glow with visible light. If the burner is set to a higher temperature, it starts to glow a dull red. At a still higher setting, it glows a brighter orange-red (shorter wavelength). At even higher temperatures, which cannot be reached with ordinary stoves, metal can appear brilliant yellow or even blue-white.

We can use these ideas to come up with a rough sort of "thermometer" for measuring the temperatures of stars. The Sun and stars emit energy that approximates that from a blackbody, so it is possible to estimate temperatures by noting the spectrum of waves they give off. Since many stars give off most of their waves in visible light, the color of light that dominates a star's appearance is a rough indicator of its temperature. If one star looks red, and another looks blue, which one has the higher temperature? Since blue is the shorter wavelength color, it is the sign of a hotter star. (By the way, note that the temperatures we associate with different colors are not the same as the ones artists use. In art, red is often called a "hot" color and blue a "cool" color, while in nature, it's the other way around.)

We can develop a more precise star thermometer by measuring how much energy a star gives off at each wavelength and constructing diagrams like Figure 4.8. The graph of the energy emitted by a blackbody at each wavelength is called its *blackbody curve*. The location of the peak (or maximum) in the blackbody curve of each star can tell us its temperature. The temperature at the surface of the Sun, which is where the radiation that we see is emitted, turns out to be 5800 K. (Throughout this text we use the Kelvin or absolute temperature scale. On this scale, water freezes at 273 K and boils at 373 K. All molecular motion ceases at 0 K. The various temperature scales are described in Appendix 5.) There are stars cooler than the Sun and stars hotter than the Sun; there are even some stars so hot that most of their energy is given off at ultraviolet wavelengths.

The wavelength at which a blackbody emits its maximum energy can be calculated according to the equation

$$\lambda_{\max} = 3 \times \frac{10^6}{T}$$

where the wavelength is in nanometers and the temperature is in Kelvins. This relationship is called **Wien's law.** For the Sun, the wavelength at which the maximum en-

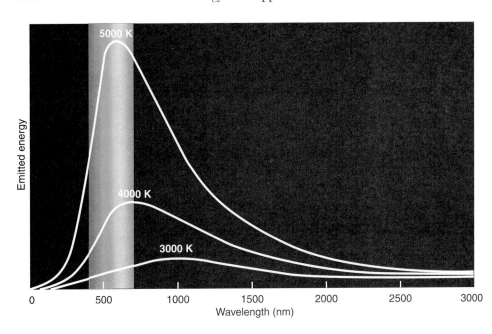

Figure 4.8
Energy emitted at different wavelengths for blackbodies at three different temperatures. At hotter temperatures, more energy is emitted at all wavelengths. The peak amount of energy is radiated at shorter wavelengths for higher temperatures (Wien's law).

ergy is emitted is 520 nm, which is near the middle of that portion of the electromagnetic spectrum called visible light. Characteristic temperatures of other astronomical objects, and the wavelengths at which they emit most of their energy, are listed in Table 4.1.

We can also describe our observation that hotter objects radiate more energy at all wavelengths in a mathematical form. If we sum up the contributions from all parts of the electromagnetic spectrum, we obtain the total energy emitted by a blackbody. That total energy, emitted per second per square meter by a blackbody at a temperature T, is proportional to the fourth power of its absolute temperature. This relationship is known as the **Stefan-Boltzmann law,** which can be written in the form of an equation as

$$E = \sigma T^4$$

where E stands for the energy and σ is a constant number.

Notice how impressive this result is. Increasing the temperature of a star would have a tremendous effect on its ability to generate electromagnetic energy. If the Sun, for example, were twice as hot as it is today—that is, if it had a temperature of 11,600 K—it would radiate 2^4, or 16, times more energy than it does now.

4.3

Spectroscopy in Astronomy

As we hinted at the beginning of the chapter, electromagnetic radiation carries a tremendous amount of information about the nature of stars and other astronomical objects. To extract this information, however, astronomers must be able to study the amounts of energy we receive at different wavelengths of light (and other radiation) in fine detail. Let's examine how we can do this and what we can learn.

Optical Properties of Light

Light and other forms of electromagnetic energy exhibit certain behaviors that are important to the design of telescopes and other instruments. For example, light is *reflected* from a surface. If the surface is smooth and shiny, as in a mirror, the direction of the reflected light beam can be accurately calculated from a knowledge of the shape of the reflecting surface. Light is also bent, or *refracted,* when it passes from one kind of transparent medium into another, say from the air into a glass lens.

The reflection and refraction of light are the basic properties that make possible all optical instruments—from eyeglasses to giant astronomical telescopes. Such instruments are generally combinations of glass lenses, which bend light according to the principles of refraction, and curved mirrors, which depend on the properties of reflection. Small optical devices, such as eyeglasses or binoculars, generally use lenses, while large telescopes depend almost entirely on mirrors for their main optical elements. In Chapter 5 we will discuss a number of astronomical instruments and their uses. For now, we turn to another behavior of light, one that is essential for making a spectrometer.

When light passes from one transparent medium to another, an interesting effect occurs in addition to simple refraction. Because the bending of the beam depends on the wavelength of the light as well as the properties of the medium, different wavelengths or colors of light are bent by different amounts, and separated. This phenomenon is called **dispersion.**

Figure 4.7 shows how light can be separated into different colors with a *prism*—a piece of glass in the shape of a triangle. Upon entering one face of the prism, light is refracted once, the violet light more than the red, and upon leaving the opposite face, the light is bent again and further dispersed. If the light leaving the prism is focused on a screen, the different wavelengths or colors that compose white light are lined up side by side (Figure 4.9) as in a rainbow, which is formed by the dispersion of light through raindrops (see the "Making Connections" box). Since this array of colors is a spectrum, the instrument used to disperse the light and form the spectrum is called a **spectrometer.**

The Value of Stellar Spectra

If the spectrum of the white light from the Sun and stars were simply a continuous rainbow of colors, astronomers, once they had learned a star's approximate temperature, as described earlier, would have little interest in the detailed study of its spectrum. When Newton, who first described the laws of refraction and dispersion in optics, observed the solar spectrum, all he could see was a continuous band of colors.

Figure 4.9
When white light passes through a prism, its dispersion forms a continuous spectrum of all the colors.

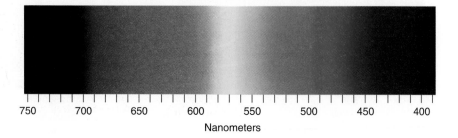

| | | | | | | | |
|750|700|650|600|550|500|450|400|

Nanometers

Figure 4.10
The visible spectrum of the Sun. The spectrum is crossed by dark lines produced by atoms in the solar atmosphere that absorb light at certain wavelengths. (National Solar Observatory/National Optical Astronomy Observatories)

However, in 1802 William Wollaston built an improved spectrometer that included a lens to focus the Sun's spectrum on a screen. With this device Wollaston saw that the colors were not spread out uniformly, but that instead some ranges of color were missing, appearing as dark bands in the solar spectrum. He mistakenly attributed these lines to natural boundaries between the colors. In 1815 the German physicist Joseph Fraunhofer, upon a more careful examination of the solar spectrum, found about 600 such dark lines (Figure 4.10), which pretty quickly ruled out the boundary hypothesis.

These physicists called the missing colors *dark lines* because the rainbow-colored (continuous) spectrum as they viewed it was interrupted at various wavelengths, and in their spectrometers such interruptions looked like dark lines. Later, researchers found that similar dark lines could be produced in the spectra (*spectra* is the plural of spectrum) of artificial light sources by passing their light through various apparently transparent substances (usually containers with just a bit of thin gas in them).

These gases turned out not to be transparent at *all* colors; they were quite opaque at a few sharply defined wavelengths. Something in each gas had to be absorbing just a few colors of light and no others. Interestingly, gases made of different elements all did this, but each element absorbed a different set of colors.

What would happen if there were no continuous spectrum for these gases to remove light from? What if, instead, we heated the same gases until they were hot enough to glow with their own light? (Bear in mind that the gas containers used in these experiments held gases in pretty low concentrations, so collisions among the molecules or atoms were rare.) When the gases were heated, a spectrometer revealed no continuous spectrum, but several separate *bright lines*. That is, these gases emitted light only at certain discrete wavelengths or colors. Interestingly, the colors at which they emitted when they were heated were the very same colors at which they had absorbed when a continuous source was behind them.

When the gas was pure hydrogen it would absorb or emit one pattern of colors; when it was sodium, it would absorb or emit a different pattern. A mixture of hydrogen and sodium emitted both sets of spectral lines. Thus scientists began to see that different substances showed dis-

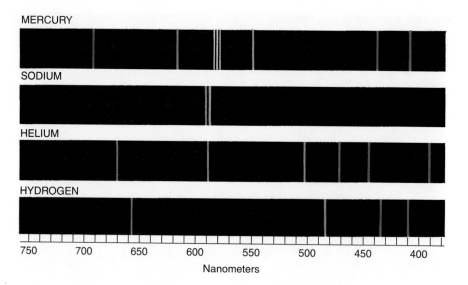

Figure 4.11
The line spectra produced by several different kinds of hot gas. Each type of gas (each element) produces its own unique pattern of lines, so the composition of a gas can be identified by its spectrum.

tinctive *spectral signatures* by which their presence could be detected (Figure 4.11). Just as your signature allows the bank to identify you, the unique pattern of colors for each type of atom can help us identify which element or elements are in the gas.

Types of Spectra

We can distinguish, then, among the three different types of spectra. A **continuous spectrum** (formed when a dense collection of gas or a solid gives off radiation) is an array of all wavelengths or colors of the rainbow. A continuous spectrum can serve as a backdrop from which the atoms of a much less dense gas can absorb light. A dark line, or **absorption line spectrum,** consists of a series or pattern of dark lines—missing colors—superimposed upon the continuous spectrum of a source. A bright line, or **emission line spectrum,** appears as a pattern or series of bright lines; it consists of light in which only certain discrete wavelengths are present.

Each particular chemical element or compound, when in the gaseous form, produces its own characteristic pattern of spectral lines—its spectral signature. No two patterns are alike. In other words, each particular gas can absorb or emit only certain wavelengths of light peculiar to that gas. The temperature and other conditions determine whether the lines are bright or dark, but the wavelengths are the same in either case, and it is the precise pattern of wavelengths that makes the signature of each element unique. Liquids and solids can also generate spectral lines or bands, but they are broader and less well defined, and hence more difficult to interpret.

The dark lines in the solar spectrum thus give evidence of certain chemical elements between us and the Sun, absorbing those wavelengths of light. Since the space between us and the Sun is pretty empty, astronomers quickly realized that the atoms doing the absorbing must be in a thin atmosphere of cooler gas around the Sun. Re-

markably, this outer atmosphere is not all that different from the rest of the Sun, just thinner and cooler. Thus we can use what we learn about its composition as a pretty good indicator of what the whole Sun is made of. Similarly, we can use the presence of absorption and emission lines to analyze the composition of other stars and clouds of gas in space.

Such analysis of spectra is the key to modern astrophysics. Only in this way can we "sample" the stars, which are too far away for us to have much hope of visiting. Even if we could visit them, they are so hot that it would be very difficult to find graduate students willing to go take a sample. But there is no need to ask for such volunteers: encoded in the electromagnetic radiation from celestial objects is clear information about the chemical makeup of these objects.

In 1860, the German physicist Gustav Kirchhoff became the first person to use spectroscopy to identify an element in the Sun when he found the spectral signature of sodium gas. In the years that followed, astronomers found many other chemical elements in the Sun and stars. In fact, the element helium was found first in the Sun from its spectrum, and only later identified on Earth. (The word helium comes from *helios*, the Greek name for the Sun.) But only in the 20th century, with the development of a model for the atom, did scientists learn why such dark and bright spectral lines exist. We therefore turn next to a closer examination of the atoms that make up all matter.

4.4

The Structure of the Atom

The idea that matter is composed of tiny particles called atoms is at least 25 centuries old. It took until the 20th century, however, for scientists to invent instruments that permitted them to probe inside an atom and find that it is not, as had been thought, hard and indivisible. Instead, the atom is a complex structure composed of still smaller particles.

Probing the Atom

The first of these smaller particles was discovered by British physicist J. J. Thomson in 1897. Named the electron, this particle is negatively charged. (It is the flow of these particles that produces currents of electricity, whether in lightning bolts or in the wires leading to your hair dryer.) Since an atom in its normal state is electrically neutral, each electron in an atom must be balanced by the same amount of positive charge.

The next problem was to determine where in the atom the positive and negative charges are located. In 1911 British physicist Ernest Rutherford devised an experiment that provided part of the answer to this question. He bombarded an extremely thin piece of gold foil, only about 400 atoms thick, with a beam of alpha particles emitted from a radioactive material (Figure 4.12). We now know that alpha particles are helium atoms that have lost all of their electrons and are thus positively charged. Most of the alpha particles passed through the gold foil just as if it and the atoms composing it were nearly empty space. About 1 in 8000 of the alpha particles, however, completely reversed direction and bounced backward from the foil. Rutherford wrote, "It was quite the most incredible event that has ever happened to me in my life. It was almost as incredible as if you fired a 15-inch shell at a piece of tissue paper and it came back and hit you."

The only way to account for the alpha particles that reversed direction when they hit the gold foil was to assume that nearly all of the mass as well as all of the positive charge in each individual gold atom was concentrated in a tiny **nucleus.** When an alpha particle strikes a nu-

cleus, it reverses direction, much as a cue ball reverses direction when it hits another billiard ball. Rutherford's model placed the other type of charge—the negative electrons—in orbit around this nucleus.

Rutherford's model required that the electrons be in motion. Since positive and negative charges attract each other, stationary electrons would fall into the nucleus. Also, since both the electrons and the nucleus are extremely small, most of the atom is empty, which is why nearly all of Rutherford's alpha particles were able to pass right through the gold foil without colliding with anything. Rutherford's model was a very successful explanation of the experiments he conducted, although eventually scientists would discover that the nucleus has structure as well.

The Atomic Nucleus

The simplest possible atom (and the most common one in the Sun and stars) is hydrogen; the nucleus of ordinary hydrogen contains a single positively charged particle called a *proton.* Moving around this proton is a single *electron.* The mass of an electron is nearly 2000 times smaller than the mass of a proton, but the electron carries an amount of charge exactly equal to that of the proton but opposite in sign (Figure 4.13). Opposite charges attract each other, so it is electromagnetic force that holds the proton and electron together, just as gravity is the force that keeps the planets in orbit around the Sun.

There are, of course, other types of atoms. Helium, for example, is the second most abundant element in the Sun. Helium has two protons in its nucleus, instead of the single proton that characterizes hydrogen. In addition, the helium nucleus contains two *neutrons,* particles with a mass comparable to that of the proton but with no electric charge. Moving around this nucleus are two electrons, so

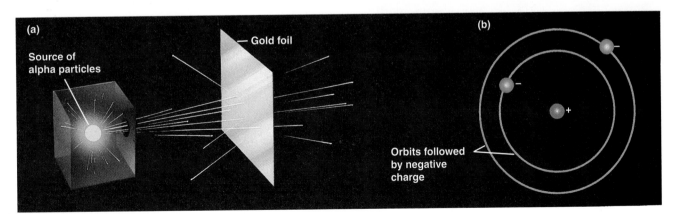

Figure 4.12
(a) When Rutherford allowed alpha particles from a radioactive source to strike a target of gold foil, he found that some of them rebounded back in the direction from which they came.
(b) From this experiment, he concluded that the atom must be constructed like a miniature solar system, with the positive charge concentrated in the nucleus, and the negative charge orbiting in the large volume around the nucleus.

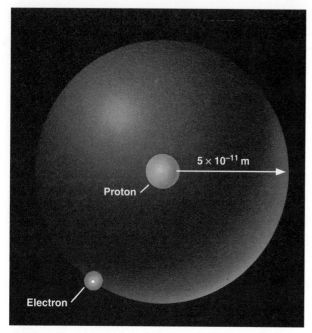

Figure 4.13
Schematic diagram of a hydrogen atom in its lowest energy state, also called the ground state. The proton and electron have equal but opposite charges, which exert an electromagnetic force that binds the hydrogen atom together.

In the diagram: 5×10^{-11} m, Proton, Electron.

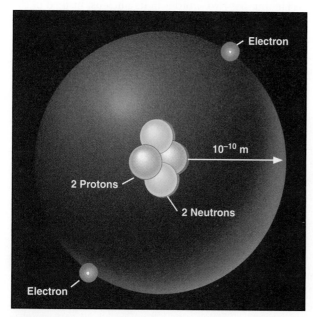

Figure 4.14
Schematic diagram of a helium atom in its lowest energy state. Two protons are present in the nucleus of all helium atoms. In the most common variety of helium, the nucleus also contains two neutrons, which have nearly the same mass as the proton but carry no charge. Two electrons can be seen orbiting the nucleus.

In the diagram: Electron, 10^{-10} m, 2 Protons, 2 Neutrons, Electron.

the total net charge of the helium atom is zero (Figure 4.14).

From this description of hydrogen and helium, perhaps you have guessed the pattern for building up all the elements (different types of atoms) that we find in the universe. The type of element is determined by the *number of protons* in the nucleus of the atom. Any atom with 6 protons is the element carbon, with 8 protons is oxygen, with 26 is iron, and with 92 is uranium. On Earth, a typical atom has the same number of electrons as protons, and these electrons follow complex orbital patterns around the nucleus. (Deep inside stars, however, it is so hot that the electrons get loose from the nucleus.)

Although the number of neutrons in the nucleus is usually approximately equal to the number of protons, the number of neutrons is not necessarily the same for all atoms of a given element. For example, most hydrogen atoms contain no neutrons at all. There are, however, hydrogen atoms that contain one proton and one neutron, and others that contain one proton and two neutrons. The various types of hydrogen nuclei are called **isotopes** of hydrogen (Figure 4.15), and other elements have isotopes as well. You can think of isotopes as siblings, closely related but with different characteristics and behaviors.

The Bohr Atom

There is one serious problem with Rutherford's model for atoms. Maxwell's theory of electromagnetic radiation says that when electrons change either their speed or direction of motion, they must emit energy. Since orbiting electrons constantly change their direction of motion, they should emit a constant stream of energy. According to Maxwell's theory, all electrons should spiral into the nucleus of the atom as they lose energy. And this collapse should happen very quickly—in about 10^{-16} seconds.

It was the Danish physicist Niels Bohr (1885–1962) who solved the mystery of how electrons remain in orbit.

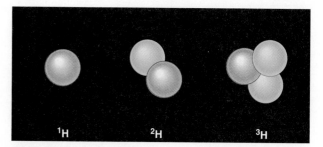

^{1}H ^{2}H ^{3}H

Figure 4.15
Schematic diagram of the nuclei of hydrogen isotopes. A single proton in the nucleus defines the atom to be hydrogen, but there may be zero, one, or two neutrons. The most common isotope is the one with only a single proton. A hydrogen nucleus with one neutron is called deuterium; one with two neutrons is called tritium. The symbol for hydrogen is H; the superscript at left tracks the total number of particles (protons plus neutrons) in the nucleus.

He was trying to develop a model of the atom that would also explain certain regularities observed in the spectrum of hydrogen (see Section 4.5). He suggested that the spectrum of hydrogen can be understood if we assume that only orbits of certain sizes are possible for the electron. Bohr further assumed that as long as the electron moves only in one of these allowed orbits, it radiates no energy. Its energy would change only if it moved from one orbit to another.

This suggestion, in the words of science historian Abraham Pais, was "one of the most audacious hypotheses ever introduced in physics." If something equivalent were at work in the everyday world, you might find that, as you went for a walk after astronomy class, nature permitted you to walk 2 steps per minute, 5 steps per minute, and 12 steps per minute, but no speeds in between. No matter how you tried to move your legs, only *certain* walking speeds would be permitted. And to make things more bizarre, it would take no effort to walk at one of the allowed speeds, but be difficult to change from one speed to another. Luckily, no such rules apply at the macroscopic level of human behavior, but at the microscopic level of the atom, experiment after experiment has confirmed the validity of Bohr's idea. Bohr's suggestions became one of the foundations of the new (and much more sophisticated) model of the subatomic world called quantum mechanics.

In Bohr's model, if the electron moves from one orbit to another closer to the atomic nucleus, it must give up some energy in the form of electromagnetic radiation. If, on the other hand, the electron goes from a smaller orbit to one farther from the nucleus, it requires some additional energy. One way to obtain the necessary energy is to absorb electromagnetic radiation that may be streaming past the atom from an outside source.

A key feature of Bohr's model is that each of the permitted electron orbits around a given atom has a certain energy value. To move from one orbit to another (which will have *its* own specific energy value) requires a change in the electron's energy—a change determined by the difference between the two energy values. If the electron goes to a lower level, the energy difference will be given off; if the electron goes to a higher level, the energy difference must be obtained from somewhere else. Each jump (or transition) to a different level has a fixed and definite energy change associated with it.

A crude analogy for this situation might be life in a tower of luxury apartments where the rent is determined by the quality of the view. Such a building has certain definite numbered levels or floors on which apartments are located. No one can live on floor 5.37 or 22.5. Second, the rent gets higher as you go up. If you want to exchange an apartment on the 20th floor for one on the 2nd floor, you will not owe as much rent, and you will get a refund from the landlord. But if you want to move from the 3rd floor to the 25th floor, you'd better be prepared to find some extra resources, because your rent will increase. In the atom, too, the "cheapest" place for an electron to live is the lowest possible level, and energy is required to move to a higher level.

Here we have one of the situations where it is easier to think of electromagnetic radiation as **photons** than as waves, see page 86. As electrons move from one level to another they give off or absorb little packets of energy. When an electron moves to a higher level, it absorbs a photon of just the right energy (provided one is available). When it moves to a lower level, it emits a photon with the exact amount of energy it no longer needs in its "lower-cost" living situation.

The photon and wave perspectives must be equivalent: light is light, no matter how we look at it. Thus each photon carries a certain amount of energy that is proportional to the frequency of the wave it represents. The constant of proportionality, h, called Planck's constant, is named for Max Planck, the German physicist who was one of the originators of the quantum theory. If metric units are used (that is, if energy is measured in joules and frequency in cycles or waves per second), Planck's constant has the value $h = 6.626 \times 10^{-20}$ joule/s. As we saw, the highest-energy photons are the high-frequency gamma rays; those of lowest energy are the lowest-frequency radio waves.

4.5

Formation of Spectral Lines

The Hydrogen Spectrum

Now let's use Bohr's model of the atom to understand how spectral lines are formed. Suppose a beam of white light (which consists of photons of all wavelengths) shines through a gas of atomic hydrogen. It turns out that a photon of wavelength 656 nm has just the right energy to raise an electron in a hydrogen atom from the second to the third orbit. Thus, as all the photons of different energies (waves of different wavelengths) stream by the hydrogen atoms, photons with *this* particular wavelength can be absorbed by those atoms whose electrons are orbiting on the second level. When they are absorbed, the electrons on the second level will now be on the third level, and a number of the photons of this wavelength and energy will be missing from the general stream.

Other photons will have the right energies to raise electrons from the second to the fourth orbit, or from the first to the fifth orbit, and so on. Only photons with exactly these correct energies can be absorbed. All of the other photons will stream past the atoms untouched. Thus, the hydrogen atoms absorb light only at certain wavelengths and produce dark lines in the spectrum we see.

Now suppose we have a container of hydrogen gas through which a whole series of photons is passing, allowing many electrons to move up to higher levels. Next we turn off the light source. These electrons then "fall" back

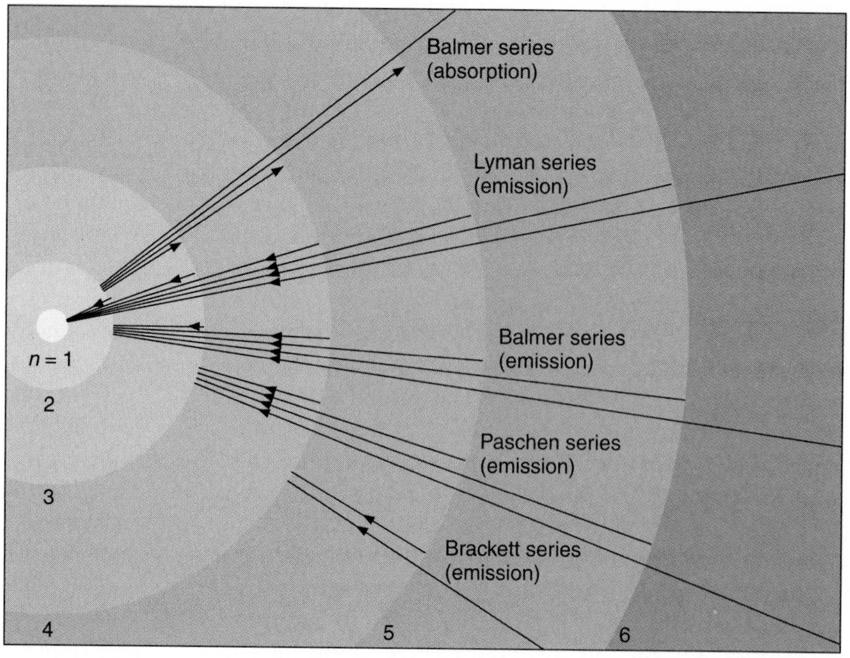

Figure 4.16
Emission and absorption of photons by a hydrogen atom according to the Bohr model. Several different series of spectral lines are shown, corresponding to transitions of electrons from or to certain allowed energy levels. Each series of lines that terminates on a specific inner orbit is named for the physicist who studied it.

down from larger to smaller orbits and emit photons of light—but, again, only light of those energies or wavelengths that correspond to the energy difference between permissible orbits. The orbital changes of hydrogen electrons giving rise to spectral lines are shown in Figure 4.16. (Note that in a real hydrogen atom the electron orbits are not as evenly spaced as those shown in this diagram.)

Similar pictures can be drawn for atoms other than hydrogen. However, since these other atoms ordinarily have more than one electron each, the orbits of their electrons are much more complicated, and the spectra are more complex as well. For our purposes, the key difference is this: each type of atom has its own unique pattern of electron orbits, and no two sets of orbits are exactly alike. This means that each type of atom has its own unique set of spectral lines, produced by electrons moving between its unique set of orbits. If we can learn the lines that go with each element by studying atoms here on Earth, we can use this knowledge to identify the elements in celestial bodies. In this way, astronomers today know the chemical makeup of objects so distant that their light started on its way to us long before the Earth had even formed.

Energy Levels and Excitation

Bohr's model of the hydrogen atom was a great step forward in our understanding of the atom. However, we know today that atoms cannot be represented by quite so simple a picture. Even the concept of sharply defined electron orbits is not really correct. One of the most interesting (and puzzling) results of quantum mechanics is that subatomic particles such as electrons sometimes behave like waves (just as waves of electromagnetic radiation sometimes behave like particles). Since an electron has

wave-like characteristics, its position is much harder to pinpoint than you might expect from the Bohr model. We can only estimate the *probability* that it will follow a particular orbit. (If these ideas are starting to sound a little strange, we assure you that they also sounded strange to physicists when first being proposed. But remember that nature, especially at the subatomic level, is under no obligation to fit "common sense" human perceptions, which developed in the world of much larger-scale phenomena.)

Luckily, it turns out that the most likely orbits for electrons fall in a fairly narrow range when compared to the size of an atom. This means we can still retain the concept that only certain discrete energies are allowable for an atom. These energies, called **energy levels,** can be thought of as representing certain average distances of the electron's possible orbits around the atomic nucleus.

Ordinarily, an atom is in the state of lowest possible energy, its **ground state.** In the Bohr model, the ground state corresponds to the electron's being in the innermost orbit. An atom can absorb energy, which raises it to a higher energy level (corresponding, in the Bohr picture, to an electron's movement to a larger orbit). The atom is then said to be in an **excited state.** Generally, an atom remains excited for only a very brief time. After a short interval, typically a hundred-millionth of a second or so, it drops back down to its ground state, with the simultaneous emission of light. The atom may return to its lowest state in one jump, or it may make the transition in steps of two or more jumps, stopping at intermediate levels on the way down. With each jump it emits a photon of the wavelength that corresponds to the energy difference between the levels at the beginning and end of that jump.

An energy-level diagram for a hydrogen atom and several possible atomic transitions are shown in Figure

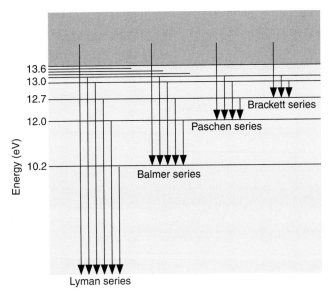

Figure 4.17
Energy-level diagram for hydrogen. As you get to higher and higher energy levels, they get more and more crowded together, approaching a limit. The shaded region represents energies at which the atom is ionized.

We mentioned that atoms that have absorbed specific photons from a passing beam of white light and have thus become excited generally de-excite themselves and emit that light again in a very short time. You might therefore wonder why *dark* spectral lines are ever produced. In other words, why doesn't this re-emitted light quickly "fill in" the darker absorption lines?

Imagine a beam of white light coming toward you through some cooler gas. Some of the re-emitted light is actually returned to the beam of white light you see, but this fills in the absorption lines only to a slight extent. The reason is that the atoms in the gas re-emit light in mostly random directions, and only a small fraction of the re-emitted light is in the direction (toward you) of the original beam. In a star, much of the re-emitted light actually goes in directions leading back into the star, which does observers outside the star no good whatsoever.

Atoms in a hot gas are moving at high speeds and continually colliding with one another and with any loose electrons. They can be excited and de-excited by these collisions, as well as by absorbing and emitting light. The speed of atoms in a gas depends on the temperature. When the temperature is higher, so are the speed and energy of the collisions. The hotter the gas, therefore, the more likely that electrons will occupy the outermost orbits, which correspond to the highest energy levels. This means that the level where electrons *start* their upward jumps in a gas can serve as an indicator of how hot that gas is. In this way, the absorption lines in a spectrum can also give astronomers information about the temperature of the regions where the lines originate. Figure 4.18 summarizes the different kinds of spectra.

4.17; compare this figure with the Bohr model, shown in Figure 4.16. The transitions to or from the ground state, called the Lyman series of lines, result in the emission or absorption of ultraviolet photons. But the transitions to or from the first excited state, called the Balmer series, produce emission or absorption in visible light. In fact it was to explain this Balmer series that Bohr first suggested his model of the atom.

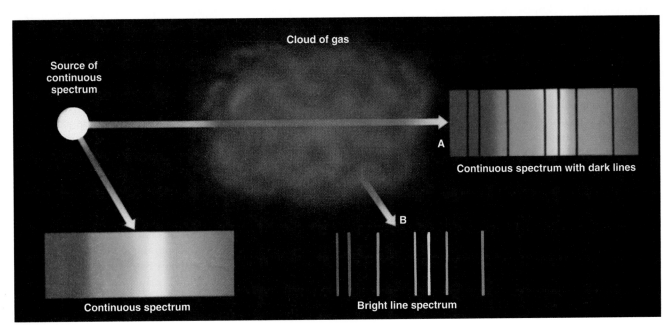

Figure 4.18
Production of bright and dark spectral lines. The atoms in the gas cloud produce absorption lines in the continuous spectrum of the white light source when viewed from direction A, but they produce emission lines (of the light they re-emit) when viewed from direction B.

Ionization

We have described how certain discrete amounts of energy can be absorbed by an atom, raising it to an excited state and moving one of its electrons farther from its nucleus. If enough energy is absorbed, the electron can be completely removed from the atom. The atom is then said to be **ionized.** The *minimum* amount of energy required to remove one electron from an atom in its ground state is called its ionization energy, or ionization potential.

Still greater amounts of energy must be absorbed by the now-ionized atom (called an **ion**) to remove an additional electron deeper in the structure of the atom. Successively greater energies are needed to remove the third, fourth, fifth, and so on, electrons from the atom. If enough energy is available (in the form of very short-wavelength photons or from a collision with a very fast-moving electron or another atom), an atom can become completely ionized, losing all of its electrons. A hydrogen atom, having only one electron to lose, can be ionized only once; a helium atom can be ionized twice, and an oxygen atom up to eight times. When we examine regions of the cosmos where there is a great deal of radiation, such as the neighborhoods where hot young stars have recently formed, we see a lot of ionization going on.

An atom that has become ionized has lost a negative charge—which was carried away by the electron—and thus is left with a net positive charge. It therefore feels a strong attraction for any free electron. Eventually, one or more electrons will be captured, and the atom will become neutral (or ionized to one less degree) again. During the electron capture process, the atom emits one or more photons, depending on whether the electron is captured at once to the lowest energy level of the atom, or whether it stops at one or more intermediate levels on its way to the ground state.

Just as the excitation of an atom can result from a collision with another atom, ion, or electron (collisions with electrons are usually most important), so ionization can also result from collisions. The rate at which such collisional ionizations occur depends on the speeds of the atoms and hence on the temperature of the gas.

The rate at which ions and electrons recombine also depends on their relative speeds—that is, on the temperature. In addition, it depends on the density of the gas: the higher the density, the greater the chance for recapture, because the different kinds of particles are crowded more closely together. From a knowledge of the temperature and density of a gas, it is possible to calculate the fraction of atoms that have been ionized once, twice, and so on. In the Sun, for example, we find that most of the hydrogen and helium atoms in its atmosphere are neutral, whereas most of the calcium atoms, as well as many other heavier atoms, are ionized once.

The energy levels of an ionized atom are entirely different from those of the same atom when it is neutral. Each time an electron is removed from the atom, the energy levels of the ion, and thus the wavelengths of the spectral lines it can produce, change. This helps astronomers tell the different ions of a given element apart. Ionized hydrogen, having no electron, can produce no absorption lines.

4.6

The Doppler Effect

Although the last two sections contain many new ideas, we hope that you have seen one major idea emerge. Astronomers can learn about the elements in stars and galaxies by decoding the information in the spectral lines. There is a complicating factor in learning how to decode the message of starlight, however. If a star is moving toward or away from us, its lines will be in a slightly different place in the spectrum from where they would be in a star at rest. And most objects in the universe do have some motion relative to the Sun.

In 1842 Christian Doppler pointed out that if a light source is approaching or receding from the observer, the light waves will be, respectively, crowded more closely together or spread out. (This is actually true for all kinds of waves, including sound: Doppler first measured the effect of motion on waves by hiring a group of musicians to play on an open railroad car as it was moving along the track.) The general principle, now known as the **Doppler effect,** is illustrated in Figure 4.19.

In part (a) of the figure, the light source (S) is stationary with respect to the observer. The source gives off a series of waves, whose crests we have labeled 1, 2, 3, and 4. The light waves spread out evenly in all directions, like the ripples from a splash in a pond. The crests are separated by a distance, λ, where λ is the wavelength. The observer, who happens to be located in the direction of the bottom of the page, sees the light waves coming nice and evenly, one wavelength apart. Observers located anywhere else would see the same thing.

On the other hand, if the source is moving with respect to the observer, as in part (b), the situation is more complicated. Between the time that one crest is emitted and the next is ready to come out, the source has moved a bit—in our case toward the bottom of the page and toward our original observer. By moving, the source has decreased the distance between crests—squeezed the crests together a bit, you might say.

In part (b) we show the source in four positions, S_1, S_2, S_3, and S_4, each corresponding to the emission of one wave crest. To our original observer, now labeled observer A, the waves seem to follow one another more closely, at a decreased wavelength and thus increased frequency. (Remember, since all light waves travel at the speed of light, motion cannot affect the speed. This means that as the wavelength decreases, the frequency must increase. If the waves are shorter, more will be able to fit during each second.)

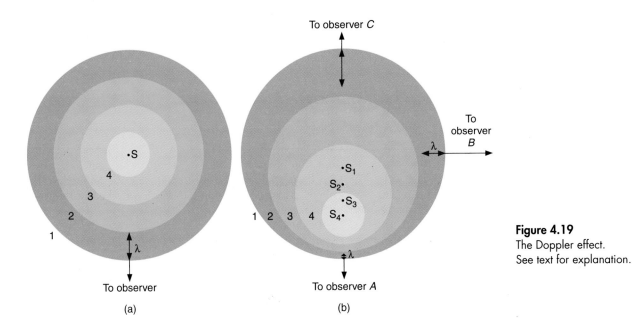

Figure 4.19
The Doppler effect.
See text for explanation.

But the situation is not the same for other observers. Let's look at the situation from the point of view of observer C, located opposite observer A in the figure. For her, the source is moving away from her location. As a result, the waves are not squeezed together, but instead are spread out by the motion of the source. The crests arrive with an increased wavelength and decreased frequency. To observer B, in a direction at right angles to the motion of the source, no effect is observed. The wavelength and frequency remain the same as they were in part (a) of the figure.

We can see that the Doppler effect is produced only by a motion toward or away from the observer, a motion called **radial velocity.** Observers between A and B and between B and C would observe, respectively, some shortening or lengthening (increase or decrease in frequency) of the light waves for that part of the motion of the source that is along their line of sight.

You have heard the Doppler effect with sound waves in everyday life. When a train whistle or police siren approaches you and then moves away, you will notice a decrease in the pitch (or frequency) of the sound waves. They have changed from slightly more frequent than at rest when coming toward you, to slightly less frequent than at rest when moving away from you. (A nice example of this change in the sound of a train whistle can be heard at the end of the Beach Boys song "Caroline, No" on their album *Pet Sounds.*)

When the source of waves moves toward you, the wavelengths decrease a bit. If the waves involved are visible light, then the colors of the light change slightly. As wavelengths decrease, they shift a bit toward the blue end of the spectrum: astronomers call this a *blueshift.* (Since the end of the spectrum is really violet, the term should probably be violetshift, but blue is a more common color.) When the source moves away from you, and wavelengths get longer, we call the change in colors a *redshift.* Because the Doppler effect was first used with visible light in astronomy, the terms blueshift and redshift became well established. Today astronomers use these words to describe changes in radio wavelengths or x-ray wavelengths as comfortably as they use them to describe changes in visible light.

The greater the motion toward us or away from us, the greater the Doppler shift. If the relative motion is entirely along the line of sight, the formula for the Doppler shift of light is

$$\frac{\Delta\lambda}{\lambda} = \frac{v}{c}$$

where λ is the wavelength emitted by the source, $\Delta\lambda$ is the difference between λ and the wavelength measured by the observer, c is the speed of light, and v is the relative velocity (speed) of the observer and the source in the line of sight. The variable v is counted as positive if the velocity is one of recession, and negative if it is one of approach. Solving this equation for the velocity, we find

$$v = c\,\frac{\Delta\lambda}{\lambda}$$

If a star approaches or recedes from us, the wavelengths of light in its continuous spectrum appear shortened or lengthened, respectively, as do those of the dark lines. However, unless its speed is tens of thousands of kilometers per second, the star does not appear noticeably bluer or redder than normal. The Doppler shift is thus not easily detected in a continuous spectrum and cannot be measured accurately in such a spectrum. On the other hand, the wavelengths of the absorption lines can be measured accurately, and their Doppler shift is relatively simple to detect.

The Rainbow

Rainbows are an excellent illustration of refraction of sunlight. You have a good chance of seeing a rainbow anytime you are between the Sun and a rain shower, and this situation is illustrated in Figure 4A. The raindrops can act like little prisms and break white light into the spectrum of colors. Suppose that a ray of sunlight encounters a raindrop and passes into it. The light changes direction—is refracted (Figure 4B)—when it passes from air to water; the blue light is refracted more than the red. Some of the light is then reflected at the backside of the drop and re-emerges from the front, where it is again refracted. As a result, the white light is spread out into a rainbow of colors.

Note that in Figure 4B, violet light lies above the red light after it emerges from the raindrop. When you look at a

rainbow, however, it is the red light that is higher in the sky. Why? Look again at Figure 4A. If the observer looks at a raindrop that is high in the sky, the violet light passes over her head, while the red light enters her eye. Similarly, if the observer looks at a raindrop that is low in the sky, the violet light reaches her eye and the drop appears violet, while the red light from that same drop strikes the ground and is not seen. Colors of intermediate wavelengths are refracted to the eye by drops that are intermediate in altitude between the drops that appear violet and the ones that appear red. Thus, a single rainbow always has red on the outside and violet on the inside.

For an even simpler example of refraction, put a pencil at a slanted angle in a glass of water. What do you see? Can you offer an explanation?

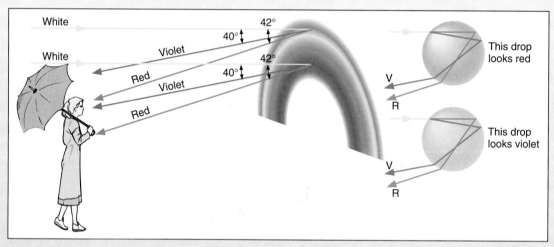

Figure 4A
This diagram shows how light from the Sun, which is located behind the observer, can be refracted by raindrops to produce a rainbow.

Figure 4B
A diagram showing the path of light passing through a raindrop. Refraction separates white light into its component colors.

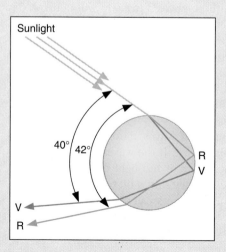

This may sound like a horrible note on which to end the chapter. If all the stars are moving, and motion changes the wavelength of each spectral line, won't this be a disaster for astronomers trying to figure out what elements are present in the stars? After all, it is the precise wavelength (or color) that tells astronomers which lines belong to which element. And we first measure these wavelengths in containers of gas in our laboratories, which are not moving. If every line is now shifted by motion to a different wavelength (color), how can we be sure which lines and which elements we are looking at in a new star whose speed we do not know?

This situation sounds worse than it really is. Astronomers rarely judge the presence of an element in an astronomical object by a single line. It is the *pattern* of lines unique to hydrogen or calcium that enables us to determine that those elements are part of the star or galaxy we are observing. The Doppler effect does not change the pattern of lines from a given element—it only shifts the pattern slightly toward redder or bluer wavelengths. The shifted pattern is still pretty easy to recognize. And best of all, when we do recognize a familiar element's pattern, we get a bonus: the amount the pattern is shifted now tells us the motion of the object in our line of sight.

Summary

4.1 James Clerk Maxwell showed that whenever charged particles change their motion, as they do in every atom and molecule, they give off waves of energy. Light is one form of this **electromagnetic radiation.** The **wavelength** of light determines the color of visible radiation. Wavelength (λ) is related to **frequency** (f) and the speed of light (c) by the equation $c = \lambda f$. Electromagnetic radiation sometimes behaves like waves, but at other times it behaves as if it were a little packet of energy, called a **photon.** The **apparent brightness** of a source of electromagnetic energy decreases with increasing distance from that source in proportion to the square of the distance, a relationship known as the **inverse-square law.**

4.2 The **electromagnetic spectrum** consists of **gamma rays, x rays,** and **ultraviolet radiation** (all forms of electromagnetic radiation with wavelengths shorter than that of visible light), **visible light,** and **infrared, microwave,** and longer-wave **radio** radiation (the last three with wavelengths longer than that of light). Many of these wavelengths cannot penetrate the layers of the Earth's atmosphere and must be observed from space. The emission of electromagnetic radiation is intimately connected to the temperature of the source. The higher the temperature of a **blackbody,** the shorter the wavelength at which the maximum amount of electromagnetic radiation is emitted. The mathematical equation describing this relationship ($\lambda_{max} = 3 \times 10^6/T$) is known as **Wien's law.** The total energy emitted per square meter increases with increasing temperature. The relationship between emitted energy and temperature ($E = \sigma T^4$) is known as the **Stefan-Boltzmann law.**

4.3 A **spectrometer** is a device that forms a spectrum, often utilizing the optical phenomenon of **dispersion.**

The light from an astronomical source can consist of a **continuous spectrum,** a bright line or **emission line spectrum,** or a dark line or **absorption line spectrum.** Because each element leaves its spectral "signature" in the pattern of lines we observe, spectral analyses reveal the composition of the Sun and stars.

4.4 Atoms consist of a **nucleus** containing one or more positively charged protons. All atoms except hydrogen also contain one or more neutrons in the nucleus. Negatively charged electrons orbit the nucleus. The number of protons defines the element (hydrogen, helium, and so on) of the atom. Nuclei with the same number of protons but different numbers of neutrons are different **isotopes** of the same element. According to the Bohr model of the atom, when an electron moves from one orbit to another closer to the atomic nucleus, a photon is emitted and a spectral emission line is formed. Absorption lines are formed when an electron moves to an orbit farther from the nucleus. Since each atom has its own characteristic set of orbits, each is associated with a unique pattern of spectral lines.

4.5 An atom in its lowest **energy level** is said to be in the **ground state.** If an electron is in an orbit other than the least energetic one possible, the atom is said to be **excited.** If an atom has lost one or more electrons, it is called an **ion** and is said to be **ionized.**

4.6 If an atom is moving toward us when an electron changes orbits and produces a spectral line, we see that line shifted slightly toward the blue of its normal wavelength in a spectrum. If the atom is moving away, we see the line shifted toward the red. This shift is known as the **Doppler effect** and can be used to measure the **radial velocities** of distant objects by the formula $v = c(\Delta\lambda/\lambda)$.

1. What distinguishes one type of electromagnetic radiation from another? What are the main categories (or bands) of the electromagnetic spectrum?

2. What is a wave? Use the terms "wavelength" and "frequency" in your definition.

3. What is a blackbody? Is this textbook a blackbody? Why or why not? How does the energy emitted by a blackbody depend on its temperature?

4. Where in an atom would you expect to find electrons? Protons? Neutrons?

5. Explain how emission lines and absorption lines are formed.

6. Explain how the Doppler effect works for sound waves, and give some familiar examples.

7. What kind of motion for a star does *not* produce a Doppler effect? Explain why.

8. Describe how Bohr's model used the work of Rutherford and Maxwell. Why was Bohr's model considered a radical notion?

9. Make a list of some of the many practical consequences of Maxwell's theory of electromagnetic waves (television would be one example). Put a check mark next to all the electromagnetic wave technology that you use during the course of a typical day.

10. Suppose the Sun radiates like a blackbody. Explain how you would calculate the total amount of energy radiated into space by the Sun each second. What information about the Sun would you need to make this calculation?

11. What type of electromagnetic radiation is best suited to observing
  a. a star with a temperature of 5800 K?
  b. a gas heated to a temperature of 1 million K?
  c. a person on a dark night?

12. Why is it dangerous to be exposed to x rays but not (or at least way less) dangerous to be exposed to radio waves?

13. Go outside on a clear night and look carefully at the brightest stars. Some should look red and others blue. The primary factor determining the color of a star is its temperature. Which is hotter, a blue star or a red one? Explain.

14. Water faucets are often labeled with a red dot for hot water and a blue dot for cold. Given Wien's law, does this labeling make sense?

15. The planet Jupiter appears yellowish, while Mars is red. Does this mean that Mars is cooler than Jupiter? Explain your answer.

16. Suppose you are standing at the exact center of a park surrounded by a circular road. An ambulance drives completely around this road, with siren blaring. How does the pitch of the siren change as it circles around you?

17. How could you measure the Earth's orbital speed by photographing the spectrum of a star at various times throughout the year? (*Hint:* Suppose the star lies in the plane of the Earth's orbit.)

18. "Tidal waves," or tsunamis, are waves caused by earthquakes that travel rapidly through the ocean. If tsunamis travel at the speed of 600 km/h and approach a shore at the rate of one wave crest every 15 min, what would be the distance between those wave crests at sea?

19. How many times brighter or fainter would a star appear if it were moved to
  a. twice its present distance?
  b. ten times its present distance?
  c. half its present distance?

20. Two stars with identical diameters are the same distance away. One has a temperature of 5800 K; the other has a temperature of 2900 K. Which is brighter? How much brighter is it?

21. If the emitted infrared radiation from Pluto has a wavelength of maximum intensity at 50,000 nm (50 μm), what is the temperature of Pluto (assuming it behaves like a black body)?

22. What is the temperature of a star whose maximum light is emitted at a wavelength of 290 nm?

23. Suppose that a spectral line of some element, normally at 500 nm, is observed in the spectrum of a star to be at 500.1 nm. How fast is the star moving toward or away from the Earth?

Suggestions for Further Reading

Augensen, H. and Woodbury, J. "The Electromagnetic Spectrum" in *Astronomy*, June 1982, p. 6.

Bova, B. *The Beauty of Light*. 1988, Wiley. A readable introduction to all aspects of the production and decoding of light by a science writer.

Connes, P. "How Light Is Analyzed" in *Scientific American*, Sep. 1968.

Darling, D. "Spectral Visions: The Long Wavelengths" in *Astronomy*, Aug. 1984, p. 16; "The Short Wavelengths" in *Astronomy*, Sep. 1984, p. 14.

Gingerich, O. "Unlocking the Chemical Secrets of the Cosmos" in *Sky & Telescope*, July 1981, p. 13.

Hearnshaw, J. *The Analysis of Starlight*. 1986, Cambridge U. Press. A history of spectroscopy.

Newman, J. "James Clerk Maxwell" in *Scientific American*, June 1955.

Sobel, M. *Light*. 1987, U. of Chicago Press. An excellent nontechnical introduction to all aspects of light.

Stencil, R. et al. "Astronomical Spectroscopy" in *Astronomy*, June 1978, p. 6.

The 3.6-m Canada-France-Hawaii telescope on Mauna Kea, Hawaii, at an altitude of 4200 m (about 14,000 ft). (Charles Kaminski, University of Hawaii)

CHAPTER 5

Astronomical Instruments

Thinking Ahead

When you look at the night sky far from city lights on a camping trip, there seem to be an overwhelming number of stars up there. But in reality, only about 6000 are visible to the unaided eye. The light from most stars is so weak by the time it reaches Earth that it cannot register on human vision. How can we learn about the vast majority of objects in the universe that our unaided eyes simply cannot see?

A s we saw in the last chapter, we are almost entirely dependent on electromagnetic radiation from space to learn about the universe. Fortunately, objects in the cosmos send us radiation at all wavelengths, from gamma rays to radio waves. Unfortunately for astronomical observations, the atmosphere blocks much of the electromagnetic spectrum from reaching the surface.

There are, however, two regions in the spectrum where the Earth's atmosphere is transparent and through which we can observe (see Figure 4.5). One of these is called the *optical (or visible light) window,* and it includes wavelengths from about 300 nm or 0.3 μm up to about 30 μm. The other is the *radio window,* which includes radio waves that range in length from about 1 mm to about 20 m.

Until recently, astronomers could work only with the radiation in these windows, and all telescopes were designed to receive these bands of the spectrum. Today, however, we can also study radiation in parts of the spectrum where the atmosphere is opaque; to observe in these regions, astronomers must put telescopes in orbit.

Beyond the [faintest stars the eye can see] you will behold through the telescope a host of other stars, which escape the unassisted sight, so numerous as to be almost beyond belief . . .

Galileo Galilei in *Siderius Nuncius,* 1610, reporting on his first observations of the night sky with a telescope.

(a)

(b)

Figure 5.1

Two surviving pretelescopic observatories. (a) The Jantar Mantar, built in 1724 by Maharaja Jai Singh in Delhi, India. (b) Seventeenth-century bronze instruments from the old Chinese imperial observatory, Beijing. (David Morrison)

5.1

Telescopes

Many ancient cultures constructed special sites for observing celestial objects, especially the Sun, the Moon, and the planets (Figure 5.1). At these ancient *observatories* they could measure the positions of objects in the sky, mostly to keep track of time and their calendars. Many of these ancient observatories had religious and ritual functions as well. Telescopes are a relatively late addition to the instruments housed within observatories, becoming important only in the past 300 years. But their introduction revolutionized our understanding of the universe.

The Astronomical Telescope

Galileo built the first telescopes used for astronomical observations in 1610. As we have seen, these were simple tubes that could be held in a person's hand, but even so they got Galileo into a lot of trouble with Church authorities. Astronomical telescopes have come a long way since Galileo's time. Now they tend to be huge devices, constructed at costs of tens of millions of dollars.

The key to understanding telescopes is to realize that celestial objects, such as planets, stars, and galaxies, send much more light to Earth than any human eye (with its tiny opening) can catch. There is plenty of starlight to go around: if you have ever watched the stars with a group of friends, you know that each of you can see each of the stars. And if there were a thousand more people watching,

each of them would also catch a bit of each star's light. But as far as you are concerned, the light not shining into your eye is wasted. It would be great if some of this "wasted" light could also be captured and brought to your eye. This is precisely what a telescope does.

The most important functions of a telescope are to *collect* the faint light from an astronomical source, and to *focus* all the light it collects into an image. Most objects of interest to astronomers are extremely faint; the more light we can collect, the better we can study such objects. You might think of a telescope as a big light-gathering "bucket": just as a bigger bucket allows you to collect more rainwater when it starts to pour, so a bigger telescope can collect more of the light that "rains down" from the night sky. (Bear in mind that, while we use the word "light" in this paragraph, there are types of telescopes that can collect not just visible light, but each of the various forms of electromagnetic radiation. We want no light-chauvinists among our readers!)

In telescopes that collect radiation in the optical window, the light is gathered by a lens or mirror (which we will discuss in a moment). In other types of telescopes, the collecting devices may not look exactly like the lenses or mirrors you know, but they serve the same function. In all types of telescopes, the light-gathering ability is determined by the diameter, or **aperture,** of the device functioning as the light-gathering bucket.

In order to study astronomical objects, it is useful to form an *image* of a source of radiation. The image can then be detected, recorded, measured, reproduced, and analyzed in a host of ways. When telescopes were first perfected, astronomers simply viewed images with their eyes, but this was very inefficient and did not provide a very good long-term record. Today astronomers rarely look through the larger astronomical telescopes. The image is recorded electronically and stored in computers, and this permanent record becomes the object of later detailed study.

Formation of an Image by a Lens or a Mirror

Whether or not you wear glasses, you see the world through lenses; they are key elements of your eyes. A lens is a transparent piece of material that bends parallel rays of light passing through it, bringing them to the same focus. Figure 5.2 shows how a simple lens forms an image. If the curvatures of its surfaces are just right, two parallel rays of light (say from a star) are bent, or refracted, in such a way that they converge toward a point, called the **focus** of the lens. At the focus, an image of the light source appears. The distance of the focus, or image, behind the lens is called the **focal length** of the lens.

As you look at Figure 5.2, you may ask why two rays of light from a star would be parallel to each other. After all, if you draw a picture of a star, shining in all directions, the rays of light coming from the star don't look parallel at all. But you have to remember that the stars (and other astronomical objects) are all outrageously far away. By the time the rays of light headed our way actually arrive at Earth, they are—for all practical purposes—parallel to each other. Put another way, any rays that were not more or less parallel to the ones pointed at Earth are now heading in some very different direction in the universe.

To view the image formed by the primary lens in a telescope, we can install another lens called an eyepiece. This lens can magnify the image if appropriate; stars, for example, are points of light, and magnifying them makes little difference. But the image of a planet or galaxy,

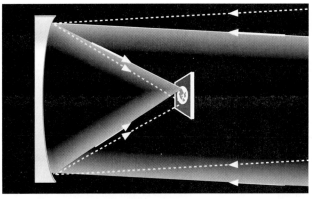

Figure 5.3

Formation of an image by a concave mirror. Here we see rays from two sides of an object (solid lines and dashed lines). Each set of parallel rays is reflected by the curved surface of the mirror, bringing them to a focus at points in front of the mirror.

which has structure, can often benefit from magnification, just as a tiny image in a magazine can sometimes look clearer through a magnifying glass. By installing different eyepieces, we can change the magnification of the image. But remember, these days the eyepiece of a telescope is usually replaced by a camera or electronic light detector (discussed shortly).

Rays of light can also be focused to form an image with a concave mirror—one curved like the inner surface of a sphere (Figure 5.3) and coated with silver or aluminum to make it highly reflecting. If the mirror has the correct concave shape, all parallel rays are reflected back through the same point, the focus of the mirror. Thus images are produced by a mirror exactly as they are by a lens.

Many people, when thinking of a telescope, picture a long tube with a large glass lens at one end. This design is called a **refracting telescope.** Galileo's telescopes were refractors, as are binoculars or opera glasses today (Figure 5.4.) But refractors (and their lenses) are not very good for most astronomical applications. In large sizes their lenses are difficult to support (since light must pass through them, they can be held at the edge only, not from behind). Also, making a large piece of glass that does not have flaws in it (such as bubbles) is costly and difficult. Most astronomical telescopes (both amateur and professional) use a mirror rather than a lens as their primary optical part; these are called **reflecting telescopes.** Since reflecting telescopes are standard in modern astronomy, we will refer to mirrors in most of the discussion that follows. However, similar statements can also be made about telescopes that use a lens as their primary optical element.

The reflecting telescope was conceived by James Gregory in 1663, and the first successful model was built by Newton in 1668. The concave mirror is placed at the bottom of a tube or open framework. The mirror reflects the light back up the tube to form an image near the front end at a location called the *prime focus.* The image can be

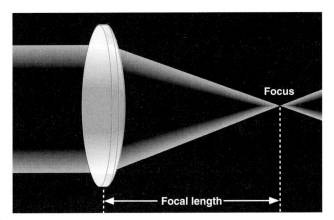

Figure 5.2

Formation of an image by a simple convex lens. Parallel rays from a star are bent by the lens to a single focus.

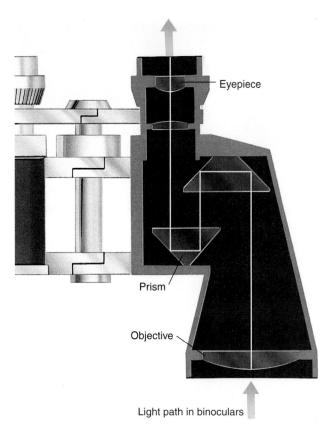

Figure 5.4
Binoculars are common examples of refracting telescopes. The main light-collecting element in each side is the initial (objective) lens; after following a path through several prisms (used to shorten the length of the instrument), the light is viewed through a magnifying eyepiece. This makes the image both brighter and larger, helping you see what is happening on stage or on the field even if you can't afford really good tickets.

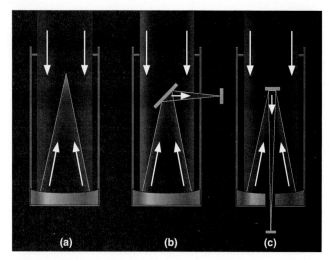

Figure 5.5
Various focus arrangements for reflecting telescopes. (a) Prime focus. (b) Newtonian focus. (c) Cassegrain focus.

observed at the prime focus, or, alternatively, various systems of auxiliary mirrors can intercept the light and bring it to a focus at more convenient locations. Two such arrangements are shown in Figure 5.5. In the Newtonian mode, a small secondary mirror is angled to reflect the light off to the side of the telescope tube. In the Cassegrain mode, the secondary mirror reflects the light back down the tube, through a small hole in the primary mirror (which doesn't hurt that mirror's ability to reflect in any significant way), and to a focus below the telescope tube.

Properties of Images

The *brightness* of an image is a measure of the amount of light energy that is concentrated into a unit area—say, a square millimeter—of the image. This is a crucial issue for viewing dim stars at the edge of our Galaxy, or dim galaxies very far from us in the cosmos. Astronomers also want more light from all objects so they can break up the light into the spectrum we discussed in the last chapter and look for absorption or emission lines. The brightness

of the image depends on the amount of light focused into it by the primary mirror. Doubling the aperture (size) of the mirror increases its area by a factor of 2×2 or 4 and thus results in images that are four times brighter. For example, astronomers recently completed the world's largest optical telescope on a volcanic peak called Mauna Kea in Hawaii; its 10-m mirror is twice the aperture of the older Hale telescope at Palomar and thus can gather four times as much light from any object.

Resolution refers to the fineness of detail present in the image. As you can imagine, astronomers are always eager to make out more detail in the images they study, whether following the weather on Jupiter or trying to peer into the violent heart of a distant galaxy. The larger the telescope aperture, the sharper the image will be. However, additional limits on resolution are imposed by fluctuations in the Earth's atmosphere, which always introduce a certain amount of blurring in telescopes on the ground. It is important to place observatories at locations where this atmospheric distortion is minimal. (You have seen these fluctuations as the "twinkling" of stars seen from Earth. In space the light of the stars is steady.)

The resolution of an astronomical image is measured by the angular size of a point source, such as a star. This size is expressed in seconds of arc, or arcsec, where 1 **arcsec** is 1/3600 degree. One arcsec is the apparent or angular diameter of a quarter at a distance of 50 km. During the past century, a resolution of about 1 arcsec has been considered the standard in astronomy, achieved at major observatory sites only when the atmosphere is unusually steady.

In the 1990s, however, astronomers are raising their standards. At the best observing locations, such as Mauna Kea in Hawaii, it is possible to achieve resolutions of 0.3 arcsec. The Hubble Space Telescope has a resolution of 0.1 arcsec. New techniques on the ground now allow astronomers to reduce the blurring of images due to turbulence in the atmosphere (see opening figure in Chapter 17). One such technique, called *adaptive optics,* makes very quick changes in the shape of a flexible telescope mirror to compensate for changes in the Earth's atmosphere. And, as we will see, radio astronomers have built instruments with resolution better than 0.001 arcsec—the width of a quarter at 50,000 km.

The Complete Telescope

Now that we have seen how an image is produced, we can understand the operation of a telescope. Since its main purpose is to collect light from faint sources, astronomers usually want a telescope with as large an aperture as possible. Thus they must obtain a large disk of glass, which is laboriously ground and polished (and coated with a reflective layer) to produce a concave mirror of high optical quality. The shape of the mirror must be just right so that all parallel rays of light, no matter where they hit the mirror's surface, come to a single focus.

Once we have a good primary mirror, it must be mounted in such a way that the entire telescope can be pointed toward any object in the sky. When your mirror is 5 m wide, it has an enormous weight; the mirror alone in the telescope on Mount Palomar weighs 14.5 tons. Thus you need a complex support structure in order to move it quickly and smoothly from orientation to orientation (Figure 5.6). In addition, since the Earth is rotating, the telescope must have a motorized drive system that moves it backwards at exactly the same rate the Earth is moving forward, so it can continue to point at the object being observed. Sophisticated instruments must be built to analyze and record the light collected at the telescope's focus. And all this machinery must be housed in a dome to protect it from the elements.

Figure 5.6
The Hale telescope on Palomar Mountain has a complex mounting structure that enables the telescope (in the open "tube" pointing upwards in this photo) to swing easily into any position. The people in the bottom foreground can give you a sense of scale. (California Institute of Technology)

ASTRONOMY BASICS

How Astronomers Use Telescopes

In the popular view, an astronomer is a person who spends most nights in a cold observatory peering through a telescope. But this picture, which may have had some validity in previous centuries, is not very accurate today. Most astronomers do not live at observatories, but near the universities or laboratories where they work. A typical astronomer might spend a total of only a week or so each year observing at the telescope, and the rest of the time measuring or analyzing the data acquired during that week. Many astronomers use only radio telescopes or space experiments, which work just as well during the day. Still others work at purely theoretical problems (often using high-speed computers) and never observe at a telescope of any kind.

Even when astronomers are observing with a large telescope, they seldom peer through it. Electronic detectors are used to record the data permanently for detailed analysis after the observations are completed. At some observatories, it is now possible to conduct observations remotely, with the astronomer sitting at an office computer terminal—which can be thousands of miles away from the telescope.

Time on the major instruments used by astronomers is at a premium, and an observatory director will typically receive many more requests for telescope time than can be accommodated during the year. Astronomers must therefore write a convincing *proposal* explaining how they would like to use the telescope and why their observations will be important to the progress of astronomy. A committee of astronomers is then asked to judge and rank the proposals, and time is assigned to those with the most merit. Even if your proposal is among the high-rated ones, you may have to wait many months for your turn. And if the skies are cloudy on the nights you have been assigned, it may be more than a year before you get another chance.

5.2

Optical Detectors and Instruments

The main role of a telescope—as we have seen—is to collect as much of the radiation as possible from a given object. But once the radiation has been captured, it must be *detected* and *measured.* The first detector used for astronomical observations was the human eye, but it suffers from being connected to an imperfect recording and retrieving device, the human brain. Photography and modern electronic detectors have eliminated the foibles and quirks of human memory by making a permanent record of the information from the cosmos.

In addition, the eye also has a very short *integration time*—it takes only a fraction of a second to add light energy together before sending the image to the brain. One important advantage of photography is that the light from astronomical objects can be collected by the detector over longer periods of time; this technique is called "taking a long exposure." Exposures of several hours are required to detect very faint objects in the cosmos.

We can divide the way astronomers use large telescopes into three basic categories. The first is *imaging*—photographing or otherwise recording the appearance of a small portion of the sky. The second is making an accurate measurement of the *brightness* and *color* of objects. The last is *spectroscopy*—the measurement of the spectra of astronomical sources. All three of these require similar detectors to record and measure the properties of light.

Photographic and Electronic Detectors

Throughout most of the 20th century, photographic plates and film served as the prime astronomical detectors, whether for direct imaging or for photographing spectra. To photograph an astronomical object, its image is allowed to fall on a light-sensitive chemical coating that, when developed, provides a record of the image—one that can be measured, studied, enlarged, published, and inspected by many individuals. When the coating is applied to a sheet of glass, astronomers call the record a *photographic plate* (Figure 5.7). At observatories around the world, vast collections of such plates preserve what the sky has looked like during the past 100 years.

Figure 5.7
Astronomer Eleanor Helin inspects a 14- × 14-in. glass plate taken at the Palomar Observatory's 48-in. Schmidt telescope. She is searching for trails left by asteroids. (Kritan Lattu, World Space Foundation)

Figure 5.8
A CCD chip, showing the array of light-sensitive picture elements (called pixels that can be read off by a computer and used to construct a digital image. These light detectors, chilled to very low temperatures, are far more sensitive to light than the chemicals used in photography. (National Optical Astronomy Observatories)

Photographic film and plates are superb devices for collecting a large amount of information; they do, however, have serious limitations. Perhaps the most important is that they are inefficient; only about 1 percent of the light actually falling on the film or plate contributes to the image, with the rest wasted as far as our record is concerned. For this and other reasons, astronomy has moved from the photographic to the electronic detection of light.

The most important electronic detectors are called **charge-coupled devices,** abbreviated **CCDs** (Figure 5.8), in which photons of radiation generate charged particles (electrons) that are then stored on the detector and counted at the end of the exposure. The number of charged particles is proportional to the brightness of the astronomical source.

CCDs offer several advantages over photography. For example, they record as much as 60 to 70 percent of all the photons that strike them. This means we can record much fainter objects with CCDs. They also provide more accurate measurements of the brightness of astronomical objects than do photographic films and plates. (CCD technology is now widely used in video camcorders.)

Infrared Observations

Observing the universe in the infrared band of the spectrum presents some additional challenges. The infrared extends from wavelengths near 1 μm, which is about the longwave sensitivity limit of both CCDs and photography, out to 100 μm or longer. Recall from Chapter 4 that infrared radiation is heat radiation. The main challenge to infrared astronomers is to distinguish the tiny amount of heat that reaches the Earth from stars and galaxies, from the much greater heat radiated by telescopes and our planet's atmosphere.

Typical temperatures on the Earth's surface are near 300 K, and the atmosphere through which observations are made is only a little cooler. According to Wien's law (see Section 4.2), the telescope, the observatory, and even the sky are radiating infrared energy with a peak wavelength of about 10 μm. To infrared eyes, everything on Earth is brightly aglow. The challenge is to detect faint cosmic sources against this sea of light. The infrared astronomer must always contend with the situation that a visible-light observer would face if working in broad daylight with a telescope and optics lined with bright fluorescent lights.

The first step in solving this problem is to protect the infrared detector from nearby radiation, just as you would shield photographic film from bright daylight. Since anything warm radiates infrared energy, the detector must be isolated in very cold surroundings; often it is held near absolute zero (1 to 3 K) by immersing it in liquid helium. The second step is to reduce the radiation emitted by the telescope structure and optics. All the optics must be kept very clean, since every bit of dust is an infrared source. The telescope is designed so that the infrared radiation from its own structure does not reach the detector.

Observing in other parts of the electromagnetic spectrum can present other challenges to the astronomer.

What kind of mirror can you use to reflect such penetrating radiation as x rays and gamma rays, for example? And, as we saw, many kinds of waves from space can be detected only from observatories in orbit, which are costly to launch and sometimes impossible to repair. We will return to these topics later in the chapter.

Spectroscopy

We have discussed detectors as if they were always used to record an image of a portion of the sky, but they can also be used to record a spectrum. Spectroscopy is one of the astronomer's most powerful tools, and more than half the time spent on most large telescopes is used for spectroscopy.

The many different wavelengths present in light can be separated by passing it through a prism to form a spectrum. A *spectrometer* is an instrument designed to record the spectrum of a light source. The design of a simple spectrometer is illustrated in Figure 5.9. Light from the source (actually, the image of a source produced by the telescope) enters the instrument through a small hole or narrow slit and is then collimated (made into a beam of parallel rays) by a lens. The light then passes through a prism, producing a spectrum: different wavelengths leave the prism in different directions because of dispersion. A second lens placed behind the prism focuses the many different images of the slit or entrance hole on the CCD or other detecting device. This collection of images (spread out by color) is the desired spectrum.

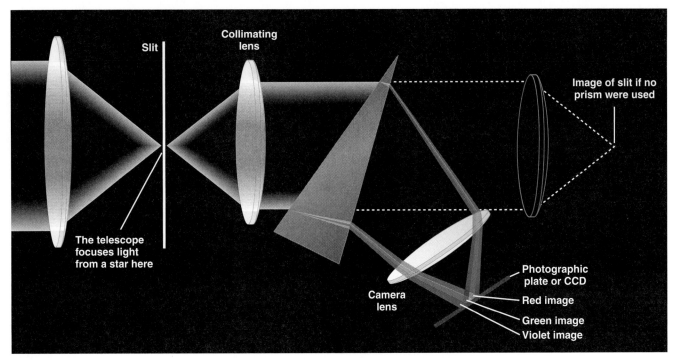

Figure 5.9

Design of a simple prism spectrometer for astronomy. The light from the telescope is focused on a slit, so the light of one object at a time enters the spectrometer. A prism (or grating) disperses the light into a spectrum, which is then photographed or recorded electronically.

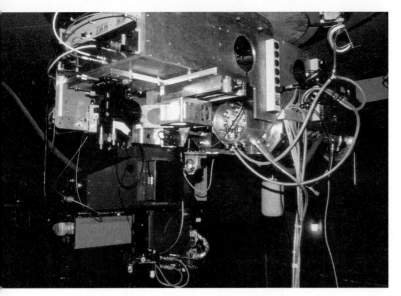

Figure 5.10
Modern telescope instrumentation is complex. This is a photograph of an infrared spectrometer at the Cassegrain focus of the 3-m NASA Infrared Telescope Facility on Mauna Kea. The spectrometer is operated remotely by the astronomer, who sits in a heated control room and records the data on a computer disk.
(University of Hawaii photo by Alan Tokunaga)

Spectrometers can be complex and impressive instruments (Figure 5.10). In practice, it is more common to use a different dispersing piece of optics called a *grating* to form the spectrum in an astronomical spectrometer instead of a prism. A grating is a piece of glass with thousands of tiny grooves in its surface; the grooves cause the light waves to interfere with each other (slightly differently for each wavelength) with the result that the light spreads out into a spectrum.

5.3

Optical and Infrared Observatories

The Astronomical Observatory

A modern observatory is often a collection of astronomical telescopes (Figure 5.11). Since the end of the 19th century, astronomers have realized that the best observatory sites are on mountains far from the lights and pollution of cities. Although a number of urban observatories remain, especially in the large cities of Europe, they have all become administrative centers or museums. The real action takes place far away, often on desert mountains or isolated peaks in the Atlantic and Pacific Oceans.

Typically, the astronomer travels for hours by plane or car in order to reach the observatory. Sometimes the trip is halfway around the world, as for the British and French astronomers who operate large telescopes in Hawaii. At the observatory are living quarters, computers, electronic and machine shops, and of course the telescopes themselves. A large observatory today requires a supporting staff of 20 to 100 persons in addition to the astronomers.

The performance of an optical telescope is determined not only by the size of its mirror but also by its location. The world's largest telescopes are found in remote sites such as the Andes mountains of Chile, the Canary Islands in the Atlantic Ocean, and Mauna Kea in Hawaii, a

Figure 5.11
The Kitt Peak National Observatory near Tucson, Arizona. (National Optical Astronomy Observatories)

George Ellery Hale: Master Telescope Builder

All of the major research telescopes in the world today were built within this century. The giant among early telescope builders was George Ellery Hale (1868–1938). Not once but four times he initiated projects that led to construction of what was at the time the world's largest telescope. He was a master at finding and winning over wealthy benefactors to underwrite the construction of these new instruments.

Hale's training and early research were in solar physics. In 1892, at age 24, he was named Associate Professor of Astral Physics and director of the astronomical observatory at the University of Chicago. At that time, the largest telescope in the world was the 36-in. refractor at the Lick Observatory near San Jose, California. Taking advantage of an existing glass blank for a 40-in. telescope, Hale set out to raise the money for a larger telescope than the one at Lick. One prospective donor was Charles T. Yerkes, who, among other things, ran the trolley system in Chicago.

Hale wrote to Yerkes, encouraging him to support construction of the giant telescope by saying, ". . . the donor could have no more enduring monument. It is certain that Mr. Lick's name would not have been nearly so widely known today were it not for the famous observatory established as a result of his munificence." Yerkes agreed, and the new telescope was completed in May 1897; it still remains the largest refractor in the world.

Even before the completion of the Yerkes refractor, Hale was not only dreaming of building a still-larger telescope, but also taking concrete steps to achieve that goal. In the 1890s there was a major controversy about the relative quality of refracting and reflecting telescopes. Hale realized that 40 in. was close to the maximum feasible aperture for refracting telescopes. If telescopes with apertures a factor of 2 or more

George Ellery Hale (Caltech)

larger than that were to be built, they would have to be reflecting telescopes.

Using funds borrowed from his own family, Hale set out to construct a 60-in. reflector. For a site, he left the Midwest for the much better conditions on Mount Wilson, at that time a wilderness peak above the small city of Los Angeles. In 1904, at the age of 36, Hale received funds from the Carnegie Foundation to establish the Mount Wilson Observatory. The 60-in. mirror was placed in its mount in December 1908.

Two years earlier, in 1906, Hale had already approached John D. Hooker, who had made his fortune in hardware and steel pipe, with a proposal to build a 100-in. telescope. The technological risks were substantial. The 60-in. telescope was not yet complete, and the utility of large reflectors for astronomy had yet to be demonstrated. George Ellery Hale's brother called him "the greatest gambler in the world." Once again, Hale was successful in obtaining funds, and the 100-in. telescope was completed in November 1917. (It was with this telescope that Edwin Hubble was able to establish that the spiral nebulae were separate islands of stars—or galaxies—quite removed from our own Milky Way; see Chapter 26.)

Hale was not through dreaming. In 1926 he wrote an article in *Harper's Magazine* about the scientific value of a still-larger telescope. This article came to the attention of the Rockefeller Foundation, which granted $6 million (a very large sum in those days) for the construction of a 200-in. telescope. Hale died in 1938, but the 200-in. (5-m) telescope on Palomar Mountain was dedicated ten years later and is now named in Hale's honor.

The 40-in. (1-m) refracting telescope of the Yerkes Observatory. (Yerkes Observatory)

Figure 5.12
Cerro Paranal, a mountain summit 2.7 km above sea level in Chile's Atacama desert, will be the site of the European Southern Observatory's Very Large Telescope. This photograph, taken in late 1990, shows the beginnings of the construction and vividly illustrates that astronomers prefer high and dry sites for their instruments. (ESO)

mountain that is 13,700 ft (4200 m) high. What makes these places such good sites for astronomy?

The Earth's atmosphere, so vital to life, is the biggest headache to the observational astronomer. In at least four ways the air imposes limitations on the usefulness of telescopes:

1. The most obvious limitation is weather—clouds, wind, rain, and the like. At the best sites the weather is clear as much as 75 percent of the time.
2. Even on a clear night, the atmosphere filters out a certain amount of starlight, especially in the infrared, where the absorption is due primarily to water vapor. Astronomers therefore prefer dry sites, generally found at the highest altitudes.
3. The sky should also be dark. Near cities the air scatters the glare from lights, producing an illumination that hides the faintest stars and limits the distances that can be probed by telescopes. Observatories are best located at least a hundred miles from the nearest large city.
4. Finally, the air is often unsteady; light passing through this turbulent air is disturbed, resulting in blurred star images. Astronomers call these effects "bad **seeing.**" When seeing is bad, images of celestial objects are distorted by the constant twisting and bending of light rays by turbulent air.

The best observatory sites combine dark skies and good seeing. For infrared work, sites should also be high and dry, usually above 3000 m in elevation. And the need for as many clear nights as possible usually means placing observatories in desert regions at latitudes between about 20° and 35°. Sites that meet these criteria are found along the western mountains that border the North and South American continents from California to Chile, in the desert mountains of Spain, in the U.S. in Arizona and New Mexico, and on island mountaintops in Hawaii and the Canaries (Figure 5.12).

Major New Telescopes

Most of the great telescopes constructed during the first half of the 20th century, including the 100-in. (2.5-m) and 200-in. (5-m) telescopes (see "Voyagers in Astronomy" box) were built in the Northern Hemisphere with private funds and were available only to a small group of astronomers working in the institutions that owned them. Following the success of the 200-in. telescope, astronomers from universities in the East and Midwest began a campaign for a national observatory to provide comparable facilities for the rest of the astronomical community. The National Science Foundation eventually agreed, and Kitt Peak National Observatory in Arizona began operations in 1960 (Figure 5.11).

A few years later, a second U.S. national observatory was established at Cerro Tololo in the Andes mountains of Chile to provide access to the skies of the Southern Hemisphere. Meanwhile, a consortium of European nations founded the European Southern Observatory in the Andes, while the British and Australians undertook the construction of major facilities in Australia. The largest telescopes at these observatories were about 4 m in aperture. Still larger (6 m) was a telescope built by what was then called the Soviet Union, but this instrument was much less productive due to a poorer site and long history of technical problems.

The kinds of research that astronomers are working on today require even bigger "light buckets." In the 1990s, construction of ground-based telescopes is proceeding at an unprecedented pace. Worldwide, more than 25 telescopes with apertures of 3 m or more are either op-

Aperture (m)	Telescope Name	Location	Status
16.4	VLT (four 8.2-m telescopes)	Cerro Paranal, Chile†	First light of first telescope 1997
10.0	Keck I	Mauna Kea, HI, USA	Completed 1993
10.0	Keck II	Mauna Kea, HI, USA	Completed 1996
9.9	Hobby–Eberly (SST)	Mount Locke, TX, USA	First light 1997
8.3	Subaru (Pleiades)	Mauna Kea, HI, USA	First light 1998
8.0	Gemini (North)	Mauna Kea, HI, USA*	First light 1998
8.0	Gemini (South)	Cerro Panchon, Chile*	First light 2000
6.5	Multi-Mirror (MMT)	Mount Hopkins, AZ, USA	First light 1997
6.5	Magellan	Las Campanas, Chile	First light 1997
6.0	Large Alt–Azimuth	Mount Pastukhov, Russia	Completed 1976
5.0	Hale	Mount Palomar, CA, USA	Completed 1948
4.2	William Herschel	Canary Islands, Spain	Completed 1987
4.0	NOAO Cerro Tololo	Cerro Tololo, Chile*	Completed 1974
3.9	Anglo–Australian (AAT)	Siding Spring, Australia	Completed 1975
3.8	NOAO Mayall	Kitt Peak, AZ, USA*	Completed 1973
3.8	United Kingdom Infrared (UKIRT)	Mauna Kea, HI, USA	Completed 1979
3.6	Canada–France–Hawaii (CFHT)	Mauna Kea, HI, USA	Completed 1979
3.6	ESO	Cerro La Silla, Chile†	Completed 1976
3.6	ESO New Technology	Cerro La Silla, Chile†	Completed 1989
3.5	Max Planck Institut	Calar Alto, Spain	Completed 1983
3.5	WIYN	Kitt Peak, AZ, USA*	Completed 1993
3.5	Astrophysical Research Corp.	Apache Point, NM, USA	Completed 1993
3.0	Shane (Lick Observatory)	Mount Hamilton, CA, USA	Completed 1959
3.0	NASA Infrared (IRTF)	Mauna Kea, HI, USA	Completed 1979

* Part of the U.S. National Optical Astronomy Observatories.
† Part of the European Southern Observatory.

erating or under construction (Table 5.1). New technologies are being used to construct instruments with roughly twice the aperture of the 200-in. telescope. These instruments have thinner, lighter mirrors and more compact mechanical designs.

The twin 10-m Keck telescopes on Mauna Kea (Figure 5.13) are the first of these new-technology instruments and are currently the largest optical telescopes in the world. Operated by the University of California and the California Institute of Technology, the Keck telescopes were built mostly with private rather than government funds, returning to an older tradition in astronomical observatories. Instead of constructing a single primary mirror 10 m in diameter, each Keck telescope achieves its

Figure 5.13
The W. M. Keck Observatory is located on Hawaii's Mauna Kea—acknowledged as the world's best site for astronomical observations. The white dome houses the first Keck telescope, which has a primary mirror 10 m in diameter. A twin of this telescope, named Keck II, is housed in the second dome, still showing its red primer paint in this May 1994 photo that was taken during construction.
(Leonard Nakahashi, Keck Observatory)

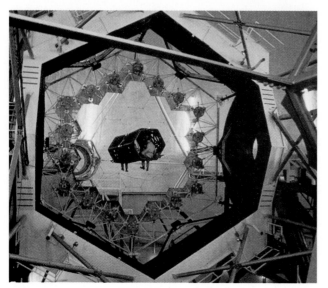

Figure 5.14
A close-up of the 10-m Keck telescope's mirror as it was being assembled. The finished mirror is composed of 36 hexagonal sections. In this view, 18 sections have been installed. To get a sense of the mirror's size, note the person working at the center of the mirror assembly. (California Association for Research in Astronomy)

large aperture by combining the light from 36 separate hexagonal mirrors, each 1.8 m in width (Figure 5.14). Computer-controlled actuators constantly adjust these 36 mirrors so that the overall reflecting surface maintains just the right shape to collect and focus the light with high accuracy.

Several additional large telescopes are under construction by various governments or consortia of universities around the world. They will all use single mirrors with apertures of 7.5 to 8.4 m. By the turn of the millennium, visitors to Mauna Kea, for example, can expect to see new 8-m telescopes operated by Japan and the U.S. National

Observatories. The largest instrument of all, consisting of four 8-m telescopes operated together, is under construction on Cerro Paranal, an isolated peak in Chile's Atacama desert. This instrument, called the Very Large Telescope (VLT), is being built by a consortium of European nations as part of their European Southern Observatory (Figures 5.12 and 5.15).

Each of these large new instruments is designed to operate in the infrared as well as the visible part of the spectrum. Since they are located on high, dry sites, the atmosphere is relatively transparent most of the time for wavelengths from the visible out to about 30 μm. Because atmospheric seeing is generally better at infrared than at visible wavelengths, the highest-resolution imaging with these instruments is likely to be done in the infrared, at wavelengths near 2 μm, where resolution approaching 0.1 arcsec may be achievable. Such new telescopes will also be equipped with systems for adaptive optics, as discussed in Section 5.1.

5.4
Radio Telescopes

Origins of Radio Astronomy

In 1931 Karl G. Jansky of the Bell Telephone Laboratories was experimenting with antennas for long-range radio communication when he encountered some mysterious static—radio radiation coming from an unknown source (Figure 5.16). He discovered that this radiation came in strongest about 4 min earlier on each successive day and correctly concluded that since the Earth's sidereal rotation period is 4 min shorter than a solar day (see Section 3.3), the radiation must be originating from some region fixed on the celestial sphere. Subsequent investigation showed that the source of the radiation was part of the Milky Way.

Figure 5.15
An artist's conception of the Very Large Telescope, now being built in Chile. Four individual telescopes with mirrors 8.2 m in diameter will work together, creating a total light-collecting area equivalent to a 16-m telescope. The Very Large Telescope should be able to detect a firefly 10,000 km away. The first of the four telescopes should be ready in 1998, and the entire assembly by 2002. (Jean Quebatte, European Southern Observatory)

Figure 5.16
The rotating radio antenna used by Jansky in his serendipitous discovery of radio radiation from the Milky Way. (Bell Laboratories)

In 1936 Grote Reber, an amateur astronomer and radio ham, built from galvanized iron and wood the first antenna specifically designed to receive these cosmic radio waves (Figure 5.17). Over the years, Reber built several such antennas and used them to carry out pioneering surveys of the sky for celestial radio sources; he remained active in radio astronomy for more than 30 years. During the first decade, he worked practically alone, because professional astronomers had not yet recognized the vast potential of radio astronomy. Many of the objects that Reber discovered became subjects of intensive investigation years later.

Figure 5.17
Grote Reber, an amateur astronomer and electronics expert, built the first telescope specifically designed to observe radio waves from space. From 1937 until after the end of World War II, he was the world's only active radio astronomer. (Ohio State University)

Detection of Radio Energy from Space

It is important to understand that radio waves cannot be "heard"; they are not the sound waves you hear coming out of the radio receiver in your home or car. Like light, radio waves are a form of electromagnetic radiation; but unlike light, we cannot detect them with our senses—we must rely on electronic equipment to pick them up. In commercial radio broadcasting, we encode sound information (music or the voice of a newscaster) into the radio waves, which must be decoded at the other end and turned back into sound by speakers or headphones. There are two familiar ways to *modulate* (encode information into) radio waves: by changing the amplitude of the waves, called *amplitude modulation* or AM radio, and by changing the frequency of the waves, called *frequency modulation* or FM.

The radio waves we receive from space do not, of course, have rock and roll or other program information encoded in them (although astronomers would love to tune in on a radio broadcast from some distant civilization; see Epilogue). If cosmic radio signals were translated into sound, they would sound like the static you hear when dialing between stations. Nevertheless, there is information in the radio waves we receive—information that can tell us about the chemistry and physical conditions in the sources of the waves.

Just as vibrating charged particles can produce electromagnetic waves (see Chapter 4), electromagnetic waves, in turn, can make charged particles move up and down. Radio waves can produce a current in conductors of electricity such as metals. An *antenna* is such a conductor; it intercepts radio waves, which induce a feeble current in it. The current is then amplified in a radio receiver until it is strong enough to measure or record. Like your television or radio set, receivers can be tuned to select a single frequency (channel). In astronomy today, however, it is more common to use sophisticated data-processing techniques that allow thousands of separate frequency bands to be detected simultaneously. Thus the astronomical radio receiver operates much like a spectrometer on an optical telescope, giving information about how much radiation we receive at each wavelength or frequency. The radio signals, after computer processing, are typically recorded on magnetic disk or tape for further analysis.

Radio Telescopes

Radio waves are reflected by conducting surfaces just as light is reflected from an optically shiny surface, and according to the same laws of optics. A radio-reflecting telescope consists of a concave metal reflector (called a *dish*), analogous to a telescope mirror. The radio waves collected by the dish are reflected to a focus, where they can then be directed to a receiver and analyzed. Because humans are such visual creatures, however, radio astronomers often construct an artificial picture of the radio sources they observe. Figure 5.18 shows such a radio image of a distant galaxy, where radio telescopes reveal vast jets and regions of radio emission that are completely invisible in photographs taken with light.

Radio astronomy is a young field compared with optical astronomy, but it has experienced tremendous growth in recent decades. Several European nations lacking high-quality sites for optical telescopes have instead chosen to concentrate their astronomical efforts in the radio part of the spectrum; these include the Netherlands, Britain, and Germany. The world's largest radio reflectors that can be pointed to any direction have apertures of 100 m. One of these is at the Max Planck Institute for Radio Astronomy, located near Bonn, Germany (Figure 5.19); the other is at the U.S. National Radio Astronomy Observatory in West

Figure 5.19
The 100-m radio telescope near Bonn, Germany. (Max Planck Institut für Radioastronomie)

Virginia. Table 5.2 lists some of the major radio telescopes of the world.

Radio Interferometry

As we discussed earlier, a telescope's ability to show us fine detail (its resolution) depends on its aperture. But it also depends on the wavelength of the radiation the telescope is gathering. The longer the waves, the harder it is to resolve fine detail in the images or maps we make. Because radio waves have such long wavelengths, they present tremendous challenges for astronomers who need good resolution. In fact, even the largest radio telescopes

Figure 5.18
A radio image of a galaxy called Cygnus A. Colors have been added to help the eye sort out regions of different radio intensity. Red regions are the most intense, blue the least. The visible galaxy would be a small dot in the center of the image. The radio image reveals jets of expelled material (over 160,000 LY long) on either side of the galaxy. (NRAO/AUI)

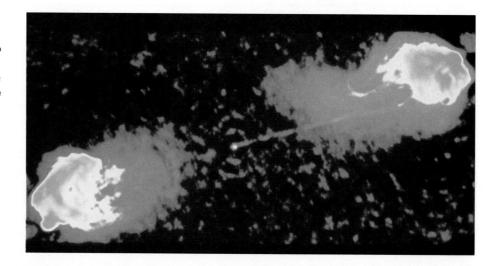

TABLE 5.2
Major Radio Observatories of the World

Observatory	Location	Description
National Astronomy and Ionospheric Center	Arecibo, Puerto Rico	305-m fixed dish
National Radio Astronomy Observatory	Green Bank, West Virginia	100-m steerable dish and 43-m steerable dish
Max Planck Institut für Radioastronomie	Bonn, Germany	100-m steerable dish
Jodrell Bank Radio Observatory	Manchester, England	76-m steerable dish and 66-m steerable dish
Goldstone Tracking Station (NASA/JPL)	Barstow, California	70-m steerable dish
Australia Tracking Station (NASA/JPL)	Tidbinbilla, Australia	70-m steerable dish
Parkes National Radio Observatory	Parkes, Australia	64-m steerable dish
MERLIN	Cambridge, England and other British sites	Network of 7 dishes (the largest of which is 32 m)
Owens Valley Radio Observatory (Caltech)	Big Pine, California	Two 27-m dishes
Very Large Array (NRAO)	Socorro, New Mexico	27-element array of 25-m dishes (36-km baseline)
Westerbork Radio Observatory	Westerbork, the Netherlands	12-element array of 25-m dishes (1.6-km baseline)
Very Long Baseline Array (NRAO)	Ten U.S. sites, Hawaii to Virgin Islands	10-element array of 25-m dishes (9000-km baseline)
IRAM	Granada, Spain	30-m steerable mm-wave dish
Nobeyama Cosmic Radio Observatory	Minamimaki-Mura, Japan	25-m steerable mm-wave dish
Australia Telescope	Several sites in Australia	8-element array (seven 22-m dishes plus Parkes)
James Clerk Maxwell Telescope	Mauna Kea, Hawaii	15-m steerable mm-wave dish
Hat Creek Radio Observatory (U. Cal.)	Cassel, California	6-element array of 5-m mm-wave dishes

on Earth—operating alone—cannot make out as much detail as the typical small optical telescope used in a college astronomy lab. To overcome this difficulty, radio astronomers have learned to sharpen their images by linking two or more radio telescopes together electronically.

Telescopes linked together in this way are called an **interferometer.** Imagine two radio dishes placed some distance apart. The radio waves from a radio source in the sky will strike one antenna a brief instant before the other, so that the two antennas receive the same waves at slightly different times and thus become "out of phase" with each other. The difference in phase between the waves detected at the two antennas can be measured electronically, and the direction to the source can be calculated from the difference. The farther apart the components of an interferometer are placed (the longer the *baseline*), the more accurately we can pinpoint the direction of the source.

The use of pairs of radio telescopes as interferometers was pioneered in England. If one of the telescopes is mounted on rails to make it movable, interferometers with different baselines can be established. The next step is to combine a large number of radio dishes into an **interferometer array.** In effect, such an array works as a large number of two-dish interferometers, all observing the same part of the sky together. Computer processing of the results permits the reconstruction of a radio image, allowing us to trace its structure more clearly.

The resolution of such an array is determined by the largest spacing between dishes, just as the resolution of an optical telescope is determined by how big the primary mirror is. The most extensive radio array is the National Radio Astronomy Observatory's Very Large Array (VLA)

near Socorro, New Mexico (see opening figure in Chapter 4. It consists of 27 movable radio telescopes, each having an aperture of 25 m, spread over a total span of about 36 km. By electronically combining the signals from all of its individual telescopes, this array permits the radio astronomer to make pictures of the sky at radio wavelengths comparable to those obtained with an optical telescope, with a resolution of about 1 arcsec.

Very Long Baseline Interferometers

Initially, interferometer arrays were limited in size by the requirement that all of the dishes be accurately wired together. The maximum dimensions were thus only a few tens of kilometers. However, larger interferometer baselines can be achieved if the telescopes do not require a physical connection. Using high-precision clocks at each site, astronomers can record the radiation coming from space at each telescope and combine the data later. Thus we have learned to build **very long baseline interferometers** with baselines all the way from California to Parkes, Australia, and from West Virginia to the Crimea in Ukraine. The resulting resolution far surpasses that of optical telescopes.

Recently the United States completed the construction of the Very Long Baseline Array (VLBA), made up of ten individual telescopes stretching from the Virgin Islands to Hawaii (Figure 5.20). The VLBA, which began operations in 1993, can form astronomical images with a resolution of 0.0001 arcsec, permitting features as small as 10 AU to be distinguished at the center of our Galaxy. The addition of one or more telescopes in space is planned as

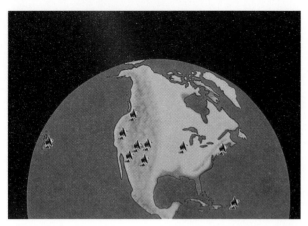

Figure 5.20
Distribution of the ten antennas that constitute the U.S. VLBA (Very Long Baseline Array). (National Radio Astronomy Observatory)

a means to increase further the capability of the ground-based VLBA beyond the limits currently set by the Earth's dimensions. Russia, in collaboration with the United States and several European nations, is developing an orbiting receiver called Radioastron that could be launched in the late 1990s to link up with the VLBA and increase its resolution still further.

Radar Astronomy

Radar is the technique of transmitting radio waves to an object in our solar system and then detecting the radio radiation that the object reflects back. The time the radio waves take to make the round trip can be measured electronically with great precision. Because we know the speed at which radio waves travel (the speed of light), we can determine the distance to the object or a particular feature on its surface (such as a mountain).

Radar observations of the Moon and planets have yielded our best knowledge of the distances to these worlds, and played an important role in navigating spacecraft throughout the solar system. In addition, as will be discussed in later chapters, radar observations have determined the rotation periods of Venus and Mercury, probed tiny Earth-approaching asteroids and the nuclei of comets, analyzed the rings of Saturn, and investigated the surfaces of Mercury, Venus, Mars, and the large satellites of Jupiter.

Any radio dish can be used as a radar telescope if it is equipped with a powerful transmitter as well as a receiver. The most spectacular facility in the world for radar astronomy is the 1000-ft (305-m) telescope at Arecibo in Puerto Rico (Figure 5.21). The Arecibo telescope is too large to be pointed directly toward different parts of the sky. Instead, it is made from a huge natural "bowl" (more than a mere dish!) formed by several hills and is lined with a reflecting material. The Arecibo bowl has a volume roughly equal to the world's annual beer consumption. A limited ability to track astronomical sources is achieved by moving the receiver system, which is suspended on cables 100 m above the surface of the bowl.

5.5

Observations Outside the Earth's Atmosphere

Since the Earth's atmosphere blocks most radiation at wavelengths shorter than visible light, we can make ultraviolet, x-ray, and gamma-ray observations only from space. Getting above the distorting effects of the atmosphere is also advantageous for many observations at visible and infrared wavelengths. The stars don't twinkle in space, and thus the amount of detail you can observe is limited only by the size of your instrument. On the other hand, it is difficult and expensive to get telescopes into space, and any needed repairs can present a major challenge. This is why astronomers continue to build telescopes for use on the ground *and* for launching into space.

Airborne and Space Infrared Telescopes

Water vapor, the main source of atmospheric interference throughout the infrared, is concentrated in the lower part of the Earth's atmosphere. For this reason a gain of even a few hundred meters in elevation can make an important difference in the quality of an observatory site. Given the limitations of high mountains, most of which attract clouds and violent storms, it was natural for astronomers

Figure 5.21
The Arecibo Observatory is part of the National Astronomy and Ionosphere Center which is operated by Cornell University under a cooperative agreement with the National Science Foundation. (National Astronomy and Ionosphere Center)

Figure 5.22
The Kuiper Airborne Observatory consists of a 0.9-m infrared telescope in a C-141 airplane operated by NASA. The research done with this telescope included discovery of the rings around the planet Uranus. (NASA)

to investigate the possibility of observing infrared waves from airplanes and ultimately from space.

The first airborne infrared observations were made in the 1960s. Based on the success of these efforts, NASA constructed and operated a 0.9-m airborne telescope from 1974 through 1995, flying regularly out of the Ames Research Center south of San Francisco. The Gerard P. Kuiper Airborne Observatory—named for a pioneer of infrared astronomy who first applied modern detectors to astronomy after World War II—flew at elevations of 12 to 13 km, above 99 percent of the atmospheric water vapor (Figure 5.22). With the retirement of the Kuiper Observatory in 1995, NASA has plans to build a larger 2.5-m infrared telescope, called the Stratospheric Observatory for Infrared Astronomy (SOFIA), to fly in a modified Boeing 747.

The next step, observations from space itself, has important advantages for infrared astronomy. First, of course, is the elimination of all interference from the atmosphere. But equally important is the opportunity to cool the entire optical system of the telescope in order to nearly eliminate infrared radiation from this source as well. If we tried to cool a telescope within the atmosphere, it would quickly become coated with condensing water vapor and other gases, making it useless. Only in the vacuum of space can optical elements be cooled to hundreds of degrees below freezing and still remain operational.

The first orbiting infrared observatory, launched in 1983, was the Infrared Astronomy Satellite (IRAS), built as a joint project among the United States, the Netherlands, and Britain (Figure 5.23). IRAS was equipped with a 0.6-m telescope cooled to a temperature of less than 10 K. For the first time the infrared sky could be seen at night, as it were, rather than through a bright foreground of atmospheric and telescope emissions. IRAS carried out

Figure 5.23
The Infrared Astronomy Satellite (IRAS), a joint project of NASA, the Netherlands, and the United Kingdom. Here the spacecraft is being tested before its 1983 launch. (NASA/JPL)

a rapid but comprehensive survey of the entire infrared sky over a ten-month period, until its liquid helium coolant was exhausted. It cataloged over 250,000 sources of infrared radiation. A second small infrared telescope called the Cosmic Background Explorer (COBE) was launched in 1990 to make very precise observations of the remnant radiation from the formation of the universe (discussed in Chapter 28).

The European Space Agency launched a new telescope, called the Infrared Space Observatory (ISO), in 1995. While its primary mirror is also 0.6 m in diameter, its instruments are significantly more advanced than those in IRAS. It is cooled to approximately 2 K and is expected to operate for about 24 months. Beyond ISO, three proposals have been made for the next generation of infrared space telescopes, but none is yet under construction. The U.S. project is called SIRTF, or Space Infrared Telescope Facility. In Europe, a telescope called Edison is being studied, with a larger aperture than SIRTF but operating at a higher temperature. The third option is a Japanese orbiting telescope. All three will, if built, take advantage of a new generation of more capable infrared detectors.

Hubble Space Telescope

In April 1990, a great leap forward (or should we say upward) in astronomy was made with the launch of the first new-generation large space observatory, the Hubble Space Telescope (HST) (Figure 5.24). This telescope has an aperture of 2.4 m, the largest put into space so far. (Its aperture was limited by the size of the payload bay in the Space Shuttle, which was its mode of transportation into space.) It was named for Edwin Hubble, the astronomer who discovered the expansion of the universe in the 1920s (Table 5.3).

HST is operated from Goddard Space Flight Center, working with the NASA-funded Space Telescope Science Institute in Baltimore. It is instrumented for direct imaging and spectroscopy in the visible and ultraviolet, but in 1997 two of its instruments will be replaced, and the HST will then also work in the near-infrared part of the spectrum. Because it is designed to be serviced by Shuttle astronauts, HST can be reconfigured and improved over time, unlike any of its smaller orbiting predecessors.

HST's mirror has been ground and polished to a remarkable de-

TABLE 5.3
The Hubble Space Telescope

Aperture	2.4 m
Resolution	0.1 arcsec
Pointing precision	0.01 arcsec
Tracking precision	0.007 arcsec
Electrical power	4000 watts
Spacecraft length	13.1 m
Spacecraft mass	11.6 tons

gree of accuracy. If we were to scale up its mirror to the size of the entire continental United States, there would be no hill or valley larger than about 6 cm in its smooth surface. Unfortunately, after it was launched, scientists discovered that the primary mirror had a slight error in its shape equal to roughly 1/50th the width of a human hair. Small as that sounds, it was enough to ensure that much of the light entering the telescope did not come to a clear focus, and that all the images were thus blurry.

The solution was to do something very similar to what we do for astronomy students with blurry vision: put corrective optics in front of their eyes. In December 1993, in one of the most exciting and difficult space missions to date, astronauts captured the orbiting telescope and brought it back into the Shuttle payload bay. There they were able successfully to install a package containing com-

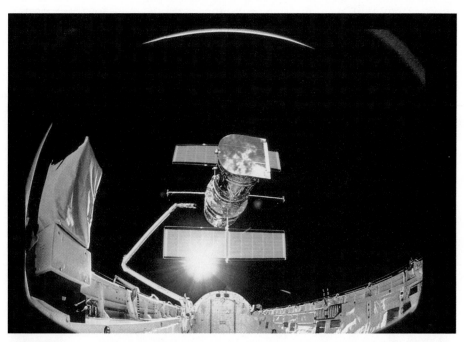

Figure 5.24
The Hubble Space Telescope being launched in April 1990 from the Space Shuttle *Discovery*. The robot arm, or Remote Manipulator System, was operated by astronomer Steven Hawley on this historic occasion. The image was taken with a special 70-mm IMAX camera system. You can see the light of the Earth reflected from the closed cover of the telescope. The orange panels are solar arrays, which capture the Sun's light to provide energy for the instruments. (IMAX/OMNIMAX image ©1990 Smithsonian Institution/Lockheed Corporation)

pensating optics as well as a new improved camera, and then to release HST back into orbit. Figure 5.25 illustrates the improvement in the clarity of the images that were returned to Earth. We have sprinkled the new images and discoveries from HST throughout this book, and invite you to enjoy them for their beauty as well as for their information content.

Now that it's been refurbished, HST is the premier astronomical instrument in the world, used by hundreds of astronomers from around the globe. It is also the most expensive telescope in the world, with an annual operating cost of about a quarter of a billion dollars—ten times as much as any ground-based observatory. In spite of its great success, cutbacks in the NASA budget make it unlikely that other space observatories of this scale will be orbited in the foreseeable future, and no successor to HST is currently planned.

High-Energy Observatories

As we have seen, ultraviolet, x-ray, and gamma-ray observations must be made from space. Such observations first became possible in 1946 with V2 rockets captured from the Germans. The U.S. Naval Research Laboratory instrumented these rockets for a series of pioneering flights, used initially to detect ultraviolet radiation from the Sun. Since then, many other rockets have been launched to make x-ray and ultraviolet observations of the Sun, and later of other celestial objects as well. But such rocket flights allowed the instruments only a brief period of time above the Earth's atmosphere before they fell or parachuted back to the ground. The next step was to put instruments in orbit around the Earth.

Since the 1960s, a number of satellite observatories have been launched to carry out astronomical observations at short wavelengths (high energies). These have included the orbiting solar observatories, the orbiting astronomical observatories, and the high-energy astronomy observatories. The U.S.-built x-ray satellite Uhuru (Swahili for "freedom"), launched in 1970 from Kenya, discovered about 100 x-ray sources, some of which were found to emit higher-energy gamma rays as well. Another highly successful space observatory has been the International Ultraviolet Explorer, a joint U.S.–European effort to observe spectra of objects that give off ultraviolet light.

X-ray astronomy came into its own in 1978 with the launching of the Einstein Observatory. Einstein was the

(a)

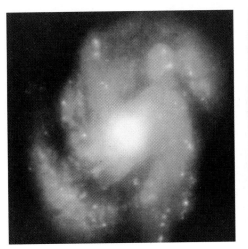

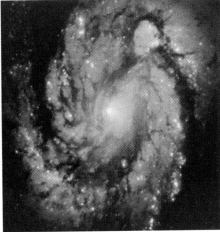

(b)

Figure 5.25

(a) Repairing the Hubble Space Telescope. Among those working in the Shuttle payload bay is astronomer Jeffrey Hoffman, at bottom. (b) The improvement in HST images after the repair mission. Both images show the central regions of the galaxy M100. (NASA/STScI)

first x-ray telescope capable of forming an image of a source, and it had a sensitivity a thousand times greater than anything preceding it. This is equivalent to changing from a small amateur telescope to the 5-m Hale reflector on Palomar.

The primary x-ray observatory of the 1990s was a German satellite (with British and U.S. collabora-

(continued on page 127)

Choosing Your Own Telescope

If the astronomy course you are taking whets your appetite for exploring the sky further, you may be thinking about buying your own telescope. The good news is that many excellent amateur telescopes are now on the market, and prices are much more reasonable than they were 20 years ago. The bad news is that a good telescope, like a good camera or video recorder, is still not inexpensive, and requires some research to find the model that is best for you. The best sources of information about personal telescopes are the two popular-level magazines published for amateur astronomers, *Astronomy* and *Sky & Telescope*. Both carry regular articles of advice, reviews, and ads from reputable telescope dealers around the country.

In some ways, choosing a telescope is like choosing a car: personal preference can play a major role in the decision. A certain shape or brand name may be important to some car buyers, but no big deal to others. In the same way, some of the factors determining which telescope is right for you depend on personal preferences:

- Will you be setting up the telescope in one place and leaving it there, or do you want an instrument that is portable and can come with you on camping trips? Does it have to be carried some distance by a person, or can it get to its destination in a car?

- Do you want to observe the sky with your eyes only, or do you want to take photographs? (Long-exposure photography, for example, requires a good clock-drive to turn your telescope in compensation for the Earth's rotation.)

- What sorts of objects will you be observing? Are you interested primarily in comets, planets, star clusters, or galaxies? Or do you want to observe all kinds of celestial sights?

You may not know the answers to some of these questions yet. For this reason, our number-one recommendation is that you "test-drive" some telescopes first. Most communities have amateur astronomy clubs that sponsor star parties open to the public. The members of these clubs often know a lot about telescopes and can share their ideas with you. Your instructor may know where the nearest amateur astronomy club meets; many of the clubs in the country are also listed once a year in *Sky & Telescope* and *Astronomy* magazines.

Furthermore, you may already have an instrument like a telescope at home (or have access to one through a relative or friend). Many amateur astronomers recommend starting your survey of the sky with a good pair of binoculars. These are easily carried around and can show you many objects not visible (or clear) to the unaided eye.

When you are ready to purchase a telescope, you might find the following thoughts useful:

- The key characteristic of a telescope is the aperture of the main mirror or lens; when someone says they have a 6-in. or 8-in. telescope, they mean the diameter of the collecting surface. The larger the aperture, the more light you can gather and the fainter the objects you can see or photograph.

- Telescopes of a given aperture that use lenses (refractors) are typically more expensive than those using mirrors (reflectors). This is because both sides of a lens must be polished to great accuracy. And, because the light passes *through* the lens, it must be made of high-quality glass throughout. In contrast, only the front surface of a mirror must be accurately polished.

- Magnification is *not* one of the criteria on which to base your choice of a telescope. Since, as we discussed in the main text, the magnification of the image is done by a smaller eyepiece, the magnification can be changed by changing eyepieces. However, a telescope will magnify not only the astronomical object you are viewing, but also the turbulence of the Earth's atmosphere. If the magnification is too high, your image will shimmer and shake and be difficult to view. A good telescope will come with a variety of eyepieces that stay within the range of useful magnifications.

- The mount of a telescope is one of its most critical elements. Because a telescope shows a tiny field of view which is magnified significantly, even the smallest vibration or jarring of the telescope can move the object you are viewing around, or out of your field of view. A sturdy and stable mount is essential for serious viewing or photography (although it clearly affects how portable your telescope can be).

- A telescope requires some practice to set up and use effectively. Don't expect everything to go perfectly on your first try. Take some time to read the instructions. If there is one near you, use your local amateur astronomy club as a resource.

tion) called the Roentgensatellit, named for Wilhelm Roentgen, the scientist who discovered x rays (Figure 5.26). More commonly called simply Rosat, it had a mass of 2.4 tons and was launched into orbit in June 1990 (two months after HST). It operated for about two years, producing both an x-ray survey of the entire sky and a series of more detailed images of individual sources. In its survey mode, Rosat was able to catalog more than 100,000 x-ray sources. Its x-ray telescope had both higher sensitivity and greater resolving power than the Einstein satellite.

Toward the end of the decade, NASA plans to launch a larger x-ray satellite called the Advanced X-Ray Astrophysics Facility (AXAF), designed for imaging with still greater sensitivity and resolution.

Gamma rays of cosmic origin were first discovered by the Vela satellites, launched by the U.S. Department of Defense to carry out worldwide surveillance for possible explosions of nuclear bombs, which emit gamma rays. In 1967 the Vela system detected bursts of gamma radiation that investigation showed could not originate from within the solar system. Such bursts were also detected by Uhuru and other orbiting observatories equipped with detectors of high-energy radiation. Each of these mysterious bursts lasts anywhere from about a tenth of a second to several seconds.

NASA launched the Compton Gamma Ray Observatory in April 1991 in order to conduct an all-sky survey for gamma-ray sources, as well as to study individual objects in detail at a variety of energies. The spacecraft was named for American physicist Arthur Holly Compton, an early 20th-century pioneer in the laboratory study of gamma rays. With a mass of more than 16 tons, the Compton Observatory is one of the largest scientific payloads ever launched into space (Figure 5.27). One of the specific aims of the Compton Observatory is to learn more about the gamma-ray bursts. It has detected over a thousand of these bursts already (as of the end of 1995), and

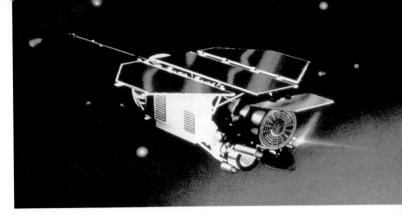

Figure 5.26
Painting of the Rosat spacecraft, which was placed into an almost circular orbit 585 km above the Earth. It carried the largest x-ray telescope built so far. (Max Planck Institute for Extraterrestrial Physics)

learned a great deal about them. Nevertheless, their origin remains a mystery.

The worldwide reassessment of government priorities in the 1990s is having an impact on astronomy. Both NASA and its European counterpart, the European Space Agency (ESA), have scaled back their commitments to space science, while the Russian space program, once the most ambitious of any nation, has been devastated by budgetary problems. In the United States, the earlier idea of maintaining several large, long-lived space observatories like HST, with their high operating costs, no longer seems politically feasible.

In the late 1990s NASA's emphasis is expected to be on smaller, cheaper, faster missions rather than on the "great observatories in space." AXAF has been reduced in size and cost, while SIRTF has been repeatedly delayed due to lack of funds. Most astronomers hope that these trends will be reversed, but at this writing it appears that space astronomy will be progressing at a rather slow pace as we approach the next century.

Figure 5.27
The deployment of the Compton Gamma Ray Observatory in April 1991 from the Space Shuttle *Atlantis*. When the satellite's antenna would not open, two of the astronauts took an unscheduled space walk to free it. Here the observatory is shown high above Africa just before it was released into orbit. (NASA)

Summary

5.1 The astronomical telescope collects light and forms an image, using a convex lens in a **refractor** or a concave mirror in a **reflector** to bring it to a **focus**. The distance from the lens or mirror to the focus is called the **focal length**. The diameter, or **aperture,** of the telescope determines the brightness and **resolution** of the image. Resolution is usually expressed in **arcsec** units.

5.2 Optical detectors include the eye, photographic film or plates, and the **charge-coupled device (CCD)** detector. Detectors sensitive to infrared radiation must be cooled to very low temperatures. A spectrometer can be built using either a prism or grating to disperse the light into a spectrum.

5.3 Telescopes are housed in domes and controlled by computers. Observatory sites must be carefully chosen for clear weather, dark skies, low water vapor, and excellent atmospheric **seeing** (low atmospheric turbulence). In the 1990s, a new generation of instruments is being constructed, including the 10-m Keck telescope at Mauna Kea and the four 8-m instruments that constitute the European Very Large Telescope in Chile.

5.4 In the 1930s, radio astronomy was pioneered by Jansky and Reber. A radio telescope is basically a radio antenna (often a large parabolic dish) connected to a receiver. Significantly enhanced resolution can be obtained by **interferometry,** including the development of **interferometer arrays** like the 27-element VLA. Expanding to **very long baseline interferometers,** radio astronomers can achieve resolutions as good as 0.0001 arcsec. **Radar** astronomy involves transmitting as well as receiving. The largest radar telescope is the 305-m bowl at Arecibo.

5.5 Infrared observations are made with telescopes in aircraft and in space, as well as from ground-based facilities on dry mountain peaks. IRAS and COBE provided the first all-sky surveys, succeeded by the European ISO mission. Ultraviolet, x-ray, and gamma-ray observations must be made from above the atmosphere. Many orbiting observatories have been flown to observe in these bands in the last few decades. The largest aperture instrument in space is the Hubble Space Telescope (HST), whose flawed mirror was repaired in 1993. HST can observe in both visible and ultraviolet light with greater resolution than is possible from the ground. The Compton Gamma Ray Observatory (GRO) is the most sophisticated instrument for observing high-energy gamma rays ever built. Budget problems are affecting plans for more ambitious space observatories.

Review Questions

1. Name the two spectral windows through which electromagnetic radiation reaches the surface of the Earth, and describe the largest aperture telescope currently in use for each window.

2. List the six bands into which we commonly divide the electromagnetic spectrum, and list the largest-aperture telescope currently in use in each band.

3. When astronomers discuss the apertures of their telescopes, they say bigger is better. Explain why.

4. What are the properties of an image, and what factors determine each?

5. Compare the eye, photographic film, and CCDs as detectors for light. What are the advantages and disadvantages of each?

6. Radio and radar observations are often made with the same antenna, but otherwise they are very different techniques. Compare and contrast radio and radar astronomy in terms of the equipment needed, the methods used, and the kind of results obtained.

7. Why do astronomers place telescopes in Earth orbit? What are the advantages for different spectral regions?

8. What was the problem with the Hubble Space Telescope and how was it solved?

9. Describe the techniques radio astronomers use to obtain a resolution comparable to what astronomers working with visible light can achieve.

Thought Questions

10. What happens to the image produced by a lens if the lens is "stopped down" with an iris diaphragm—a device that covers its periphery?

11. What would be the properties of an ideal astronomical detector? How closely do the actual properties of a CCD approach this ideal?

12. Fifty years ago, the astronomers on the staff of Mount Wilson and Palomar Observatories each received about 60 nights per year for their observing programs. Today an astronomer feels fortunate to get ten nights per year on a large telescope. Can you suggest some reasons for this change?

13. The largest observatory complex in the world is on Mauna Kea in Hawaii, at an altitude of 4.2 km. This is by no means the tallest mountain on Earth. What are some factors astronomers consider when selecting an observatory site? Don't forget practical ones. Should astronomers, for example, consider building an observatory on Mount McKinley (Denali) or Mount Everest?

14. Another site recently developed for astronomy is the Antarctic plateau. Discuss its advantages and disadvantages.

15. Suppose you are looking for sites for an optical observatory, an infrared observatory, and an x-ray observatory. What are the main criteria of excellence for each? What sites are actually considered the best today?

16. Radio astronomy involves wavelengths much longer than those of visible light, while many orbiting observatories have probed the universe for radiation of very short wavelengths. What sorts of objects and physical conditions would you expect to be associated with radiation emissions of very long and very short wavelengths?

17. The dean of a university located near the ocean proposes building an infrared telescope right on campus and operating it in a nice heated dome so astronomers will be comfortable on cold winter nights. Criticize this proposal, giving your reasoning.

Problems

18. The resolution of a radio interferometer is proportional to the maximum spacing of the antennas. The VLA, with a maximum antenna separation of 36 km, achieves a resolution of 1 arcsec at a wavelength of 6 cm.

    a. What is the resolution of a very long baseline interferometer at this wavelength if the antennas are separated by 3600 km?

    b. What is the resolution of the VLBA, with antennas on opposite sides of the Earth?

19. A typical large telescope today requires about $4 million per year to operate. What is the cost per night to use such a telescope? Per hour? Per minute? Can you think of any other activities that have such a high associated cost, in dollars per hour?

20. The HST cost about $1.7 billion for construction and $300 million dollars for its Shuttle launch, and it costs $250 million per year to operate. If the telescope lasts a total of ten years, what is the cost per year? Per day? If the telescope can be used just 30 percent of the time for actual observations, what is the cost per hour and per minute for the astronomer's observing time on this instrument?

Suggestions for Further Reading

Bunge, R. "Dawn of a New Era: Big Scopes" in *Astronomy*, Aug. 1993, p. 49.

Chaisson, E. *The Hubble Wars.* 1994, HarperCollins. Controversial story of the building of the HST.

Cohen, M. *In Quest of Telescopes.* 1980, Cambridge U. Press. What it's like to observe with large astronomical telescopes.

Cornell, J. and Carr, J., eds. *Infinite Vistas: New Tools for Astronomy.* 1985, Scribners. Essays on new telescopes and projects.

Dyer, A. "ROSAT's Penetrating X-ray Visions" in *Astronomy*, June 1991, p. 42. Summary of the work of the European x-ray satellite.

Florence, R. *The Perfect Machine: Building the Palomar Telescope.* 1994, Harper Perennial. Highly readable account of how the 200-in. telescope was built.

Fugate, R. and Wild, W. "Untwinkling the Stars" in *Sky & Telescope*, May 1994, p. 24; June 1994, p. 20. Discusses adaptive optics.

Gehrels, N. et al. "The Compton Gamma-ray Observatory" in *Scientific American*, Dec. 1993, p. 68.

Janesick, J. and Blouke, M. "Sky on a Chip: The Fabulous CCD" in *Sky & Telescope*, Sep. 1987, p. 238.

Jastrow, R. and Baliunas, S. "Mount Wilson: America's Observatory" in *Sky & Telescope*, Mar. 1993, p. 18. A history and current review.

Krisciunas, K. "Science with the Keck Telescope" in *Sky & Telescope*, Sep. 1994, p. 20. What it's like to use the telescope, and what work is being done with it.

Krisciunas, K. *Astronomical Centers of the World.* 1988, Cambridge U. Press. History of and guide to major observatories.

Preston, R. *First Light.* 1987, Atlantic Monthly Press. Superbly written popular book on the work being done at the 200-in. telescope on Mount Palomar.

Shore, L. "VLA: The Telescope That Never Sleeps" in *Astronomy*, Aug. 1987, p. 15.

Sinnott, R. and Nyren, K. "The World's Largest Telescopes" in *Sky & Telescope*, July 1993, p. 27. Several pages of data tables.

Stephens, S. "Telescopes that Fly" in *Astronomy*, Nov. 1994, p. 46. What it was like to fly and do research aboard the Kuiper.

Tucker, W. and Tucker, K. *Cosmic Inquirers.* 1986, Harvard U. Press. An excellent introduction to the VLA, Einstein x-ray observatory, and other major instruments.

Many of the major observatories and space missions have sites on the World Wide Web that you can visit, often with information and images you can download. See Appendix 1 for a listing.

This Hubble Space Telescope view is the clearest image of Mars ever taken from the distance of the Earth (103 million km at the time the photo was taken). The northern polar cap is visible at the top. Ascraeus Mons, a 25-km high volcano, can be seen peeking through the morning clouds at the very left. Valles Marineris, a rift valley as long as the United States is wide, is clearly visible at the lower left. (P. James, S. Lee, and STScI/NASA)

CHAPTER 6

Other Worlds: An Introduction to the Solar System

Thinking Ahead

Surrounding the Sun is a family of worlds large and small: nine planets, dozens of moons, and smaller pieces too numerous to count. It is a complex system, with a wide range of conditions on the various worlds. Yet the system has some remarkable regularities to it. Most objects circle the Sun in the same direction and in the same plane, for example. What caused the orbits of all these worlds to share a common motion?

During the past 30 years we have learned more about the solar system than any other field of astronomy. In addition to gathering information with powerful new telescopes at observatories on the ground and in orbit, we have been able to send spacecraft directly to many members of the planetary system. (Planetary astronomy is the only branch of our science where we can—at least vicariously—travel to the objects we want to study.) With evocative names like Voyager, Pioneer, and Mariner, our instrumented robot explorers have flown past or around every planet except Pluto, returning images and data that have dazzled both astronomers and the public. In the process, we have also investigated dozens of fascinating moons, four ring systems, two asteroids, and one comet.

Some of our probes have penetrated the atmospheres of Venus, Mars, and Jupiter, while others have landed on the sur-

In all of the history of mankind, there will be only one generation that will be the first to explore the Solar System, one generation for which, in childhood, the planets are distant and indistinct discs moving through the night sky, and for which, in old age, the planets are *places,* diverse new worlds in the course of exploration.

Carl Sagan in *The Cosmic Connection* (1973, Doubleday)

Figure 6.1
The Viking 2 lander on the surface of Mars. The reddish color is the result of iron oxides in the soil. (NASA)

faces of Venus, Mars, and the Moon (Figure 6.1). Humans have even set foot on the Moon and returned samples of its surface soil for analysis in the laboratory.

The planet chapters in today's textbooks are nothing like the ones earlier generations of students were reading. We can now envision the members of the solar system as other worlds like our own, each with its own chemical and geological history, and each with unique sights that interplanetary tourists may someday even visit. Some have called these last few decades the golden age of planetary exploration, in analogy with another golden age of exploration—the period starting in the 15th century, when great sailing ships plied the Earth's oceans, and humanity began to get familiar with our own planet's surface.

In this chapter we examine our planetary system in general, introducing the idea of *comparative planetology*—studying how the planets work by comparing them with each other. We want to get to know the planets not only for their own sakes, but also to see what they can tell us about the origins and evolution of the entire solar system. Then in the next seven chapters, we will discuss how scientists have learned so much about the planets, and describe the better-known members of the solar system in some detail, returning to the theme of origins and comparative planetology in Chapter 13.

6.1

Overview of Our Planetary System

The solar system[1] consists of the Sun and many smaller objects: the planets, their satellites and rings, and such "debris" as asteroids, comets, and dust. Decades of obser-

[1] The generic term for a group of planets and other bodies circling a star is *planetary system*. Ours is called the solar system, because our Sun is sometimes called Sol. Strictly speaking, then, there is only one solar system; planets circling other stars are planetary systems.

TABLE 6.1
Mass of Members of the Solar System

Object	Percentage of Total Mass
Sun	99.80
Jupiter	0.10
Comets	0.05
All other planets	0.04
Satellites and rings	0.00005
Asteroids	0.000002
Cosmic dust	0.0000001

vation and spacecraft exploration have revealed that most of these objects formed together with the Sun about 4.5 billion years ago. They represent aggregations of material that condensed from an enormous cloud of gas and dust. The central part of this cloud became the Sun, and a small fraction of the material in the outer parts eventually formed the other objects.

An Inventory

The Sun (Figure 6.2), a rather ordinary star, is by far the most massive member of the solar system, as shown in Table 6.1. It is an enormous ball of incandescent gas, 1.4 million km in diameter. Its surface temperature is about 6000 K, with a central temperature of millions of degrees. Since it is a star, the Sun will be discussed later in this book. For the present, let us examine the planets and smaller objects that make up our planetary system.

Table 6.1 also shows that most of the material of the planets is actually concentrated in the largest one, Jupiter. The mass of the other eight added together is only about 40 percent that of Jupiter. Besides the Earth, five other

Figure 6.2
An eruption on the Sun, with the Earth shown to scale. In this false-color image, we see the Sun in the high-energy radiation given off by ionized helium atoms. Clouds of helium gas have blasted some 560,000 km from the surface layers of the Sun. (NASA)

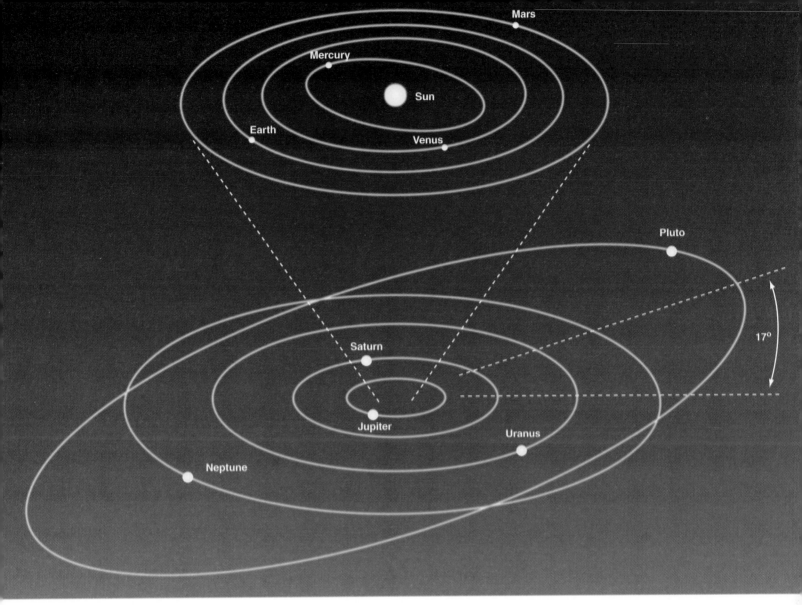

Figure 6.3
All the planets except Pluto orbit the Sun in roughly the same plane.

planets (Mercury, Venus, Mars, Jupiter, and Saturn) were known to the ancients, while three (Uranus, Neptune, and Pluto) were discovered since the invention of the telescope.

The nine planets all revolve in the same direction around the equator of the Sun (Figure 6.3). They orbit in approximately the same plane, like cars traveling on concentric tracks on a giant, flat racing course. However, their motion is not random, as we have seen in earlier chapters. Each planet follows a nearly circular orbit about the Sun, obeying the laws of motion discovered by Galileo, Kepler, and Newton.

Each of the planets also rotates about an axis running through it, and in most cases the direction of rotation is the same as that of revolution about the Sun. The exception is Venus, which rotates very slowly backward (that is,

in a retrograde direction). Uranus and Pluto also have strange rotations, each spinning about an axis tipped nearly on its side.

The four planets closest to the Sun (Mercury through Mars) are called the inner or **terrestrial planets;** often the Moon is also discussed as a part of this group, bringing the total of terrestrial bodies to five. (We generally call the Earth's satellite the *Moon,* with a capital M, and the other satellites *moons,* with lowercase m's.) The terrestrial planets are relatively small worlds, warmed by their proximity to the Sun and composed primarily of rock and metal. All of them have solid surfaces that bear the records of geological forces in the form of craters, mountains, and volcanoes (Figure 6.4).

The next four planets (Jupiter through Neptune) are much larger and are composed primarily of lighter ices,

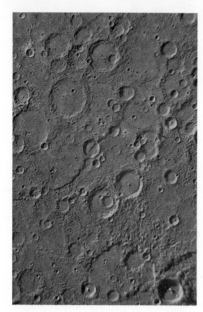

Figure 6.4
The surface of Mercury. The pockmarked face of a terrestrial world, more typical of the inner planets than the watery surface of the Earth. This image, taken with the Mariner 10 spacecraft, shows a region more than 400 km wide. (NASA)

liquids, and gases. We call these four the jovian (after "Jove," another name for Jupiter in mythology) or **giant planets,** a name they richly deserve (Figure 6.5). Over 1400 Earths could fit inside Jupiter, for example. These planets do not present any solid surface on which future explorers might land. They are more like vast, ball-shaped oceans with much smaller, solid cores at their centers.

At the outer edge of the system lies little Pluto, which is neither a terrestrial nor a jovian planet; it is similar to the satellites of the outer planets and may have a common origin with them. Table 6.2 summarizes some of the main facts about these nine planets.

The Titius-Bode Rule

For a long time scientists looked for evidence of regularity in the spacing of the orbits of the planets. Look at Table 6.2. You will see that the four terrestrial planets have orbits that are almost evenly spaced, with semimajor axes of 0.4, 0.7, 1.0, and 1.5 AU. In contrast, the distances between the giant planets are much greater. There is an especially large space between the orbits of Mars and Jupiter.

In the 18th century, astronomers thought they had discovered just such a relationship, called the Titius-Bode rule, after the two men who independently pointed it out. The semimajor axes of the orbits of the planets from Mercury through Uranus are matched quite well by the series of numbers formed by carrying out the following instructions: First write down the numbers 0, 3, 6, 12, . . . , with each number in the sequence (after the first) being double the previous one. Then add 4 to each number and divide the result by 10. Compare your results to the distances from the Sun shown in Table 6.2; the sequence looks pretty good at first, until we discover that there is no planet at the distance indicated by the fifth number in the sequence, 2.8 AU. The fact that a planet was "missing" spurred astronomers to look for something in that orbit, and led (as we shall see) to the discovery of a belt of asteroids between Mars and Jupiter.

On the other hand, note that this rule breaks down completely for the two outer planets, Neptune and Pluto. It also fails to predict the spacing of each planet exactly, serving only as a rough guide. Astronomers today are not sure what to make of the rule. Most agree that the Titius-

Figure 6.5
A to-scale montage showing the four giant planets (as photographed by the Voyager cameras) and the Earth. (NASA)

EARTH

TABLE 6.2
The Planets

Name	Distance from Sun (AU)*	Revolution Period (yrs)	Diameter (km)	Mass (10^{23} kg)	Density (g/cm³)
Mercury	0.39	0.24	4,878	3.3	5.4
Venus	0.72	0.62	12,102	48.7	5.3
Earth	1.00	1.00	12,756	59.8	5.5
Mars	1.52	1.88	6,787	6.4	3.9
Jupiter	5.20	11.86	142,984	18,991	1.3
Saturn	9.54	29.46	120,536	5,686	0.7
Uranus	19.18	84.07	51,118	866	1.2
Neptune	30.06	164.82	49,660	1,030	1.6
Pluto	39.44	248.60	2,200	0.01	2.1

* An AU (or astronomical unit) is the distance from the Earth to the Sun.

Bode rule is an interesting piece of numerology, but that it really doesn't tell us anything fundamental about why the planets are spaced as they are.

Smaller Members of the Solar System

Most of the planets are accompanied by one or more satellites or moons; only Mercury and Venus are alone. There are 61 known satellites (see Appendix 11), and undoubtedly many other very small ones remain undiscovered. The largest of these satellites are as big as small planets and just as interesting; these include, in addition to our Moon, the four largest moons of Jupiter (called the Galilean satellites) and one each of the moons of Saturn and Neptune (confusingly named Titan and Triton).

Each of the jovian planets also has a system of rings, made up of countless small bodies ranging in size from mountains down to mere grains of dust, all in orbit about the equator of the planet. The bright rings of Saturn are the widest and by far the easiest to see; they are among the most beautiful sights in the solar system (Figure 6.6). But all of the four ring systems are interesting to scientists because of the complicated forms they can assume under the gravitational pull of the satellites that also orbit these giant planets.

The solar system has many other, less-conspicuous members. **Asteroids** are rocky and metallic bodies that orbit the Sun like little planets, mostly in the space between Mars and Jupiter (Figure 6.7). Most asteroids are remnants of the initial population of the solar system that existed before the planets themselves formed. Some of the smallest satellites of the planets, such as the moons of Mars, are very likely captured asteroids.

Another class of small bodies is composed in large part of frozen gases such as water, carbon dioxide, carbon

Figure 6.6
Voyager 1 image of a crescent Saturn and its complex system of rings, taken from a distance of about 1.5 million km. Notice that you can see the edge of the planet through the rings at left.
(NASA/JPL)

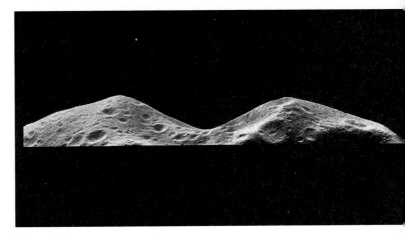

Figure 6.7
This image showing part of Ida (taken with the Galileo spacecraft in 1993) is the most detailed close-up of an asteroid we presently have. The spacecraft was 2480 km from the asteroid, about 46 s after its closest approach, when the image was taken. Ida is approximately 50 km in its longest dimension.
(NASA/JPL)

Names in the Solar System

We humans just don't feel comfortable until something has a name. Types of butterflies, new elements, and mountains of Venus all need names for us to feel we are acquainted with them. How do we give names to objects and features in the solar system?

Planets and satellites are named after gods and heroes in Greek and Roman mythology (with a few exceptions among the moons of Uranus that are drawn from English literature). When William Herschel, a German immigrant to England, first discovered the planet we now call Uranus, he wanted to name it Georgium Sidus (George's star) after King George III of his adopted country. This caused such an outcry among astronomers in other countries, however, that the classical tradition was resumed and has been maintained ever since. Luckily, there were a lot of minor gods in the ancient pantheon, so there are plenty of names left for the many small moons we are discovering around the giant planets. (Appendix 8 lists all of the known satellites; how many names do you recognize?)

Comets are named after their discoverers (giving an extra incentive to comet hunters), while asteroids are named by their discoverers after just about anyone or anything they want. Recently, asteroid names have been used to recognize people who have made significant contributions to astronomy, including two of the authors of this book.

That was pretty much all the naming that was needed while our study of the solar system was confined to Earth. But now our robot probes have surveyed and photographed a number of worlds in great detail, and on each world there are many features that suddenly need names. To make sure that naming things in space remains multinational, rational, and somewhat dignified, astronomers have given the responsibility of approving names to a special committee (including author Morrison) of the International Astronomical Union, the body that includes scientists from every country that does astronomy.

This committee has developed a set of rules for naming features on other worlds. For example, many newly discovered features on Venus are being named for women who have made significant contributions to human knowledge and welfare. Volcanic features on Jupiter's moon Io, which is in a constant state of volcanic activity, are being named after gods of fire and thunder from the mythologies of many cultures. Craters on Mercury now commemorate famous novelists, playwrights, artists, and composers. On Saturn's satellite Tethys, all the features have been named after characters in Homer's great epic poem, *The Odyssey*. As we explore further, it may well turn out that there are more places in the solar system needing names than Earth history can provide. Perhaps by then, explorers and settlers on these worlds will be ready to develop their own names for the places that (if only for a while) they will call home.

---

monoxide, and methyl alcohol; these are called **comets.** Comets also are remnants from the formation of the solar system, but they were formed and continue (with rare exceptions) to orbit the Sun in distant, cooler regions—stored in a sort of cosmic deep freeze.

Finally, there are countless grains of rock and dust, which we call simply **cosmic dust,** scattered throughout the solar system. When these particles collide with the Earth's atmosphere (as millions do each day), they burn up, producing brief flashes of light that we call shooting stars or **meteors.** Occasionally, some larger chunk of material from space survives the passage through the atmosphere to become a **meteorite,** a piece of the solar system that has landed on Earth. (You can see meteorites on display in many natural history museums and can purchase them from gem and mineral dealers.)

A Scale Model of the Solar System

Astronomy often deals with dimensions and distances that far exceed our ordinary experience. What does 1.4 billion

kilometers—the distance from the Sun to Saturn—really mean to anyone? Sometimes it helps to visualize the system in terms of a scale model.

In our imagination, let us build a scale model of the solar system, adopting a scale factor of 1 billion (10^9)—that is, reducing the actual solar system by dividing every dimension by a factor of 10^9. The Earth then has a diameter of 1.3 cm, about the size of a grape. The Moon is a pea orbiting this grape at a distance of 40 cm, or a bit over a foot away. The Earth-Moon system fits into an attaché case.

In this model the Sun is nearly 1.5 m in diameter, about the average height of an adult, and is found at a distance of 150 m—about one city block. Jupiter is five blocks away from the Sun, and its diameter is 15 cm, about the size of a very large grapefruit. Saturn is 10 blocks from the Sun, Uranus 20 blocks, and Neptune and Pluto each 30 blocks (Pluto's eccentric orbit has currently brought it a bit closer than Neptune to the Sun). Most of the satellites of the outer solar system are the sizes of various kinds of seeds orbiting the grapefruit, oranges, and lemons that represent the outer planets.

In our scale model, a human is reduced to the dimensions of a single atom, and cars and spacecraft to the size of molecules. Sending the Voyager spacecraft to Neptune involves navigating a single molecule from the Earth-grape toward a lemon 5 km away with an accuracy equivalent to the width of a thread in a spider's web.

That's the solar system; what about the nearest stars? If we keep the same scale, the closest stars would be tens of thousands of kilometers away. If you build the model in the city where you live, you would have to place the representations of these stars on the other side of the Earth or beyond.

6.2

Composition and Structure of Planets

The fact that there are two distinct kinds of planets leads us to believe that they formed under different conditions. Certainly their compositions are dominated by different elements. Let us look at each type in more detail.

The Giant Planets

The two largest planets, Jupiter and Saturn, have nearly the same composition as the Sun itself, consisting primarily of the two elements hydrogen and helium, in the proportion of about 75 percent hydrogen and 25 percent helium (as measured by the mass of each). On Earth, both hydrogen and helium are gases, so Jupiter and Saturn are sometimes called gas planets. But this name is misleading. Jupiter and Saturn are so large that the gas is compressed in their interior until the hydrogen becomes a liquid. Since the bulk of both planets consists of compressed, liquified hydrogen, we should really call them liquid planets (Figure 6.8).

Under the force of gravity, the heavier elements tend to sink toward the inner parts of a liquid or gaseous planet. Both Jupiter and Saturn have cores composed of rock, metal, and ice, but we cannot see these regions directly. When we look down from above, all we see is the atmosphere with its swirling clouds.

Uranus and Neptune are much smaller than Jupiter and Saturn, but each also has a core of metal, rock, and ice. Interestingly, the cores of all four of the giant planets have about the same mass—roughly ten times the entire mass of the Earth. We suspect that these cores formed early on and subsequently attracted the surrounding atmospheres of hydrogen and helium. Uranus and Neptune were less efficient at attracting this gas, so they remained smaller, with much less hydrogen and helium in proportion to their cores.

Chemically, each of these planets is dominated by hydrogen and its many compounds. Nearly all of the oxygen present is chemically combined with hydrogen to form

Figure 6.8
Hubble Space Telescope view of Jupiter taken in May 1994 from 670 million km away. The dark spot is the shadow of the moon Io. Eleven Earths could fit side by side across the diameter of Jupiter. (H. Weaver and T. Smith, STScI/NASA)

H_2O (water). Chemists call such a hydrogen-dominated composition *reduced*. Throughout the outer solar system, we find abundant water (mostly in the form of ice) and reducing chemistry.

The Terrestrial Planets

The terrestrial planets are quite different from the giants. In addition to being much smaller, they are all composed primarily of rocks and metals, which are made of elements that are less common in the universe as a whole. The most abundant rocks, called *silicates*, are made of silicon and oxygen, while the most common metal is iron. Judging from their densities (see Table 6.2), Mercury has the highest proportion of metals (which are denser), and the Moon has the lowest. (To refresh your memory about the definition of density, see Section 2.2.)

Earth, Venus, and Mars all have roughly similar bulk compositions: about one-third of their mass consists of iron-nickel or iron-sulfur combinations, while two-thirds is made of silicates. Because they contain very little hydrogen, these planets display a wide variety of oxygen compounds (such as the silicate minerals of their crusts). Their chemistry is said to be *oxidized*.

When we look at the internal structure of each of the terrestrial planets, we find that the densest metals are in a central core, with the lighter silicates near the surface. If these planets were liquid like the giant planets, we could understand this effect as the result of the sinking of heavier materials due to the pull of gravity. Thus we believe that the terrestrial planets must once have been much hotter and therefore substantially molten.

Differentiation is the name given to the process by which a planet organizes its interior into layers of different compositions and densities. Once the planet becomes molten, the heavier metal sinks to form a core, while the lightest minerals float to the surface to form a crust. Later, when the planet cools, this layered structure is preserved. In order for a rocky planet to differentiate, it must be heated to the melting point of rocks, typically above 1300 K. (Temperature was defined in Chapter 4; see Appendix 5 for a discussion of temperature scales.)

Moons, Asteroids, and Comets

Chemically and structurally, the Earth's Moon is like the terrestrial planets, but it is an exception among satellites. Most of the moons in the solar system are found far from the Sun, and they have compositions similar to the cores of the giant planets around which they orbit. The three largest satellites, Ganymede and Callisto in the jovian system and Titan in the saturnian system, are composed half of frozen water, half of rocks and metals. These moons differentiated easily during formation since their interior temperatures had to rise only to the melting point of ice, rather than to that of rock, in order for much of their substance to become liquid. Today these satellites have cores of rock and metal, with upper layers and crusts of very cold and thus very hard ice (Figure 6.9).

The asteroids and comets, as well as the smallest moons, were probably never heated to the melting point. They therefore retain their original composition and structure. This is one of their main attractions for astronomers: they represent relatively unmodified material dating back to the formation of the solar system. In a sense, they are ancient "fossils" helping us to learn about a time whose traces have long been erased on larger worlds.

Temperatures: Going to Extremes

Generally speaking, the farther a planet or satellite is from the Sun, the cooler it is. The planets are heated by the radiant energy of the Sun, which declines as the square of the distance from the Sun. You know how rapidly the heating effect of a fireplace or an outdoor "radiant heater" diminishes as you walk away from it. Mercury, the closest planet to the Sun, has a blistering surface temperature of more than 500 K (230°C), as hot as the cleaning cycle on an electric oven, while the surface temperature on Pluto is only about 50 K, colder than liquid air.

Mathematically, the temperatures (expressed on the Kelvin scale) decrease approximately in proportion to the square root of the distance from the Sun. Pluto is 30 AU from the Sun, or 100 times the distance of Mercury. Thus Pluto's temperature is less than that of Mercury by the square root of 100, or a factor of 10—from 500 K to 50 K.

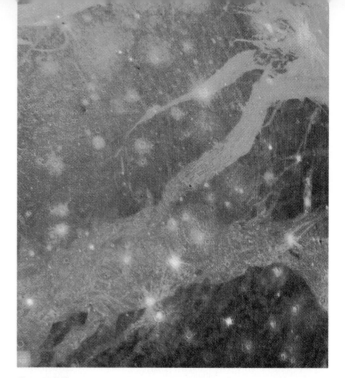

Figure 6.9
The surface of Jupiter's satellite Ganymede, as imaged by the camera system aboard Voyager. (NASA/JPL)

Because of this strong dependence of temperature on distance from the Sun, we find that the Earth is the only planet where surface temperatures generally lie between the freezing and boiling points of water. Our planet is the only one with liquid water on its surface, and therefore (as far as we know) the only one to support life. All the other worlds in the solar system are too hot or too cold, or else lack air and water like our Moon (see Astronomy Basics box).

The giant planets don't follow the simple rule that temperature declines with distance from the Sun. In fact, it is difficult to define a surface temperature at all for these huge balls of gas and liquid. In the atmospheres of the giant planets, temperature increases with depth. While the cloud tips are frigid, their interiors can be much hotter than the surface of Mercury. The situation is further complicated by the presence of substantial internal heat sources on Jupiter, Saturn, and Neptune. Jupiter, which has the largest internal heat source, generates as much energy within its interior as it absorbs from the Sun.

There's No Place Like Home

In the classic film *The Wizard of Oz*, Dorothy, our heroine, concludes after her many adventures in "alien" environments that "there's no place like home." The same can be said of the worlds with which we share the solar system. There are many fascinating places, large and small, that we might like to visit, but on none of them could human beings survive without a great deal of artificial assistance.

A thick atmosphere keeps the surface temperature on our neighbor Venus at a sizzling 700 K (near 900°F), hot enough to melt several metals. And the pressure of its mostly carbon dioxide atmosphere equals that found almost a kilometer underwater on Earth. Mars, on the other hand, has temperatures generally below freezing, with air (also mostly CO_2) so thin, it's like that found at an altitude of 30 km (100,000 ft) in the Earth's atmosphere. And the red planet is so dry that it has not had any rain for billions of years.

In the outer layers of the jovian planets, it is neither warm enough nor solid enough for human habitation. Any bases we build in the systems of the giant planets may well have to be in space or on one of their satellites—none of which is particularly hospitable to a luxury hotel with swimming pool and palm trees. Or perhaps we will find warmer havens deep inside the clouds of Jupiter, or in a possible ocean under the frozen ice of its moon Europa.

All of this suggests that we had better take good care of the Earth, as it is the only site where large numbers of our species can survive unaided. Recent human activity may be reducing the habitability of our planet by adding pollutants to the atmosphere or reducing the fraction of the Earth's surface where vegetation can thrive and continue to supply us with oxygen. In a solar system that seems unready to receive us, making the Earth less hospitable to life may be a grave mistake.

Figure 6.10
The Hubble Space Telescope took this sequence of images of Jupiter in summer 1994, when the fragments of Comet Shoemaker-Levy 9 collided with the giant planet. Here we see the site hit by fragment G, from five minutes to five days after impact. Several of the dust clouds generated by the collisions became larger than the Earth. (STScI/NASA)

Geological Activity

The crusts of all of the terrestrial planets, as well as of the larger satellites, have been modified over their histories by both internal and external forces. Externally, each has been battered by a rain of projectiles from space, leaving surfaces pockmarked by impact craters of all sizes (see Figure 6.4). We have good evidence that this bombardment was far greater in the early history of the solar system, but it certainly continues to this day. The collision of more than 20 large pieces of Comet Shoemaker-Levy 9 with Jupiter in the summer of 1994 is one recent example of this process (Figure 6.10).

At the same time, internal forces on the planets have buckled and twisted their crusts, built up mountain ranges, erupted as volcanoes, and generally reshaped the surfaces in what we call *geological activity*. (Since the prefix *geo* means "Earth," this is a bit of an "Earth-chauvinist" term, but it is so widely used, your authors bow to tradition.) Among the terrestrial planets, the Earth and Venus have experienced the greatest level of geological activity over their histories, although some of the satellites in the outer solar system are also surprisingly active. In contrast, our own Moon is a dead world where geological activity ceased billions of years ago.

We can understand the reason for the different levels of activity on the planets when we note that geological activity is the result of a hot interior; the forces of volcanism and mountain building are driven by heat escaping from the cores of the planets. As we will see, each of the planets was heated at the time of its birth, and this primordial heat initially powered extensive volcanic activity, even on our Moon. But small objects like the Moon soon cooled off. The larger the planet or satellite, the more likely it is to retain its internal heat, and therefore the more we expect to see surface evidence of continuing geological activity. The effect is similar to our own experience with a baked potato: the larger the potato, the more slowly it cools. If we want a potato to cool quickly, we cut it into small pieces.

For the most part, the history of volcanic activity seen on the terrestrial planets conforms to the predictions of this simple theory. The Moon, smallest of these objects, is a geologically dead world. Although we know much less about Mercury, it seems likely that this planet, too, ceased most volcanic activity about the same time that the Moon did. Mars represents an intermediate case: it has been much more active than the Moon, but less so than the Earth. Earth and Venus, the largest terrestrial planets, still have molten interiors and high levels of geological activity even today, some 4.5 billion years after their birth.

The level of geologicial activity also plays an important role in determining how well we can still "read" the history of external bombardment on these worlds. On Earth, for example, all but the most recent impact craters have been erased by surface activity (and the effects of weather), while the cratered surface of the Moon bears silent witness to billions of years of impacts.

Dating Planetary Surfaces

How do we know the age of the surfaces we see on planets and satellites? If a world *has* a surface (as opposed to being mostly gas and liquid), we have developed some techniques for estimating how long ago that surface solidified. Note that if a world is active, the age of a particular surface layer may tell us little about how long ago its entire world formed. On several bodies (including the Earth), vast outpourings of molten rock or water have erased evidence of earlier epochs, and present us only with a relatively young surface for investigation.

Counting the Craters

If a planet has had little erosion or internal activity, such as Mercury or the Moon, it is possible to estimate the age of its surface by counting the number of impact craters. This technique works because the evidence shows that the rate at which impacts have occurred in the solar system has been roughly constant for several billion years. Thus, in the absence of forces to eliminate craters, the number of craters is simply proportional to the length of time the surface has been exposed. Chunks of material fell and still fall from space randomly, so we have to be sure to do our crater counting over a large enough sample of the surface to eliminate chance clusters of many craters, or regions that happen to be emptier. Since spacecraft have given us close-up views of many worlds too far away for crater counting by telescope, this technique has been successfully applied to planets, satellites, and even to two asteroids (Figure 6.11).

But bear in mind that crater counts can only tell us the time since the surface experienced a major change. On worlds with strong internal activity, that age may be shorter than a geologist might like. Estimating ages from crater counts is a little like walking along a sidewalk in a snowstorm after the snow has been falling steadily for a day or more. You may notice that in front of one house the snow is deep, while next door the sidewalk may be almost clear. Do you conclude that less snow has fallen in front of Ms. Jones' house than Mr. Smith's? Of course not. Instead, you conclude that Jones has recently swept the walk clean, while Smith has not. Similarly, the numbers of craters indicate how long since a planetary surface was last "swept clean" by lava flows or material ejected from a nearby large impact.

Thus the number of craters provides an important clue to the evolution of different worlds. On a given planet, the more heavily cratered terrain will always be older (according to our definition of age in the preceding paragraph). And if we can calibrate the relationship between crater numbers and ages—using what we have learned about the Moon, from which we can also gather samples (see following paragraph)—we may be able to

Figure 6.11
This image of the Moon's surface was taken during the Apollo 16 mission and shows craters of many different sizes. (NASA)

estimate the ages of surfaces on other cratered planets, such as Mars or Mercury.

Radioactive Rocks

Another way to trace the history of a solid world is to determine the ages of individual rocks. After samples were brought back from the Moon by the Apollo astronauts, the techniques that had been developed to date rocks on Earth could also be applied to establish a chronology for the Moon. In addition, a few samples of material from the Moon, Mars, and the large asteroid Vesta have fallen to Earth as meteorites and can also be examined directly (see Chapter 13).

The ages of rocks can be measured using the properties of natural **radioactivity.** Around the turn of the century, physicists began to understand that some atomic nuclei are not stable, but can spontaneously split apart or *decay* into smaller nuclei. The process of radioactive decay involves the emission of particles such as electrons, or of radiation in the form of gamma rays (see Chapter 4).

For any given nucleus, it is not possible to say when the decay process might happen. It is random in nature, just as the throw of dice is: as gamblers have all too often found out, it is impossible to say just when dice will come up 7 or 11. But for a very large number of dice tosses, you can calculate the odds that 7 or 11 will come up. Similarly, if you have a very large number of radioactive atoms of one type (say, uranium), there is a specific time period, called its **half-life,** during which the chances are fifty-fifty that decay will occur for any of the nuclei.

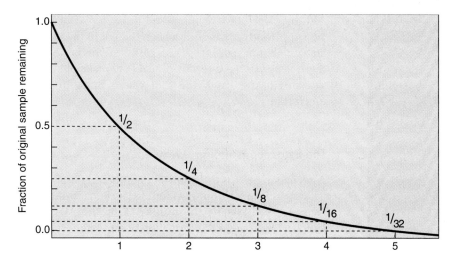

Figure 6.12
Radioactive decay. This graph shows the amount of a radioactive sample that remains after several half-lives have passed. After one half-life, half the sample is left; after two half-lives, one-half of the remainder (or 1/4) is left; and after three half lives, one-half of that (or 1/8) is left.

A particular nucleus may last a shorter or longer time than its half-life, but in a large sample almost exactly half will have decayed after a time equal to one half-life. Half of those remaining will have decayed in two half-lives, leaving only one-half of a half—or one-quarter—of the original sample (Figure 6.12). So if you had a gram of pure radioactive nuclei whose half-life was 100 years, then after 200 years you would have only 1/4 gram, and after 300 years only 1/8 gram, and so forth.

Thus, radioactive elements provide accurate nuclear clocks. The presence of the nucleus that the element decays into is a signal to scientists that decay has taken place. By comparing how much of a radioactive element is left in a rock with how much we have of the element it decays into, we can learn how long the decay process has been going on and hence how long ago the rock formed. Table 6.3 summarizes the decay reactions used most often to date lunar and terrestrial rocks. Typically, more than one element is used to date a rock, and the process is carried out for several different mineral grains in the same rock to eliminate uncertainties. The final age, checked by the agreement of the different methods, is usually accurate to

within a few percentage points—that is, to a few tens of millions of years in a rock several billion years old.

This method works only if both the radioactive element and its decay product remain trapped in the same location of a rock sample. A radioactive age, therefore, is the time during which the rock has remained undisturbed, so that both the radioactive element and its decay product are still present in each mineral grain. In most cases it can be thought of as the *solidification age* of the rock—the time since it cooled from the molten state. As we will see, for the lunar samples these ages range from about 3.3 to 4.4 billion years—substantially older than most of the rocks on the Earth.

By the way, note that the decay of radioactive nuclei generally releases energy in the form of heat. While the energy from a single nucleus is not very large (in human terms), the enormous numbers of radioactive nuclei in a planet or satellite (especially early on) can be a significant source of internal energy for that world.

Radioactive dating techniques can be used to determine the date of a planet's formation, as well as the solidification age of individual rocks. The results of this technique agree for both Earth and Moon: each was formed 4.5 billion years ago. Similar results are obtained when we date primitive meteorites—they too formed about 4.5 billion years ago, a date we can therefore associate with the beginning of the solar system.

6.4

Origin of the Solar System

Much of astronomy is motivated by a desire to understand the origin of things—to find at least partial answers to the age-old questions of where the universe, the Sun, the Earth, and we ourselves came from. Each planet and satellite is a fascinating place that may stimulate our imaginations as we try to picture what it would be like to visit there. But collectively the members of the solar system

TABLE 6.3
Radioactive Decay Reactions Used to Date Rocks

Parent	Daughter	Half-Life
		(billion yr)
Samarium-147*	Neodymium-143	106
Rubidium-87	Strontium-87	48.8
Thorium-232	Lead-208	14.0
Uranium-238	Lead-206	4.47
Potassium-40	Argon-40	1.31

* The number after each element is its atomic weight, equal to the number of protons plus neutrons in its nucleus. This specifies the *isotope* of the element; different isotopes of the same element differ in the number of neutrons.

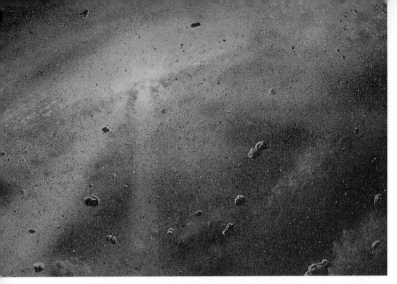

Figure 6.13
Artist's conception of the solar nebula, the flattened cloud of gas and dust from which our planetary system formed. Icy and rocky planetesimals can be seen in the foreground. The bright center is where the Sun is forming. (Art by William Hartmann)

also preserve patterns that can tell us about the formation of the system. As we begin our exploration of the planets, we want to introduce our modern picture of how the solar system formed. We will return to the question of origins in more detail in Chapter 13.

Looking for Patterns

One way to approach the question of origins is to look for regularities among the planets. We found, for example, that all the planets revolve in the same direction around the Sun, and that they lie in nearly the same flat plane. The Sun also spins in the same direction about its own axis. Astronomers interpret this pattern as evidence that the Sun and planets formed together from a spinning system of gas and dust that we call the **solar nebula** (Figure 6.13).

The composition of the planets gives another clue. Spectroscopic analysis (see Chapter 4) allows us to determine what elements are present in the Sun and the planets. The Sun has the same hydrogen-dominated composition as Jupiter and Saturn, and therefore appears to have been formed from the same reservoir of material. Each of the terrestrial planets and satellites is deficient to some degree in the light gases and in the various ices that form out of the common elements oxygen, carbon, and nitrogen. Instead, we see mostly the rarer heavy elements like iron and silicon. This pattern suggests that the processes that led to planet formation in the inner solar system allowed most of the lighter materials to escape, leaving just a residue of heavy stuff as available building blocks.

The reason for this is not hard to guess, bearing in mind the heat of the Sun. The inner planets are made of rock and metal, which require a high temperature to evaporate, and they contain very little ice or gas, which evaporate more easily. Farther from the Sun, the planets

and their satellites are composed in large part of ice and gas. Apparently it has always been too hot in the inner solar system for most ices and gases to condense, while these materials were readily available to make planets and satellites farther from the Sun where it has remained much colder.

The Evidence from Far Away

A second approach to understanding the origin of the solar system is to look outward for evidence that other, similar systems of planets are forming elsewhere. We cannot look back in time to the formation of our own system, but there are many stars in space that are much younger than the Sun. In these systems, the process of planet formation might still be accessible to direct observation. One of the important astronomical discoveries of recent years is the presence of just such *planetary nurseries.* What we actually observe are other "solar nebulas" or *circumstellar disks*—flattened, spinning systems of gas and dust surrounding young stars. These collapsing, spinning disks are analogues of our own solar system's initial stages of formation billions of years ago. But because the stars are so far away, astronomers have not yet been able to observe the actual formation of planets in these circumstellar disks.

Building Planets

One approach is to use theoretical calculations to build simulated "model solar nebulas" on a large computer to see how solid bodies might form from the gas and dust in these disks as they cool. These models show that material begins to coalesce by first forming smaller objects, precursors of the planets, which we call **planetesimals** (see Figure 6.13).

Fast computers can now be used to simulate the way that millions of planetesimals, each no larger than 100 km in diameter, might have gathered together under their mutual gravity to form the planets we see today. We are beginning to understand that this process was a violent one, with planetesimals crashing into each other and sometimes even disrupting the growing planets themselves. As a consequence of these violent impacts (as well as the heat from radioactive materials in them), all of the planets were heated to above their melting points, and they differentiated to acquire their present internal structures.

The process of impacts and collisions was complex and often apparently random. While the solar nebula model can explain many of the regularities we find in the solar system, the random collisions of massive collections of planetesimals could be the reason we have some exceptions to the "rules" of solar system behavior. For example, why do the planets Uranus and Pluto spin on their sides? Why does Venus spin slowly and in the opposite direction from all the other planets? Why does the composition of

Carl Sagan: Solar System Advocate

Arguably the best-known astronomer in the world, the late Carl Sagan devoted most of his professional career to studying the planets, and much of his considerable energy to raising public awareness of what we can learn from exploring the solar system. Born in Brooklyn, New York, in 1934, Sagan became interested in astronomy as a youngster, but also credits science fiction stories for sustaining and encouraging his fascination with what's out there. After earning both his bachelor and PhD degrees from the University of Chicago, he worked as a research associate at the University of California and as an assistant professor at Harvard before becoming professor of astronomy at Cornell University in 1968.

In the early 1960s, when many scientists still thought that Venus might turn out to be a hospitable place, Sagan demonstrated that the atmosphere of our neighborhood planet could act like a giant greenhouse, keeping the heat in and raising the temperature enormously. He showed that the seasonal changes astronomers had seen on Mars were caused not by vegetation, but by wind-blown dust. He was a member of the scientific teams for many of the robotic missions that have explored the solar system, and was instrumental in getting NASA to put a message-bearing plaque aboard the Pioneer spacecraft, as well as audio–video records on the Voyagers.

To encourage public interest and public support of planetary exploration, Sagan helped found The Planetary Society, now the largest space-interest organization in the world. And he was a tireless and eloquent exponent of the need to study the solar system close-up, and of the value of learning about other worlds in order to take better care of our own.

Sagan did experimental work simulating conditions on the early Earth and demonstrated how some of life's fundamental building blocks might have formed from the primordial "soup" of natural compounds on our planet. He and his students have built "Jupiter boxes" to demonstrate how conditions in the giant planet's atmosphere may in some ways resemble what it was like on Earth before life began. More recently, he and his colleagues made models showing that the consequences of nuclear war for the Earth would be even more devastating than anyone had thought (the *nuclear winter* hypothesis), and demonstrating some of the serious consequences of continued pollution of our atmosphere.

Sagan was perhaps best known, however, as a brilliant popularizer of astronomy and the author of many books on science, including the best-selling *Cosmos*, the Pulitzer-prize-winning *Dragons of Eden*, and the recent *Pale Blue Dot*, an evocative tribute to solar system exploration. His intriguing science fiction novel, *Contact*, now a successful film, is still recommended by many science instructors as a scenario for making contact with life elsewhere that is much more reasonable than most science fiction.

An appearance on the *Tonight Show* in 1973 (where his lucid explanations caught the attention of then-host Johnny Carson, an amateur astronomer) began a long series of television appearances that culminated in Sagan's 13-part public television series, *Cosmos*. It has been seen by an estimated 500 million people in 60 countries, and has become the most-watched series in the history of public broadcasting. Sagan, by the way, was quick to point out that not once in the episodes of *Cosmos* did he say "billions and billions," a phrase now indelibly associated with him (although he did use it in other media appearances over the years). While a few astronomers scoffed at a scientist who spent so much time in the public eye, it is probably fair to say that Sagan's enthusiasm and skill as an explainer won more friends for the science of astronomy than anyone or anything else in this century.

Carl Sagan

the Moon resemble the Earth in many ways, and yet exhibit substantial differences? Although we cannot know for sure, the answers to such questions may well lie in enormous collisions that took place in the solar system long before life on Earth began.

Today, some 4.5 billion years after it began, the solar system is a much less violent place. But as we will see, some planetesimals have continued to interact and collide, and their fragments move about the solar system as roving "transients" that can "make trouble" for the established members of the Sun's family, such as our own Earth. In Chapter 7 we will examine some of the devastation caused by impacts from space in the past, and some of the dangers that may face us in the future.

What might other planetary systems be like, given these ideas about the formation of our own? While planets of some sort are always expected to form from a circumstellar disk like the solar nebula, they would not necessarily become organized into a system that looks just like our own solar system. There might be two planets as large as Jupiter, or none at all. There might be an analogue of the Earth—a smallish planet in just the right place to have liquid water on its surface—or there might not. The first discoveries of planets around other stars—made in 1995 and 1996—are quite different from our own solar system. But this may be a selection effect, since astronomers don't yet have the technology to detect planets like the Earth or Venus in other systems.

Summary

6.1 Our solar system consists of the Sun, nine planets, 61 known satellites, and a host of smaller objects. The planets can be divided into two groups: the inner **terrestrial planets** and the outer **giant planets;** Pluto does not fit into either category. The giant planets are composed mostly of liquids and gases. Smaller members of the solar system include **asteroids,** rocky and metallic objects found mostly between Mars and Jupiter; **comets,** made mostly of frozen gases, which for the most part orbit far from the Sun; and countless smaller grains of **cosmic dust.** When a meteor survives its passage through our atmosphere and falls to Earth, we call it a **meteorite.**

6.2 The giant planets have dense cores roughly ten times the mass of the Earth, surrounded by layers of hydrogen and helium. The terrestrial planets consist mostly of rocks and metals. They were once molten, allowing their structures to **differentiate** (i.e., their denser materials sank to the center). The Earth's moon resembles the terrestrial planets in composition, but most of the other satellites—which orbit the giant planets—have a large quantity of frozen ices within them. Generally, worlds closer to the Sun have higher temperatures, although three of the giant planets have internal heat sources. The surfaces of terrestrial planets have been modified by impacts from space, and by varying degrees of geological activity.

6.3 The ages of the surfaces of objects in the solar system can be estimated by counting craters: on a given world, a more heavily cratered region will always be older. We can also use samples of rocks with **radioactive** elements in them to obtain the time since the layer in which the rock formed last solidified. The **half-life** of a radioactive element is the time it takes for half the sample to decay; thus we determine how many half-lives have passed by how much of a sample remains the radioactive element and how much has become the decay product.

6.4 Regularities among the planets lead astronomers to hypothesize that the Sun and the planets formed together in a giant, spinning cloud of gas and dust called the **solar nebula.** Recent observations show tantalizingly similar circumstellar disks around other stars. Within the solar nebula, material first coalesced into **planetesimals;** many of these gathered together to make the planets and satellites. The remainder can still be seen as comets and asteroids.

Review Questions

1. List some reasons that the study of the planets has progressed more in the last few decades than has any other branch of astronomy.

2. Imagine that you are a travel agent in the next century. An eccentric billionaire asks you to arrange a "Guinness Book of Solar System Records" kind of tour. Where would you direct him to find the following (use this chapter and Appendices 7 and 8):

   a. the least-dense planet
   b. the densest planet
   c. the largest moon in the solar system
   d. the planet at whose surface you would weigh the most (*Hint:* weight is directly proportional to surface gravity)
   e. the smallest planet
   f. the planet that takes the longest time to rotate
   g. the planet that takes the least time to rotate
   h. the planet whose diameter is the closest to Earth's
   i. the moon that takes the shortest time to orbit its planet
   j. the densest satellite
   k. the moon that orbits the farthest from its planet
   l. the most massive moon
   m. the most spectacular ring system

3. How do terrestrial and giant planets differ from each other? List as many ways as you can think of.

4. Why are there so many craters on the Moon and so few on the Earth?

5. How do asteroids and comets differ?

6. How and why is the Earth's moon different from the larger satellites of the giant planets?

7. Where would you look for some "original" planetesimals left over from the formation of our solar system?

8. Describe how we use radioactive elements and their decay products to find the age of a rock sample? Is this necessarily the age of the entire world from which the sample comes? Explain.

9. What was it like in the solar nebula? Why did the Sun form at its center?

Thought Questions

10. Using some of the astronomical resources in your college library, find five names of features on each of three other worlds that are named after people. In a sentence or two, describe who each of these people was.

11. Explain why the planet Venus is differentiated, but asteroid Fraknoi, a very boring and small member of the asteroid belt, is not.

12. Would you expect as many impact craters per unit area on the surface of Venus as on the surface of Mars? Why or why not?

13. Interview a sample of 20 people who are *not* taking an astronomy class and ask them if they can name a living astronomer. What percent of those interviewed were able to name one? Typically, the two astronomers the public knows these days are Carl Sagan and Stephen Hawking. Why are they better known than most astronomers? How would your results have differed if you had asked the same people to name a movie star or a professional football player?

Problems

14. Looking at Appendices 7 and 8, find the satellite whose diameter is the largest fraction of the diameter of the planet it orbits.

15. Use the rules given in the chapter for calculating the spacing of each planet with the Titius-Bode "law." Now compare your answers to the listing of the semimajor axes of each planet. What is the ratio of your calculated value to the actual value in each case?

16. Barnard's Star, the second closest star to us, is about 56 trillion (5.6×10^{12}) km away. Calculate how far it would be using the scale model in Section 6.1.

17. A radioactive nucleus has a half-life of 5×10^8 years. Assuming a sample of rock (say in an asteroid) solidified right after the solar system formed, approximately what fraction of the radioactive element should be left in the rock today?

Suggestions for Further Reading

Badash, L. "The Age of the Earth Debate" in *Scientific American*, Aug. 1989.

Barnes-Svarney, P. "The Chronology of Planetary Bombardments" in *Astronomy*, July 1988, p. 21.

Beatty, J. et al. *The New Solar System*, 3rd ed. 1990, Sky Publishing. A compendium of articles on planetary astronomy by leading scientists in the field. Occasionally technical.

Hartmann, W. "Piecing Together Earth's Early History" in *Astronomy*, June 1989, p. 24.

Kross, J. "What's in a Name?" in *Sky & Telescope*, May 1995, p. 28. Discusses the naming of features in the solar system.

Miller, R. and Hartmann, W. *The Grand Tour: A Traveler's Guide to the Solar System*, 2nd ed. 1993, Workman. Lavishly illustrated beginners' primer.

Morrison, D. *Exploring Planetary Worlds*. 1993, Scientific American Library/W. H. Freeman. Clear, up-to-date, nontechnical survey.

Morrison, D. and Owen, T. *The Planetary System*, 2nd ed. 1996, Addison-Wesley. A thorough introduction to our modern knowledge.

Nicks, O. *Far Travelers: The Exploring Machines*. 1985, NASA Special Publication 480, U.S. Government Printing Office. Describes lunar and planetary spacecraft.

O'Dell, R. "Exploring the Orion Nebula" in *Sky & Telescope*, Dec. 1994, p. 20. Discusses planetary nurseries.

Sagan, C. *Pale Blue Dot*. 1994, Random House. Eloquent and impressive introduction to planetary exploration.

Sheehan, W. *Worlds in the Sky*. 1992, U. of Arizona Press. History of how we came to understand the planets.

Using **REDSHIFT** ™

1. Tour the solar system by turning on only the planets. Set the location to heliocentric, longitude to 270°, latitude to 90°, and the height to 100 AU. Step in time and turn on Draw Paths. How long does the slowest planet take to complete one orbit? In what year will Pluto again be the farthest planet from the Sun?

Now set the latitude to 0°, keeping the height set to 100 AU, and resume motion. Which planet has the most inclined orbit? Do most of the planets seem to share the plane of Jupiter's orbit? Pluto's orbit crosses Neptune's orbit in this view. Why won't they collide?

2. View thousands of asteroids by turning them on and the stars and deep-sky objects off and setting the location to heliocentric with longitude 270°, latitude 90°, and height 20 AU. How are the asteroids distributed? Do you notice anything unusual about the motion of the outermost asteroids?

Set the range of asteroid sizes from 100–1000 to 0–10. What new feature of the asteroid distribution do you see? Now set the latitude to 0°. What do the orbits of the planets and asteroids say about their origins?

Molten lava fountaining from a volcanic vent in Hawaii—
testament to the dynamic nature of the Earth's crust.
(G. Briggs/USGS)

CHAPTER 7

Earth as a Planet

Thinking Ahead

Airless worlds in our solar system seem peppered with craters large and small. The Earth, on the other hand, has few craters, but a thick atmosphere and much surface activity. We believe that in the past the Earth, too, had many craters that have since been erased by forces in its crust and atmosphere. What can the persistent cratering on so many other worlds tell us about the history of our own planet?

As our first step in exploring the solar system in more detail, we turn to the most familiar planet, our own Earth. The first humans to see the Earth as a blue sphere floating in the blackness of space were the *Apollo 8* astronauts who made the first voyage around the Moon in 1988. For many people, the historic visits to the Moon and the images showing our world as a small, distant globe represent a pivotal moment in human history, when it became difficult for educated human beings to avoid viewing our world with a global perspective.

In previous chapters we discussed the size and orbit of the Earth, and we described terrestrial phenomena, such as seasons and tides, that are related to the motions of the Earth. In this chapter we examine the composition and structure of our planet with its envelope of ocean and atmosphere. We ask how that envelope and the surface of our planet came to be the way they are today. Because we live on it, we know the Earth much better than any other planet, and we can perform experiments on its surface that we cannot yet do elsewhere with any regularity. This direct knowledge helps us use the Earth as a reference point for understanding other worlds where we are mostly limited to remote observations.

> By coming face to face with alternative fates of worlds more or less like our own, we have begun to better understand the Earth. Every one of these worlds is lovely and instructive. But, so far as we know, they are also, every one of them, desolate and barren. Out there, there are no "better places." So far, at least.
>
> Carl Sagan in *Pale Blue Dot* (1994, Random House)

Figure 7.1
The Earth from space, taken by the *Apollo 17* astronauts in December 1972. This is one of the rare images of a full Earth taken during the Apollo program; most images show only some of the Earth's disk in sunlight. (NASA)

7.1

The Global Perspective

Basic Properties

The Earth is a medium-sized planet, with a diameter of approximately 13,000 km (Figure 7.1). As one of the terrestrial planets, it is composed primarily of a few heavy elements, such as iron, silicon, and oxygen—very different from the composition of the Sun and stars, which are dominated by the light elements hydrogen and helium. The Earth's orbit is nearly circular, and it is close enough to the Sun to support liquid water on its surface. It is the only planet in our solar system that is neither too hot nor too cold, but "just right" for the development of life as we know it. Some of the basic properties of the Earth are summarized in Table 7.1.

TABLE 7.1
Some Properties of the Earth

Semimajor axis	1.00 AU
Period	1.00 year
Mass	5.98×10^{24} kg
Diameter	12,756 km
Escape velocity	11.2 km/s
Rotation period	23h 56m 4s
Surface area	5.1×10^8 km^2
Atmospheric pressure	1.00 bar

Earth's Interior

The interior of a planet—even our own Earth—is difficult to study, and its composition and structure must be determined indirectly. Our only direct experience is with the outermost skin of the Earth's crust, a layer no more than a few kilometers deep. It is important to remember that in many ways we know less about our own planet 5 km beneath our feet than we do about the surfaces of Venus and Mars.

The Earth is composed largely of metal and silicate rock (see Section 6.2). Most of this material is in a solid state, but some of it is hot enough to be molten. The structure of the material in the Earth's interior has been probed in considerable detail by measuring the transmission of **seismic waves** through the Earth. These are waves produced in the material of our planet by natural earthquakes or by impacts or explosions.

Seismic waves travel through a planet rather like sound waves through a struck bell. Just as the sound frequencies vary depending on what material the bell is made of and how it is constructed, so a planet's response depends on its composition and structure. By monitoring the seismic waves in different locations, scientists can learn about the layers the waves have traveled through. Some of these vibrations travel along the surface; others pass directly through the interior. Seismic studies have shown that our planet's interior consists of several distinct layers with different compositions, illustrated in Figure 7.2.

The top layer is the **crust,** the part of the Earth we know best (Figure 7.3). The crust under the oceans, which cover 55 percent of the surface, is typically about 6 km thick and is composed of volcanic rocks called **basalts.** Produced by the cooling of volcanic lava, basalts are made primarily of the elements silicon, oxygen, iron, aluminum, and magnesium. The continental crust, which covers 45 percent of the surface, is from 20 to 70 km thick and is predominantly composed of a different volcanic class of silicates called **granites.** These crustal rocks typically have densities of about 3 g/cm^3. The crust is the easiest layer for geologists to study, but it makes up only about 0.3 percent of the total mass of the Earth.

The largest part of the solid Earth, called the **mantle,** stretches from the base of the crust down to a depth of 2900 km. The mantle is more-or-less solid, but at the temperatures and pressures found there, the mantle rock can deform and flow slowly. The density in the mantle in-

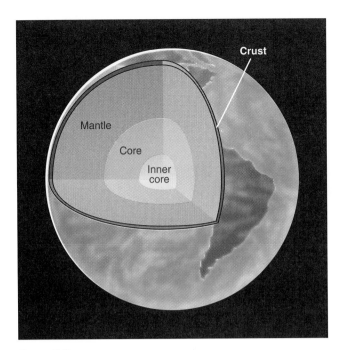

Figure 7.2
The interior structure of the Earth.

creases downward from about 3.5 g/cm^3 to more than 5 g/cm^3 as a result of the compression produced by the weight of overlying material. Samples of upper-mantle material are occasionally ejected from volcanoes, permitting a detailed analysis of its chemistry.

Figure 7.3
This computer-generated globe shows the surface of the Earth's crust with the oceans removed, as determined from satellite images and ocean floor radar mapping. The amounts of vegetation on the continents, as recorded by the instruments aboard the NOAA satellites, are indicated by color, ranging from light brown for the least vegetation, through yellow and orange, to green for the most. (J. Frawley and NASA)

Beginning at a depth of 2900 km, we encounter the dense metallic **core** of the Earth. With a diameter of 7000 km, it is substantially larger than the entire planet Mercury. The outer core is liquid. The innermost part of the core (about 2400 km in diameter) is extremely dense and probably solid. In addition to iron, the core probably also contains substantial quantities of nickel and sulfur.

The separation of the Earth into layers of different densities is an example of differentiation (see Chapter 6). The fact that the Earth is differentiated is evidence that it was once warm enough for the mantle rocks to melt, permitting the heavier metals to sink to the center and form the dense core.

Magnetic Field and Magnetosphere

We can find additional clues about the Earth's interior from its magnetic field. Our planet behaves in some ways as if there were a giant bar magnet inside, approximately aligned with the rotational poles of the Earth. This magnetic field is actually generated by moving material in the Earth's liquid metallic core. As the liquid metal inside the Earth circulates, it sets up a circulating electric current. And, as discussed in Chapter 4, whenever charged particles are moving, they set up a magnetic field.

The Earth's magnetic field extends into surrounding space. When a charged particle encounters a magnetic field, it begins to spiral around it and becomes trapped in the magnetic zone. Above the Earth's atmosphere, our field is able to trap small quantities of electrons and other atomic particles. This region, called the Earth's **magnetosphere,** is defined as the zone within which the Earth's magnetic field dominates over the weak interplanetary field that extends outward from the Sun (Figure 7.4).

Where do the charged particles trapped in our magnetosphere come from? They flow from the hot surface of the Sun; astronomers call this flow the *solar wind* (see Chapter 14), and it not only provides particles for the Earth's magnetic field to trap, but also elongates our field in the direction pointing away from the Sun. Typically, the Earth's magnetosphere extends about 60,000 km, or ten Earth radii, in the direction of the Sun. But pointing away from the Sun, the magnetic field is shaped like a wind sock and can reach as far as the orbit of the Moon.

The magnetosphere was discovered in 1958 by instruments on the first U.S. Earth satellite, Explorer 1, which recorded the ions (charged particles) trapped in its inner part. The regions of high-energy ions in the magnetosphere are often called the *Van Allen belts* in recognition of the University of Iowa professor who built the scientific instrumentation for Explorer 1 and correctly interpreted the satellite measurements. Since 1958, hundreds of spacecraft have explored various regions of the magnetosphere. In Chapter 14 we will discuss how the magnetic field and Van Allen belt of the Earth give rise to the brilliant curtains of light called *aurorae,* visible in our atmosphere near the poles.

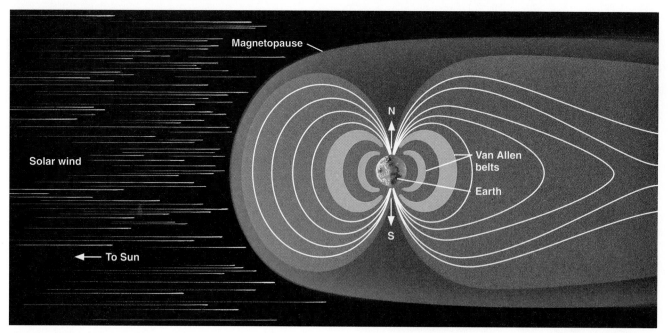

Figure 7.4
A cross-sectional view of the Earth's magnetosphere as revealed by numerous spacecraft missions.

The Crust of the Earth

Let us now examine our planet's outer layers in more detail. The Earth's crust is a dynamic place. Volcanic eruptions, erosion, and large-scale movements of the continents constantly rework the surface of our planet. Geologically, ours is the most active planet. Many of the geological processes described in this section have taken place on other planets as well, usually in their distant pasts. But, as we will see in Chapter 11, some of the satellites (moons) do have impressive activity. For example, Jupiter's satellite Io has a remarkable number of active volcanoes.

Composition of the Crust

The Earth's crust is largely made up of oceanic basalts and continental granites. These are both examples of **igneous** rock, the term used for any rock that has cooled from a molten state. All volcanically produced rock is igneous (Figure 7.5).

Two other kinds of rock are familiar to us on the Earth, although it turns out that neither is common on other planets. **Sedimentary** rocks are made of fragments of igneous rocks or living organisms, deposited by wind or water and cemented together without melting. On Earth, these include the common sandstones, shales, and limestones. **Metamorphic** rocks are produced when high temperature or pressure alters igneous or sedimentary rocks physically or chemically (the word *metamorphic* means changed in form). Metamorphic rocks are pro-

duced on Earth because geological activity carries surface rocks to considerable depths and then brings them back up to the surface. Without such activity, these rocks would not exist.

There is a fourth very important category of rock that can tell us much about the early history of the planetary system. This is **primitive** rock, which has largely escaped chemical modification by heating. Primitive rock represents the original material out of which the planetary system was made. There is no primitive material left on the

Figure 7.5
The formation of igneous rock as liquid lava cools and freezes. This is a lava flow from a basaltic eruption in Hawaii. (David Morrison)

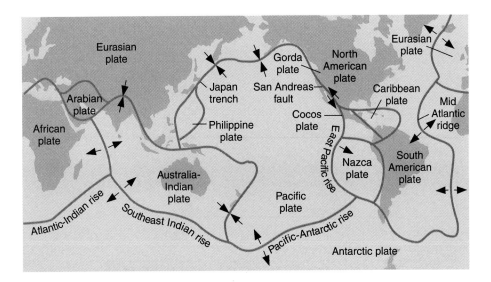

Figure 7.6
The major plates into which the crust of the Earth is divided. Arrows indicate the motion of the plates.

Earth because the entire planet was strongly heated early in its history. To find primitive rocks, we must look to smaller objects such as comets, asteroids, and small planetary satellites.

A block of marble on Earth is composed of materials that have gone through all four of these states. Beginning as primitive material before the Earth was born, it was heated in the early Earth to form igneous rock, subsequently eroded and redeposited (perhaps many times) to form sedimentary rock, and finally transformed several kilometers below the Earth's surface into the hard white metamorphic stone we see today.

Plate Tectonics

Geology is the study of the Earth's crust, particularly the processes that have shaped its surface throughout history. Heat escaping from the interior provides the energy for the formation of mountains, valleys, volcanoes, and even the continents and ocean basins themselves. But not until the middle of the 20th century did geologists succeed in understanding how these landforms are created.

Plate tectonics is a theory that explains how slow motions within the mantle of the Earth move large segments of the crust, resulting in a gradual drifting of the continents as well as the formation of mountains and other large-scale geologic features. It is a concept as basic to geology as evolution by natural selection is to biology, or gravity is to understanding the orbits of planets.

The Earth's crust and upper mantle (to a depth of about 60 km) are divided into about a dozen major plates that fit together like the pieces of a jigsaw puzzle (Figure 7.6). But these plates are also capable of moving slowly with respect to each other. In some places, such as the Atlantic Ocean, the plates are moving apart, while in others they are being forced together. The power to move the plates is provided by slow **convection** of the mantle, a process by which heat escapes from the interior through

the upward flow of warmer material and the slow sinking of cooler material. (Convection, in which energy is transported from a warm region such as the interior of the Earth to a cooler region such as the upper mantle, is a process we encounter often in astronomy.)

As the plates slowly move, they come into contact with each other and cause dramatic changes to the Earth's crust over time. Four basic kinds of interactions between crustal plates are possible at their boundaries: (1) they can pull apart; (2) one plate can burrow under another; (3) they can slide alongside each other; or (4) they can jam together. Each of these activities is important in determining the geology of the Earth.

Rift and Subduction Zones

Plates pull apart from each other along **rift zones,** such as the Mid-Atlantic Ridge, driven by upwelling currents in the mantle (Figure 7.7a). A few rift zones are found on land, the best known being the central African rift, an area where the African continent is slowly breaking apart. Most rift zones, however, are in the oceans. New material rises from below to fill the space between the receding plates; this is basaltic lava, the kind of igneous rock that forms most of the ocean basins.

From a knowledge of how the sea floor is spreading, we can calculate the average age of the oceanic crust. About 60,000 km of active rifts have been identified, with average separation rates of about 4 cm per year. The new area added to the Earth each year is about 2 km^2, enough to renew the entire oceanic crust in a little more than 100 million years. This is a very short interval in geological time, less than 3 percent of the age of the Earth. The present ocean basins thus turn out to be among the youngest features on our planet.

When two plates come together, one plate is often forced down beneath another in what is called a **subduction zone** (Figure 7.7b). Generally, the thicker continen-

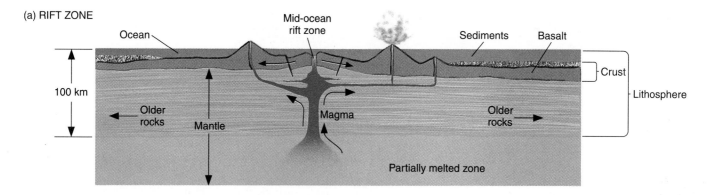

(a) RIFT ZONE

Ocean

Mid-ocean
rift zone

Sediments   Basalt

Crust

Lithosphere

100 km

Older
rocks

Mantle

Magma

Older
rocks

Partially melted zone

(b) SUBDUCTION ZONE

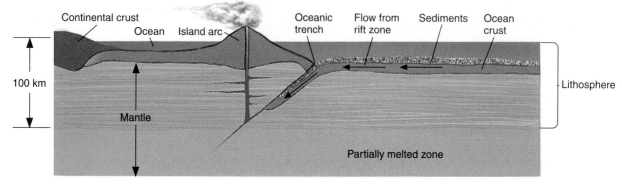

Continental crust

Ocean   Island arc

Oceanic
trench

Flow from
rift zone

Sediments   Ocean
crust

Lithosphere

100 km

Mantle

Partially melted zone

Figure 7.7
Rift zones (a) and subduction zones (b) in the Earth's crust.

tal masses cannot be subducted, but the thin oceanic
plates can be rather readily thrust down into the upper
mantle. Often a subduction zone is marked by an ocean
trench, a fine example being the deep Japan Trench along
the coast of Asia. The subducted plate is forced down into
regions of high pressure and temperature, eventually
melting several hundred kilometers below the surface. Its
material is recycled back into a downward-flowing convection current, ultimately balancing the flow of material that
rises along rift zones. The amount of crust destroyed at
subduction zones is approximately equal to the amount
formed at rift zones.

All along the subduction zone, earthquakes and volcanoes mark the death throes of the plate. Some of the most
destructive earthquakes in history have taken place along
subduction zones. These include the 1923 Yokohama
earthquake and fire, which killed 100,000 people, and the
1976 earthquake that leveled the Chinese city of Tang-
shan and resulted in nearly half a million deaths.

Fault Zones and Mountain Building

Along much of their lengths, the crustal plates slide parallel to each other. These plate boundaries are marked by
cracks or **faults.** Along active fault zones, the motion of
one plate with respect to the other is several centimeters
per year, about the same as the spreading rates along rifts.

One of the most famous faults is the San Andreas
Fault, lying on the boundary between the Pacific Plate

and the North American Plate. This fault runs from the
Gulf of California to the Pacific Ocean northwest of San
Francisco (Figure 7.8). The Pacific Plate, to the west, is
moving northward, carrying Los Angeles, San Diego, and

Figure 7.8
The San Andreas fault, a very active region where one crustal
plate is sliding sideways with respect to the other. The fault is
marked by the valley running up the center of the photo; the dark
line to the left is tumbleweeds piled along a fence. (USGS)

parts of the southern California coast with it. In several million years, Los Angeles will be an island off the coast of San Francisco.

Unfortunately for us, the motion along most fault zones does not take place smoothly. The creeping motion of the plates against each other builds up stresses in the crust that are released in sudden, violent slippages, generating earthquakes. Since the average motion of the plates is constant, the longer the interval between earthquakes, the greater the stress and the larger the energy released when the surface finally moves.

For example, the part of the San Andreas Fault near the central California town of Parkfield has slipped every 22 years or so during the past century, moving an average of about 1 m each time. In contrast, the average interval between major earthquakes in the Los Angeles region is about 140 years, and the average motion is about 7 m. The last time the San Andreas slipped in this area was in 1857; tension has been building ever since, and sometime soon it is bound to be released. Sensitive instruments placed within the Los Angeles Basin show that the basin is actually contracting in size as these tremendous pressures mount beneath the surface.

When two continental masses are brought together by the motion of the crustal plates, they are forced against each other under great pressure. The Earth buckles and folds, forcing some rock deep below the surface and raising other folds to heights of many kilometers. This is the way most of the mountain ranges on Earth were formed; the Alps, for example, are a result of the African plate's bumping into Europe. As we will see, however, quite different processes produced the mountains on other planets.

Once a mountain range is formed by upthrusting of the crust, its rocks are subject to the erosional force of water and ice. The sharp peaks and serrated edges characteristic of our most beautiful mountains (Figure 7.9) have little to do with the forces that initially make mountains. Instead they result from the processes that tear them down. Ice is an especially effective sculptor of rock. In a planet without moving ice or running water, mountains remain smooth and dull.

Volcanoes

Volcanoes mark locations where molten rock (called **magma**) rises to the surface. One example is the mid-ocean ridges, long undersea mountain ranges formed by hot material rising from the Earth's mantle at plate boundaries. A second major kind of volcanic activity is associated with subduction zones, and volcanoes sometimes also appear in regions where continental plates are colliding. In each case, the volcanic activity gives us a way to sample some of the material from deeper within our planet.

Another location for volcanic activity on our planet is found above so-called mantle hot spots, areas far from

Figure 7.9
The Alps, a young region of the Earth's crust where sharp mountain peaks are being sculpted by glaciers. (David Morrison)

plate boundaries where heat is nevertheless rising from the interior of the Earth. Perhaps the best-known such hot spot is under the island of Hawaii, where it currently supplies the magma to maintain three active volcanoes, two on land and one under the ocean (Figure 7.10). The Hawaiian hot spot has been active for at least 100 million years. As the Earth's plates have moved during that time, the hot spot has generated a 3500-km-long chain of volcanic islands. The tallest Hawaiian volcanoes are among the largest individual mountains on Earth, more than 100

Figure 7.10
Mauna Loa, a large volcano in Hawaii, during eruption. The stars of the Southern Cross are visible to the left of the volcanic plume. (Dale P. Cruikshank)

Alfred Wegener: Catching the Drift of Plate Tectonics

When studying maps or globes of the Earth, many students notice that the coast of North and South America, with only minor adjustments, could fit pretty well against the coast of Europe and Africa. It seems as if these great land masses could once have been together and then were somehow torn apart. The same idea had occurred to a number of scientists (including Francis Bacon as early as 1620). But not until the 20th century could such a proposal be more than speculation. The man who made the case for continental drift in 1912 was a German meteorologist and astronomer named Alfred Wegener.

Born in Berlin in 1880, Wegener was from an early age fascinated by Greenland, the world's largest island, which he dreamed of exploring. He studied at the universities in Heidelberg, Innsbruck, and Berlin, receiving a doctorate in astronomy by re-examining 13th-century astronomical tables. But his interests turned more and more toward the Earth, particularly its weather. He carried out experiments using kites and balloons, becoming so accomplished that he and his brother set a world record in 1906 by flying for 52 hours in a balloon.

Wegener first conceived of continental drift in 1910, while examining a world map in an atlas. But it took two years for him to assemble sufficient data to propose the idea in public; he published the results of his work in book form in 1915. Wegener's evidence went far beyond the congruence in the shapes of the continents. He proposed that the similarities between unusual fossils found only in certain parts of South America and Africa indicated that these two continents were together at one time. He also showed that resemblances among living animal species on different continents could best be explained by assuming that the continents were once

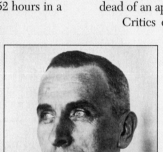

Alfred Wegener (1880–1930). (Historical Pictures Service, Chicago)

connected in a supercontinent he called Pangea (from Greek elements meaning "all land").

Wegener's suggestion first met with a hostile reaction from scientists. Although he had marshaled an impressive list of arguments for his theory, he was missing a *mechanism*: no one could explain *how* solid continents could drift over thousands of miles. A few scientists were sufficiently impressed by Wegener's work to continue searching for additional evidence, but many found the notion of moving continents too different from the accepted view of the Earth's surface to take seriously. Developing an understanding of the mechanism (plate tectonics) would take decades of further progress in geology, oceanography, and geophysics.

Wegener was disappointed in the reception to his suggestion, but continued his research on the weather, the causes of craters on the Moon, and the exploration of Greenland. In 1924 he was appointed to a special meteorology and geophysics professorship created especially for him at the University of Graz. Four years later, on his fourth expedition to his beloved Greenland, he celebrated his 50th birthday with colleagues and then set off towards a different camp on the island. He never made it; he was found a few days later, dead of an apparent heart attack.

Critics of science often point to the resistance to the continental drift hypothesis as an example of the flawed way that scientists regard new ideas. (Many people advancing crackpot theories have claimed that they are being unjustly ridiculed, just as Wegener was.) But we think there is a more positive light in which to view the story of Wegener's suggestion. Scientists in his day maintained a skeptical attitude because they needed more evidence and a clear mechanism that would fit what they understood about nature. Once the evidence and the mechanism were clear, Wegener's hypothesis quickly became the centerpiece of our new view of a dynamic Earth.

---

km in diameter and rising 9 km above the ocean floor. As we saw in Chapter 5, one of these volcanic mountains, the now-dormant Mauna Kea, has become one of the world's great sites for doing astronomy.

Not all volcanic eruptions produce mountains. If the lava is very fluid and flows rapidly from long cracks, it spreads out to form lava plains. The largest known terrestrial eruptions, such as those that produced the Snake River Basalts in the northwestern United States, or the Deccan Plains in India, are of this type. As we will see later, similar lava plains are found on the Moon and the other terrestrial planets.

7.3

The Earth's Atmosphere

We live at the bottom of the ocean of air that envelops our planet. The atmosphere, weighing down upon the Earth's surface under the force of gravity, exerts a pressure at sea level which scientists define as 1 **bar.** A bar of pressure means that each square centimeter of the Earth's surface has a weight of 1.03 kg pressing down on it. Humans have evolved to exist best at this pressure; make the pressure a lot lower or higher and we do not function well at all.

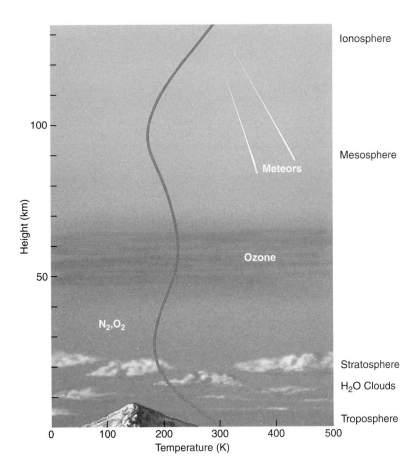

Figure 7.11

The structure of the Earth's atmosphere. Height increases up the left side of the diagram, and the names of the different layers are shown at the right. The curving red line shows the temperature; see scale at bottom of figure.

The total mass of our atmosphere is about 5×10^{18} kg; while this sounds like a big number, it is only about a millionth of the total mass of the Earth. The atmosphere represents a smaller fraction of the Earth than the fraction of your mass represented by the hair on your head.

Structure of the Atmosphere

The structure of the atmosphere is illustrated in Figure 7.11. Most of the atmosphere is concentrated near the surface of the Earth, within about the bottom 10 km where clouds form and airplanes fly. Within this region, called the **troposphere,** warm air, heated by the surface, rises and is replaced by descending currents of cooler air—another example of convection. This circulation generates clouds and other manifestations of weather. Within the troposphere, temperature drops rapidly with increasing elevation to values near 50°C below freezing at its upper boundary, where the **stratosphere** begins. Most of the stratosphere, which extends to about 80 km above the surface, is cold and free of clouds.

Near the top of the stratosphere is a layer of **ozone** **(O_3),** a heavy form of oxygen with three atoms per molecule instead of the usual two. Since ozone is a good absorber of ultraviolet light, it protects the surface from some of the Sun's dangerous ultraviolet radiation, making it possible for life to exist on land. Because ozone is essential to our survival, we react with justifiable concern to

evidence that atmospheric ozone is being destroyed. Just such evidence has been accumulating; one especially important agent of this destruction is a series of industrial chemicals called CFCs (chlorofluorocarbons).

Each year a large ozone hole forms above the Antarctic continent, and by now the ozone loss has progressed into the temperate zones. The production of CFCs has been banned by international agreement, but these chemicals are destroyed so slowly that we can expect increasing loss of ozone well into the next century.

At heights above 100 km, the atmosphere is so thin that orbiting satellites can pass right through it with very little friction. At these elevations, individual atoms can occasionally escape completely from the gravitational field of the Earth. Thus there is a continuous, slow leaking of atmosphere—especially of lightweight atoms, which move faster than heavy ones. The Earth's atmosphere cannot, for example, hold on to hydrogen or helium, which escape into space.

Atmospheric Composition and Origin

At the Earth's surface, the atmosphere consists of 78 percent nitrogen (N_2), 21 percent oxygen (O_2), and 1 percent argon (A), with traces of water vapor (H_2O), carbon dioxide (CO_2), and other gases. Variable amounts of dust particles and water droplets are also found suspended in the air.

Figure 7.12
A large tropical storm marked by clouds swirling around a low-pressure region, photographed by astronauts orbiting the Earth. (NASA)

A complete census of the Earth's volatile materials, however, should look at more than the gas now present. Volatile materials are those that evaporate at a relatively low temperature. If the Earth were just a little bit warmer, some materials that are now liquid or solid might become part of the atmosphere. Suppose, for example, that our planet were heated to above the boiling point of water (100°C, or 373 K); that's a large change for humans, but not a very large change compared to the range of possible temperatures in the universe. At 100°C the oceans would boil, and this water vapor would become a part of the atmosphere.

To estimate how much water vapor would be released, we note that there is enough water to cover the entire Earth to a depth of about 3000 m. Since the pressure exerted by 10 m of water is about equal to 1 bar, the average pressure at the ocean floor is about 300 bars. Since water weighs the same whether in liquid or vapor form, if the oceans boiled away the atmospheric pressure of the water would still be 300 bars. Water would therefore greatly dominate the Earth's atmosphere, with nitrogen and oxygen reduced to the status of trace constituents.

On a warmer Earth another source of additional atmosphere would be found in the sedimentary carbonate rocks of the crust. These are minerals that contain abundant carbon dioxide, which if released by heating would generate about 70 bars of CO_2, far more than the current CO_2 pressure of only 0.0005 bar. Thus the atmosphere of a warm Earth would be dominated by water vapor and carbon dioxide, with a surface pressure close to 400 bars.

There are several lines of evidence that convince scientists that the composition of the Earth's atmosphere today is not the same as it was originally. We examine this question in more detail in the next section. For now, just note that we do not know the source of the Earth's original atmosphere. Today we see that CO_2, H_2O, sulfur dioxide (SO_2), and other gases are released from deeper within the Earth through the action of volcanoes. Much of this apparently new gas, however, is probably recycled material that has been subducted through plate tectonics.

Three possibilities exist for the original source of the Earth's atmosphere and oceans (since the early Earth was warmer, we can think of today's oceans as liquified early atmosphere, as just discussed): (1) the atmosphere could have been formed with the rest of the Earth, as it accumulated from the debris left over from the formation of the Sun; (2) it could have been released from the interior through volcanic activity, subsequent to the formation of the Earth; or (3) the atmosphere may have been derived from impacts by comets or other icy materials from the outer parts of the solar system. Current evidence favors the comet hypothesis, but all three mechanisms may have contributed.

Weather and Climate

All planets with atmospheres have *weather*, which is just a name we give to the circulation of the atmosphere. The energy that powers the weather is derived primarily from the sunlight that heats the surface. As the planet rotates, and as the slower seasonal changes cause variations in the amount of sunlight striking different parts of the Earth, the atmosphere and oceans try to redistribute the heat from warmer to cooler areas. Weather on any planet represents the response of its atmosphere to changing inputs of energy from the Sun (Figure 7.12).

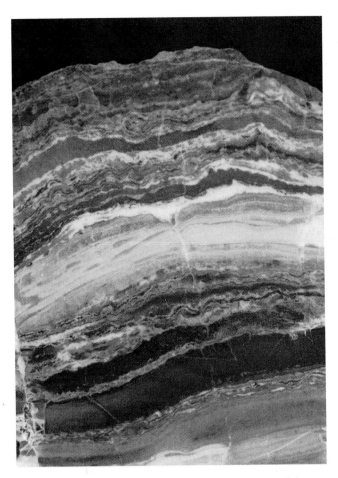

Figure 7.13
Cross section of stromatolites, both fossil and contemporary. The layered, dome-like structures are mats of sediment trapped in shallow waters by large numbers of blue-green bacteria that can photosynthesize. Such colonies of microorganisms date back more than 3 billion years. (NASA/ARC, courtesy of David DesMarais)

7.4

Life and Chemical Evolution

Earth is the only inhabited planet in the solar system, as we know from the investigation of other planets by spacecraft. The origin and development of life is an important part of the story of our planet. Life arose early in Earth's history, and living organisms have been interacting with their environment for billions of years. We recognize that life forms have evolved to adapt themselves to the environment on Earth, and we are now beginning to realize that the Earth itself has been changed in important ways by the presence of living matter.

The Origin of Life

The record of the birth of life on Earth has been lost in the restless motions of the crust. By the time the oldest surviving rocks were formed, about 3.8 billion years ago, life already existed. At 3.5 billion years ago, life had achieved the sophistication to build large colonies called stromatolites (Figure 7.13), a form so successful that stromatolites still grow on Earth today. But little crust survives from these ancient times, and abundant fossils have been produced only during the past 600 million years—less than 15 percent of our planet's history.

Climate is a term used to refer to the effects of the atmosphere through decades and centuries. Changes in climate (as opposed to the random variations in weather from one year to the next) are often difficult to detect over short time periods, but as they accumulate their effect can be devastating. Modern farming is especially sensitive to temperature and rainfall; for example, calculations indicate that a drop of only 2°C throughout the growing season would cut the wheat production by half in Canada and the United States.

The best-documented changes in the Earth's climate are the great ice ages, which have periodically lowered the temperature of the Northern Hemisphere over the past million years or so. The last ice age, which ended about 14,000 years ago, lasted some 20,000 years. At its height, the ice was almost 2 km thick over Boston and stretched as far south as New York City. Today we are in a relatively warm period, interpreted by many scientists as a fairly short-lived interglacial interval between major ice ages.

Scientists think that the ice ages are primarily the result of changes in the tilt of the Earth's rotational axis, produced by the gravitational effects of the other planets. This idea of an astronomical cause of climate changes was first proposed in 1920 by the Serbian scientist Milutin Milankovich. As we will see in Chapter 9, there is also evidence of periodic climate changes on Mars, and modern calculations suggest that these also have their origin in slow changes of that planet's orbit and rotational axis.

Any theory of the origin of life must therefore be partly speculative, since there is little direct evidence to go on. All we really know is that the atmosphere of the early Earth, unlike today, contained abundant carbon dioxide but no oxygen gas. In the absence of oxygen, many complex chemical reactions are possible that lead to the production of amino acids, proteins, and other chemical building blocks of life. It now seems relatively certain that life arose from these building blocks very early in Earth's history.

For tens of millions of years after its formation, life (perhaps little more than large molecules, like the viruses of today) probably existed in warm, nutrient-rich seas, living off accumulated organic chemicals. Eventually, however, as this easily accessible food became depleted, life began the long evolutionary road that led to the vast numbers of different organisms on Earth today. As it did so, life began to influence the chemical evolution of the atmosphere.

Evolution of the Atmosphere

One of the key steps in the evolution of life on Earth was the development of plants, a very successful life form that takes in carbon dioxide from the environment and releases oxygen as a waste product. Since the energy for making new plant material from chemical building blocks comes from sunlight, we call the process *photosynthesis.*

Studies of the chemistry of ancient rocks show that Earth's atmosphere lacked free oxygen until about two billion years ago, in spite of the presence of plants releasing oxygen by photosynthesis. Apparently chemical reactions with the crust removed the oxygen gas as quickly as it formed. Slowly, however, the increasing evolutionary sophistication of life led to a growth in plant population and thus increased oxygen production. At the same time, increased geological activity resulted in heavy erosion that buried most of the plant carbon before it could recombine with oxygen to form CO_2.

Free oxygen began accumulating in the atmosphere about two billion years ago, and the increased amount of this gas led to the formation of the Earth's ozone layer, which protects the surface from lethal solar ultraviolet light. Before that, it was unthinkable for life to venture outside the protective oceans, and so the land masses of Earth were barren. The presence of oxygen and hence ozone thus allowed the colonization of the land. It also made possible a tremendous proliferation of animals, who lived by taking in and using the organic materials produced by plants as their own energy source.

As animals evolved in an environment increasingly rich in oxygen, they were able to develop techniques for breathing oxygen directly from the atmosphere. We humans take it for granted that plenty of free oxygen is available in the Earth's atmosphere to release energy from the food we take in. Although it may seem funny to think of it

this way, we are life forms that have evolved to breathe in the waste product of plants.

On a planetary scale, one of the most important consequences of life has been a decrease in atmospheric carbon dioxide. In the absence of life, Earth would probably have a much more massive atmosphere dominated by CO_2 like that of the planet Venus. But life in combination with high levels of geological activity has effectively stripped us of most of this gas.

The Greenhouse Effect and Global Warming

We have a special interest in the carbon dioxide content of the atmosphere because of the role this gas plays in retaining heat from the Sun through a process called the **greenhouse effect.** To understand how the greenhouse effect works, consider the fate of the sunlight that strikes the surface of the Earth. It is absorbed, heats the surface layers, and is then re-emitted as infrared or heat radiation (Figure 7.14). However, the CO_2 in our atmosphere, which is transparent to visible light, is largely opaque to infrared energy. As a result, it acts like a blanket, trapping the heat in the atmosphere and impeding its flow back to space. Over time this extra heat means that the Earth's surface regions get warmer than they would have without the effect of the CO_2. The more CO_2 in our atmosphere, the higher the temperature at which the Earth's surface reaches a new balance.

The greenhouse effect in a planetary atmosphere is similar to the heating of a gardener's greenhouse, or the inside of a car left out in the Sun with the windows rolled

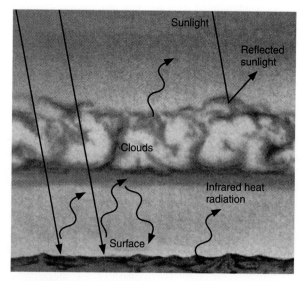

Figure 7.14

How the greenhouse effect works. Sunlight that penetrates to the Earth's lower atmosphere and surface is reradiated as infrared or heat radiation, which is trapped by the atmosphere. The result is a higher surface temperature for our planet.

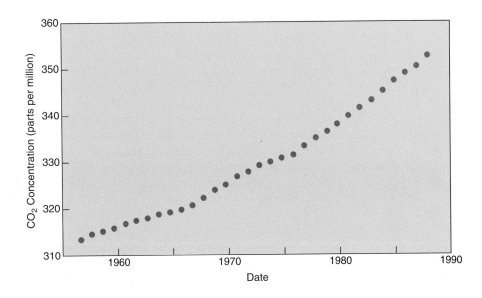

Figure 7.15
The increase of atmospheric carbon dioxide over time. The amount of CO_2 is expected to double by the middle of the 21st century. (Adapted from data obtained at the Mauna Loa Observatory of NOAA)

up. In these examples, the window glass plays the role of carbon dioxide, letting sunlight in but reducing the outward flow of heat radiation. As a result, a greenhouse or car interior winds up much hotter than would be expected from the heating of sunlight alone. On Earth, the current greenhouse effect elevates the surface temperature by about 23°C. Without this greenhouse effect, the average surface temperature would be well below the freezing point, and the Earth would be locked in a global ice age.

That's the good news; the bad news is that the greenhouse effect is increasing. Modern industrial society depends on energy extracted from burning fossil fuels. As these ancient coal and oil deposits are oxidized, additional carbon dioxide is released into the atmosphere. The problem is increased by the widespread destruction of tropical forests, which we depend on to extract CO_2 from the atmosphere and replenish our supply of O_2. So far in this century, the amount of CO_2 in the atmosphere has increased by about 25 percent, and it is continuing to rise at 0.5 percent per year.

By early in the 21st century, the CO_2 level is predicted to reach twice the value it had before the industrial revolution (Figure 7.15). The consequences of such an increase for the Earth's surface and atmosphere (and the creatures who live there) are likely to be complex. Many groups of scientists are now studying the effects of such *global warming* with elaborate computer models. At the present time we are not sure what all the effects will be, since the Earth's surface is an extremely complicated system, and the specific results of increased temperature will surely vary from one location to another.

Already some global warming is apparent. In North America and Europe, summer temperatures throughout the 1980s and 1990s reached record highs. Because they are superimposed on the usual year-to-year fluctuations in weather, these effects are not easy to sort out, but most scientists are convinced that global warming due to an enhanced greenhouse effect is a reality. The increasing

greenhouse effect comes at a particularly difficult time: our planet is already in an unusually warm interglacial period. We are rapidly entering unknown territory where human activities are contributing to the highest temperatures on Earth in more than 50 million years.

7.5

Cosmic Influences on the Evolution of Earth

Where Are the Craters on Earth?

In discussing the Earth's geology in Section 7.2, we dealt only with the effects of internal forces, expressed through the processes of plate tectonics and volcanism. In contrast, on the Moon we see primarily craters, produced by the impacts of interplanetary debris (as described in Chapter 6). Why do we not see more evidence on Earth of the kinds of impact craters that are so prominent on the Moon and other planets?

It is not possible that the Earth escaped being struck by the interplanetary debris that has pockmarked the Moon. From a cosmic perspective, the Moon is almost next door, and it is unlikely that if our closest neighbor was bombarded by projectiles from space over vast periods of time, we could somehow have been immune. Our atmosphere does make small pieces of cosmic debris burn up (which we see as meteors—commonly called shooting stars). But the layers of air provide no shield against the large impacts that form craters several kilometers or more in diameter; nor is there any other known way to avoid these cosmic events.

In the course of its history, the Earth must therefore have been cratered as heavily as the Moon. The difference is that on the Earth these craters are destroyed by our active geology before they can accumulate. As plate tectonics constantly renews our crust, evidence of past

Figure 7.16
Aftermath of the Tunguska explosion. This photograph, taken 21 years after the blast, shows a part of the forest that was devastated by the 15-megaton explosion. (Novosty)

cratering events is destroyed. Only in the past few decades have geologists succeeded in identifying the eroded remnants of many old impact craters. Even more recent is our realization that these impacts may have had an important influence on the evolution of life.

Recent Impacts

The collision of interplanetary debris with the Earth is not a hypothetical idea: evidence of a number of relatively recent impacts can be found on our planet's surface. The best-studied historic collision took place on June 30, 1908, near the Tunguska River in Siberia. In this desolate region, a remarkable explosion took place in the atmosphere about 8 km above the surface. The shock wave flattened more than 1000 km² of forest. Herds of reindeer and other animals were killed, and a man at a trading post 80 km from the blast was thrown from his chair and knocked unconscious (Figure 7.16). The blast wave spread around the world, recorded by instruments designed to measure changes in atmospheric pressure.

Despite this violence, no craters were formed by the Tunguska explosion. Shattered by atmospheric pressure, a stony projectile with a mass of approximately 100,000 tons disintegrated to create a blast equivalent to a 15-megaton nuclear bomb[1]. Had it been smaller or more fragile, the impacting body would have dissipated its energy at high

[1] A megaton is the explosive energy released by a million tons of TNT— not a million pounds or kilograms, but a million tons!

altitude and probably attracted no attention. If it had been larger or made of stronger material (such as metal), it would have penetrated all the way to the surface and formed a crater. Instead, only the shock of the explosion reached the surface, but the devastation it left behind in Siberia bore witness to the power of such impacts. Imagine if the same rocky impactor had exploded over New York City in 1908; history books might today record it as one of the most deadly events in human history.

The most recent impact to produce a substantial crater took place 50,000 years ago in Arizona. The projectile in this case was a lump of iron about 50 m in diameter (we will discuss how the explosion of a small chunk of material can make a much bigger crater in Chapter 8). The crater, called Meteor Crater and now a major tourist attraction, is about a mile across and has all of the features associated with similar-sized lunar impact craters (Figure 7.17). Meteor Crater is one of the few impact features on the Earth that remain relatively intact; other, older craters are so eroded that only a trained eye can distinguish them. Nevertheless, more than 150 have been identified as further evidence that impacts continue to happen to our planet over the millennia.

Extinction of the Dinosaurs

The impact that produced Meteor Crater would have been dramatic indeed to any humans who witnessed it, since the energy release was also equivalent to a 15-megaton nuclear bomb. But such explosions are devastating only in their local areas; they have no *global* consequences. Much larger (and rarer) impacts, however, can disturb the ecological balance of the entire planet and thus have a major influence on the course of evolution.

The best-documented large impact took place 65 million years ago, at the end of what is now called the Cretaceous era of geological history. This point in the history of life on Earth was marked by a **mass extinction** in which more than half of the species on our planet died out. While there are a dozen or more mass extinctions in the geological record, this particular event has always intrigued paleontologists because it marks the end of the dinosaur age. For tens of millions of years these great creatures had flourished. Then they suddenly disappeared (along with many other species), and thereafter the mammals began the development and diversification that ultimately led to the readers (and authors) of this book.

The object that collided with the Earth at the end of the Cretaceous era struck what is now the Yucatan peninsula of Mexico. Its mass must have been more than a trillion tons (Figure 7.18), because scientists have found a worldwide layer of sediment deposited from the dust cloud that enveloped the planet after its impact. First identified in 1980, this sediment layer is rich in the rare metal iridium and other elements that are relatively abundant in asteroids and comets but very rare in the Earth's crust. Even diluted by the terrestrial material excavated

Figure 7.17
Meteor Crater in Arizona, a 50,000-year-old impact scar. While impact craters are common on less-active bodies such as the Moon, this is one of the very few well-preserved craters on the Earth. (Meteor Crater, Northern Arizona)

by the explosion from the crater it made, this cosmic component is easily identified in layers dating back 65 million years around the globe. In addition, the sediment contains many minerals characteristic of the temperatures and pressures of a gigantic explosion.

The impact released energy equivalent to five billion Hiroshima-sized nuclear bombs, excavating a crater 200 km across and deep enough to penetrate through the Earth's crust. This large crater, named Chicxulub for a small town near its center, has subsequently been buried in sediment, but its outlines can still be identified (Figure 7.19). The explosion that created the Chicxulub crater lifted about *100 trillion tons* of dust into the atmosphere, determined by measuring the thickness of the sediment layer that formed when this dust settled to the surface.

Such a quantity of airborne material would have blocked sunlight completely, plunging the Earth into a period of cold and darkness that lasted several months. Other worldwide effects included large-scale fires (started by the hot debris from the explosion) that destroyed most of the planet's forests and grasslands, and a long period in which rainwater around the globe was acidic. Presumably these environmental effects, rather than the explosion itself, were responsible for the mass extinction, including the death of the dinosaurs.

A demonstration of such effects took place in July 1994, when pieces of a comet called Shoemaker-Levy 9 impacted the planet Jupiter. Although Jupiter has no solid

Figure 7.18
Artist's impression of the impact of a 20-km asteroid with the Earth. This is the approximate magnitude of the event that led to mass extinctions 65 million years ago. (Painting by Don Davis)

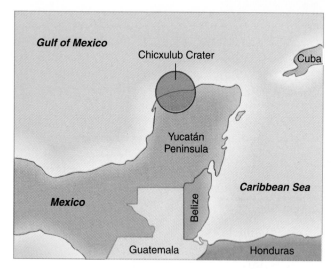

Figure 7.19
Map showing the location of the buried crater from the impact event 65 million years ago on Mexico's Yucatan peninsula. (NASA)

surface, the comet pieces themselves contained a great deal of dusty material. The comet fragments exploded as they penetrated the giant planet's atmosphere, and vast clouds of dark material were seen to expand over great distances (see Chapter 12). In Figure 17.20 a computer graphic shows what it might have looked like if the largest fragment of the comet (roughly 1 km across) had struck Detroit instead.

Impacts and the Evolution of Life

Several other mass extinctions in the geological record have been tentatively identified with large impacts, although none is so dramatic as the blast that destroyed the dinosaurs. But even without such specific documentation, it is now clear that impacts of this size do occur on all planets, and that their effects can be catastrophic for life. A catastrophe for one group of living things, however, may create opportunities for another group. Following each

Figure 7.20
Computer simulation by astronomer John Spencer at the Lowell Observatory, showing what the Earth might have looked like 2 hours after impact if the largest fragment of Comet Shoemaker-Levy 9 had hit near Detroit, Michigan. A photograph of the debris cloud from the fragment explosion on Jupiter was superimposed onto a global view of the Earth. If the roughly 1-km-wide comet fragment had actually hit the Earth, the debris would have included a mixture of vaporized Earth solids, as well as the material from the comet. Nevertheless, this image makes clear why large impacts can cause planet-wide effects.

mass extinction, there is a sudden evolutionary burst as new species develop to fill the ecological niches opened by the event.

Impacts by comets and asteroids represent the only mechanisms we know of that could cause truly global catastrophes and seriously influence the evolution of life all over the planet. According to some estimates, the *majority* of all species extinctions may be due to such impacts. As paleontologist Stephen Jay Gould of Harvard has noted, such a perspective fundamentally changes our view of biological evolution. The central issues for the survival of a species must now include more than just its success in competing with other species and adapting to slowly changing environments, as envisioned by Darwinian natural selection. Of at least equal importance is its ability to survive random global ecological catastrophes due to impacts.

Still earlier in its history, the Earth was subject to even larger impacts from the leftover debris of planet formation. We know the Moon was repeatedly struck by objects more than 100 km in diameter, a thousand times more massive than the object that wiped out most terrestrial life 65 million years ago. The Earth must have experienced similar large impacts during its first 700 million years of existence. Some of them were probably violent enough to strip the planet of most of its atmosphere and to boil away its oceans. Such events would sterilize the planet, utterly destroying any life that had begun. Life may have formed and been wiped out several times before our own ancestors took hold sometime about 4 billion years ago.

Today, then, we think of the Earth as a target in a cosmic shooting gallery, subject to random violent events that were unsuspected a few decades ago. We owe our existence to such events. Had the impact 65 million years ago not redirected the course of evolution, mammals might never have become the dominant large animals they are today (and intelligent dinosaur students might be reading this text). But scientists have begun to ask an even more fundamental question. If impacts with comets and asteroids had not occurred throughout our planet's history, could biological evolution in more stable environments have produced the wondrous diversity of life that populates the Earth today?

Summary

7.1 The Earth is the prototype terrestrial planet. Its interior composition and structure are probed using **seismic waves.** Such studies reveal that we have a metal **core** and silicate **mantle.** The outer layer or **crust** consists primarily of oceanic **basalt** and continental **granite.** A global magnetic field, generated in the core, produces the Earth's **magnetosphere,** which can trap charged atomic particles.

7.2 Terrestrial rocks can be classified as **igneous, sedimentary,** or **metamorphic.** A fourth type, **primitive rock,** is not found on the Earth. Our planet's geology is dominated by **plate tectonics,** in which crustal plates move slowly in response to mantle **convection.** The surface expression of plate tectonics includes continental drift, recycling of the ocean floor, mountain building, **rift zones, subduction zones, faults,** earthquakes, and volcanic eruptions of **magma** from the interior.

7.3 The atmosphere has a surface pressure of 1 **bar** and is composed primarily of N_2 and O_2, plus such important trace gases as H_2O, CO_2, and **ozone (O_3).** Its structure consists of the **troposphere,** the **stratosphere,** and tenuous higher regions. Atmospheric oxygen is the product of life forms on Earth, especially plants. Atmospheric circulation (weather) is driven by seasonally changing deposition of sunlight. Longer-term climatic variations, such as the ice ages, are probably due to changes in the planet's orbit and axial tilt.

7.4 Life originated on Earth at a time when the atmosphere lacked O_2 and consisted mostly of CO_2. Later photosynthesis gave rise to free oxygen and ozone. Carbon dioxide in the atmosphere heats the surface through the **greenhouse effect;** today, increasing atmospheric CO_2 is leading to the global warming of our planet.

7.5 Earth, like the Moon and other planets, has been influenced by the impacts of cosmic debris, including such small recent examples as Meteor Crater and the Tunguska explosion. Larger past impacts are responsible for at least some **mass extinctions,** including the large extinction 65 million years ago that ended the Cretaceous period. Such impacts have probably played an important role in the evolution of life.

Review Questions

1. What are the Earth's core and mantle made of? Explain how we know.

2. Describe the differences among primitive, igneous, sedimentary, and metamorphic rock, and relate these differences to their origins.

3. Explain briefly how the following phenomena happen on Earth, relating your answers to the theory of plate tectonics:
  a. earthquakes
  b. continental drift
  c. mountain building
  d. volcanic eruptions
  e. creation of the Hawaiian island chain

4. Describe three ways in which the presence of life has affected the composition of the Earth's atmosphere.

5. How do impacts by comets and asteroids influence the Earth's geology, its atmosphere, and the evolution of life?

6. Why are there so many impact craters on our neighbor world, the Moon, and so few on the Earth?

Thought Questions

7. If you wanted to live where the chances of a destructive earthquake were small, would you pick a location near a fault zone, near a mid-ocean ridge, near a subduction zone, or on a volcanic island such as Hawaii? What are the relative risks of earthquakes at each of these locations?

8. If all life were destroyed on Earth, would new life eventually form to take its place? Explain how conditions would have to change for life to start again on our planet.

9. Why will a decrease in the Earth's ozone be harmful to life?

10. Why are we concerned about the increases in CO_2 and other gases that cause the greenhouse effect in the Earth's atmosphere? What steps could we take in the future to reduce the levels of CO_2 in our atmosphere? What factors stand in the way of taking the steps you suggest (you may include technological, economic, and political factors in your answer).

11. Is there evidence of climate change in your area over the past century? How would you distinguish a true climate change from the random variations in weather that take place from one year to the next?

Problems

12. How might the history of the Earth have been different if it were closer to the Sun? Farther from the Sun?

13. What fractions of the volume of the Earth are occupied by the core, the mantle, and the crust?

14. Suppose that the next slippage along the San Andreas Fault in southern California takes place in the year 2000, and that it completely relieves the accumulated strain in this region. How much slippage is required for this to occur?

15. Measurements using Earth satellites have shown that Europe and North America are moving apart by 4 m per century due to plate tectonics. As the continents separate, new ocean floor is created along the Mid-Atlantic Rift. If the rift is 5000 km long, what is the total area of new ocean floor created in the Atlantic each century? Each year?

16. Over the entire Earth there are 60,000 km of active rift zones, with average separation rates of 4 m per century. How much area of new ocean crust is created each year over the entire planet? (This area is approximately equal to the amount of ocean crust that is subducted, since the total area of the oceans remains about the same.)

17. With the information from Problem 16, you can calculate the average age of the ocean floor. First find the total area of ocean floor (equal to about 60 percent of the surface area of the Earth). Then compare this with the area created (or destroyed) each year. The average lifetime is simply the ratio of these numbers—the total area of ocean crust compared to the amount created (or destroyed) each year.

18. What is the volume of new oceanic basalt added to the Earth's crust each year? Assume that the thickness of the

new crust is 5 km, that there are 60,000 km of active rifts, and that the average speed of plate motion is 4 cm per year. What fraction of the Earth's entire volume does this annual addition of new material represent?

19. The sea-level pressure of the atmosphere (1 bar) corresponds to 10^4 kg of mass above each square meter of the Earth's surface. Calculate the total mass of the atmosphere in kilograms and in tons (1 ton equals 1000 kg). Then compute the total mass of ocean, given that the oceans would be 3000 m deep if water covered the globe uniformly. (Note: 1 m^3 of water has a mass of 1 ton.) Compare the mass of atmosphere and the mass of the oceans with the total mass of the Earth to determine what percentage of our planet is represented by the atmosphere and oceans.

20. Suppose a major impact that produces a mass extinction takes place on the Earth once every 5 million years. Further suppose that if such an event occurred today, you and most other humans would be killed (this would be true even if the human species as a whole survived). Such impact events are random, and one could take place at any time. Calculate the probability that such an impact will occur within the next 50 years (within your lifetime). This is equal to the probability that you will be killed by this means, rather than dying from an auto accident or heart disease or some other "natural" cause. How do the risks of dying from the impact of an asteroid or comet compare with other risks we are concerned about? (*Hint:* To find the risk, go to the library and look up the annual number of deaths from a particular cause in a particular country, and then divide by the population of that country.)

Suggestions for Further Reading

Allegre, C. and Schneider, S. "The Evolution of the Earth" in *Scientific American,* Oct. 1994, p. 66.

Broadhurst, L. "Earth's Atmosphere: Terrestrial or Extraterrestrial" in *Astronomy,* Jan. 1992, p. 38.

Cattermole, P. and Moore, P. *The Story of the Earth.* 1985, Cambridge U. Press.

Hartmann, W. "Piecing Together Earth's Early History" in *Astronomy,* June 1989, p. 24.

Hartmann, W. and Miller, R. *The History of the Earth.* 1993, Workman. Lavishly illustrated chronicle of our planet's evolution.

Heppenheimer, T. "Journey to the Center of the Earth" in *Discover,* Nov. 1987, p. 86.

Jones, P. and Wigley, T. "Global Warming Trends" in *Scientific American,* Aug. 1990, p. 84. Three hundred years of data analyzed.

Lewis, J. *Rain of Fire and Ice: The Very Real Threat of Comet and Asteroid Bombardment.* 1995, Addison-Wesley. Popular overview of impacts.

Murphy, J. and Nance, R. "Mountain Belts and the Supercontinent Cycle" in *Scientific American,* Apr. 1992, p. 84. There has been more than one supercontinent through geologic time.

Orgel, L. "The Origin of Life on Earth" in *Scientific American,* Oct. 1994, p. 77.

Schneider, S. and Londer, R. *The Coevolution of Climate and Life.* 1984, Sierra Club Books.

Weiner, J. *Planet Earth.* 1986, Bantam. From the fine public TV series.

Apollo 11 astronaut Buzz Aldrin on the surface of the Moon. Because there is no atmosphere, ocean, or geological activity on the Moon today, the footprints you see in this image will likely be preserved in the lunar soil for millions of years. (NASA)

CHAPTER 8

Cratered Worlds: The Moon and Mercury

Thinking Ahead

The Moon is the only other world human beings have ever visited. What is it like to stand on the surface of our natural satellite? And what can we learn from going there and bringing home pieces of a different world?

We begin our discussion of the planets as worlds with two relatively simple objects: the Moon and Mercury. Unlike the Earth, the Moon is geologically dead, a place that has exhausted its internal energy sources. Because its airless surface preserves the events that happened long ago, the Moon provides us with a window on the earlier epochs of solar system history. The planet Mercury is in many ways similar to the Moon, which is why the two are discussed together. Both are relatively small, and are lacking in atmospheres, deficient in geological activity, and dominated by the effects of impact cratering from outside. Still, the processes that have molded their surfaces have acted on many other members of the planetary system as well.

This is one small step for a man, one giant leap for mankind.

What astronaut Neil Armstrong *meant* to say when he became the first human being to walk on another world. (In his nervousness, he left out the word "a," making the statement a bit less effective.)

Figure 8.1

that faces the Earth with telescopic resolution of about 1 km, but lunar geology hardly existed as a scientific subject. All that changed beginning in the early 1960s.

Initially the Russians took the lead in lunar exploration with Luna 3, which returned the first photos of the lunar far side in 1959, and then with Luna 9, which landed on the surface in 1966 and transmitted pictures and other data to Earth. However, these efforts were overshadowed on July 20, 1969, when the first American astronaut set foot on the Moon.

Table 8.2 summarizes the nine Apollo flights—six that landed and three others that circled the Moon but did not land. The initial landings were on flat plains selected out of safety considerations, but with increasing experience and confidence, NASA targeted the last three missions to more geologically interesting locales. The level of scientific exploration also increased with each mission, as the astronauts spent longer times on the Moon and carried more elaborate equipment. Finally, on the last Apollo Moon landing, NASA included one scientist, geologist Jack Schmitt, among the astronauts (Figure 8.2).

In addition to landing on the lunar surface and studying it at close hand, the Apollo missions accomplished three objectives of major importance for lunar science. First, the astronauts collected nearly 400 kg of samples for detailed laboratory analysis on Earth (Figure 8.3). These

8.1

General Properties of the Moon

Some Basic Facts

The Moon (Figure 8.1) has only 1/80 the mass of the Earth and about 1/6 of the Earth's surface gravity—too low to retain an atmosphere. If, early in its history, the Moon expelled an atmosphere from its hot interior or collected a temporary envelope of gases from impacting comets, such an atmosphere was lost before it could leave any recognizable evidence of its short existence. Any sign of water is similarly absent. Indeed, the Moon is dramatically deficient in a wide range of *volatiles,* those elements and compounds that evaporate at relatively low temperatures. Some of the Moon's properties are summarized in Table 8.1, along with comparative values for Mercury.

Exploration of the Moon

Most of what we know about the Moon today derives from the U.S. Apollo program, which sent nine spacecraft with people on board to our satellite between 1968 and 1972, landing 12 astronauts on its surface (see photograph that opens this chapter). Before the era of spacecraft studies, astronomers had mapped the side of the Moon

TABLE 8.1
Properties of the Moon and Mercury

	Moon	Mercury
Mass (Earth = 1)	0.0123	0.055
Diameter (km)	3476	4878
Density (g/cm³)	3.3	5.4
Surface gravity (Earth = 1)	0.17	0.38
Escape velocity (km/s)	2.4	4.3
Rotation period (days)	27.3	58.65
Surface area (Earth = 1)	0.27	0.38

TABLE 8.2
Apollo Flights to the Moon

Flight	Date	Landing Site	Main Accomplishment
Apollo 8	Dec. 1968	—	First humans to fly around the Moon.
Apollo 10	May 1969	—	First rendezvous in lunar orbit.
Apollo 11	July 1969	Mare Tranquillitatis	First human landing on the Moon; 22 kg of samples returned.
Apollo 12	Nov. 1969	Oceanus Procellarum	First ALSEP; visit to Surveyor 3.
Apollo 13	Apr.1970	—	Landing aborted due to explosion in Command Module.
Apollo 14	Jan. 1971	Mare Nubium	First "rickshaw" on the Moon.
Apollo 15	July 1971	Imbrium/Hadley	First "rover"; visit to Hadley Rille; astronauts traveled 24 km.
Apollo 16	Apr. 1972	Descartes	First landing in highlands; 95 kg of samples returned.
Apollo 17	Dec. 1972	Taurus Mountains	Geologist among the crew; 111 kg of samples returned.

samples, still being studied, have probably revealed more about the Moon and its history than all other lunar studies combined. Second, each Apollo landing after the first one deployed an Apollo Lunar Surface Experiment Package (ALSEP), which continued to operate for years after the astronauts departed. (The ALSEPs were turned off by NASA in 1978 as a cost-cutting measure.) Third, the orbiting Apollo Command Modules carried a wide range of instruments to photograph and analyze the lunar surface from above.

The last human left the Moon in December 1972, just a little more than three years after Neil Armstrong took his "giant leap for mankind." The program of lunar exploration was cut off in midstride due to political and economic pressures. It had cost just about $100 per American, spread over ten years—the equivalent of one large pizza per person per year. Yet for many people, the Moon landings were one of the central events in 20th-century history.

The giant Apollo rockets built to travel to the Moon have been left to rust on the lawns of NASA centers at Cape Canaveral and Houston (Figure 8.4). Today no nation on Earth has the capability of returning to the Moon; NASA estimates that it would require more than a decade to mount such an effort, and the financial and political will to do so are lacking. Having reached our nearest neighbor in space, humans retreated back to our own planet. It is hard to guess how long it will be before humans again venture out into the solar system.

Figure 8.2
Geologist (and later U.S. Senator) Jack Schmitt in front of a large boulder in the Litrow Valley at the edge of the lunar highlands. The "rover" vehicle the astronauts used for driving around on the Moon is seen at left. Note how black the sky is on the airless Moon. (NASA)

Figure 8.3
Lunar samples collected in the Apollo project are analyzed and stored in NASA facilities at the Johnson Space Center in Houston, Texas. Here a technician is examining a rock sample using gloves in a sealed environment to avoid contaminating the sample. (NASA)

Figure 8.4
One of the unused Saturn 5 rockets built to go to the Moon, now a tourist attraction at NASA's Johnson Space Center in Houston. (NASA)

The scientific legacy of Apollo remains, however, as discussed in the following sections of this chapter. The Apollo results have also been recently supplemented by two small scientific orbiters, one launched by Japan and the other, Clementine, built and operated by the U.S. mil-

itary as a by-product of the "star wars" space defense effort (Figure 8.5). NASA is itself launching a small orbiter called Lunar Prospector in 1997.

Composition and Structure of the Moon

The composition of the Moon is not the same as that of the Earth. With an average density of only 3.3 g/cm³, the Moon must be made almost entirely of lighter silicate rock. Compared to the Earth, it is *depleted* in iron and other metals. We also know from the study of lunar samples that water and other volatiles are absent. It is as if the Moon were composed of the same basic silicates as the Earth's mantle and crust, with the core metals and the volatiles selectively removed. The differences in composition between Earth and Moon provide important clues concerning the origin of the Moon, a topic we return to later in this chapter.

Probes of the interior carried out with seismometers taken to the Moon as part of the Apollo program confirm the absence of a large metal core. The Moon also lacks a global magnetic field like that of the Earth, a result consistent with the theory that such a field must be generated by motions in a liquid metal core. Not only is the metal lacking, but in addition the Moon is cold and has a solid

Figure 8.5
This mosaic of approximately 1500 images taken with the Clementine spacecraft shows the south pole region of the Moon. Don't let the round border of the image fool you; you are only looking at a small portion of the Moon. For a sense of scale, note that the largest visible impact feature, the Schrodinger Basin at the lower right, is 320 km in diameter. Clementine gave astronomers their first close-up look at the Moon's polar regions. The south pole is of particular interest because there may be a layer of ice in the permanent shadow zone (center). (USGS)

interior. The level of seismic activity is very low; the total energy released by moonquakes is 100 billion times less than that of earthquakes on our planet.

Figure 8.6
The old, heavily cratered lunar highlands make up 83 percent of the Moon's surface. The width of this image is 250 km. (NASA)

8.2

The Lunar Surface

General Appearance

If you look at the Moon through a telescope, you see that it is covered by impact craters of all sizes. However, none of these craters or other topographic features is large enough to be seen without optical aid. The most conspicuous of the Moon's surface features—those that can be seen with the unaided eye and that make up the feature often called "the man in the Moon"—are vast splotches of darker material of volcanic origin.

Centuries ago, the early lunar observers thought that the Moon had continents and oceans, and that it was a possible abode of life. Early names given to the lunar "seas"—Mare Nubium (Sea of Clouds), Mare Tranquillitatis (Tranquil Sea), and so on—are still in use today. In contrast, the "land" areas between the seas do not have names. Thousands of individual craters have been named, however, mostly for great scientists and philosophers. Among the most prominent are those named for Plato, Copernicus, Tycho, and Kepler. Galileo has only a very small crater, however, reflecting his low standing among the Roman Catholic scientists who made some of the first lunar maps.

We know today, however, that the resemblance of lunar features to terrestrial ones is superficial. Even when they look somewhat similar, the origins of lunar features such as craters and mountains may be very different from their terrestrial counterparts. The Moon's relative lack of internal activity, together with the absence of air and water, make most of its geological history unlike anything we know on Earth. And much of the lunar surface is older than the rocks of the Earth's crust, as we learned when the first lunar samples were analyzed in the laboratory.

Ages of Lunar Rocks

In order to trace the detailed history of the Moon or of any planet, we must have a way to determine the ages of individual rocks. Once lunar samples were brought back by the Apollo astronauts, the radioactive dating techniques that had been developed to date rocks on Earth (see Section 6.3) were applied to establish a chronology for our satellite.

The results of these measurements gave solidification ages ranging from about 3.3 to 4.4 billion years for individual Moon samples—substantially older than most of the rocks on the Earth. As we saw in Chapter 6, it now seems clear that both the Earth and the Moon were formed 4.5 billion years ago.

Geological Features

Most (83 percent) of the surface of the Moon is heavily cratered and consists of relatively light-colored silicate rocks called anorthosites. These regions are known as the lunar **highlands.** With ages of more than 4.0 billion years, the highlands are the oldest surviving part of the lunar crust. They represent material that solidified on the crust of the cooling Moon like slag floating on the top of a smelter. Because they formed so early in lunar history, the highlands are also extremely heavily cratered, bearing the scars of billions of years of impacts by interplanetary debris (Figure 8.6).

The most prominent lunar features are the so-called seas, still called **maria** (Latin for "seas") by scientists. Maria is the plural term, by the way; the singular is **mare** (pronounced "mah-ray"). These dark, round plains (Figure 8.7), which are much less cratered than the highlands, cover just 17 percent of the lunar surface, mostly on the side facing the Earth. Today we know that the maria are not ocean basins at all. They are volcanic plains, laid down in eruptions billions of years ago and partly filling huge depressions called *impact basins*, which were produced by collisions of large chunks of material with the Moon relatively early in its history (Figure 8.8).

The lunar maria are all composed of basalt, very similar in composition to the oceanic crust of the Earth or to the lavas erupted by many terrestrial volcanoes. A series of large eruptions on the Moon between 3.3 and 3.8 billion years ago (dated from laboratory measurements of returned samples) formed smooth flows typically a few meters thick but extending over distances of hundreds of kilometers. Eventually, these lava flows filled in the lowest parts of the basins to form the mare surfaces we see today.

Volcanic activity may have begun very early in the Moon's history, although most evidence of the first half-

Figure 8.7
About 17 percent of the Moon's surface consists of the maria—flat plains of basaltic lava. This view of Mare Imbrium also shows numerous secondary craters, and evidence of material ejected from the large crater Copernicus on the upper horizon. Copernicus is an impact crater almost 100 km in diameter that was formed after the lava in Imbrium had already been deposited. (NASA)

billion years is lost. What we do know is that the major mare volcanism, which involved the release of highly fluid magma from hundreds of kilometers below the surface, ended about 3.3 billion years ago (Figure 8.9). After that the Moon's interior cooled, and by 3 billion years ago all volcanic activity had ceased. Since then our satellite has been a geologically dead world, changing only slowly as the result of random impacts.

The major lunar mountains are all the result of the Moon's history of impacts. Long, arc-shaped ranges that border the maria are debris ejected from the impacts that formed these giant basins. These mountains have low, rounded profiles that resemble old, eroded mountains on Earth (Figure 8.10). But appearances can be deceiving. The mountains of the Moon have not been eroded, except for the effects of meteoritic impacts. They are rounded because that is the way they formed, and there has been no water or ice to carve them into cliffs and sharp peaks.

On the Lunar Surface

> The surface is fine and powdery. I can pick it up loosely with my toe. But I can see the footprints of my boots and the treads in the fine sandy particles.
>
> —Neil Armstrong, *Apollo 11* astronaut, immediately after stepping onto the Moon for the first time

The surface of the Moon is buried under a fine-grained soil of tiny, shattered rock fragments. The dark

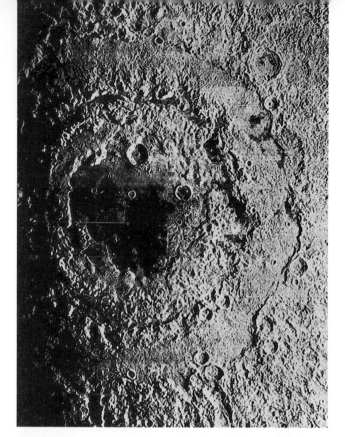

Figure 8.8
The youngest of the large lunar impact basins is Orientale, formed 3.8 billion years ago. Its outer ring is about 1000 km in diameter, roughly the distance between New York City and Detroit, Michigan. Unlike most of the other basins, Orientale has not been filled in with lava flows, so it retains its striking "bull's-eye" appearance. It is located on the edge of the Moon as seen from Earth; some authors have speculated that if it were facing us, the giant feature might have been seen as a kind of "evil eye" in more superstitious times. (NASA)

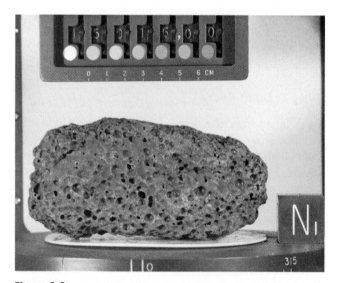

Figure 8.9
Sample of basalt from the mare surface. The gas bubbles are characteristic of rock formed from lava. (NASA)

Figure 8.10
Mt. Hadley on the edge of Mare Imbrium, photographed by the *Apollo 15* astronauts. Note the smooth contours of the lunar mountains, which have not been sculpted by water or ice. (NASA)

basaltic dust of the lunar maria was kicked up by every astronaut footstep, and this dust eventually worked its way into all of the astronauts' equipment. The upper layers of the surface are porous, consisting of loosely packed dust into which their boots sank several centimeters (Figure 8.11). This lunar dust, like so much else on the Moon, is the product of impacts. Each cratering event, large or small, breaks up the rock of the lunar surface and scatters the fragments. Ultimately, billions of years of impacts reduce much of the surface layer to particles about the size of dust or sand.

In the absence of any air, the lunar surface experiences much greater temperature extremes than the surface of the Earth, even though the Earth is virtually the same distance from the Sun. Near local noon, when the Sun is highest in the sky, the temperature of the dark lunar soil rises to the boiling point of water. During the long lunar night (which lasts two Earth weeks) it drops to

about 100 K (−173°C). The extreme cooling is a result not only of the absence of air, but also of the porous nature of the dusty soil, which cools more rapidly than would solid rock.

Impact Craters

The Moon provides an important benchmark for understanding the history of the planetary system. Most solid bodies show the effects of impacts, often extending back to the era when a great deal of debris from the formation process was still present. On the Earth, this long history has been erased by our active geology. On the Moon, in contrast, most of the impact history is preserved. If we can understand what has happened on the Moon, we may be able to extrapolate this knowledge to the other cratered planets and satellites.

Volcanic Versus Impact Origin of Craters

Until the middle of the 20th century, scientists did not generally recognize that lunar craters were the result of impacts. Since impact craters are extremely rare on Earth, geologists did not consider them to be the major feature of lunar geology. They reasoned (perhaps unconsciously) that since the craters we have on Earth are volcanic, the lunar craters must have a similar origin.

One of the first geologists to argue for an impact origin of lunar craters was Grove K. Gilbert (1843–1918), a scientist with the U.S. Geological Survey in the 1890s (Figure 8.12). He pointed out that the large lunar craters, which are mountain-rimmed, circular features with floors generally below the level of the surrounding plains, are

Figure 8.11
Apollo photo of an astronaut's bootprint in the lunar soil. (NASA)

Figure 8.12
Grove K. Gilbert, a founding member of the U.S. Geological Survey, was among the first scientists to recognize the importance of impact cratering on the Earth and Moon. (USGS)

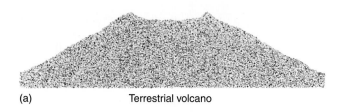

(a)   Terrestrial volcano

(b)   Lunar impact crater

Figure 8.13
Profiles of typical terrestrial volcanic craters (a) and typical lunar impact craters (b) are quite different from each other.

different in form from terrestrial volcanic craters. Volcanic craters are smaller, deeper, and almost always occur at the tops of volcanic mountains (Figure 8.13). Gilbert therefore concluded that the lunar craters were not volcanic. His careful arguments, although decades ahead of their time, laid the foundations for the modern science of lunar geology.

Gilbert believed that the lunar craters were of impact origin, but he still had difficulty explaining why all of them were circular. Impacts of the sort we are most familiar with, such as those made by throwing a stone into a sandbox, only make circular features when falling straight down; otherwise, when the projectile comes in at an angle, the outline of the crater is more elliptical.

The solution to this problem lies in the speed with which projectiles approach the Earth or Moon. Attracted by the gravity of the larger body, the incoming chunk strikes with at least escape velocity, which is 11 km/s for the Earth and 2.4 km/s (5400 mi/h) for the Moon. To this escape velocity is added whatever speed the projectile already had with respect to the Earth or Moon, typically 10 km/s or more.

At these speeds the energy of impact is so great, it leads to an *explosion* that excavates a much larger volume of material than mere impact can, and does so in a more-or-less symmetrical way. Photographs of bomb and shell craters on Earth confirm that explosion craters are always essentially circular. This recognition of the similarity between impact craters and explosion craters removed the last important objection to the impact theory for the origin of lunar craters.

The Cratering Process

Let's consider how an impact at these high speeds produces a crater. When such a fast projectile strikes a planet, it penetrates two or three times its own diameter before stopping. During these few seconds its energy of motion is transferred into a shock wave, which spreads through the target body, and into heat, which vaporizes most of the projectile and some of the surrounding target. The shock wave fractures the rock of the target, while the hot silicate vapor generates an explosion not too different from that of a nuclear bomb detonated at ground level (Figure 8.14). The size of the excavated crater depends primarily on the speed of impact, but generally it is about 10 to 15 times the diameter of the projectile.

An impact explosion of the sort described above leads to a characteristic kind of crater, as illustrated in Figure 8.15. The central cavity is initially bowl-shaped (*crater* comes from the Greek word for bowl), but the gravitational rebound of the crust partially fills it in, producing a

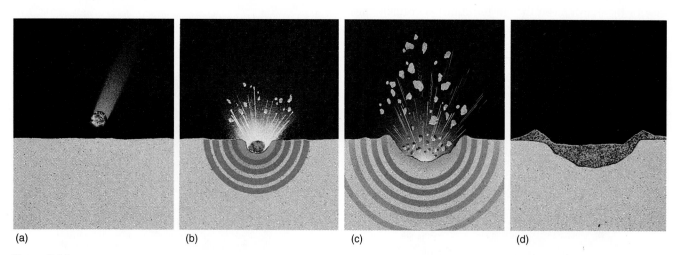

(a)   (b)   (c)   (d)

Figure 8.14
Stages in the formation of an impact crater. (a) The impact occurs. (b) The projectile vaporizes and a shock wave spreads through the lunar rock. (c) Ejecta are thrown out of the crater.
(d) Most of the ejected material falls back to fill the crater and form an ejecta blanket.

Figure 8.15
King Crater on the far side of the Moon, a fairly recent lunar crater 75 km in diameter, shows most of the features associated with large impact structures. (NASA)

flat floor and sometimes creating a central peak. Around the rim, landslides create a series of terraces.

The rim of the crater is turned up by the force of the explosion, so it rises above both the floor and the adjacent terrain. Surrounding the rim is an *ejecta blanket* consisting of material thrown out by the explosion. This debris falls back to create a rough, hilly apron, typically about as wide as the crater diameter. Additional, higher-speed ejecta fall at greater distances from the crater, often digging small *secondary craters* where they strike the surface (see Figure 8.7).

Some of these streams of ejecta can extend from the crater for hundreds or even thousands of kilometers, creating on the Moon the bright *crater rays* that are so prominent in photos taken near full phase (see "Seeing for Yourself" box). The brightest lunar crater rays are associated with large young craters such as Kepler and Tycho.

Using Crater Counts

As discussed in Chapter 6, if a world has had little erosion or internal activity, like the Moon during the past 3 billion years, it is possible to use the number of impact craters on its surface to estimate the age of that surface. By "age" we mean the time since a major disturbance (such as the volcanic eruptions that produced the lunar maria) occurred.

We cannot directly measure the rate at which craters are being formed in the vicinity of the Earth and Moon, since (as we saw in Chapter 7) the average interval between large crater-forming impacts is longer than the span of human history. Remember that Meteor Crater in Arizona is 50,000 years old. However, the cratering rate can be estimated from the number of craters on the lunar maria, or it can be calculated from the number of potential projectiles (asteroids and comets) present in the

solar system today, as will be discussed in Chapter 11. Fortunately, both lines of reasoning lead to about the same answers.

For the Moon these calculations indicate that a crater 1 km in diameter should be produced about every 200,000 years, a 10-km crater every few million years, and one or two 100-km craters every billion years. Craters made by comets and asteroids appear to contribute about equally to these statistics. For the land area of the Earth, the numbers are about 20 times greater, primarily because of its larger surface area.

If these cratering rates have stayed the same over the millennia, we can figure out how long it must have taken to make all the craters we see crowded onto the lunar maria: our calculations show that it would have taken several billion years. This result is similar to the age determined for the maria from radioactive dating—3.3 to 3.8 billion years old.

The fact that these two calculations agree suggests that our assumption was right: comets and asteroids in approximately their current numbers have been impacting planetary surfaces for at least 3.8 billion years. Calculations carried out for other planetary bodies (and their satellites) indicate that they also have been subject to about the same number of interplanetary impacts during this time.

Earlier than 3.8 billion years ago, however, we have good reason to believe that the impact rates must have been a great deal higher. This becomes immediately evident when we compare the numbers of craters on the lunar maria with those on the highlands. Typically, there are ten times more craters on the highlands than on a similar area of maria. Yet the radioactive dating of highland samples showed that they are only a little older than the maria, typically 4.2 billion years rather than 3.8 billion years. If the rate of impacts had been constant, the highlands would have had to be at least ten times older. They would thus have had to form 38 billion years ago—long before the entire universe began.

The Moon, therefore, must have experienced a period of *heavy bombardment* earlier than 3.8 billion years ago, as illustrated in Figure 8.16. This heavy bombardment produced most of the craters we see today in the highlands. Was this a local event, or did similar high impact rates apply throughout the solar system? A partial answer is provided by Voyager spacecraft pictures of the satellites of Jupiter and Saturn, which reveal some surfaces as heavily cratered as the lunar highlands. Because it is impossible to accumulate this many craters *at the present rate* of impacts within the lifetime of the solar system, there must have been a period of high bombardment in the outer solar system as well.

Since we do not have samples of other planets to date in the laboratory, we do not know for sure that the heavy bombardments elsewhere coincided exactly with events on the Moon. However, for most purposes it does not matter whether the heavy bombardments were really si-

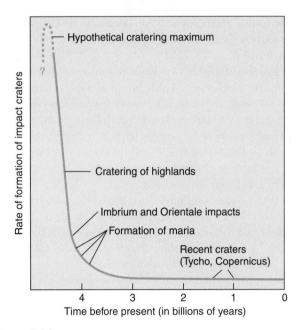

Figure 8.16
How the number of craters being made on the Moon's surface has varied with time over the past 4.3 billion years.

multaneous. The main point is that they all occurred a long time ago, so that any very heavily cratered surface can be assigned an age going back to the first billion years of solar system history. Thus the measurement of craters on planets and satellites throughout the solar system allows scientists to establish a rough chronology for planetary evolution, even without samples. We will make use of these ideas in the following chapters as we discuss the interpretation of spacecraft photos of other planets.

8.4

The Origin of the Moon

Since the Moon is our closest celestial neighbor, the question of where it came from has intrigued philosophers and scientists for centuries. In our time, however, understanding the origin of the Moon has proved to be extremely difficult for planetary scientists. Part of the problem is simply that we know so much about our satellite. The great wealth of data, particularly on the details of the elemental and isotopic composition of the Moon, presents a challenge to any simple theory of lunar origins. Put in simple terms, our problem is that the Moon is both tantalizingly similar to the Earth, *and* frustratingly different.

Theories for the Origin of the Moon

Most of the various theories for the Moon's origin follow one of three general ideas:

1. The fission theory—that the Moon was once part of the Earth but separated from it early in their history.

2. The sister theory—that the Moon formed together with (but independently of) the Earth, as we believe many satellites of the outer planets formed.

3. The capture theory—that the Moon formed elsewhere in the solar system and was captured by the Earth.

Unfortunately, there are fundamental problems with each of these ideas. Perhaps the easiest theory to reject is the capture theory. Its primary drawback is that no one knows of any way that the early Earth could have captured a large satellite from elsewhere. One body approaching another cannot go into orbit around it without a substantial loss of energy; this is the reason that spacecraft destined to orbit other planets are equipped with retrorockets. Further, if such a capture did take place, it would be into a very eccentric orbit rather than the nearly circular orbit the Moon occupies today. And finally, there are too many compositional similarities between the Earth and the Moon, particularly an identical fraction of the major isotopes of oxygen, to justify seeking a completely independent origin.

The fission theory, which states that the Moon separated from the Earth, was suggested in the late 19th century by George Darwin, whom we profiled in Chapter 3. More modern dynamical calculations have shown that the sort of spontaneous fission or splitting imagined by Darwin is impossible. Further, it is difficult to understand how a Moon made out of terrestrial material in this way could have developed the many distinctive chemical differences now known to characterize our satellite.

Scientists have therefore been left with the sister theory—that the Moon formed alongside the Earth—or with some modification of the fission theory that finds a more acceptable way for the lunar material to have separated from the Earth.

The Giant Impact Theory

Scientists have recently put forward a new theory for the origin of the Moon, in part as a response to the problems of lunar chemistry posed by the sister theory, and in part as a reflection of interest among planetary scientists in the role of impacts in planetary evolution. There is increasing evidence that large projectiles—objects of essentially planetary mass—were orbiting in the inner solar system at the time of the terrestrial planets' formation. What if one of these very large projectiles had struck the Earth?

The *giant impact theory* envisions the Earth being struck obliquely by an object approximately one-tenth the Earth's mass, about the size of Mars. This is very nearly the largest impact the Earth could experience without being shattered. Computer calculations (Figure 8.17) show that such an impact would disrupt much of the Earth by penetrating into the core, ejecting a vast amount of material into space, and releasing almost enough energy to break the planet apart.

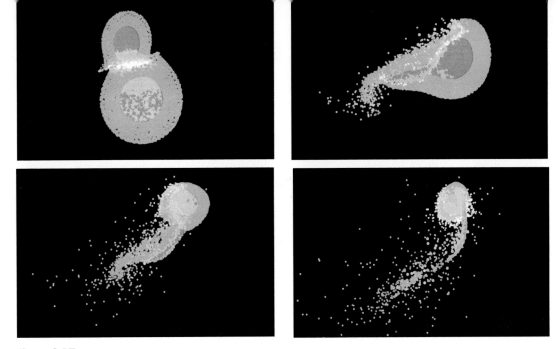

Figure 8.17
Four frames from one of a series of computer simulations showing an oblique giant impact on the Earth. Iron is shown as spheres of blue and green. Rock is represented by red, pink, yellow, and white spheres, in order of increasing temperature. In this particular simulation, the impactor has 1/5 of the mass in the system, and the Earth has 4/5. After the impact, a plume of material is ejected into space. It is this plume that, under the right circumstances, can condense to form the Moon. Note that the plume eventually has much more pink (rock) than blue or green (iron), explaining why the Moon resembles the mantle of the Earth. (Courtesy of A. G. W. Cameron, Center for Astrophysics)

The computer simulations show that material totaling several percent of the Earth's mass could be ejected in such an impact. Most of this material would be from the mantles of the Earth and the impacting body, and it would initially be ejected at high speed in the form of hot vapor. Perhaps—and the calculations are not really clear on this point—much of this hot vapor could condense into a ring of material orbiting around the Earth. Alternatively, it might form a sort of superheated silicate atmosphere rotating rapidly around the planet. Either way, the theory suggests that this ejected material ultimately condensed to form the Moon.

While we do not have any current way of showing that the giant impact hypothesis is the correct model of the Moon's origin, it does offer potential solutions to most of the major problems raised by the chemistry of the Moon. First, since the Moon's raw material is derived from the mantles of the Earth and the projectile, the absence of metals is easily understood. Second, most of the volatile elements would have been lost during the high-temperature phase following the impact. Yet by making the Moon primarily of terrestrial mantle material, it is also possible to understand similarities such as identical abundances of various oxygen isotopes.

8.5

Mercury

The planet Mercury is similar to the Moon in many ways. Like the Moon, it has no atmosphere and its surface is heavily cratered. As described later in this chapter, it also shares with the Moon a violent birth history.

Mercury's Orbit

Mercury is the nearest to the Sun of the nine planets and, in accordance with Kepler's third law, it has the shortest period of revolution about the Sun (88 of our days) and the highest average orbital speed (48 km/s). It is appropriately named for the fleet-footed messenger god of the Romans. Because Mercury remains close to the Sun, it can be difficult to pick out in the sky. As you might imagine, it's best seen when its eccentric orbit takes it as far from the Sun as possible.

The semimajor axis of Mercury's orbit—that is, the planet's average distance from the Sun—is 58 million km or 0.39 AU. However, because its orbit has the high eccentricity of 0.206, Mercury's actual distance from the Sun varies from 46 million km at perihelion to 70 million km at aphelion. (If some of our terms are new to you, you may want to look at Chapter 2 or the glossary at the end of the book.) Pluto is the only planet with a more eccentric orbit than Mercury. Furthermore, the 7° inclination of Mercury's orbit to the plane of the ecliptic is also greater than that of any other planet except Pluto.

Composition and Structure

Table 8.1 compares some basic properties of Mercury with those of the Moon. Mercury's mass is 1/18 that of the Earth; again Pluto is the only planet with a smaller mass. Mercury is also the second smallest of the planets, having a diameter of only about 4880 km, less than half that of the Earth. Mercury's density is 5.4 g/cm³, much greater than the density of the Moon, indicating that the composition of these two objects differs substantially.

What a Difference a Day Makes

Mercury rotates three times for each two orbits around the Sun. It is the only planet that exhibits this relationship, and there are some interesting consequences for any observers who might someday be stationed on the surface of Mercury.

Here on Earth, we take it for granted that days are much shorter than years. Therefore the two ways of defining the local "day"—how long the planet takes to rotate, and how long the Sun takes to return to the same position in the sky (see Chapter 3)—are the same for most practical purposes. But this is not the case on Mercury. While Mercury rotates in 59 Earth days, the time for the Sun to return to the same place in Mercury's sky turns out to be two Mercury years or 176 Earth days. So if one day at noon a Mercury explorer suggests to her companion that they should meet at noon the next day, this could mean a very long time apart!

To make things even more interesting, recall that Mercury has an eccentric orbit, meaning that its distance from the Sun varies significantly during each mercurian year. By Kepler's law, the planet moves fastest in its orbit when closest to the Sun. Let's examine how this affects the way we would see the Sun in the sky if we were standing on the surface of Mercury in the center of a giant basin that astronomers call Caloris (Figure 8.21).

At the location of Caloris, Mercury is most distant from the Sun at sunrise; this means the rising Sun looks smaller in the sky (although still more than twice the size it appears from the Earth). As the Sun rises higher and higher, it gets bigger and bigger; Mercury is now getting closer to the Sun in its eccentric orbit. At the same time, the apparent motion of the Sun slows down as Mercury's faster motion in orbit begins to catch up with its rotation.

At noon the Sun is now three times larger than it looks from Earth, and hangs almost motionless in the sky. As the afternoon wears on, the Sun eventually looks smaller and smaller again, and moves faster and faster in the sky. At sunset, a full Mercury-year after sunrise, the Sun is back to its smallest apparent size as it dips out of sight. (Sunrise and sunset are much more sudden on Mercury, since there is no atmosphere to bend or scatter the rays of sunlight.)

Astronomers call locations like the Caloris basin the "hot longitudes" on Mercury, because the Sun is closest to the planet at noon, just when it is lingering overhead for many Earth days. This makes them the hottest places on the planet.

We bring all this up not because the exact details of this scenario are so important, but to illustrate how many of the things we take for granted on Earth are not the same on other worlds. As we've mentioned before, one of the best things about taking an astronomy class should be ridding you forever of any *Earth chauvinism* you might have. The way things are on our planet is just one of the many ways nature can arrange reality.

Its composition, in fact, is one of the most interesting things about Mercury; it is unique among the planets. Mercury's high density tells us that it must be composed largely of heavier materials such as metals. The most likely models for Mercury's interior suggest a metallic iron-nickel core amounting to 60 percent of the total mass, with the rest of the planet made up primarily of silicates. The core has a diameter of 3500 km and extends to within 700 km of the surface (Figure 8.18). We could think of Mercury as a metal ball the size of the Moon surrounded by a rocky crust 700 km thick. Unlike the Moon, however, Mercury also has a weak magnetic field. The existence of this field is consistent with a large metal core, and it suggests that at least part of the core must be liquid in order to generate the observed field.

The escape velocity is too low and the surface temperature too high for Mercury to retain any substantial atmosphere. In 1985, however, an extremely thin atmosphere of sodium was detected spectroscopically. The atoms in this atmosphere must be lost to space all the time and must therefore be regenerated by some process. The most likely explanation is that high-speed particles from the solar wind are regularly colliding with Mercury's surface and "chipping off" atoms of sodium.

Mercury's Strange Rotation

Visual studies of Mercury's indistinct surface markings were once thought to indicate that the planet kept one face to the Sun (as the Moon does to the Earth). Thus for many years it was widely believed that Mercury's rotation period was equal to its revolution period of 88 days, making one side perpetually hot, while the other was always cold.

Radar observations of Mercury in the mid-1960s, however, showed conclusively that Mercury does not keep one side fixed to the Sun. Recall from Section 5.4 that the frequency of a transmitted radar pulse can be controlled precisely. Any motion of the target then introduces a measurable change in this frequency from the Doppler effect

Figure 8.18
The interior of Mercury is dominated by a metallic core about the same size as our Moon. (University of Arizona, courtesy Robert Strom)

(see Section 4.6). Such target motions can include rotation: if a planet is turning, one side seems to be approaching the Earth while the other is moving away from it. The result is to spread or broaden the precise transmitted frequency into a range of frequencies in the reflected signal (Figure 8.19). The degree of broadening provides an exact measurement of the rotation rate of the planet.

Mercury's sidereal period of rotation (how long it takes to turn with respect to the distant stars) was found to be about 59 days. Shortly after this period was discovered by radar astronomers, the Italian physicist Giuseppe

Colombo pointed out that this is very nearly two-thirds of the planet's period of revolution, and subsequently astronomers found theoretical reasons (connected with the tidal forces discussed in Chapter 3) for expecting that Mercury can rotate stably with a period exactly two-thirds that of its revolution—58.65 days.

Mercury is very hot on its daylight side, but because it has no appreciable atmosphere, it gets surprisingly cold during the long nights. The temperature on the surface climbs to 700 K at noontime. After sunset, however, the temperature drops, reaching 100 K just before dawn. The range in temperature on Mercury is thus 600 K, more than on any other planet.

The Surface of Mercury

The first close-up look at Mercury came in 1974, when the U.S. spacecraft Mariner 10 passed 9500 km from the surface of the planet and transmitted more than 2000 photographs to Earth, revealing details with a resolution down to 150 m. Mercury strongly resembles the Moon in appearance (Figures 8.20 and 8.21). It is covered with thousands of craters and larger basins up to 1300 km in diameter. Some of the brighter craters are rayed, like Tycho and Copernicus on the Moon, and many have central peaks. There are also scarps (cliffs) over a kilometer high and hundreds of kilometers long, as well as ridges and plains.

Most of the mercurian features have been named in commemoration of artists, writers, composers, and other contributors to the arts and humanities, in contrast with the scientists commemorated on the Moon. Among the most prominent craters are Bach, Shakespeare, Tolstoy, Mozart, and Goethe.

Mercury's larger basins resemble the lunar maria in both size and appearance. They show evidence of flooding

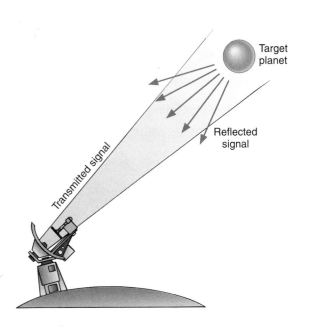

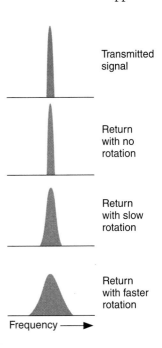

Figure 8.19
The use of Doppler radar to measure the rotation of a planet.

Figure 8.20
The planet Mercury as photographed by Mariner 10 in 1974.
(Mosaic courtesy of the Astrogeology Team, USGS, Flagstaff)

Figure 8.21
Part of the Caloris basin on Mercury, 1300 km in diameter, can be seen at the bottom. This partially flooded impact basin is the largest structural feature on Mercury seen by Mariner 10. Compare photo with the Orientale basin on the Moon (Figure 8.7). (NASA/JPL)

from lava, although the flows do not have the distinctive dark color that characterizes the lunar maria. Since Mercury is so difficult to study telescopically, and has been visited by only one spacecraft, we actually know very little about the chemistry of its surface or the details of its geological history.

There is no evidence of plate tectonics on Mercury. Distinctive long scarps, however, can sometimes be seen cutting across craters; this means the scarps must have formed later than the craters (Figure 8.22). These long cliffs appear to have their origin in the slight compression of Mercury's crust. Apparently, at some point in its history this planet shrank, wrinkling the crust, and it must have done so after most of the craters on its surface had already formed. If the standard cratering chronology applies to Mercury, this shrinkage must have taken place during the past 4 billion years. If we understood the interior structure and composition of the planet better, we could probably calculate how internal changes in temperature might have led to this global compression, which has no counterpart on the Moon or the other terrestrial planets.

Although it has become something of a forgotten planet in the two decades since the Mariner 10 flybys, as-

tronomers who study Mercury continue to be surprised. For example, enhanced radar reflectivity from both poles of the planet was discovered in 1992, indicating the presence of ice just beneath the surface. Even though the temperature very near the poles remains low enough for ice to survive, the discovery of these icy polar caps was unexpected, since they seem to imply that the polar regions of Mercury have remained below freezing since the formation of the planet.

Figure 8.22
Discovery Scarp on Mercury. This long cliff, nearly 1 km high and more than 100 km long, cuts across several craters. We conclude that the compression that made "wrinkles" like this in the planet's surface must have taken place after the craters were formed. (NASA/JPL)

The Origin of Mercury

The problem with understanding how Mercury formed is the reverse of that posed by the composition of the Moon. We have seen that, unlike the Moon, Mercury is composed mostly of metal. However, Mercury should have formed with about the same ratio of metal to silicate as that found on the Earth or Venus. How did it lose so much of its rocky material?

The most probable explanation for Mercury's silicate loss may be similar to the explanation for the Moon's lack of a metal core. Mercury is likely to have experienced several giant impacts very early in its youth, and one or more of these may have torn away a fraction of its mantle and crust, leaving a body dominated by its iron core.

Today astronomers recognize that the early solar system was a chaotic place, with the final stages of planet formation characterized by impacts of great violence. Some bodies of planetary mass may have been destroyed, while others could have fragmented and then reformed, perhaps more than once. Both the Moon and Mercury, with their strange compositions, bear testimony to the catastrophes that must have characterized the solar system during its youth.

Observing the Moon

The Moon is one of the most beautiful sights in the sky, and it is the only object close enough to reveal its topography to us without a visit from a spacecraft. A fairly small amateur telescope easily shows craters or mountains on the Moon as small as a few kilometers across.

Even as seen through a good pair of binoculars, the appearance of the Moon's surface changes dramatically with its phase (see Chapter 3 for an explanation of the phases). At full phase it shows almost no topographic detail, and you must look closely to see more than a few craters. This is because the sunlight illuminates the surface straight on, and in this flat lighting no shadows are cast. Much more revealing is the view near first or last quarter, when sunlight streams in from the side and topographic features cast sharp shadows. It is almost always more rewarding to study a planetary surface under such *oblique lighting*, when the maximum information about the surface relief can be obtained.

The flat lighting at full phase does, however, accentuate brightness contrasts on the Moon, such as those between the maria and highlands. Notice on the accompanying photo that several of the large mare craters seem to be surrounded by aprons of white material, and that the light streaks or *rays* that can stretch for hundreds of kilometers across the surface are clearly visible. These lighter features are ejecta, splashed out from the crater-forming impact.

By the way, there is no danger in looking at the Moon with binoculars or telescopes. The reflected sunlight is never bright enough to hurt you. In fact, the sunlit surface of the Moon has about the same brightness as a sunlit landscape on Earth. So explore away. Several guides to observing the Moon are listed in the recommendations for further reading at the end of this chapter.

One interesting thing about the Moon that you can see without binoculars or telescopes is popularly called "the new Moon in the old Moon's arms." Look at the Moon when it is a thin crescent and you can often make out the faint circle of the entire lunar disk, even though the sunlight shines on only a narrow section. The rest of the disk is illuminated not by sunlight but by earthlight—reflected light from the Earth. The light of the full Earth on the Moon is about 50 times brighter than that of the full Moon shining on the Earth.

The appearance of the Moon at different phases. Illumination from the side brings craters and other topographic features into sharp relief. At full phase (right) there are no shadows and it is more difficult to see such features. However, the flat lighting at full phase brings out some classes of surface features, such as the bright rays of ejecta that stretch out from a few large young craters. (Lick Observatory)

Summary

8.1 Most of what we know about the Moon derives from the Apollo program, including 400 kg of lunar samples still being intensively studied. The Moon has 1/80 the mass of the Earth and is severely depleted in both metals and volatile materials. It is made almost entirely of silicates like those in the Earth's mantle and crust.

8.2 Lunar rocks can be dated by their radioactivity. Like the Earth, the Moon was formed 4.5 billion years ago. The heavily cratered **highlands** are made of rocks more than 4 billion years old. The darker volcanic plains of the **maria** were erupted between 3.3 and 3.8 billion years ago. Generally, the surface is dominated by impacts, including continuing small impacts that produce its fine-grained soil.

8.3 A century ago Gilbert suggested that the lunar craters were of impact origin, but the cratering process was not well understood until more recently. High-speed impacts produce explosions and excavate craters with raised rims, ejecta blankets, and often central peaks. Cratering rates have been roughly constant for the past 3 billion years, but earlier were much greater. Crater counts can be used to derive approximate ages for geologic features on the Moon and other planets.

8.4 There are three standard theories for the origin of the Moon: the fission theory; the sister theory; and the capture theory. All have problems, and recently they have been supplanted by the giant impact theory, which ascribes the origin of the Moon to the impact 4.5 billion years ago of a Mars-sized projectile with the Earth.

8.5 Mercury is the nearest planet to the Sun, and the fastest moving. Mercury is similar to the Moon in having a heavily cratered surface and no atmosphere, but it differs in having a very large metal core. Early in its evolution, it apparently lost part of its silicate mantle, probably due to one or more giant impacts. Long scarps on its surface testify to a global compression of Mercury's crust during the past 4 billion years.

Review Questions

1. What is the composition of the Moon, and how does it compare to the composition of the Earth? Of Mercury?

2. Outline the main events in the Moon's geological history.

3. Explain how high-speed impacts form circular craters. How can this explanation account for the characteristic features of impact craters?

4. Explain the evidence for a period of heavy bombardment on the Moon about 4 billion years ago. What might have been the source of this early flux of impacting debris?

5. How did our exploration of the Moon differ from that of Mercury (and the other planets)?

6. Summarize the four main theories for the origin of the Moon. Give pros and cons for each.

7. What do current ideas about the origins of the Moon and Mercury have in common?

Thought Questions

8. One of the primary scientific objectives of the Apollo program was the return of lunar material.
 a. Why was this so important?
 b. What can be learned from samples?
 c. Are they still of value now?

9. Apollo astronaut David Scott dropped a hammer and a feather together on the Moon, and both reached the ground at the same time. What are the two distinct advantages that this experiment on the Moon had over the same experiment as performed by Galileo on the Earth?

10. Galileo thought the lunar maria were seas of water. If you had no better telescope than the one he had, could you prove that they are not composed of water?

11. Why did it take so long for geologists to recognize that the lunar craters had an impact origin rather than a volcanic one?

12. How might a crater made by the impact of a comet with the Moon differ from a crater made by the impact of an asteroid?

13. Why are the lunar mountains smoothly rounded rather than having sharp, pointed peaks (as they were almost always depicted in science fiction illustrations and films before the first lunar landings)?

14. The lunar highlands have about ten times more craters on a given area than do the maria. Does this mean that the highlands are ten times older? Explain your reasoning.

15. Give several reasons that Mercury would be a particularly unpleasant place to live.

16. If, in the remote future, we establish a base on Mercury, keeping track of time will be a challenge. Discuss how to define a year on Mercury, and the two ways to define a day. Can you come up with ways that humans raised on Earth might deal with time cycles on Mercury?

17. The Moon has too little iron, Mercury too much. How

can both of these anomalies be the result of giant impacts? Explain how the same process can yield such apparently contradictory results.

18. Some scientists are urging a return to the Moon in the next decade or two. List some reasons (astronomical and otherwise) that it may be useful for humanity to return to the Moon, and some reasons against making such an effort.

Problems

19. The Moon was once closer to the Earth than it is now. When it was at half its present distance, how long was its period of revolution? (See Chapter 2.)

20. Astronomers believe that the deposit of lava in the giant mare basins did not happen in one flow, but in many different eruptions spanning some time. Indeed, in any one mare we find a variety of rock ages, typically spanning about 100 million years. The individual lava flows as seen in Hadley Rill by the *Apollo 15* astronauts were about 4 m thick. Estimate the average time interval between the beginnings of successive lava flows if the total depth of the lava in the mare is 2 km.

21. The Moon requires about one month (0.08 year) to orbit the Earth. Its distance from us is about 400,000 km (0.0027 AU). Use Kepler's third law, as modified by Newton, to calculate the mass of the Earth relative to the Sun.

Suggestions for Further Reading

On the Moon

Chaikin, A. *A Man on the Moon*. 1994, Viking Press. A well-reviewed history of manned lunar exploration.

Coco, M. "Staging a Moon Shot" in *Astronomy*, Aug. 1992, p. 62. Describes photographing the Moon.

Hockey, T. *The Book of the Moon*. 1986, Prentice Hall. A basic primer on many aspects of the Moon.

Kitt, M. *The Moon: An Observing Guide for Backyard Telescopes*. 1992, Kalmbach. Eighty-page illustrated primer for beginners.

MacRobert, A. "Close-up of an Alien World" in *Sky & Telescope*, July 1984, p. 29. An observing guide.

Moore, P. *The Moon*. 1980, Rand McNally. A reference atlas and guide.

Morrison, D. and Owen, T. "Our Ancient Neighbor the Moon" in *Mercury*, May/June 1988, p. 66; July/Aug. 1988, p. 98. An overview.

Murray, C. and Cox, C. *Apollo: The Race to the Moon*. 1989, Simon & Schuster. Popular-level account, from many interviews.

Price, F. *The Moon Observer's Handbook*. 1989, Cambridge U. Press. Comprehensive observing guide.

Ryder, G. "Apollo's Gift: The Moon" in *Astronomy*, July 1994, p. 40. Good evolutionary history of the Moon.

Schmitt, H. "Exploring Taurus–Littrow: Apollo 17" in *National Geographic*, Sep. 1973.

Taylor, G. "The Scientific Legacy of Apollo" in *Scientific American*, July 1994, p. 40.

On Mercury

Chapman, C. "Mercury's Heart of Iron" in *Astronomy*, Nov. 1988, p. 22. Good introduction to our modern view of the planet.

Cordell, B. "Mercury: The World Closest to the Sun" in *Mercury*, Sep./Oct. 1984, p. 136. An overview.

Gingerich, O. "How Astronomers Finally Captured Mercury" in *Sky & Telescope*, Sep. 1983, p. 203. A history of early observations.

Strom, R. "Mercury: The Forgotten Planet" in *Sky & Telescope*, Sep. 1990, p. 256.

Strom, R. *Mercury: The Elusive Planet*. 1987, Smithsonian Institution Press. Best book on the inner planet.

See also some of the World Wide Web sites for the planets cited in Appendix 1.

Using REDSHIFT ™

1. Locate the crater Copernicus on the Moon image with magnification set to 30. Lock on the Moon and *step* once per day. Watch the terminator (day/night dividing line) travel across the lunar surface. For the month ahead determine when sunrise, noon (when the Sun is overhead as seen at Copernicus), and sunset occur. If you can, view Copernicus at these times through a telescope and describe the appearance of the crater under these lighting conditions.

2. To compare Mercury's rotation and revolution, set the *Follow Planet* panel to Mercury, inward view, with a height of 25,000 units. This effectively allows you to view the planet as seen from the Sun. Turn on the stars and their proper names. *Step time* until Mercury appears near a labeled star. Note the Julian date and *step* until Mercury appears near the labeled star again. The difference in days is Mercury's sidereal revolution period.

To watch Mercury's rotation, set the height on the *Follow Planet* panel to about 10,000 units. Pick a large, easily identifiable feature on the surface and *step time* until this feature rotates around once.

Heavily eroded canyonlands on Mars. This Viking photograph, taken from Mars orbit, looks down on a small part of the giant Valles Marineris complex and shows an area about 60 km across. It is one of a series of images taken in 1976 that has been processed by the staff of the U.S. Geological Survey in Flagstaff, Arizona, to bring out some of the marvelous detail in each photograph. (NASA/USGS, courtesy of Alfred McEwen)

CHAPTER 9

Earth-Like Planets: Venus and Mars

Thinking Ahead

For decades science fiction writers have conjured up visions of Venus and Mars as more-or-less Earth-like worlds where humans will someday visit and (perhaps) come face-to-face with alien life-forms. Yet we have discovered that neither planet is very similar to our own, and it seems unlikely that either could sustain life-forms like those that populate the Earth. Why is this? How did it happen that the three largest terrestrial planets have turned out to be so different from each other?

The Moon and Mercury are geologically dead. In contrast, the larger terrestrial planets—the Earth, Venus, and Mars—are more active and interesting worlds. We have already discussed the Earth, and we now turn to Venus and Mars. These are the nearest planets and the most accessible to spacecraft. Not surprisingly, the greater part of the effort of the United States and the Soviet Union (now Russia) in planetary exploration has been devoted to these fascinating worlds. In this chapter we discuss some of the results of more than three decades of scientific exploration of Mars and Venus.

One deduction from this thin air [on Mars] we must be careful not to make—that because it is thin, it is incapable of supporting intelligent life. That beings constituted physically as we are would find it a most uncomfortable habitat is pretty certain. But lungs are not wedded to logic, as public speeches show, and there is nothing in the world or beyond it to prevent, so far as we know, a being with gills, for example, from being a most superior person.

Percival Lowell, from *Mars* (1895).

The Nearest Planets: An Overview

Mars and Venus are among the brightest objects in the night sky. The average distance of Mars from the Sun is 227 million km (1.52 AU), but with an orbital eccentricity of 0.093, its distance varies by 42 million km as it goes around the Sun. Venus' orbit, on the other hand, is very nearly circular, at a distance of 108 million km (0.72 AU) from the Sun.

Venus has the distinction of approaching the Earth more closely than does any other planet: at its nearest it is only 40 million km away. Like Mercury, Venus sometimes appears as an "evening star" and sometimes as a "morning star." Under favorable circumstances, Venus is so bright that it even casts a visible shadow. The closest Mars ever gets to Earth is about 56 million km.

Appearance

Seen from the ground through a telescope, Mars is both tantalizing and disappointing (Figure 9.1). The planet is distinctly red in color, due (as we now know) to the presence of iron oxides in its soil, which may account for its association with war (and bloodletting) in the legends of many early cultures. The best resolution obtainable from telescopes on the ground is about 100 km, or about the same as what we can see on the Moon with the unaided

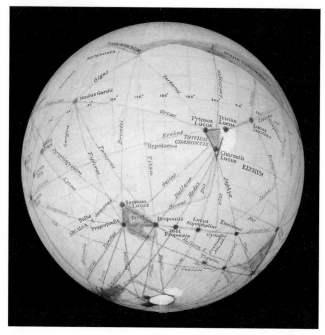

Figure 9.2
One of the remarkable globes of Mars prepared by Percival Lowell, showing a network of dozens of canals, oases, and triangular water reservoirs that he claimed were visible on the red planet. (Lowell Observatory)

eye. At this resolution, however, no hint of topographic structure can be detected—no mountains, no valleys, not even impact craters. On the other hand, the bright polar ice caps can easily be seen from the ground, together with dusky surface markings that gradually change in outline and intensity from season to season. Of all the planets, only Mars has a surface that can be clearly made out from the Earth, and this surface exhibits changes that bespeak a dynamic atmosphere.

For a few decades around the turn of the 20th century, some astronomers believed they saw evidence of an intelligent civilization on Mars. The controversy began in 1877, when the Italian astronomer Giovanni Schiaparelli announced the he could see long, faint, straight lines on Mars that he called *canale,* or channels. In English-speaking countries, the term was mistakenly translated as canals, implying an artificial origin.

Even before Schiaparelli's observations, astronomers had seen the bright caps of ice in Mars' polar regions change size with the seasons, and had seen seasonal changes in the dark surface features as well. With a little imagination, it was not difficult to picture the canals as irrigation ditches bringing water from the melting polar ice to the parched deserts of the red planet.

Until his death in 1916, the most effective proponent of intelligent life on Mars was Percival Lowell, self-made American astronomer and member of the wealthy Lowell family of Boston (see "Voyagers in Astronomy" box). An effective author and public speaker, Lowell made what seemed to the public to be a convincing case for intelli-

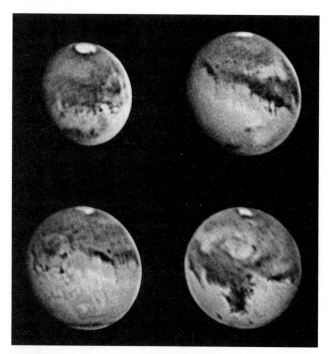

Figure 9.1
These are among the best Earth-based photos of Mars, taken in 1988 when the planet was exceptionally close to the Earth. The polar caps and dark surface markings are evident, but not the topographic features. (Steve Larson, University of Arizona)

Percival Lowell: Dreaming of an Inhabited Mars

Percival Lowell (1855–1916) was born into the well-known and well-to-do Massachusetts family about whom John Bossidy made the famous toast:

> And this is good old Boston,
> the home of the bean and the cod,
> where the Lowells talk to the Cabots
> and the Cabots talk only to God.

Percival's brother Lawrence became president of Harvard University, and his sister Amy, who liked to dress in men's clothing and smoke cigars, became a distinguished poet. Percival was already interested in astronomy as a boy: he made observations of Mars at age 13. His undergraduate thesis at Harvard dealt with the origin of the solar system, but he did not pursue this interest immediately. Instead he entered the family business and traveled extensively in Asia. In 1892, however, he decided to dedicate himself to carrying on Schiaparelli's work and solving the mysteries of the martian canals.

In 1894, with the help of astronomers at Harvard but using his own funds, Lowell built an observatory on a high plateau in Flagstaff, Arizona, where he hoped the seeing would be clear enough to show him Mars in unprecedented detail. He and his assistants quickly accumulated a tremendous number of drawings and maps, purporting to show a vast network of martian canals (see Figure 9.2). He elaborated his ideas about the inhabitants of the red planet in several books, including *Mars* (1895) and *Mars and Its Canals* (1906), and in hundreds of articles and speeches. As Lowell put it:

> . . . A mind of no mean order would seem to have presided over the system we see—a mind certainly of considerably more comprehensiveness than that which presides over the various departments of our own public works. Party politics, at all events,

Percival Lowell in about 1910, observing with his 24-inch telescope at Flagstaff, Arizona.
(Lowell Observatory)

have had no part in them; for the system is planet-wide . . . Certainly what we see hints at the existence of beings who are in advance of, not behind us, in the journey of life.

Lowell's views captured the public imagination and inspired many novels and stories, the most famous of which was H. G. Wells' *War of the Worlds* (1897). In this famous "invasion" novel, the thirsty inhabitants of a dying planet Mars (based entirely on Lowell's ideas) come to conquer the Earth with advanced technology.

Although the Lowell Observatory first became famous for its work on the martian canals, both Lowell and the observatory eventually turned to other projects as well. As we will see in Chapter 11, Lowell became interested in the search for a ninth (and then undiscovered) planet in the solar system. In 1930 Pluto was found at the Lowell Observatory, and it is not a coincidence that the name selected for the new planet starts with Lowell's initials. It was also at the Lowell Observatory that the first measurements were made of the great speed at which galaxies are moving away from us, observations that would ultimately lead to our modern view of an expanding universe.

Lowell continued to live at his observatory, marrying at age 53 and publishing extensively. He relished the debate his claims about Mars caused far more than the astronomers on the other side, who often complained that Lowell's work was making planetary astronomy a less-respectable field. At the same time, the public fascination with the planets fueled by Lowell's work (and its interpreters) may, several generations later, have helped fan support for the space program and the many missions whose results grace the pages of our text.

gent Martians, who he believed had constructed the huge canals to preserve their existence in the face of a deteriorating climate (Figure 9.2).

The argument for Martians, however, hinged on the reality of the canals, a matter that remained in serious dispute among astronomers. The "canal" markings were always difficult to study, glimpsed only occasionally because atmospheric conditions caused the tiny image of Mars to shimmer in the telescope. Lowell saw them everywhere, but many other observers could not see them at all, and remained unconvinced of their existence. When larger telescopes failed to confirm the presence of canals, the skeptics felt vindicated.

Astronomers had given up the idea of martian canals by the 1930s, although it persisted in the public consciousness until the first spacecraft photographs clearly showed that none existed. Now it is generally accepted that the canals were an optical illusion, the result of the human mind's tendency to see order in random features glimpsed dimly at the limits of the eye's resolution. When we see small, dim dots of surface markings, our minds tend to connect those dots into straight lines.

Figure 9.3
Venus as photographed by the Pioneer Venus orbiter. This ultraviolet image shows upper-atmosphere cloud structure that would be invisible at visible wavelengths, but not a glimpse of the planet's cloud-shrouded surface. (NASA/ARC)

Venus, in contrast to Mars, exhibits little of interest through even the largest telescope. Galileo discovered that Venus displays a full range of phases as it revolves about the Sun, but the planet's actual surface is not visible because it is shrouded by dense clouds that reflect about 70 percent of the sunlight that falls on them. These clouds help make Venus one of the brightest objects in the sky, but they also frustrate efforts to study the underlying surface, even with cameras aboard spacecraft in orbit around the planet (Figure 9.3).

Rotation of the Planets

Astronomers can determine the rotation period of Mars with great accuracy by watching the motion of permanent surface markings; its sidereal day is 24 h 37 m 23 s, just a little greater than the rotation period of the Earth. This high precision is not obtained by watching Mars for a single rotation, but by noting how many turns it makes over a long period of time. Good observations of Mars date back more than 200 years, a period during which tens of thousands of martian days have passed. As a result, the rotation period is now known to within a few hundredths of a second.

The rotational axis of Mars has a tilt of about 25°, very similar to the tilt of the Earth's axis. Thus Mars experiences seasons very much like those on Earth. Because of the longer martian year (almost two Earth years), however, each season there lasts about six of our months.

Since no surface detail can be seen on Venus, its rotation period can only be found with radar. The first radar observations of the planet's rotation were made in the early 1960s. Surprisingly, they showed Venus to rotate from east to west—the reverse direction from the rotation of most other planets—in a period of about 250 days.

Subsequently, topographical surface features were identified on the planet that showed up in the reflected radar signals. The rotation period of Venus, precisely determined from the motion of such features across its disk, is 243.08 days retrograde (east to west). The rotation period of Venus, by far the longest in the solar system, is thus about 19 days longer than its period of revolution about the Sun. The length of a day on Venus—the time the Sun takes to return to the same place in the sky—turns out to be 116.67 Earth days. Although we do not know the reason for Venus' slow backwards rotation, we can guess that it may have suffered one or more extremely powerful collisions during the formation process of the solar system (see Chapter 8).

Basic Properties of Venus and Mars

Before discussing each planet individually, let us compare some of their basic properties with each other and with the Earth (Table 9.1). Venus is in many ways the Earth's twin, with a mass 0.82 times the mass of the Earth and an almost identical density. The level of geological activity is also relatively high, creating one of the most geologically complex and diverse surfaces in the solar system. On the other hand, Venus' atmosphere is much more massive than that of the Earth, with a surface pressure nearly 100 times greater than ours. Its surface is also remarkably hot, with a temperature of 730 K (over 850°F, hotter than the self-cleaning cycle of your oven). One of the major challenges presented by Venus is to understand why the atmosphere and surface environment of this twin have diverged so sharply from those of our own planet.

Mars, by contrast, is rather small, with a mass only 0.11 times the mass of the Earth. It is larger than either the Moon or Mercury, however, and unlike them it retains

TABLE 9.1
Properties of Earth, Venus, and Mars

	Earth	Venus	Mars
Semimajor axis (AU)	1.00	0.72	1.52
Period (year)	1.00	0.61	1.88
Mass (Earth = 1)	1.00	0.82	0.11
Diameter (km)	12,756	12,102	6,790
Density (g/cm³)	5.5	5.3	3.9
Surface gravity (Earth = 1)	1.00	0.91	0.38
Escape velocity (km/s)	11.2	10.4	5.0
Rotation period (hours or days)	23.9 h	−243 d	24.6 h
Surface area (Earth = 1)	1.00	0.90	0.28
Atmospheric pressure (bar)	1.00	90	0.007

a thin atmosphere. Mars is also large enough to have supported considerable geological activity. But the most fascinating thing about Mars is that it probably once had a thick atmosphere and seas of liquid water; there is even a chance that some sort of life flourished there in the distant past.

9.2

The Geology of Venus

Since Venus has about the same size and composition as the Earth, we might expect its geology to be similar. This is partly true, but Venus does not exhibit the same kind of plate tectonics as the Earth, and we will see that its lack of erosion results in a very different surface appearance.

Spacecraft Exploration of Venus

More spacecraft have been sent to Venus than to any other planet. Although the 1962 U.S. Mariner 2 flyby was the first, the Soviet Union launched most of the missions to Venus, which served as the major focus of their planetary exploration program. The early Soviet Venera entry probes were crushed by the high pressure of the atmosphere before they could reach the surface, but in 1970

Venera 7 became the first probe to land and broadcast data from the surface of Venus. It lasted for 23 minutes before succumbing to the high surface temperature. Additional Venera probes and landers followed, photographing the surface and analyzing the atmosphere and soil. In 1985 two instrumented balloons were deployed into the planet's atmosphere by the Soviet VEGA flyby missions, which then continued on to intercept Comet Halley.

However, a better understanding of Venus required astronomers to make a global study of its surface, a task made very difficult by the perpetual cloud layers surrounding the planet. The problem, however, resembles the challenge facing air traffic controllers on many an evening at San Francisco airport. Often, the weather is so foggy that they can't locate the incoming planes visually. The solution is similar in both cases: use a radar instrument to probe through the obscuring layer.

The first crude global radar map was made by the U.S. Pioneer 12 orbiter in the late 1970s, followed by better maps from the twin Soviet Venera 15 and 16 radar orbiters in the early 1980s. However, most of our information on the geology of Venus is derived from the U.S. Magellan spacecraft, which mapped Venus from orbit between 1991 and 1993 (Figure 9.4). With its powerful *imaging radar*, Magellan was able to study the surface at a resolution of 100 m, much higher than that of previous

Figure 9.4
The Magellan radar orbiter spacecraft being launched from the Space Shuttle in 1989. (NASA)

Figure 9.5

Two globes of the same face of Venus, constructed from Magellan radar data. Both views show Aphrodite Terra, the largest continental area on Venus. At left, it can be identified by its brightness (the result of rougher terrain). The reddish colors have been added to simulate the red light filtering down to the surface of Venus through the vast cloud layers. At right, the same image (but with contrast slightly improved) was color-coded to resemble the Earth: continental highlands are brown, and the lava basins are blue. (JPL/USGS)

missions, yielding our first detailed look at the surface of our sister planet. (The Magellan spacecraft actually returned more data to Earth than all previous planetary missions combined; each 100 min of data transmission from the spacecraft provided enough information—if translated into characters—to fill two 30-volume encyclopedias.)

Consider for a moment how good a resolution of 100 m really is. It means the radar images from Venus can show anything on the surface larger than a football field. Suddenly a whole host of topographic features on Venus became accessible to our view. As you look at the radar images throughout this chapter, bear in mind that these are constructed from radar reflections, not from visible-light photographs. On these images, for example, bright features are an indication of rough terrain, while darker regions are smoother. Colors are sometimes added artificially to simulate the light that filters through the Venus clouds, or to show altitude differences (Figure 9.5).

Probing Through the Clouds

The radar maps of Venus reveal a planet that looks much the way the Earth might look if our planet's surface were not constantly being modified by erosion and deposition of sediment. Since there is no water or ice on Venus, and the surface wind speeds are very low, almost nothing ob-

scures or erases the complex geological features produced by widespread crustal forces and volcanic eruptions. Having finally penetrated below the clouds of Venus, we find its surface to be naked, revealing the history of hundreds of millions of years of geological activity. Venus is in many ways a geologist's dream planet.

About 75 percent of the surface of Venus consists of lowland lava plains. Superficially, these resemble the basaltic ocean basins of the Earth, but they were not produced in quite the same way. There is no evidence of subduction zones (see Chapter 7) on Venus, indicating that, unlike the Earth, this planet never experienced plate tectonics. Although convection in its mantle generated great stresses in the crust of Venus, it was never able to initiate plate motion.

The formation of the lava plains of Venus resembled that of the lunar maria, which were also the result of widespread lava eruptions without the crustal spreading associated with plate tectonics. The venusian lava plains are much younger than the lunar plains, however, and they have been distorted by stresses in the crust, as described shortly. In contrast, the lunar plains have languished almost unchanged since their formation more than 3 billion years ago.

Rising above the lowland lava plains are individual mountains and mountain ranges, as well as two full-scale continents. The largest continental area on Venus, called

Aphrodite, is about the size of Africa (you can see it stand out in Figure 9.5). Aphrodite stretches along the equator for about one-third of the way around the planet. Next in size is the northern highland region Ishtar, which is about the size of Australia. Ishtar contains the highest region on the planet, the Maxwell Mountains, which rise about 11 km above the surrounding lowlands. (The Maxwell Mountains are the only feature on Venus named after a man; they commemorate James Clerk Maxwell, whose theory of electromagnetism led to the invention of radar.)

Craters and the Age of the Surface

One of the first questions astronomers addressed with the high-resolution Magellan images was the age of the surface of Venus. Remember that the age of a surface is rarely the age of the world it is on; a young age merely implies an active geology that can resurface the world in short order. As described in the previous chapter, the age can be derived from counting impact craters (Figure 9.6a is an example of what these look like on the Venus radar images). The more densely cratered the surface, the greater its age. The largest crater on Venus (called Mead) is 275 km in diameter, slightly larger than the largest known terrestrial crater (Chicxulub) but much smaller than the lunar impact basins.

You might think that the thick atmosphere of Venus would protect the surface from impacts, burning up the projectiles long before they could reach the surface. But this is only the case for smaller projectiles. The effect of the atmosphere is readily seen in the crater statistics, which show very few craters less than 10 km in diameter, indicating that projectiles smaller than about 1 km (the size that typically produces a 10-km crater) were stopped by the atmosphere.

Those craters with diameters from 10 to 30 km are frequently distorted, apparently because the incoming projectile broke apart and exploded in the atmosphere before it could strike the ground. There are also examples of multiple craters that occurred when the projectile broke into several pieces before striking the surface (Figure 9.6b). But if we limit ourselves to impacts that produce craters with diameters of 30 km or greater, crater counts are as useful on Venus for measuring surface age as they are on airless bodies such as the Moon.

The numbers of craters on the plains of Venus are typically only about 15 percent of the lunar mare values, indicating a surface age only about 15 percent as great, or about 500 to 600 million years old. These results indicate that Venus is indeed a planet with persistent geological activity, intermediate between that of the Earth's ocean basins (which are younger and more active) and that of its continents (which are older and less active).

Almost all of the craters look fresh, with little degradation or filling in by either lava or windblown dust. This is one way we know that the rates of erosion or sediment

(a)

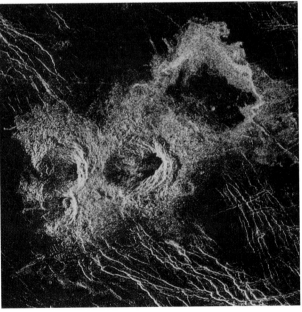

(b)

Figure 9.6

(a) Large impact craters in the Lavinia region of Venus. Because they are rough, the crater rims and ejecta appear brighter on these radar images than do the smoother surrounding lava plains. The largest of these craters has a diameter of 50 km. (b) Crater Stein, named after writer Gertrude Stein. The triple impact was caused by the breaking apart of the incoming asteroid during its passage through the thick atmosphere of Venus. We estimate the projectile had an initial diameter of between 1 and 2 km. (NASA/JPL)

Figure 9.7
This computer-generated view of two large volcanoes on the surface of Venus was assembled from Magellan radar data. In the center foreground is Sapas Mons (named for a Phoenician goddess), which is about 400 km across at its base, and 1.5 km high. The image shows a perspective located 4 km above the terrain and 527 km in front of Sapas. Lava flows extending for hundreds of kilometers are visible in the foreground. In the background you can see Maat Mons, 8 km high. The colors have been added from the Russian Venera lander images. The vertical scale has been exaggerated ten times to make the mountains more clearly visible. In real life the mountains would appear to hug the ground much more closely if viewed from over 500 km away. (NASA/JPL)

The largest individual volcano on Venus, called Sif Mons, is about 500 km across and 3 km high—broader but lower than the Hawaiian volcano Mauna Loa. At its top is a volcanic crater or caldera about 40 km across, and its slopes show individual lava flows up to 500 km long. Thousands of smaller volcanoes dot the surface, down to the limit of visibility of the Magellan images, which corresponds to cones or domes about the size of a shopping-mall parking lot. Most of these seem similar to terrestrial volcanoes. (Figure 9.7 shows a computer-generated oblique view of two of the larger volcanoes.)

Much more striking are circular, flat-topped volcanoes informally known as "pancake domes" (Figure 9.8). The larger examples are about 25 km in diameter and 2 km high, with steep ramparts around their rims. These volcanoes, which have no terrestrial counterpart, appear to be the product of very thick, viscous lava flowing out evenly in all directions.

All of this volcanism is the result of eruption of lava onto the surface of the planet. But the hot magma rising from the interior of a planet does not always make it to the surface. On both the Earth and Venus, this magma can collect to produce bulges in the crust. Many of the granite mountain ranges on Earth, such as the Sierra

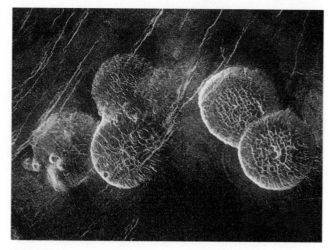

Figure 9.8
Pancake-shaped volcanoes on Venus. These remarkable circular domes, each about 25 km across and about 2 km tall, are the result of eruptions of highly viscous lava. (NASA/JPL)

deposition are very low. We have the impression that most of the venusian plains were resurfaced by large-scale volcanic activity a few hundred million years ago, and that relatively little has happened since. We see a similar situation in the lunar maria, which were all formed within a relatively short interval of time (geologically speaking), but on Venus this period of widespread volcanic activity was much more recent. As planetary scientists study the Magellan data, they are becoming more and more convinced that Venus experienced some sort of planet-wide volcanic convulsion about 500 million years ago.

Volcanoes on Venus

It appears that, like the Earth, Venus is a planet that has experienced widespread volcanism, and we can see many different types of volcanic features. In the lowland plains, volcanic eruptions are the principal way the surface is renewed, with large flows of highly fluid lava destroying old craters and generating a fresh surface every 500 million years or so. In addition, there are numerous younger volcanic mountains and other structures associated with surface hot spots—places where convection in the planet's mantle transports the interior heat to the surface.

Nevada in California, involve subsurface collections of magma that push up on the material above them.

Such subsurface features are common on Venus, and in the absence of plate tectonics or surface erosion, they are much more visible than on the Earth. Their characteristic visible expression is a large circular or oval feature called a *corona* (plural: *coronae*) (Figure 9.9). Coronae are typically several hundred kilometers across, with slightly raised interiors surrounded by a depressed ring or moat. They are a unique feature of venusian geology, not seen on any other planet or satellite in the solar system.

Tectonic Activity

The mantle convection currents on Venus do more than bring magma to the surface. As on the Earth, these convection currents exert forces on the crust. Although full-scale plate motion has not occurred on Venus, its crust is constantly subjected to pushing and stretching. These forces are called **tectonic,** and the geological features that result from such forces are called *tectonic features*.

The geology of Venus is dominated by tectonic stresses resulting from mantle convection. On the lowland plains, tectonic forces have broken the lava surface to create remarkable patterns of ridges and cracks (Figure 9.10). In a few places the crust has even torn apart to generate rift valleys. The circular features associated with

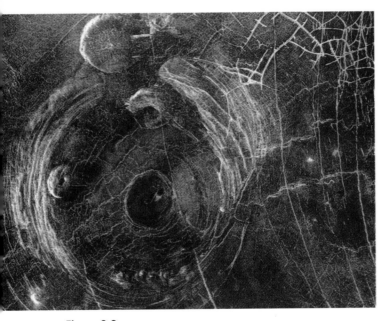

Figure 9.9
Fotla Corona, located in the plains to the south of Aphrodite Terra. This circular feature, about 200 km across, is a region where hot material from deeper inside Venus is pushing up on the surface. Curved fracture patterns clearly show where the material beneath has put stress on the surface. A number of pancake and dome volcanoes are also visible. Fotla was a Celtic fertility goddess. Some students see a resemblance between this corona and Miss Piggy of the Muppets (her right ear is the pancake volcano in the upper center of the image). (NASA/JPL)

Figure 9.10
This region of the Lakshmi Plains on Venus has been fractured by tectonic forces to produce a "cross-hatched" grid of cracks and ridges. Be sure to notice the fainter linear features that run perpendicular to the brighter ones! This image shows a region 40 km across. Lakshmi was a Hindu goddess of prosperity. (NASA/JPL)

coronae are tectonic ridges and cracks, and most of the mountains of Venus also owe their existence to tectonic forces.

The Ishtar continent, which has the highest elevations on Venus, is perhaps the most dramatic product of these tectonic forces. In many ways Ishtar and its high Maxwell Mountains resemble the Tibetan Plateau and Himalayan Mountains on the Earth. Both are the product of compression of the crust, and both are maintained by the continuing forces of mantle convection. Figure 9.11 illustrates the steep flanks of the Ishtar plateau, where folded mountains rise from the lowland plains to the south. These features look much more like the mountains of the Earth than anything seen on the Moon, Mars, or other planets.

On the Surface

The successful Soviet Venera landers found themselves on an extraordinarily inhospitable planet, with a surface pressure of 90 bars and a temperature hot enough to melt lead and zinc. Despite these unpleasant conditions, the spacecraft were able to photograph their surroundings and collect surface samples for chemical analysis before their instruments gave out. They found that the rock in the landing areas is igneous, primarily basalts. Examples of the Venera photographs are shown in Figure 9.12. Each picture shows a flat, desolate landscape with a variety of

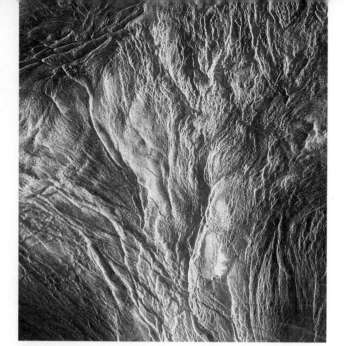

Figure 9.11
This Magellan radar image shows folded mountains with individual peaks rising about 2 km high. This region of the Danu mountains on the Ishtar continent of Venus may be similar in origin to the Himalayan Mountains on the Earth. Danu was a goddess in the mythology of ancient Ireland. (NASA/JPL)

TABLE 9.2
Atmospheric Compositions of Earth, Venus, and Mars (in %)

Gas	Earth	Venus	Mars
Carbon dioxide (CO_2)	0.03	96	95.3
Nitrogen (N_2)	78.1	3.5	2.7
Argon (Ar)	0.93	0.006	1.6
Oxygen (O_2)	21.0	0.003	0.15
Neon (Ne)	0.002	0.001	0.0003

rocks, some of which may be ejecta from impacts. Other areas show flat, layered lava flows.

The Sun cannot shine directly through the heavy, opaque clouds, but the surface is fairly well lit by diffused light. The illumination is about the same as that on Earth under a very heavy overcast, but with a strong red tint because the massive atmosphere blocks shorter-wavelength colors of light. The weather at the bottom of this deep atmosphere remains perpetually hot and dry, with calm winds. Because of the heavy blanket of clouds and atmosphere, one spot on the surface of Venus is similar to any other as far as weather is concerned.

9.3

The Massive Atmosphere of Venus

Composition and Structure

The most abundant gas in the atmosphere of Venus is carbon dioxide (CO_2), which accounts for 96 percent of the atmosphere. The second most abundant gas is nitrogen. The predominance of carbon dioxide over nitrogen is not surprising when you recall that the Earth's atmosphere would also be mostly carbon dioxide if this gas were not locked up in marine sediments (see Section 7.3). Table 9.2 compares the compositions of the atmospheres of Venus, Mars, and the Earth. Expressed in this way, as percentages, the proportions of major gases are very similar for Venus and Mars, but in total quantity their atmospheres are dramatically different. With its surface pressure of 90 bars, the venusian atmosphere is more than 10,000 times more massive than its martian counterpart.

In addition to these gases, we have found sulfur dioxide (SO_2) in Venus' middle atmosphere, which, as we will see, plays a role in the chemistry of the venusian clouds. Overall, the atmosphere of Venus is very dry; absence of water is one of the important ways that Venus differs from the Earth.

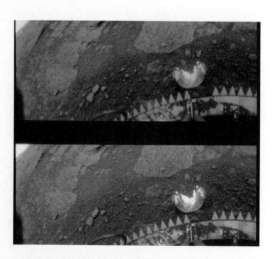

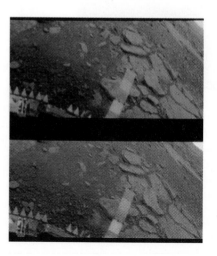

Figure 9.12
Views of the surface of Venus from the Venera 13 spacecraft. The upper images show the scene as recorded by the spacecraft cameras; everything looks orange because the thick atmosphere of Venus absorbs the bluer colors of light. Computer processing has transformed the bottom images to show what the scenes would look like under Earth lighting conditions. The horizon is visible in the upper corner of each image. (USSR Academy of Science and Carle Pieters, Brown University)

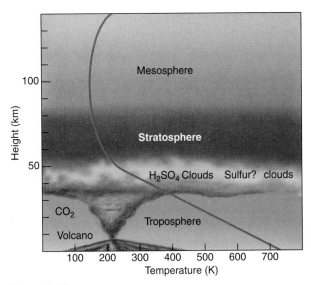

Figure 9.13
The layers of the massive atmosphere of Venus, based on data from the Pioneer and Venera entry probes. Height is measured along the left axis; the bottom scale shows temperatures, and the red line allows you to read off the temperature at each height. Notice how steeply the temperature rises below the clouds from the greenhouse effect.

The venusian atmosphere (Figure 9.13) has a huge troposphere that extends up to at least 50 km above the surface. Within the troposphere, the gas is heated from below and circulates slowly, rising near the equator and descending over the poles. With no rapid rotation to break up this flow, the atmospheric circulation is highly stable. In addition, the very size and mass of the atmosphere maintains stability. Being at the base of the atmosphere of Venus is something like being a kilometer or more below the ocean surface on the Earth. There, also, the mass of water evens out temperature variations and results in a uniform environment.

In the upper troposphere, between 30 and 60 km above the surface, there is a thick cloud layer composed primarily of sulfuric acid droplets. Sulfuric acid (H_2SO_4) is formed from the chemical combination of sulfur dioxide (SO_2) and water (H_2O). In the atmosphere of the Earth, sulfur dioxide is one of the primary gases emitted by volcanoes, but it is quickly diluted and washed out by rainfall. In the dry atmosphere of Venus, this unpleasant substance is apparently stable. Below 30 km the venusian atmosphere is clear of clouds.

The atmospheric balloons released by the VEGA spacecraft in 1985 floated for 46 h in the middle of Venus' cloud layer, at a height of 53 km. At that altitude, its instruments reported the conditions to be almost Earth-like. The pressure there is 0.5 bar and the temperature a comfortable 305 K, just a little warmer than the room in which you are reading this book. Were it not for the absence of oxygen and the nasty sulfuric acid clouds, this would not be a bad place to visit. Certainly it beats conditions on the *surface* of our "sister" planet.

Surface Temperature

The high surface temperature of Venus was discovered by radio astronomers in the late 1950s and confirmed by Mariner 2 observations and by the early Venera probes. It was not easy to understand, however, how this planet could be so hot; although Venus is somewhat closer to the Sun than is the Earth, its surface is hundreds of degrees hotter than you might expect from the extra sunlight it receives. Scientists wondered what could be heating the surface of Venus to a temperature above 700 K. The answer turned out to be the *greenhouse effect.*

The greenhouse effect works on Venus just as it does on the Earth (see Section 7.3). But since Venus has so much more CO_2—almost a million times more—the effect is much stronger. Sunlight that diffuses through the atmosphere of Venus heats the surface, but the CO_2 acts as a blanket, making it very difficult for the infrared radiation to leak back into space. As a result, the surface heats up until eventually it is emitting enough to balance the energy it receives from the Sun.

Has Venus always had such a massive atmosphere and high surface temperature, or might it have evolved to such conditions from a climate that was once more nearly Earth-like? The answer to this question is of particular interest to us as we look at the increasing levels of CO_2 in the Earth's atmosphere, as discussed in Section 7.4. Are we in any danger of transforming our own planet into a hellish place like Venus?

Let us try to reconstruct the possible evolution of Venus from an Earth-like beginning to its present state. Imagine that Venus began with moderate temperatures, water oceans, and much of its CO_2 dissolved in the ocean or chemically combined with the surface rocks, as is the case on Earth today. Then allow for modest additional heating, by a small rise in the energy output of the Sun, for example, or by an increase in atmospheric CO_2. It turns out that even a small amount of extra heat can lead to increased evaporation from the oceans and the release of gas from surface rocks.

This in turn means a further increase in the atmospheric CO_2 and H_2O, amplifying the greenhouse effect and hence leading to still more heating and the release of further CO_2 and H_2O. Unless some other processes intervene, the temperature thus continues to rise. Such a situation is called the **runaway greenhouse effect.**

The runaway greenhouse is not just a larger greenhouse effect; it is a process whereby an atmosphere evolves from a state in which the greenhouse effect is small, such as on the Earth, to one where it is a major factor, as we see today on Venus. Once the larger greenhouse conditions develop, the planet establishes a new, much hotter equilibrium near its surface. Reversing the situation is difficult if not impossible.

Note that in our scenario, if large bodies of water are available, the runaway greenhouse leads to their evaporation, creating an atmosphere of hot water vapor, which itself is a major contributor to the greenhouse effect. Water vapor in the atmosphere is not stable in the presence of solar ultraviolet light, however. It tends to break the molecules of H_2O into their constituent parts—oxygen and hydrogen. The light element hydrogen can escape from the atmospheres of the terrestrial planets, leaving the oxygen behind to combine chemically with surface rock. The loss of water is therefore an *irreversible* process; once the water is gone, it cannot be restored. There is evidence that this is exactly what happened to the water once present on Venus.

While we do not know the point at which a greenhouse effect turns into a runaway greenhouse effect, Venus stands as clear testament to the fact that a planet cannot continue heating indefinitely without a major effect on its oceans and atmosphere. It is a conclusion that we and our descendants will surely want to pay close attention to.

9.4

The Geology of Mars

Spacecraft Exploration of Mars

The U.S. Mariner 4 spacecraft flew by Mars in 1965 and radioed 22 photos to Earth. These pictures showed an apparently bleak planet with abundant impact craters. In those days craters were unexpected, and perhaps people who were romantically inclined still hoped to see canals. In any case, the Mariner 4 results represented something of a shock to the public, and newspaper headlines announced that Mars was a "dead planet."

In 1971 Mariner 9 became the first spacecraft to orbit another planet, mapping the entire surface of Mars at a resolution of about 1 km and discovering a great variety of geological features—including volcanoes, huge canyons, intricate layers on the polar caps, and channels that appeared to have been cut by running water—that the previous flybys had missed. Geologically, Mars didn't look so dead after all.

Mariner 9 set the stage for the Viking spacecraft, which were among the most ambitious and successful of all planetary missions. Two Viking *orbiters* surveyed the planet and served to relay communications for two *landers* on the surface of Mars. After an exciting and

sometimes frustrating search for a safe landing spot, the Viking 1 lander touched down on the surface of Chryse Planitia (the Plains of Gold) on July 20, 1976, exactly seven years after Neil Armstrong's historic first step on the Moon. Two months later, Viking 2 landed with equal success in another plain farther north, called Utopia. Most of the information about Mars in this chapter is derived from the work of the four Viking spacecraft (Figure 9.14).

The Viking mission continued until November 5, 1982, when Viking 1 sent its last message after more than six Earth-years on the surface. Even then, the mission was terminated by a human programming error on Earth, not by failure of the robot spacecraft on Mars. In 1981, NASA transferred ownership of the Viking 1 lander to the National Air and Space Museum—the only museum exhibit located on another planet!

No major missions were launched to Mars after Viking, in spite of continuing scientific interest in the exploration of the red planet. However, several smaller Mars probes are planned for the late 1990s, beginning with the 1996 launches of two U.S. spacecraft (Mars Pathfinder and Mars Global Surveyor) and one from Russia (Mars '96).

These are the forerunners of a number of scientific missions that will extend into the next century, including a series of spacecraft to return selected samples of martian rocks and soil to Earth, beginning in 2005. The ultimate long-term goal is human flight to Mars, leading perhaps to permanent settlements. These ambitious piloted missions are beyond the current space budgets of the United States and other space-faring nations, but Mars is well within our reach if we choose to travel there in the next century.

Figure 9.14

This globe is a photographic mosaic assembled by the mapmakers at the U.S. Geological Survey from many Mars images taken with the Viking cameras. Crossing the center of the planet is a 5000-km-long rift valley called Mariner Valley (a name it received when first discovered by the Mariner 9 spacecraft). Three giant volcanoes can be seen at the left edge of the planet. Compare this to Lowell's globe in Figure 9.2. (JPL/USGS)

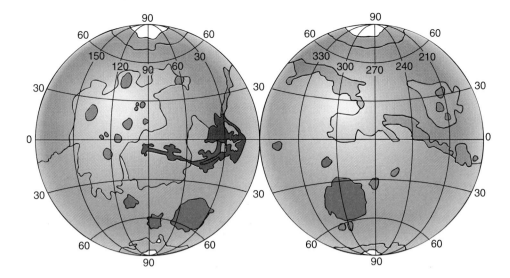

Global Properties

Mars has a diameter of 6790 km, just over half the diameter of the Earth, giving it a total surface area very nearly equal to the *continental* area of our planet. Its density of 3.9 g/cm^3 suggests a composition consisting primarily of silicates but with the possibility of a significant metal core. The planet has no detectable magnetic field, suggesting that it has no liquid conducting material in its core.

Like the Earth, Moon, and Venus, the surface of Mars has continental or highland areas as well as lower volcanic plains. Approximately half the planet, lying primarily in the southern hemisphere, consists of heavily cratered higher-elevation terrain. The other half, which is primarily in the north, contains younger, lightly cratered volcanic plains at an average elevation about 4 km lower than the highlands.

This division into older highlands and younger lowland plains seems to be characteristic of all the terrestrial planets except Mercury. The uplands are shown in light blue and the lowlands in yellow in Figure 9.15, a map of the major geological areas of Mars, many of which are described in more detail shortly. Mars has a few large impact basins in the old southern highlands, shown as dark blue in Figure 9.15. The largest basin, called Hellas, is approximately 1800 km in diameter and 6 km deep, about the size of the largest basin on the Moon.

One of the main features of martian geology is an impressive uplifted continent the size of North America. This is the 10-km-high Tharsis bulge, a volcanically active region crowned by four great volcanoes that rise another 15 km into the martian sky (the bulge is pink and the volcanoes are red in Figure 9.15).

Volcanoes on Mars

The lowland plains of Mars look very much like the lunar maria, and they have about the same number of impact craters. Like the lunar maria, they probably formed between 3 and 4 billion years ago. Apparently Mars experienced extensive volcanic activity at about the same time the Moon did, producing similar basaltic lavas.

The largest volcanic mountains of Mars are found in the Tharsis area (you can see three of them in Figure 9.14), although many smaller volcanoes dot much of the surface of the younger, northern half of the planet. The most dramatic volcano on Mars is Olympus Mons (Mount Olympus), with a diameter of more than 500 km and a summit that towers 25 km above the surrounding plains (Figure 9.16). The volume of this immense volcano is nearly 100 times greater than that of Mauna Loa in Hawaii.

The Viking imagery permits a detailed search for impact craters on the slopes of these volcanoes. Many of the

Figure 9.16
The largest volcano on Mars, and probably the largest in the solar system, is Olympus Mons, illustrated in this Viking orbiter photograph. Placed on Earth, the base of the Olympus Mons would completely cover the state of Missouri; the caldera, the circular opening at the top, is 65 km across, about the size of Los Angeles. Note the extensive clouds over the lower slopes of the volcano. (NASA/JPL)

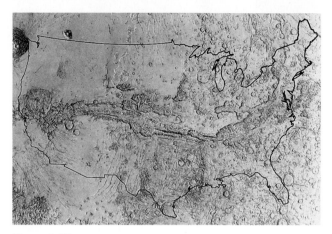

Figure 9.17
The outline of the United States is superimposed to scale on a mosaic of photographs showing the Valles Marineris canyons on Mars. You can see that if it were on Earth, the system would stretch from Los Angeles to Washington, D.C. Two volcanoes are visible in the upper left, which is the direction of the Tharsis bulge. (Stephen Meszaros/NASA)

volcanoes show fair numbers of such craters, suggesting that they ceased activity a billion years or more ago. However, Olympus Mons has very, very few impact craters. Its present surface cannot be more than about 100 million years old; it may even be much younger. Some of the fresh-looking lava flows might have been formed a hundred years ago, or a thousand, or a million; but geologically speaking they are quite young. This leads geologists to the conclusion that Olympus Mons probably remains intermittently active today.

Cracks and Canyons

The Tharsis bulge has many interesting geological features in addition to its huge volcanoes. In this part of the planet the surface itself has bulged upward, forced by great pressures from below and resulting in extensive tectonic cracking of the crust. None of these cracks shows any evidence of sliding motion such as that associated with most faults on Earth, however. Like Venus, Mars never reached the stage of plate tectonics. This may be a result of Mars' lower mass: the planet may not have had enough interior heat and pressure to set up the long-lasting tectonic activity we still see on Earth and Venus today.

Among the most spectacular tectonic features on Mars are the great canyons called the Valles Marineris, which extend for about 5000 km (nearly a quarter of the way around Mars) along the slopes of the Tharsis bulge (Figure 9.17). The main canyon is about 7 km deep and up to 100 km wide. It is so large that the Grand Canyon of the Colorado River would fit comfortably into one of its side canyons. Figure 9.18 and the opening image for this chapter show spectacular, detailed views of parts of Valles Marineris.

The term *canyon* is somewhat misleading, because the Valles Marineris canyons have no outlets and were not cut by running water. They are basically tectonic cracks, produced by the same crustal tensions that caused the Tharsis uplift. However, water is believed to have played a later role in shaping the canyons, primarily by seeping from deep springs and undercutting the cliffs. This undercutting led to landslides, gradually widening the original cracks into the great valleys we see today.

Figure 9.18
This mosaic of Viking orbiter images shows one section of the Valles Marineris canyon system. The canyon walls are about 100 km apart here. Look carefully and you can see enormous landslides whose debris is piled up underneath the cliff walls. (NASA/USGS)

While the Tharsis bulge and Valles Marineris are impressive, in general we see fewer tectonic structures on Mars than on Venus. In part, this may reflect a lower general level of geological activity, as would be expected for a smaller planet. But it is also possible that evidence of widespread faulting has been buried by wind-deposited sediment over much of Mars. Like the Earth, Mars may have hidden part of its geological history under a cloak of soil.

The View from the Surface

Viking 1 landed at a latitude of 22° N on a 3-billion-year-old windswept plain near the lowest point of a broad basin. Its desolate but strangely beautiful surroundings included numerous angular rocks, some more than 1 m across, interspersed with dune-like deposits of fine-grained, reddish soil. On the horizon, the low profiles of several distant impact craters could be seen (Figure 9.19a). At the Viking 2 site in Utopia, at latitude 48° N, the surface was somewhat similar but with a substantially greater number of rocks (Figure 9.19b). At both sites it seems that winds, sometimes blowing up to 100 km/h or more, have stripped the surface of some of its loose, fine covering to leave the rocks exposed.

Each lander peered at its surroundings through color stereo cameras, sniffed the atmosphere with a variety of analytical instruments, and poked at nearby rocks and soil with its mechanical arm. As part of its mission of searching for martian life, each lander collected soil samples and brought them on board for analysis, as described in more detail shortly. Viking found that the soil consisted of clays and iron oxides, as had long been expected from the red color of the planet (Figure 9.20).

Figure 9.20
The surface of Mars as recorded by the Viking 2 cameras. In this image, produced from a black-and-white original, the colors Viking recorded in less-detailed images have been added to this wonderfully detailed scene using computer-assisted techniques. You can see two trenches dug by the Viking collecting scoop.
(Image processing by Mary Dale-Bannister)

Each lander also carried a weather station to measure temperature, pressure, and wind. As expected, temperatures vary much more on Mars than on Earth, owing to the absence of moderating oceans and clouds. Typically, the summer maximum was 240 K (−33°C), dropping to 190 K (−83°C) at the same location just before dawn. The lowest air temperatures, measured farther north by Viking 2, were about 173 K (−100°C). During the winter, Viking 2 also photographed water frost deposits on the ground (Figure 9.21). We make a point of saying "water frost" here, because at some locations on Mars it gets cold

(a)

(b)

Figure 9.19
The martian surface as photographed by Viking. (a) The Viking 1 landing site in Chryse.
(b) The Viking 2 landing site in Utopia. (NASA/JPL)

Figure 9.21
Surface frost photographed at the Viking 2 landing site during late winter. (NASA/JPL)

enough for carbon dioxide (dry ice) to freeze out of the atmosphere as well.

Most of the winds measured at the Viking sites were low to moderate, only a few kilometers per hour. However, Mars is capable of great windstorms that can shroud the entire planet in dust (one greeted the Mariner 9 probe when it first arrived in 1971). During such storms, the Sun was greatly dimmed at the Viking sites, and the sky turned a dark red color.

Martian Samples

Much of what we know of the Moon, including the circumstances of its origin, comes from studies of lunar sam-

Figure 9.22
One of the SNC meteorites, a fragment of basalt ejected from Mars. (NASA/JSC)

ples, but spacecraft have not yet returned martian samples to Earth for laboratory analysis. It is with great interest, therefore, that scientists have concluded that samples of martian material are already here on Earth, available for study. About a dozen of these martian rocks exist, all members of a rare class of meteorites (rocks that fall to Earth from space) called *SNC meteorites* (Figure 9.22). The most obvious special characteristic of this small group is that they are volcanic *basalts;* most of them are also relatively *young,* with ages of about 1.3 billion years. We know from details of their composition that they are not from the Moon, and in any case there was no lunar volcanic activity as recently as 1.3 billion years ago. Theory suggests that it would be impossible for ejecta from impacts on Venus to escape through its thick atmosphere. By process of elimination, the only reasonable origin seems to be Mars, where the Tharsis volcanoes were certainly active at that time.

The martian origin of the SNC meteorites has been confirmed by analysis of tiny gas bubbles trapped inside several of them. These bubbles match the atmospheric properties of Mars as measured directly by Viking. Apparently some atmospheric gas was trapped in the rock by the shock of the impact that ejected it from Mars and started it on its way toward Earth.

One of the most exciting results from analysis of SNC meteorites has been the discovery of both water and organic (carbon-based) compounds, which suggest to a few scientists—although this suggestion is rather speculative—that Mars may once have had oceans and even life on its surface. As we will see in a moment, there is other evidence for the presence of flowing water on Mars in the remote past.

9.5

Martian Polar Caps, Atmosphere, and Climate

The atmosphere of Mars today has an average surface pressure of only 0.007 bar, less than 1 percent that of the Earth. (This is how thin the air is about 30 km above the Earth's surface.) Martian air is composed primarily of carbon dioxide (95 percent), with about 3 percent nitrogen and 2 percent argon. The proportions of different gases are similar to those in the atmosphere of Venus (Table 9.2), but a lot less of each gas is found on Mars.

Clouds on Mars

Several types of clouds can form in the martian atmosphere. First there are dust clouds, raised by winds, which can sometimes grow to cover a large fraction of the surface. Second are water-ice clouds similar to those on Earth. These often form around mountains, just as happens on our planet (Figure 9.16). Finally, the CO_2 of the atmosphere can itself condense at high altitudes to form hazes of dry ice crystals. The CO_2 clouds have no counterpart on Earth, since on our planet, temperatures never drop low enough (down to about 150 K) for this gas to condense.

Although the atmosphere contains water vapor, and occasional clouds of water ice can form, *liquid* water is not stable under present conditions on Mars. Part of the problem is the low temperatures on the planet. But even if the temperature on a sunny summer day rises above the freezing point, liquid water still cannot exist. At a pressure of less than 0.006 bars, only the solid and vapor forms are possible. In effect, the boiling point is as low or lower than the freezing point, and water changes directly from solid to vapor without an intermediate liquid state.

The Polar Caps

Through a telescope the most prominent surface features on Mars are the bright polar caps, which change with the seasons (see the opening image for Chapter 6). These seasonal caps are similar to the seasonal snow-cover on Earth. We do not usually think of the winter snow in northern latitudes as a part of our polar caps, but seen from space, the thin snow merges with the Earth's thick, permanent ice caps to create an impression much like that seen on Mars.

The seasonal caps on Mars are composed not of ordinary snow, but of frozen CO_2 (dry ice). These deposits condense directly from the atmosphere when the surface temperature drops below about 150 K. The caps develop during the cold martian winters, extending down to about latitude 50° by the start of spring.

Quite distinct from these thin seasonal caps of CO_2 are the permanent, or residual, caps that are always present near the poles (Figure 9.23). As the seasonal cap retreats during spring and early summer, it reveals the brighter, thicker cap beneath. The southern permanent cap has a diameter of 350 km and is composed of frozen carbon dioxide deposits together with an unknown thickness of water ice. Throughout the southern summer, it remains at the freezing point of CO_2, 150 K, and this cold reservoir is thick enough to survive the summer heat intact.

The northern permanent cap is different. It is much larger, never shrinking below a diameter of 1000 km, and is composed of water ice. Summer temperatures in the north are too high for the frozen CO_2 to be retained. We do not know the thickness of the water ice in either

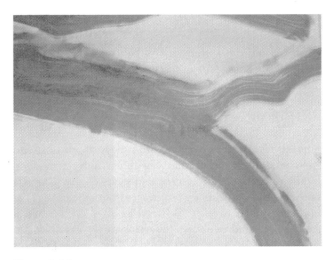

Figure 9.24
Detail of the southern polar cap showing terracing on exposed slopes. These terraces preserve a record of past climate variations. The width of this frame is about 50 km. (NASA/JPL)

cap, but it may be as much as several kilometers. In any case, the polar caps represent a huge reservoir of water compared with the very small amounts of water vapor in the atmosphere. The two caps are different because Mars' distance from the Sun varies substantially during the course of its year. This means seasons (and temperatures) on Mars are affected by both the tilt of its axis and its distance from the Sun.

The Viking images also show some distinctive terrain surrounding the permanent polar caps, as well as in ice-free areas within the caps themselves (Figure 9.24). At

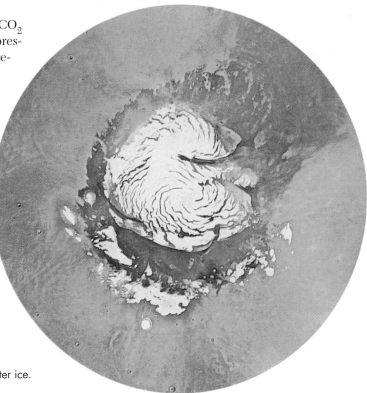

Figure 9.23
Viking orbiter photograph of the residual north polar cap of Mars, which is about 1000 km across and composed of water ice. (NASA/USGS)

latitudes above 80° in both hemispheres, the surface consists of recent, layered sedimentary deposits that entirely cover the older cratered ground below. Individual layers are typically a few tens of meters in thickness, marked by alternating light and dark bands of sediment. Probably the material in the polar deposits is dust carried by wind from the equatorial regions of Mars.

What do these terraced layers tell us about Mars? Some cyclic process is laying down deposits of dust and ice over long periods of time. The time scales represented by the polar layers are tens of thousands of years. Apparently the martian climate experiences periodic changes at intervals similar to those between ice ages on the Earth. Calculations indicate that the causes are probably also similar: gravitational perturbations from other planets produce variations in Mars' orbit and tilt as the great clockwork of the solar system goes through its paces.

Channels and Floods

Although no liquid water exists on Mars today, the Viking data have revealed fascinating evidence that rain once fell and rivers once flowed on the red planet. Two kinds of geological features appear to be remnants of ancient watercourses. We will examine each of them in turn, but note for the record that both types are far too narrow to have been the "canals" Percival Lowell imagined seeing on Mars.

In the highland equatorial plains, the Viking spacecraft photographed multitudes of small, sinuous (twisting) channels—typically a few meters deep, some tens of me-

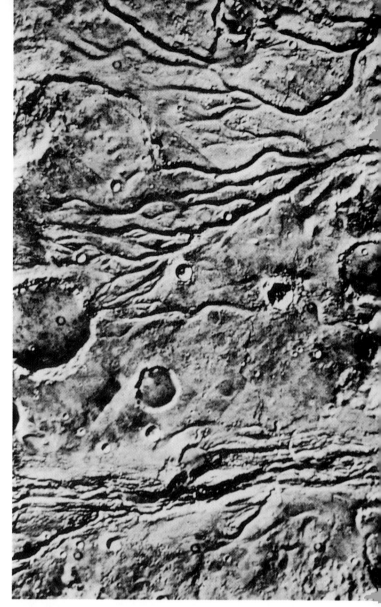

Figure 9.26
Large outflow channels photographed by Viking. These features appear to have been formed in the distant past from massive floods of water. The width of this image is about 150 km. (NASA/JPL)

ters wide, and perhaps 10 or 20 km long (Figure 9.25). They are called *runoff channels* because their form looks like what geologists would expect from troughs that carried the surface runoff of ancient rainstorms. If they were indeed made by rain, these runoff channels seem to be telling us that the planet had a very different climate long ago.

How can we tell when rain might have last fallen on Mars? Crater counts show that this part of the planet is more cratered than the lunar maria but less so than the lunar highlands. Thus the runoff channels are probably older than the lunar maria, presumably at least 3.9 billion years old.

The other water-related features we see are *outflow channels* (Figure 9.26), much larger than the older runoff channels. The largest of these, which drain into the

Figure 9.25
Runoff channels in the old martian highlands, interpreted as the valleys of ancient rain-fed rivers. The width of this image is about 200 km. (NASA/from Mars Digital Image Map, processing by Brian Fessler, LPI)

Chryse basin where Viking 1 landed, are 10 km or more in width and hundreds of kilometers long. Many features of these outflow channels have convinced geologists that they were carved by huge volumes of running water, far too great to be produced by ordinary rainfall. Where did such floodwater come from on Mars?

As far as we can tell, the regions where the outflow channels begin contained abundant water frozen in the soil as permafrost. Some local source of heating must have released this water, leading to a period of rapid and catastrophic flooding. Perhaps this heating was associated with the formation of the volcanic plains on Mars, which date back to roughly the same time as the outflow channels.

The older runoff channels, however, require one crucial thing that is absent on the Mars we see today: an atmosphere thick enough to sustain liquid water.

Climate Change on Mars

The evidence discussed so far suggests that billions of years ago martian temperatures must have been warmer, rain must have fallen, and the atmosphere must have been much more substantial than it is today. What could have changed the climate on Mars so dramatically over the millennia?

We presume that, like the Earth and Venus, Mars formed with a much thicker atmosphere than it has now, and that this atmosphere maintained a higher surface temperature thanks to the greenhouse effect. But Mars is a smaller planet, and its lower gravity means that gases could escape more easily than from Earth and Venus. As more and more of the atmosphere escaped into space, the temperature on the surface gradually fell.

Eventually Mars became so cold that any water froze out of the atmosphere, further reducing its ability to retain heat—a sort of runaway refrigerator effect, just the opposite of the runaway greenhouse effect that occurred on Venus. The result is the cold, dry Mars we see today. Probably this loss of atmosphere took place within less than a billion years; based on the absence of runoff channels in the northern plains, it seems that rain has not fallen on Mars for at least 3 billion years.

The Search for Life on Mars

If there was running water on Mars in the past, as the channels imply, perhaps there was life as well; could life, in some form, remain in the martian soil today? Testing this possibility, however unlikely, was one of the primary objectives of the Viking landers (Figure 9.27).

These landers carried, in addition to the instruments already discussed, miniature biological laboratories to test for microorganisms in the martian soil. They looked for evidence of *respiration* by living animals, *absorption of nutrients* offered to organisms that might be present, and an *exchange of gases* between the soil and its surroundings

for any reason whatsoever. In various tests, martian soil was scooped up by the spacecraft's long arm and placed into the experimental chambers, where it was isolated and incubated in contact with a variety of gases, radioactive isotopes, and nutrients to see what would happen. A fourth instrument pulverized the soil and analyzed it carefully to determine what organic (carbon-bearing) material it contained.

The Viking experiments were so sensitive that, had one of the spacecraft landed anywhere on Earth (with the possible exception of Antarctica), it would easily have detected life. But, to the disappointment of many scientists and members of the public, no life was detected on Mars. The soil tests for absorption of nutrients and gas exchange did show some activity, but this was most likely caused by chemical reactions that began as the soil was heated and had nothing to do with life. In fact, these experiments showed that martian soil seems much more chemically active than terrestrial soils because of its exposure to solar ultraviolet radiation (Mars has no ozone layer).

The organic chemistry experiment showed no trace of organic material, which is apparently destroyed by the sterilizing effect of this ultraviolet light. While the possibility of martian life has not been eliminated, most experts consider it negligible. Although Mars has the most Earth-like environment of any planet in the solar system, the sad fact is nobody seems to be home.

9.6

Divergent Planetary Evolution

Venus, Mars, and our own planet Earth form a remarkably diverse triad of worlds. Although all three orbit at about the same distance from the Sun, and all apparently started with about the same chemical mix of silicates and metals, their evolutionary paths have diverged. As a result, Venus became hot and dry, Mars became cold and dry, and only the Earth ended up with what we consider a hospitable climate.

We have discussed the runaway greenhouse effect on Venus and the runaway refrigerator effect on Mars. But we do not understand exactly what started these two planets down these evolutionary paths. Was the Earth ever in danger of a similar fate? Or might it still be diverted onto one of these paths, perhaps due to stress on the atmosphere generated by human pollutants? One of the reasons for studying Venus and Mars is to seek insight into these questions.

Some people have even suggested that if we understood the evolution of Mars and Venus better, we could possibly reverse their evolution and restore more Earth-like environments. This process, which is highly speculative, is called *terraforming*. While it seems unlikely that humans could ever make either Mars or Venus into a replica of the Earth, considering such possibilities is a

Figure 9.27
Engineering model of the Viking lander in a Mars simulation laboratory. The boom picks up soil and rocks for analysis. The two cylinders at the top of the lander are survey cameras. Below the right-hand camera is one of three rocket engines used during the final soft landing on Mars. Jim Martin (foreground) was the project manager for Viking. (NASA/JPL)

useful part of our more general quest to understand the delicate environmental balance that distinguishes our planet from its two neighbors.

In Chapter 13 we will return to the comparative study of the terrestrial planets and their divergent evolutionary histories.

Summary

9.1 Venus, the nearest planet, is a great disappointment through the telescope, due to its impenetrable cloud cover. Mars is more tantalizing, with dark markings and polar caps. Early in the 20th century it was widely believed that the "canals" of Mars indicated intelligent life there. Mars has only 11 percent the mass of the Earth, but Venus is nearly our twin in size and mass. Mars rotates in 24 hours and has seasons like the Earth; Venus has a retrograde rotation period of 243 days. Both planets have been extensively explored by spacecraft.

9.2 Venus has been mapped by radar; its crust consists of 75 percent lowland lava plains, numerous volcanic features including strange pancake domes, and many large coronae, which are the expression of subsurface volcanism. The planet has been modified by widespread **tecton-**

ics driven by mantle convection, forming complex patterns of ridges and cracks and building high continental regions such as Ishtar. The surface is extraordinarily inhospitable, with pressure of 90 bars and temperature of 730 K, but several Russian Venera landers investigated it successfully.

9.3 The atmosphere of Venus is 96 percent CO_2. Thick clouds at altitudes of 30 to 60 km are made of sulfuric acid, and a CO_2 greenhouse effect maintains the high surface temperature. Venus presumably reached its current state from more Earth-like initial conditions as a result of a **runaway greenhouse effect,** which included the loss of large quantities of water.

9.4 Mars has heavily cratered highlands in its southern hemisphere but younger, lower volcanic plains over much of its northern half. The Tharsis bulge, as big as North America, includes several huge volcanoes; Olympus Mons is 25 km high and 500 km in diameter. The Valles Marineris canyons are tectonic features widened by erosion. The Viking landers revealed barren, windswept plains at Chryse and Utopia. Currently there is great interest in the SNC meteorites, which are samples of the martian crust.

9.5 The martian atmosphere has a surface pressure of less than 0.01 bar and is 95 percent CO_2. There are dust clouds, water clouds, and carbon dioxide (dry ice) clouds. Liquid water is not possible, but there may be subsurface permafrost. Seasonal polar caps are made of dry ice, but the northern residual cap is water ice. Evidence of a very different climate in the past is found in water erosion features, both runoff channels and outflow channels, the latter carved by catastrophic floods. The Viking landers searched for martian life in 1976, with negative results, but life might have flourished long ago.

9.6 Earth, Venus, and Mars have diverged in their evolution. We need to understand why if we are to protect the environment of the Earth.

Review Questions

1. List several ways that Venus, Earth, and Mars are similar, and several ways that they are different.

2. Although neither Venus nor Mars has oceans today, both are considered to have "continents" somewhat like those on the Earth. Describe the continents on each planet. What fraction of the surface area do they take up on each world?

3. Compare the current atmospheres of Earth, Venus, and Mars in terms of:
   a. composition
   b. thickness (and pressure at the surface)
   c. greenhouse effect

4. How might Venus' atmosphere have evolved to its present state through a runaway greenhouse effect?

5. The Magellan mission has taught us the most about Venus, while the Viking mission has been the source of the largest amount of information about Mars. Compare the instruments on each mission, and describe how each helped us understand the planet it surveyed.

6. Describe the current atmosphere on Mars. What evidence suggests that it must have been different in the past?

7. Explain the "runaway refrigerator effect" and the role it may have played in the evolution of Mars.

Thought Questions

8. What are the advantages of using radar imaging rather than ordinary cameras to study the topography of Venus? What are the relative advantages of these two approaches to mapping the Earth or Mars?

9. Venus and Earth are nearly the same size and distance from the Sun. What are the main differences in the geology of the two planets? What might be some of the reasons for these differences?

10. Why is there so much more carbon dioxide in the atmosphere of Venus than in that of the Earth? Why so much more than on Mars?

11. Compare Mars with Mercury and the Moon in terms of overall properties. What are the main similarities and differences?

12. Contrast the mountains on Mars and Venus with those on Earth and the Moon.

13. We believe that all of the terrestrial planets had similar histories when it comes to impacts from space. Explain how this idea can be used to date the formation of the martian highlands, the martian basins, and the Tharsis volcanoes. How certain are the ages derived for these features?

14. Is it likely that life ever existed on either Venus or Mars? Justify your answer.

15. Suppose that, decades from now, astronauts are sent to Mars and Venus. In each case, describe what kind of protective gear they would have to carry, and what their chances for survival would be if their spacesuits ruptured.

16. We believe that Venus, Earth, and Mars all started with a significant supply of water. Explain where that water is now for each planet.

17. Assume you are a travel agent in the next century, planning a tour of the "great natural tourist sights" of Mars. Assuming you have good transportation available, and money is no object, what places on Mars would you include on your tour? Give a brief justification of each choice.

Problems

18. At its nearest approach, Venus comes within about 40 million km of the Earth. How distant is it at its farthest?

19. If you weigh 150 lb on the surface of the Earth, how much would you weigh on Venus? On Mars?

20. Calculate the relative *land* areas—that is, the amounts of the surface not covered by liquids—of the Earth, Moon, Venus, and Mars. (*Note:* 70 percent of the Earth is covered with water.)

21. Where is the water on Mars? Try to estimate how much might be present in various forms, such as in the polar caps (using the dimensions given in the text) or in subsurface permafrost (assuming various permafrost thicknesses, from 1 to 10 km, and a concentration of ice in the permafrost of 10 percent by volume).

Suggestions for Further Reading

On Venus

Cattermole, P. *Venus: The Geological Story.* 1994, Johns Hopkins Press. Clear, terse introduction to Venus after Magellan.

Cooper, H. *The Evening Star: Venus Observed.* 1993, Farrar, Straus, and Giroux. Excellent primer on Venus exploration, with a focus on the Magellan mission.

Grinspoon, D. "Venus Unveiled" in *The Sciences,* July/Aug. 1993, p. 20.

Saunders, S. "Venus: A Hellish Place Next Door" in *Astronomy,* Mar. 1990, p. 18.

Stofan, E. "The New Face of Venus" in *Sky & Telescope,* Aug. 1993, p. 22. Excellent review.

On Mars

Carr, M. "The Surface of Mars: A Post-Viking View" in *Mercury,* Jan./Feb. 1983, p. 2.

Carroll, M. "Digging Deeper for Life on Mars" in *Astronomy,* Apr. 1988, p. 6.

Cooper, H. *The Search for Life on Mars.* 1980, Holt Rinehart & Winston. The Viking program described by a well-known journalist.

Hartmann, W. "What's New on Mars" in *Sky & Telescope,* May 1989, p. 471.

Hoyt, W. *Lowell and Mars.* 1976, U. of Arizona Press. Probably the best study of the canal controversy.

Kargel, J. and Strom, R. "The Ice Ages of Mars" in *Astronomy,* Dec. 1992, p. 40.

McKay, C. "Did Mars Once Have Oceans?" in *Astronomy,* Sep. 1993, p. 27.

Robinson, M. "Surveying Scars of Ancient Martian Floods" in *Astronomy,* Oct. 1989, p. 38.

Wilford, J. *Mars Beckons.* 1990, Random House. *A New York Times* reporter looks at Mars past, present, and future.

On Both

Kasting, J. et al. "How Climate Evolved on the Terrestrial Planets" in *Scientific American,* Feb. 1988.

Morrison, D. and Owen, T. *The Planetary System,* 2nd ed. 1996, Addison-Wesley. A review of modern planetary astronomy.

Sagan, C. *Pale Blue Dot.* 1994, Random House. Fine book on planetary exploration.

Mars Explorer and *Venus Explorer* are two pieces of software that enable the user to "walk around" on a grand map of each planet. (Virtual Reality Laboratories: 1-800-829-VRLI) See also some of the planet World Wide Web sites listed in Appendix 1.

1. Use the *Follow Planet* panel to watch Mars every 30 days.

From your study can you explain why the seasonal part of Mars' south polar cap is larger than that of the northern polar cap?

2. Change the view of Mars in the *Follow Planet* panel to above the planet and the height to 200,000.

Measure the orbital periods of Phobos and Deimos.

Knowing that the Moon–Earth distance is 41 times the Phobos–Mars distance and 16 times the Deimos–Mars distance, determine the mass of Mars compared to Earth's using Kepler's laws of orbital motion.

The four jovian planets to scale, in a composite Voyager image. (NASA/JPL)

CHAPTER **10**

The Giant Planets

Thinking Ahead

Are you a surface chauvinist? You should be: you grew up on a planet with a solid surface and probably take it for granted that a world must have a surface. Yet, in the outer solar system, four planets do not have solid surfaces. You cannot land *on* Jupiter, but only *in* it. What would it be like to visit such a world?

W hat do we learn about the Earth by studying the planets? Humility.

Andrew Ingersoll in discussing the Voyager missions in 1986.

B eyond Mars and the asteroid belt we encounter a new region of the solar system: the realm of the giants. Here planets are much larger, the distances between them are vastly increased, and they are accompanied by extensive systems of satellites and rings. From many perspectives the outer solar system is where the action is, and the giant planets are the more important members of the Sun's family. When compared to these outer giants, the little cinders of rock and metal that orbit closer to the Sun can seem like insignificant afterthoughts.

Exploring the Outer Planets

Five planets—Jupiter, Saturn, Uranus, Neptune, and Pluto—are found in the outer solar system. The first four of these are the giant or jovian planets, the subject of this chapter. In contrast, Pluto is a small world, by far the smallest in the planetary system. Pluto is physically similar to the satellites of the giant planets, and is therefore discussed with them in Chapter 11.

Composition and Chemistry

As we saw in Chapter 6, most of the mass in the planetary system is found in the outer solar system. Jupiter alone exceeds the mass of all the other planets combined (Figure 10.1). The chemistry of the outer solar system is also different, with hydrogen rather than oxygen dominating. When the planets and the Sun were forming, it was cooler in the parts of the solar nebula farther from the Sun. This allowed water ice and other volatile compounds to condense, whereas these materials remained as gas (and eventually dissipated) in the inner solar system.

As a result, the formative stages of the outer planets included a significantly greater mass of solid material to begin with. This meant that the core bodies for the giant

Figure 10.1
This Hubble Space Telescope image of Jupiter, taken on Feb. 13, 1995, from a distance of 961 million km, shows a remarkable amount of detail in the turbulent atmosphere of the giant planet. The Great Red Spot is visible to the lower right; just below and to the left of it are three white oval storms being swept eastward by prevailing winds while the Red Spot moves westward. The Hubble Space Telescope now permits astronomers to make routine "weather reports" from Jupiter. (R. Beebe, A. Simon, and NASA)

TABLE 10.1 Abundances in the Outer Solar Nebula

Material	Percent (by mass)
Hydrogen (H$_2$)	75
Helium (He)	24
Water (H$_2$O)	0.6
Methane (CH$_4$)	0.4
Ammonia (NH$_3$)	0.1
Rock (incl. metal)	0.3

planets had more gravity to capture the gases hydrogen and helium, which were abundant in the solar nebula. Hence, the outer planets were able to grow significantly bigger than those closer to the Sun. Table 10.1 lists the main materials that were available in the outer solar nebula to form a planet, based on the elements we observe making up the Sun.

With so much hydrogen available, the chemistry of the outer solar system became *reducing* (see Chapter 6). Most of the oxygen chemically combined with hydrogen to make H$_2$O, and was thus unavailable to form many oxidized compounds with other elements. As a result, the compounds detected in the atmospheres of the giant planets are hydrogen-based gases such as methane (CH$_4$) and ammonia (NH$_3$), or more complex hydrocarbons (combinations of hydrogen and carbon) such as ethane (C$_2$H$_6$) and acetylene (C$_2$H$_2$).

Exploration of the Outer Solar System

Six spacecraft, five from the United States and one from Europe, have penetrated beyond the asteroid belt. The challenges of probing so far away from Earth are considerable. Flight times to the outer planets are measured in years to decades, rather than the few months required to reach Venus or Mars. Spacecraft must be highly reliable, and also capable of a fair degree of independence and autonomy. This is because, even at the speed of light, messages between Earth and the spacecraft take several hours to arrive. If a problem develops near Saturn, for example, the spacecraft computer must deal with it directly. To wait hours for the alarm to reach Earth and instructions to be routed back to the spacecraft could spell disaster.

These spacecraft also must carry their own electrical energy sources, since sunlight is too weak to supply energy through solar cells. Heaters are required to keep instruments at proper operating temperatures, and spacecraft must have powerful radio transmitters and large antennas if their precious data are to be transmitted to receivers on Earth a billion kilometers or more away. Table 10.2 summarizes the encounter dates for the spacecraft missions to the outer solar system.

The first spacecraft to the outer solar system were Pioneers 10 and 11, launched in 1972 and 1973 as pathfind-

TABLE 10.2
Missions to the Outer Solar System

Planet	Spacecraft	Encounter Date
Jupiter	Pioneer 10	Dec 73
	Pioneer 11	Dec 74
	Voyager 1	Mar 79
	Voyager 2	Jul 79
	Ulysses*	Dec 91
	Galileo	Dec 95
Saturn	Pioneer 11	Sep 79
	Voyager 1	Nov 80
	Voyager 2	Aug 81
	Cassini	Jul 04
Uranus	Voyager 2	Jan 86
Neptune	Voyager 2	Aug 89

* The Ulysses spacecraft was designed to study the polar regions of the Sun. To get away from the ecliptic, it needed a gravity boost from flying by Jupiter.

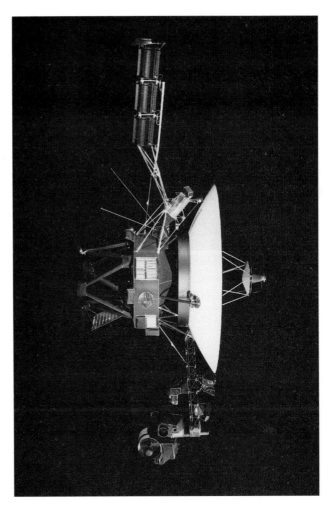

Figure 10.2
The Voyager spacecraft. The camera (at bottom) is on an arm that swivels. Below the white main antenna for sending data back to Earth, you can see a gold disk; this is the cover for an audio-video recording of the sights and sounds of Earth, included for the remote possibility that some other civilization might find the capsule someday. The entire spacecraft weighs about a ton on Earth. (NASA/JPL)

ers to Jupiter. One of their main objectives was simply to determine whether a spacecraft could actually navigate through the belt of asteroids that lies beyond Mars without colliding with small particles in the belt. Another was to measure the radiation hazards in the enormous magnetosphere of Jupiter (see Chapter 7 for a discussion of magnetospheres). Both spacecraft passed through the asteroids without incident, but the energetic particles in Jupiter's magnetic field nearly wiped out their electronics, providing information necessary for the safe design of subsequent missions.

Pioneer 10 flew past Jupiter in 1974, after which it sped outward toward the limits of the solar system. Pioneer 11 undertook a more ambitious program, using the gravity of Jupiter during its 1975 encounter to divert it toward Saturn, which it reached in 1979. The two small craft eventually became the first human-built objects to leave the realm of the planets.

The most productive scientific missions to the outer solar system were Voyagers 1 and 2, launched in 1977 (Figure 10.2). The Voyagers each carried 11 scientific instruments, including cameras and spectrometers, as well as devices to measure the characteristics of planetary magnetospheres. Voyager 1 reached Jupiter in 1979 and used a gravity assist from that planet to take it on to Saturn in 1980 (see Figure 2.12). The second Voyager arrived at Jupiter four months later and then followed a different path to complete a grand tour of the outer planets, reaching Saturn in 1981, Uranus in 1986, and Neptune in 1989. Most of the information in this chapter and in Chapter 11 is derived from the Voyager missions.

The trajectory of the Voyagers was made possible by the approximate alignment of the four giant planets on the same side of the Sun. About once every 175 years, these planets are in a position that allows a single spacecraft to visit them all by using gravity-assisted flybys to adjust its course for each next encounter. We are fortunate that this opportunity was seized. Because of the alignment, every planet in the outer solar system except Pluto has been visited by spacecraft; without it, our basic reconnaissance of the planetary system would probably have waited until well into the next century.

All of the above missions were *flybys* of the giant planets: one quick look and the spacecraft moved on. For a more detailed study, we must send spacecraft that can go into orbit around a planet: this is the mission of Galileo, launched to Jupiter in 1989, and Cassini, which will be sent to Saturn in 1997.

The Galileo Mission

The Galileo spacecraft arrived at Jupiter in December 1995, beginning its investigations by deploying an entry probe into the planet for the first direct studies of its

Engineering and Space Science: Teaching an Old Spacecraft New Tricks

By the time Voyager 2 arrived at Neptune in 1989, almost exactly 12 years after its launch, the spacecraft was beginning to show almost human signs of old age. The arm on which the camera and other instruments were located was "arthritic"—it could no longer move easily in all directions. The communications system was "hard-of-hearing"—part of its radio receiver was not operating anymore. The "brains" had significant "memory loss"—some of the computer memory on board had failed. And the whole spacecraft was beginning to run out of energy—its radioactive generators had begun showing serious signs of wear.

To make things even more of a challenge, the craft's mission at Neptune was in many ways the most difficult of all four flybys. For example, since sunlight at Neptune is 900 times weaker than at Earth, the on-board camera had to take longer exposures in this light-starved environment. This was a non-trivial requirement, given that the spacecraft was shooting by Neptune at ten times the speed of a rifle bullet, or 42,000 mi/h (at this speed, it could have gone around the Earth's equator in only 36 min).

The solution was to swivel the camera backwards at a rate that would exactly compensate for the forward motion of the spacecraft. But since the arm on which the camera was mounted could not move with the ease and grace it once had (and because instructions from Earth took about 4 hr to reach Neptune), engineers had to preprogram the ship's computer to execute an incredibly complex series of maneuvers for each image. The beautiful illustrations you see in this chapter are a testament to their ingenuity.

Another problem that confronted the mission scientists was the sheer distance of the craft from its controllers on Earth. Voyager communicates its results (including a pattern of dark and light dots that make up each image) by means of its on-board radio transmitter. By the time the radio beam arrived at the Earth in 1989, it had traveled (and spread out over) about 4.8 billion km. The power that reached us was approximately 10^{-16} watts, or 20 billion times less power than it takes to operate a digital watch. Thirty-eight different antennas on four continents were used by NASA to collect the faint signals from the spacecraft and decode the precious information they contained.

---

gaseous outer layers. Like a flaming meteor, the 339-kg probe plunged at a shallow angle into the jovian atmosphere, traveling at a speed of 50 km/s (more than 160,000 km/h, fast enough to fly from New York to San Francisco in 100 seconds).

Atmospheric friction slowed the probe within two minutes, reaching a maximum deceleration of nearly 230 g's and producing temperatures at the front of its heat shield as high as 15,000°C. This high speed was a result of the strong gravitational attraction of Jupiter, which presents the most difficult spacecraft entry condition anywhere in the solar system (except for the Sun itself). As the probe speed dropped to 2500 km/h, the remains of the glowing heat shield were jettisoned, and a parachute was deployed to lower the instrumented probe spacecraft more gently into the atmosphere (Figure 10.3).

While the probe descended into the clouds of Jupiter, the Galileo main spacecraft was passing overhead at about half the Earth–Moon distance. The data from the probe instruments were sent to the orbiter over a tiny 23-watt radio transmitter, to be recorded and later relayed to the Earth. The probe made its measurements during its 57-min descent, covering about 200 km in depth while being carried another 500 km sideways by the strong jovian winds. The transmitter then switched off due to increasing temperature as the probe reached hotter layers of the jovian atmosphere. All of these events were controlled by the computers on the probe and orbiter spacecraft without any intervention from controllers on the Earth. A few minutes later the polyester parachute melted, and within a few hours the main aluminum and titanium structure of the probe vaporized to became a part of Jupiter itself.

About two hours after the receipt of the final probe data, Galileo fired its retro-rockets to slow the spacecraft so it could be captured into orbit around the planet. An additional rocket firing in April 1996 completed the orbital insertion and started the spacecraft on a two-year tour of Jupiter's main satellites. Repeated close flybys of these big moons allow us to take images and make other observations at resolutions as much as 100 times greater than the best achieved by Voyager. Although the spacecraft's main antenna has been inoperable, mission engineers have been using the backup antenna and some clever data compression techniques to get most of the information from the mission back to Earth.

The similar Cassini mission to Saturn is under development as a cooperative venture between NASA and the European Space Agency. Cassini is planned for launch in

Figure 10.3
Artist's impression of the Galileo Probe descending into the clouds on its parachute just after the protective heat shield separated. The Probe made its measurements of Jupiter on December 7, 1995. (NASA/ARC)

1997, and arrival at Saturn in 2004. It will also deploy an entry probe, but not into Saturn; instead it will explore the atmosphere of Saturn's moon Titan before beginning its orbital tour of Saturn and its satellites. (Titan, a giant moon with a planet-like atmosphere, will be discussed more thoroughly in Chapter 11.)

10.2

The Jovian Planets

Let us now examine the four jovian planets in more detail. Our approach is not just to catalog their characteristics, but to *compare* them with each other, noting their similarities and differences, and attempting to relate their properties to their differing masses and distances from the Sun.

Basic Characteristics

The giant planets are much farther from the Sun than we are. The median distance of Jupiter from the Sun is 778 million km, 5.2 times the Earth's distance (or 5.2 AU); thus it takes just under 12 years to circle the Sun. Saturn is about twice as far away as Jupiter and takes very nearly the standard human generation of 30 years to complete one revolution. Its slow cycle around the sky provided the longest natural time interval available to ancient peoples.

As we saw in Chapter 2, Uranus and Neptune were not discovered until larger telescopes came into regular use. Uranus orbits at 19 AU with a period of 84 years, while Neptune at 30 AU requires 165 years for each circuit of the Sun. Not until 2010 will a full Neptune "year" have passed since its discovery in 1845.

Jupiter and Saturn have many similarities in composition and internal structure, although Jupiter is nearly four times more massive. In contrast, Uranus and Neptune are smaller worlds that differ in composition and structure from their larger siblings. Some of the main properties of these four planets are summarized in Table 10.3.

Jupiter, the giant among giants, has enough mass to make 318 Earths. Its diameter is about 11 times the Earth's (and about one-tenth the Sun's). Jupiter's average density is 1.3 g/cm^3, much lower than that of any of the terrestrial planets. (Recall that water has a density of 1 g/cm^3.) Jupiter's material is spread out over a volume so large that more than 1400 Earths could fit within it.

Saturn's mass is 95 times that of the Earth, and its average density is only 0.7 g/cm^3—the lowest of any planet. Since this is less than the density of water, Saturn would be light enough to float if a bathtub large enough to contain it existed (but imagine the ring it would leave).

Uranus and Neptune each have a mass about 15 times that of the Earth and hence only 5 percent as great as Jupiter. Their densities of 1.2 g/cm^3 and 1.6 g/cm^3, respectively, are much higher than that of Saturn, in spite of their smaller mass and weaker gravity. This tells us that their composition must be fundamentally different, consisting for the most part of heavier materials than hydro-

TABLE 10.3
Basic Properties of the Jovian Planets

Planet	Distance (AU)	Period (years)	Diameter (km)	Mass (Earth = 1)	Density (g/cm^3)	Rotation (hours)
Jupiter	5.2	11.9	142,800	318	1.3	9.9
Saturn	9.5	29.5	120,540	95	0.7	10.7
Uranus	19.2	84.1	51,200	14	1.2	17.2
Neptune	30.1	164.8	49,500	17	1.6	16.1

gen and helium, the primary constituents of Jupiter and Saturn. We will return to the question of their composition shortly.

Appearance and Rotation

When we look at the giant planets, we see only their atmospheres, composed primarily of hydrogen and helium gas (Figure 10.4). The uppermost cloud deck of Jupiter and Saturn, the part we see when looking down at these planets from above, is composed of ammonia crystals. On Neptune the upper cloud deck is made of methane. On Uranus we see no obvious cloud deck at all, but only a deep and featureless haze.

Seen through a telescope, Jupiter is a colorful and dynamic planet. Distinct details in its cloud patterns allow us to determine the rotation rate of its atmosphere at the cloud level, although such an apparent atmospheric rotation may have little to do with the spin of the underlying planet. Much more fundamental is the rotation of the mantle and core; these can be determined by periodic variations in radio waves coming from Jupiter, which are controlled by its magnetic field. Since the magnetic field originates deep inside the planet, it shares the rotation of the interior. This rotation period of 9^h 56^m gives Jupiter the shortest "day" of any planet.

In the same way, we can measure that the underlying rotation period of Saturn is 10^h 40^m. Uranus and Neptune have slightly longer rotation periods of about 17^h, also determined from the rotation of their magnetic fields.

Figure 10.4
Jupiter, as photographed by Voyager. The banded structure on the clouds represents strong east-west winds in the atmosphere. The colors in this image are the true colors of Jupiter, more subtle than the exaggerated colors in the early Voyager photo releases seen in Figure 10.8. (NASA/JPL, courtesy of Charles Avis)

Figure 10.5
A computer simulation showing the orbits of Uranus' satellites and its ring system, as it would have been seen by an observer aboard the Voyager 2 spacecraft. (JPL)

Remember from Chapters 3 and 9 that the Earth and Mars have seasons because their axes of spin, instead of "standing up straight," are tilted relative to the Sun's orbital plane. This means that, as the Earth revolves around the Sun, sometimes one hemisphere and sometimes the other "leans into" the Sun. What are the seasons like for the giant planets?

The axis of rotation of Jupiter is tilted by only 3°, so there are no seasons to speak of. Saturn, however, does have seasons, since its axis of rotation is inclined at 27° to the perpendicular to its orbit. Neptune has about the same tilt as Saturn (29°); therefore, it experiences similar seasons (only more slowly).

The strangest seasons of all are on Uranus, which has an axis of rotation tilted by 98° with respect to the north direction. Practically speaking, we can say that Uranus orbits on its side, and its ring and satellite system follows the same pattern, orbiting about Uranus in the plane of the planet's equator. This meant that when Voyager 2 approached Uranus in the plane of the planets' orbits, it was aimed like a well-thrown dart into the giant bull's-eye of the Uranus system (Figure 10.5).

We don't know what caused Uranus and only Uranus to be tipped over like this, but one possibility is a collision with a large planetary body when our system was first forming. Whatever the cause, this unusual tilt creates dramatic seasons. When Voyager 2 arrived at Uranus, its south pole was facing directly into the Sun. The southern hemisphere was having a roughly 21-year sunlit summer, while during that same period the northern hemisphere was plunged into darkness. For the next 21-year season (Figure 10.6), the Sun shines on Uranus' equator, and both hemispheres go through cycles of light and dark. Then there are 21 years of an illuminated northern hemisphere and a dark southern hemisphere. If you were to install a floating platform at the south pole of Uranus, for example, it would experience 42 years of light and 42 years of darkness. Any future astronauts crazy enough to set up camp there could spend most of their lives without ever seeing the Sun.

Composition and Structure

Although we cannot see *into* the giant planets, astronomers are confident that the interiors of Jupiter and Saturn are composed primarily of hydrogen and helium. Of course, these gases have been measured only in their atmospheres, but calculations first carried out 50 years ago by astronomer Rupert Wildt have shown that these two light gases are the only possible materials out of which a planet with the observed masses and densities of Jupiter and Saturn could be constructed.

Having said that much, we should add that the *precise* internal structure of the two planets is not so easy to predict. This is mainly because the planets are so big that the hydrogen and helium in their centers becomes tremendously compressed, and behaves in ways that these gases can never behave on Earth. The best theoretical models we have of Jupiter's structure predict a central pressure of over 100 million bars and a central density of about 31 g/cm^3. (By contrast, the Earth's core has a cen-

tral pressure of 4 million bars and a central density of 17 g/cm^3.)

At the pressures inside the giant planets, familiar materials can take on strange forms. At depths of only a few thousand kilometers below Jupiter and Saturn's visible clouds, pressures become so large that hydrogen changes from a gaseous to a liquid state. Still deeper, this liquid hydrogen is further compressed and begins to act like a metal, something it never does on Earth. (In a metal, electrons are not firmly attached to their parent nuclei, but can wander around. This is why metals are such good conductors of electricity.) On Jupiter, the greater part of the interior is liquid metallic hydrogen. Because Saturn is less massive, it has only a small volume of metallic hydrogen, but most of its interior is liquid. Uranus and Neptune are probably too small to reach internal pressures sufficient to liquify hydrogen. We will return to the question of the metallic hydrogen layers when we examine the magnetic fields of the giant planets.

Each of these planets has a core composed of heavier materials, as demonstrated by detailed analyses of their gravitational fields. Presumably these cores are the original rock-and-ice bodies that formed before the capture of gas from the surrounding nebula (see Chapters 6 and 13). The cores exist at pressures of tens of millions of bars. While scientists speak of the giant planet cores being composed of rock and ice, we can be sure that neither rock nor ice assumes any familiar forms at such pressures and the accompanying high temperatures. What is really meant by "rock" is any material made up primarily of iron, silicon, and oxygen. The term "ice" denotes materials composed primarily of the elements carbon, nitrogen, and oxygen in combination with hydrogen.

Figure 10.7 illustrates the interior structures of the four jovian planets. It appears that all four have similar cores of rock and ice. On Jupiter and Saturn, the cores constitute only a few percent of the total mass, consistent with the initial composition of raw materials shown in Table 10.1. However, most of the mass of Uranus and Neptune resides in these cores, demonstrating that the two planets were unable to attract massive quantities of hydrogen and helium when they were first forming.

Internal Heat Sources

Because of their large sizes, all of the giant planets were strongly heated during their formation by the collapse of surrounding material onto their cores. Jupiter, being the largest, became by far the hottest. Some of this *primordial* heat can still remain inside such large planets. In addition, it is possible for giant, largely gaseous planets to *generate* heat after formation by slowly contracting. (With so large a mass, even a minuscule amount of shrinking can generate significant heat.) The effect of these internal energy sources is to raise the temperatures in the interiors and atmospheres of the planets to values higher than

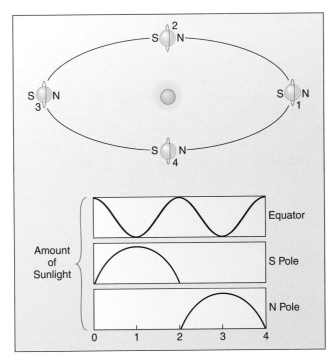

Figure 10.6
The strange seasons on Uranus. (a) The orbit of Uranus as seen from above. At the time Voyager 2 arrived (position 1), the south pole was facing the Sun. As we move counterclockwise in the diagram, we see the planet 21 years later at each step. (b) The amount of sunlight seen at the poles and the equator of Uranus over the course of its 84-year revolution around the Sun.

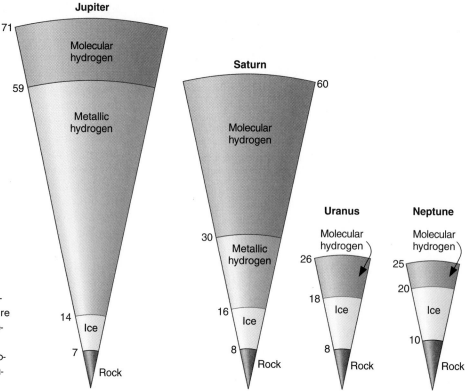

Figure 10.7
Internal structures of the four jovian planets, drawn to scale. Jupiter and Saturn are composed primarily of hydrogen and helium, but Uranus and Neptune consist in large part of compounds of carbon, nitrogen, and oxygen. (The numbers show radius in thousands of km.)

those we would expect from the heating effect of the Sun alone.

Jupiter has the largest internal energy source, amounting to 4×10^{17} W (i.e., it is glowing with the equivalent of 4 million billion 100-W light bulbs). This is about the same as the total solar energy absorbed by Jupiter. The atmosphere of Jupiter is therefore something of a cross between a normal planetary atmosphere (like the Earth's), which obtains most of its energy from the Sun, and the atmosphere of a star, which is entirely heated by an internal energy source. Most of the internal energy of Jupiter is primordial heat, left over from the formation of the planet 4.5 billion years ago.

Saturn has an internal energy source about half as large as that of Jupiter, which means (since its mass is only about one-quarter as great) that it is producing twice as much energy per kilogram of material as does Jupiter. Since Saturn is expected to have much less primordial heat, there must be another source at work generating most of this 2×10^{17} W of power. This source is believed to be the separation of helium from hydrogen in the interior. In the liquid hydrogen mantle, the heavier helium forms drops that sink toward the core, releasing gravitational energy. In effect, Saturn is still differentiating.

Uranus and Neptune are different. Neptune has a small internal energy source, while Uranus does not emit a measurable amount of internal heat. As a result, these two planets have almost the same temperature, in spite of Neptune's greater distance from the Sun. No one knows why these two planets differ in their internal heat.

10.3

Atmospheres of the Jovian Planets

The atmospheres of the jovian planets are the parts we can observe or measure directly. Since these planets have no solid surfaces, their atmospheres are more representative of their general compositions than is the case with the terrestrial planets. These atmospheres also present us with some of the most dramatic examples of weather patterns in the solar system. As we will see, storms on these planets can grow bigger than the entire planet Earth.

Atmospheric Composition

When sunlight reflects from the atmospheres of the giant planets, the atmospheric gases leave their "fingerprints" in the spectra (as we saw in Chapter 4). Spectroscopic observations of the jovian planets began in the 19th century, but for a long time astronomers were not able to interpret the spectra they observed. As late as the 1930s, the most prominent features photographed in these spectra remained unidentified. Then, better spectra revealed the presence of methane (CH_4) and ammonia (NH_3) in the atmospheres of Jupiter and Saturn.

At first we thought that methane and ammonia might be the main constituents of these atmospheres, but now we know that hydrogen and helium are actually the dominant gases. The confusion arose because neither hydrogen nor helium possesses easily detected spectral features. It was not until the Voyager spacecraft measured the far-infrared spectra of Jupiter and Saturn that a reliable abundance for the elusive helium could be found. The compositions of the two atmospheres are generally similar, except that on Saturn there is less helium—the result of the precipitation of helium that contributes to Saturn's internal energy source. The most precise measurements of composition were made on Jupiter by the Galileo entry probe in December 1995, and as a result, we know the elements in the jovian atmosphere even better than those in the Sun.

Clouds and Atmospheric Structure

The clouds of Jupiter are among the most spectacular sights in the solar system, much beloved by makers of science-fiction films. They range in color from white to orange to red to brown, swirling and twisting in a constantly changing kaleidoscope of patterns (Figure 10.8). Saturn shows similar but much more subdued cloud activity: instead of vivid colors, its clouds have a nearly uniform butterscotch hue (Figure 10.9).

Different gases freeze at different temperatures. At the temperatures and pressures of the upper atmospheres of Jupiter and Saturn, methane remains a gas, but ammonia can condense, just as water vapor condenses in the Earth's atmosphere, to produce clouds. The primary clouds that we see around these planets, whether from a spacecraft or through a telescope, are composed of frozen ammonia crystals. The ammonia cloud deck marks the upper edge of the planets' tropospheres; above it lies the

Figure 10.9
Saturn and its rings, photographed by Voyager. The clouds of Saturn are less colorful than those of Jupiter, but the structure and motions of the atmosphere are similar. (NASA/JPL)

cold stratosphere. (These layers were defined, for the Earth, in Chapter 7.)

Figure 10.10 is a diagram showing the structure and clouds in the atmospheres of all four jovian planets. On both Jupiter and Saturn, the temperature near the cloud tops is about 140 K (only a little cooler than the polar caps of Mars). On Jupiter this cloud level is at a pressure of about 0.1 bar, but on Saturn it occurs lower in the atmosphere, at about 1 bar. Because the ammonia clouds lie so much deeper on Saturn, they are more difficult to see, and the overall appearance of the planet is much more bland than is Jupiter.

Figure 10.8
Jupiter's complex and colorful clouds. Also shown in this Voyager photo are two of Jupiter's satellites: Io (left) and Europa (right). The colors have been somewhat exaggerated to emphasize the differences. (NASA/JPL)

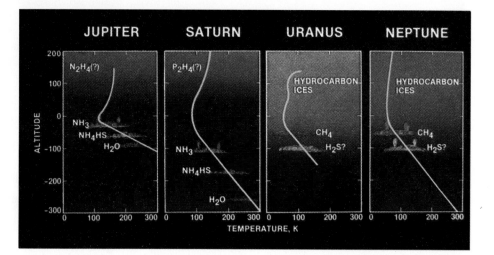

Figure 10.10

The structure of the atmosphere and clouds for the jovian planets. At each altitude, the yellow line enables you to read the temperature at the scale on the bottom. The altitude of the main cloud decks is also shown. (NASA/JPL)

Within the tropospheres of these planets, the temperature and pressure both increase with depth. Through breaks in the ammonia clouds, we can see tantalizing glimpses of other cloud layers that exist in these deeper regions of the atmosphere—regions that were sampled directly by the Galileo probe. Below the thin ammonia clouds the probe descended through a clear region, but at a pressure of about 2 bars it entered a deck of clouds thought to be composed of ammonium hydrosulfide (NH_4SH). The ammonium hydrosulfide clouds may also contain some sulfur particles, which color them a darker yellow or brown.

As it descended to a pressure of 5 bars, the Galileo probe should have passed into a region of frozen water clouds, then below that into clouds of liquid water droplets perhaps similar to the common clouds of the terrestrial troposphere. At least this is what scientists expected. But the probe saw no water clouds, and it measured a surprisingly low abundance of water vapor in the atmosphere. We now understand that the probe happened to descend through an unusually dry, cloud-free region of the atmosphere—the most prominent hole in the clouds visible on the entire planet. Andrew Ingersoll of Caltech, who believes that water clouds are common over most of the planet, has called this entry site the "desert" of Jupiter.

The probe continued to make measurements to a pressure of 22 bars, but found no other cloud layers before its instruments stopped working. It also detected lightning storms, but only at great distances, suggesting that the probe itself was in a region of clear weather.

Above the visible ammonia clouds in Jupiter's atmosphere, we find a region that is clear and cold, reaching a minimum temperature near 120 K. At still higher altitudes temperatures rise again, just as they do in the upper atmosphere of the Earth, because here the molecules absorb ultraviolet light from the Sun. This input of energy causes chemical reactions in the air, a process we call **photochemistry.** In Jupiter's upper atmosphere, photochemical reactions create a variety of fairly complex compounds of hydrogen and carbon that form a thin layer of smog far above the visible clouds. We show this smog as a fuzzy orange region in Figure 10.10; however this thin layer does not block our view of the clouds beneath it.

One puzzle of the main jovian cloud deck that we must mention is the mystery of its colors. The ammonia-condensation clouds identified on the planet should be white, like water clouds on Earth, yet we see beautiful and complex patterns of pastel red, orange, and brown. Some additional chemical or chemicals must be present to create such colors, but we do not know what they are. Various photochemically produced organic compounds have been suggested, as well as sulfur and red phosphorus. But there are no firm identifications, nor any immediate prospects of solving this mystery.

The atmospheric structure on Saturn is similar to that of Jupiter. Temperatures are somewhat colder, however, and the atmosphere is more extended due to Saturn's lower surface gravity. Thus the layers are stretched out over a longer distance (as you can see in Figure 10.10). Overall though, the same atmospheric regions, condensation clouds, and photochemical reactions that we see on Jupiter should be present on Saturn.

Unlike Jupiter and Saturn, Uranus is almost entirely featureless as seen at wavelengths that range from the ultraviolet to the infrared (see its rather boring image in the opening collage for this chapter). Calculations indicate that the basic atmospheric structure of Uranus should resemble that of Jupiter and Saturn, although its upper clouds (at the 1-bar pressure level) are composed of methane (CH_4) rather than ammonia. However, the absence of an internal heat source suppresses up-and-down movement and leads to a very stable atmosphere with little visible structure. In addition, the troposphere is hidden from our view by a deep, cold, hazy stratosphere.

Neptune differs dramatically from Uranus in its appearance (Figure 10.11), although the basic atmospheric temperatures are almost identical. The upper clouds are composed of methane, which forms a thin cloud deck near the top of the troposphere at a temperature of 70 K and a pressure of 1.5 bar. Most of the atmosphere above this level is clear and transparent, with less haze than is found on Uranus. The scattering of sunlight by gas molecules lends Neptune a pale blue color similar to that of the Earth's atmosphere. Another cloud layer, perhaps composed of hydrogen sulfide ice particles, exists below the methane clouds at a pressure of 3 bar.

An important difference between the atmospheres of Uranus and Neptune is the presence on Neptune of convection currents from the interior, powered by the planet's internal heat source. These currents carry warm gas above the 1.5-bar cloud level, forming additional clouds at elevations about 75 km higher. These high-altitude clouds can be seen in the Voyager images: they form bright white patterns against the blue planet beneath. They can even cast distinct shadows on the methane cloud tops, permitting their altitudes to be calculated (see Figure 10.12, a remarkable close-up of Neptune's outer layers that could never have been obtained from Earth).

Winds and Weather

Thick atmospheres like the ones on the jovian planets have many regions of high pressure (where there is more air) and low pressure (where there is less). Just as it does on Earth, air flows between these regions, setting up wind patterns that are then distorted by the rotation of the planet. By observing the changing cloud patterns on the jovian planets, we can measure wind speeds and track the circulation of their atmospheres.

Figure 10.12
High clouds in the atmosphere of Neptune. These bright narrow cirrus clouds are made of methane ice crystals injected into the lower stratosphere. From the shadows they cast on the thicker cloud layer below, we can measure that they are about 75 km higher. (NASA/JPL)

The atmospheric motions we see on these planets are fundamentally different from those of the terrestrial planets. For one thing, the giants spin faster and their rapid rotation tends to smear out the circulation of the air into horizontal (east-west) patterns parallel to the equator.

In addition, there is no solid surface below the atmosphere against which the circulation patterns can rub and lose energy. And, as we have seen, on all the giants except Uranus, heat from the inside contributes about as much energy to the atmosphere as sunlight from the outside. This means that deep currents of rising hot air and falling cooler air circulate throughout the atmospheres of the planets in the vertical direction; these are called convection currents. All three of these factors help explain some of the features we see in the Voyager and Hubble Space Telescope images.

The main features of Jupiter's visible clouds (see Figures 10.1 and 10.4), for example, are alternating dark and light bands that stretch around the planet parallel to the equator. These bands are semipermanent features, although they shift in intensity and position from year to year. Consistent with the small tilt of Jupiter, the pattern does not change with the seasons.

More fundamental than these bands are underlying east-west wind patterns in the atmosphere, which do not appear to change at all, even over many decades. These are displayed in Figure 10.13, which shows maps of Jupiter and Saturn from the Voyager images, and indicates how strong the winds are at each latitude. At Jupiter's equator, a jet stream flows eastward with a speed of about 90 m/s (300 km/h), similar to the speed of jet streams in the Earth's upper atmosphere. At higher latitudes there are alternating east- and west-moving

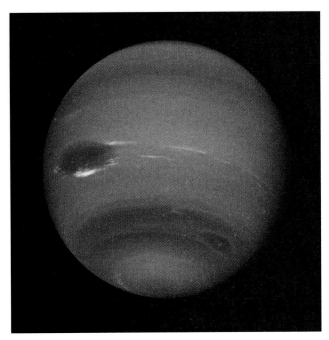

Figure 10.11
The planet Neptune as photographed in 1989 by Voyager 2. The blue color, exaggerated here with computer processing, is caused by the scattering of sunlight in the upper atmosphere. (NASA/JPL)

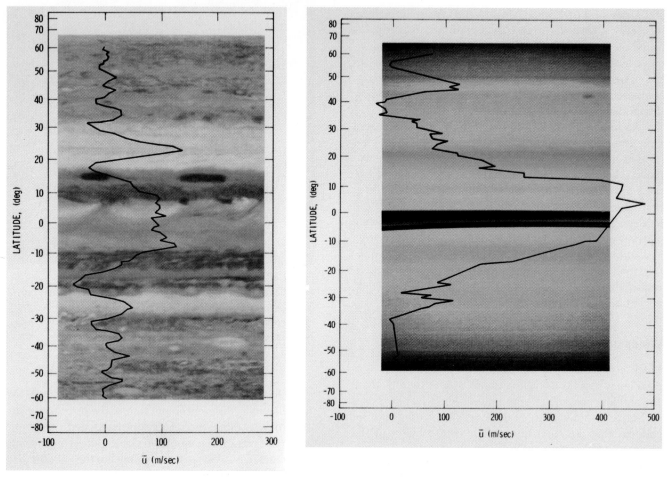

Figure 10.13

East-west winds on Jupiter (left) and Saturn (right) as measured by Voyager. Latitude is shown in the left column of each chart, with the equator being 0°. The dark line shows the wind speed at each latitude (measured relative to the planet's core). Notice that Saturn has much stronger winds at the equator. (NASA/JPL)

streams, with each hemisphere an almost perfect mirror image of the other. Saturn shows a similar pattern, but with a much stronger equatorial stream flowing at a speed of 1300 km/h.

Generally, the light zones on Jupiter are regions of upwelling air capped by white ammonia cirrus clouds. They apparently represent the tops of upward-moving convection currents. The darker belts are regions where the cooler atmosphere moves downward, completing the convection cycle; they are darker because fewer ammonia clouds mean we can see deeper into the atmosphere, perhaps down to the ammonium hydrosulfide clouds.

In spite of the strange seasons induced by the 98° tilt of its axis, Uranus' basic circulation is parallel with its equator, as is the case on Jupiter and Saturn. The mass of the atmosphere and its capacity to store heat are so great that the alternating 40-year periods of sunlight and darkness have little effect; in fact, Voyager measurements show that the atmospheric temperatures are a few degrees higher on the dark winter side than on the hemisphere facing the Sun. Here is another indication that the behavior of such giant planet atmospheres is a complex

problem: we simply do not understand the seasonal effects on Uranus in detail.

Neptune's weather is characterized by strong east-west winds generally similar to those observed on Jupiter and Saturn. The highest wind speeds near its equator reach 2100 km/h (600 m/s), nearly twice as fast as the peak winds on Saturn. The Neptune equatorial jet stream actually approaches supersonic speeds (speeds higher than the speed of sound in Neptune's air).

Storms

Superimposed on the regular atmospheric circulation patterns just described are many local disturbances—weather systems or storms, to borrow the term we use on Earth. The most prominent of these are large oval high-pressure regions on both Jupiter and Neptune.

The largest and most famous "storm" on Jupiter is the Great Red Spot, a reddish oval in the southern hemisphere that was almost 30,000 km long when the Voyagers flew by—big enough to hold more than two Earths side by side (Figure 10.14). First seen 300 years ago, the Red

Figure 10.14
The Great Red Spot of Jupiter as seen during the Voyager spacecraft flyby, with the Earth superimposed to give a sense of scale. Below and to the right of the Red Spot is one of the white ovals, which are similar but smaller high-pressure features. The colors on the Jupiter image have been exaggerated so astronomers (and astronomy students) can study their differences more effectively. See Figures 10.1 and 10.4 to get a better sense of the colors your eye would actually see near Jupiter. (S. Meszaros and the Astronomical Society of the Pacific)

Spot has changed in size but has not disappeared; it is clearly much longer-lived than storms in our own atmosphere. It also differs from terrestrial storms in being a high-pressure region; on our planet such storms are regions of lower pressure. The Red Spot's counterclockwise rotation has a period of six days. Three similar but smaller disturbances formed on Jupiter about 1940; they look like white ovals. These are "only" about 10,000 km across (they are clearly visible to the left of the Red Spot in Figure 10.1).

We don't know what causes the Great Red Spot or the white ovals, but we do understand how they can last so long once they form. On Earth, the lifetime of a large oceanic hurricane or typhoon is typically a few weeks, or even less when it moves over the continents and encounters friction with the land. Jupiter has no solid surface to slow down an atmospheric disturbance, and furthermore the sheer size of the disturbances lends them stability. We can calculate that on a planet with no solid surface, the lifetime of anything as large as the Red Spot should be measured in centuries, while lifetimes for the white ovals should be measured in decades. These time scales are consistent with the observed lifetimes of jovian storms.

Despite Neptune's smaller size and different cloud composition, the 1989 Voyager images showed that it had an atmospheric feature surprisingly similar to the jovian Great Red Spot. Neptune's Great Dark Spot (Figure 10.15) was nearly 10,000 km long. On both planets, the giant storms formed at latitude 20° S, had the same shape,

and took up about the same fraction of the planet's diameter. The Great Dark Spot rotated with a period of 17 days, versus about 6 days for the Great Red Spot.

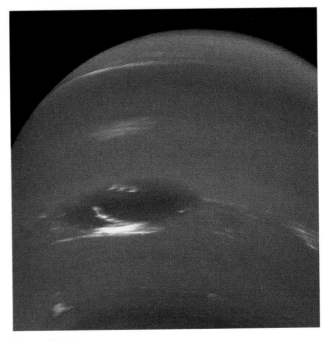

Figure 10.15
The Great Dark Spot of Neptune as seen by Voyager in 1989. The spot was accompanied by streamers of bright methane cirrus clouds that form around it but at a higher altitude. (NASA/JPL)

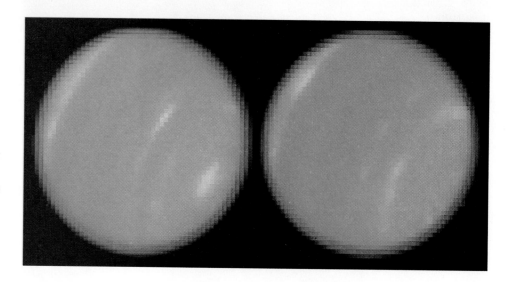

Figure 10.16
Two images of Neptune taken with the Hubble Space Telescope half-a-Neptune-day apart show both sides of the planet. Detailed examination showed no trace of the Great Dark Spot so prominent on the Voyager images. (D. Crisp, H. Hammel, and NASA)

When the repaired Hubble Space Telescope examined Neptune in June 1994, however, astronomers could find no trace of the Great Dark Spot (Figure 10.16) on their images. It is not clear whether the Spot has disappeared or merely faded in brightness, as Jupiter's Red Spot also does over time. In November 1994, a new dark spot was seen in Neptune's northern hemisphere, but it had faded somewhat by 1995, indicating that such large storm systems may form and dissipate more quickly on Neptune than on Jupiter.

Large storms on Saturn are much rarer and appear to be connected with the seasons that result from the large tilt of its axis. Approximately once every 30 years observers have seen outbreaks of spots in the equatorial regions of Saturn, with the most recent such outbreak occurring in 1990. As photographed by the Hubble Space Telescope about a month after its appearance (Figure 10.17), this storm had spread most of the way around the planet and grew to be over ten times the size of the Earth. It resembled some of the turbulent cloud structure that accompanies the jovian Great Red Spot. These seasonal storms on Saturn appear to be large bubbles or plumes of warmer air that rise into the stratosphere and then gradually dissipate in the strong equatorial jet streams.

10.4
Magnetospheres

The largest features of the giant planets are their *magnetospheres*. Like the magnetosphere of the Earth, these regions are defined as the cavities within which the planet's own magnetic field dominates over the general interplanetary magnetic field. When a planet like Jupiter has a strong magnetic field, its zone of magnetic influence can, as we will see, extend over enormous distances.

Inside a planet's magnetosphere, charged particles (either from the Sun or local sources) are trapped; as a result they can be accelerated to high energies. These physical processes turn out to be milder versions of what astronomers find in many distant objects, from the remnants of dead stars to the puzzling distant powerhouses we call quasars. One reason to study the magnetospheres of the giant planets and the Earth is that they provide nearby accessible analogues of more energetic and challenging cosmic processes.

Jupiter's Magnetic Field

In the late 1950s, astronomers discovered that Jupiter was a source of radio waves that got more intense at longer rather than at shorter wavelengths—just the reverse of what is expected from *thermal* radiation (radiation caused by the normal vibrations of particles within all matter; see Chapter 4). Such behavior is typical, however, of the radi-

Figure 10.17
A huge storm on Saturn as photographed by the Hubble Space Telescope in November 1990. The colors have been significantly exaggerated to bring out subtle details. Note similarities to the structure visible on Jupiter in Figure 10.14. (NASA)

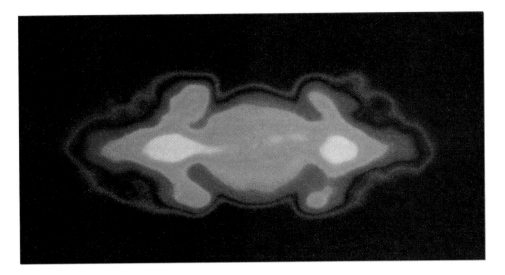

Figure 10.18
An image of Jupiter in radio waves, made with the Very Large Array. We see part of the magnetosphere, brightest in the middle because the largest number of charged particles are in the equatorial zone of Jupiter. The planet itself is slightly smaller than the circular green region in the center. Different colors are used to indicate different intensities of synchrotron radiation. (I. dePater/ NRAO)

ation emitted when high-speed electrons are accelerated by a magnetic field. We call this **synchrotron radiation** because it was first observed on Earth in particle accelerators, called synchrotrons. This was our first hint that Jupiter must have a strong magnetic field.

Later observations showed that the radio energy originates from a region surrounding Jupiter with a diameter several times that of the planet itself (Figure 10.18). The evidence suggested, therefore, that a vast number of charged atomic particles must be circulating around Jupiter, spiraling around the lines of force of a magnetic field associated with the planet. This is just what we observe happening, but on a smaller scale, in the Van Allen belt around the Earth (see Chapter 7).

We now understand that the jovian magnetosphere is one of the largest features in the solar system. It is actually much larger than the Sun and completely envelops the innermost satellites of Jupiter. The total *mass* of the ions and electrons in the giant magnetosphere, however, is less than the mass of the Great Pyramid of Giza in Egypt.

The Pioneer and Voyager spacecraft allowed astronomers to map and study the magnetic field and magnetosphere of Jupiter directly (Figure 10.19). They found the magnetic field at Jupiter's cloudtops to be from 20 to 30 times as strong as the Earth's field. Because of Jupiter's great size, moreover, its total magnetic energy is enormous compared with the Earth's.

Figure 10.19
The magnetosphere of Jupiter as mapped out by the Pioneer and Voyager missions. Blue lines are representative magnetic field lines; here they are shown only in two dimensions, but in reality they surround Jupiter on all sides. The yellow ellipses are the orbits of the Galilean satellites. Note the orbit of the satellite Io (inside the orange donut), which is a local supplier of charged particles for the system. (Computer visualization by John Spencer, Lowell Observatory)

James Van Allen:
Several Planets Under His Belt

The career of physicist James Van Allen has spanned the birth and growth of the Space Age; in fact he played a major role in its development. Born in Iowa in 1914, Van Allen received his PhD from the University of Iowa. He then worked for several research institutions and served in the Navy during World War II. In 1951 he was appointed Professor of Physics at the University of Iowa, where he has remained ever since.

After the war Van Allen and his collaborators began using the V2 rockets captured from the Germans, as well as rockets built in the United States, to explore cosmic radiation in the Earth's outer atmosphere. Van Allen designed a technique, nicknamed "the rockoon," in which a small rocket is lifted up by a balloon before being launched, enabling it to reach a higher altitude.

Over dinner one night in 1950, Van Allen and several colleagues came up with the idea of the International Geophysical Year (IGY), an opportunity for scientists around the world to coordinate their investigations of the physics of the Earth, especially research done at high altitudes. Stretched to 18 months, the IGY took place between July 1957 and December 1958. In 1955, the United States and the Soviet Union each committed themselves to launching an Earth-orbiting satellite during IGY, a competition that began what came to be known as the Space Race.

When the Soviet Union won the race by launching Sputnik 1 in October 1957, the U.S. government spurred its scientists and engineers to even greater efforts to get something into space and thereby maintain the country's prestige. However, the main U.S. satellite program, called Vanguard, was running into all sorts of difficulties, and each of its early launches crashed or exploded. In the meantime, a second team of rocket engineers and scientists had quietly been working on a military launch-vehicle called Jupiter-C. Van Allen spearheaded the design of the instruments aboard the small satellite, called Explorer I, that this rocket would carry. On January 31, 1958, Explorer I became the first U.S. satellite in space.

Unlike Sputnik, Explorer I was equipped to make scientific measurements. Van Allen had placed an instrument aboard Explorer I that could detect high-energy charged particles above the atmosphere, and it was this instrument that discovered what we now call the Van Allen belts around the Earth. This was the first scientific discovery of the space program and made Van Allen's name a household word around the world.

Van Allen and his colleagues continued making measurements of the magnetic and radiation environment around our planet with more sophisticated spacecraft and instruments. When Pioneers 10 and 11 made the first exploratory surveys of the Jupiter and Saturn environments, they again carried instruments designed by Van Allen. Some scientists therefore refer to the charged-particle zones around those planets as Van Allen belts as well. (Once, when Van Allen was giving a lecture at the University of Arizona, the graduate students in planetary science asked him if he would leave his belt at the school. It is now proudly displayed as the "Van Allen belt.")

Van Allen has continued to be a strong supporter of space science, and has been an eloquent senior spokesman for the American scientific community, warning NASA not to put all its efforts into human exploration but to remember how productive and (relatively) inexpensive its robot probes and instruments have always been in helping us learn more about the solar system.

James Van Allen
(Photo by D. Hokanson)

The axis of Jupiter's magnetic field (the line that connects magnetic north and south) is not aligned exactly with the axis of rotation of the planet; rather, it is tipped at some 10°. The magnetic axis also does not pass exactly through the planet's center but is offset by about 18,000 km. Furthermore, the jovian field has the opposite polarity from the Earth's current value (i.e., magnetic north is in the southern hemisphere). However, the Earth's field is known to reverse polarity from time to time, and the same may be true of Jupiter's field. The reasons for many of these phenomena are not well understood at all.

When we measure the amount of synchrotron radiation coming from Jupiter, it becomes clear that far more particles are spiraling in the giant's planet's magnetic field than can be supplied by the Sun or by Jupiter's upper atmosphere (the main sources of the particles in the Earth's Van Allen belt). Where do the other particles around Jupiter come from?

As we will see in the next chapter, Io, one of Jupiter's large moons, turns out to be the most volcanically active world in the solar system. Voyager revealed that Io's volcanic vents are erupting large quantities of sulfur and sulfur dioxide into the space surrounding it. While most of this material falls back to the surface (as an eerie kind of snow), it is estimated that about 10 tons per second is lost to the magnetosphere. Expelled atoms of oxygen and sulfur (stripped of some electrons) form a donut-shaped reservoir of charged particles surrounding Jupiter approx-

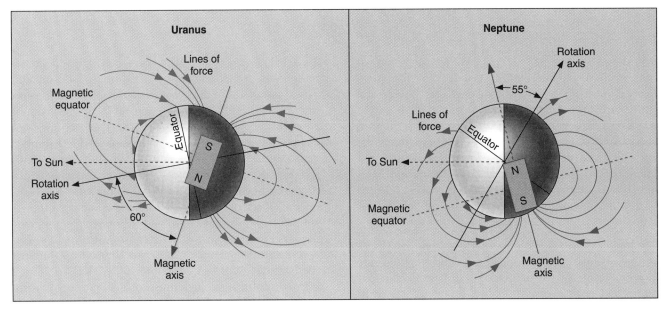

Figure 10.20
The magnetic fields of Uranus and Neptune as revealed by Voyager; note the large (and puzzling) offset from the center of the planet, and the tilt with respect to the planet's axis of rotation.

imately at Io's orbit, at a distance of five Jupiter radii from the planet (see Figure 10.19).

The energetic ions in this donut and its surroundings would be very dangerous to both spacecraft and humans if any future astronauts should ever think of venturing close to Jupiter. When these ions strike a solid surface, they generate lethal x rays. On or near Io, a human (even in a spacesuit) could survive for only a few minutes. Special shielding would be required for spacecraft electronics, and no spacecraft has been built that could last more than a few hours in this environment. Thus an Io lander or orbiter is far beyond our present capabilities.

The Magnetic Fields of the Other Giants

Saturn does not emit strong synchrotron radiation, since its magnetosphere has a very poor supply of charged particles. This is because electrons near Saturn tend to collide with the planet's large ring system (see Chapter 11) and thus get removed from "active duty" in the magnetosphere. Saturn does have a substantial magnetic field, however, as discovered by Pioneer 11 and the Voyagers. Unlike the Earth and Jupiter, Saturn's field is almost perfectly aligned with its rotation axis.

The magnetic field of Uranus was not discovered until the Voyager flyby in 1986. The strength of the field is comparable to that of Saturn, about what would be expected from the size of the planet. However, the orientation of Uranus' magnetic field is very different. Like Jupiter's field, it is offset from the center of the planet, but to a greater degree (by about one-third of the planet's radius). In addition, the magnetic field of Uranus is tilted by a remarkable 60° with respect to the axis of rotation (Figure 10.20).

Neptune's magnetic field was discovered in 1989, when the Voyager 2 spacecraft reached this distant world. Its configuration is similar to that of Uranus, with a magnetic axis tilted by 55° from the rotational axis. The offset of the neptunian field is the greatest of any planet, amounting to nearly half the planet's radius.

We presume that the magnetic fields of the outer planets are generated in much the same way that the field of the Earth is. All of these planets spin rapidly, supplying a ready source of energy to power their internal magnetic generators. We've seen that Jupiter and Saturn have large interior regions of metallic liquid hydrogen that can act like the liquid iron core of the Earth—electrons in a metal circulate in vast currents, which are the source of the magnetic field.

With Uranus and Neptune, the situation is a bit less clear. The metallic region needed for a magnetic field may not be near the center of either planet, but rather in their hydrogen-water mantles. This could account for the large offset of the fields from the centers of the planets. The detailed mechanisms, however, are still not very well understood.

The magnetospheres of the other three jovian planets are generally similar to that of Jupiter. The physical dimensions of Saturn's magnetosphere are about one-third as great, and those of Uranus and Neptune are smaller still, approximately proportional to the sizes of the planets themselves.

As in the case of Jupiter, the source of some of the particles in the magnetospheres of the other giants can be found in their satellites. Each of the giants is surrounded by a variety of worlds, some quite small, others larger than Mercury. We turn to these accompanying worlds in the next chapter.

Summary

10.1 The outer solar system contains the four jovian (or giant) planets: Jupiter, Saturn, Uranus, and Neptune. Their chemistry is generally reducing, and Jupiter and Saturn have overall compositions similar to that of the Sun. These planets have been explored by the Pioneer, Voyager, and Galileo spacecraft. Voyager 2, perhaps the most successful of all space-science missions, explored Jupiter (1979), Saturn (1981), Uranus (1986), and Neptune (1989)—a grand tour of the jovian planets. Galileo successfully deployed a probe into the atmosphere of Jupiter in 1995.

10.2 Jupiter is 318 times more massive than the Earth. Saturn is about 25 percent, and Uranus and Neptune only 5 percent, as massive as Jupiter. All four have deep atmospheres and opaque clouds, and all rotate quickly (with periods from 10 to 17 hours). Jupiter and Saturn have extensive mantles of liquid hydrogen. Uranus and Neptune are depleted in hydrogen and helium relative to Jupiter and Saturn (and the Sun). Each jovian planet has a core of "ice" and "rock" of about 10 Earth masses. Jupiter, Saturn, and Neptune have major internal heat sources, obtaining as much (or more) energy from their interiors as by radiation from the Sun. Uranus has no measurable internal heat.

10.3 The four jovian planets have generally similar atmospheres, composed mostly of hydrogen and helium. The atmospheres of the jovian planets contain small quantities of methane (CH_4) and ammonia (NH_3) gas, both of which also condense to form clouds. Deeper (invisible) cloud layers consist of H_2O and possibly NH_4SH (Jupiter and Saturn) and H_2S (Neptune). In the upper atmospheres, hydrocarbons and other trace compounds are produced by **photochemistry.** We do not know what colors the clouds of Jupiter. Atmospheric motions on the giant planets are dominated by east-west circulation. Jupiter displays the most active cloud patterns, with Neptune second. Saturn is generally bland, and Uranus is featureless (perhaps due to its lack of an internal heat source). Large storms (oval-shaped high-pressure systems such as the Great Red Spot on Jupiter and the Great Dark Spot on Neptune) can be found in some of the giant planet atmospheres.

10.4 The jovian planets have substantial magnetic fields, approximately in proportion to their sizes. Within their large magnetospheres, trapped ions and electrons are accelerated to high energies and emit **synchrotron radiation.** Jupiter has the most active magnetosphere, partly because of the ions provided by the volcanoes of Io. The magnetic fields of Jupiter, Uranus, and Neptune are offset from the center of each planet.

Review Questions

1. Describe the differences in the chemical makeup of the inner and outer parts of the solar system. What is the relationship between what the planets are made of and the temperature where they formed?

2. How did the giant planets grow to be so large?

3. Why is it difficult to drop a probe like Galileo's *gently* into the atmosphere of Jupiter? How did the Galileo engineers solve this problem?

4. What are the visible clouds on the four jovian planets composed of, and why are they different from each other?

5. Describe the seasons on the planet Uranus.

6. Compare the atmospheric circulation (weather) for the four jovian planets.

7. What are the main atmospheric heat sources of each of the jovian planets?

8. Compare the magnetic fields and magnetospheres of the four jovian planets.

Thought Questions

9. Jupiter is denser than water, yet composed for the most part of two light gases, hydrogen and helium. What makes Jupiter so dense?

10. Would you expect to find free oxygen gas in the atmospheres of the giant planets? Why or why not?

11. Why would a tourist brochure (of the future) describing the most dramatic natural sights of the jovian planets have to be revised more often than one for the terrestrial planets?

12. The water clouds believed to be present on Jupiter and Saturn exist at temperatures and pressures similar to those in the clouds of the terrestrial atmosphere. What would it be like to visit such a location on Jupiter or Saturn? In what ways would the environment differ from that in the clouds of Earth?

13. Describe the different processes that lead to substantial internal heat sources for Jupiter and Saturn. Since these two objects generate much of their energy internally, should they be called stars instead of planets? Justify your answer.

14. Use sources in your college library to read up on the Galileo mission. What technical problems occurred between the mission launch and the arrival of the craft in the jovian system, and how did the mission engineers deal with them? (Good sources of information include *Astronomy* and *Sky & Telescope* magazines.)

Problems

15. Calculate how many Earths would fit into the volumes of Saturn, Uranus, and Neptune.

16. As the Voyager spacecraft penetrated into the outer solar system, the illumination from the Sun declined. Relative to the situation at Earth, how bright is the sunlight at each of the jovian planets?

17. Jupiter's Great Red Spot rotates in six days and has a circumference equivalent to a circle with radius 10,000 km.

Neptune's Great Dark Spot was one-fourth as large and rotated in 17 days. For each, calculate the wind speeds at the outer edges of the spots.

18. The ions in the inner parts of the jovian magnetosphere rotate with the same period as Jupiter. Calculate how fast they are moving at the orbit of Io (see Appendix 11). Will these ions strike Io from behind or in front as it moves about Jupiter?

Suggestions for Further Reading

Beebe, R. *Jupiter: The Giant Planet.* 1994, Smithsonian Institution Press. Clear, nontechnical summary.

Beebe, R. "Queen of the Giant Storms" in *Sky & Telescope,* Oct. 1990, p. 359. Excellent review of the Red Spot.

Burgess, E. *Far Encounter: The Neptune System.* 1992, Columbia U. Press. Survey of Voyager flyby by a veteran science journalist.

Cowling, T. "Big Blue: The Twin Worlds of Uranus and Neptune" in *Astronomy,* Oct. 1990, p. 42. Nice long review of the two planets.

Davis, J. *Flyby.* 1987, Atheneum. About the Voyager mission, with emphasis on Uranus.

Gore, R. "Neptune: Voyager's Last Picture Show" in *National Geographic,* Aug. 1990, p. 35.

Littmann, M. *Planets Beyond: Discovering the Outer Solar System,* 2nd ed. 1989, J. Wiley. Good introduction to Uranus, Neptune, and Pluto.

Littmann, M. "The Triumphant Grand Tour of Voyager 2" in *Astronomy,* Dec. 1988, p. 34.

Miner, E. *Uranus.* 1990, Ellis Horwood/Simon and Schuster. Definitive introduction to the planet.

Montoya, E. and Fimmel, R. "Pioneers in Space: The Story of the Pioneer Missions" in *Mercury,* May/June 1988, p. 81.

Morrison, D. *Voyages to Saturn.* 1982, NASA-SP 451, U.S. Government Printing Office. Well-illustrated guide to the Voyager missions.

Sagan, C. *Pale Blue Dot.* 1994, Random House. Chapters 6 through 9 explore the outer solar system.

Smith, B. "Voyage of the Century" in *National Geographic,* Aug. 1990, p. 48. Beautiful summary of the Voyager mission to all four outer planets.

Washburn, M. *Distant Encounters: The Exploration of Jupiter and Saturn.* 1983, Harcourt, Brace, Jovanovich. A science reporter covers the first Voyager encounters.

Images of and data about the giant planets can be found at many sites on the World Wide Web. There are also sites devoted to specific missions, such as Galileo. See Appendix 1 for a listing of some of these.

Using **REDSHIFT** ™

1. Check out the unusual rotation of Uranus by clicking on the planet in the *Follow Planet* panel and choosing the inward view. Follow the motion of its moons by sliding the distance pointer to its maximum setting.

2. Adrastea orbits Jupiter at about the same distance Miranda does Uranus. Measure each satellite's orbital period in hours. The ratio of the square of Miranda's period to that of Adrastea will tell you the ratio of Jupiter's mass to Uranus.

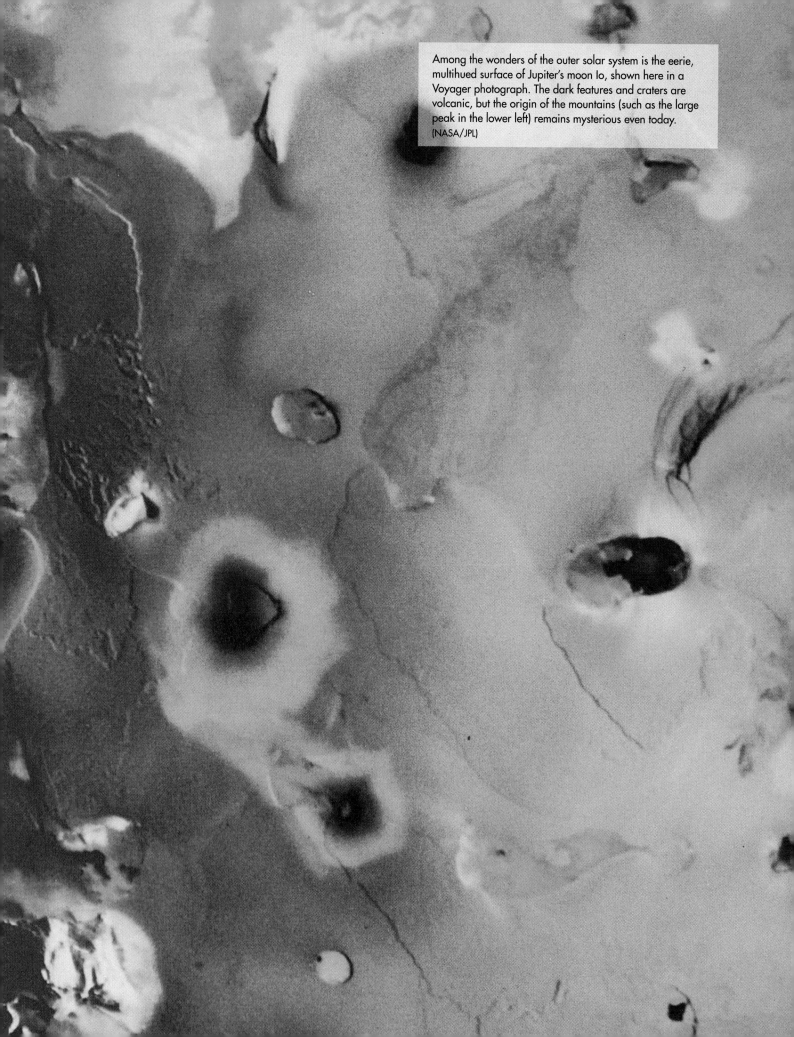

Among the wonders of the outer solar system is the eerie, multihued surface of Jupiter's moon Io, shown here in a Voyager photograph. The dark features and craters are volcanic, but the origin of the mountains (such as the large peak in the lower left) remains mysterious even today. (NASA/JPL)

CHAPTER 11

Rings, Moons, and Pluto

Thinking Ahead

On an active world, a map is always temporary. The boundaries of towns change, new roads are built, rivers slowly alter their channels. Over geological time, even the Earth's mountains, seas, and continents change shape and form. But what kind of map would be worth making of a world where the surface landforms changed substantially in only a decade? Such a world can be found orbiting the planet Jupiter.

Our imaginations always fall short of anticipating the beauty we find in nature.

Geologist Laurence Soderblom, discussing the 1989 Voyager encounter with Neptune's moons.

All of the giant planets are accompanied by moons or satellites (astronomers use the two terms interchangeably), which orbit about them like planets in a miniature solar system. Over 60 satellites are known in the outer solar system, too many to discuss individually in any detail, but we list their characteristics in Appendix 8. Astronomers suspect that other small satellites will be found in the future.

In this chapter we focus our attention on a few of the more interesting satellites, and on the tiny planet Pluto, itself smaller than a number of the outer planets' moons. Then we examine the ring systems that gird each of the four giant planets. The processes we see at work in these rings have applications ranging from the formation of planetary systems to the spiral structure of galaxies.

In large part, this chapter continues the remarkable saga of the two Voyager spacecraft that explored the outer solar system between 1979 and 1989. Before Voyager, many of the satellites we will be discussing were mere points of light in our telescopes.

Suddenly, in less than a decade, we had close-up images (and much non-visual information as well); they then became individual *worlds*, each with unique features and peculiarities.

As discussed in Chapter 10, during 1996 the attention of scientists again focused on the jovian system with the arrival of the Galileo spacecraft. This billion-dollar space robot began orbiting Jupiter in December 1995, to make repeated close flybys of all four of Jupiter's large satellites, photographing them and making other measurements of their surfaces with a resolution much higher than what Voyager was capable of during its brief flybys. Keep a sharp eye out for newspaper and TV stories that describe how the data returned from Galileo extend and modify our picture of these four worlds.

Figure 11.1

A montage assembled from individual Voyager images showing Jupiter and its four large Galilean satellites. The image is meant to show a spacecraft perspective on the way in to Jupiter, so the size of each moon is an accident of perspective. In order of increasing size, we see Io, Europa, Ganymede, and Callisto. (NASA/JPL)

11.1

Ring and Satellite Systems

General Properties

As discussed in Chapter 6, the rings and satellites of the outer solar system have compositions different from objects in the inner solar system. We should expect this, since they formed in regions of lower temperature, cool enough so that large quantities of water ice were available as building materials. Most of these objects also contain dark, organic compounds mixed with their ice and rock. This dark, primitive material often means that the objects reflect very little light. Don't be surprised, therefore, to find many objects in the ring and satellite systems that are both icy and black.

Most of the satellites in the outer solar system are in direct or regular orbits; that is, they revolve about their parent planet in an east-to-west direction and in the plane of the planet's equator. Ring systems are also in direct revolution in the planet's equatorial plane. Such objects probably formed at about the same time as the planet by processes similar to those that formed the planets in orbit around the Sun (see Chapter 13).

In addition to the regular satellites, irregular satellites orbit in a *retrograde* (west-to-east) direction or else have orbits of high eccentricity (more elliptical than circular) or high inclination (moving in and out of the planet's equatorial plane). These are usually smaller satellites, located relatively far from their planet; they were probably formed far away and subsequently captured by the planet they now orbit.

The Jupiter System

Jupiter has 16 satellites and a faint ring. The satellites include four large moons—Callisto, Ganymede, Europa, and Io—(Figure 11.1) discovered in 1610 by Galileo and therefore often called the Galilean satellites. The smaller

of these, Europa and Io, are about the size of our Moon, while the larger, Ganymede and Callisto, are bigger than the planet Mercury. We will discuss the Galilean satellites individually in a moment, and try to understand the interesting differences among them.

The other 12 jovian moons are much smaller. We can divide them conveniently into three groups of four each. The inner four all circle the planet inside the orbit of Io; one of these, Amalthea, has been known for about a century, but the other three were discovered by Voyager. The outer satellites consist of four in direct but highly inclined orbits, and four farther out in retrograde orbits. These eight are believed to be captured objects. The two groupings may indicate two parent bodies that were broken up in a collision early in the history of the jovian system. The eight outer satellites are dark and apparently primitive (that is, they have not undergone much chemical change since the formation of the solar system).

The Saturn System

Saturn has 19 known satellites in addition to a magnificent set of rings (Figure 11.2). The largest of the satellites, Titan, is almost as big as Ganymede in the jovian system, and it is the only satellite with a substantial atmosphere. We discuss it in detail in the next section. Saturn has six other regular satellites with diameters between 400 and 1600 km, and a handful of small moons orbiting in or near

Figure 11.2
A montage of Voyager images of Saturn and several of its satellites. The satellites are not to scale: giant Titan is the tiny orange ball toward the back. The moon in the foreground is Enceladus. The color of Saturn is a bit intensified by the image processing. (NASA/JPL)

the rings. There are also two distant irregular satellites, one of which has a retrograde orbit.

The rings of Saturn, one of the most impressive sights in the solar system, are broad and flat, with only a few gaps. They are not solid, but rather a huge collection of icy pieces, all orbiting the equator of Saturn in a traffic pattern that makes the Los Angeles freeway system look simple. Individual ring particles are composed of water ice and are typically the size of ping-pong, tennis, and basketballs.

The Uranus System

The regular ring and satellite system of Uranus is tilted at 98°, just like the planet itself. It consists of 11 rings and 15 regular satellites. The five largest satellites are similar in size to the six regular satellites of Saturn, with diameters from 500 to 1600 km, while the ten smaller satellites and the ring particles are very dark, reflecting only a few percent of the sunlight that strikes them.

Discovered in 1977, the rings of Uranus are narrow ribbons of material with broad gaps in between—fundamentally different from the broad rings of Saturn. Astronomers suppose that the ring particles are confined to these narrow paths by the gravitational effects of numerous small satellites, most of which we have not yet glimpsed.

The Neptune System

Neptune has eight satellites: six regular satellites close to the planet, and two irregular satellites farther out. The

most interesting of these is Triton, a relatively large moon in a retrograde orbit. Triton has an atmosphere, and active volcanic eruptions were discovered there by Voyager in its 1989 flyby. We compare it in detail with the planet Pluto, which it resembles, later in this chapter.

The rings of Neptune are narrow and faint. Like those of Uranus, they are composed of dark materials and are thus not easy to see.

11.2

The Galilean Satellites and Titan

In this section we discuss the five largest satellites of the outer solar system. Table 11.1 summarizes their properties, with the Moon also listed for comparative purposes.

Callisto, a Garden-Variety Icy Satellite

We begin our discussion of the Galilean satellites with Callisto, not because it is remarkable, but because it is not. Callisto is a simple, large, icy satellite, an object with which other, more peculiar objects can be compared.

Callisto is the outermost Galilean satellite. Its distance from Jupiter is about 2 million km, and it takes 17 days to orbit the planet. Like our own Moon, Callisto's period of rotation is the same as its period of revolution, so it always keeps the same face toward Jupiter. Its noontime surface temperature is a chilly 130 K (about 140°C below freezing).

Callisto has a diameter of 4820 km, almost the same as that of the planet Mercury. However, its mass (and

TABLE 11.1
The Largest Satellites

Name	Diameter (km)	Mass (Moon = 1)	Density (g/cm³)	Reflectivity (%)
Moon	3476	1.0	3.3	12
Callisto	4820	1.5	1.8	20
Ganymede	5270	2.0	1.9	40
Europa	3130	0.7	3.0	70
Io	3640	1.2	3.5	60
Titan	5150	1.9	1.9	20

therefore its density) is only one-third as great, immediately telling us that Callisto has far less of the rocky, metallic materials found in the inner planets; instead it must be mostly ice. This satellite can therefore show us how the geology of an icy object compares with one made primarily of rock.

Like other icy bodies of the outer solar system, Callisto has differentiated: the heavier materials have sunk toward the center. Actually, an icy body can differentiate much more easily than a rocky one can, since the melting temperature of ice is so low. Only a little heating will soften the ice and get the process started, as the rock and metal sink to the center and the slushy ice floats to the surface. We therefore surmise that Callisto consists of a rocky, metallic core about the size of the Moon, surrounded by a mantle and crust of water ice.

The surface of Callisto, like the highlands of the Moon, is covered with impact craters (Figure 11.3). The survival of these craters tells us that an icy object can form and retain impact craters in its surface. Their large numbers show that Callisto, like both the Moon and Mercury, has experienced little if any geological activity for billions of years. The craters on Callisto further demonstrate that the early period of heavy bombardment we saw on worlds nearer the Sun also peppered the outer solar system with impact scars.

In form, the smaller icy craters of Callisto look very much like the rocky craters of the Moon and Mercury. Ice on Callisto does not deform or flow like ice in glaciers on the Earth. At the frigid temperatures of the outer solar system, ice is nearly as hard as rock, and it behaves similarly.

Ganymede, the Largest Satellite

Ganymede, the largest satellite in the solar system, is also cratered, but less so than Callisto (Figure 11.4). About one-third of its surface seems to be darker and as heavily cratered as Callisto; the rest formed later, after the end of the heavy bombardment period. Judging from crater counts, this younger terrain on Ganymede is probably 3 to 4 billion years old, like the lunar maria or the martian volcanic plains. It is the result of tectonic forces, great move-

ments of the moon's outer layers. As Ganymede's crust cracked, many of the craters were flooded with water from the interior, just as the lunar maria were flooded with lava. These same forces formed extensive parallel ice-mountain ridges on Ganymede (Figure 11.5). The mountainous terrain, with its ridges evenly spaced about 15 km apart, covers more than one-quarter of the surface.

Why did Ganymede experience volcanic activity and consequent resurfacing during the first billion years after its formation, while Callisto did not? Ganymede is a larger world, with more radioactive materials inside to provide a source of internal heat. Apparently, even a small difference in size and internal heating can lead to a significant difference in the evolution of two worlds.

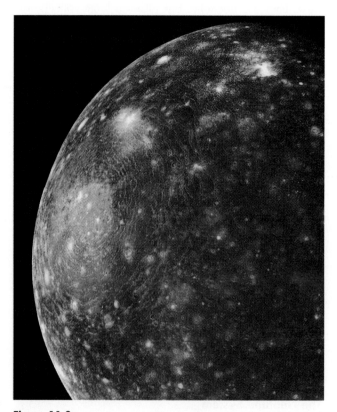

Figure 11.3
Callisto, Jupiter's outermost large satellite, shows a heavily cratered surface of dark ice in this Voyager image. (NASA/JPL)

Figure 11.4
Jupiter's largest satellite, Ganymede. (NASA/JPL)

Europa, the Smoothest Satellite

Europa and Io, the two inner Galilean satellites, are not icy worlds like most of the satellites of the outer planets. With densities and sizes similar to our Moon, they instead appear to be predominantly rocky objects. How did they fail to acquire the ice that must have been plentiful at the time of their formation? The most probable cause is Jupiter itself, which became hot and radiated a great deal of infrared energy during its first few million years. Temperatures therefore rose, causing the ice to evaporate and leaving Europa and Io with compositions resembling bodies in the inner solar system.

In spite of its mainly rocky composition, Europa (Figure 11.6) has an ice-covered surface. In this way it is like the Earth, which also has global oceans of water, except that most of Europa's ocean may be frozen. There are very few impact craters, indicating that the surface of Europa has been capable of an impressive degree of self-renewal. Additional indications of continuing internal activity are an extensive and complex network of cracks in its icy crust.

Europa's interior could be heated by the gravitational stress of Jupiter, in a milder version of what Io experiences (see following discussion). Some scientists even speculate that beneath the ice, Europa may have a layer of liquid water. Since the Voyager spacecraft did not pass as close to Europa as to the other Galilean satellites, we

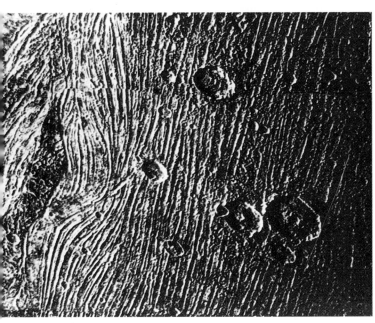

Figure 11.5
This remarkable Voyager close-up of Ganymede's surface is about 300 km wide. You can see the parallel ridges of mountains, about 15 km across, made of ice that, at the cold temperatures of the outer solar system, acts in many ways like rock. (NASA/JPL)

Figure 11.6
A mosaic of Voyager images showing part of Jupiter's satellite Europa. Note the complicated system of cracks and low ridges in the surface. (USGS, Flagstaff/NASA)

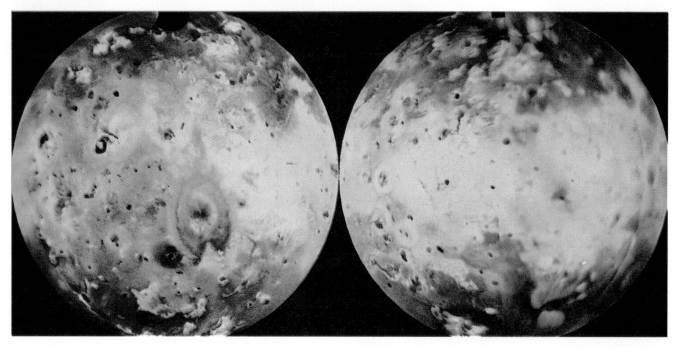

Figure 11.7
A composite of Voyager images assembled by the U.S. Geologic Survey to show both sides of the volcanically active moon Io. The orange colored deposits are sulfur snow; the white color is sulfur dioxide. (Carl Sagan once quipped that Io looks like it desperately needs a shot of penicillin.) (NASA/USGS)

know less about it. One of the highest priorities of the Galileo mission is to learn more about this enigmatic world.

Io, a Volcanic Satellite

Io, the innermost of Jupiter's Galilean satellites, turned out to be the most dramatic world visited by the Voyager spacecraft. Io is a close twin of our Moon, with nearly the same size and density. We might therefore expect it to have a similar history. Its appearance, as photographed from space, tells another story, however (Figure 11.7). Rather than a dead, cratered world, Io turns out to be the most volcanically active place in the solar system.

Io's active volcanism was discovered with the Voyager spacecraft instruments. Eight volcanoes were actually seen erupting when Voyager 1 passed in March 1979, and six of these were still active four months later when Voyager 2 passed (Figure 11.8). These eruptions consisted of graceful plumes that extended hundreds of kilometers up into space. The erupted material was not lava or steam or carbon dioxide, all of which are vented by terrestrial volcanoes, but instead was sulfur and sulfur dioxide.

As the rising plumes of Io's active volcanoes cool, the sulfur and sulfur dioxide recondense as solid particles, which fall back to the surface in colorful "snowfalls" that extend as much as 1000 km from the vent. Because of Io's low gravity, some of the material in the plumes can actually escape into space to produce a huge "donut" of es-

caped particles. Io's surface also has a variety of volcanic hot spots, one of them a sort of "lava lake" 200 km in diameter, with "lava" composed of liquid sulfur.

How can Io maintain this remarkable level of volcanism in spite of its small size? The answer lies in gravitational heating of the satellite by Jupiter. Io is about the same distance from Jupiter as our Moon is from the Earth. Yet Jupiter is more than 300 times more massive than the Earth, and its gravity pulls the satellite into an elongated shape with a several-kilometer-high bulge extending toward Jupiter. If Io always kept exactly the same face turned toward Jupiter, this bulge would not generate heat. However, due to the perturbing effects of Europa and Ganymede, Io's orbit is not exactly circular. In its slightly eccentric orbit, Io twists back and forth with respect to Jupiter, at the same time moving nearer and farther from the planet on each revolution. The twisting and flexing heats Io, much as repeated flexing of a wire coathanger heats the wire.

After billions of years, constant flexing and heating have taken their toll on Io, driving away water, carbon dioxide, and other volatiles. The inside of the moon is entirely melted, and the crust itself is constantly recycled by volcanic activity. In a sense, Io is turning itself inside out. If the volcanism observed by Voyager is typical (and we have no reason to believe it isn't) then we estimate that approximately 100 billion tons of material are expelled through the volcanic vents of Io each year!

Io was thoroughly mapped during the Voyager flybys; today you can even buy a U.S. Geologic Survey map of its

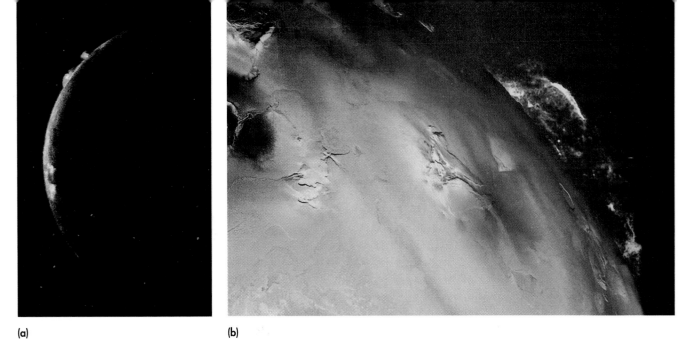

(a)   **(b)**

Figure 11.8
Two views of erupting volcanoes on Io, taken by the Voyager spacecraft. (a) Crescent Io with two bluish eruption plumes visible. (b) The large plume rising about 250 km above Io's surface (seen in white above the edge of the moon) erupted from the center of the enormous hoofprint-shaped volcanic feature that takes up most of this image. It is named Pele, after the Hawaiian goddess of fire, and is about the size of the state of Alaska. (NASA/JPL)

surface. But future explorers, and even the scientists directing the Galileo spacecraft, will not be able to rely on the details of those maps. Io is one world whose face will keep changing as time goes on.

Titan, a Satellite with Atmosphere

Titan, discovered in 1655 by the Dutch astronomer Christian Huygens, is the largest satellite of Saturn. It was the first satellite discovered since Galileo saw the four moons of Jupiter that bear his name. Titan has roughly the same diameter, mass, and density as Callisto. Presumably it also has a similar composition—about half ice and half rock. What is remarkable about this satellite, however, is the presence of a substantial atmosphere, discovered by Gerard Kuiper in 1944 (Figure 11.9). He noticed that spectra of Titan showed absorption features indicating the presence of substantial amounts of methane gas.

The 1980 Voyager flyby of Titan determined that the atmospheric surface pressure on this satellite is 1.6 bars, higher than that on any other satellite, and even higher than that of the terrestrial planets Mars and Earth. The atmospheric composition is primarily nitrogen, an important way in which Titan's atmosphere resembles Earth's.

A variety of additional compounds have been detected in Titan's upper atmosphere, including carbon monoxide (CO), hydrocarbons (compounds of hydrogen and carbon) such as methane (CH_4), ethane (C_2H_6), and propane (C_3H_8), and nitrogen compounds such as hydrogen cyanide (HCN), cyanogen (C_2N_2), and cyanoacetylene (HC_3N). These indicate an active chemistry in which

Figure 11.9
To the Voyager cameras, Titan looked like a big fuzzy orange ball (see Figure 10.2). In this enhanced close-up view, we see the main orange haze layer that blocks our view of the surface, but also the blue haze that lies higher in the moon's atmosphere. (NASA/JPL)

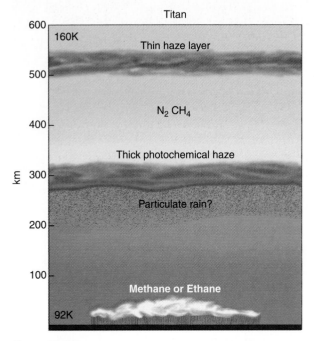

Figure 11.10
A chart of the structure of Titan's atmosphere. Heights above the surface are given on the left border. Titan has the most Earth-like atmosphere in the solar system, although it is much colder than our planet. (NASA/JPL)

sunlight interacts with atmospheric nitrogen and methane to create a rich mix of organic molecules. This is very much the way we believe organic compounds were formed on the Earth when it still had its original atmosphere. The discovery of HCN on Titan is particularly interesting, since this molecule is the starting point for formation of some of the components of DNA, the fundamental genetic molecule essential to life on Earth.

Titan has multiple layers of clouds (Figure 11.10). The lowest are within the bottom 10 km of the atmosphere; these are condensation clouds probably composed of methane. Much higher is a dark reddish haze or smog, consisting of complex organic chemicals created by the energy of sunlight falling on the molecules of Titan's atmosphere. Formed at an altitude of several hundred kilometers, this aerosol slowly settles downward to the surface, where it may have built up a deep layer of tar-like organic chemicals.

Much to the disappointment of the Voyager scientists, the spacecraft revealed that the reddish haze completely blankets Titan, hiding its surface from our view. Still the Voyager data allow us to make models of what it might be like to land on this intriguing world. For example, we know Titan's surface temperature is about 90 K, held uniform by the blanketing atmosphere. At such a low temperature, methane plays the same role that water does on Earth: the gas is only a minor constituent of the atmosphere, but it can exist in gas, liquid, or solid form. Methane condenses to form clouds, and there may be

seas of liquid methane and ethane, as well as solid methane ice.

Organic compounds are chemically stable in Titan's cold environment, unlike those on the warmer, oxidizing Earth. Titan's surface, therefore, probably records a chemical history that goes back billions of years. Many scientists believe that this satellite will provide more insights into the early history of Earth's atmosphere, and even into the origin of life, than any other object in the solar system.

11.3

Triton and Pluto

In this section we discuss two apparently similar objects: Neptune's satellite Triton and the planet Pluto. Seeking to understand the ways in which these two worlds are similar to each other is an example of the comparative approach to studying astronomy. Pluto is the only planet not visited by a spacecraft; therefore comparing it with Triton, which was studied by Voyager in 1989, proves especially useful.

Triton and Its Volcanoes

Triton has a diameter of 2720 km and a density of 2.1 g/cm^3, indicating a probable composition of 75 percent rock mixed with 25 percent water ice. Measurements indicate that Triton's surface has the coldest temperature of any of the worlds our robot representatives have visited. Because its reflectivity is so high (about 80 percent), Triton reflects most of the solar energy falling on it, resulting in a surface temperature between 35 and 40 K.

The surface material of Triton is made of frozen water, nitrogen, methane, and carbon monoxide. Although methane and nitrogen exist as gas in most of the solar system, they are frozen at Triton's temperatures. Only a small quantity of nitrogen vapor persists to form an atmosphere. Although the surface pressure of this atmosphere is only 16 millionths of a bar, this is sufficient to support thin haze or cloud layers.

Triton's surface, like that of many other satellites in the outer solar system, reveals a long history of geological evolution (Figure 11.11). While some impact craters are found, so are many regions that have been flooded by the local version of "lava" (perhaps water or water–ammonia mixtures). The evidence for such low-temperature volcanism includes a number of frozen "lava lakes" more than 100 km across (Figure 11.12). There are also mysterious regions of jumbled or mountainous terrain.

The Voyager flyby of Triton took place at a time when the satellite's southern pole was tipped toward the Sun, allowing this part of the surface to enjoy a period of relative warmth. (Remember that warm on Triton is still outrageously colder than anything we experience on Earth.) A polar cap covers much of its southern hemisphere, apparently evaporating along the northern edge. This polar cap

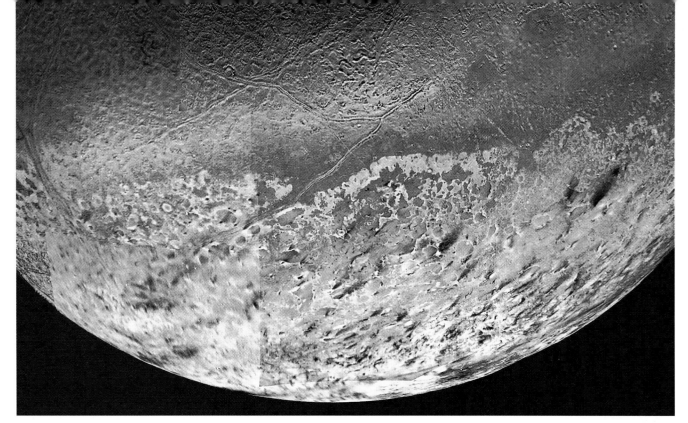

Figure 11.11

A mosaic of Voyager images of Triton, showing a wide range of surface features. The pinkish area at the bottom is Triton's large southern polar cap. The south pole of Triton currently faces the Sun, and the slight heating effect is driving some of the material northward, where it is colder. (NASA/Image processing at USGS)

may consist of frozen nitrogen that was deposited during the previous winter. Remarkably, the evaporation of this polar cap generates geysers or volcanic plumes of nitrogen gas that make fountains about 10 km high. These plumes differ from the volcanic plumes of Io in their composition, and also in that they derive their energy from sunlight warming the surface rather than from internal heat.

Discovery of Pluto

Now we turn to Pluto, a small planet that resembles Triton in many ways, and is quite different from the giant planets in the outer solar system. Pluto was discovered through a careful, systematic search, unlike Neptune, whose position was calculated from gravitational theory. Nevertheless, the history of the search for Pluto began with indications that Uranus and Neptune had slight departures from their predicted orbits. Early in the 20th century several astronomers—most notably Percival Lowell, then at the peak of his fame as an advocate of intelligent life on Mars (see Chapter 9)—became interested in this problem.

At the time Lowell made his calculations, Neptune had moved such a short distance since its discovery that it could not be used effectively to search for perturbations by an unknown ninth planet. Therefore Lowell and his

Figure 11.12

Old flooded "lava lakes" on Triton can be seen in this Voyager close-up. Now frozen solid, these features, 100 to 200 km across, date from an earlier period of volcanic activity which may have poured water onto Triton's surface. (NASA/JPL)

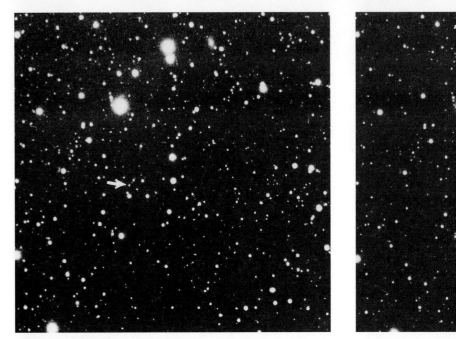

Figure 11.13
Portions of the two photographs on which Clyde Tombaugh discovered Pluto in 1930. The left one was taken on January 23 and the right one on January 29. Note that Pluto, indicated by an arrow, has moved among the stars during those six nights. Note also that if we hadn't put an arrow next to it, you probably would never have spotted the dot that moved. (Lowell Observatory)

contemporaries based their calculations primarily on the tiny remaining irregularities in the motion of Uranus. Lowell's computations indicated two possible locations for a perturbing planet; the more likely of the two was in the constellation of Gemini. He predicted a mass for the planet intermediate between those of the Earth and Neptune (his calculations gave about 6.6 Earth masses). Other astronomers, however, obtained other solutions from the tiny orbital irregularities, including one model that indicated *two* exterior planets.

At his Arizona observatory, Lowell searched without success for the unknown planet from 1906 until his death in 1916. Subsequently, his brother donated to the observatory a 33-cm photographic telescope that could record a 12° × 14° area of the sky on a single photograph. The new camera went into operation in 1929.

In February 1930, Clyde Tombaugh, comparing photographs made on January 23 and 29 of that year (Figure 11.13), found an object whose motion appeared to be about right for a planet far beyond the orbit of Neptune. It was within 6° of the position Lowell had predicted for the unknown planet. The new planet was named for Pluto, the god of the underworld; the choice of this name, among hundreds suggested, may have been helped by the fact that the first two letters were Percival Lowell's initials.

Although at the time the discovery of Pluto appeared to be a vindication of gravitational theory similar to the triumph of Adams and Leverrier in predicting the position of Neptune, we now know that Lowell's calculations were wrong. When the mass of Pluto was finally measured, it was found to be much less than that of the Moon. It could not possibly have exerted any measurable pull on either Uranus or Neptune. To the degree that the orbital discrepancies of these two planets are real, they must be due to some other cause. A survey of the entire sky in 1983 by the Infrared Astronomical Satellite (IRAS), however, has revealed no hidden "Planet X," and today it is generally accepted that the supposed perturbations of Uranus and Neptune are not, and never were, real.

Pluto's Motion and Satellite

Pluto's orbit has the highest inclination to the ecliptic (17°) of any planet, and also the largest eccentricity (0.248). Its average distance from the Sun is 40 AU, or 5.9 billion km, but its perihelion distance is under 4.5 billion km, within the orbit of Neptune. As it happens, Pluto is presently closer to us than is Neptune, and it will remain so until 1999. Even though the orbits of these two planets cross, there is no danger of collision because of the high inclination of Pluto's orbit.

Pluto completes its orbital revolution in a period of 248.6 years; since its discovery in 1930, it has traversed less than one-quarter of its long path around the Sun. But almost 50 years after that discovery, astronomers found that it is not alone in its distant travels.

In 1978, astronomer James Christy of the U.S. Naval Observatory found that Pluto has a satellite, which was named Charon after the boatman who ferried the dead into the realm of Pluto in mythology. The exact nature of

Clyde Tombaugh: From the Farm to Fame

Clyde Tombaugh discovered Pluto when he was 24 years old, and his position as staff assistant at the Lowell Observatory was his first paying job. Tombaugh had been born on a farm in Illinois, but when he was 16 his family moved to Kansas. There, with his uncle's encouragement, he observed the sky through a telescope the family had ordered from the Sears catalog. He later constructed a larger telescope on his own, and devoted his nights (when he wasn't too tired from farmwork) to making detailed sketches of the planets.

In 1928, after a hailstorm ruined the crop, Tombaugh decided he needed a job to help support his family. Although he had only a high school education, he thought of becoming a telescope builder. He sent his planet sketches to the Lowell Observatory, seeking advice about whether such a career choice was realistic. By a wonderful twist of fate, his query arrived just when the Lowell astronomers realized that a renewed search for a ninth planet would require a very patient and dedicated observer.

The large photographic plates (pieces of glass with photographic emulsion on them) that Tombaugh was hired to take at night and search during the day contained an average of about 160,000 star images each. How to find Pluto among them? The technique involved taking two photographs about a week apart. During that week, a planet would move a tiny bit, while the stars remained in the same place relative to each other. A new instrument called a "blink comparator" could quickly alternate the two images in an eyepiece. The stars, be-

Clyde Tombaugh at the Lowell Observatory at the time of his discovery of Pluto in 1930.
(Lowell Observatory)

ing in the same position on the two plates, would not appear to change as the two images were "blinked." But a moving object would appear to wiggle back and forth as the plates were alternated.

After examining more than 2 million stars (and many false alarms), Tombaugh found his planet on February 18, 1930. The astronomers at the observatory checked his results carefully, and the find was announced on March 13, the 149th anniversary of the discovery of Uranus. Congratulations and requests for interviews poured in from around the world. Visitors descended on the observatory by the scores, wanting to see the place where the first new planet in almost a century had been discovered, as well as the person who had discovered it.

In 1932 Tombaugh took leave from Lowell, where he had continued to search and blink, to get a college degree. Eventually he received a master's degree in astronomy, and taught navigation for the Navy during World War II. In 1955, after working to develop a rocket-tracking telescope, he became a professor at New Mexico State University, where he helped found the astronomy department. He is still alive today, giving lectures on astronomy and defending Pluto against a few astronomers who have unkindly suggested that Pluto is too small to be a "real" planet.

---

its orbit was not confirmed until 1985, when the system had turned enough that the satellite and the planet began to go behind each other as seen from the Earth. These observations showed that the satellite was in a retrograde orbit and had a diameter of about 1200 km, more than half the size of Pluto itself. This makes Charon the satellite whose size is the largest fraction of its parent planet. Seen from Pluto, Charon would be as large as eight Moons on Earth, side by side in the sky.

Furthermore, Pluto orbits "on its side" like Uranus, and Charon circles in its equatorial plane around its rotation axis. The gravitational interaction between the two worlds has locked them in the tightest of embraces. Not only does Charon take the same time to rotate and revolve, thus keeping the same face toward Pluto, but this period is the same length as Pluto's day. This means that if you were on the Charon-facing side of Pluto, you would see the moon hover in the same place all the time, while

from Pluto's opposite hemisphere Charon will never be visible.

The Nature of Pluto

Pluto has not been visited by spacecraft, and it is so faint that the world's largest telescopes are required to study it. The diameter of Pluto is 2190 km, only 60 percent that of the Moon. From the diameter and mass, we find a density of 2.1 g/cm^3, suggesting a mixture of rocky materials and water ice in about the same proportions as in Triton.

Pluto's surface is highly reflective, and its spectrum demonstrates the presence of frozen methane, carbon monoxide, and nitrogen (Figure 11.14). The surface temperature ranges from about 50 K near aphelion to 60 K near perihelion, resulting in partial evaporation of the methane and nitrogen ice to generate an atmosphere. Pluto is now in its warmest period, near perihelion, and its

Figure 11.14

(Top) An image of Pluto and its satellite Charon, taken with the Faint Object Camera on the Hubble Space Telescope soon after the repair mission. The clarity of the image allowed astronomers to measure the diameters of the two worlds to an accuracy of 1%. (R. Albrecht, ESA/ESO & NASA)   (Bottom) A global map of the two hemispheres of Pluto, constructed using computer processing techniques from images taken during Pluto's full 6.4-day rotation period. The tile pattern is an artifact of the image enhancement techniques. Pluto was over 4.8 billion km from Earth at the time the images were taken (Summer 1994). A number of bright and dark features are visible, including a prominent northern polar cap. (A. Stern, M. Buie, NASA & ESA)

atmosphere is accordingly near maximum size and density. Observations of a distant star seen through that thin atmosphere suggest that the surface pressure is about a ten-thousandth of the Earth's. Because Pluto is a few degrees warmer than Triton, its atmospheric pressure is about ten times greater.

The Origin of Pluto

We have a long way to go to understand how Pluto arrived at its unusual place in the outer solar system. Neither its path around the Sun nor its size resembles the jovian planets. The presence of a moon as large as Charon is also a mystery. As we have seen, Pluto is much more similar to a moon such as Triton than to a planet such as Saturn. One thought is that Pluto and Charon are merely the largest examples of a family of distant asteroid- or comet-like bodies. As we will discuss in Chapter 12, we have recently discovered some smaller siblings in that family.

It is interesting to note that Triton is the only large satellite in the solar system with a retrograde orbit. The most distant of Neptune's satellites, Nereid, the one beyond Triton, has an orbit that is direct but very large and highly elliptical. Some astronomers speculate that the strange orbits of Pluto, Charon, Nereid, and Triton are the result of violent collisions during the early history of the outer solar system.

11.4

Planetary Rings

All four of the giant planets have rings, with each ring system consisting of billions of small particles or moonlets orbiting close to their planet. Each of these rings displays a complicated structure that seems related to interactions between the ring particles and the larger satellites. How-

TABLE 11.2
Properties of Ring Systems

Planet	Outer Radius (km)	Outer Radius (R_{planet})	Mass (kg)	Reflectivity (%)
Jupiter	128,000	1.8	$10^{10}(?)$	?
Saturn	140,000	2.3	10^{19}	60
Uranus	51,000	2.2	10^{14}	5
Neptune	63,000	2.5	10^{12}	5

ever, the four ring systems are very different from each other in mass, structure, and composition (Table 11.2).

Saturn's large ring system is made up of icy particles spread out into several vast, flat rings containing a great deal of fine structure. The Uranus and Neptune ring systems, on the other hand, are nearly the reverse of Saturn's; they consist of dark particles confined to a few narrow rings with broad empty gaps in between. Jupiter's ring and at least one of Saturn's are merely transient dust bands, constantly renewed by erosion of dust grains from small satellites. In this section we focus on the two most massive ring systems, those of Saturn and Uranus.

What Causes Rings?

A ring is a collection of vast numbers of particles, each obeying Kepler's laws as it follows its own orbit around the planet. Thus the inner particles revolve faster than those farther out, and the ring as a whole does not rotate as a solid body. In fact, it is better not to think of a ring *rotating* at all, but rather to consider the *revolution* of its individual moonlets.

If the ring particles were widely spaced, they would move independently, like separate small satellites. However, in the main rings of Saturn and Uranus the particles are close enough to each other to exert mutual gravitational influence, and occasionally even to rub together or bounce off each other in low-speed collisions. Because of these interactions, we see phenomena such as waves that move across the rings the way water waves move over the surface of the ocean.

There are two basic theories of how such rings come to be. First is the breakup theory, which suggests that the rings are the remains of a shattered satellite. The second theory, which takes the reverse perspective, suggests that the rings are made of particles that were unable to come together to form a satellite in the first place. In either theory, an important role is played by the gravitation of the planet. Close to the planet, gravitational forces (like those that cause tides on the Earth) can tear bodies apart or inhibit loose particles from coming together. The rings of both Saturn and Uranus lie within the region in which large satellites are probably not stable against such forces (Figure 11.15).

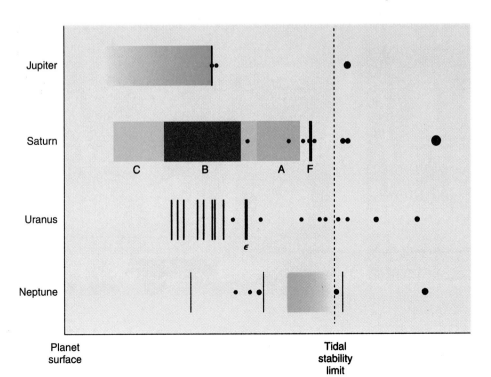

Figure 11.15
A diagram showing the location of the ring systems of the four giant planets. The left axis represents the planet's surface. The dotted vertical line is the limit inside which gravitational forces can break up satellites. Notice that each planet's system is drawn to a different scale, so that this stability limit lines up for all four of them. The black dots are the inner satellites of each planet on the same scale. Notice only really small bodies survive inside the stability limit.

In the breakup theory of ring formation, we can imagine a satellite or even a passing comet coming too close and being torn apart. The fragments then remain in orbit as one or more rings. A more likely variant of this idea suggests that a small satellite close to the planet might have broken apart in a collision, with the fragments dispersing into a disk. We do not know which explanation holds for the rings, although many scientists have concluded that at least a few of the rings are relatively young and must therefore be the result of breakup.

Rings of Saturn

The rings of Saturn are one of the most beautiful sights in the solar system (Figure 11.16). From outer to inner, the three brightest rings are labeled the A, B, and C Rings. In Table 11.3 the dimensions of the rings are given in both kilometers and in units of the radius of Saturn, R_S. The B Ring is the brightest and has the most closely packed particles, whereas the A and C Rings are translucent. The total mass of the B Ring, which is probably close to the mass of the entire ring system, is about equal to that of an icy satellite 250 km in diameter. Between the A and B Rings is a wide gap named the Cassini Division after Gian Domenico Cassini, who first glimpsed it in 1675.

The rings of Saturn (which you can see very clearly in Figure 10.9) are very broad and very thin. The width of the main rings is 70,000 km, yet their thickness is only about 20 m. If we made a scale model of the rings out of paper the thickness of the pages in this book, we would have to make them 1 km across—about eight city blocks. On this scale, Saturn itself would loom as high as an 80-story building.

TABLE 11.3
Rings of Saturn

Ring Name	Outer Edge (R_S)	Outer Edge (km)	Width (km)
F	2.324	140,180	90
A	2.267	136,780	14,600
Cassini Division	2.025	122,170	4,590
B	1.949	117,580	25,580
C	1.525	92,000	17,490

The ring particles are composed primarily of water ice, and they range from grains the size of sand up to house-sized boulders. An insider's view of the rings would probably resemble a bright cloud of floating snowflakes and hailstones, with a few snowballs and larger objects, many of these loose aggregates of smaller particles (Figure 11.17).

In addition to the broad A, B, and C Rings, Saturn has a handful of very narrow rings, no more than 100 km wide. The most substantial of these, which lies just outside the A Ring, is called the F Ring. In general, Saturn's narrow rings resemble the rings of Uranus and Neptune.

Rings of Uranus and Neptune

The rings of Uranus are narrow and black, making them almost invisible from the Earth. The nine main rings were discovered in 1977 from observations made of a star as Uranus passed in front of it. We call the passage of one astronomical object in front of another an **occultation.** During the 1977 occultation, we merely expected the star's light to disappear as the planet moved across it. But

Figure 11.16
Voyager image of Saturn's rings as seen from underneath. Since you are looking up at sunlight filtering through the rings, here the brightest parts of the ring system are the gaps and the least dense regions. The denser B Ring is not transmitting much light. (NASA/JPL)

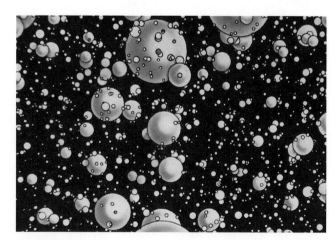

Figure 11.17
An artist's idealized version of the sizes of typical particles inside the A Ring of Saturn. The ball at the center is roughly the size of an adult's head. In the real rings, the particles are not spherical but are likely to have irregular shapes. (NASA/JPL)

Astronomy and Future Tourism:
The Seven Wonders of the Solar System

Imagine it is the year 2075 and humanity has been exploring the solar system for over a century. It is now possible to visit many of the worlds we have been discussing in these chapters—not cheap, you understand, but possible. And suppose you are in charge of planning the itinerary for a once-in-a-lifetime tour of the most spectacular sights in all of these worlds. (By "sights" here, we don't mean a whole planet or moon, but something specific on or around it.)

Ancient writers enjoyed making lists of the seven great wonders of the world, such as the pyramids and the hanging gardens of Babylon. What would you include on your list of seven great solar system wonders? Make your own list first and then take a look at one possible answer, on the following page.

continued

---

the star dimmed briefly several times before Uranus even reached it, as each narrow ring passed between the star and the telescope. Thus the rings were mapped out in detail even though they could not be seen or photographed

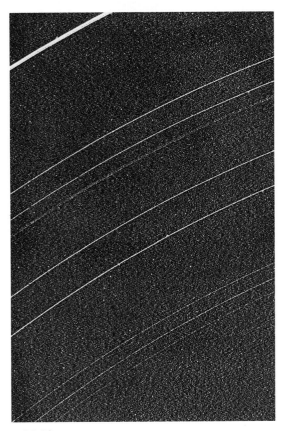

Figure 11.18
Voyager photograph of the narrow rings of Uranus. Note that we had to expose the image for a long time to get a glimpse of these dark rings; you can see the grainy structure of "noise" in the electronics of the camera. (NASA/JPL)

directly. When Voyager approached Uranus in 1986, it was able to study the rings at close range; the spacecraft also photographed two new rings (Figure 11.18).

The outermost and most massive of the rings of Uranus is called the Epsilon Ring. It is only about 100 km wide and probably no more than 100 m thick (similar to the F Ring of Saturn). The Epsilon Ring circles Uranus at a distance of 51,000 km, about twice the radius of Uranus. This ring probably contains as much mass as all of Uranus' other ten rings combined; most of them are narrow ribbons less than 10 km wide—just the reverse of the broad rings of Saturn.

The individual particles in the uranian rings are nearly as black as lumps of coal. While astronomers do not understand the composition of this material in detail, it seems to consist in large part of black carbon and hydrocarbon compounds. Organic material of this sort is rather common in the outer solar system. Many of the asteroids and comets (discussed in Chapter 12) also are composed of dark, tar-like materials. In the case of Uranus, its ten small inner satellites have a similar composition, suggesting that one or more satellites might have broken up to make the rings.

The rings of Neptune are generally similar to those of Uranus but even more tenuous (Figure 11.19). There are only four of them, and the particles are not uniformly distributed along their lengths. Because these rings are so difficult to investigate from the Earth, it will probably be a long time before we understand them very well.

Satellite-Ring Interactions

Much of our current fascination with planetary rings is a result of the intricate structures discovered by the Voyager spacecraft. We now understand that most of these structures owe their existence to the gravitational effect of

Just to get an argument going, here is one list of favorite solar system sights. How does it compare to yours?

1. Olympus Mons, the giant volcano on Mars.
2. The enormous lava channel on Venus.
3. The Pele volcano on Io.
4. The kinky F Ring around Saturn.
5. The Red Spot storm on Jupiter.
6. The Caloris impact basin on Mercury.
7. The human footprints on the Earth's Moon, showing signs of intelligence in the solar system.

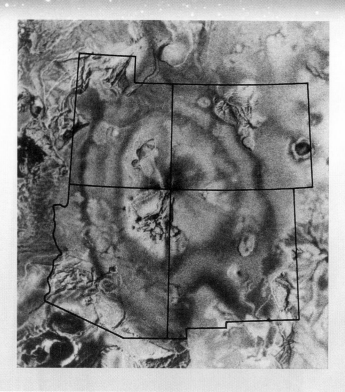

The volcanic mound called Pele on Io, with the outlines of the states of Arizona, New Mexico, Colorado, and Utah, drawn in to scale. (Stephen Meszaros, NASA)

Figure 11.19
A long exposure of the rings of Neptune, as photographed by Voyager in 1989. Note the two denser regions of the outer ring. (JPL/NASA)

satellites. Without satellites the rings would be flat and featureless. Indeed, without satellites there would probably be no rings at all, since left to themselves, thin disks of matter gradually spread and dissipate.

Most of the gaps in Saturn's rings, and also the location of the outer edge of the A Ring, result from gravitational resonances with small inner satellites. A **resonance** takes place when two objects have orbital periods that are exact ratios of each other, such as one-to-two or one-to-three. For example, any particle in the gap at the inner side of the Cassini Division of Saturn's rings would have a period equal to one-half that of Saturn's satellite Mimas. Such a particle would be nearest Mimas in the same part of its orbit every second revolution. The repeated gravitational tugs of Mimas, acting always in the same direction, would perturb it, forcing it into a new orbit outside the gap and therefore no longer representing a resonance.

One of the most interesting rings of Saturn is the narrow F Ring, which contains several apparent ringlets within its 90-km width (Figure 11.20). In places the F Ring breaks up into two or three parallel strands that sometimes show bends or kinks. Most of the rings of Uranus and Neptune are also narrow ribbons like the F Ring of Saturn. Clearly, the gravity of some objects must be keeping the particles in these rings from spreading out.

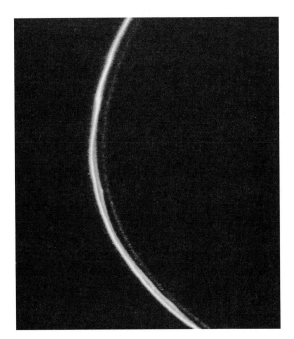

Figure 11.20
Voyager close-up of the narrow but complex F Ring of Saturn, showing some of the kinky strands. (NASA/JPL)

The best theory is that the rings are controlled gravitationally by small satellites orbiting very close to them. This certainly seems to be the case for Saturn's F Ring: Voyager showed that it is bounded by the orbits of two satellites, now called Pandora and Prometheus. These two small objects (each about 100 km in diameter) are referred to as "shepherd satellites" since their gravitation serves to "shepherd" the ring particles and keep them confined to a narrow ribbon. A similar situation applies to the Epsilon Ring of Uranus, which is shepherded by the satellites Cordelia and Ophelia. These two shepherds, each about 50 km in diameter, orbit about 2000 km inside and outside the ring.

Theoretical calculations suggest that the other narrow rings in the uranian and neptunian systems should also be controlled by shepherd satellites, but none has been located. The calculated diameter for such shepherds—about 10 km—is just at the limit of detectability for the Voyager cameras, so it is impossible to say if they are present or not. Given all the narrow rings we see, many scientists hope to find another, more satisfactory explanation.

The main rings of Saturn include several narrow gaps that are clearly the result of satellites, which clear lanes in a manner similar to that of the shepherd satellites. As each small satellite moves through its gap, it produces waves in the surrounding ring material like the wake left by a moving ship. One of these satellites, called Pan, was discovered in 1991 from its wake; calculations based on the waves it produced pinpointed its location, and when the appropriate ten-year-old Voyager picture was examined, there it was!

Studies of planetary rings, and of the gravitational interactions between rings and small satellites, have come a long way in the past few years, but many problems in understanding these complex phenomena remain unsolved. Without additional spacecraft missions to the outer planets, however, it may be difficult to obtain the data necessary to test the new dynamical theories now being developed. NASA and ESA (the European Space Agency) are developing a new mission to Saturn called Cassini, which is scheduled to begin orbiting the planet in 2004. The Cassini mission (which will also drop a probe into the atmosphere of Titan) should provide a wealth of new information on the rings.

Summary

11.1 The four jovian planets are accompanied by impressive systems of satellites and rings. Jupiter has 16 satellites, Saturn 19, Uranus 15, and Neptune 8, many discovered by Voyager. Most of these satellites are composed in part of water ice, with varying amounts of rock and dark primitive material mixed in. Of the four ring systems, Saturn's is the largest; Uranus and Neptune have narrow rings of dark material; and Jupiter has a tenuous ring of dust.

11.2 The largest satellites are Ganymede, Callisto, and Titan, all differentiated objects which are about half water ice. Callisto has an ancient cratered surface while Ganymede shows evidence of having once experienced tectonic and volcanic activity. Voyager was not able to see Titan's surface because it is unique among jovian moons in having a thick atmosphere (surface pressure of 1.6 bar) with an opaque haze layer. Io and Europa are denser and smaller, each about the size of the Moon. Io is the most volcanically active object in the solar system; its eruptions of sulfur and sulfur compounds are powered by tidal interactions with Jupiter.

11.3 Triton, Neptune's retrograde satellite, and Pluto resemble each other in size, composition, and temperature. Triton has a very thin atmosphere, and we can see eruptions of nitrogen when surface layers of its extensive polar cap evaporate while facing the Sun. The planet Pluto, discovered by Clyde Tombaugh in 1930, has never been visited by spacecraft. Observations indicate the presence of frozen methane, carbon monoxide, and nitrogen on its surface, and a thin (perhaps temporary) atmosphere. Pluto's moon Charon has a diameter more than half of Pluto's.

11.4 Rings are composed of vast numbers of individual particles (ranging in size) orbiting so close to a planet that its gravitational forces could have broken larger pieces apart or kept pieces from gathering together. Saturn's rings are broad, flat, and nearly continuous, except for a handful of gaps. The particles are mostly water ice, with typical dimensions of a few cm. The rings of Uranus are narrow ribbons separated by wide gaps and contain much less mass. They were discovered in 1977 during a stellar **occultation.** Neptune's rings are similar, but contain even less material. Much of the complex structure of the rings is due to waves and **resonances** induced by satellites within the rings or orbiting outside them. Voyager found that several thin rings were accompanied by shepherd moons that kept the ring material confined.

Review Questions

1. What are the satellites of the outer planets made of, and why is their composition different from that of the Moon?

2. Compare the geology of Callisto, Ganymede, and Titan. Why do we think Titan has an atmosphere while the other two large satellites do not?

3. Explain the energy source that powers the volcanoes of Io.

4. Compare the properties of Titan's atmosphere with those of the Earth's atmosphere.

5. How was Pluto discovered?

6. How are Triton and Pluto similar?

7. Describe and compare the rings of Saturn and Uranus.

8. Three possibilities were suggested in the text for the origin of Saturn's rings. List them, and briefly summarize the arguments in favor of each.

Thought Questions

9. Why do you think the outer planets have such extensive systems of rings and satellites, while the inner planets do not?

10. Which would have the longer orbital period, a satellite 1 million km from the center of Jupiter, or a satellite 1 million km from the center of Earth? Why?

11. Ganymede and Callisto were the first icy objects to be studied from a geological point of view. Summarize the main differences between their geology and that of the rocky terrestrial planets.

12. Compare the properties of the volcanoes on Io with those of terrestrial volcanoes. Give at least two similarities and two differences.

13. Would you expect to find more impact craters on Io or Callisto? Why?

14. Where did the nitrogen in Titan's atmosphere come from? Compare its origin with that of the nitrogen in our atmosphere.

15. Do you think there are many impact craters on the surface of Titan? Why or why not?

16. Why do you suppose the rings of Saturn are made of bright particles, whereas the particles in the rings of Uranus and Neptune are black?

17. Suppose you miraculously removed all of Saturn's satellites. What would happen to its rings?

Problems

18. Saturn's A, B, and C Rings extend about 75,000 to 137,000 km from the center of the planet. Use Kepler's third law to calculate the difference between how long the inner edge and the outer edge of the three-ring system take to revolve about the planet.

19. Use the information in Appendix 8 to calculate what you would weigh on Titan, Io, and Uranus' satellite Miranda.

20. Occultations of stars by the rings of Uranus have yielded resolutions of 10 km in determining ring structure. What angular resolution (in arcsec) would a space telescope have to achieve to obtain equal resolution from Earth orbit? How close to Uranus would a spacecraft have to get to obtain equal resolution with a camera having an angular resolution of 2 arcsec? To solve this problem, you need the "small-angle formula" to relate angular and linear size in the sky. It is usually written as

$$\frac{\text{angular diameter}}{206{,}265} = \frac{\text{linear diameter}}{\text{distance}}$$

where angular diameter is expressed in arcsec.

Suggestions for Further Readings

Binzel, R. "Pluto" in *Scientific American,* June 1990.

Burnham, R. "At the Edge of Night: Pluto and Charon" in *Astronomy,* Jan. 1994, p. 41.

Elliot, J. and Kerr, R. *Rings: Discoveries from Galileo to Voyager.* 1985, MIT Press.

Elliot, J. et al. "Discovering the Rings of Uranus" in *Sky & Telescope,* June 1977, p. 412.

Esposito, L. "The Changing Shape of Planetary Rings" in *Astronomy,* Sep. 1987, p. 6.

Harrington, R. and Harrington, B. "The Discovery of Pluto's Moon" in *Mercury,* Jan./Feb. 1979, p. 1.

Hartmann, W. "View from Io" in *Astronomy,* May 1981, p. 17.

Hoyt, W. *Planets X and Pluto.* 1980, U. of Arizona Press. Superb history of the search for Pluto and other outer planets.

Johnson, T. et al. "The Moons of Uranus" in *Scientific American,* Apr. 1987.

Levy, D. *Clyde Tombaugh: Discoverer of Pluto.* 1991, U. of Arizona Press.

Morrison, D. "An Enigma Called Io" in *Sky & Telescope,* Mar. 1985, p. 198.

Rothery, D. *Satellites of the Outer Planets: Worlds in Their Own Right.* 1992, Oxford U. Press.

Sobel, D. "Secrets of the Rings" in *Discover,* Apr. 1994, p. 86. Discusses the outer planet ring systems.

Talcott, R. "The Violent Volcanoes of Io" in *Astronomy,* May 1993, p. 41.

Tombaugh, C. "The Discovery of Pluto" in *Mercury,* May/June 1986, p. 66, and Jul./Aug. 1986, p. 98.

See also the books on the giant planets recommended in Chapter 10, and the books on the planetary system in general listed in Chapter 6.

Using REDSHIFT ™

1. Observe the satellites of Jupiter by centering and locking on the planet. Set the *Zoom Panel* to level 8.

Which satellites move fastest? Why do the orbits appear to be ovals?

Step time and, using a water-based pen, mark the farthest each of the satellites moves away from Jupiter on the right side of the monitor. Measure the distance from the planet's center to the marks and the time it takes each satellite to return to the marked positions. For each satellite, square the period and divide it by the cube of the orbital radius.

Do these numbers have a trend?

What do they say about Kepler's laws of motion (Chapter 2)?

2. Center Saturn and lock on it. Set the *Zoom Factor* to 1000 and turn off the satellites. Set the date to January 1985 and step time.

What happens to Saturn's rings? You can better understand the ring behavior by setting the *location* to heliocentric, latitude to 0°, and height to 30 AU. Center on the Sun, set Saturn's magnification to 1000, and step time.

Comet Halley photographed from Australia in 1985. This is a composite of three images, one each in red, green, and blue light. During the time they were taken, the comet moved among the stars. The telescope was moved to keep the image of the comet steady, causing the stars to become colored streaks in the background. (Anglo-Australian Observatory)

CHAPTER 12

Comets and Asteroids: Debris of the Solar System

Thinking Ahead

Hundreds of smaller members of the solar system, called asteroids and comets, are known to have crossed the Earth's orbit in the past, and many others will do so in the centuries ahead. What could we do if one of these bodies (that had significant mass) were predicted to cross the Earth's orbit at exactly the place our planet then happened to be?

To understand the early history of life on Earth, scientists need fossils from as long ago as possible. To piece together the early history of the solar system, we need some cosmic fossils— pieces formed when our system was very young. When it comes to answering our questions about origins, the planets themselves are largely mute. Melted, battered by giant impacts, twisted by tectonic forces, they retain little evidence of their births. Reconstructing their early history is almost as difficult as determining the circumstances of human birth by looking at an adult. Instead, we must turn to clues provided by the surviving remnants of the creation process—ancient but smaller bodies with which we share our cosmic neighborhood.

We have rather arbitrarily divided these bodies into two categories. **Asteroids** are small, rocky objects containing little volatile (easily evaporated) material. They differ from the terrestrial planets primarily in size, and are sometimes even called minor plan-

The day will yet come when posterity will be amazed that we remained ignorant of things that will to them seem so plain. . . . Men will someday be able to demonstrate in what regions comets have their paths [and] what is their size and constitution. Let us be satisfied with what we have discovered, and leave a little truth for our descendants to find out.

Lucius Annaeus Seneca, in *Natural Questions* (about 63 A.D.)

ets. **Comets** are small, icy pieces containing frozen water and other volatile materials (but with solid grains mixed in). Comets are most easily noticed when they get near the Sun and form a visible (but temporary) atmosphere and an extended tail of gas and dust. Recently astronomers have begun finding some objects that seem intermediate between the two types, but our categories are still a handy way to remember them.

In this chapter we discuss the asteroids and comets, and in Chapter 13 we continue the story of the smaller members of our system by looking at meteorites, which are samples of cosmic material that survive their fall to the surface of the Earth.

Figure 12.1

Time exposure showing trails left by asteroids (arrows). The telescope moved backwards to compensate for the rotation of the Earth, thus the stars remained fixed. But in the time the image was taken, the asteroids moved relative to the stars and left a trail. (Yerkes Observatory)

12.1

Asteroids

The Discovery and Motions of the Asteroids

The orbits of most of the asteroids lie between those of Mars and Jupiter, in a region called the **asteroid belt.** The asteroids are too small to be seen without a telescope; hence the first of them was not discovered until the beginning of the 19th century. At that time, astronomers were hunting for an additional planet they thought should exist in the large gap between the orbits of Mars and Jupiter. Some German astronomers even organized themselves into a group they called "the celestial police" to search for this "missing" member of the solar system.

In January 1801, the Sicilian astronomer Giovanni Piazzi thought he had found this missing planet when he discovered the first asteroid, which he named Ceres, orbiting at 2.8 AU from the Sun. However, his discovery was followed the next year by that of another little planet in a similar orbit, and two more were found in 1804 and 1807. Clearly, there was not a single missing planet between Mars and Jupiter, but rather a whole group of objects, each no more than 1000 km across. By 1890 more than 300 had been discovered by sharp-eyed observers. In that year, Max Wolf of Heidelberg introduced astronomical photography to the search for asteroids, greatly accelerating the discovery of additional objects (Figure 12.1). More than 10,000 asteroids now have well-determined orbits.

Asteroids are given both a number (corresponding to the order of discovery) and a name. Originally, the names of asteroids were chosen from goddesses in Greek and Roman mythology. These, however, were soon used up. After exhausting other female names (including those of wives, friends, flowers, cities, colleges, pets, and the like), astronomers have recently turned to the names of colleagues whom they wish to honor for contributions to the field. For example, asteroids 2410 and 4859 are named Morrison and Fraknoi, respectively, for two of the authors of this text.

It would be a formidable task to discover, determine orbits for, and catalog all the asteroids bright enough to be photographed with modern telescopes. Nevertheless, the total number of such objects can be estimated by systematically sampling regions of the sky. These studies indicate that there are approximately a million (10^6) asteroids with diameters greater than 1 km.

The largest asteroid is Ceres, with a diameter just under 1000 km. Two, Pallas and Vesta, have diameters near 500 km, and about 15 more are larger than 250 km (Table 12.1). The number of asteroids increases rapidly with decreasing size; there are about 100 times more objects 10 km across than there are 100 km across. The total mass in all the asteroids, which probably represent just a tiny fraction of the original asteroid population, is less than that of the Moon. The orbits of the four largest asteroids are illustrated in Figure 2.9.

The asteroids all revolve about the Sun in the same west-to-east direction as the planets, and most of their orbits lie near the plane in which the Earth and other planets circle. We define the asteroid belt as the region that contains all asteroids with semimajor axes (see Chapter 2) in the range 2.2 to 3.3 AU. Asteroids at these distances take 3.3 to 6 years to orbit the Sun (Figure 12.2). Although more than 75 percent of the known asteroids are in the main belt, they are not closely spaced. The volume of the belt is actually very large, and the typical spacing between objects (down to 1 km in size) is several million kilometers. (This is fortunate for spacecraft like Galileo

TABLE 12.1
The Largest Asteroids

Name	Year of Discovery	Semimajor Axis (AU)	Diameter (km)	Class*
Ceres	1801	2.77	940	C
Pallas	1802	2.77	540	C
Vesta	1807	2.36	510	†
Hygeia	1849	3.14	410	C
Interamnia	1910	3.06	310	C
Davida	1903	3.18	310	C
Cybele	1861	3.43	280	C
Europa	1868	3.10	280	C
Sylvia	1866	3.48	275	C
Juno	1804	2.67	265	S
Psyche	1852	2.92	265	M
Patientia	1899	3.07	260	C
Euphrosyne	1854	3.15	250	C

* C = carbonaceous; S = stony; M = metallic.

† Vesta has a very unusual (once thought unique) basaltic surface.

and Cassini that need to travel through the belt without a collision.)

In 1917, the Japanese astronomer Kiyotsuga Hirayama found that a number of the asteroids fall into *families* or groups with similar orbital characteristics. He hypothesized that each family may have resulted from an explosion of a larger body or, more likely, from the collision of two bodies. Slight differences in the speed with which the various fragments left the scene account for the small spread in orbits now observed for the different as-

teroids in a given family. Several dozen such families exist, and observations have shown that individual members of the larger ones are physically similar, just as we would expect if they were fragments of a common parent. The existence of these families testifies to the fact that asteroids must have collided frequently during the history of the solar system.

Composition and Classification

Asteroids are as different as black and white. The majority are very dark, with reflectivities of only 3 to 4 percent, like a lump of coal. However, another large group has typical reflectivities of 15 to 20 percent, a little greater than that of the Moon, and still others have reflectivities as high as 60 percent. To understand more about these differences, astronomers can study the spectra of the light reflected from asteroids for clues about their composition.

The dark asteroids are revealed from spectral studies to be *primitive* bodies (chemically unchanged since the beginning of the solar system) composed of silicates mixed with dark, organic carbon compounds. Two of the largest asteroids, Ceres and Pallas, are primitive, as are almost all of the objects in the outer third of the belt. Most of the primitive asteroids are classed as C asteroids, where C stands for carbonaceous or carbon-rich, but several other classes of primitive objects with different minerals have also been identified.

The second most populous asteroid group is that of the S asteroids, where S stands for a stony composition. Here the dark carbon compounds are missing, resulting in higher reflectivities and clearer spectral signatures of silicate minerals. It appears that most of the S-type asteroids are also primitive.

Asteroids of a third class, much less numerous than those of the first two, are composed primarily of metal

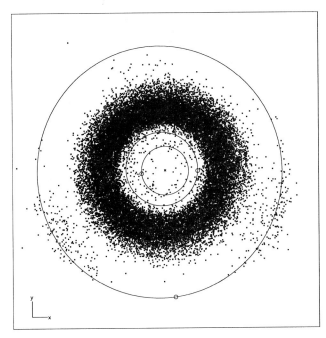

Figure 12.2
The positions of more than 6000 asteroids in February 1990, seen from above. Also shown are the planets Earth, Mars, and Jupiter. (Edward Bowell, Lowell Observatory)

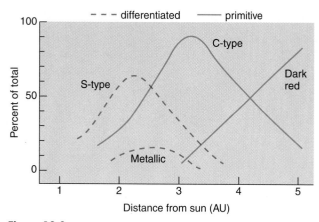

Figure 12.3
Asteroids of different composition are distributed at different distances from the Sun. (Adapted from work carried out by D. Tholen and J. Gradie, University of Hawaii, and E. Tedesco, JPL)

and are called M asteroids. Spectroscopically, the identification of metal is difficult, but for at least the largest M asteroid, Psyche, this identification has been confirmed by radar. Since a metal asteroid, like an airplane or ship, is a much better reflector of radar than is a stony object, Psyche seems bright when we aim a radar beam at it.

How did such metal asteroids come to be? We suspect that each came from a parent body large enough for its molten interior to settle out or differentiate. The heavier metals sank to the center. When this parent body shattered in a later collision, the fragments from the core were rich in metals. There is enough metal in even a 1-km M-type asteroid to supply the world with iron and most other industrial metals for the foreseeable future, if only we could bring one safely to Earth. The iron and nickel in one of the world's largest mines, in Sudbury, Canada, originated in just such a collision with a metallic asteroid more than a billion years ago.

In addition to the M asteroids, a few other asteroids show signs of early heating and differentiation. These have basaltic surfaces like the volcanic plains of the Moon and Mars; the large asteroid Vesta (discussed in a moment) is in the latter category. Why only a small percentage of the total number of asteroids have these characteristics, we do not know.

One of the most interesting products of asteroid compositional studies has been the discovery that different classes of asteroids are grouped together at different distances from the Sun (Figure 12.3). Apparently the asteroids are still located near their birthplaces, and by tracing how their compositions vary with distance from the Sun, we can reconstruct some of the properties of the solar nebula from which they originally formed.

Vesta: A Volcanic Asteroid

Vesta is one of the most interesting of the asteroids. It orbits the Sun with a semimajor axis of 2.4 AU, and its rela-

tively high reflectivity of almost 30 percent makes it the brightest of the main belt objects, visible to the unaided eye if you know just where to look. But its real claim to fame is the fact that its surface is covered with basalt, indicating that Vesta was once volcanically active in spite of its small size (about 500 km in diameter).

You may be surprised to learn that samples of Vesta's surface are available for direct study in the laboratory; hence we know a great deal about this asteroid. Scientists have long suspected that meteorites (chunks of solar system material that survived their trip to the Earth's surface; see Chapter 13) come from the asteroids, but there is generally no way to identify the particular source of a given meteorite that strikes the Earth. However, because of Vesta's unusual surface composition, its identification seems fairly firm.

The meteorites believed to come from Vesta are a group of about 30 basaltic meteorites very similar in composition (Figure 12.4). Chemical analysis has shown that they cannot have come from the Earth, Moon, or Mars. On the other hand, their spectra (measured in the laboratory) perfectly match the spectrum of Vesta obtained telescopically. The age of the lava flows from which these meteorites derived has been measured at 4.4 to 4.5 billion years, very soon after the formation of the solar system. This age is consistent with what we might expect for Vesta; whatever process heated such a small object was probably intense and short-lived. Evidence for large lava flows of different colors on the surface of Vesta can be inferred from photographs of this asteroid made in 1995 by the Hubble Space Telescope (Figure 12.5).

Asteroids Up Close

The Hubble photo of Vesta represents the best we can see from the Earth, but spacecraft give us much closer views of asteroids. On the way to its 1995 encounter with

Figure 12.4
Photograph of one of the meteorites believed to be a volcanic fragment from the crust of asteroid Vesta. (NASA)

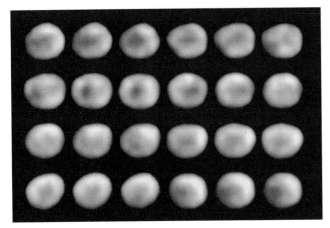

Figure 12.5
A series of 24 images, taken with the Hubble Space Telescope, which show the full 5.34-hour rotation of the asteroid Vesta. Details as small as 80 km across can be resolved. (B. Zellner, Georgia Southern University, and NASA)

Figure 12.6
Galileo spacecraft image of the small main-belt asteroid, Gaspra, from a distance of 1600 km, with a resolution of about 100 m. The color is highly exaggerated to bring out subtle differences in surface composition. The dimensions of Gaspra are approximately 16 × 11 × 10 km. (NASA/JPL)

Jupiter, the Galileo spacecraft was targeted to fly close to two main-belt asteroids called Gaspra and Ida. Gaspra, the first target, is a member of the Flora family of asteroids, and therefore probably a fragment from the collision that formed that family. It is classed as an S-type asteroid, and astronomers knew from variations in its brightness that it had a rotation period of 7 hours.

The Galileo camera revealed that Gaspra is 16 km long and highly irregular, as befits a fragment from a catastrophic collision (Figure 12.6). The detailed images have allowed us to count the craters on Gaspra, and to estimate the time its surface has been exposed to collisions. From the rather sparse number of craters, the Galileo scientists have concluded that Gaspra is only about 200 million years old (that is, the collision that formed Gaspra and other members of the Flora family took place about 200 million years ago). Calculations suggest that an asteroid the size of Gaspra can expect another catastrophic collision sometime in the next billion years, at which point it will be disrupted to form another generation of still-smaller fragments.

The second target was Ida, a larger S-type asteroid 56 km in length (Figure 12.7). It is much more heavily cratered than Gaspra, suggesting that its surface has been exposed to small impacts for a longer time period, probably more than a billion years. The greatest surprise of the Galileo flyby of Ida, and the most important result scientifically, was the discovery of a satellite—named Dactyl—in orbit about the asteroid.

Although only 1.5 km in diameter, smaller than many college campuses, Dactyl provides scientists with something otherwise beyond their reach—a measurement of the mass and density of Ida using Kepler's laws. The satellite's distance of about 100 km, and its orbital period of about 24 hours, indicate that Ida has a density approximately 2.5 g/cm^3, which matches the density of primitive rather than differential rocks. Ida is primitive in composi-

Figure 12.7
Asteroid Ida and its moon Dactyl photographed by the Galileo spacecraft in 1993. Irregularly shaped Ida is 56 km in its longest dimension, while Dactyl is about 1.5 km across. (NASA/JPL)

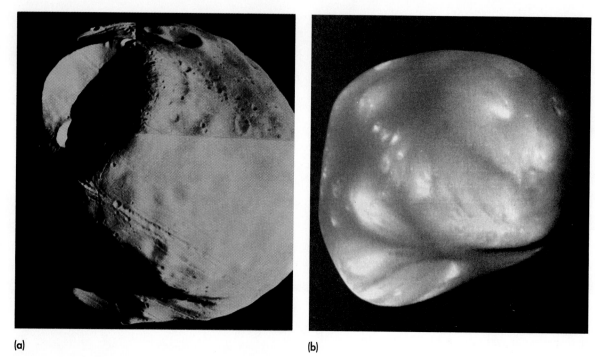

(a)   (b)

Figure 12.8
The satellites of Mars, which are probably captured asteroids, are shown here as photographed by the Viking cameras. (a) Phobos has a 10-km-wide impact crater named Stickney, in honor of the wife of Asaph Hall, who discovered the satellites. (b) Deimos, seen here at full phase, resembles a whale from this perspective. (NASA/JPL, courtesy of Peter Thomas, Cornell University)

tion. By inference, the other S-type asteroids are also primitive, a point that had been in dispute among scientists before the discovery of Dactyl.

Phobos and Deimos, two satellites of Mars, are probably captured asteroids. They were first studied at close range by the Viking orbiters in 1977. Both are rather irregular, somewhat elongated, and heavily cratered (Figure 12.8). Their largest dimensions are about 25 km and 13 km, respectively. Both are dark brownish-gray, and spectral analysis suggests that they are made of primitive materials similar to those that compose most asteroids.

In 1996, NASA launched a spacecraft that will match orbit with an asteroid rather than flying past at high speed. Its target is Eros, an asteroid about the size of Gaspra. The mission, called NEAR (Near Earth Asteroid Rendezvous) will arrive in 1999 and spend about a year mapping Eros in detail and determining its surface composition and density with high precision.

12.2

Asteroids Far and Near

Not all of the asteroids are in the main asteroid belt. In this section, we consider some of the special groups of asteroids that stray outside the belt's boundaries.

The Trojans

The Trojan asteroids are objects located far beyond the main belt, orbiting the Sun at 5.2 AU, about the same distance as Jupiter. The gravity of the giant planet makes most orbits of asteroids near it unstable, but calculations first carried out in the 18th century show that there are two points in the orbit of Jupiter near which an asteroid can remain almost indefinitely. These two points make equilateral triangles with Jupiter and the Sun (Figure 12.9). Between 1906 and 1908, four such asteroids were found; the number has now increased to several hundred. These asteroids are named for the Homeric heroes from the *Iliad* and are collectively called the Trojans.

Measurements of the reflectivities and spectra of the Trojans show that they are dark, primitive objects like those in the outer part of the asteroid belt. They appear faint because they are so dark and far away, but actually the larger Trojans are quite sizable. Four of them—Hektor, Diomedes, Agamemnon, and Patroclus—have diameters between 150 and 200 km.

In 1990 the first asteroids in Trojan-type orbits were discovered in association with Mars. Like Mars, these asteroids have semimajor axes of 1.5 AU. Other planets, including the Earth, may have their own collection of small Trojan asteroids.

Asteroids in the Outer Solar System

There may be many asteroids with orbits that carry them far beyond Jupiter, but they are difficult to detect and only a few have been discovered. The largest of these mysterious objects is Chiron, with a path that carries it from just inside the orbit of Saturn at its closest approach to the Sun, out to almost the distance of Uranus. The diameter of Chiron is estimated to be about 200 km. In 1992 a still-more-distant object named Pholus was discovered, with an orbit that takes it 33 AU from the Sun, beyond the orbit of Neptune. Pholus has the reddest surface of any object in the solar system, indicating a strange (and still-unknown) surface composition. As more objects are discovered in these distant reaches, astronomers have decided that they too will be given the names of Centaurs from classical mythology; this is because the Centaurs were half human, half horse, and these new objects display some of the properties of both asteroids and comets.

In 1988, as it was reaching its closest approach to the Sun, Chiron was seen to brighten by about a factor of two. This is not the sort of behavior we expect from self-respecting asteroids. Astronomers realized that Chiron (and perhaps the other Centaurs) likely contained abundant volatile materials such as water ice or carbon monoxide ice. As the object gets closer to the Sun, these materials begin to evaporate and develop bright reflective atmospheres of gas and dust. As we will see in the next section, this is precisely the behavior we associate with *comets*. But Chiron and Pholus are much larger than the nuclei of any known comet. What a spectacular show they would put on if either were diverted into the inner solar system! Chiron has been shown to be in an unstable orbit, so this may yet happen sometime in the distant future.

Earth-Approaching Asteroids

Not just astronomers, but many readers of this text, political leaders, and even military planners have an interest in the asteroids with orbits that come close to or cross the orbit of the Earth. Some of these briefly become the closest celestial objects to us. In 1989 a 200-m object passed within 800,000 km of the Earth, and in 1994 a tiny 10-m object was picked up passing just 105,000 km away. Some of these objects have collided with the Earth in the past, and others will continue to do so, as we saw in Chapter 7. Together with any comets that come close to our planet, they are known collectively as **Near-Earth Objects (NEOs)**.

At present fewer than 250 NEOs have been located, although the total population of such objects with diameters larger than 1 km is calculated to be approximately 2000. There will likely be many more that are smaller yet, but the smaller they are, the more difficult they are to find. Searches for additional members of this group result in the discovery of several new objects each month.

The orbits of Earth-approaching asteroids are unstable. These objects will meet one of two fates: either they will impact one of the terrestrial planets, or they will be ejected gravitationally from the inner solar system due to a near-encounter with a planet. The probabilities of these two outcomes are about the same. The time scale for impact or ejection is only about 100 million years, very short compared with the age of the solar system. Calculations show that approximately one-quarter of the current Earth-approaching asteroids will end up crashing into the Earth itself. The larger of these impacts will generate environmental catastrophes for our planet, as described in Section 7.5 (where we also discuss possible responses to

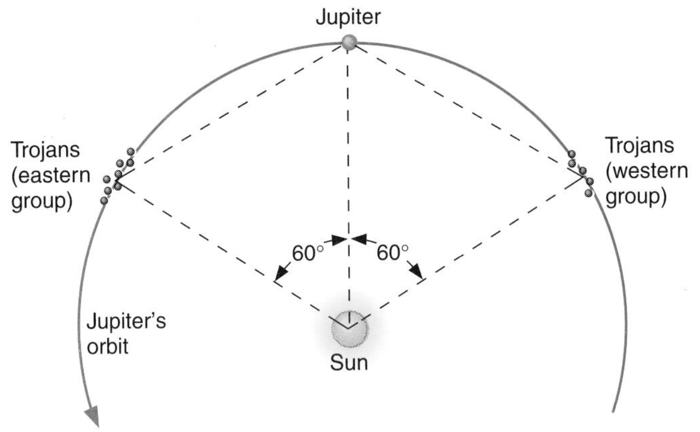

(a)

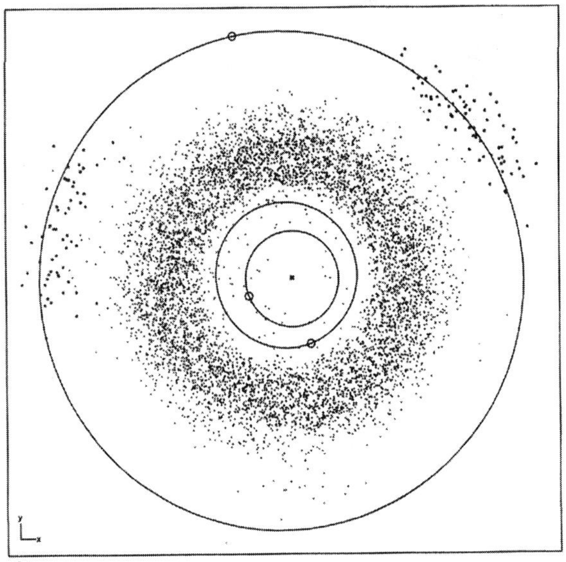

(b)

Figure 12.9
(a) Locations of the Trojan points of the orbit of Jupiter. (b) Plot of the actual positions of the 132 known Trojan asteroids in February 1990. (Part b courtesy of Edward Bowell, Lowell Observatory)

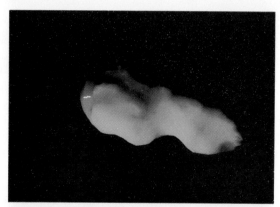

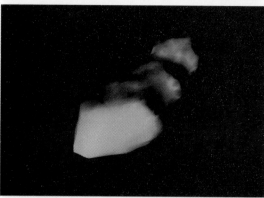

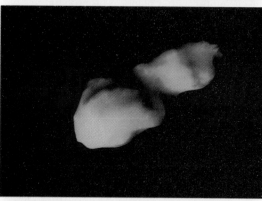

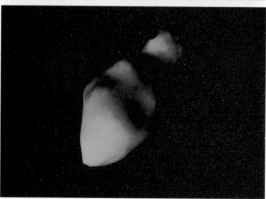

Figure 12.10
Reconstruction of the size and shape of asteroid Toutatis obtained from radar data collected during its close flyby of Earth in December 1992. Toutatis appears to consist of two irregular, lumpy bodies rotating in contact with each other. Its maximum dimension is about 5 km. (NASA/JPL, courtesy of Steven Ostro)

such a threat). This is a strong argument for additional study of NEOs; at the very least, we should have a complete catalog of the larger objects and their orbits in order to anticipate any collisions.

If the current population of Earth-approaching asteroids will be removed by impact or ejection in 10^8 years, there must be a continuing source of new objects. Some of them probably come from the asteroid belt, where collisions between asteroids can eject fragments into Earth-crossing orbits. Others may be dead comets that have exhausted their volatile materials (see next section). So far we have no way of distinguishing between asteroidal fragments and the solid remnants of former comets.

If an Earth-approaching asteroid comes close enough, it can be studied with radar as well as with optical telescopes. Radar can even be used to produce an image of the asteroid. Steven Ostro and his colleagues at the NASA Jet Propulsion Laboratory have succeeded in imaging several such objects with sufficient resolution to establish their shapes and sizes. Their first target was a tiny object called Castalia, which they studied in 1989 using the 300-m Arecibo radar in Puerto Rico. Although Castalia is only 400 m long, the radar images show that it is a double object consisting of two nearly equal lumps in a dumbbell configuration.

The second successfully imaged asteroid was the larger object Toutatis, which approached to within 3 million km of the Earth in 1992 — less than ten times the distance to the Moon. Several radar images of Toutatis are shown in Figure 12.10, obtained with the NASA radar system in California. It too appears to be a double object, consisting of two irregular lumps, with diameters of 3 km and 2 km, that rotate in contact with each other. Scientists were surprised to see such double objects, and they now wonder if this shape is common among small asteroids.

12.3

The "Long-Haired" Comets

Appearance of Comets

Comets have been observed from the earliest times; accounts of spectacular comets are found in the histories of virtually all ancient civilizations. A typical comet is not spectacular, however, but has the appearance of a rather faint, diffuse spot of light somewhat smaller than the Moon and many times less brilliant (Figure 12.11). Recall that comets are chunks of icy material that develop an atmosphere as they get closer to the Sun. Later there may be a very faint, nebulous **tail** extending several degrees away from the main body of the comet.

Like the Moon and planets, comets appear to wander among the stars, slowly shifting their positions in the sky from night to night. Unlike the planets, however, most comets appear at unpredictable times, perhaps explaining why they have frequently inspired fear and superstition.

Edmund Halley:
Astronomy's Renaissance Man

Halley, a brilliant astronomer who made contributions in many fields of science and statistics, was by all accounts a generous, warm, outgoing man. In this he was quite the opposite of his good friend Isaac Newton, whose great work, the *Principia* (see Chapter 2), Halley encouraged, edited, and helped pay to publish. Halley himself published his first scientific paper at age 20, while still in college. As a result, he was given a royal commission to go to Saint Helena (a remote island off the coast of Africa where Napoleon would later be exiled) to make the first telescopic survey of the southern sky. After returning, he received the equivalent of a master's degree and was elected to the prestigious Royal Society in England, all at the age of 22.

In addition to his work on comets, Halley was the first astronomer to recognize that the so-called "fixed" stars move relative to each other, by noting that several bright stars had changed their positions since the ancient Greek catalogs published by Ptolemy. He wrote a paper on the possibility of an infinite universe, proposed that some stars may be variable, and discussed the nature and size of *nebulae* (glowing cloud-like structures visible in telescopes). While in Saint Helena, he observed the planet Mercury going across the face of the Sun, and developed the mathematics of how such *transits* could be used to establish the scale of the solar system.

Edmund Halley (1656–1742).
(Yerkes Observatory, University of Chicago)

In other fields, Halley published the first table of human life expectancies (the precursor of life-insurance statistics); wrote papers on monsoons, trade winds, and tides (charting the tides in the English Channel for the first time); laid the foundations for the systematic study of the Earth's magnetic field; studied evaporation and how inland waters become salty; and even designed an underwater diving bell. He served as a British diplomat, advising the emperor of Austria and squiring the future czar of Russia around England (avidly discussing, we are told, both the importance of science and the quality of local brandy).

For a brief time Halley was a deputy comptroller of the British Mint (where he uncovered some local graft) and served as a Navy captain for scientific voyages (on one of which he had to deal with a mutiny). In 1703 he became a professor of geometry at Oxford, and in 1720 was appointed Astronomer Royal of England. He continued observing the Earth and the sky and publishing his ideas for another 20 years, until death claimed him at age 85.

They typically remain visible for periods that vary from a few days to a few months.

Today we recognize comets as the best-preserved, most primitive material available in the solar system. Stored in the deep freeze of space until some cosmic accident hurls them our way, these icy objects provide us with unique access to the initial material from which the planets formed 4.5 billion years ago.

Comet Orbits

The study of comets as members of the solar system dates from the time of Isaac Newton, who first suggested that their orbits were extremely elongated ellipses. Newton's colleague Edmund Halley (see "Voyagers in Astronomy" box) developed these ideas, and in 1705 he published calculations of 24 cometary orbits. In particular, he noted that the orbits of the bright comets appearing in the years 1531, 1607, and 1682 were so similar that the three could well be the same comet, returning to perihelion at average intervals of 76 years. If so, he predicted that the object should next return about 1758. When the comet ap-

Figure 12.11
Comet Halley in the spring of 1986 had the appearance typical of a moderately bright comet with a tail a few degrees long. It is shown here rising above Mauna Kea Observatory in Hawaii.
(William Golisch, University of Hawaii)

TABLE 12.2
Some Well-Known Comets

Name	Period	Significance
Great Comet of 1577	Long	Tycho Brahe showed it was beyond the Moon.
Great Comet of 1843	Long	Brightest recorded comet; visible in daytime.
Daylight Comet of 1910	Long	Brightest comet of 20th century (so far).
West	Long	Nucleus broke into pieces in 1976.
Hyakutake	Long	Passed within 15 million km of Earth in 1996.
Hale-Bopp	Long	Predicted to be very bright in April 1997.
Swift-Tuttle	133 years	Parent comet of Perseid meteor shower.
Halley	76 years	First comet found to be periodic; explored by spacecraft in 1986.
Biela	6.7 years	Broke up in 1846 and not seen again.
Giacobini-Zinner	6.5 years	First comet to be explored by spacecraft (1985).
Encke	3.3 years	Shortest known period.
Shoemaker-Levy 9	Changed	Broke into pieces in 1992; collided with Jupiter in 1994.

peared as predicted, it was given the name Comet Halley (rhymes with valley) in honor of the man who first recognized it as a permanent member of our solar system.

Comet Halley has been observed and recorded on every passage near the Sun at intervals from 74 to 79 years since 239 B.C. The period varies somewhat because of orbital changes produced by the pull of the jovian planets. In 1910 the Earth was brushed by the comet's tail, causing much needless public concern. Comet Halley last appeared in our skies in 1986, when it was met by several spacecraft that gave us a wealth of information about its makeup; it will return in 2061.

Observational records exist for about a thousand comets. Today, new comets are discovered at an average rate of five to ten per year. Most never become conspicuous and are visible only on photographs made with large telescopes. Every few years, however, a comet may appear that is bright enough to be seen easily with the unaided eye; Table 12.2 lists some well-known comets. The brightest recent comet, called Hyakutake, appeared in March 1996, and at this writing another (called Hale-Bopp) is expected in April 1997. These are the first comets easily visible to Northern Hemisphere observers since Comet West in 1976.

The Comet's Nucleus

When we look at a comet, all we see is its transient atmosphere of gas and dust, illuminated by sunlight. Since the escape velocity from such small bodies is very low, the atmosphere is rapidly escaping all the time; it must be replenished by new material, which has to come from somewhere. The source is the small, solid **nucleus** inside, usually hidden by the glow from the much larger atmosphere surrounding it. The nucleus is the *real* comet, the fragment of ancient material responsible for the atmosphere and the tail (Figure 12.12).

The modern theory of the physical and chemical nature of comets was first proposed by Harvard astronomer Fred L. Whipple in 1950 (Figure 12.13). Before

Whipple's work, many astronomers thought that a comet's nucleus might be a loose aggregation of solids of meteoritic nature, a sort of orbiting "gravel bank." Whipple proposed instead that the nucleus is a solid object a few kilometers across, composed in substantial part of water ice (but with other ices as well) mixed with silicate grains and dust. This proposal became known as the "dirty snowball" model.

The water vapor and other volatiles that escape from the nucleus when it is heated can be detected telescopically in the comet's head and tail. We are somewhat less certain of the non-icy component of the nucleus, however. No large fragments of solid matter from a comet have ever survived passage through the Earth's atmosphere to be studied as meteorites. Some very fine, microscopic grains of comet dust have been collected in the Earth's upper atmosphere, however, and have been studied in the

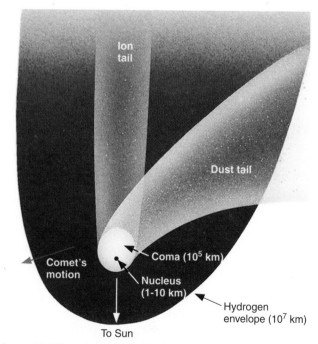

Figure 12.12
Schematic illustration of the parts of a comet.

Figure 12.13
Fred Whipple, the father of modern comet studies, in 1957.
(Smithsonian Astrophysical Observatory)

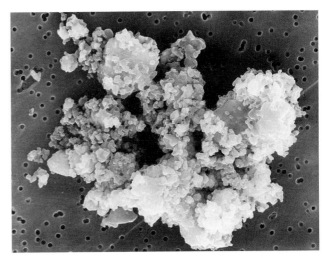

Figure 12.14
A particle that is believed to be a tiny fragment of cometary dust, collected in the upper atmosphere of the Earth. (Donald Brownlee, University of Washington)

the European Space Agency to allow them to target their Giotto spacecraft for an even closer encounter on March 14, 1986, just 605 km from the comet's nucleus. Giotto measured the dimensions of Comet Halley to be 6 km by 10 km, and obtained a beautiful photograph of the nucleus itself (Figure 12.15).

laboratory (Figure 12.14). The spacecraft that encountered Comet Halley in March 1986 also carried dust detectors. From these various investigations it seems that much of the "dirt" in the dirty snowball consists of tiny bits of dark, primitive hydrocarbons and silicates, rather like the material thought to be present on the dark, primitive asteroids.

Since the nuclei of comets are small and dark, they are difficult to study. Just measuring their diameters has been a problem. Our only direct measurements of a comet's nucleus were obtained in 1986, when three spacecraft swept past Comet Halley at close range. The Soviet VEGA 1 and VEGA 2 were the first to arrive, on March 6 and 9, 1986. Each plunged deeply into the inner atmosphere and dust cloud of the comet, passing within about 8000 km of the nucleus. Both VEGA craft were severely damaged by dust impacts, losing most of their solar cells and suffering the loss of several instruments at the time of closest approach.

However, the battering they received was not in vain. The trajectory data for the VEGA craft were provided to

Figure 12.15
Composite photograph of the black, irregularly shaped nucleus of Comet Halley. This historic image, obtained by the Giotto spacecraft at a distance of about 1000 km, has a resolution of better than 1 km. (Max Planck Institut für Aeronomie, and Ball Aerospace Corporation, courtesy of Harold Reitsema)

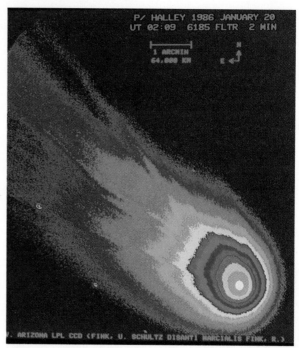

Figure 12.16
Two versions of the same photograph of the head of Comet Halley taken January 20, 1986. In the right image different colors represent different values of the brightness of the comet's coma. False color is frequently used in modern image processing to display and to study the wide range in brightness that is recorded in images obtained with electronic charge-coupled device (CCD) cameras. (University of Arizona, courtesy of Uwe Fink)

The Comet's Atmosphere

The spectacular activity that allows us to see comets is caused by the evaporation of cometary ices heated by sunlight. Beyond the asteroid belt, where comets spend most of their time, these ices are solidly frozen. But as a comet approaches the Sun, it begins to warm up. If water (H_2O) is the dominant ice, significant quantities vaporize as temperatures rise toward 200 K, which happens somewhat beyond the orbit of Mars. The evaporating H_2O in turn releases the dust that was mixed with the ice. Since the comet's nucleus is so small, its gravity cannot hold back either the gas or the dust, both of which flow away into space at speeds of about 1 km/s.

The comet continues to absorb energy as it approaches the Sun. A great deal of this energy goes into the evaporation of its ice, as well as into heating the surface. However, observations of many comets indicate that the evaporation is not uniform, and that most of the gas is released in sudden spurts, perhaps confined to a few areas of the surface. Such jets were observed directly on the surface of Comet Halley by the spacecraft that photographed it in 1986; the jets turned out to resemble volcanic plumes or geysers (Figure 12.15). Most of the comet's surface is apparently inactive, with the ice buried under a layer of black silicates and carbon compounds.

The atmosphere of a comet is composed of the gas released from the nucleus, together with the dust and other solid material being carried along with it. Expanding at a speed of about 1 km/s, the atmosphere can reach an enormous size. The diameter of a comet's head is usually as large as Jupiter, and it often approaches 1 million km (Figure 12.16).

The composition of the gas is primarily H_2O (about 80 percent in the case of Comet Halley), plus a few percent each of carbon dioxide (CO_2) and carbon monoxide (CO), along with small quantities of many additional gases, including hydrocarbons. Ultraviolet light from the Sun can break up the molecules of H_2O into oxygen and lighter hydrogen, which easily escapes. Modern telescopes have revealed huge hydrogen clouds, up to tens of millions of kilometers across, around comets. In fact, for a brief period in the spring of 1986, the hydrogen cloud around Comet Halley was the largest object in our solar system.

Many comets develop tails as they approach the Sun. A comet's tail is an extension of its atmosphere, consisting of the same gas and dust that make up its head. As early as the 16th century, observers realized that comet tails always point away from the Sun (Figure 12.17), not back along the comet's orbit. Newton proposed that comet tails are formed by a repulsive force of sunlight driving particles away from the head, an idea close to our modern view. In addition to sunlight, however, we now know that cometary gas is also repulsed by streams of ions (charged particles) emitted by the Sun (Figure 12.18).

Figure 12.17
Orientation of a typical comet tail as the comet passes perihelion.

12.4

Origin and Evolution of Comets

The Oort Comet Cloud

Although comets are part of the solar system, observations show that they come initially from very great distances. By following their orbits, we can calculate that the aphelia (points farthest from the Sun) of new comets typically have values near 50,000 AU (more than a thousand times farther than Pluto). This clustering of aphelion distances was first noted by Dutch astronomer Jan Oort, who in 1950 proposed a scheme for the origin of the comets that is still accepted today.

It is possible to calculate that a star's *gravitational sphere of influence*—the distance within which it can exert sufficient gravitation to hold onto orbiting objects—is about one-third of the distance to the nearest other stars. Stars in the vicinity of the Sun are spaced in such a way that the Sun's sphere of influence extends only a little beyond 50,000 AU, or about 1 LY. At such distances, objects in orbit about the Sun can be perturbed by the gravitation of passing stars. Oort suggested, therefore, that the new comets were objects orbiting the Sun with aphelia near the edge of its sphere of influence, and that nearby stars disturbed their orbits, eventually bringing them close to the Sun where we can see them. This reservoir of ancient icy objects from which the new comets are derived is now called the **Oort comet cloud.**

Astronomers estimate that there are about a trillion (10^{12}) comets in the Oort cloud. In addition, we estimate that about ten times this number of potential comets could be orbiting the Sun in the volume of space *between* the planets and the Oort cloud at 50,000 AU. These objects remain undiscovered because their orbits are too stable to permit any of them to be deflected inward close to the Sun. The total number of cometary objects could thus be on the order of 10 trillion (10^{13}), a very large number indeed.

What is the mass represented by 10^{13} comets? We can make an estimate if we assume something about typical comet sizes and masses. Let us suppose that the nucleus of Comet Halley is typical. Its observed volume is about 600 km^3. If the primary constituent is water ice with a density of about 1 g/cm^3, the total mass of the nucleus must be about 6×10^{14} kg, or 10^{-10} Earth masses.

Figure 12.18
Comet Mrkos, photographed in 1957 with the Schmidt telescope at Palomar Observatory. Note the smoother tail of dust curving to the right as individual dust particles spread out along the comet's orbit, and the straight ion tail pushed outward from the Sun by its wind of charged particles. (Caltech/Palomar Observatory)

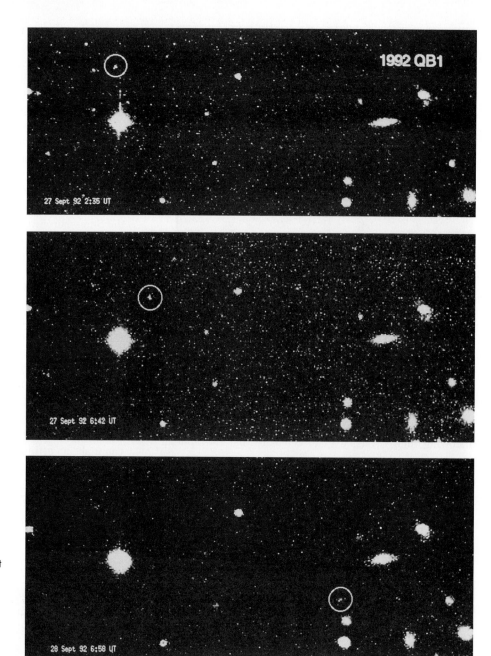

Figure 12.19
Three images of object 1992QB, taken in September 1992 with the 3.5-m New Technology Telescope at the European Southern Observatory. The object, circled on each image as it moved among the stars, is the first identified member of the Kuiper belt.
(European Southern Observatory)

The corresponding mass for all the comets is then about 1000 Earth masses—greater than the mass of all the planets put together. Therefore, cometary material could be the most important constituent of the solar system after the Sun itself.

The Oort cloud spawns comets because they are deflected into the inner solar system by passing stars. In addition, astronomers have located a second source of comets just beyond the orbit of Pluto. The relatively nearby reservoir of cometary material, in the form of a flattened disk, is called the **Kuiper belt** after the Dutch-American astronomer who first suggested its existence. Astronomers have suspected for a number of years that some comets originate in this belt, but not until 1992 was the first member of this group of objects discovered within the belt itself. It turns out to be an exceedingly

faint object (designated 1992QB), with a diameter of less than 100 km (bigger than the kinds of comets we see come inward, but much smaller than planets). It has an orbit beyond Pluto and a period of revolution about the Sun of nearly 300 years (Figure 12.19). More than 30 such objects had been found by 1996, and data from the Hubble Space Telescope suggest that hundreds of additional objects can be picked up with long-exposure photographs. These are probably icy objects, but they are too faint for us to determine their sizes or chemical properties.

The Fate of Comets

Any comet we see today will have spent nearly its entire existence in the Oort cloud or Kuiper belt, at a tempera-

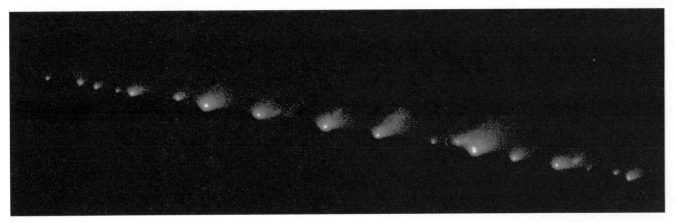

Figure 12.20
Comet Shoemaker-Levy 9 as photographed by the Hubble Space Telescope in May 1994, just two months before impact with Jupiter. Some 21 distinct objects can be seen stretching over a distance of 1.1 million km. (NASA/STScI)

ture near absolute zero. But once a comet enters the inner solar system, its previously uneventful life history begins to accelerate. It may, of course, survive its initial passage near the Sun and return to the cold reaches of space where it spent the previous 4.5 billion years. At the other extreme, it may impact the Sun, or come so close that it is destroyed on its first perihelion passage. Frequently, however, the new comet does not come that close to the Sun, but instead interacts with one or more of the planets.

A comet coming within the gravitational influence of a planet has three possible fates. It can (1) impact the planet, ending the story at once; (2) be speeded up and ejected, leaving the solar system forever; or (3) be perturbed into an orbit with a shorter period. In the last case, its fate is sealed. Each time it approaches the Sun, it loses part of its material and also has a significant chance of collision with a planet. Once the comet is in a short-period orbit, its lifetime starts being measured in thousands, not billions, of years.

A few comets end their lives catastrophically. The brightest comet of the past half-century, Comet West, broke into four pieces in 1976. Even more spectacular was the fate of a faint comet called Shoemaker-Levy 9, which broke into about 20 pieces (Figure 12.20) when it passed close to Jupiter in July 1992. The fragments of Shoemaker-Levy were actually captured into a very elongated, two-year orbit around Jupiter, more than doubling the number of known jovian satellites. Even more remarkable, however, was the fate of these new satellites, which astronomers soon realized would crash into Jupiter in July 1994, releasing energy equivalent to millions of megatons of TNT.

As each cometary fragment streaked into the jovian atmosphere at a speed of 60 km/s, it disintegrated and exploded, launching a hot fireball that carried the comet dust as well as atmospheric gases to high altitudes. These fireballs were clearly visible in profile, with the actual point of impact just beyond the jovian horizon as viewed

from the Earth (Figure 12.21). As each explosive plume fell back into Jupiter, a region of the upper atmosphere larger than the Earth was heated to incandescence and glowed brilliantly for about 15 min, as seen with infrared-sensitive telescopes. Dark clouds of debris settled into the stratosphere of Jupiter, producing long-lived "bruises" (each larger than the Earth) that could be easily seen through even small telescopes (Figure 12.22). Millions of people all over the world peered at Jupiter through telescopes or followed the event via television and the internet. Astronomers will be analyzing data from these impacts for many years, learning more about both the planet and the comet fragments in the process.

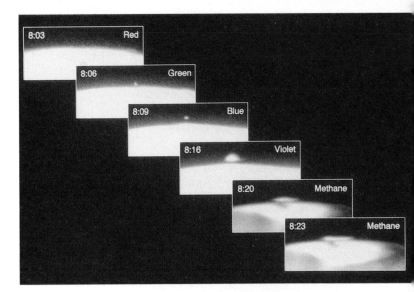

Figure 12.21
This sequence of images shows the last fragment of Comet Shoemaker-Levy 9 to hit Jupiter. Although the initial impact is on the edge of Jupiter as seen from Earth, the giant planet's rotation soon brings the fireball caused by the impact into view. The time of each observation and the filter through which it was taken are shown in each box. (NASA/STScI)

Comet Hunting as a Hobby

When amateur astronomer David Levy, the co-discoverer of Comet Shoemaker-Levy 9, found his *first* comet, he had already spent 917 fruitless hours searching through the dark night sky. But the discovery of that first comet only whetted his appetite. Since then he has found 6 others on his own, and 12 more working with others. Despite this impressive record, he ranks only fourth in the record books for number of comet discoveries. But David is still young and hopes to break that record someday.

All around the world, dedicated amateur observers spend countless nights scanning the sky for new comets. Astronomy is one of the very few fields of science where amateurs can still make a meaningful contribution, and the discovery of a comet is one of the most exciting ways they can establish their place in astronomical history. Don Machholz, a California amateur (and comet hunter) who has been making a study of comet discoveries, reports that between 1975 and 1995, 38 percent of all comets discovered were found by amateurs. Those 20 years yielded 67 comets for amateurs, or almost four per year. That might sound pretty encouraging to new comet hunters, until they learn that the average number of hours the typical amateur spent searching for a comet before finding one was about 420. Clearly, this is not an activity for impatient personalities.

What should you do if you think you have found a new comet? First, you must check your object's location in an atlas of the sky, to make sure it really is a comet. Since the first sighting of a comet usually occurs when it is still far from the Sun and before it sports a significant tail, it will look like a small, fuzzy patch. But through most amateur telescopes, so will *nebulae*, clouds of cosmic gas and dust, and *galaxies*, distant groupings of stars. Next you must check that you have not come across a comet that is already known, in which case you will only get a pat on the back instead of fame and glory. Then you must reobserve or rephotograph it some time later to see if its motion in the sky is appropriate for comets. Often comet hunters who think they have made a discovery get another comet hunter elsewhere in the country to confirm it. But if everything checks out, the place to contact is the Central Bureau for Astronomical Telegrams at the Harvard-Smithsonian Center for Astrophysics in Cambridge, Massachusetts. If your discovery is confirmed, they will send the news out to astronomers and observatories around the world.

Amateur astronomer David Levy.
(Andrew Fraknoi)

For comets that do not meet so dramatic an end, measurements of the amount of gas and dust in their atmospheres permit us to estimate the total losses during one orbit. Typical loss rates are up to a million tons per day from an active comet near the Sun, adding up to some tens of millions of tons per orbit. At that rate, a typical comet will be gone after a few thousand orbits. This will probably be the fate of Comet Halley in the long run.

Figure 12.22
One group of features resulting from the impact of Comet Shoemaker-Levy 9 with Jupiter, seen with the Hubble Space Telescope 105 minutes after the impact that produced the dark rings (the compact black dot came from another fragment). The inner edge of the diffuse outer ring is about the same size as the Earth. Later, the winds on Jupiter blended these features into a broad spot that remained visible for more than a month. (NASA/STScI)

Summary

The solar system includes many bodies that are much smaller than the planets and their larger satellites. The rocky ones are called **asteroids,** and the icy ones are called **comets.**

12.1 Ceres is the largest asteroid; about 15 are larger than 250 km, and 100,000 are larger than 1 km. Most are in the **asteroid belt,** between 2.2 and 3.3 AU from the Sun. The presence of asteroid families in the belt indicates that many asteroids are the remnants of asteroid collisions and fragmentation. The asteroids include both primitive and differentiated objects. Most asteroids are classed as C-type, meaning they are composed of carbonaceous materials. Dominating the inner belt are S-type (stony) asteroids, with a few M-type (metallic) ones.

Vesta is rare in having a volcanic (basaltic) surface. We now have spacecraft images of the two asteroids Gaspra and Ida, plus Ida's moon Dactyl.

12.2 The Trojan asteroids are dark, primitive objects gathered into two swarms controlled by Jupiter's gravity at 5.2 AU from the Sun. Of great interest are the Earth-approaching asteroids, called **Near-Earth Objects (NEOs),** estimated to number about 2000 (down to 1 km in diameter). These are on unstable orbits, and on time scales of 10^8 years, they will either impact one of the terrestrial planets or be ejected. Most of them probably come from the asteroid belt, but some may be dead comets. Radar images have been constructed of Castalia and Toutatis; both turn out to be double objects.

12.3 Edmund Halley first showed that some comets are on closed orbits and return periodically to the Sun. The heart of a comet is its **nucleus,** a few kilometers in diameter and composed of volatiles (primarily H_2O) and solids (including both silicates and carbonaceous materials). Whipple first suggested this "dirty snowball" model in 1950; it has been confirmed by spacecraft studies of Comet Halley. As the nucleus approaches the Sun, its volatiles evaporate (perhaps in localized jets or explosions) to form the comet's head or atmosphere, which escapes at about 1 km/s. The atmosphere streams away from the Sun to form a long **tail.**

12.4 Oort proposed in 1950 that comets are derived from what we now call the **Oort comet cloud,** which surrounds the Sun out to about 50,000 AU (near the limit of the Sun's gravitational sphere of influence) and contains between 10^{12} and 10^{13} comets. Some comets can be found in the **Kuiper belt,** a region of cometary pieces beyond the orbit of Pluto. Comets are primitive bodies left over from the formation of the outer solar system. Once a comet is diverted into the inner solar system, it typically survives no more than about 1000 perihelion passages before losing all its volatiles. Some comets die spectacular deaths: Shoemaker-Levy 9, for example, broke into 20 pieces before colliding with Jupiter in 1994.

Review Questions

1. Why are asteroids and comets important to our understanding of solar system history?

2. Describe the main differences between C-type and S-type asteroids.

3. Compare asteroids of the asteroid belt with the Earth-approaching asteroids. What are the main differences between the two groups?

4. Describe the nucleus of a typical comet, and compare it with an asteroid of similar size.

5. Describe the origin and eventual fate of the comets we see from the Earth.

6. What evidence do we have for the existence of the Kuiper belt of comets?

7. How did Comet Shoemaker-Levy 9 end its life in 1994?

Thought Questions

8. Give at least two reasons that today's astronomers are so interested in the discovery of additional Earth-approaching asteroids.

9. Suppose you were designing a spacecraft that would match course with an asteroid and follow along its orbit. What sorts of instruments would you put on board, and what would you like to learn?

10. Suppose you were designing a spacecraft that would match course with a comet and follow along its orbit. What sorts of instruments would you put on board, and what would you like to learn?

11. Suppose a comet were discovered approaching the Sun, one whose orbit would cause it to collide with the Earth 20 months later, after perihelion passage. (This is approximately the situation described in the science-fiction novel *Lucifer's Hammer* by Larry Niven and Jerry Pournelle.) What could we do? Would there be any way to protect ourselves from a catastrophe?

Problems

12. What is the period of revolution about the Sun for an asteroid with a semimajor axis of 3 AU in the middle of the asteroid belt?

13. What is the period of revolution for a comet with aphelion at 5 AU and perihelion at the orbit of the Earth?

14. Suppose the Oort comet cloud contains 10^{12} comets with an average diameter of 10 km each. Calculate the mass of a comet 10 km in diameter, assuming it is composed mostly of water ice with a density of 1 g/cm^3. Next calculate the total mass of the comet cloud. Finally, compare this mass with those of the Earth and Jupiter.

15. The calculation in Problem 14 refers to the known Oort cloud, the source for the comets we see. If, as some astronomers suspect, there are ten times this many cometary objects in the solar system, how does the total mass of cometary matter compare with the total mass of the planets?

16. If the Oort comet cloud contains 10^{12} comets, and 10 new comets are discovered each year, what percentage of the comets have been used up since the beginning of the solar system?

On Asteroids

Burnham, R. "Here's Looking at Ida" in *Astronomy,* Apr. 1994, p. 38.

Cunningham, C. "The Captive Asteroids" in *Astronomy,* June 1992, p. 41. Describes the Trojans.

Durda, D. "All in the Family" in *Astronomy,* Feb. 1993, p. 36. Discusses asteroid families.

Hartmann, W. "Vesta: A World of Its Own" in *Astronomy,* Feb. 1983, p. 6.

Kowal, C. *Asteroids: Their Nature and Utilization.* 1988, Ellis Horwood/Wiley. Introductory book by the astronomer who discovered Chiron.

McFadden, L. and Chapman, C. "Near-Earth Objects: Interplanetary Fugitives" in *Astronomy,* Aug. 1992, p. 30.

Morrison, D. "The Spaceguard Survey: Protecting the Earth from Cosmic Impacts" in *Mercury,* Sep./Oct. 1992, p. 103.

Ostro, S. "Radar Reveals a Double Asteroid" in *Astronomy,* Apr. 1990, p. 38.

On Comets

Benningfield, D. "Where Do Comets Come From?" in *Astronomy,* Sep. 1990, p. 28.

Bortle, J. "A Halley Chronicle" in *Astronomy,* Oct. 1985, p. 98. A chronology of each pass.

Brandt, J. and Chapman, R. *Rendezvous in Space.* 1992, W. H. Freeman. Introduction by two leading comet experts.

Gore, R. "Halley's Comet '86: Much More Than Met the Eye" in *National Geographic,* Dec. 1986, p. 758.

Levy, D. *The Quest for Comets.* 1994, Plenum Press. Personal story of comet discovery and comet science by an amateur astronomer who has found many comets.

Levy, D. "How to Discover a Comet" in *Astronomy,* Dec. 1987, p. 74.

Sagan, C. and Druyan, A. *Comet.* 1986, Random House. Very good presentation of historical material and human connections.

Sky & Telescope, March 1987, was a special issue about what we learned from Halley's Comet in 1986.

Spencer, J. and Mitton, J., eds. *The Great Comet Crash: The Collision of Comet Shoemaker-Levy 9 and Jupiter.* 1995, Cambridge U. Press. Good, non-technical summary.

Stern, A. "Chiron: Interloper from the Kuiper Disk?" in *Astronomy,* Aug. 1994, p. 26.

Weissman, P. "Comets at the Solar System's Edge" in *Sky & Telescope,* Jan. 1993, p. 26.

Using **REDSHIFT**™

1. Click on *Objects Filter* under *Display*. View the different families of asteroids by clicking on their names in the *Asteroids panel*. Set the *Location* to heliocentric, latitude to 90°, and height to 20 AU. *Step time* to follow the asteroids' orbits.

What sets the Trojans apart from the others?

What are the Mars crossers?

Why are the Apollos special?

2. Display only the Hilda group of asteroids. Set the *Location* to heliocentric, latitude to 90°, and height to 20 AU. Do all the Hildas seem to orbit in about the same period?

Watch one Hilda by increasing the minimum diameter in the Asteroid panel until only one asteroid remains. Set the time step to 1 year and count how many times Jupiter and the Hilda go around the Sun between times when they roughly line up. Look up resonance in Redshift *Dictionary of Astronomy.*

Which resonance does the Hilda asteroid follow?

3. Turn the comets and planets on. Set the *Location* to heliocentric, latitude to 0°, height to 100 AU, and center the Sun. Compare the view with that for latitude 90°.

How can you distinguish between long and short period comets in these views?

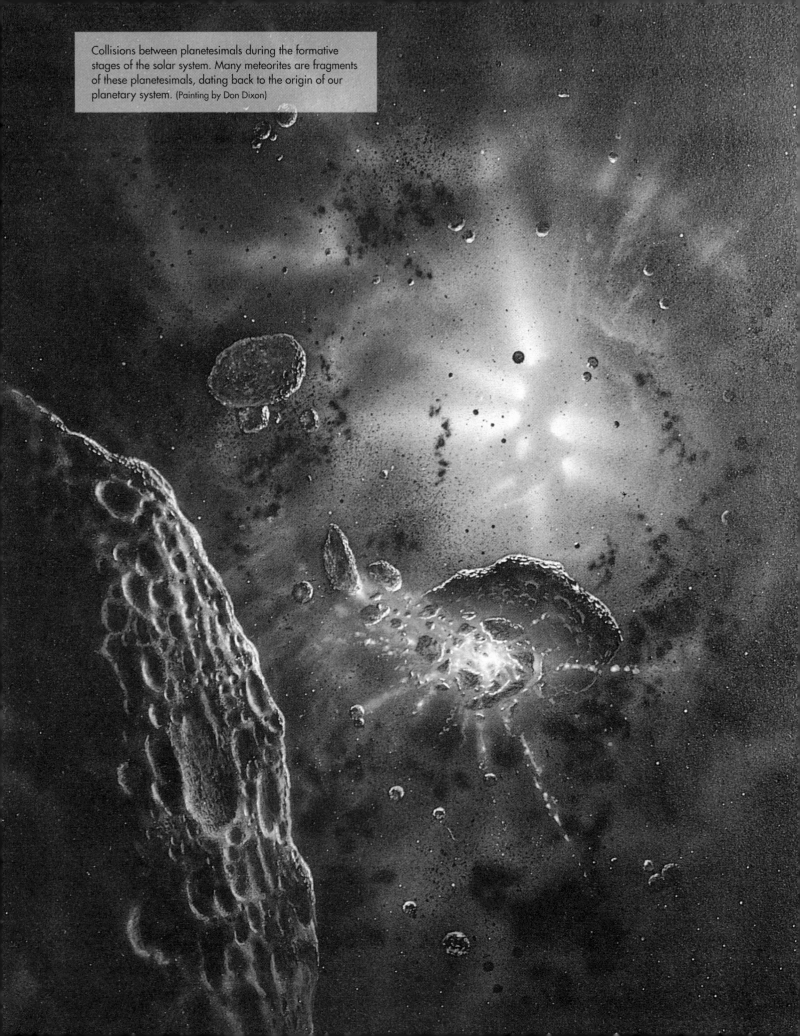

Collisions between planetesimals during the formative stages of the solar system. Many meteorites are fragments of these planetesimals, dating back to the origin of our planetary system. (Painting by Don Dixon)

CHAPTER 13

Cosmic Samples and the Origin of the Solar System

Thinking Ahead

Imagine you are a scientist examining a sample of rock that had fallen from space a few days earlier, and you find within it some of the chemical building blocks of life. How could you determine if those "organic" materials were extraterrestrial or merely the result of earthly contamination?

> I could more easily believe that two Yankee professors would lie than that stones would fall from the heavens.
>
> Thomas Jefferson, responding to a report of a Connecticut meteorite fall in 1807.

We conclude our survey of the solar system with a discussion of its origin and evolution. Some of these ideas were introduced in Chapter 6; we now return to them, using the information we have learned about individual planets and smaller objects. But another crucial way that astronomers learn about the ancient history of the solar system is by examining samples of primitive matter, the debris of the processes that formed the solar system some 4.5 billion years ago. Unlike the Apollo Moon rocks, these samples of cosmic material come to us free of charge — they literally fall from the sky. We call this material cosmic dust and meteorites.

269

Figure 13.1
Two meteors can be seen against a starry background that includes the Milky Way. (P. Parviainen, Polar Image)

13.1

Meteors

As we saw in the last chapter, the ices in comets evaporate when they get close to the Sun, spraying millions of tons of rock and dust into the inner solar system. The Earth is surrounded by this material, sweeping through thin clouds and streams of dust in its orbit about the Sun. As each of the larger dust or rock particles enters the Earth's atmosphere, it creates a brief fiery trail; this is often called a shooting star, but it is properly known as a meteor.

Observing Meteors

Meteors are the result of solid particles that enter the Earth's atmosphere from interplanetary space (Figure 13.1). Since they move at speeds of many kilometers per second, friction with the air vaporizes them at altitudes between 80 and 130 km. The resulting flashes of light fade out within a few seconds. These "shooting stars" got their name because at night their luminous vapors look like stars moving rapidly across the sky. To be visible, a meteor must be within about 200 km of the observer. On a typical dark, moonless night, an alert observer can see

half-a-dozen meteors per hour. Over the entire Earth, the total number of meteors bright enough to be visible must total about 25 million per day.

The typical bright meteor is produced by a particle with a mass of less than 1 gram—no larger than a pea. The light you see comes from the much larger region of glowing gas surrounding this little grain of interplanetary material. Because of its high speed, the energy in a pea-sized meteor is as great as that of an artillery shell fired on Earth, but it is all dispersed high in the Earth's atmosphere.

If a particle the size of a golf ball strikes our atmosphere, it produces a much brighter trail called a *fireball* (Figure 13.2). A piece as large as a bowling ball has a fair chance of surviving its fiery entry if its approach speed is not too high. The total mass of meteoric material entering the Earth's atmosphere is estimated to be about 100 tons per day (which seems like a lot, if you imagine it all falling on your home, but is not very significant when spread out over the whole planet).

Meteor Showers

Many—perhaps most—of the meteors that strike the Earth can be associated with specific comets. Some of these periodic comets still return to our view; others have long ago fallen apart, leaving only a trail of dust behind them. The many dust particles from a given comet retain approximately the orbit of their parent, continuing to move together through space (spreading out over the orbit with time). When the Earth, in its travels around the

Figure 13.2
When a larger piece strikes the Earth's atmosphere, it can make a bright fireball. Note that during the long exposure needed to catch a fireball, the Earth's motion has made the stars into streaks. (Photo © 1986 John Sanford)

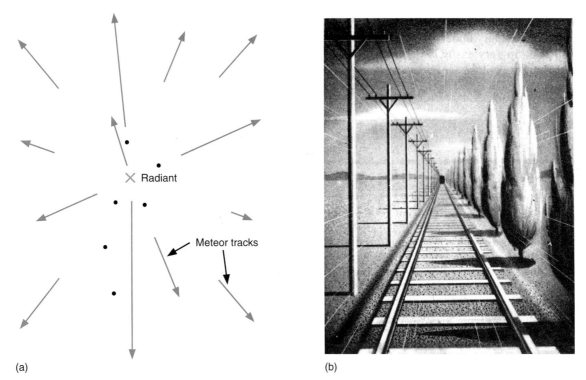

(a)   (b)

Figure 13.3
The radiant of a meteor shower. The tracks of the meteors converge to a point in the distance, just as long, straight railroad tracks appear to do.

Sun, crosses such a dust stream, we see a sudden burst of meteor activity that usually lasts several hours; these events are called **meteor showers.**

The dust particles that produce meteor showers are moving together in space before they encounter the Earth. Thus, as we look up at the atmosphere, their parallel paths seem to diverge from a place in the sky called the *radiant.* This is the direction in space from which the meteor stream is moving at us, just as long railroad tracks seem to diverge from a single spot on the horizon (Figure 13.3). Meteor showers are often designated by the con-

stellation in which this radiant is located: for example, the Perseid meteor shower has its radiant in the constellation of Perseus. The characteristics of some of the more famous meteor showers are summarized in Table 13.1.

The meteoric dust is not always evenly distributed along the orbit of the comet, so that during some years more meteors are seen when the Earth intersects the dust stream, and in other years fewer. A very clumpy distribution is associated with the Leonid meteors, which in 1833 and again in 1866 (after an interval of 33 years—the period of the comet) yielded the most spectacular showers

TABLE 13.1
Major Annual Meteor Showers

Shower Name	Date of Maximum	Associated Comet	Comet's Period (years)
Quadrantid	January 3	Unknown	—
Lyrid	April 21	Thatcher	415
Eta Aquarid	May 4	Halley	76
Delta Aquarid	July 30	Unknown	—
Perseid	August 11	Swift-Tuttle	105
Draconid	October 9	Giacobini-Zinner	7
Orionid	October 20	Halley	76
Taurid	October 31	Encke	3
Leonid	November 16	Tempel-Tuttle	33
Geminid	December 13	Phaethon*	1.4

* Phaethon is an Earth-approaching asteroid, not a comet.

Figure 13.4
A woodcut of the great meteor shower or storm of 1833, shown here with considerable artistic license.

(sometimes called meteor storms) ever recorded (Figure 13.4). The last good Leonid shower was on November 17, 1966, when in some southwestern states up to 100 meteors could be observed per second. Some scientists and engineers are concerned about the effect a meteor storm might have on space stations in orbit around the Earth, and in 1994 NASA postponed a Shuttle flight because of concern over a predicted meteor storm.

The most dependable meteor shower at present is the Perseid shower, which appears each year for about three nights near August 11. In the absence of bright moonlight, you can see meteors at a rate of about one per minute during a typical Perseid shower. Astronomers estimate that the total combined mass of the particles in the Perseid swarm is nearly a billion tons; the comet that gave rise to the particles, called Swift-Tuttle, must originally have had at least that much mass. However, if its initial mass was comparable to the mass measured for Comet Halley, then Swift-Tuttle would have contained several hundred billion tons, suggesting that only a very small fraction of the original cometary material survives in the meteor stream.

No shower meteor has ever survived its flight through the atmosphere and been recovered for laboratory analysis. However, there are other ways to investigate the nature of these particles and thereby to gain additional in-

sight into the comets from which they are derived. Analysis of the flight paths of meteors shows that most of them are very light or porous, with densities typically less than 1.0 g/cm^3. If you placed a fist-sized lump of meteor material on a table in the Earth's gravity, it might well fall apart under its own weight.

Such light particles break up very easily in the atmosphere, accounting for the failure of even relatively large shower meteors to reach the ground. Comet dust is apparently fluffy, rather inconsequential stuff, as we can also infer from the tiny comet particles recovered in the Earth's atmosphere (Figure 12.14). This fluff, by its very nature, cannot reach the Earth's surface intact. However, more substantial fragments, from asteroids, do make it into our laboratories, as we will see in the next section.

13.2

Meteorites: Stones from Heaven

Any fragment of interplanetary debris that survives its fiery plunge through the Earth's atmosphere is called a **meteorite.** Meteorites fall only very rarely in any one locality, but over the entire Earth hundreds fall each year. These rocks from the sky carry a remarkable record of the formation and early history of the solar system.

Extraterrestrial Origin of Meteorites

While occasional meteorites have been found throughout history, their extraterrestrial origin was not accepted by scientists until the beginning of the 19th century. Before that, these strange stones were either ignored or considered to have a supernatural origin.

The falls of the earliest recovered meteorites are lost in the fog of mythology. A number of religious texts speak of stones from heaven, which sometimes arrived at opportune moments to smite the enemies of the authors of those texts. At least one sacred meteorite has survived in the form of the Ka'aba, the holy black stone in Mecca that is revered by Islam as a relic from the time of the Patriarchs.

The modern scientific history of the meteorites begins in the late 18th century, when a few scientists suggested that some strange-looking stones were of such peculiar composition and structure that they were probably not of terrestrial origin. The general acceptance that indeed "stones fall from the sky" occurred after the French physicist Jean Baptiste Biot investigated a well-observed fall in 1803, in which many meteoritic stones were found, still warm, on the ground.

Meteorites sometimes fall in groups or showers. Such a fall may result when a group of pieces has been moving together in space before colliding with the Earth, but more likely the different stones are fragments of a single larger object that broke up during its violent passage through the atmosphere. It is important to remember that

Observing a meteor shower is one of the easiest and most enjoyable astronomy activities for beginners. The best thing about it is that you don't need a telescope or binoculars—in fact, they would positively get in your way! What you do need is a site far from city lights, with an unobstructed view of as much sky as possible. While the short bright lines in the sky made by individual meteors *could* in theory be traced back to a radiant point (as shown in Figure 13.3), the quick blips of light that represent the end of the meteor could happen anywhere above you.

So the trick of observing meteor showers is *not* to restrict your field of view, but to lie back and scan the sky alertly. Try to select a good shower (see the list in Table 13.1) and a night when the Moon will not be bright at the time you are observing. The Moon, street lights, car headlamps, and bright flashlights will all get in the way of your seeing the faint meteor streaks.

You will also see more meteors after midnight, when you are on the hemisphere of the Earth that faces *forward*—in the direction of the Earth's revolution around the Sun. Before midnight, you are observing from the "back side" of the Earth, and the only meteors you see will be those that traveled fast enough to catch up with the Earth's orbital motion.

When you've gotten away from all the lights, give your eyes about 15 minutes to get "dark adapted"—that is, for the pupil of your eye to open up as much as possible. (This adaptation is the same thing that happens in a dark movie theater. When you first enter, you can't see a thing; but eventually, as your pupils open wider, you can see pretty clearly by the faint light of the screen.)

Seasoned meteor observers find a hill or open field and make sure to bring warm clothing, a blanket, and a thermos of hot coffee or chocolate with them. (It's also nice to take along someone with whom you enjoy sitting in the dark.) Don't expect to see fireworks or a laser show—meteor showers are subtle phenomena, best approached with a patience that reflects the fact that some of the dust you are watching burn up may first have been gathered into its parent comet more than 4.5 billion years ago, as the solar system was just forming.

---

such a *shower of meteorites* has nothing to do with a *meteor shower*. No meteorites have ever been recovered in association with meteor showers. Whatever the ultimate source of the meteorites, they do not appear to come from the comets or their associated particle streams.

Meteorite Falls and Finds

Meteorites are found in two ways. First, sometimes bright meteors (fireballs) are observed to penetrate the atmosphere to very low altitudes. If we search the area beneath the point where the fireball burned out, we may find one or more remnants that reached the ground. Observed *meteorite falls*, in other words, may lead to the recovery of fallen meteorites.

Second, people sometimes discover unusual-looking rocks that turn out to be meteoritic; these are termed *meteorite finds*. Now that the public has become meteorite-conscious, many unusual fragments, not all of which turn out to be from space, are sent to experts each year. Some scientists divide these objects into two categories: "meteorites" and "meteorwrongs." Outside Antarctica (see next paragraph), genuine meteorites turn up at an average rate of 25 or so per year. Most of these end up in natural history museums (Figure 13.5) or specialized meteoritical laboratories throughout the world.

Since the 1980s, a new source from the Antarctic has dramatically increased our knowledge of meteorites. More than ten thousand meteorites have been recovered

Figure 13.5
This 1906 photo shows a 15-ton iron meteorite found in the Willamette Valley in Oregon. Although known to Native Americans in the area, it was "discovered" by an enterprising local farmer in 1902, who proceded to steal it and put it on display. It was eventually purchased for the American Museum of Natural History and is now on display in the Hayden Planetarium in New York City, the largest iron meteorite in the U.S. (J.O. Wheeler; courtesy Department of Library Services, American Museum of Natural History)

Figure 13.6
An iron meteorite lying on the Antarctic ice just before being added to our collection. (NASA)

TABLE 13.2
Frequency of Occurrence of Meteorite Classes

Class	Falls	Finds	Antarctic
	(%)	(%)	(%)
Primitive stones	88	51	85
Differentiated stones	8	1	12
Irons	3	42	2
Stony-irons	1	5	1

ground for some time and been subject to weathering. The most scientifically valuable stones are those collected immediately after they fall, or the Antarctic samples preserved in a nearly pristine state by the ice.

Table 13.2 summarizes the frequencies of occurrence of the different classes of meteorites among the fall, find, and Antarctic categories.

Ages and Compositions of Meteorites

It was not until the ages of meteorites were measured, and techniques were developed for the detailed analysis of their compositions, that scientists appreciated their true significance. The meteorites include the oldest and most primitive materials available for direct study in the laboratory. The ages of stony meteorites can be determined from the careful measurement of radioactive isotopes and their decay products, as described in Chapter 6. Almost all meteorites have radioactive ages between 4.48 and 4.56 billion years, as old as any ages we have mea-

(text cont. on page 276)

from the Antarctic ice, as a result of the low precipitation and peculiar motion of the ice in some parts of that continent (Figure 13.6). Meteorites that fall in regions where ice accumulates are buried and then carried slowly, with the motion of the ice, to other areas where the ice is gradually worn away. After thousands of years, the rock again finds itself on the surface, along with other meteorites carried to these same locations. The ice thus concentrates the meteorites that have fallen both in a large area and over a long period of time.

Meteorite Classification

The meteorites in our collections have a wide range of compositions and histories, but traditionally they have been placed into three broad classes. First are the **irons,** composed of nearly pure metallic nickel–iron. Second are the **stones,** the term used for any silicate or rocky meteorite. Third are the much rarer **stony-irons,** made (as the name implies) of mixtures of stone and metallic iron (Figure 13.7).

Of these three types, the irons and stony-irons are the most obviously extraterrestrial in origin because of their metallic content. Pure iron almost never occurs naturally on Earth; it is generally found here as an oxide (chemically combined with oxygen) or other mineral ore. Therefore, if you ever come across a chunk of metallic iron, it is sure to be either man-made or a meteorite.

The stones are much more common than the irons but more difficult to recognize. Often laboratory analysis is required to demonstrate that a particular sample is really of extraterrestrial origin, especially if it has lain on the

Figure 13.7
A variety of meteorite types. At the top left is a piece of the Allende carbonaceous meteorite, with white inclusions that may date back to before the formation of the solar nebula. At the top right is a fragment of the iron meteorite responsible for the formation of Meteor Crater in Arizona (see Chapter 12). At bottom left is a piece of the Imilac stony-iron meteorite, a beautiful mixture of green olivine crystals and metallic iron. At bottom right is part of the Mern primitive stony meteorite. (Chip Clark)

Some Striking Meteorites

Although meteorites fall regularly onto the Earth's surface, few of them have much of an impact (!) on human civilization. There is so much water and uninhabited land on our planet that rocks from space typically fall where no one even sees them come down. But given the number of meteorites that land each year, you may not be surprised that a few *have* struck buildings, cars, and even people.

In November 1982, for example, Robert and Wanda Donahue of Wethersfield, Connecticut, were watching "M°A°S°H" on television when a 6-lb meteorite came thundering through their roof, making a hole in the living room ceiling. After bouncing back up into the attic, it finally came to rest under their dining room table.

Eighteen-year-old Michelle Knapp of Peekskill, New York, got quite a surprise one morning in October 1992. She had just purchased her very first car, her grandmother's 1980 Chevy Malibu. But she awoke to find its rear end mangled and a crater in the family driveway—thanks to a 3-lb meteorite. Michelle was not sure whether to be devastated by the loss of her car or thrilled by all the media attention.

In June 1994, José Martín and his wife were driving from Madrid, Spain, to a golfing vacation when a fist-sized meteorite crashed through the windshield, bounced off the dashboard, broke José's little finger, and then landed in the car's back seat. Before Martín, the most recent person known to have been struck by a meteorite was Annie Hodges of Sylacauga, Alabama. In November 1954, she was napping on a couch when a meteorite came through the roof, bounced off a large radio set, and hit her first on the arm and then on the leg.

In a recent compilation, amateur astronomer and meteorite historian Christopher Spratt found 61 recorded instances between 1790 and 1990 of meteorites striking human habitats or machinery (and we presume there have been other cases that have gone unreported or unrecorded). Although his compilation makes for an impressive list, two centuries are a lot of time and the Earth is a big place. The chance of you or your property being hit by a meteorite is so small that it is certainly not worth losing sleep over.

John McAuliffe, deputy chief of the Wethersfield Fire Department, examines the hole made by a meteorite that crashed into the home of Robert and Wanda Donahue on November 8, 1982. McAuliffe is now chief of the department. (© 1982, 1992 Dan Haar/ *The Hartford Courant*)

sured in the solar system. The few exceptions are igneous rocks that are ejecta from cratering events on the Moon or Mars.

The average age for all of the old meteorites, calculated using the best data and most accurate values now available for radioactive half-lives, is 4.54 billion years, with an uncertainty of less than 0.1 billion years. This value is taken to represent the *age of the solar system*—the time since the first solids condensed and began to form into larger bodies.

The traditional classification of meteorites into irons, stones, and stony-irons is easy to use because it is obvious from inspection which category a meteorite falls into (although it may be much more difficult to distinguish a meteoritic stone from a terrestrial rock). Much more significant, however, is the distinction between *primitive* and *differentiated* meteorites.

The differentiated meteorites are fragments of larger parent bodies that were molten before they broke up, allowing the denser materials (such as metals) to sink to their centers. Like many rocks on Earth, they have been subject to a degree of chemical reshuffling, with the different materials becoming sorted out according to density. Differentiated meteorites include the irons, which come from the metal cores of their parent bodies, and the stony-irons, which probably originate in regions between a metal core and a stony mantle.

For information on the *earliest* history of the solar system, however, we turn to the primitive meteorites—those made of materials that have not been subject to great heat or pressure since their formation. The fiery passage of the meteorite through the air takes place so rapidly that the interior (below a burned crust a few millimeters thick) never even becomes hot; hence a fragment of primitive interplanetary debris is still primitive (in our sense of the word) after it lands on the Earth. What are the parent bodies of these primitive meteorites?

A comparison of their spectra indicates that the parent bodies of primitive meteorites are almost certainly asteroids. Recall from Chapter 12 that asteroids are believed to be fragments left over from the formation process of the solar system; thus it makes sense that they should be the parent bodies of the primitive meteorites.

The great majority of the meteorites that reach the Earth are primitive stones. Many of them are composed of light-colored gray silicates with some metallic grains mixed in, but there is also an important group of darker stones called **carbonaceous meteorites.** As their name suggests, these meteorites contain carbon, as well as various complex organic compounds—chemicals based on carbon, which on Earth are the simple building blocks of life. In addition, some of them also contain chemically bound water, and many are depleted in metallic iron. The carbonaceous meteorites probably originate among the dark, carbonaceous asteroids concentrated in the outer part of the asteroid belt.

The Allende and Murchison Primitive Meteorites

Next to the tiny dust particles from comets, the carbonaceous meteorites are the most primitive materials available for laboratory study. Two large carbonaceous meteorites that fell within a few months of each other have proved particularly valuable in probing the birth of the solar system. The Allende meteorite fell in Mexico (see Figure 13.7) and the Murchison meteorite in Australia, both in 1969. Like other meteorites, Allende and Murchison are named for the towns near which they fell.

The Murchison meteorite is best known for the variety of organic chemicals it has yielded. Most of the carbon compounds in carbonaceous meteorites are complex, tar-like substances that defy exact analysis. Murchison also contains 16 amino acids (the building blocks of proteins), 11 of which are rare on Earth. The most remarkable thing about them is that they include equal numbers with right-handed and left-handed molecular symmetry. Amino acids can have either kind of symmetry, but all life on Earth has evolved using only the left-handed versions to make proteins. The presence of both kinds of amino acids clearly demonstrates that the ones in the meteorite had a nonterrestrial origin!

The presence of these naturally occurring amino acids and other complex organic compounds in Murchison shows that a great deal of interesting chemistry must have taken place when the solar system formed. Perhaps some of the molecular building blocks of life on Earth were delivered by primitive meteorites and comets. This is an interesting idea, because our planet was probably much too hot for any organic materials to survive its earliest history. But after the Earth's surface cooled, the asteroid and comet fragments that pelted it could have refreshed its supply of organic materials.

The Allende meteorite is a rich source of information on the formation of the solar system because it contains many individual grains with varied chemical histories. As much as 10 percent of the material in Allende appears to be older than the solar system. In examining these materials, we may be looking at some interstellar dust grains (see Chapter 19) formed by previous generations of stars that were not destroyed in the processes that gave rise to our own system. We will return to some of these ideas when we examine how stars form (in Chapter 20).

13.3

Formation of the Solar System

The comets, asteroids, and meteorites are surviving remnants from the processes that formed the solar system. (The planets and the Sun, of course, also are the products of the formation process, although the material in them has undergone a wide range of changes.) We are now ready to put together the information from the previous chapters in order to discuss what is known about the origin of the solar system.

Observational Constraints

There are certain basic properties of the planetary system that any viable theory of its formation must explain. These may be summarized under three categories: motion constraints, chemical constraints, and age constraints. We call them constraints because they place restrictions on our theories; unless a theory can explain the observed facts, it will not survive in the competitive marketplace of ideas that characterizes the endeavor of science.

Motion Constraints

We saw that the planets all revolve around the Sun in the same direction and approximately in the plane of the Sun's own rotation. In addition, most of the planets rotate in the same direction they revolve and most of the satellites also move in counterclockwise orbits (when seen from the north). With the exception of the comets, the motions of the system members define a disk or frisbee

shape. On the other hand, a full theory must be prepared to deal with the exceptions to these trends, such as the retrograde rotation of Venus.

Chemical Constraints

Jupiter and Saturn have approximately the same composition—dominated by hydrogen and helium—that the Sun and stars do. Each of the other members of the planetary system is, to some degree, lacking in the light elements. A careful examination of the composition of solid solar-system objects shows a striking progression from the metal-rich inner planets, through those made predominantly of rocky materials, out to objects with ice-dominated compositions in the outer solar system. The comets in the Oort cloud and the Kuiper belt are also icy objects, whereas the asteroids (Figure 13.8) represent a transitional rocky composition with abundant dark, carbon-rich material.

As we saw in Chapter 6, this general chemical pattern can be interpreted as a temperature sequence, with the inner parts of the system strongly depleted in materials that could not condense (form a solid) at the high temperatures found near the Sun. Again, however, there are important exceptions to the general pattern. For example, it

Figure 13.8
In the 1990s the Galileo spacecraft sent back the first close-up images of asteroids (to add to our pictorial inventory of the solar system). This sequence of images was taken as the spacecraft approached the asteroid Ida in 1993 (from bottom right to top left). The asteroid rotates clockwise in 4 hours and 39 minutes when viewed from above, and the 14 images cover about one rotation. (JPL/NASA)

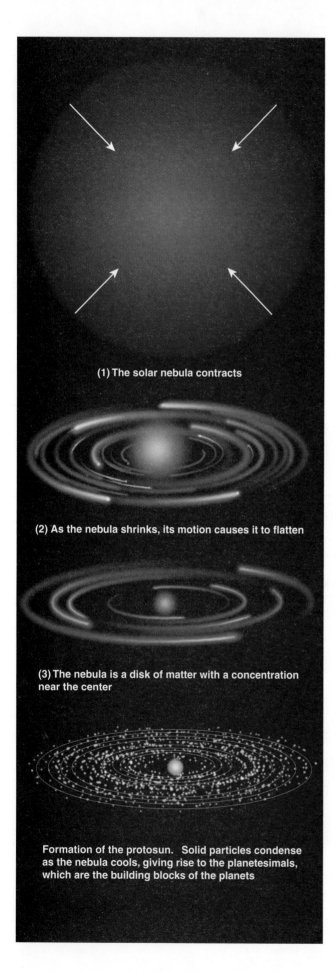

(1) The solar nebula contracts

(2) As the nebula shrinks, its motion causes it to flatten

(3) The nebula is a disk of matter with a concentration near the center

Formation of the protosun. Solid particles condense as the nebula cools, giving rise to the planetesimals, which are the building blocks of the planets

Figure 13.9
Steps in the formation of the solar system from the solar nebula. As the nebula shrinks, its rotation causes it to flatten into a disk. Much of the material is concentrated in the hot center, which will ultimately become a star. Away from the center, solid particles can condense as the nebula cools, giving rise to planetesimals, the building blocks of the planets and satellites.

is difficult to explain the presence of water on Earth and Mars if these planets formed in a region where the temperature was too hot for ice to condense, unless the ice or water was brought in later from cooler regions.

Age Constraints

Radioactive dating demonstrates that some rocks on the surface of the Earth have been present for at least 3.8 billion years, and that certain lunar samples are 4.4 billion years old. In addition, the primitive meteorites all have radioactive ages near 4.5 billion years. The age of these unaltered building blocks is considered the age of the planetary system. The similarity of the measured ages tells us that planets formed and their crusts cooled within a few hundred million years (at most) of the beginning of the solar system. Further, detailed examination of primitive meteorites indicates that they are made primarily from material that condensed or coagulated out of a hot gas; few identifiable fragments or grains appear to have survived from before this hot-vapor stage 4.5 billion years ago.

The Solar Nebula

All of the foregoing constraints are consistent with the general idea introduced in Chapter 6—that the solar system formed 4.5 billion years ago out of a rotating cloud of hot dust and vapor called the *solar nebula,* with an initial composition similar to that of the Sun today. As the solar nebula collapsed under its own gravity, material fell toward the center, where things became more and more concentrated and hot. Increasing temperatures in the shrinking nebula vaporized most of the solid material that was originally present.

At the same time, the collapsing nebula began to rotate faster and faster through the conservation of *angular momentum* (see Chapter 2). One effect of this increased spin was to make it easier for material to fall in at the poles than at the equator (where the spinning tends to push things outward). Thus, more material fell inward along the axis of rotation, eventually giving the nebula a disk shape (Figure 13.9). The existence of this disk-shaped rotating nebula explains the primary motions in the solar system just discussed. Observations of very young stars (described in Chapter 20) support these ideas; a number of such stars are surrounded by flattened disks of the type we are describing here.

Picture the solar nebula at the end of the collapse phase, when it was at its hottest. With no more gravitational energy (from particles falling in) to heat it, most of

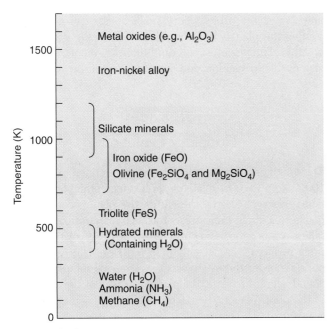

Figure 13.10
The chemical condensation sequence in the solar nebula. The scale at left shows temperature; at right are the materials that would condense out at each temperature under the conditions expected to prevail in the nebula. (Adapted from a diagram published by John Lewis, University of Arizona)

the nebula began to cool. The material in the center, however, where it was hottest and most crowded, formed a star (see Chapters 15 and 20) that was able to maintain high temperatures in its immediate neighborhood.

The temperature within the nebula therefore decreased with increasing distance from the Sun, much as the planets' temperatures vary with position today. As the nebula cooled, the gases interacted chemically to produce compounds; eventually these compounds condensed into liquid droplets or solid grains. This is similar to the process by which raindrops on Earth condense from moist air in a cloud as it rises over a mountain. Let's look in more detail at how material condensed at different places in the maturing nebula.

In the 1970s, geochemists proposed a detailed theory of what substances could condense at what temperatures in the solar nebula. We call the results of their calculations the *chemical condensation sequence* (Figure 13.10). The first materials to form grains were the metals and various rock-forming silicates. As the temperature dropped, these were joined throughout much of the solar nebula by sulfur compounds and by carbon- and water-rich silicates, such as those now found abundantly among the asteroids. However, in the inner parts of the nebula, the temperature never dropped low enough for such materials as ice or carbonaceous organic compounds to condense, so they are lacking on the innermost planets.

Far from the Sun, cooler temperatures allowed the oxygen to combine with hydrogen and condense in the form of water (H_2O) ice. Beyond the orbit of Saturn, car-

bon and nitrogen combined with hydrogen to make ices such as methane (CH_4) and ammonia (NH_3). This chemical condensation sequence explains the basic chemical composition differences among various regions of the solar system.

Formation of the Terrestrial Planets

The grains that condensed in the solar nebula rather quickly joined into larger and larger aggregates, until most of the solid material was in the form of *planetesimals* a few kilometers to a few tens of kilometers in diameter (see Section 6.4). These planetesimals then became the building blocks for the planets. Some of them still survive today as the comets and asteroids. Others have left their imprint on the cratered surfaces of many of the worlds we studied in earlier chapters. A substantial step up in size is required, however, to go from planetesimals to planets.

Some planetesimals were large enough to attract their neighbors gravitationally and thus to grow by the process called **accretion.** While the intermediate steps are not well understood, ultimately several dozen centers of accretion seem to have grown in the inner solar system. Each of these attracted surrounding planetesimals until it had acquired a mass similar to that of Mercury or Mars. At this stage we may think of these objects as *protoplanets*—"not quite ready for prime time" planets.

Each of these protoplanets continued to grow by the accretion of planetesimals. When planetesimals were colliding with other planetesimals, the energies involved were not necessarily large enough to cause either body to melt. Now, however, every incoming planetesimal was accelerated by the much larger gravity of the protoplanet. This means it struck with great energy—sufficient to melt both the projectile and a part of the impact area. Soon the entire protoplanet was heated to above the melting temperature of rocks. The result was planetary differentiation, with heavier metals sinking toward the core and lighter silicates rising toward the surface. As they were heated, the protoplanets lost some of their more volatile constituents (the lighter gases), leaving more of the heavier elements and compounds behind.

Formation of the Giant Planets

In the outer solar system, where the building blocks included ices as well as silicates, the protoplanets grew to be much larger, with masses ten times that of the Earth. These protoplanets of the outer solar system were so large that they were able to attract and hold the surrounding gas. As the hydrogen and helium rapidly collapsed onto their cores, the giant planets were heated by the energy of contraction. Such heating, as we will see in Chapter 15, also begins the process of forming stars. But these giant planets were far too small to raise their central temperatures and pressures to the point where nuclear reactions could begin (and it is such reactions that give us our definition of a star). After glowing dull red for a few thousand years, the giant planets gradually cooled to their present state.

The collapse of nebular gas onto the cores of the giant planets explains how these objects came to have about the same hydrogen-rich composition as the Sun itself. The process was most efficient for Jupiter and Saturn; hence their compositions are most nearly "cosmic." Much less gas was captured by Uranus and Neptune, which is why these two planets have compositions dominated by the icy and rocky building blocks that made up their large cores, rather than by hydrogen and helium.

Further Evolution of the System

All of the just-described processes, from the collapse of the solar nebula to the formation of protoplanets, took place within a few million years. However, the story of the formation of the solar system is not complete at this stage; there were many planetesimals and other debris that did not initially accumulate to form the planets. What was their fate?

Recall from Chapter 12 that we have evidence that the comets visible to us are merely the tip of the cosmic iceberg (if you'll pardon the pun) as far as the full comet population is concerned. Most comets are believed to be in the Oort cloud, far from the region of the planets. Our models show that these icy pieces probably formed near the present orbits of Uranus and Neptune, but were ejected from their initial orbits by the gravitational influence of the giant planets. Additional comets are in the Kuiper belt, which stretches beyond the orbit of Neptune.

In the inner parts of the system, remnant planetesimals and perhaps several dozen protoplanets continued to whiz about. Collisions among these objects were inevitable. Giant impacts at this stage probably stripped Mercury of part of its mantle and crust, reversed the rotation of Venus, and broke the Earth apart to create the Moon.

Smaller-scale impacts also added mass to the inner protoplanets. Because the gravity of the giant planets can "stir up" the orbits of the planetesimals, the material impacting on the inner protoplanets could have come from almost anywhere within the solar system. In contrast to the previous stage of accretion, therefore, this new material did not represent just a narrow range of compositions (from "local" planetesimals). Much of the debris striking the inner planets was ice-rich material that had condensed in the outer part of the solar nebula. As this comet-like bombardment progressed, the Earth accumulated the water and various organic compounds that would later be critical to the formation of life. Mars and Venus probably also acquired water and organic materials from the same source.

Gradually, as the planets swept up or ejected the remaining debris, most of the leftover planetesimals disappeared. In the region between Mars and giant Jupiter,

however, stable orbits where small bodies can avoid impacting the planets or being ejected from the system are possible. The planetesimals (and their fragments) that survive in this special location are what we call the asteroids.

Planetary Evolution

The era of giant impacts was probably confined to the first 100 million years of solar system history, ending by 4.4 billion years ago. Shortly thereafter, the planets cooled and began to assume their present aspects. Up until about 4.0 billion years ago they continued to acquire volatiles, and their surfaces were heavily cratered from the remaining debris that hit them. However, as external influences declined, all of the terrestrial planets, as well as the satellites of the outer planets, began to follow their own evolutionary courses. The nature of this evolution depended on the composition of each object, its mass, and its distance from the Sun.

Geological Activity

We have seen a wide range in the level of geological activity on the terrestrial planets and icy satellites. Internal sources of such activity (as opposed to pummelling from above) require energy, either in the form of primordial heat left over from the formation of a planet or from the decay of radioactive elements in the interior. The larger the planet or satellite, the more likely it is to retain its internal heat, and the more slowly it cools—this is the "baked potato effect" mentioned in Chapter 6. Therefore we are more likely to see evidence of continuing geological activity on the surface of larger worlds (Figure 13.11). Jupiter's satellite Io is an interesting exception to this rule; we saw that it has an unusual source of heat—the gravitational flexing of its interior by the tidal pull of Jupiter.

The Moon, the smallest of the terrestrial worlds, was internally active until about 3.3 billion years ago, when its major volcanism ceased. Since that time its mantle has cooled and become solid, and today even internal seismic activity has declined to almost zero. The Moon is a geologically dead world. Although we know much less about Mercury, it seems likely that this planet, too, ceased most volcanic activity about the same time the Moon did.

Mars represents an intermediate case, and it has been much more active than the Moon. The southern hemisphere crust had formed by 4 billion years ago, and the northern hemisphere volcanic plains seem to be contemporary with the lunar maria. However, the Tharsis bulge formed somewhat later, and activity in the large Tharsis volcanoes has apparently continued intermittently to the present.

Earth and Venus are the largest and most active terrestrial planets. Our planet experiences global plate tec-

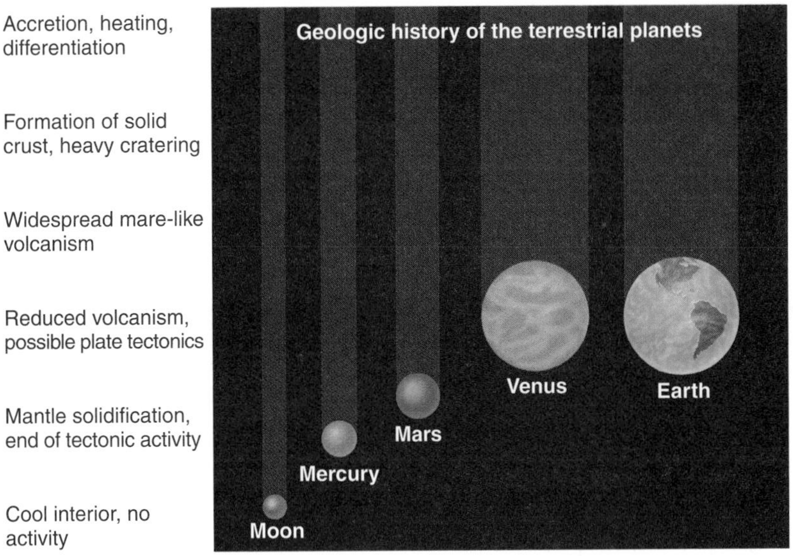

Geologic history of the terrestrial planets

Accretion, heating, differentiation

Formation of solid crust, heavy cratering

Widespread mare-like volcanism

Reduced volcanism, possible plate tectonics

Mantle solidification, end of tectonic activity

Cool interior, no activity

Moon
Mercury
Mars
Venus
Earth

Figure 13.11
Stages in the geological history of a terrestrial planet (time increases downward along the left axis). Each planet is shown roughly in its present stage. The smaller the planet, the more quickly it passes through these stages.

tonics driven by mantle convection; as a result, our surface is continually reworked, and most of the Earth's surface material is less than 200 million years old. Venus has generally similar levels of volcanic activity, but unlike the Earth, it has not experienced plate tectonics. Most of its surface appears to be roughly 600 million years old. A better understanding of the geological differences between Venus and Earth is a high priority for planetary geologists now that we have a detailed radar map of our sister planet.

The geological evolution of the icy satellites has been distinct from that of the terrestrial planets. The energy sources were the same, but the materials are different. Here we see evidence of low-temperature volcanism, with the silicate lava replaced by sulfur and sulfur compounds on Io, and by water and other ices on the colder and more distant moons.

Elevation Differences

The mountains on the terrestrial planets have very different origins. On the Moon and Mercury, the major mountains are ejecta thrown up by the large basin-forming impacts that took place billions of years ago. Most large mountains on Mars are volcanoes, produced by repeated eruptions of lava from the same vents. There are

similar (but smaller) volcanoes on Earth and Venus. However, the highest mountains on Earth and Venus are the result of compression and uplift of the surface. On the Earth, this crustal compression results from collisions of one continental plate with another.

It is interesting to compare the maximum heights of the volcanoes on Earth, Venus, and Mars (Figure 13.12). On Venus and Earth, the maximum elevation differences between these mountains and their surroundings are about 10 km. Olympus Mons, in contrast, towers 26 km above its surroundings, and nearly 30 km above the lowest elevation areas on Mars (Figure 13.13).

One reason Olympus Mons is so much larger than its terrestrial counterparts is that the crustal plates on Earth never stop long enough to let a really large volcano grow. Instead, the moving plate creates a long row of volcanoes like the Hawaiian Islands. On Mars (and perhaps Venus) the crust remains stationary with respect to the underlying hot spot, and so the volcano can continue to grow for hundreds of millions of years.

A second difference relates to the force of gravity on the three planets. The surface gravity on Venus is nearly the same as that on Earth, but on Mars it is only about one-third as great. In order for a mountain to survive, its internal strength must be great enough to support its weight against the force of gravity. Volcanic rocks have

Olympus Mons

Mauna Loa   Mt. Everest   Maxwell Mountains

Sea level

Figure 13.12
Comparison of the highest mountains on Mars, Venus, and Earth. Mountains can rise taller on Mars because the surface gravity is less and there are no moving plates. The vertical scale is exaggerated by a factor of 3 to make comparison easier.

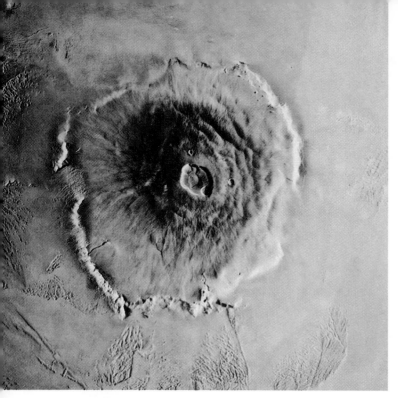

Figure 13.13
The giant martian volcano Olympus Mons is seen here from above on this composite image created from many Viking orbiter photographs. The volcano is nearly 500 km wide at its base, and 25 km high. (NASA/USGS, courtesy of Alfred McEwen)

known strengths, and we can calculate that on Earth 10 km is about the limit. For instance, when new lava is added to the top of Mauna Loa in Hawaii, the mountain slumps downward under its own weight. The same height limit applies on Venus, where the force of gravity is the same as ours. On Mars, however, with its lesser surface gravity, much greater elevation differences can be supported; which means that Olympus Mons is more than twice as high as the tallest mountains of Venus or Earth.

The same kind of calculation that determines the limiting height of a mountain can be used to ascertain the largest body that can have an irregular shape. Gravity, if it can, pulls all objects into the most "efficient" shape (where all the outside points are equally distant from the center). All of the planets and larger satellites are nearly spherical, due to the force of their own gravity. But the smaller the object, the greater is the departure from spherical shape that the strength of its rocks can support. For silicate bodies, the limiting diameter is about 400 km; larger objects will always be approximately spherical, while smaller ones can have almost any shape (as we see when on photographs of asteroids such as Figure 13.8).

Atmospheres

The atmospheres of the planets were formed by a combination of gas escaping from their interiors and the impacts of volatile-rich debris from the outer solar system. Each of the terrestrial planets must have originally had similar atmospheres, but Mercury was too small and too hot to retain its gas. The Moon probably never had an atmosphere, since the material composing it was depleted in volatiles.

The predominant volatile gas on the terrestrial planets is now carbon dioxide (CO_2), but initially there were probably also hydrogen-containing gases. In this more chemically reduced environment, there should have been large amounts of carbon monoxide (CO) and perhaps also traces of ammonia (NH_3) and methane (CH_4). Ultraviolet light from the Sun destroyed the reducing gases, however, and most of the hydrogen escaped, leaving behind the oxidized atmospheres we see today on Earth, Venus, and Mars.

The fate of water was different for each of these three planets, depending on size and distance from the Sun. Early in its history, Mars apparently had a thick atmosphere with abundant liquid water, but it could not retain those conditions. As we saw in Chapter 9, the CO_2 necessary for a substantial greenhouse effect was lost, the temperature dropped, and eventually the water froze. On Venus the reverse process took place, with a runaway greenhouse effect leading to the permanent loss of water. Only the Earth managed to maintain the delicate balance that permits liquid water to persist.

With the water gone, Venus and Mars each ended up with an atmosphere about 96 percent carbon dioxide and a few percent nitrogen. On Earth, the presence first of water and then of life led to a very different kind of atmosphere. The CO_2 was removed to be deposited in marine sediment, while proliferation of photosynthetic life eventually led to the release of enough oxygen to overcome natural chemical reactions that eliminate this gas from the atmosphere. As a result, we on Earth find ourselves with a great deficiency of CO_2 and with the only planetary atmosphere containing free oxygen.

In the outer solar system, Titan is the only satellite with a substantial atmosphere. Apparently this object contained sufficient volatiles—such as ammonia, methane, and nitrogen—to form an atmosphere. It was also large and cold enough to retain these gases. Thus, as we saw in Chapter 11, today Titan's atmosphere consists primarily of nitrogen. Compared with those on the inner planets, temperatures on Titan are too low for either carbon dioxide or water to be in vapor form. With these two common volatiles frozen solid, it is perhaps not too surprising that nitrogen has ended up the primary atmospheric constituent.

We see that nature, starting with one set of chemical building blocks, can fashion a wide range of final atmospheres appropriate to the conditions and history of each world. The atmosphere we have on Earth is the result of many eons of evolution and adaptation. And, as we saw, it can be changed by the actions of the life-forms that inhabit the planet.

Conclusion

We now come to the end of our study of the planetary system. Although we have learned a great deal about the other planets during the past few years of spacecraft exploration, much remains unknown. Now that we have adopted the comparative approach to study these different worlds and the processes that form them and their environments, we see how limited our knowledge really is. We do not understand even our own planet well enough to assess the consequences of the changes humans are inadvertently inflicting on the atmosphere and oceans. The exploration of the solar system is one of the greatest human adventures, and it will continue, in part so that we can learn to appreciate and protect our own planet.

Summary

13.1 When a fragment of interplanetary dust strikes the Earth's atmosphere, it burns up to create a **meteor.** Streams of dust particles traveling through space together produce **meteor showers,** in which we see meteors diverging from a spot in the sky called the radiant of the shower. Many meteor showers recur each year and are associated with particular comets.

13.2 **Meteorites** are the debris from space (mostly asteroid fragments) that survive to reach the surface of the Earth. Meteorites are called finds or falls according to how they are found; the most productive source today is the Antarctic ice cap. Meteorites are classified as **irons, stony-irons,** or **stones** according to their composition. Most stones are primitive objects, dated to the origin of the solar system 4.5 billion years ago. The most primitive are the **carbonaceous meteorites,** such as Murchison and Allende. These can contain a number of organic molecules.

13.3 A theory of solar system formation must take into account motion constraints, chemical constraints, and age constraints. Meteorites, comets, and asteroids are survivors of the solar nebula out of which the solar system formed. The solar nebula formed by the collapse of an interstellar cloud of gas and dust, which contracted (conserving its angular momentum) to form the protosun surrounded by a thin, spinning disk of dust and hot vapor. Condensation in the disk led to the formation of planetesimals, which became the building blocks of the planets. **Accretion** of infalling materials heated the planets, leading to their differentiation. The giant planets were also able to attract and hold gas from the solar nebula. After a few million years of violent impacts, most of the debris was swept up or ejected, leaving only the asteroids and cometary remnants surviving to the present. All of these violent events terminated by about 4.4 billion years ago.

13.4 After their common beginning, each of the planets evolved on its own path. The different possible outcomes are illustrated by comparison of the terrestrial planets (Earth, Venus, Mars, Mercury, and the Moon). All are rocky, differentiated objects. The level of geological activity is proportional to mass: greatest for Earth and Venus, less for Mars, and absent for the Moon and Mercury. Mountains can result from impacts, volcanism, or uplift. Whatever their origin, higher mountains can be supported on smaller planets where the surface gravity is less. All the terrestrial planets may have acquired atmospheric volatiles from comet impacts. The Moon and Mercury lost their atmospheres; most volatiles on Mars are frozen due to its greater distance from the Sun; and Venus retained CO_2 but lost H_2O when it developed a massive greenhouse effect. Only Earth still has liquid H_2O and hence can support life.

Review Questions

1. A friend of yours sees a meteor shower and becomes concerned that the stars he knows (such as the ones in the Big Dipper) could self-destruct like shooting stars. How would you put his mind at rest?

2. In what ways are meteorites different from meteors? What is the probable origin of each?

3. How is Comet Halley connected to meteor showers?

4. What do we mean by primitive material? How can we tell if a meteorite is primitive?

5. Describe the solar nebula, and outline the sequence of events within the nebula that gave rise to the planetesimals.

6. Why do the giant planets and their satellites have compositions different from those of the terrestrial planets?

7. Explain the role of impacts in planetary evolution, including both giant impacts and more modest ones.

8. Why are some planets more geologically active than others?

9. Summarize the origins and evolutions of the atmospheres of Venus, Earth, and Mars.

10. What sorts of methods do scientists use to distinguish a meteorite from terrestrial material?

11. Explain why we do not believe meteorites come from comets, or meteors from asteroids.

12. Why do iron meteorites represent a much higher percentage of finds than of falls.

13. Why is it more useful to classify meteorites according to whether they are primitive or differentiated, rather than whether they are stones, irons, or stony-irons?

14. Which meteorites are the most useful for defining the age of the solar system? Why?

15. Suppose a new primitive meteorite is discovered (sometime after it falls in a field of soybeans), and analysis reveals that it contains a trace of amino acids, all of which show the same rotational symmetry (unlike the Murchison meteorite). What might you conclude from this finding?

16. How do we know when the solar system formed? Usually we say that the solar system is about 4.5 billion years old. To what does this age correspond?

17. We have seen how Mars can support greater elevation differences than can Earth or Venus. According to the same arguments, the Moon should have higher mountains than any of the other terrestrial planets, yet we know it does not. What is wrong with applying the same line of reasoning to the mountains on the Moon?

Problems

18. Consider the differentiated meteorites. We think the irons are from the cores, the stony-irons from the interfaces between mantles and cores, and the stones from the mantles of their differentiated parent bodies. If these parent bodies were like the Earth, what fraction of the meteorites would you expect to consist of irons, stony-irons, and stones? Is this consistent with the observed numbers of each?

19. Estimate the maximum height of the mountains on a hypothetical planet similar to the Earth but with twice the surface gravity of our planet.

20. The angular momentum of an object is proportional to the square of its size divided by its period of rotation (D^2/P). If angular momentum is conserved, then any change in size must be compensated for by a proportional change in period, in order to keep D^2/P a constant. Suppose that the solar nebula began with a diameter of 10,000 AU and a rotation period of 1 million years. What would be its rotation period when it had shrunk to the size of Pluto's orbit? Of Jupiter's orbit? Of the Earth's orbit?

Suggestions for Further Reading

On Meteors

Bagnall, P. "Watching Halley's Debris" in *Astronomy,* May 1992, p. 78. Describes the two meteor showers caused by Comet Halley.

Bone, N. *Meteors.* 1993, Sky Publishing. A good manual and information source.

Kronk, G. "Meteor Showers" in *Mercury,* Nov./Dec. 1988, p. 162.

Taubes, G. "U-2 Mission: To Catch the Dust of Comets" in *Discover,* Oct. 1983, p. 74.

On Meteorites

Chaikin, A. "A Stone's Throw from the Planets" in *Sky & Telescope,* Feb. 1983, p. 122. Discusses meteorites from the Moon and Mars.

Dodd, R. *Thunderstones and Shooting Stars: The Meaning of Meteorites.* 1986, Harvard U. Press. A clear guide to the science of meteorites.

Hutchinson, R. *The Search for Our Beginnings.* 1983, Oxford U. Press. Focuses on the role of meteorites in helping us understand the solar system.

McSween, H. *Meteorites and Their Parent Planets.* 1987, Cambridge U. Press. A good general introduction by a geologist.

Norton, O. *Rocks from Space.* 1994, Mountain Press Publishing. A guide for amateurs, with information on science, folklore, and collecting.

Spratt, C. and Stephens, S. "Against All Odds: Meteorites That Have Struck [the Earth]" in *Mercury,* Mar./Apr. 1992, p. 50.

On the Evolution of the Solar System

See the readings recommended in Chapter 6, plus:

Greeley, R. *Planetary Landscapes,* 2nd ed. 1994, Chapman and Hall. A geological survey of the solar system.

Kastings, J. "How Climate Evolved on the Terrestrial Planets" in *Scientific American,* Feb. 1988, p. 90.

McSween, H. *Stardust to Planets.* 1993, St. Martin's Press. Essays on a number of current topics in planetary astronomy.

Rubin, A. "Microscopic Astronomy" in *Sky & Telescope,* July 1995, p. 24. Discusses the study of solar system history from analysis of samples.

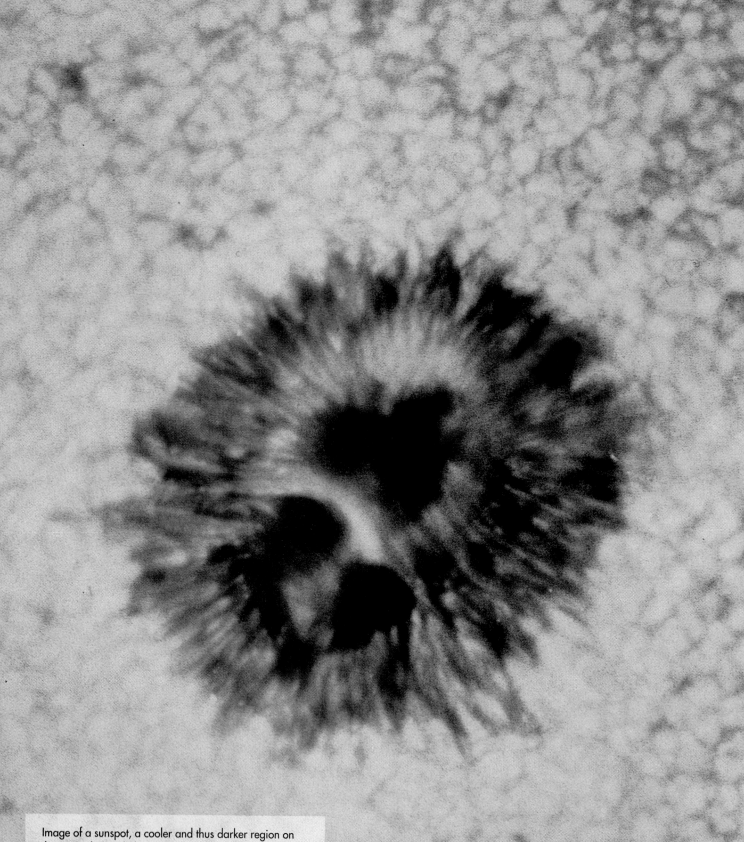

Image of a sunspot, a cooler and thus darker region on the Sun, taken with the McMath-Pierce Solar Telescope. The spot has a dark central region (called the umbra) surrounded by a less-dark region (the penumbra). Note that you can see the granulation on the Sun's surface. (W. Livingston, National Solar Observatories)

CHAPTER 14

The Sun: A Garden-Variety Star

Thinking Ahead

Is the Sun's energy output constant? Since life on Earth ultimately depends on the energy from the Sun, this is a question of personal importance to those of us who share the Sun's cosmic neighborhood. What would happen, for example, if the Sun's energy output decreased just a little bit?

We begin our voyage toward the realm of the stars by considering *our* star, the Sun. It is, by many measures, a rather ordinary star—not unusually hot or cold, old or young, large or small. Indeed, we are lucky that the Sun is typical. By studying it, we learn much that helps us understand the stars in general. Just as studies of the Earth help us understand observations of the more distant planets, so too the Sun serves as a guide to interpreting the messages contained in the faint light we receive from distant stars.

In this chapter we describe what the Sun looks like, how it changes with time, and how those changes affect the Earth. The next chapter will discuss the Sun's interior and the way it produces the energy we see as sunlight. Some of the basic characteristics of the Sun are listed in Table 14.1, although many of the terms in that table may be unfamiliar to you until you've read further in this and the next few chapters.

The adventure of the Sun is the great natural drama by which we live, and not to have joy in it and awe of it, not to share in it, is to close a dull door on nature's sustaining and poetic spirit.

Henry Beston, from "Midwinter" in *The Outermost House* (1928)

TABLE 14.1
Solar Data

Datum	How Found	Value
Mean distance	Radar reflection from planets	1 AU (149,597,892 km)
Maximum distance from Earth		1.521×10^8 km
Minimum distance from Earth		1.471×10^8 km
Mass	Orbit of Earth	333,400 Earth masses (1.99×10^{30} kg)
Mean angular diameter	Direct measure	31'59".3
Diameter of photosphere	Angular size and distance	109.3 times Earth diameter (1.39×10^6 km)
Mean density	Mass/volume	1.41 g/cm^3
Gravitational acceleration at photosphere (surface gravity)	$\dfrac{GM}{R^2}$	27.9 times Earth surface gravity 273 m/s^2
Solar constant	Measure with instrument sensitive to radiation at all wavelengths	1370 watts/m^2
Luminosity	Solar constant times area of spherical surface 1 AU in radius	3.8×10^{26} watts
Spectral class	Spectrum	G2 V
Effective temperature	Derived from luminosity and radius of Sun	5800 K
Visual magnitude		
Apparent	Visible flux	−26.7
Absolute	Apparent magnitude and distance	+4.8
Rotation period at equator	Sunspots and Doppler shift in spectra taken at the edge of the Sun	24^{d}16^h
Inclination of equator to ecliptic	Motions of sunspots	7°10'.5

14.1

Outer Layers of the Sun

The Sun, like all stars, is an enormous ball of extremely hot gas, shining under its own power. To sense how large a ball it is, note that the Sun has enough volume (takes up enough space) to hold about 1.3 million Earths. Most of the Sun's regions are shielded from our view by its outside layers, just as the Earth shows only its surface and atmosphere to our direct observation. The Sun has no solid surface, but it does have an extensive atmosphere—and this is the only part of the Sun we can actually see.

Astronomers divide the Sun's atmosphere into different regions—the photosphere, the chromosphere, and the corona—that we will examine in more detail shortly.

The Composition of the Sun

But first, let us ask what the Sun's outer layers are made of. As explained in Chapter 4, we can use the *absorption line spectrum* in a diffuse gas to determine what elements are present. More than 60 of the elements known on the Earth have now been identified in spectra of the Sun's atmosphere. Those not identified in the Sun either do not produce lines in the observable part of the spectrum, or are so rare in the universe that they are not expected to produce lines strong enough to be measurable. Most of the elements found in the Sun are in the form of atoms, but more than 18 types of molecules, including water vapor and carbon monoxide, have been identified in the

light from the Sun's cooler regions, such as the sunspots (see Section 14.2).

The atoms and molecules contained in the Sun are all in the form of gases; the Sun is so hot that no matter can survive as a liquid or a solid. Still, the Sun's gravity can compress the gases deep inside to a much higher density than the gases on Earth. The density is highest in the center of the Sun, and decreases outward.

Although the Sun contains the same elements as the Earth, their relative proportions are quite different. About 73 percent of the Sun's mass is hydrogen, and another 25 percent is helium. All the other chemical elements (including those we know and love, such as carbon, oxygen, and nitrogen) make up only 2 percent of our star. The ten most abundant gases in the solar photosphere are listed in Table 14.2. Examine that table, and notice how different the composition of the Sun's outer layer is from the Earth's crust, where we live. (In our planet's crust, the three most abundant elements are oxygen, silicon, and aluminum.)

As we will see when we turn to the study of other stars, the composition of the Sun is much more consistent with that of the surrounding universe than is the crust of the Earth. (The reasons for the Earth's oddity can be explained by the early evolution of the solar system, as described in Chapters 6 and 13.)

The Solar Photosphere

The Earth's air is generally transparent. But on a smoggy day it can become opaque, which prevents us from seeing

TABLE 14.2		
Abundance of Elements in the Sun		
Element	Percentage by Number of Atoms	Percentage by Mass
Hydrogen	92.0	73.4
Helium	7.8	25.0
Carbon	0.02	0.20
Nitrogen	0.008	0.09
Oxygen	0.06	0.8
Neon	0.01	0.16
Magnesium	0.003	0.06
Silicon	0.004	0.09
Sulfur	0.002	0.05
Iron	0.003	0.14

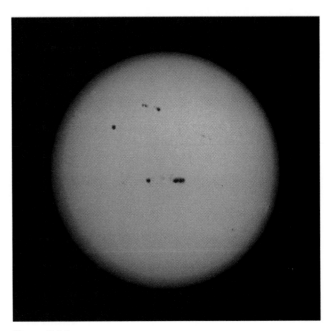

Figure 14.1
This photograph showing several sunspots on the photosphere—the visible surface of the Sun—was taken through a filter that lets only a particular set of wavelengths through. (National Solar Observatory/National Optical Astronomy Observatories)

through it past a certain point. Something similar happens in the Sun. Its outer atmosphere is transparent, allowing us to look a short distance into deeper layers. But when our line of sight reaches thicker layers that are aglow with light, we can no longer see through them. The **photosphere** (Figure 14.1) is where the Sun becomes opaque to our view.

Beneath the photosphere, the gas—like all of the Sun's layers—absorbs and re-emits photons of energy that were produced in the solar interior. But these photons are then quickly "grabbed" by other atoms and so remain trapped inside the Sun where we cannot see them. Only photons absorbed and then re-emitted in the photosphere can escape from the Sun and reach the Earth.

We have found that the solar atmosphere changes from almost perfectly transparent to almost completely opaque in a distance of just over 400 km; it is this thin region that we call the photosphere. When astronomers speak of the "diameter" of the Sun, they mean the size of the region surrounded by the photosphere.

The photosphere only looks sharp from a distance. If you were falling into the Sun, you would not feel any surface but would just sense a gradual increase in the density of the gas surrounding you. It is much the same as falling through a cloud while skydiving. The cloud looks as if it

has a sharp surface, but you do not feel a surface as you fall into it. (One big difference between these two scenarios, however, is temperature. The Sun is so hot that you would be vaporized long before you reached the photosphere. Skydiving in the Earth's atmosphere is much safer!)

From analysis of the light received from the Sun, astronomers can calculate how the temperature, density, and pressure of the gases vary through the photosphere. The average characteristics of the photosphere are listed in Table 14.3, where zero depth has been set arbitrarily at the point with the lowest temperature.

Look at how drastically things can change through such a thin layer. For example, over a distance of a little more than 300 km, the pressure and density increase by a factor of 10, while the temperature climbs from 4500 to

TABLE 14.3				
Properties of the Solar Photosphere				
Depth in Atmosphere (km)	Percent of Light Emerging from That Depth	Temperature (K)	Pressure (bars)	Density (g/cm³)
0	99.5	4465	6.8×10^{-3}	2.3×10^{-8}
100	97	4780	1.7×10^{-2}	5.4×10^{-8}
200	89	5180	3.9×10^{-2}	1.2×10^{-7}
250	80	5455	5.8×10^{-2}	1.6×10^{-7}
300	64	5840	8.3×10^{-2}	2.1×10^{-7}
350	37	6420	1.2×10^{-1}	2.7×10^{-7}
375	18	6910	1.4×10^{-1}	3.0×10^{-7}
400	4	7610	1.6×10^{-1}	3.1×10^{-7}

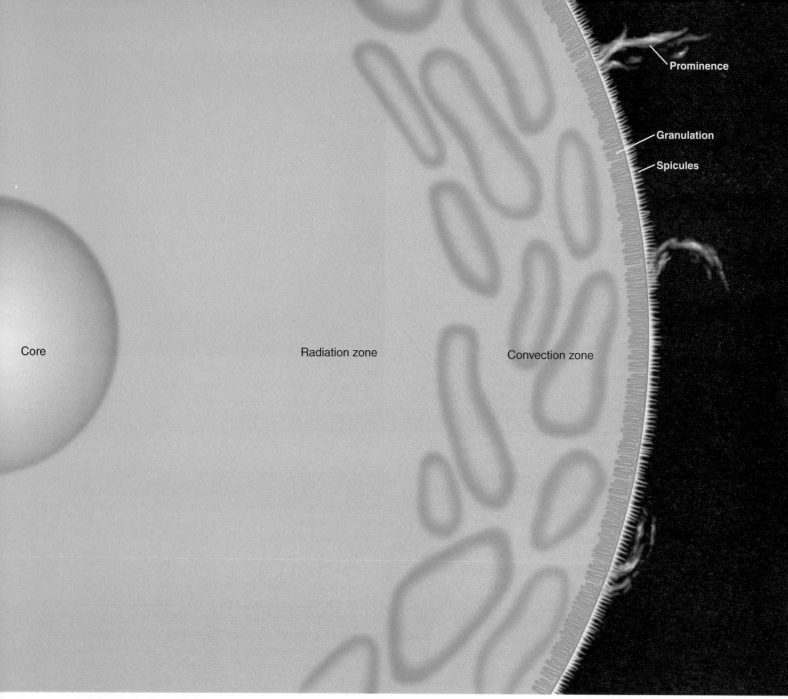

Core

Radiation zone

Convection zone

Prominence

Granulation

Spicules

Figure 14.2
The structure of the Sun. Note that not all the layers are to scale. Some of the outer regions of the Sun are so thin, they would actually not be visible on an accurate drawing, so we have exaggerated their width.

6000 K. At a depth of 400 km, only 4 percent of the photons escape from the Sun. The remainder are absorbed and re-emitted again, and through this process they gradually work their way outward, eventually reaching a region of low density with too few atoms to absorb the photons again. By the time they get to the top of the photosphere, 99.5 percent of the photons are escaping from the Sun, and these make up most of the sunlight we detect here on Earth.

Note that the atmosphere of the Sun is not a very dense layer compared to the air in the room where you are reading this book. At a typical point in the photosphere, the pressure is less than 10 percent of the Earth's pressure at sea level, and the density is about one ten-thousandth of the Earth's atmospheric density at sea level.

The Chromosphere

The Sun's outer gases extend far beyond the photosphere (Figure 14.2). Because they are transparent to most visible radiation and emit only a small amount of light, these outer layers are difficult to observe. The region of the Sun's atmosphere that lies immediately above the photosphere is called the **chromosphere.** Until this century, the chromosphere was visible only when the photosphere was concealed by the Moon during a total solar eclipse (see Section 3.7). In the 17th century, several observers described what appeared to them as a narrow red "streak" or "fringe" around the edge of the Moon during a brief instant after the Sun's photosphere had been covered. The name chromosphere, from the Greek for "colored sphere," was given to this red streak.

Observations made during eclipses show that the chromospheric spectrum consists of bright *emission* lines, indicating that this layer is composed of hot, transparent gases emitting light at discrete wavelengths. The reddish color of the chromosphere arises from one of the strongest emission lines in the visible part of its spectrum—the bright red line caused by hydrogen, the element that, as we have already seen, dominates the composition of the Sun.

In 1868, observations of the chromospheric spectrum revealed a yellow emission line that did not correspond to any previously known element on Earth. Scientists quickly realized they had found a new element, and named it *helium* (after *helios*, the Greek word for "Sun"). It took until 1895 for helium to be discovered on our planet. Today, students are probably most familiar with it as the light gas used to blow up balloons, although, as we have seen, it turned out to be the second most abundant element in the universe.

The chromosphere is about 2000 to 3000 km thick. The density of the chromospheric gases decreases as we go up, but the temperature *increases* from 4500 K at the photosphere to 10,000 K or so in the lower levels of the chromosphere. This temperature increase is a surprise, since throughout the rest of the Sun the temperature steadily decreases from the center outward.

The Transition Region

The increase in temperature does not stop with the chromosphere, however. There is a thin region in the solar atmosphere where the temperature changes from 10,000 K, typical of the chromosphere, to nearly a million degrees, characteristic of the corona. Appropriately, this part of the Sun is called the **transition region.** It can vary in thickness: in small areas on the Sun, this region may be only a few tens of kilometers thick.

The transition region does not, however, form a smooth shell around the Sun at a fixed height above the photosphere. Figure 14.3 shows that the chromosphere contains many jet-like spikes of gas, called *spicules*, rising vertically through it. When viewed near the *limb* (or edge) of the Sun, so many are seen in projection that they give the effect of a forest. They consist of gas jets moving upward at about 30 km/s and rising to heights of 5000 to 20,000 km above the photosphere. Individual spicules last

Figure 14.3
Solar spicules, photographed in the light of the red Balmer line of hydrogen, which is referred to as Hα. (National Solar Observatory/National Optical Astronomy Observatories)

only 10 min or so. Solar astronomers believe that the transition region may be wrapped around the spicules like a thin cloak.

Figure 14.4 summarizes how the temperature of the solar atmosphere changes from the photosphere to the corona.

The Corona

The outermost part of the Sun's atmosphere is the **corona.** Like the chromosphere, the corona was first observed during total eclipses (Figure 14.5). Unlike the chromosphere, the corona has been known for many centuries: it was referred to by the Roman historian Plutarch and was discussed in some detail by Kepler. The corona extends millions of kilometers above the photosphere and emits half as much light as the full moon. The reason we don't see this light is the overpowering brilliance of the photosphere. Just as bright city lights make it difficult to see faint starlight, so too the intense light from the photosphere hides the faint light from the corona. While the best time to see the corona is during a total solar eclipse, its brighter parts can be photographed with special instruments even at other times.

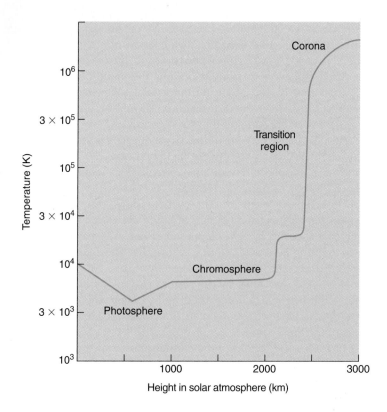

Figure 14.4
Temperatures in the solar atmosphere as a function of height above the photosphere. Note the very rapid increase in temperature over a very short distance in the transition region between the chromosphere and the corona.

Studies of its spectrum show the corona to be very low in density. At the bottom of the corona there are about 10^9 atoms per cubic centimeter, compared with about 10^{16} atoms per cubic centimeter in the upper pho-

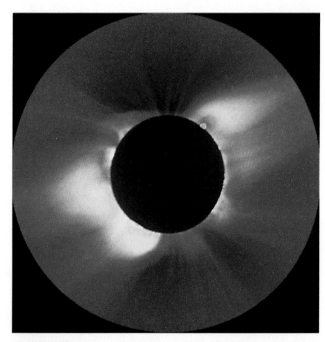

Figure 14.5
Image of the Sun taken during the solar eclipse on March 18, 1988. Since the light from the brilliant surface (photosphere) of the Sun is blocked by the Moon, it is possible to see the tenuous outer atmosphere of the Sun which is called the corona.
(High Altitude Observatory/NCAR)

tosphere, and 10^{19} molecules per cubic centimeter at sea level in the Earth's atmosphere. The corona thins out very rapidly at greater heights, where it corresponds to a high vacuum by laboratory standards.

The corona is very hot. We know this because the spectral lines produced in it come from highly ionized atoms of iron, nickel, argon, calcium, and other elements. For example, astronomers observe lines of iron whose atoms have lost 16 electrons in the corona. Such a high degree of ionization requires a temperature of millions of degrees Kelvin. Indeed, because we don't routinely deal with such hot gases on Earth, for a while astronomers were unsure what caused the lines in the corona's spectrum. They thought the lines might come from a new element, which they named *coronium*. But in the 1930s this hypothetical element became unnecessary, because astronomers discovered that known elements with many electrons missing could account for the coronal spectrum.

Because of its high temperature, most of the spectral lines emitted by the corona lie in the far ultraviolet region (at wavelengths shorter than about 100 nm), and the corona is also very bright at x-ray wavelengths (Figure 14.6). Images and data from instruments in space have helped considerably to unravel the mechanisms at work in the corona.

How is it possible to heat the outer layers of the Sun's atmosphere to much higher temperatures than those of the photosphere? Observations indicate that magnetic fields play a major role. The Sun, like several of the planets, is a giant magnet: it has a complex magnetic field that can have powerful effects on a thin, electrically charged gas like the corona. Magnetic fields from below appar-

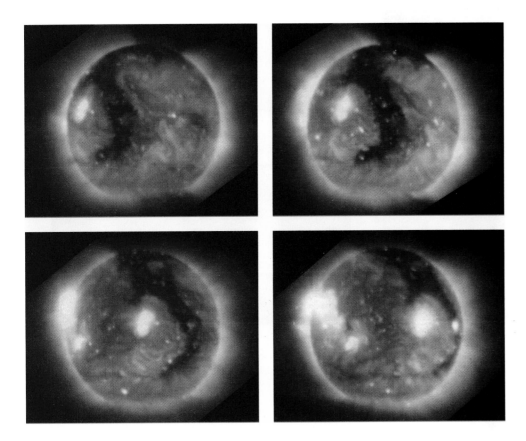

Figure 14.6

X-ray images of the Sun taken from the Skylab satellite show hot coronal gas. The corona is patchy, with bright spots indicating regions where hot gas is concentrated. The long dark area with no x-ray emission is called a coronal hole. In these regions, hot gas streams away from the solar surface out through the solar system. This stream of particles is called the solar wind. The four successive images of the Sun clearly show how the positions of solar features change as the Sun rotates.

(Courtesy Leon Golub, SAO)

ently store energy and carry it to the chromosphere and corona, where it is converted to electrical currents; these in turn heat the gases of the Sun's outer atmosphere. The precise way in which magnetic energy is converted to heat energy is not yet understood, and explaining this process in detail is one of the major challenges facing solar astronomers.

The Solar Wind

One of the most remarkable discoveries about the Sun's atmosphere is that it produces a stream of charged particles (mainly protons and electrons) that we call the **solar wind.** These particles flow outward from the Sun at a speed of about 400 km/s (about 900,000 mi/h!). The solar wind exists because the gases in the corona are so hot and moving so rapidly that they cannot be held back by solar gravity. (This phenomenon was actually discovered by its effects on the charged tails of comets; in a sense, we can see the comet tails blow in the solar breeze the way windsocks flutter on Earth.)

By means of the solar wind, material actually leaves the Sun and travels outward into the solar system. Although the solar wind material is very, very rarified, the Sun has an enormous surface area. Astronomers estimate that the Sun is losing about 10 million tons of material each year. While this amount of lost mass seems large by Earth standards, it is completely insignificant for the Sun.

In visible light photographs the solar corona appears fairly uniform and smooth. X-ray pictures, however, show that the corona has loops, plumes, and both bright and dark regions (Figure 14.7). Sometimes large regions of the corona are relatively cool and quiet. These **coronal holes** are places of extremely low density and are usually (but not always) found in the polar regions of the Sun. They cause the empty spaces seen on some of the eclipse photographs of the solar corona.

Measurements show that hot coronal gas is present mainly where magnetic fields have trapped and concentrated it. The coronal holes lie between these concentrations of gas. The solar wind comes predominantly from these coronal holes, streaming through them into space unhindered by magnetic fields.

At the surface of the Earth, we are protected from the solar wind by our atmosphere and by the Earth's magnetic field (see Chapter 7). However, our magnetic field lines come into the Earth at the North and South Poles. Here particles from the solar wind can follow the field down to our upper atmosphere. As the Sun's particles strike atoms and molecules of air, they cause them to glow, producing beautiful curtains of light called the **aurora** (or the northern and southern lights). The most spectacular auroras occur at altitudes of 75 to 150 km. Strong activity on the Sun can trigger "gusts" in the solar wind, which, in turn, can lead to truly dramatic auroral displays (Figure 14.8).

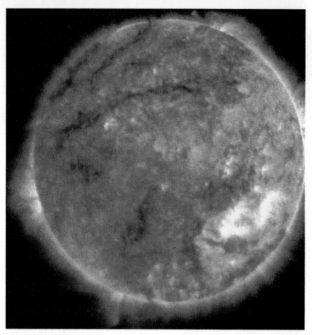

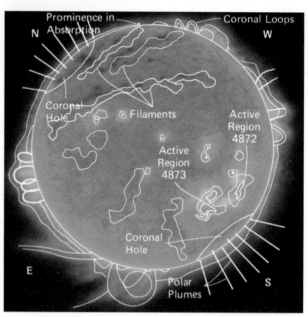

Figure 14.7
An x-ray/extreme ultraviolet image of the Sun taken from a sounding rocket 100 miles above White Sands Missile Range on October 23, 1987. The features shown in the photograph are identified in the drawing. (Art Walker/Stanford University and NASA)

The Active Sun Introduced

Overall, the Sun is quite stable, and we rely on that stability to maintain a life-sustaining environment on Earth. The surface of the Sun is not at all quiet, however. It is a seething, bubbling cauldron of hot gas. Areas that are

Figure 14.8
A spectacular auroral display in the Earth's atmosphere. We see aurorae when charged particles from the magnetosphere hit the atmosphere near the poles (where the magnetic field lines bring the particles to interact with the molecules of air). See Figure 7.4. (John Warden/Alaska Stock Images)

darker and cooler than the rest of the surface come and go. Vast plumes of gas are ejected into the chromosphere and corona. Occasionally, there are even giant explosions on the Sun that have major effects on the Earth.

Photospheric Granulation

Observations show that the photosphere has a mottled appearance resembling grains of rice spilled on a dark tablecloth. This structure of the photosphere is now generally called **granulation** (Figure 14.9). Typically, granules are 700 to 1000 km in diameter. They appear as bright areas surrounded by narrow, darker regions, and individual granules persist for only 5 to 10 min.

The motions of the granules can be studied by examining the Doppler shifts (see Section 4.6) in the spectra of gases just above them. The bright granules are columns of hotter gases rising from below the photosphere. As rising gas reaches the photosphere, it spreads out and sinks down again into the darker regions between the granules. The centers of the granules are hotter than the intergranular regions by 50 to 100 K, and the vertical motions of gases in the granules have speeds of 2 or 3 km/s.

The granules, then, are the tops of *convection* currents of gases rising through the photosphere. We discussed convection when we talked about the layers of the Earth's interior. It is the rising of a hot liquid or gas, transferring energy from the hotter layers underneath to the cooler layers above. As the gas cools, it falls to lower layers and the area looks darker from above. (You can see the same sort of hot currents rising in a pot of boiling water as you make more coffee in order to stay awake while read-

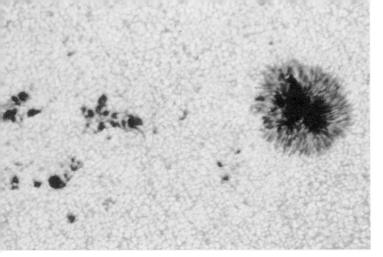

Figure 14.9

Solar granulation in the vicinity of sunspots. Each small, bright region is a rising column of hotter gas about 1000 km across. Cooler gas descends in the darker regions between the granules. The larger dark regions are sunspots. (National Solar Observatory/National Optical Astronomy Observatories)

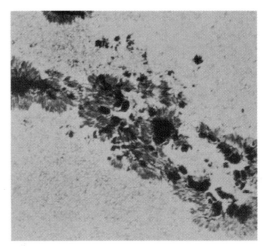

Figure 14.10

This excellent photograph of a group of sunspots and solar granulation was taken through a green filter. (National Solar Observatory/National Optical Astronomy Observatories)

ing your astronomy book.) Granulation is one piece of evidence that the Sun's interior must be supplying heat to the layers above.

Sunspots

The most conspicuous of the photospheric features are the **sunspots** (Figure 14.10). Occasionally, these spots are large enough to be visible to the unaided eye.

Sunspots are darker than the photosphere in which they are embedded because the gases in sunspots are as much as 1500 K cooler than the surrounding gases. Sunspots are nevertheless hotter than the surfaces of many stars. If they could be removed from the Sun, they would shine brightly. They appear dark only in contrast with the hotter, brighter surrounding photosphere.

Individual sunspots have lifetimes that range from a few hours to a few months. If a spot lasts and develops, it usually consists of two parts: an inner darker core, the *umbra,* and a surrounding less-dark region, the *penumbra.* Many spots become much larger than the Earth, and a few have reached diameters of 50,000 km. Frequently, spots occur in groups of 2 to 20 or more. The largest groups are very complex and may have over 100 spots. Like storms on the Earth, sunspots are not fixed in position but they drift slowly compared with solar rotation, which carries them across the visible disk of the Sun.

By recording the apparent motions of the sunspots as the turning Sun carried them across its disk, Galileo demonstrated that the Sun rotates on its axis (Figure 14.11). He found that the rotation period of the Sun is a little less than one month. Modern measurements show that the Sun's rotation period is about 25 days at the equator, 28 days at latitude 40°, and 36 days at latitude 80°, in a west-to-east direction (like the orbital motions of the planets). Note that the Sun, being composed of gas, does not have to rotate rigidly the way a solid body does.

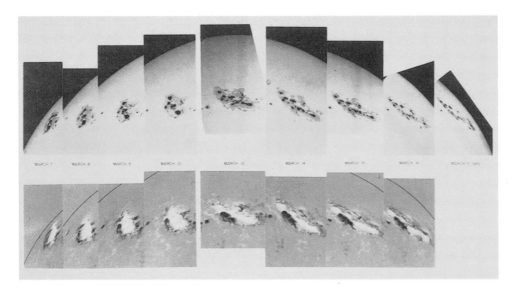

Figure 14.11

Photographs of the Sun's surface, showing a large group of sunspots. The series of exposures follows the rotation of sunspots across the visible hemisphere of the Sun. The top sequence shows the Sun in ordinary light; the bottom sequence shows emission from the chromosphere. (National Solar Observatory/National Optical Astronomy Observatories)

Figure 14.12

A comparison of the number of sunspots and magnetic activity on the active (left) and quiet (right) Sun. The computer-generated images use yellow to indicate positive or north polarity, and dark blue for negative or south polarity. In the image of the active Sun, note that pairs of sunspots have opposite polarity. Note also that the polarity of the leading spot is different in the upper and lower hemispheres. At solar minimum there are no large sunspots, and the magnetic fields are weak. (National Solar Observatory/National Optical Astronomy Observatories)

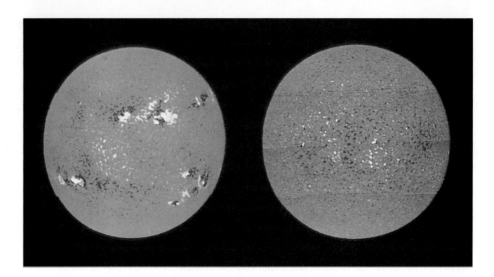

14.3

The Sunspot Cycle

Between 1826 and 1850, Heinrich Schwabe, a German pharmacist and amateur astronomer, kept daily records of the number of sunspots. His diligence paid off, and he was able to conclude that the number varied systematically in cycles of about ten years. Since Schwabe's work, the existence of such a **sunspot cycle** has been clearly established. Although individual spots are short-lived, the total number visible on the Sun at any one time is likely to be very much greater at certain times—the periods of sunspot maximum—than at other times—the periods of sunspot minimum (Figure 14.12).

Sunspot maxima have occurred at an average interval of 11.1 years, but the intervals between successive maxima have ranged from as little as 8 years (from 1830 to 1838) to as long as 16 years (from 1888 to 1904). During sunspot maxima, more than 100 spots can often be seen at once. During sunspot minima, sometimes no spots are visible. Activity was last near maximum in 1990 and 1991.

Magnetism in the Solar Cycle

The sunspot cycle is closely related to magnetism in the Sun, and it is the Sun's changing magnetic field that provides the driving force for many aspects of solar activity.

The solar magnetic field is measured using a property of atoms called the *Zeeman effect*. Recall from Chapter 4 that an atom has many energy levels, and that spectral lines are formed when electrons shift from one level to another. If each energy level is precisely defined, then the difference between them is also quite precise. As an electron changes levels, the result is a sharp, narrow spectral line (either an absorption or emission line, depending on whether the electron's energy increases or decreases in the transition).

In the presence of a strong magnetic field, however, each energy level is separated into several levels very close to one another. The separation of the levels is proportional to the strength of the field. As a result, spectral lines formed in the presence of a field are not single lines, but a series of very closely spaced lines corresponding to the subdivision of the atomic energy levels. This splitting of lines in the presence of a magnetic field is termed the Zeeman effect.

Measurements of the Zeeman effect in the spectra of the light from sunspot regions (Figure 14.13) show them to have strong magnetic fields. Whenever sunspots are observed in pairs, or in groups containing two principal spots, one of the spots usually has the magnetic polarity of a north-seeking magnetic pole, and the other has the opposite polarity. Moreover, during a given cycle the leading spots of pairs (or leading principal spots of groups) in the northern hemisphere all tend to have the same polarity, while those in the southern hemisphere all tend to have the opposite polarity.

During the next sunspot cycle, however, the polarity of the leading spots is reversed in each hemisphere. For example, if during one cycle the leading spots in the northern hemisphere all had the polarity of a north-seeking pole, the leading spots in the southern hemisphere would have the polarity of a south-seeking pole. During the next cycle, then, the leading spots in the northern hemisphere would have south-seeking polarity, while those of the southern hemisphere would have north-seeking polarity. We see, therefore, that the sunspot cycle does not repeat itself in regard to magnetic polarity until two maxima have passed. The solar activity cycle, fundamentally a magnetic cycle, therefore, averages 22 years in length, not 11.

Magnetic fields hold the key to explaining why sunspots are cooler and darker than the regions without strong magnetic fields. The forces produced by the magnetic field resist the motions of the bubbling columns of

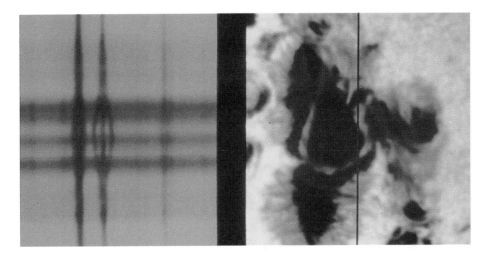

Figure 14.13

These photographs show how magnetic fields in sunspots are measured by means of the Zeeman effect. The vertical black line in the picture at right indicates the position of the spectrograph slit through which light passed in order to obtain the spectrum in the picture at left. Note that the strongest spectral line in the left-hand picture is split into three components.

(National Optical Astronomy Observatories)

rising hot gases. Since these columns carry most of the heat from inside the Sun to the surface by means of convection, less heating occurs where there are strong magnetic fields. As a result, these regions are seen as darker, cooler sunspots.

14.4

Activity Above the Photosphere

To see those portions of the Sun lying directly above the photosphere, we should observe at wavelengths where the photospheric gases are especially opaque—for example, the strong absorption lines of elements such as hydrogen and calcium. Photons emitted at these wavelengths cannot escape directly from the photosphere, but rather are re-absorbed by it. Any calcium and hydrogen *emission* must come from the thinner, hotter layers of the solar atmosphere that lie well above the photosphere. Since many aspects of solar activity involve regions of gas in the chromosphere, these strong spectral lines provide a means to monitor such activity. Astronomers routinely photograph the Sun through monochromatic filters that pass light only at these special wavelengths (Figure 14.14).

Plages and Prominences

Pictures taken through filters that pass only the light of either calcium or hydrogen show bright "clouds" in the chromosphere around sunspots; these bright regions are known as **plages** (Figure 14.15). The plages are not really clouds of any particular element, but are regions of higher temperature and density within the chromosphere.

Moving higher into the Sun's atmosphere, we come to the spectacular phenomena called **prominences** (Figures 14.16 and 14.17). Eclipse observers see prominences as red, flame-like protuberances rising above the eclipsed Sun and reaching high into the corona. Some, the quiescent prominences, are graceful loops that can remain nearly stable for many hours or even days. They may extend to heights of tens of thousands of kilometers above the solar surface. Others can move upward or have arches that surge slowly back and forth.

The relatively rare eruptive prominences appear to send matter upward into the corona at speeds up to 700 km/s, and the most active surge prominences may move as fast as 1300 km/s (almost 3 million mi/h). Some eruptive prominences have reached heights of over 1 million km above the photosphere; the Earth would be completely lost inside one of these awesome displays.

Prominences usually originate near regions of sunspot activity and lie on the boundary between regions of opposite magnetic polarity. Eruptive prominences appear to result from sudden changes in the magnetic fields.

(text cont. on page 299)

Figure 14.14
Image of the Sun taken on March 18, 1990, with a filter that transmits only the light of the spectral line produced by singly ionized calcium. (National Solar Observatory/National Optical Astronomy Observatories)

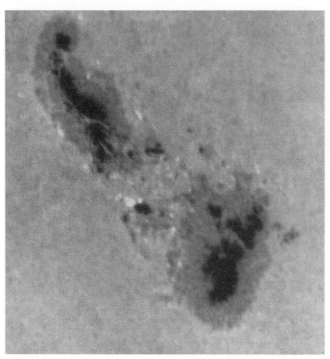
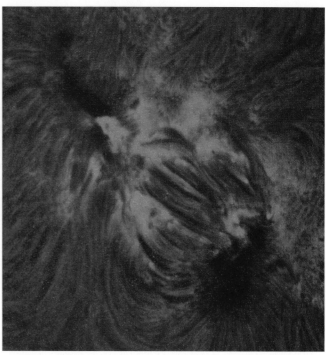

Figure 14.15

Two pictures showing the same area on the Sun—a large sunspot group photographed in 1981 at the Big Bear Solar Observatory. The image on the left was taken in normal (white) light and shows the photosphere. The image on the right, taken in the light of the red Balmer line of hydrogen, shows bright "clouds" or plages in the upper solar atmosphere (chromosphere) in the regions around sunspots. (Big Bear Solar Observatory, California Institute of Technology)

Figure 14.16

An eruptive prominence as seen in the light of the red line of the hydrogen Balmer series (called Hα). This picture was taken on June 20, 1989. (National Solar Observatory/National Optical Astronomy Observatories)

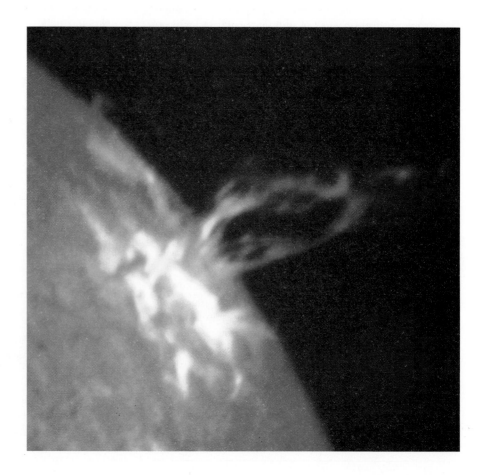

Flares

The most awesome event on the surface of the Sun is a rapid eruption called a **solar flare.** A typical flare lasts for 5 to 10 min and releases a total amount of energy equivalent to that of perhaps a million hydrogen bombs. The largest flares last for several hours and emit enough energy to power the entire United States at its current rate of electrical consumption for 100,000 years. Near sunspot maximum, small flares occur several times per day, and major ones may occur every few weeks.

Flares are often observed in the red light of hydrogen (Figure 14.18 is a dramatic example), but the visible emission is only a tiny fraction of the energy released when a solar flare explodes. At the moment of the explosion, the matter associated with the flare is heated to temperatures as high as 10 million K. At such high temperatures, a flood of x-ray and ultraviolet radiation is emitted.

There is evidence that flares occur when magnetic fields pointing in opposite directions release energy by interacting with and destroying each other—much as a stretched rubber band releases energy when it breaks. The gases at the solar surface are in constant churning motion, and this motion frequently brings oppositely directed fields together. If the magnetic interactions cover a large volume in the solar corona, then in addition to the flare of electromagnetic radiation, immense quantities of coronal material—mainly protons and electrons—may be ejected at high speeds (500–1000 km/s) into interplanetary space. Such *coronal mass ejections* can affect the Earth in several ways. (See "Making Connections" box.)

Active Regions

Sunspots, flares, and bright regions in the chromosphere and corona tend to occur together on the Sun. That is, they all tend to have similar longitudes and latitudes, but they are located at different heights in the atmosphere. A place on the Sun where

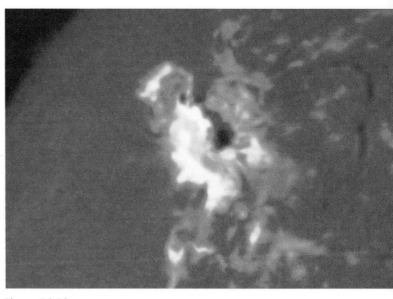

Figure 14.18
A giant solar flare as seen in the red light of hydrogen. This flare occurred on March 10, 1989. (National Solar Observatory/National Optical Astronomy Observatories)

these phenomena are seen is called an **active region** (Figure 14.19).

Astronomers do not know what causes an active region to form, but they do know that such regions are found in areas with strong magnetic fields. Therefore, in developing models to explain the solar cycle, astronomers have tried to understand why the Sun's magnetic field changes in strength and polarity in a nearly regular way. Calculations show that rotation and convection just below the solar surface can distort magnetic fields, which causes them to grow and then decay, regenerating with opposite polarity approximately every 11 years.

This calculated change in the strength and polarity of magnetic fields is just what we actually see over the course of the solar cycle. The calculations also show that as the fields grow stronger near solar maximum, they float from the interior of the Sun toward its surface in the form of loops. When a loop emerges

(text cont. on page 301)

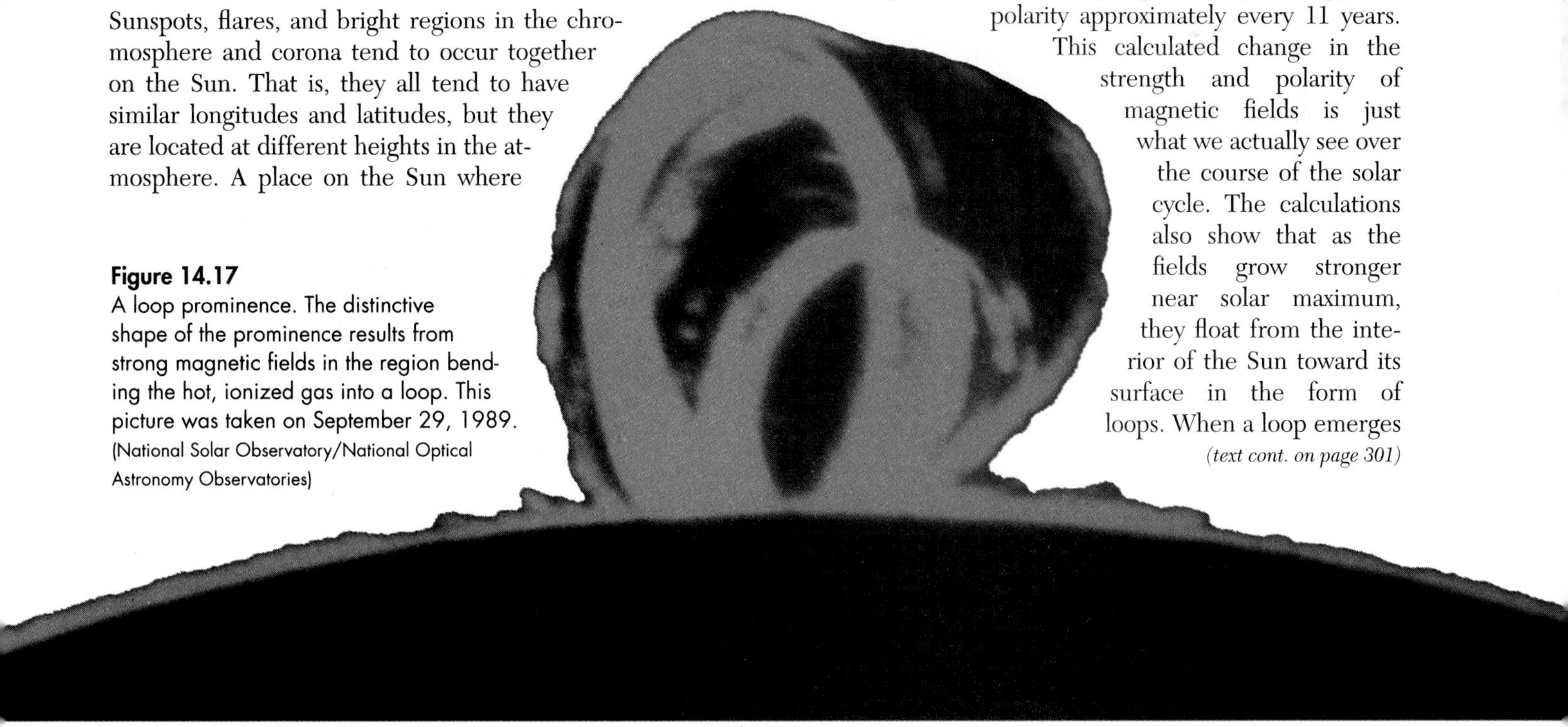

Figure 14.17
A loop prominence. The distinctive shape of the prominence results from strong magnetic fields in the region bending the hot, ionized gas into a loop. This picture was taken on September 29, 1989. (National Solar Observatory/National Optical Astronomy Observatories)

MAKING CONNECTIONS

Solar Flares and Their Effects on Earth

The most obvious effect of coronal mass ejections on the Earth is the appearance of a brilliant aurora. In March 1989 a gigantic flare, which occurred as the Sun approached maximum activity, was accompanied by a coronal mass ejection that produced an aurora visible as far south as Arizona (latitude 32°). (Recall that auroras occur preferentially near the Earth's magnetic poles because the charged particles from the Sun tend to flow down into the Earth's atmosphere along our magnetic field, which comes into the Earth near the poles.)

The changes in the Earth's magnetic field caused by interactions with the sudden onslaught of charged particles from the Sun in turn generate changing electric currents. The effects are most noticeable in long power lines, but solar flares can even cause power station components to burn out. As a result of the solar explosion in March 1989, parts of Montreal and Quebec Province in Canada were without power for up to 9 hours. Other effects occurred as well. For example, because of the electrical interference, people found their automatic garage doors opening and closing for no apparent reason. Charged particles that reach the Earth from the Sun can also overload telephone circuits.

The short-wavelength radiation produced during solar flares heats the outer atmosphere of the Earth. In 1981 a very large solar flare occurred while the Space Shuttle *Columbia* was in orbit. The astronauts aboard found that the flare, which lasted for 3 hours, increased the temperature of the Earth's atmosphere at an altitude of 260 km from its normal value of 1200 K up to 2200 K. When the outer atmosphere is heated, it also expands, so that it reaches farther into space. As a consequence, friction between the atmosphere and spacecraft increases, dragging satellites to lower altitudes. At the time of the March 1989 flare, the system responsible for tracking some 19,000 objects orbiting the Earth temporarily lost track of 11,000 of them because their orbits were changed by the expansion of the Earth's atmosphere. During solar maximum, a number of satellites are brought to such a low altitude that they are destroyed by friction with the atmosphere.

The level of solar activity is a critical factor in calculating the lifetimes and orbits of satellites in near-Earth orbit. Flares could also be life-threatening to astronauts on a voyage to Mars. Obviously, it would be extremely valuable to predict both the overall level of solar activity and the occurrence of individual flares. Solar astronomers are working very hard to learn how to make reliable predictions, but accurate forecasts of solar "weather" are proving to be a goal as elusive as reliable forecasts of the weather on Earth.

A solar flare can result in the ejection of vast numbers of particles at speeds of thousands of kilometers per hour. (National Solar Observatory/National Optical Astronomy Observatories)

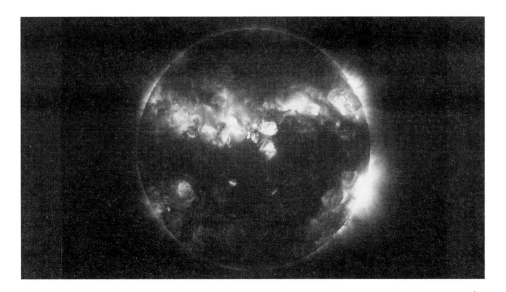

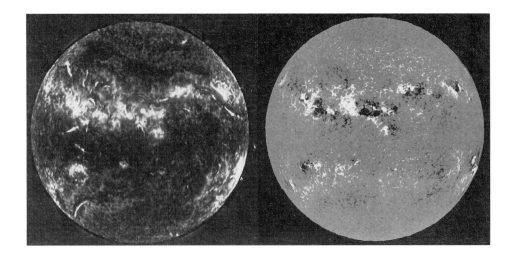

Figure 14.19
Three images of the Sun taken at the same time but at different wavelengths. The top image shows x-ray emission, which originates high in the hotter chromosphere and the corona. Image at lower left, taken in the light of a strong line of neutral helium, shows where chromospheric emission is strong. In image at lower right, white and black spots are regions of strong magnetic field; white and black correspond to opposite magnetic polarities. You can see that x-ray emission, chromospheric emission, and strong magnetic fields tend to occur in the same locations, called active regions. (NASA/National Solar Observatory/National Optical Astronomy Observatories)

from the solar surface, it creates an active region. The "ends" of the loop, where they penetrate the solar surface, have different polarities. This idea of magnetic loops offers a natural explanation of why the leading and trailing sunspots in an active region have opposite polarity.

14.5

Is the Sun a Variable Star?

The Sun is one of the few truly constant objects in our daily lives. It rises faithfully at a time that can be precisely calculated. Each day it deposits a constant amount of energy on the Earth, warming it and sustaining life. But is the Sun truly constant, day by day, year by year, millennium by millennium? Or does its energy output vary? Do these variations ever become large enough to affect the Earth or its climate?

Variations in the total amount of energy emitted by the Sun, if any do exist, must be subtle. The presence of life on Earth demonstrates that there have been no recent major changes in its climate. We do know, however, our planet has experienced ice ages, separated by relatively warm interglacial periods. Do these changes have anything to do with the Sun?

Variations in the Number of Sunspots

The most obvious solar variation is the number of sunspots. Considerable evidence shows that between the years 1645 and 1715, the number of sunspots was much lower than it is now. This interval of extremely low activity was first noted by Gustav Spörer in 1887, and then by E. W. Maunder in 1890; it is now called the **Maunder Minimum.** The incidence of sunspots over the past four centuries is shown in Figure 14.20. According to the data presented in this figure, sunspot numbers were also somewhat lower during the first part of the 19th century than

(text cont. on page 303)

Art Walker: Doing Astronomy in Space

Since the Sun's corona is a rich source of high-energy photons, observations at x-ray wavelengths are a good way to learn more about our star. Unfortunately, x rays do not penetrate the Earth's atmosphere, so such observations must be done from rockets and satellites. Art Walker has been one of the pioneers in designing and flying instruments in space to learn more about the Sun. He is also one of only a handful of African-Americans actively doing research in astronomy and is thus a role model for minority students around the country.

Walker received a PhD in physics from the University of Illinois, supported in his graduate training by the Air Force reserve officer program. After receiving his degree, he began active service at the Air Force Weapons laboratory, which was beginning a space research program in the physics of the Sun. (See "Making Connections" to find out why the military is interested in the Sun.) Using a specially designed spectroscope aboard an Air Force satellite, he was able to take one of the first x-ray spectra of the Sun and identify several new emission lines in those spectra. Such measurements allowed him and his co-workers to probe the temperature, composition, and structure of the corona, and get a much better sense of what the Sun's atmosphere is like.

More recently, he has been working with a novel design for x-ray telescopes that makes them similar to some of the telescopes astronomers use on the ground. These instruments allow scientists to take very high-resolution x-ray images of the active Sun; they are also relatively easy to build and can be flown aboard sounding rockets whose paths are spectacular arcs traveling 100 or more miles above the Earth's surface.

In an interview, Walker emphasized one of the key challenges of doing astronomy from space: "In the laboratory, if you build an experiment and it doesn't quite work right, you can always tinker with it until you get the instrumentation to operate. In the case of space observations, you have to build the instrument anticipating all the things that might go wrong and eliminating each possibility of failure. Once the instrument is launched, it is out of your control, and you just have to hope that you have anticipated everything."

After working for the Aerospace Corporation, Walker became a Professor at Stanford University, where he has also served as Dean of the Graduate College. Perhaps his best-known student was Sally Ride, who received her PhD in astrophysics and went on to become America's first woman in space. Walker has chaired the Astronomy Advisory Committee for the National Science Foundation, and also served on the Presidential Commission that investigated the accident that destroyed the Shuttle *Challenger.*

Dr. Art Walker

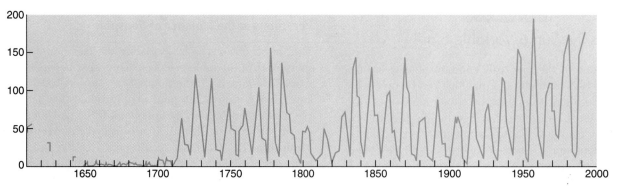

Figure 14.20
The relative numbers of sunspots as a function of time. Note the absence of sunspots from 1645 to 1715.

they are now; this period is called the Little Maunder Minimum.

When the number of sunspots is high, the Sun is active in various other ways as well, and this activity affects the Earth directly. As we have seen, auroras are caused by the impact of charged particles from the Sun on the Earth's magnetosphere. Energetic charged particles are much more likely to be ejected by the Sun when it is active and the sunspot number is high. There is a strong correlation between the number of sunspots and the frequency of auroral displays. Historical accounts indicate that auroral activity was abnormally low throughout the several decades of the Maunder Minimum.

The best quantitative evidence of long-term (over several decades) variations in the level of solar activity comes from studies of the radioactive isotope carbon-14. The Earth is constantly bombarded by cosmic rays — high-energy charged particles that include protons and nuclei of heavier elements. The rate at which cosmic rays from sources outside the solar system reach the upper atmosphere depends on the level of solar activity. When the Sun is active, charged particles streaming out into the solar system carry the Sun's strong magnetic field with them. This magnetic field shields the Earth from incoming cosmic rays. At times of low activity, when the Sun's magnetic field is weak, cosmic rays reach the Earth in larger numbers.

When the energetic cosmic-ray particles impact the upper atmosphere, they produce several different radioactive isotopes (see Section 4.4). One such isotope is carbon-14, which is produced when nitrogen is struck by high-energy cosmic rays. The rate of carbon-14 production is higher when the activity of the Sun is lower and the solar magnetic field does not shield the Earth from bombardment by cosmic rays.

Some of the radioactive carbon is contained in carbon dioxide molecules, which are ultimately incorporated into trees through photosynthesis. By measuring the amount of radioactive carbon in tree rings, we can estimate the historical levels of solar activity. Correlations with visual estimates of sunspot numbers over the past 300 years indicate that the carbon-14 estimates of solar activity are indeed valid. Because it takes an average of about ten years for a carbon dioxide molecule to be absorbed from the atmosphere or ocean into plants, this technique cannot provide data on the 11-year solar cycle. It can be used, however, to look for long-term (over several decades) changes in the level of solar activity.

Estimates of the amount of carbon-14 in tree rings now extend continuously back about 8000 years. Variations in solar activity levels have occurred throughout this period, and at different times the Sun has been both more and less active than it is now. The measurements confirm that the amount of carbon-14 was unusually high, with solar activity correspondingly low, during both the Maunder Minimum and the Little Maunder Minimum. Activity was also low from 1410 to 1530, and from 1280 to 1340. Between about 1100 and 1250, the level of solar activity may have been even higher than it is now.

Solar Variability and the Earth's Climate

Variations in the overall level of the Sun's activity seem to be well established. Did these variations have any direct impact on the Earth or its climate? The Maunder Minimum was a time of exceptionally low temperatures in Europe — so low that this period is described as a Little Ice Age. The river Thames in London froze at least 11 times during the 17th century, ice appeared in the oceans off the coasts of southeast England, and low summer temperatures led to short growing seasons and poor harvests (Figure 14.21). The global climate also appears to have been unusually cool from 1400 to 1510; this period was one of low solar activity as well.

The most obvious way in which the Sun might be linked to the Earth's climate is through variations in solar luminosity. If the Sun puts out less energy, then logically we might expect the Earth to become colder. Does the Sun become cooler at times of low activity, as would be required to account for the Little Ice Age?

The relationship between solar luminosity and activity level was determined only recently from very precise measurements made by a satellite orbiting the Earth. These observations show that the Sun's luminosity varies as it rotates, typically by about 0.1 percent. These changes are connected with changes in the number of sunspots in the sense that the Sun is slightly dimmer when more sunspots are present.

Surprisingly, measurements over the entire solar cycle indicate that we receive the maximum amount of radiation from the Sun at times of solar maximum. This is just the opposite of what we expect from the results of satellite measurements over a single rotation period. The source of the extra radiation at sunspot maximum is not known. Nevertheless, the evidence indicates that the total energy emitted by the Sun is greater when it is more active.

These observations support the idea that the Maunder Minimum was indeed associated with the Little Ice Age. The unusually cold temperatures at that time would have required approximately a 1 percent drop in solar luminosity, larger than the variations we see today.

We can get additional clues about whether a variation in luminosity of as much as 1 percent is likely to occur over several centuries by looking at activity cycles in stars that are similar to the Sun. It turns out that the Sun's current variations (just 0.1 percent of the total energy emitted) are unusually small. The luminosity of most stars varies by 0.3 percent and some by as much as 1 percent. Therefore, it seems likely that the Sun too might at times vary by larger amounts than it does currently, and that the energy output at the time of the Maunder Minimum might indeed have been low enough to cause unusually cold temperatures on Earth. The observations of higher variability in the energy output of most other solar-type

Figure 14.21
During the "Little Ice Age" in Europe, bodies of water froze over every winter. This painting by Robert van den Hoecke from 1649 is entitled "Skating in the Town Moat of Brussels."
(Kunsthistorisches Museum, Vienna; photo by Erich Lessing, Art Resources)

stars also suggests that the stability of solar behavior—and the stability of the Earth's climate—over the past three centuries may be unusual.

Very recently, scientists have reported that climate changes correlating with the subtle variation in the Sun's luminosity occur during a solar activity cycle. Over the course of the four most recent solar cycles, the average temperature in the Earth's middle and lower atmosphere has varied according to the Sun's cycle. In some places, the changes can be detected at the Earth's surface. For example, Charleston, South Carolina, appears to be several degrees chillier at solar maximum.

People who build models of the Earth's atmosphere have been trying to understand how small changes in the Sun's luminosity might cause noticeable changes in the temperature of the Earth. For example, we observe that the ultraviolet radiation from the Sun is about 1 percent higher at solar maximum. Models of our atmosphere indicate that this extra radiation can create more ozone (see Chapter 7) in the stratosphere. Ozone can absorb more sunlight and increase the temperature of the stratosphere. This in turn can affect the patterns of the winds aloft, modify the paths of storms, and change the temperature in lower layers of the Earth's atmosphere by differing amounts at different latitudes.

All of these results are very new and must still be confirmed by further observations and modeling. And phenomena on Earth, such as volcanoes, can also affect the global climate in ways we are coming to understand. But it appears that a direct link between changes in the Sun's luminosity and the Earth's climate may be on the brink of becoming firmly established.

Observing the Sun

Looking directly at the Sun is very dangerous: even a brief exposure can burn out your retina and cause severe eye damage. Looking at the Sun directly through an unfiltered telescope is even worse, because the telescope concentrates the Sun's radiation and hence damages your eye even more quickly. But there are *indirect* ways of observing the Sun that are both safe and instructive.

Not long after Galileo began training his first telescope on the sky, his assistant and disciple Benedetto Castelli came up with a way of observing sunspots, whose nature was a big controversy in those days. Castelli projected the image of the Sun made by the telescope onto a sheet of paper, where the sunspots could be sketched safely. Using this technique, Galileo was able to show that sunspots were not small, dark planets blocking sunlight, as some had suggested, but dark areas on the Sun moving around as the Sun rotated. You can try this experiment for yourself.

Set up a small telescope on a steady mount so that it points at the Sun. Remember, you have to do this by trial and error; NEVER LOOK THROUGH THE TELESCOPE AT THE SUN to see how well you're doing! Instead, watch the telescope's shadow; it will be smallest when the telescope is pointing directly at the Sun. Then you'll be able to see an image of the Sun on a sheet of white cardboard or other light surface. Sketch the sunspots if there are any. Repeat the observation over the course of a week or two, and, orienting your sketches the same way, trace how the spots have moved over the face of the Sun.

Putting a cardboard collar around the telescope makes the image easier to see by casting a shadow on it and thus making the sphere of the Sun stand out. If you observe the Sun for a while, don't be surprised if its image moves across and off the paper over time. This is caused by the turning of the Earth, which makes the Sun appear to rise and set. Also, don't be upset if you don't see any sunspots, especially if you are observing during the solar activity minimum.

Looking through a telescope at the Sun is dangerous, but you can always view the Sun safely with a small telescope by projecting its image on a sheet of white cardboard.

Summary

14.1 The Sun, our star, is surrounded by a number of layers that make up the solar atmosphere. They are, in order of increasing distance from the center of the Sun: the **photosphere,** with a temperature that ranges from 4500 K to about 6800 K; the **chromosphere,** with a typical temperature of 10^4 K; the **transition region,** a zone that may be only a few kilometers thick where the temperature increases rapidly from 10^4 K to 10^6 K; and the **corona,** with temperatures of a few million degrees K. **Solar wind** particles stream out into the solar system through **coronal holes.** Hydrogen and helium together make up 98 percent of the mass of the Sun, whose composition is much more characteristic of the universe at large than is a planet like the Earth. The Sun rotates more rapidly at its equator, where the rotation period is about 25 days, than near the poles, where the period is slightly greater than 36 days.

14.2 The Sun's surface is mottled with upwelling currents seen as hot, bright **granules. Sunspots** are dark regions where the temperature is up to 1500 K cooler than in the surrounding photosphere.

14.3 The number of visible sunspots varies according to a **sunspot cycle** that averages 11 years in length. Spots frequently occur in pairs. During a given 11-year cycle, all leading spots in the northern hemisphere have the same magnetic polarity, while all leading spots in the southern hemisphere have the opposite polarity. In the subsequent 11-year cycle, the polarity reverses. For this reason, the magnetic activity cycle of the Sun is often said to last for 22 years.

14.4 Sunspots, **solar flares, prominences,** and bright regions, including **plages,** tend to occur in **active regions**—that is, in places on the Sun with the same latitude and longitude but at different heights in the atmosphere.

14.5 There is evidence for long-term (100 years or more) variations in the level of solar activity and the number of sunspots. For example, the number of sunspots was unusually low from 1645 to 1715, a period now called the **Maunder Minimum.** It appears that the Earth tends to be cooler when the number of sunspots is unusually low for several decades.

Review Questions

1. Describe the main differences between the composition of the Earth and that of the Sun.

2. Make a sketch of the Sun's atmosphere showing the locations of the photosphere, chromosphere, and corona. What is the approximate temperature of each of these regions?

3. Why do sunspots look dark?

4. What is the Zeeman effect, and what does it tell us about the Sun?

5. Describe three different types of solar activity.

6. How does activity on the Sun affect the Earth?

7. Which aspects of the Sun's activity cycle have a period of about 11 years? Which vary during intervals of about 22 years?

8. Summarize the evidence indicating that over several decades or more there have been variations in the level of solar activity.

Thought Questions

9. The astronomer William Herschel (1738–1822) proposed that the Sun has a cool interior and is inhabited. Give at least one good argument against this idea.

10. How might you convince a misguided friend that the Sun is not hollow?

11. Suppose you took two photographs of the Sun, one in light at a wavelength centered on a strong absorption line, and the other at a wavelength region in the continuous spectrum away from strong lines. In which photograph would you observe deeper, hotter layers? Why?

12. If the rotation period of the Sun is determined by observing the apparent motions of sunspots, must any correction be made for the orbital motion of the Earth? If so, ex-

plain what the correction is and how it arises. If not, explain why the Earth's orbital revolution does not affect the observations.

13. Suppose an (extremely hypothetical) elongated sunspot forms that extends from a latitude of 30° to a latitude of 40° along a fixed line of longitude. How will the appearance of that sunspot change as the Sun rotates?

14. Suppose you live in northern Canada, and an extremely strong flare is reported on the Sun. What precautions might you take? What could compensate you for your troubles?

15. Why is it difficult to determine whether or not small changes in the amount of energy radiated by the Sun have an effect on the Earth's climate?

Problems

16. Use the data in Table 14.1 to confirm that the density of the Sun is 1.4 g/cm³. What kinds of materials have similar densities? One such material is ice. How do you know that the Sun is not made of ice?

17. Suppose you observe a major solar flare while astronauts are orbiting the Earth in the Shuttle. Use the data in Section 14.1 to calculate how long it will be before the charged particles ejected from the Sun during the flare reach the Shuttle.

18. Suppose an eruptive prominence rises at 150 km/s. If it does not change speed, how far from the photosphere will it extend after 3 hours? How does this distance compare with the diameter of the Earth?

19. From the Doppler shifts of the spectral lines in the light coming from the east and west edges of the Sun, it is found that the radial velocities of the two edges differ by about 4 km/s. Find the approximate period of rotation of the Sun.

Suggestions for Additional Reading

Akasofu, S. "The Shape of the Solar Corona" in *Sky & Telescope,* Nov. 1994, p. 24.

Bartusiak, M. "The Sunspot Syndrome" in *Discover,* Nov. 1989, p. 44.

Black, R. "Here's Looking at You, Sol" in *Air & Space,* Nov./Dec. 1986, p. 82. Describes new instruments on the ground and in space for observing the Sun.

Eddy, J. "The Case of the Missing Sunspots" in *Scientific American,* May 1977, p. 80.

Emslie, A. "Explosions in the Solar Atmosphere" in *Astronomy,* Nov. 1987, p. 18. Discusses solar flares.

Friedman, H. *Sun and Earth.* 1986, W. H. Freeman. Contains good sections about the Sun's effects on the Earth.

Giovanelli, R. *Secrets of the Sun.* 1984, Cambridge U. Press.

Golub, L. "Heating the Sun's Million-Degree Corona" in *Astronomy,* May 1993, p. 27.

Golub, L. "Solar Magnetism: A New Look" in *Astronomy,* Mar. 1981, p. 66.

Jaroff, L. "Fury on the Sun" in *Time,* July 3, 1989, p. 46. Nice introduction to solar activity and the Sun's interior.

Kippenhahn, R. *Discovering the Secrets of the Sun.* 1994, J. Wiley. Excellent modern introduction for the beginner.

Nichols, R. "Solar Max: 1980-1989" in *Sky & Telescope,* Dec. 1989, p. 601.

Noyes, R. *The Sun, Our Star.* 1982, Harvard U. Press.

Pasachoff, J. "The Sun: A Star Close-up" in *Mercury,* May/June 1991, p. 66.

Verschuur, G. "The Day the Sun Cut Loose" in *Astronomy,* Aug. 1989, p. 48. Examines a huge flare.

Wentzel, D. *The Restless Sun.* 1989, Smithsonian Institution Press. Well-written summary by a noted astronomer and educator.

The Sun, just rising above the horizon, with the silhouette of the mirror of the McMath–Pierce Solar Telescope at Kitt Peak National Observatory. This mirror tracks the Sun across the sky and directs sunlight downward into the rest of the telescope's optics. (National Optical Astronomy Observatories)

CHAPTER 15

The Sun: A Nuclear Powerhouse

Thinking Ahead

Why does the Sun shine and keep shining? It is a prodigious generator of energy, and has been so ever since first watched by human eyes. But we believe it has been shining much longer than that. How can we get a sense of just how long it has been producing energy, and thus begin to understand the mechanism that powers our star?

We do not argue with the critic who urges that the stars are not hot enough for this process [nuclear fusion]; we tell him to go and find a hotter place . . .

Arthur Eddington in *Internal Constitution of the Stars* (1926)

How much energy does the Sun radiate into space? With modern instruments, we have measured its output with great precision and found that our star puts out about 4×10^{26} watts (W). That's a very large figure; what does it mean in human terms?

The current population of our planet is about 6 billion (6×10^9) people. Suppose for a moment that all of us simultaneously turned on a thousand 100-W light bulbs. Each person on Earth would then be lit up like a Hollywood movie theater on opening night! But all those bulbs still only total 6×10^{14} W. To use as much energy as the Sun produces, we would have to find 670 billion worlds like the Earth, all doing the same stunt. In other words, the Sun is a big, big cosmic light bulb.

In considering the Sun's energy mechanism, it is not enough to determine how much energy it uses in a second. You might think of all sorts of ways to generate a huge amount of energy if you only had to do it for a second, or even for a minute. But if you had to sustain a tremendous output of energy for a year or a billion years, then you would need a steady, reliable generator.

So to understand what powers the Sun, we first have to ask how long the Sun has been shining. This is not an easy question to answer: there is nothing "built into" the structure or appearance of the Sun that allows us to pinpoint its age directly. But as we saw in Chapter 13, we have several lines of evidence indicating that the Sun formed at roughly the same time as the planetary system—about 4.5 billion years ago.

Our task, then, is to explain not only how the Sun generates so many watts now, but also what mechanism has allowed it to do so for billions of years. When geologists and biologists began to uncover clues about the great age of the Earth in the late 19th and early 20th centuries, this appeared to be an insoluble problem.

Scientists first looked at sources already familiar to them here on Earth. This seemed a reasonable strategy, since the Sun and the Earth contain the same types of atoms, albeit in different proportions. But, as we will see, none of the energy sources known at the time could explain the Sun's longevity. It was only after scientists discovered how to tap the energy stored in the nuclei of atoms that they finally identified the source of the Sun's energy.

Figure 15.1
Wood fire. (Visuals Unlimited/Doug Sokell)

consisted of a burnable material like coal or wood, our star could not produce energy at its present rate for more than a few thousand years. That's not enough time for Earth to evolve even a virus, to say nothing of a life-form as complicated as an astronomy student! Besides, geologists have found fossils in rocks that are 3.5 billion years old, so the temperature of the Earth (and the heat output of the Sun) must have been suitable to sustain life as long ago as that.

In addition, results from 20th-century spectroscopy make the coal and wood suggestions completely untenable. At the temperatures found in the Sun, nothing like solid wood or coal could survive. And the dominant elements in the Sun, as we saw in the preceding chapter, are hydrogen and helium.

ASTRONOMY BASICS

What's Watt?

Just a word about the units we are using. A watt (W) is a unit of power—that is, of energy used or given off *per unit time*. You know from everyday experience that it's not just how much energy you expend, but how long you do it. Burning 10 Calories (Cal) in 10 min takes a very different kind of exercise than burning 10 Cal in an hour. Watts tell you the *rate* at which energy is being used; for example, a 100-W bulb uses 100 joules (j) of energy every second.

And how big is a joule? A 160-lb (73-kg) astronomy instructor running at about 10 mi/h (4.4 m/s) from a football player who just flunked the midterm has a motion energy of about 700 J.

15.1

Thermal and Gravitational Energy

Nineteenth-century scientists knew of two possible sources for the Sun's energy: heat, or thermal energy, and gravitational energy. The source of heat energy most familiar to us here on Earth is the burning (the chemical term is *oxidation*) of wood, coal, gasoline, or other fuel (Figure 15.1). However, we know exactly how much energy the burning of these materials can produce, and can thus calculate that even if the immense mass of the Sun

Conservation of Energy

Other 19th-century attempts to determine what makes the Sun shine used the law of conservation of energy. Simply stated, this law says that energy cannot be created or destroyed, but can be transformed from one type to another—say from heat energy to mechanical energy. The steam engine, which was the key to the industrial revolution, is a good example. In it, the hot steam from a boiler drives the movement of a piston, converting heat energy to motion energy.

Motion can also be transformed into heat. If you clap your hands vigorously at the end of an especially good astronomy lecture, your palms become hotter. If you rub ice on the surface of a table, the heat produced by friction melts the ice.

In the 19th century, scientists considered the possibility that the source of the Sun's heat might be the mechanical motion of meteorites falling into it. Calculations show, however, that in order to produce the total amount of energy emitted by the Sun, the mass in meteorites that

Figure 15.2
British physicist William Thomson (Lord Kelvin) and German scientist Hermann von Helmholtz proposed that the contraction of the Sun under its own gravity might account for its energy. (Smithsonian Institution, courtesy AIP Emilio Segre Visual Archives)

would have to fall into it every 100 years would equal the mass of the Earth. The resulting increase in the Sun's mass would, according to Kepler's third law, change the period of the Earth's orbit by 2 s per year. Such a change would be easily measurable and is not, in fact, occurring. This source of the Sun's energy was then ruled out.

Gravitational Contraction as a Source of Energy

As an alternative, the German scientist Hermann von Helmholtz and the British physicist Lord Kelvin (Figure 15.2), in about the middle of the 19th century, proposed that the Sun might produce energy by the conversion of gravitational energy to heat. They suggested that the outer layers of the Sun might be "falling" inward because of the force of gravity. In other words, they proposed that the Sun could be shrinking in size, and staying hot and bright as a result.

To imagine what would happen, picture the outer layer of the Sun starting to fall inward. This outer layer is a gas made up of individual atoms, all moving about in random directions. Temperature, which measures the amount of stored heat energy, depends on the speed of the atoms: higher speeds equal higher temperatures. If this layer starts to fall, the atoms acquire an additional velocity because of the falling motion. As the outer layer falls inward, it also contracts, moving the atoms closer together. Collisions become more likely, and some of them transfer the velocity associated with the falling motion to other atoms, increasing their velocities and hence increasing the temperature of this layer of the Sun. Other collisions excite electrons within the atoms to higher energy

orbits. When these electrons return to their normal orbits, they emit photons, which can then escape from the Sun.

Kelvin and Helmholtz calculated that a contraction of the Sun at a rate of only about 40 m per year would be enough to produce the amount of energy that it is now radiating. Over the span of human history, the decrease in the Sun's size from such a slow contraction would be undetectable. If we assume that the Sun began its life as a large, diffuse cloud of gas, then we can calculate how much energy has been radiated by the Sun during its entire lifetime as it has contracted from a very large diameter down to its present size. That amount of energy is on the order of 10^{42} J. Since the solar luminosity is 4×10^{26} W (J/s) or about 10^{34} J per year, contraction could keep the Sun shining at its present rate for roughly 100 million years.

In the 19th century, 100 million years at first seemed plenty long enough, since the Earth was then widely thought to be much younger than this. But toward the end of that century and into the 20th, geologists and physicists showed that the Earth (and hence the Sun) is actually much older. Contraction therefore cannot be the primary source of solar energy.

In the same way, every other process suggested to explain the Sun's energy failed to account for the known facts about the Sun, including its age. Scientists were thus confronted with a puzzle of enormous proportions. Either an unknown type of energy was responsible for the most important energy source known to humanity, or estimates of the span of time that the solar system (and life on Earth) had been around had to be seriously modified. Charles Darwin, whose theory of evolution required a longer time span than the theories of the Sun seemed to offer, was discouraged by these results and continued to worry about them until his death in 1882.

It was only in the 20th century that the true source of the Sun's energy was identified. The two key pieces of information required to solve the puzzle were the structure of the nucleus of the atom, and the fact that mass can be converted into energy.

Mass, Energy, and the Special Theory of Relativity

As we have seen, energy cannot be created or destroyed, but only converted from one form to another. One of the remarkable conclusions derived by Albert Einstein when he developed the theory of relativity is that mass can be considered another form of energy and can therefore be converted to energy. This remarkable equivalence is expressed in one of the most famous equations in all of science:

$$E = mc^2$$

where E is the symbol for energy, m is the symbol for mass, and c, the constant that relates the two, is the speed of light. Note that this equation is very similar in form to

$$\text{inches} = \text{feet} \times 12 \qquad \text{or} \qquad \text{cents} = \text{dollars} \times 100$$

That is, it is a conversion formula that allows you to calculate the conversion of one thing, mass, to another, energy. The conversion factor in our case turns out to be not 12 or 100, but another constant quantity, the speed of light squared. By the way, mass does not have to *travel* at the speed of light (or the speed of light squared, which is impossible in nature) for this conversion to occur. The factor of c^2 is just the number that must be used to relate mass and energy.

Notice that this formula does not tell you *how* to convert mass into energy, just as the formula for cents does not tell you where to exchange coins for a dollar bill. The formulas merely tell you what the equivalent values are if you succeed in making the conversion. When Einstein first derived his famous formula in 1905, no one had the faintest idea how to convert mass into energy in any practical way. Einstein himself tried to discourage speculation that the conversion of atomic mass into energy would be feasible in the near future. Today, as a result of developments in nuclear physics, we convert mass into energy in power plants, nuclear weapons, and—in high-energy physics—experiments in particle accelerators.

Because c^2, the speed of light squared, is a very large quantity, the conversion of even a small amount of mass results in a very great amount of energy. For example, the complete conversion of 1 g of matter (about 1/14 oz) would produce as much energy as the combustion of 15,000 barrels of oil.

Scientists soon realized that the conversion of mass to energy is the source of the Sun's heat and light. With Einstein's equation $E = mc^2$, we can calculate that the amount of energy radiated by the Sun could be produced by the complete conversion of about 4 million tons of matter to energy inside the Sun each second. Four million tons per second sounds like a lot when compared to Earthly things, but bear in mind that the Sun is a very big reservoir of matter. In fact, we will see that the Sun contains more than enough mass to continue shining at its present rate for billions of years.

But this *still* does not tell us how mass can be converted to energy. To understand how the conversion actually occurs in the Sun, we need to explore the structure of the atom a bit further.

Elementary Particles

The fundamental components of matter are called **elementary particles.** The most familiar of these are the proton, neutron, and electron—the particles that make up ordinary atoms (see Section 4.4).

Protons, neutrons, and electrons are by no means all the particles that exist. First, for each kind of particle, there is a corresponding but opposite *antiparticle.* If the particle carries a charge, its antiparticle has the opposite charge. The antielectron is the *positron,* which has the same mass as the electron but is positively charged. Likewise, the antiproton has a negative charge. The remarkable thing about such antimatter is that when a particle comes in contact with its antiparticle, the two annihilate each other, turning into pure energy.

Since our world is made exclusively of ordinary particles of matter, any antimatter cannot survive for very long. But individual antiparticles are found in cosmic rays (particles that arrive at the top of the Earth's atmosphere from space) and can be formed in particle accelerators. In fact, when we create matter from energy in our high-energy physics labs, we always get half matter and half antimatter.

Science fiction fans may be familiar with antimatter from the *Star Trek* television series and films. The Starship *Enterprise* is propelled by the *careful* combining of matter and antimatter in the ship's engine room. According to $E = mc^2$, the complete annihilation of matter and antimatter can produce a huge amount of energy; but keeping the antimatter fuel from touching the ship before it is needed must be a big problem. No wonder that Scotty, the chief engineer in the original TV show, always looked worried!

The existence of another type of particle was originally suggested in 1933 by physicist Wolfgang Pauli. Energy seemed not to be conserved in certain types of nuclear reactions, in violation of the law of conservation of energy. Rather than accept an overthrow of one of the fundamental ideas in science, Pauli suggested that a new and so far undetected particle, named the **neutrino,** was carrying away the "missing" energy. He suggested that neutrinos were particles with zero mass that, like photons, moved with the speed of light.

Albert Einstein

For a large part of his life, Albert Einstein was one of the most recognized celebrities of his day. Strangers stopped him on the street, and people all over the world wrote asking him for endorsements, advice, and assistance. In fact, when Einstein and the great film star Charlie Chaplin met in California, they found they shared similar feelings about the loss of privacy that comes with fame. Einstein's name was a household word despite the fact that most people did not understand the ideas that had made him famous.

Einstein was born in 1879 in Ulm, Germany. Legend has it that he did not do well in school (even in arithmetic), and thousands of students have since justified a bad grade by referring to this story. Alas, like many legends, this one is not true. Records indicate that although he tended to rebel against the authoritarian teaching style in vogue in Germany at that time, Einstein was a good student.

After graduating from the Federal Polytechnic Institute in Zurich, Switzerland, Einstein at first had trouble getting a job (even as a high-school teacher), but he eventually became an examiner in the Swiss Patent Office. Working in his spare time, without the benefit of a university environment but using his superb physical intuition, he wrote four papers in 1905 that would ultimately transform the way physicists looked at the world.

One of these, which earned him the Nobel Prize in 1921, set part of the foundation of *quantum mechanics*, the rich, puzzling, and remarkable theory of the subatomic

Albert Einstein in 1905. (Permission granted by the Albert Einstein Archives, The Hebrew University of Jerusalem, Israel)

realm. But his most important paper presented the *special theory of relativity*, a re-examination of space, time, and motion that added a whole new level of sophistication to our understanding of these concepts. $E = mc^2$ was a relatively minor part of this theory, added in a later paper.

In 1916 Einstein published his *general theory of relativity*, which was, among other things, a fundamentally new description of gravity (see Chapter 23). When this theory was confirmed by measurements of the "bending of starlight" during a 1919 eclipse (the *New York Times* headline read, "Lights All Askew in the Heavens"), Einstein became world-famous.

In 1933, to escape Nazi persecution, Einstein left his professorship in Berlin and settled in the United States at the newly created Institute for Advanced Studies at Princeton. He remained there until his death in 1955, writing, lecturing, and espousing a variety of intellectual and political causes. For example, he agreed to sign a letter written by Leo Szilard and other scientists in 1939, alerting President Roosevelt to the dangers of allowing Nazi Germany to develop the atomic bomb first. In 1952 Einstein was offered the second presidency of Israel. In declining the position, he said, "I know a little about nature and hardly anything about men."

When Pauli proposed this idea, neutrinos had not yet been detected because they interact very weakly with other matter. Most of them can pass completely through a star or planet without being absorbed. The Earth is more transparent to a neutrino than the thinnest and cleanest pane of glass is to a photon of light. As we will see, this "antisocial" behavior of neutrinos makes them both frustrating and very important for scientists studying the Sun and other stars.

The elusive neutrino was finally detected in 1956, and so we know that this particle really does exist. Experiments are not, however, sufficiently precise to prove that neutrinos have exactly zero mass. If neutrinos turn out to have even a tiny mass, this fact could have interesting consequences for cosmology (see Chapters 27 and 28) and for models of the Sun's interior (see Section 15.4).

Some of the properties of the proton, electron, neutron, and neutrino are summarized in Table 15.1. (Other

TABLE 15.1
Properties of Some Elementary Particles

Particle	Mass (kg)	Charge
Proton	1.67265×10^{-27}	+ 1
Neutron	1.67495×10^{-27}	0
Electron	9.11×10^{-31}	− 1
Neutrino	0	0

subatomic particles have been produced by experiments with particle accelerators, but they do not play a role in the generation of solar energy.)

The Atomic Nucleus

The nucleus of an atom is not just a loose collection of elementary particles. Inside the nucleus, particles are held together by a very powerful force called the *strong nuclear force.* This is a short-range force, only able to act over distances about the size of the atomic nucleus. A quick thought experiment shows how important this force is. Take a look at your little finger and think of the atoms composing it. Among them is carbon, one of the basic elements of life. Focus your imagination on the nucleus of one of your carbon atoms. It contains six protons, which have a positive charge, and six neutrons, which are neutral. Thus the nucleus has a net charge of six positives; if only the electrical force were acting, the protons in this and every carbon atom would find each other very repulsive and fly apart.

The strong nuclear force is an attractive force, stronger than the electromagnetic force, and it keeps the particles of the nucleus tightly bound together. We saw earlier that if under the force of gravity a star "shrinks"—bringing its atoms closer together—gravitational energy is released. In the same way, if particles come together under the strong nuclear force, and unite to form an atomic nucleus, some of the nuclear energy is released. The energy given up in such a process is called the binding energy of the nucleus.

When such binding energy is released, the resulting nucleus has slightly less mass than the sum of the masses of the particles that came together to form it. In other words, the energy comes from the loss of mass. This slight deficiency in mass is always only a small fraction of the mass of one proton. But because each bit of lost mass can provide quite a bit of energy (remember $E = mc^2$), this nuclear energy release can be quite a potent mechanism.

The behavior of the nuclear force is more complicated than gravity, but during the 20th century, physicists have been able to investigate and establish the properties of atomic nuclei. It turns out that the binding energy is greatest for atoms with a mass near that of the iron nucleus (with a combined number of protons and neutrons equal to 56), and less for both the lighter and the heavier atoms. Iron, therefore, is the most stable element (since it gives up the most energy when it forms, it is the hardest nucleus to "undo").

What this means is that, in general, when light atomic nuclei come together to form a heavier one (up to iron), mass is lost and energy released. This joining together of atomic nuclei is called nuclear **fusion.**

Energy can also be produced by breaking up heavy atomic nuclei into lighter ones (down to iron); this process is called nuclear *fission* (see Figure 15.3). Nuclear fission was the process we learned to use first—in atomic bombs, and in nuclear reactors used to generate electric power—and it may thus be more familiar to you. It also sometimes occurs spontaneously, in natural radioactivity. But fission requires big, complex nuclei, whereas we know that the stars are made up predominantly of small, simple nuclei. So we must look to fusion to explain the Sun and the stars.

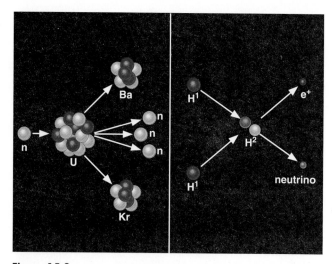

Figure 15.3
In fission, a larger nucleus breaks into two smaller components. Here a nucleus of uranium, with 92 protons and 143 neutrons, is shown undergoing fission into two smaller nuclei of barium (56 protons) and krypton (36 protons). In fusion, smaller nuclei bind together to make a larger one. We also see two nuclei of hydrogen (one proton each) fuse into a heavier hydrogen nucleus (one proton and one neutron), accompanied by the emission of a positron and a neutrino.

Nuclear Attraction Versus Electrical Repulsion

So far, this sounds like a very attractive prescription for making energy: roll some nuclei together and join them via nuclear fusion. This will cause them to lose some of their mass, which then turns into energy. However, every nucleus, even simple hydrogen, has protons in it—and protons all have positive charges. Since like charges repel via the electrical force, the closer we get two nuclei to one another, the more they repel. It's true that if we can get them within "striking distance" of the nuclear force, they will then come together with a much stronger attraction. But that striking distance is very tiny, about the size of a nucleus. How can we get nuclei close enough to participate in fusion?

The answer turns out to be heat—tremendous heat—which speeds the protons up enough to overcome the electrical forces that try to keep protons apart. Inside the Sun, as we saw, the most common element is hydrogen, whose nucleus is a single proton. Two protons can fuse only in regions where the temperature is greater than about 10 million K, and the speed of the protons averages around 1000 km/s or more. (In old-fashioned units, that's over 2 million mi/h!) Such extreme temperatures are reached only in the regions surrounding the center of the Sun, which has a temperature of 15 million K. Calculations show that nearly all of the Sun's energy is generated within about 150,000 km of its core, or within less than 10 percent of its total volume.

Even at these high temperatures, it is exceedingly difficult to force two protons to combine. On average, a proton will rebound from other protons in the Sun's crowded core for about 14 billion years, at the rate of 100 million collisions per second, before it fuses with a second proton. This is, however, only the average waiting time. Some of the enormous number of protons in the Sun's inner region are lucky and take only a few collisions to achieve a fusion reaction: they are the protons responsible for producing the energy radiated by the Sun. Since the Sun is about 4.5 billion years old, most of its protons have not yet been involved in fusion reactions.

Nuclear Reactions in the Sun's Interior

The Sun, then, taps the energy contained in the nuclei of atoms through nuclear fusion. Let's look at what happens in more detail. Deep inside the Sun, four hydrogen atoms fuse to form a helium atom. The helium atom is slightly less massive than the four hydrogen atoms that combine to form it, and that lost mass is converted to energy.

The initial steps required to form one helium nucleus from four hydrogen nuclei are shown in Figure 15.4. First, two protons combine to make a *deuterium* nucleus, which is an isotope (or version) of hydrogen that contains one proton and one neutron. In effect, one of the original protons has been converted to a neutron in the fusion reaction. Electric charge has to be conserved in nuclear reactions, and it is conserved in this one. The positive charge originally associated with one of the protons is carried away by a positron.

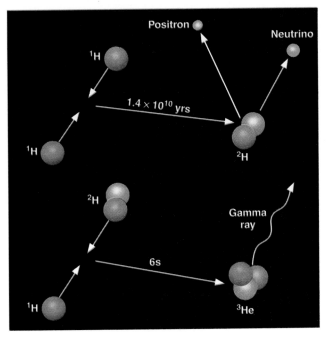

Figure 15.4
The first two steps in the process of fusing hydrogen into helium in the Sun (the p–p chain).

Since it is antimatter, this positron will instantly collide with an electron, and both will be annihilated, producing pure electromagnetic energy in the form of gamma rays. Now what happens to this gamma ray, which has been created in the center of the Sun? It finds itself in a world crammed full of fast-moving nuclei and electrons. The photon collides with particles of matter and transfers some of its energy to them. The result of this process is generally to lower the energy of the gamma-ray photon.

Such interactions happen to the gamma ray again and again and again, until—as it makes its way slowly toward the outer layers of the Sun—its energy becomes so reduced that it is no longer a gamma ray, but an x ray (recall Section 4.2). And later, as it loses more energy, it becomes an ultraviolet photon. The Sun is so full of particles (targets for the photon to hit) that it takes on the order of a million years for the average photon to emerge from the Sun's photosphere. By that time, most of the photons have given up enough energy to be ordinary light—and they are the sunlight we see coming from our star. (To be precise, each gamma-ray photon is ultimately converted into many separate lower-energy photons of sunlight.)

In addition to the positron, the fusion of two hydrogen atoms to form deuterium results in the emission of a neutrino. Because neutrinos interact so little with ordinary matter, those produced by fusion reactions near the center of the Sun travel directly to the Sun's surface and then on toward the Earth. Neutrinos move at the speed of light and they get out of the Sun only about 2 s after they are created (see Figure 15.9).

The next step in forming helium from hydrogen is to add a proton to the deuterium nucleus and create a helium nucleus that contains two protons and one neutron. In the process, some mass is again lost, and more gamma radiation is emitted. Such a nucleus is helium because an element is defined by the number of its protons; any nucleus with two protons is called helium. But this form of helium, which we call helium 3, is not the isotope we see in the Sun's atmosphere or on Earth. That helium has two neutrons and two protons and hence is called helium 4.

To produce helium 4, helium 3 combines with another just like it in the third step of fusion (illustrated in Figure 15.5). Note that two protons are left over from this step; they come out of the reaction ready to collide with other protons and thus to start step 1 all over again.

The P–P Chain

The reactions in the Sun can be described succinctly through the following nuclear formulas:

$$^1\text{H} + {}^1\text{H} \longrightarrow {}^2\text{H} + \text{e}^+ + \nu$$

$$^2\text{H} + {}^1\text{H} \longrightarrow {}^3\text{He} + \gamma$$

$$^3\text{He} + {}^3\text{He} \longrightarrow {}^4\text{He} + 2\,{}^1\text{H}$$

where the superscripts indicate the total number of neu-

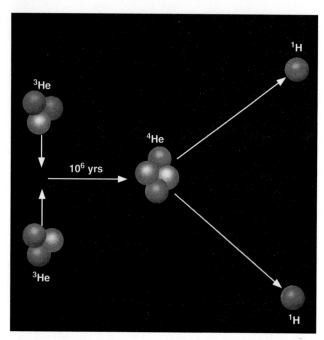

Figure 15.5
The third step in the fusion of hydrogen into helium in the Sun. Note that two of the products of the second step (see Figure 15.4) must combine before the third step becomes possible.

trons and protons in the nucleus, e^+ is the positron, ν is the neutrino, and γ indicates that gamma rays are emitted. Note that steps 1 and 2 must happen twice before step 3 can occur.

Although, as we discussed, the first step in this chain of reactions is very difficult and generally takes a long time, the other steps happen more quickly. After the deuterium nucleus is formed, it survives an average of only about 6 s before being converted to ^{3}He. About a million years after that, the ^{3}He nucleus will combine with another to form ^{4}He.

We can compute the amount of energy these reactions generate by calculating the difference in initial and final mass. The masses of hydrogen and helium atoms in the units normally used by scientists are 1.007825 u and 4.00268 u, respectively. (The unit of mass, u, is defined to be 1/12 the mass of an atom of carbon, or approximately the mass of a proton.) Here we include the mass of the entire atom, not just the nucleus, because electrons are involved as well. When hydrogen is converted to helium, two positrons are created (remember, step 1 happens twice), and these are annihilated with two free electrons, adding to the energy produced.

$$4 \times 1.007825 = 4.03130 \; u \; \text{(mass of initial hydrogen}$$
$$\text{atoms)}$$
$$- \; 4.00268 \; u \; \text{(mass of final helium atom)}$$
$$\overline{0.02862 \; u} \; \text{(mass lost in the}$$
$$\text{transformation)}$$

The mass lost, 0.02862 u, is 0.71 percent of the mass of the initial hydrogen. Thus if 1 kg of hydrogen is converted

into helium, the mass of the helium is only 0.9929 kg, and 0.0071 kg of material is converted into energy. The velocity of light is 3×10^8 m/s, so the energy released by the conversion of just 1 kg of hydrogen to helium is

$$E = 0.0071 \times (3 \times 10^8)^2$$
$$= 6.4 \times 10^{14} \; J$$

This amount, the energy released when a single kilogram of hydrogen undergoes fusion, is more than ten times the Earth's annual consumption of electricity and fossil fuels.

To produce the Sun's luminosity of 4×10^{26} W, some 600 million tons of hydrogen must be converted to helium *each second*, of which about 4 million tons turn from matter into energy. As large as these numbers are, the store of hydrogen (and thus of nuclear energy) in the Sun is still more enormous and can last a *long* time.

At the temperatures inside the Sun and in less-massive stars, most of the energy is produced by the reactions we have just described, and this set of reactions is called the **proton–proton cycle** (or sometimes, the p–p chain). It is called a cycle because the two protons produced in step 3 can fuse with other protons to initiate step 1 again. In the proton–proton cycle, protons collide directly with other protons to build into helium nuclei.

In hotter stars, another set of reactions, called the *carbon–nitrogen–oxygen (CNO) cycle*, accomplishes the same net result. In the CNO cycle, carbon and hydrogen nuclei collide to initiate a series of reactions that form nitrogen, oxygen, and ultimately helium. The nitrogen and oxygen nuclei do not survive but interact to form carbon again. Therefore, the outcome is the same as in the proton–proton cycle: four hydrogen atoms disappear, and in their place a single helium atom is created. The CNO cycle plays only a minor role in the Sun but is the main source of energy at temperatures above 15×10^6 K.

Thus we have solved the puzzle that so worried scientists at the end of the 19th century. The Sun can maintain its high temperature and energy output for billions of years through the fusion of the simplest element in the universe, hydrogen. Because most of the Sun (and, as we will see, the other stars) is made of hydrogen, it is an ideal "fuel" for powering a star. As will be discussed in the coming chapters, we can define a star as a ball of gas capable of getting its core hot enough to initiate the fusion of hydrogen. There are balls of gas that lack the mass required to do this (Jupiter is a local example); like so many hopefuls in Hollywood, they will never ever be stars.

15.3

The Interior of the Sun: Theory

Fusion of protons will occur in the center of the Sun only if the temperature exceeds 10 million K. How do we know whether the Sun is actually this hot? To determine what the interior of the Sun is like, it is necessary to resort to mathematical calculations. In effect, astronomers teach a computer everything they know about the physical processes going on in the Sun's interior. The computer then calculates the temperature and pressure at every point inside the Sun and determines what nuclear reactions, if any, are taking place. The computer can also calculate how the Sun will change with time.

After all, the Sun *must* change. In its center, the Sun is slowly depleting its supply of hydrogen and creating helium instead. Will this composition change have measurable effects? Will the Sun get hotter? Cooler? Larger? Smaller? Brighter? Fainter? Ultimately, the changes in the center could be catastrophic, since eventually all the hydrogen fuel hot enough for fusion will be exhausted. Either a new source of energy must be found, or the Sun will cease to shine. We will describe the fate of the Sun in Chapters 21 and 22. For now, let's look at some of the things we must teach the computer about the Sun in order to carry out the calculations.

The Sun Is a Gas

The Sun is so hot that the material in it is gaseous throughout. Astronomers are grateful for this because a hot gas is easier to describe mathematically than are other configurations of matter. The particles that constitute a gas are in rapid motion, frequently colliding with one another. This constant bombardment is the *pressure* of the gas (Figure 15.6). More particles within a given volume of gas produce more pressure, because the combined impact of the moving particles increases with their number. The pressure is also greater when the molecules or atoms are moving faster. Since their rate of motion is determined by the temperature of the gas, higher temperatures produce higher pressure.

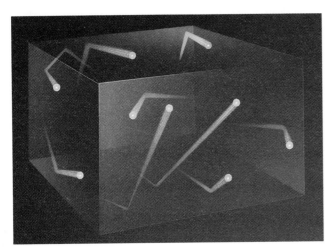

Figure 15.6
Gas pressure. The particles in a gas are in rapid motion and produce pressure through collisions with the surrounding material. Here particles are shown bombarding the sides of an imaginary container.

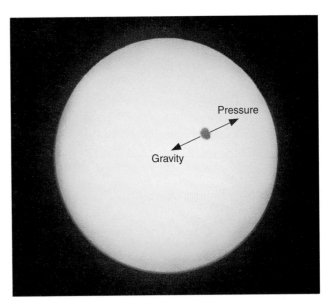

Figure 15.7
Hydrostatic equilibrium. In the interior of a star, the inward force of gravity is exactly balanced at each point by the outward force of gas pressure.

The Sun Is Stable

The Sun, like the majority of other stars, is stable: it is neither expanding nor contracting. Such a star is said to be in a condition of equilibrium. All the forces within it are balanced, so that at each point within the star the temperature, pressure, density, and so on, are maintained at constant values. We will see in Chapters 21 and 22 that even these stable stars, including the Sun, are changing as they evolve, but such evolutionary changes are so gradual that to all intents and purposes the stars are still in a state of equilibrium.

The mutual gravitational attraction between the masses of various regions within the Sun produces tremendous forces that tend to collapse the Sun toward its center. Yet we know from the history of the Earth that the Sun has been emitting approximately the same amount of energy for billions of years, and so clearly has managed to resist collapse for a very long time. The gravitational forces must therefore be counterbalanced by some other force, and that force is the pressure of the gases within the Sun (Figure 15.7). To exert enough pressure to prevent the Sun from collapsing due to the force of gravity, the gases at its center must be maintained at a temperature of 15 million K. So we can conclude that the Sun's temperature is high enough to fuse protons from the fact that it is not contracting.

If the internal pressure in a star were not great enough to balance the weight of its outer parts, the star would collapse somewhat, contracting and building up the pressure inside. If the pressure were greater than the weight of the overlying layers, the star would expand, thus decreasing the internal pressure. Expansion would stop, and equilibrium would be reached, when the pressure at every internal point again equaled the weight of the stellar layers above that point. An analogy is an inflated balloon, which will expand or contract until an equilibrium is reached between the pressure of the air inside and of that outside. This condition is called **hydrostatic equilibrium.** Stable stars are all in hydrostatic equilibrium; so are the oceans of the Earth, as well as the Earth's atmosphere. The air's own pressure keeps it from falling to the ground.

The Sun Is Not Cooling Down

Heat always flows from hotter to cooler regions. Therefore, as energy filters outward toward the surface of a star, it must be flowing from inner, hotter regions. The temperature cannot ordinarily get cooler as we go inward in a star, or energy would flow in and heat up those regions until they were at least as hot as the outer ones. We conclude that the temperature is highest at the center of a star, dropping to successively lower values toward the stellar surface. (The high temperature of the Sun's chromosphere and corona may therefore appear to be a paradox. But remember from Chapter 14 that these high temperatures are believed to be maintained by magnetic heating.)

The outward flow of energy through a star, however, robs it of its internal heat, and would result in a cooling of the interior gases if that energy were not replaced. Similarly, a hot iron begins to cool as soon as it is unplugged from its source of electric energy. Therefore a source of energy must exist within each star. In the Sun's case, we have seen that this energy source is the fusion of hydrogen to form helium.

Heat Transfer in a Star

Since the nuclear reactions that generate the Sun's energy occur deep within it, there must be ways to transport heat from the center of the Sun to its surface. There are three ways in which heat can be transported. In *conduction,* atoms or molecules pass on their energy by colliding with others nearby; this happens when the handle of a metal spoon heats up as you stir a hot cup of coffee. In *convection,* currents of warm material rise, carrying their energy with them to cooler layers; this happens when hot air from a fireplace rises in a room. In *radiation,* photons of energy move away from hot material and are absorbed by some other material to which they convey some or all of their energy; you can feel this when you put your hand close to the coils of an electric heater, allowing infrared photons to heat up your hand. Conduction and convection are both important in the interiors of planets. In stars, which are much more transparent, radiation and convection are important, while conduction can usually be ignored.

Stellar **convection** occurs as currents of hot gas flow up and down through the star (Figure 15.8). Such currents travel at moderate speeds and do not upset the con-

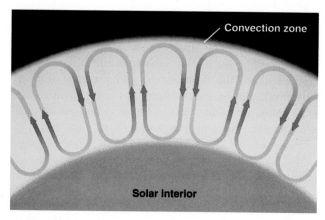

Figure 15.8
Rising convection currents carry heat from the Sun's interior to its surface, while cooler material sinks downward. Of course, nothing in a real star is as simple as diagrams in textbooks suggest.

dition of hydrostatic equilibrium. Nor do they result in a net transfer of mass either inward or outward. Nevertheless, they carry heat very efficiently outward through a star. In the Sun, convection turns out to be important in the central regions and near the surface.

Unless convection occurs, the only significant mode of energy transport through a star is by electromagnetic **radiation.** Radiation is not an efficient means of energy transport in stars, because gases in their interiors are very opaque—that is, a photon does not go far (in the Sun, typically about 0.01 m) before it is absorbed. The absorbed energy is always re-emitted, but it can be re-emitted in any direction. A photon absorbed when traveling outward in a star has almost as good a chance of being reradiated back toward the center of the star as toward its surface.

A particular quantity of energy, therefore, zigzags around in an almost random manner and takes a long time to work its way from the center of the star to the surface (Figure 15.9). As discussed earlier, in the Sun the time required is on the order of a million years. If the photons were not absorbed and re-emitted along the way, they would travel at the speed of light and could reach the surface in a little over 2 s, just as the neutrinos do.

The measure of matter's ability to absorb radiation is called its *opacity.* It should be no surprise that the Sun's gases are opaque. If they were completely transparent, we would be able to see all the way through the Sun. The processes by which atoms and ions can interrupt the flow of energy—such as by becoming ionized—were discussed in Section 4.5. In addition, individual electrons can scatter radiation helter-skelter. For a given temperature, density, and composition of a gas, all of these processes can be taken into account, and the opacity can be calculated. The computations are very complicated and thus require powerful computers, but these are now as much part of the arsenal of astronomers as are powerful telescopes.

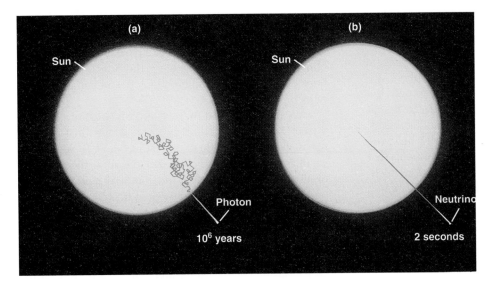

Figure 15.9
(a) Photons generated by fusion reactions in the solar interior travel only a short distance before they are absorbed. The re-emitted photons usually have lower energy and may travel in any direction. As a consequence, it takes about a million years for energy to make its way from the center of the Sun to its surface. (b) In contrast, neutrinos do not interact with matter but traverse the Sun at the speed of light, reaching the surface in only a little more than 2 s.

Model Stars

Scientists use the principles we have described to calculate what the Sun's interior is like. These physical ideas are expressed as mathematical equations that are solved to determine the values of temperature, pressure, density, and other physical quantities throughout the stellar interior. The set of solutions so obtained, based on a specific set of physical assumptions, is called a theoretical model for the interior of the Sun.

Figure 15.10 schematically illustrates the Sun's interior according to the best theoretical model. Energy is generated through fusion in the core of the Sun, which extends only about one-quarter of the way to the surface. This core contains about one-third of the total mass of the Sun, however. At the center, the temperature reaches a maximum of approximately 15 million K, and the density is nearly 150 times the density of water. The energy generated is transported toward the surface by radiation until it reaches a point about 70 percent of the distance from the center to the surface. At this point convection begins, and energy is transported the rest of the way primarily by rising columns of hot gas.

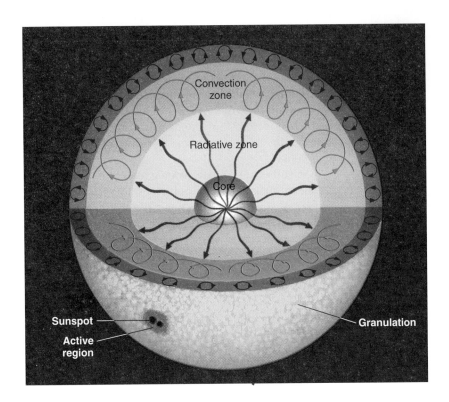

Figure 15.10
The interior structure of the Sun. Energy is generated in the core by the fusion of hydrogen to form helium. This energy is transmitted outward by radiation—that is, by the absorption and re-emission of photons. In the outermost layers, energy is transported mainly by convection.

15.4

The Interior of the Sun: Observations

Recall that when we observe the Sun's photosphere, we are not seeing very deeply into our star—certainly not into the regions where energy is generated. At first it seemed there was no way to check on the predictions of our models for the inside of the Sun. Recently, however, astronomers have devised two types of measurements that can be used to study the solar interior directly. One technique involves the analysis of tiny changes in the motion of small regions at the Sun's surface. The other relies on the measurement of the neutrinos emitted by the Sun.

Solar Pulsations

Astronomers have discovered that the Sun pulsates—that is, it alternately expands and contracts—just as your chest expands and contracts as you breathe. This pulsation is very slight, but it can be detected by measuring the *radial velocity* of the solar surface—the speed with which it moves toward or away from us. The velocity of different parts of the Sun is observed to change in a regular way, first toward the Earth, then away, then toward, and so on. Accurate measurements show that various regions on the Sun's surface with diameters ranging from 4000 to 15,000

km fluctuate back and forth this way. The pulsation cycle takes between 2.5 and 11 min, with the dominant periods lasting about 5 min (Figure 15.11). Rather than resembling a single person breathing, the Sun is more like a huge crowd of people, all breathing in and out at different rates.

We now know that these velocity fluctuations are produced by adding together millions of individual patterns of oscillation. Individual oscillations have velocities as small as 20 cm/s, and the combined sum of the velocities of all the patterns is only a few hundred meters per second. Since it takes only about 5 min to complete a full cycle from maximum to minimum velocity and back again, the change in the size of the Sun measured at any given point is no more than a few kilometers.

The remarkable thing is that these small velocity variations can be used to determine what the interior of the Sun is like. The motion of the Sun's surface is caused by waves that reach it from deep in the interior. Study of the amplitude and period of the velocity changes produced by this motion yields information about the temperature, density, and composition of the layers where the waves were generated. The situation is somewhat analogous to the use of earthquakes to infer the properties of the Earth's interior. For this reason, studies of solar oscillations are referred to as **solar seismology.**

It takes about an hour for waves to traverse the Sun, so they, like neutrinos, provide information about what the solar interior is like at the present time. In contrast, the emerging light that we measure now was generated about a million years ago in the core.

Solar seismology has yielded some important results. Measurements of solar pulsation show that convection extends inward from the surface 30 percent of the way toward the center; we have used this information in drawing Figure 15.10. The observed oscillations also indicate that the abundance of helium inside the Sun, except in the center where nuclear reactions have converted hydrogen to helium, is about the same as at the surface. That result is important to astronomers, because it means we are correct when we use the abundances of the elements measured in the solar atmosphere to construct models of the solar interior.

Scientists are now developing tools to make even better measurements of solar pulsations; these can be used to probe the structure of the Sun close to its center, where nuclear reactions are taking place. To do this, astronomers have set up a network of stations around the Earth to continuously monitor solar oscillations. This is needed because observations from any single station are interrupted by local sunset, but the Sun is always up somewhere in the world. The name given to this worldwide network of telescopes is the Global Oscillations Network Group (GONG) project (Figure 15.12). Observations began in 1995, and the goal is to continue the measurements for another 11 years in order to cover a complete solar cycle.

Figure 15.11

New observational techniques permit astronomers to measure small differences in velocity at the Sun's surface to infer what the deep solar interior is like. In this computer simulation, red shows surface regions that are moving away from the observer; blue marks regions moving toward the observer. Note that the velocity changes penetrate deep into the Sun's interior. (National Optical Astronomy Observatories)

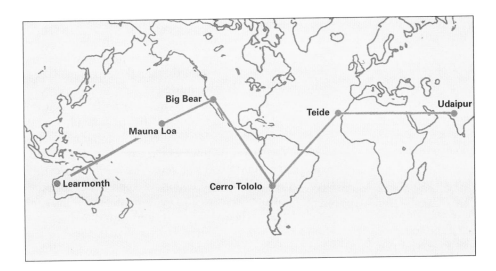

Figure 15.12
A map of the stations for the Global Oscillations Network Group (GONG) around the world. (National Optical Astronomy Observatories)

Solar Neutrinos

The second technique for obtaining information about the Sun's interior involves the detection of a few of those elusive neutrinos created during nuclear fusion. Recall from our earlier discussion that neutrinos very rarely interact with matter, and that most of the neutrinos created in the center of the Sun make their way directly out of the Sun and to the Earth at the speed of light. As far as neutrinos are concerned, the Sun is transparent. Unfortunately for those trying to "catch" some neutrinos, so is the Earth and everything on it.

About 3 percent of the total energy generated by nuclear fusion in the Sun is carried away by neutrinos. So many protons react inside the Sun's core that, scientists calculate, 35 million billion (3.5×10^{16}) solar neutrinos pass through each square meter of the Earth's surface every second. If we can devise a way to detect some of these solar neutrinos, then we can obtain information directly about what is going on in the center of the Sun.

On very, very rare occasions, a neutrino *will* interact with another atom. Several experiments have been devised to detect these interactions. The first of these uses the element chlorine, whose nucleus can be turned into a radioactive argon nucleus (an isotope called argon-37) by an interaction with a neutrino. Because the argon is radioactive, its presence can be picked up with sensitive detectors built around the tank. However, since the interaction happens so rarely, a huge amount of chlorine is needed.

Raymond Davis, Jr., and his colleagues at Brookhaven National Laboratory placed a tank containing nearly 400,000 liters of cleaning fluid (C_2Cl_4) 1.5 km beneath the Earth's surface in a gold mine at Lead, South Dakota (Figure 15.13). A mine was chosen so that the surrounding material of the Earth would keep cosmic rays from reaching the cleaning fluid and creating false signals. Calculations show that solar neutrinos should produce about one atom of argon-37 in the tank each day.

The results of this experiment, running since 1970, are that only about *one-third* as many neutrinos reach the Earth as are predicted by standard models of the solar interior. Astronomers were surprised by this result, since we thought that by now we had a pretty good understanding of both neutrinos and the Sun's interior. This demon-

Figure 15.13
The neutrino experiment operated by Raymond Davis and his colleagues one mile underground in the Homestake Gold Mine in South Dakota. The large vat contains cleaning fluid, a relatively inexpensive source of chlorine nuclei. (Brookhaven National Lab)

strates how important it is not to rest on your laurels in science, but to continue doing new experiments whenever possible.

A second experiment carried out by Japanese astronomers, who looked for the interaction of neutrinos with water, for the first time determined the direction from which the neutrinos are arriving and confirmed that they actually do come from the Sun. That's good news, but, alas, the Japanese also found fewer neutrinos than expected by models of the solar interior.

To change chlorine into argon, you need a neutrino of high energy. Calculations show that fewer than 1 percent of the neutrinos produced in the Sun have energies high enough to be detected by the chlorine, and the water experiment is likewise sensitive only to neutrinos with very high energies. The number of high-energy neutrinos that emerge from the Sun depends on the temperature of its interior. In a cooler core, the reactions that produce them occur less often. Therefore, if we lower the temperatures used in our models of the Sun and change some other details of the complex calculations that go into the models, we might conceivably be able to account for the deficiency of high-energy neutrinos.

Tinkering with the details of the calculations cannot, however, alter the predicted number of low-energy neutrinos. Nearly all of the solar energy, as well as nearly all of the solar neutrinos, are produced in the first step of the proton–proton cycle, when two protons combine to form deuterium (see Section 15.2). Since we know how much total energy is emitted by the Sun, we know very accurately how many times per second two protons combine to form deuterium. We therefore also know how many times per second the associated low-energy neutrinos are produced by this interaction. For a definitive test, scientists need experiments to look for the low-energy neutrinos associated with the proton–proton reaction.

It turns out that these low-energy neutrinos interact with the rare metallic element gallium. In the early 1990s the first results of two gallium experiments, one in Russia and the other in Italy, were reported. Our models predict that solar neutrinos, passing through a container of 60 tons of gallium, will convert about 16 atoms of normal gallium to radioactive gallium during approximately one month of operation. The results of the two gallium experiments agree: the number of low-energy neutrinos is about two-thirds the value predicted by standard solar models.

Is there something wrong with our solar models? Or, as many scientists are beginning to think, is it more likely that we are wrong about the properties of neutrinos? Physicists have shown that three types of neutrinos exist, and standard theory assumes that all three types have no mass. There is actually no laboratory evidence that the mass of a neutrino is exactly zero. If it has even a tiny mass, then it is possible for one type of neutrino to change into another type on its journey from the center of the Sun to the photosphere and on to the Earth. Fusion in the Sun produces only one type of neutrino, the so-called electron neutrino, and chlorine, water, and gallium experiments are sensitive only to this one type. If some electron neutrinos change to another type on their way from the center of the Sun to the Earth, then the experiments performed so far would have missed those neutrinos.

To test this idea, we need a new experiment that can measure all types of neutrinos. If the total number emitted by the Sun is found to be greater than the number of electron neutrinos, then we would have evidence that the mass of the neutrino is not zero. (As mentioned earlier, such a result would have important implications for other areas of astronomy as well.) One experiment of this kind, which involves heavy water (water with the hydrogen atoms replaced by deuterium), is now being built in Canada; the results should be available in the late 1990s.

Until very recently, all of the hints that neutrinos might have mass have come from measurements of solar neutrinos, over which scientists have no control. The solar measurements have inspired attempts to use laboratory accelerators to determine whether or not one type of neutrino can change into another; as we have seen, such a conversion can happen only if neutrinos have mass.

The first experiment to report success was completed at Los Alamos in 1995. This experiment was designed to produce a large quantity of one type of neutrino (technically the muon antineutrino). A detector consisting of 167 tons of baby oil was then set up 30 m away to look for a different type of neutrino (the electron antineutrino) while keeping everything else except neutrinos and antineutrinos from reaching the detector. The electron antineutrino, and no other type, produces a reaction in the baby oil that generates light. The scientists reported seeing seven flashes of light caused by electron antineutrinos. There should have been none if neutrinos have exactly zero mass. Obviously, this experiment has to be repeated, and new ones devised, before we can be sure that this early result is correct.

It will be years before we have final results of the new experiments on solar pulsations and neutrinos. Until then, we will not know which is wrong—some aspect of our solar models, or the assumption that the neutrino has no mass.

While it may seem discouraging that there are questions for which definitive answers are not yet available, science often works this way. Observations lead to the development of a model. This model then suggests a number of other measurements that can be made and predicts the outcome of those measurements. Frequently the predictions are incorrect, and the models must be modified to take into account the new measurements. And so science moves forward by successive approximations, each step providing a better and more complete description of what is actually occurring. Rather than finding the lack of final answers discouraging or frustrating, scientists find such situations exhilarating and challenging. The possibility of learning something heretofore unknown is what attracts many scientists to research.

Fusion on Earth

Wouldn't it be wonderful if we could duplicate the Sun's energy mechanism in a controlled way on Earth? (We have already duplicated it in an *uncontrolled* way in hydrogen bombs, but we hope our storehouse of these will never be used.) Fusion energy would have many advantages: it would use hydrogen (or deuterium, which is heavy hydrogen) as fuel, and there is plenty of hydrogen in the Earth's lakes and oceans. Water is much more evenly distributed around the world than is oil or uranium, meaning that a few countries would no longer hold an energy advantage over the others. And unlike fission, which leaves dangerous by-products, the nuclei that result from fusion are perfectly safe.

The problem is that, as we saw, it takes very, very high temperatures for nuclei to overcome their electrical repulsion and undergo fusion. When the first hydrogen bombs were exploded in tests in the 1950s, the "fuses" to get them hot enough were fission bombs. Interactions at such temperatures are difficult to sustain and control. Thus far, fusion experiments have generally required more energy to start them and keep them under control than is produced by the fusion itself.

Among the techniques now being tried in laboratories in the United States and around the world are keeping the hot gas under the control of powerful magnetic fields in what scientists call a "magnetic bottle," and using highly focused laser beams to compress and heat up pellets of solid deuterium. These experiments are at the forefront of our technology, and scientists estimate that it will take several decades before artificial fusion reactors become practical.

In 1989, two teams of scientists (both in Utah) made an announcement that simply astonished astronomers and physicists familiar with fusion. The two groups claimed to have achieved fusion at room temperatures, with equipment so simple it could easily be duplicated in a high-school science lab. They used an electrochemical cell in which electric current passes from one metal surface to another through a chemical solution. Both groups thought they had found evidence of deuterium fusion in their equipment—and their results, quickly dubbed "cold fusion," became a media sensation.

Alas, cold fusion did not hold up under the intense scrutiny to which all new scientific ideas and procedures are subjected. Other groups around the world could not duplicate the results. Even the original experiments failed to show many other characteristics associated with fusion. It turned out that the groups in Utah were probably seeing chemical reactions, not nuclear ones, and the initial results were discredited.

It appears that if we want to duplicate fusion on Earth we have to do what the Sun does: find a way to produce temperatures and pressures high enough to get hydrogen nuclei on intimate terms with one another. Many teams of scientists are hard at work solving the formidable technological challenges involved, although getting funding is always difficult for projects whose "payoff" is decades away. Still, perhaps by the time your children or grandchildren take an astronomy class in college, fusion will be a reality instead of a dream.

Tokamak Fusion Test Reactor (TFTR). (Princeton University, Plasma Physics Laboratory)

Summary

15.1 The Sun produces an enormous amount of energy every second. The Earth is 4.5 billion years old, so the Sun must have been shining for at least that long. Neither chemical burning nor gravitational contraction can account for the energy radiated by the Sun all this time.

15.2 Solar energy is produced by interactions of **elementary particles**—that is, protons, neutrons, electrons, and **neutrinos.** Specifically, the source of the Sun's energy is the **fusion** of hydrogen to form helium. The series of reactions required to convert hydrogen to helium is called the **proton–proton cycle.** A helium atom is about 0.71 percent less massive than the four hydrogen atoms that combine to form it, and that lost mass is converted to energy.

15.3 Even though we cannot see inside the Sun, it is possible to calculate what its interior must be like. As input for these calculations, we use what we know about the Sun. It is made entirely of hot gas. Apart from some very tiny changes, the Sun is neither expanding nor contracting (it is in **hydrostatic equilibrium**), but instead puts out energy at a constant rate. Fusion of hydrogen occurs in the center of the Sun, and the energy generated is carried to the surface by **radiation** and **convection.** A solar model describes the structure of the Sun's interior. Specifically, it describes how pressure, temperature, mass, and luminosity depend on distance from the center of the Sun.

15.4 Studies of solar oscillations (**solar seismology**) and neutrinos provide observational data about the Sun's interior. The technique of solar seismology has so far shown that the composition of the interior is much like that of the surface (except in the core, where some of the original hydrogen has been converted to helium), and that the convection zone extends 30 percent of the way from the Sun's surface to its center. Our solar models predict that we should be able to detect more neutrinos than we do. This result may indicate that the mass of the neutrino is not exactly zero.

Review Questions

1. How do we know the age of the Sun?

2. Explain how we know that the Sun's energy is not supplied either by chemical burning, as in fires here on Earth, or by gravitational contraction (shrinking).

3. What is the ultimate source of energy that makes the Sun shine?

4. How is a neutrino different from a neutron? List all the ways you can think of.

5. Describe in your own words what is meant by the statement that the Sun is in hydrostatic equilibrium.

6. Two astronomy students travel to South Dakota. One stands on the Earth's surface and enjoys some sunshine. At the same time, the other descends into the gold mine where neutrinos are detected, arriving in time to measure the creation of a new radioactive argon nucleus. Although the photon at the surface and the neutrinos in the mine arrive at the same time, they have had very different histories. Describe the differences.

7. Why do measurements of the number of neutrinos emitted by the Sun tell us about conditions deep in the solar interior?

8. Why have astronomers needed to do more than one experiment to measure neutrinos from the Sun?

Thought Questions

9. A friend who has not had the benefit of an astronomy course suggests that the Sun must be full of burning coal to shine as brightly as it does. List as many arguments as you can against this hypothesis.

10. Which of the following transformations is (are) fusion and which is (are) fission? (See Appendix 13 for a list of the elements.)

  a. helium to carbon
  b. carbon to iron
  c. uranium to lead
  d. boron to carbon
  e. oxygen to neon

11. Why is a higher temperature required to fuse hydrogen to helium by means of the CNO cycle than is required by the process that occurs in the Sun, which involves only isotopes of hydrogen and helium?

12. Explain what it means when we say that the Earth's oceans are in hydrostatic equilibrium. Now suppose you are a scuba diver. Would you expect the pressure to increase or decrease as you dive below the surface to a depth of 200 ft? Why?

13. What mechanism transfers heat away from the surface of the Moon? If the Moon is losing energy in this way, why does it not simply become colder and colder?

14. Suppose you are standing a few feet away from a bonfire on a cold fall evening. Your face begins to feel hot. What is the mechanism that transfers heat from the fire to your face? (*Hint:* Is the air between you and the fire hotter or cooler than your face?)

15. Give some everyday examples of the transport of heat by convection and by radiation.

16. Suppose the proton–proton cycle in the Sun were to slow down suddenly and generate energy at only 95 percent of its current rate. Would an observer on the Earth see an immediate decrease in the Sun's brightness? Would she immediately see a decrease in the number of neutrinos emitted by the Sun?

17. Do you think that nuclear fusion takes place in the atmospheres of stars? Why or why not?

18. Why is fission not an important energy source in the Sun?

19. Suppose a proton interacted with another proton an average of once every 100 million years rather than once every 14 billion years. In general terms, how do you think the structure of the Sun would change? (*Hint:* Think about what determines the pressure in the core of the Sun.)

20. The GONG experiment is designed to monitor the Sun's oscillations on a continuous basis. In order to save money, this experiment was designed to make use of the minimum possible number of telescopes. It turns out that if the sites are selected carefully, the Sun can be observed all but 6 percent of the time with only six observing stations. What factors have to be taken into consideration in selecting the observing sites? Can you suggest six general geographic locations that would optimize the amount of time that the Sun can be observed?

Problems

21. According to the text, a contraction of the Sun by 40 m per year would be enough to account for its current output of energy. Suppose you can measure the Sun's diameter with an accuracy of 1 percent. How long will you have to make measurements in order to determine whether or not the Sun is contracting at this rate?

22. Verify that roughly 600 million tons of hydrogen must be converted to helium in the Sun each second to explain its energy output. (*Hint:* Recall Einstein's most famous formula, and remember that for each kilogram of hydrogen, 0.0071 kg of mass is converted to energy.)

23. If an observed oscillation of the solar surface has a period of 10 min, and the average radial velocity is 1 m/s in and out, calculate the total displacement of the surface involved in this particular oscillation mode. What percent is this of the Sun's total radius?

24. Suppose you brought together a proton and an antiproton. What would happen? How much energy would result? (Take the mass of a proton to be 1.7×10^{-27} kg.)

25. Now suppose you brought together a 100-kg linebacker from the National Football League and an antiproton (ignoring the air around them). What would happen? How much energy would result?

Suggestions for Further Reading

Badash, L. "The Age of the Earth Debate" in *Scientific American*, Aug. 1989, p. 90.

Bahcall, J. "Where are the Solar Neutrinos" in *Astronomy*, Mar. 1990, p. 40.

Davies, P. "Particle Physics for Everybody" in *Sky & Telescope*, Dec. 1987, p. 582.

Fischer, D. "Closing In on the Solar Neutrino Problem" in *Sky & Telescope*, Oct. 1992, p. 378.

Goldsmith, D. *The Astronomers*. 1991, St. Martin's Press. Chapter 6 is on "Why Stars Shine."

Harvey, J. et al. "GONG: To See Inside Our Sun" in *Sky & Telescope*, Nov. 1987, p. 470.

Heppenheimer, T. *Man-made Sun: The Quest for Fusion Power*. 1984, Little Brown. Discusses our attempts to replicate the Sun's energy mechanism on Earth.

Kaler, J. *Stars*. 1992, Scientific American Library/W. H. Freeman. Good modern introduction by an astronomer on how stars work.

LoPresto, J. "Looking Inside the Sun" in *Astronomy*, Mar. 1989, p. 20. On helioseismology.

Rousseau, D. "Case Studies in Pathological Science" in *American Scientist*, Jan/Feb. 1992, p. 54. Explores the story of "cold fusion."

Sutton, C. *Spaceship Neutrino*. 1992, Cambridge U. Press. Definitive book on neutrinos.

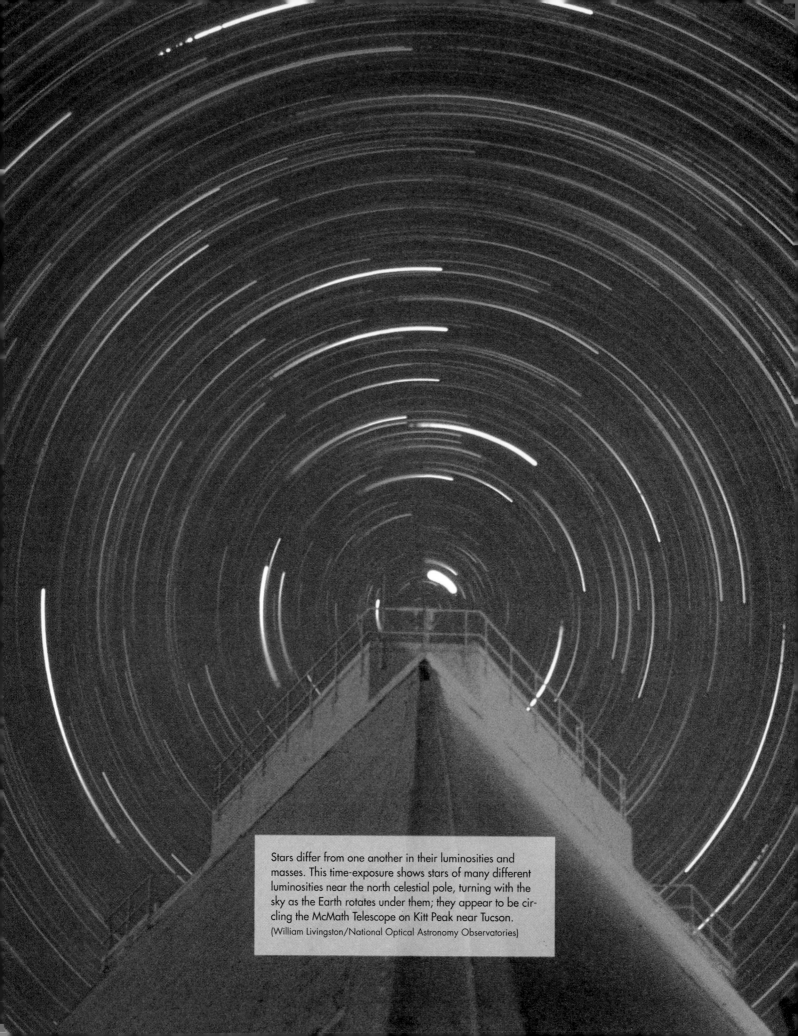

Stars differ from one another in their luminosities and masses. This time-exposure shows stars of many different luminosities near the north celestial pole, turning with the sky as the Earth rotates under them; they appear to be circling the McMath Telescope on Kitt Peak near Tucson. (William Livingston/National Optical Astronomy Observatories)

CHAPTER 16

Analyzing Starlight

Thinking Ahead

The nearest star (other than the Sun) is so far away that the fastest spacecraft the human race has built so far would take almost 100,000 years to get there. And all the other stars are even farther away. Yet we are a curious and ambitious species, and we want to know how those stars are born, live out their lives, and die. How can we learn about objects that are so remote and beyond our physical grasp?

Twinkle, twinkle little star,
I don't wonder what you are,
For by spectroscopic ken
I know that you are
hydrogen.

Anonymous, quoted by D. Bush in *Science and English Poetry* (1950).

Scientists love categories. Our scientific understanding grows when we find the right pigeonholes in which to place the objects we are studying, whether they are butterflies, human diseases, or subatomic particles. In the same way, the stars can be classified according to their properties—how big they are, how much material they contain (their mass), and how much light they give off. As we learn more about distinguishing the stars (in this and the next chapters), we will use these categories to begin assembling the pieces of the main puzzle we are interested in solving—their life stories.

We begin our voyage to the stars by describing how astronomers use the light emitted by stars to deduce what they are like. For example, we will discover that stars come in a range of sizes and masses. There are stars so large that the orbits of many of the planets in our solar system would easily fit inside them. We will see strange dying stars called white dwarfs that have been compressed until they are a thousand times denser than water. Figuring out how all the different kinds of stars fit into the big picture has been one of the most exciting and productive areas of astronomy in the last century.

Figure 16.1

A time-exposure showing the constellation Orion as it rises over Kitt Peak National Observatory. To make the photo more interesting, the photographer began with a short exposure, giving the dots at the right. Then the camera shutter was left open for a long exposure, during which the Earth's turning motion drew out the light of each star into a long line. The colors of the various stars are caused by their different temperatures. (National Optical Astronomy Observatories)

Very simple observations are enough to show that not all stars are alike. You can make some of these observations yourself, especially if you get away from city lights or use a pair of binoculars. Take a good look at the night sky. Not all the stars are the same brightness or color. Although most are too dim to excite our color vision, a few brighter stars do show distinct colors to our eyes. Some appear to be blue-white; others are obviously red. In the winter sky, a good constellation for seeing star colors is Orion, the hunter (Figure 16.1).

In Chapter 4 we discussed that in a glowing gas, color is a good indication of temperature, with red being the coolest, and blue and violet being the hottest. Yellow, the color of our Sun, is toward the middle of the visible spectrum. Since stars are giant balls of glowing gas, we can deduce that those that look blue-white are hotter than the Sun, while red ones are cooler.

Other questions about stars can be answered only by means of careful observations with large telescopes. Since the stars are suns, it should not surprise you that the same techniques, including spectroscopy, used to study the Sun can also be used to find out what stars are like.

16.1

The Brightness of Stars

Perhaps the most important characteristic of a star is its **luminosity**—the total amount of energy that it emits per second. In Chapter 15 we saw that the Sun (pictured as a big lightbulb) puts out a tremendous amount of radiation every second. (And there are stars far more luminous than the Sun out there.) Later, we will see that if we can measure how much energy a star emits, and if we also know its mass, we can calculate how long it can continue to shine before exhausting its nuclear energy and beginning to die.

Note that luminosity is how much energy a star gives off each second, *not* how much energy reaches our eyes or telescopes on Earth. Stars are very democratic in the way they emit energy: the same amount goes in every direction in space. So only a minuscule fraction of the energy given off by a star actually reaches an observer on Earth. We call the amount of a star's energy that reaches us here its **apparent brightness.** If you look at the night sky, you see a wide range of apparent brightnesses among the stars. Most stars, in fact, are so dim that you need a telescope to detect them.

If all stars were the same luminosity—if they were *standard bulbs*—we could use the difference in their apparent brightness to tell us something we very much want to know: how far away they are. Imagine you are in a big concert hall or ballroom that is dark except for a few dozen 25-watt (W) bulbs placed in fixtures around the walls. Since they are all 25-W bulbs, their luminosity is the same. But from where you are standing in one corner, they do *not* have the same apparent brightness. Those close to you look brighter (more of their light reaches your eye), while those far away look dimmer (their light has spread out more before reaching you). In the same way, if all the stars had the same luminosity, we could immediately say that the bright-looking stars were close-by and the dim-looking ones were far away.

To pin this down more precisely, recall from Chapter 4 that we know the way in which light gets dimmer with distance. The energy we receive is inversely proportional to the square of the distance. If, for example, we have two stars of the same brightness and one is twice as far away as the other, it will look four times dimmer than the closer one. If it is three times further away, it will look nine (three squared) times dimmer, and so forth.

Alas, the stars do not have the courtesy to come in one standard luminosity. (Actually, we are pretty glad about that, because having many different types of stars makes the universe a much more interesting place.) But this means that if a star looks dim in the sky, we cannot tell whether it appears dim because it has a low luminosity but is relatively close-by, or because it has a high luminosity but is very far away.

(By the way, although stars in general did not turn out to be "standard bulbs," astronomers have not given up on the idea of finding some types of objects in the cosmos that all come with the same luminosity "built-in." We will return to our quest for standard bulbs in future chapters as our voyages take us into more distant regions of the universe.)

Because the stars are not standard bulbs, we cannot just read off their luminosities from their apparent brightnesses. We must first compensate for the dimming effects of distance on light, and to do that we must know how far away they are. Distance is among the most difficult of all astronomical measurements; we will return to how it is determined in Chapter 18, after we have learned more about stars. In this chapter, we now turn to the problem of measuring apparent brightness.

The Magnitude Scale

The branch of observational astronomy that deals with the measurement of stellar brightness and luminosity is called photometry (from the Greek *photo*, "light," and *-metry*, "to measure"). As we saw in Chapter 1, astronomical photometry began with Hipparchus. Around 150 B.C. he erected an observatory on the island of Rhodes in the Mediterranean. There he prepared a catalog of nearly 1000 stars that included not only their positions but also estimates of their apparent brightness.

Hipparchus did not have photographic plates or any instruments that could measure brightness accurately, so he simply made estimates with his eye. He sorted the stars into six brightness categories, which he called **magnitudes.** He referred to the brightest stars in his catalog as first-magnitude stars, while those so faint he could barely see them were sixth-magnitude stars.

During the 19th century, astronomers attempted to make the scale more precise by establishing exactly how much the brightness of a sixth-magnitude star differs from that of a first-magnitude star; measurements showed that we receive about 100 times more light from a star of the first magnitude than from one of the sixth. Based on this measurement, astronomers then defined an accurate magnitude system in which a difference of five magnitudes corresponds exactly to a brightness ratio of 100 : 1.

This means stars that differ by about one magnitude differ in brightness by a factor of about 2.5—or, to be more precise, 2.512, which is the fifth root of 100. Thus a fifth-magnitude star gives us 2.512 times as much light as one of the sixth magnitude, a fourth-magnitude star gives 2.512 times as much light as a fifth-magnitude star, and so forth.

Today, many astronomers (and astronomy students) wish we had a less old-fashioned and cumbersome system, and many researchers have stopped using magnitudes. But tradition plays a strong role in astronomy, as it does in numerous other areas of human endeavor. Magnitudes are still used on star charts and by amateur astronomers, and news stories about astronomy may well include references to magnitudes. So even though your authors are sorely tempted to stop teaching the concept of magnitudes (and maybe start a trend), we still include it since it is commonly used.

The brightest stars, those that were traditionally referred to as first-magnitude stars, actually turned out (when measured accurately) not to be identical in brightness. For example, the brightest star in the sky, Sirius, sends us about ten times as much light as the average first-magnitude star. On the modern magnitude scale, Sirius has been assigned a magnitude of −1.5. Several of the planets appear even brighter. Venus, at its brightest, is of magnitude −4.4, while the Sun has a magnitude of −26.2.

Figure 16.2 shows the range of observed magnitudes from the brightest to the faintest, along with the actual magnitudes of several well-known objects. The important fact to remember when using magnitudes is that the system goes backwards: the *larger* the magnitude, the *fainter* the object you are observing!

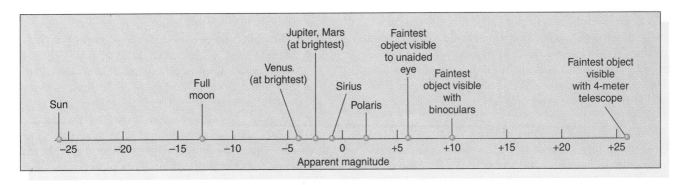

Figure 16.2
The apparent magnitudes of several well-known objects. The faintest magnitudes that can be detected by the eye, binoculars, and a large telescope are also shown.

Other Units of Brightness

Although the magnitude scale is still used for visual astronomy, it is not used at all in newer branches of the field. In radio astronomy, for example, no equivalent of the magnitude system has been defined. Rather, the radio astronomer measures the amount of energy being collected by the radio telescope, and expresses the brightness of each source in terms of the energy reaching the surface of the Earth in, for example, watts per square meter.

Similarly, most researchers in the fields of infrared, x-ray, and gamma-ray astronomy use energy units rather than magnitudes to express the results of their measurements. Nevertheless, astronomers in all fields are careful to distinguish between the total *luminosity* of the source (even when that luminosity is all in x rays) and the amount of energy that happens to reach us on Earth. After all, the luminosity is a really important characteristic that tells us a lot about the object in question, whereas the energy that reaches the Earth is an accident of cosmic geography.

In order to make the comparison among stars easy, in this text we express the luminosity of other stars in terms of the Sun's luminosity. For example, the luminosity of Sirius is 23 times that of the Sun. We use the symbol L_{Sun} to denote the Sun's luminosity; hence that of Sirius can be written as 23 L_{Sun}.

16.2

Colors of Stars

Stars are not all the same color because they are not all the same temperature. As we have seen, a hotter star has the blue colors dominating its light output, while a cool star emits most of its light energy at red wavelenths. To make such a discussion more precise, astronomers have devised quantitative methods for characterizing the color of a star and then using those colors to determine stellar temperatures.

Color Indices

In order to find the exact color of a star, astronomers normally measure the stellar brightness through filters, each of which transmits only the light from a particular narrow band of wavelengths (colors). A crude example of a filter in everyday life is a piece of red cellophane, which, when held in front of your eyes, only lets the red colors of light through.

One commonly used set of filters in astronomy measures stellar brightness at three wavelengths corresponding to ultraviolet, blue, and yellow light. The filters are given names: U (ultraviolet), B (blue), and V (visual—for yellow). These filters transmit light near the wavelengths of 360 nm, 420 nm, and 540 nm respectively. The brightness measured through each filter is usually expressed in magnitudes. The difference between any two of these magnitudes—say, between the blue and the visual magnitudes (B − V)—is called a **color index.**

If we could somehow take a star, observe it, and then move it much farther away, its apparent brightness (magnitude) would change. But since its magnitude would change by the same amount for all colors, its color index (which is two magnitudes subtracted from each other) would remain the same. This idea should make sense to you. After all, you know from everyday experience that the color of an object appears the same no matter how far away it is. Color indices of stars, therefore, provide measures of their intrinsic or true colors, and do not depend on where the observer happens to be.

The way in which color is related to the *temperatures* of stars is given by Wien's law (see Chapter 4). Hot stars emit more energy in the blue and ultraviolet part of the spectrum. Cool stars are brighter at red and infrared wavelengths. By agreement among astronomers, the ultraviolet, blue, and visual magnitudes of the UBV system are adjusted to give a color index of zero to a star with a surface temperature of about 10,000 K.

The B − V color indices of stars range from −0.4 for the bluest to more than +2.0 for the reddest. The corresponding range in surface temperature is from about 50,000 K to 2000 K. Our Sun's surface temperature is about 6000 K; its dominant color is a slightly greenish yellow; and its color index is about 0.62. It looks more yellow seen from the Earth's surface because our planet's air molecules scatter some of the green light out of the beams of sunlight that reach us, leaving more yellow behind.

16.3

The Spectra of Stars

Measuring colors is only one way of analyzing starlight. Instead of filters we can use a spectrograph to spread out the light into a spectrum (see Chapters 4 and 5). By examining that spectrum, astronomers can determine what elements make up the stars—just as they did for the Sun (see Chapter 14). As early as 1823, the German physicist Joseph Fraunhofer observed that the spectra of stars have dark lines crossing a continuous band of colors (Figure 16.3). In 1864 an English astronomer, Sir William Huggins, succeeded in identifying some of the lines in stellar spectra with those of known terrestrial elements, showing that the same chemical elements found in the Sun and planets exist in the stars (Figure 16.4). Since then, astronomers have worked hard to perfect experimental techniques for obtaining and measuring spectra, and have developed a theoretical understanding of what can be learned from spectra. Today, spectroscopic analysis is one of the cornerstones of astronomical research.

Formation of Stellar Spectra

When the spectra of different stars were first observed, astronomers found that they were not all identical. Since the dark lines are produced by the chemical elements present in the stars, astronomers first thought that the spectra differ from one another because stars are not all made of the same chemical elements. This hypothesis turned out to be wrong. *The primary reason that not all stellar spectra look alike is that stars have different temperatures.* Most stars have nearly the same composition as the Sun, with only a few exceptions.

Hydrogen, for example, is by far the most abundant element in all stars (except those at advanced stages of evolution; see Chapter 22), but lines of hydrogen are not seen in the spectra of some types of stars. In the atmospheres of the hottest stars, for example, hydrogen atoms are completely ionized. Because the electron and the proton are separated, ionized hydrogen cannot produce absorption lines.

In the atmospheres of the coolest stars, hydrogen atoms have their electrons attached and can switch energy levels to produce lines. However, practically all of the hydrogen atoms are in the lowest energy state (unexcited) in these stars, and thus can absorb only those photons able to lift an electron from that first energy level to a higher level. The photons absorbed in this way produce a series of absorption lines that lie in the ultraviolet part of the spectrum and hence cannot be studied from the Earth's surface. What this means is that if you observe the spectrum of a very hot or very cool star with a typical telescope on the surface of the Earth, the most common element in that star will not show any lines.

The hydrogen lines in the visible part of the spectrum (the Balmer lines; see Chapter 4) are strongest in stars with intermediate temperatures—not too hot and not too cool. Calculations show that the optimum temperature for producing visible hydrogen lines is about 10,000 K. At this temperature, an appreciable number of hydrogen atoms are excited to the second energy level. They can then absorb additional photons, rise to still higher levels of excitation, and produce a dark absorption line. They are less conspicuous in the spectra of both hotter and cooler stars, even though hydrogen is just about equally abundant in all the stars. Similarly, every other chemical element, in each of its possible stages of ionization, has a characteristic temperature at which it is most effective in producing absorption lines in any particular part of the spectrum.

Classification of Stellar Spectra

Because a star's temperature determines which absorption lines are present in its spectrum, we can use the spectrum to measure its surface temperature. Astronomers have identified seven principal patterns of lines organized by temperature; these are called **spectral classes.** From hottest to coldest, the seven main spectral classes are designated O, B, A, F, G, K, and M.

At this point you are probably looking at these letters with amazement and asking yourself why astronomers didn't call the spectral types A, B, C, and so on. It's a long story (partly discussed in a moment) in which tradition won out over common sense. To help them remember this crazy order of letters, generations of astronomy students have used the mnemonic "Oh be a fine girl, kiss me." (Today, when many astronomy students are women, you can easily substitute "guy" for "girl.") Other mnemonics, which we hope will not be relevant for you, include "Oh brother, astronomers frequently give killer midterms" and "Oh boy, an F grade kills me!"

Each of these spectral classes is further subdivided into ten subclasses designated by numbers. A B0 star is the hottest type of B star; a B9 star is the coolest type of B star and is only slightly hotter than an A0 star. And just one more item of vocabulary: for historical reasons, astronomers call *all* elements heavier than helium "metals," even though most of these elements do not show metallic properties. (If you are getting annoyed at the strange jargon astronomers use, just bear in mind that every field of human knowledge develops its own peculiar vocabulary.

Figure 16.4
William Huggins (1824–1910) was the first to identify the lines in the spectrum of a star other than the Sun; he also took the first spectrogram, or photograph of a stellar spectrum. (Mary Lea Shane Archives of the Lick Observatory)

TABLE 16.1
Spectral Classes for Stars

Spectral Color Class	Approximate Temperature (K)	Principal Features	Examples
O Violet	>28,000	Relatively few absorption lines. Lines of doubly ionized nitrogen, triply ionized silicon, and lines of other highly ionized atoms.	10 Lacertae
B Blue	10,000–28,000	Lines of neutral helium, singly and doubly ionized silicon, singly ionized oxygen, and magnesium. Hydrogen lines more pronounced than in O-type stars.	Rigel Spica
A Blue	7500–10,000	Strong lines of hydrogen. Lines of singly ionized magnesium, silicon, iron, titanium, calcium, and others. Lines of some neutral metals show weakly.	Sirius Vega
F Blue to white	6000–7500	Hydrogen lines weaker than in A-type stars, but still conspicuous. Lines of singly ionized calcium, iron and chromium, plus lines of neutral iron and chromium are present, as are lines of other neutral metals.	Canopus Procyon
G White to yellow	5000–6000	Lines of ionized calcium are most conspicuous spectral features. Many lines of ionized and neutral metals are present. Hydrogen lines are weaker than in F-type stars. Bands of the molecule CH are strong.	Sun Capella
K Orange to red	3500–5000	Lines of neutral metals predominate. The CH bands are still present.	Arcturus Aldebaran
M Red	<3500	Strong lines of neutral metals and molecular bands of titanium oxide dominate.	Betelgeuse Antares

Try reading a legal document these days without training in the law!)

As Figure 16.5 shows, in the hottest O stars (those with temperatures over 28,000 K), only lines of ionized helium and highly ionized atoms of other elements are conspicuous. Hydrogen lines are strongest in A stars with atmospheric temperatures of about 10,000 K. Ionized metals provide the most conspicuous lines in stars with temperatures from 6000 to 7500 K (spectral type F). In the coolest M stars (below 3500 K), absorption bands of titanium oxide and other molecules are very strong. The sequence of spectral classes is summarized in Table 16.1. The spectral class assigned to the Sun is G2.

To see how spectral classification works, let's use Figure 16.5. Suppose you have a spectrum in which the hydrogen lines are about half as strong as those viewed in an A star. Looking at the red lines in our figure, you see that the star could be either a B star or a G star. But if the spectrum also contains helium lines, then it is a B star, whereas if it contains lines of ionized iron and other metals, it must be a G star.

Much of the pioneering work on the classification of stellar spectra was carried out at Harvard University in the first decades of the 20th century. The basis for these studies was a monumental collection of nearly a million photographic spectra of stars, obtained from many years of observations made at Harvard College Observatory in Massachusetts, as well as at its remote observing stations in South America and South Africa.

Working with this data base, Annie Cannon (see

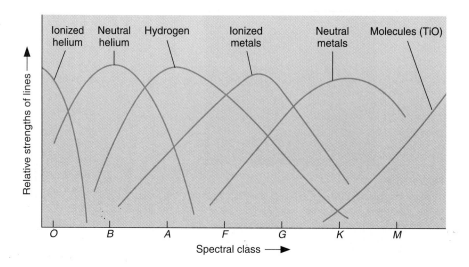

Figure 16.5
A graph showing the strength of absorption lines of different elements as we move from hot (left) to cool (right) stars along the sequence of spectral types.

Annie Cannon: Classifier of the Stars

Annie Jump Cannon was born in Delaware in 1863. In 1880 she went to Wellesley College, one of the new breed of U.S. colleges opening up to educate young women. Only five years old at the time, Wellesley had the second student physics lab in the country and provided excellent training in basic science. After college Cannon spent a decade with her parents but was very dissatisfied, longing to do scientific work. With her mother's death in 1893, she was able to return to Wellesley as a teaching assistant, and also to take courses at Radcliffe, the women's college associated with Harvard.

This was a time when the director of the Harvard Observatory, Edward C. Pickering, needed lots of help with his ambitious program of classifying stellar spectra. Pickering quickly discovered that educated young women could be hired as assistants for one-third or one-fourth the salary paid to men, and would often put up with working conditions and repetitive tasks that men with the same education would not tolerate. (We should emphasize that astronomers were not alone in reaching such conclusions about the relatively new idea of women working outside the home: women were exploited and underestimated in many fields. This is a legacy from which our society is just beginning to emerge.)

Cannon wanted to do astronomy, and she was quickly hired by Pickering to help with the classification of spectra. After a while she became so good at it that she could visually examine and determine the spectral types of several hundred stars per hour (dictating her conclusions to an assistant). She made many discoveries while investigating the Harvard photographic plates, including 300 variable stars (stars whose luminosity changes; see Chapter 18). But her main legacy is a marvelous catalog of spectral types for hundreds of thousands of stars, which serves as a foundation for much of 20th-century astronomy.

In 1911 a visiting committee of astronomers reported that "she is the one person in the world who can do this work quickly and accurately," and urged Harvard to give Cannon an official appointment in keeping with her skill and renown. Not until 1938, however, did the university appoint her an astronomer at Harvard; she was then 75 years old.

Cannon received the first honorary degree from Oxford awarded to a woman, and became the first woman to be elected an officer of the American Astronomical Society, the main professional organization of astronomers in this country. She generously donated the money from one of the major prizes she had won to found a special award for women in astronomy, now known as the Annie Jump Cannon Prize. True to form, she continued classifying stellar spectra almost to the very end of her life in 1941.

Annie Jump Cannon
(1863–1941).
(Harvard College
Observatory Archives)

"Voyagers in Astronomy" box) personally measured some 400,000 stars and assigned them to spectral classes. Because the Harvard workers did not know that temperature is the primary factor in determining the appearance of a stellar spectrum, they at first simply named the spectral classes with letters of the alphabet according to the complexity of the spectra, with type A being the simplest. It was Annie Cannon who initially glimpsed a better classification system, and focused on just a few letters from the original system. Now that we understand that spectral classes depend on the temperature of the stars, we always list the types not in order of how many lines are seen, but rather from hottest to coolest: O, B, A, F, G, K, and M.

During the 20th century, the emphasis has shifted from classification to interpretation of stellar spectra. One of the most important early contributions was the work of Cecilia Payne, then a doctoral student at Harvard. Payne's work, more than any other, proved that all stars really are composed principally of hydrogen, with the differences in their spectra resulting primarily from their different temperatures.

Effects of the Size of a Star

As we will see in Chapter 17, stars come in a wide variety of sizes. At some periods in their lives stars can expand to enormous dimensions, with correspondingly low densities and pressures in their atmospheres. Stars of such exaggerated size are called **giants.** Luckily for the astronomer who wishes to distinguish them from run-of-the-mill stars (such as our Sun), giants can be identified from a study of their spectra.

The pressure in a stellar photosphere affects its spectrum. At the very low densities that occur in the extended, tenuous photospheres of these giant stars, the pressures are also low. Ionized atoms and electrons pass close enough together to recombine much less often than they do in stars with higher pressures. Think of automobile traffic. Collisions are much more likely during rush hour, when the density of cars is high. Low-density gases, therefore, maintain a higher average degree of ionization than do high-density gases of the same temperature. Careful study of their spectra can tell which of two stars at the same temperature has higher pressure (and is thus more compressed), and which has lower pressure (and thus must be extended.)

Abundances of the Elements

Dark lines of a majority of the known chemical elements have now been identified in the spectra of the Sun and stars. Identification of the spectral lines of one of the chemical elements, in the neutral state or in one of its ionized states, immediately tells us that the star must contain that substance.

Of course, astronomy textbooks such as ours always make these things sound a bit easier than they really are. First of all, it has taken many years of careful laboratory work on Earth to determine at what precise wavelengths heated gases of each element have their spectral lines. Long books and computer listings have been compiled to show the lines that can be seen at each temperature. Second, stellar spectra often show many lines from a number of elements, and we have to be careful to sort them out correctly and beware of overlaps between lines of different elements (Figure 16.6). And third, as we saw in Chapter 4, the motion of the star changes the location of each of the lines, and therefore needs to be factored into our analysis. So, in practice, doing spectroscopy is a demanding, sometimes frustrating task that requires both training and skill. The situation has been helped considerably in recent years by automation and computers, but astronomers still need to know what they are doing when analyzing faint spectra from distant objects.

Note that the *absence* of an element's spectral lines does not necessarily mean that the element itself is absent. As we saw, the temperature and pressure in a star's atmosphere will determine what types of atoms are able to produce absorption lines. Only if the physical conditions in a star's photosphere are such that lines of an element *should* (according to calculations) be visible, can we conclude that the absence of observable spectral lines implies low abundance of the element.

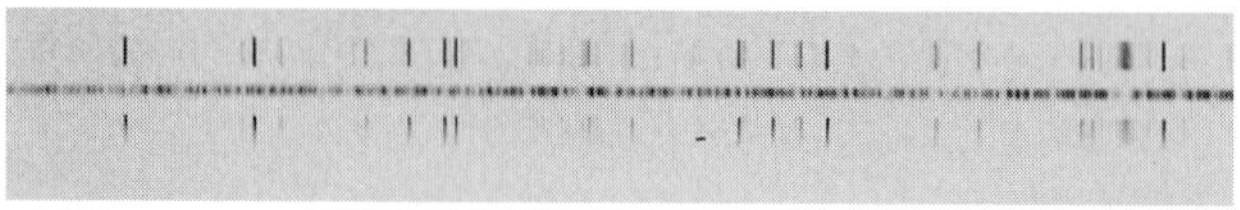

Figure 16.6
Spectrum of a variable star, showing a large number of absorption lines. The star's spectrum is the dark band crossed by white lines in the center; above and below it are some comparison lines (in this case for an iron arc near the telescope) to help astronomers measure at what wavelength each line of the star's spectrum falls. (Image courtesy of Roger Bell, University of Maryland)

Once due allowance has been made for the temperature and pressure in a star's photosphere, analyses of the amount of light subtracted from its spectrum by the absorption lines can yield information regarding the relative abundances of the various chemical elements whose lines appear. It is these kinds of detailed analyses that have shown that the relative abundances of the different chemical elements in the Sun (see Table 14.2) and in most stars are approximately the same.

Hydrogen makes up about three-quarters of the mass of most stars. Hydrogen and helium together make up from 96 to 99 percent of the mass; in some stars they amount to more than 99.9 percent. Among the 4 percent or less of "heavy elements," neon, oxygen, nitrogen, carbon, magnesium, argon, silicon, sulfur, iron, and chlorine are among the most abundant. Generally, but not invariably, the elements of lower atomic weight are more abundant than those of higher atomic weight.

Take a careful look at this list of elements. Among those contained by stars in greatest abundance are hydrogen and oxygen (which make up water); add carbon and nitrogen and you are starting to write the prescription for the chemistry of an astronomy student. We are made of elements that are common in the universe—just mixed together in a far more sophisticated form (and a much cooler environment) than is a star.

The spectrum of the Sun has been studied more thoroughly than that of any other star, and we know most about the abundances of the elements there. Appendix 13 lists how common each element is in the universe (compared to hydrogen); these estimates are based primarily on our investigation of the Sun. Some very rare elements, however, have not been detected in the Sun. Our estimate of the amounts of these elements in the universe is based on laboratory measurements of their abundance in primitive meteorites, which are considered representative of unaltered material condensed from the solar nebula (see Chapter 13).

Radial Velocity

When we measure the spectrum of a star, we determine the *wavelengths* of each of its lines. If the star is not moving with respect to the Sun, then the wavelengths corresponding to each element will be the same as those we measure in a laboratory here on Earth. But if stars are moving toward or away from us, we must consider the Doppler effect (see Section 4.6). We should see the spectral lines of moving stars shifted either toward the red end of the spectrum, if the star is moving away from us, or toward the blue (violet) end if it is moving toward us. Such motion, along an imaginary line connecting the star and the observer, is called **radial velocity** and is usually measured in kilometers per second.

William Huggins, pioneering yet again, made the first radial velocity determination of a star in 1868. He observed the Doppler shift in one of the hydrogen lines in the spectrum of Sirius and found that this star is approaching the solar system. Today, the radial velocity can be measured for any star bright enough for its spectrum to be observed. As we will see in the next chapter, radial velocity measurements of double stars are crucial in deriving stellar masses.

Proper Motion

For the sake of completeness, we should note that stars do not necessarily move directly toward or away from us. They can also move perpendicular to our line of sight; such motion is referred to as *proper motion*. If stars have some motion in this perpendicular direction, we see it as a change in the relative positions of the stars on the "dome" of the sky. However, the proper motion of a star is almost always too small to measure with much precision, even with a telescope, in a single year. For that reason we do not notice any change in the positions of the bright stars during the course of a human lifetime.

If we could live long enough, however, the changes would become obvious. For example, some 50,000 years or so from now, terrestrial observers (if humans haven't wiped themselves out by then) will find the handle of the Big Dipper unmistakably more bent than it is now (Figure 16.7).

The star with the largest proper motion is Barnard's star, our second closest neighbor, named after the American astronomer who first measured its movement. Its position in the sky changes by the width of the Moon in just two centuries (Figure 16.8). Unfortunately, Barnard's star is too faint to be seen with the unaided eye.

We measure proper motion in terms of how far the star moves in the sky in arcseconds per year. In order to convert this angular motion to a velocity, we need to know how far away the star is. If two stars are moving at the same velocity perpendicular to our line of sight, the closer one will appear to move farther across the sky in a year's time. As an analogy, imagine you are standing at the edge of a freeway. Cars will appear to whiz past you. If you watch the cars from a vantage point half a mile away, they will move much more slowly across your field of vision.

In order to know the true velocity of a star in space—that is, its total speed and the direction in which it is mov-

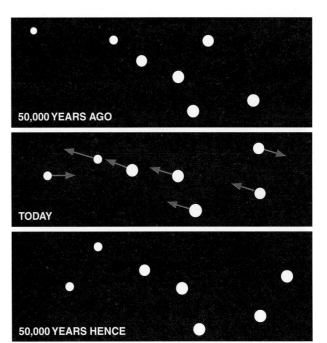

Figure 16.7
The change in the appearance of the Big Dipper due to proper motion of the stars over 100,000 years.

ing—we must know its radial velocity, proper motion, and distance.

Rotation

We can also use the Doppler effect to measure how fast a star rotates. If an object is rotating, then (unless its axis of rotation happens to be pointed exactly toward us) one of its sides is approaching us while the other is receding. This is clearly the case for the Sun or a planet; we can observe the light from either the approaching or receding edge of these nearby objects, and measure directly the Doppler shifts that arise from the rotation.

Stars, however, are so far away that they all appear as unresolved points. The best we can do is analyze the light from an entire stellar disk at once. Even so, the part of the light, including the spectral lines, that comes from the side of the star rotating towards us is shifted to shorter wavelengths, and the part from the opposite edge of the

Figure 16.8
Two photographs of Barnard's star, showing its proper motion over a period of 22 years.
(Yerkes Observatory)

Astronomy and Philanthropy

Throughout the history of astronomy, contributions from wealthy patrons of the science have made an enormous difference for building new instruments and carrying out long research projects. Edward Pickering's stellar classification project, which was to stretch over several decades, was made possible by major donations from Anna Draper, the widow of Henry Draper, a physician who was one of the most accomplished amateur astronomers of the 19th century and the first man to successfully photograph the spectrum of a star. Anna Draper gave several hundred thousand dollars (a lot more money then than today!) to Harvard Observatory; as a result, the great spectroscopic survey is still known as the Henry Draper Memorial, and many stars are still referred to by their "HD" numbers in that catalog.

In the 1870s the eccentric piano-builder and real estate magnate James Lick decided to leave some of his fortune to build the world's largest telescope. (This was actually his second plan for a memorial; fortunately he had been talked out of his first plan, which was to build the world's largest pyramid in downtown San Francisco as a monument to himself.) When, in 1887, the pier to house the telescope was finished, Lick's body was entombed in it. Atop the

James Lick
(Mary Lea Shane Archives of the Lick Observatory)

foundation rose a 36-in. refractor, for many years the main instrument at the Lick Observatory near San Jose.

As we saw in Chapter 5, the Lick telescope remained the largest in the world until 1897, when George Ellery Hale persuaded railroad millionaire Charles Yerkes to build a 40-in. telescope near Chicago.

More recently Howard Keck, whose family made their fortune in the oil business, gave $70 million dollars from his family foundation to the California Institute of Technology to help build the world's largest telescope atop the 14,000-ft peak of Mauna Kea in Hawaii (see Chapter 5). The Keck Foundation was so pleased with what is now called the Keck telescope that they gave $74 million more to build "Keck II," another 10-m reflector on the same volcanic peak.

Now if any of you reading this book become millionaires and billionaires, and astronomy has sparked your interest, do keep an astronomical instrument or project in mind as you make out your will. But frankly, private philanthropy could not possibly support the full enterprise of scientific research in astronomy. Much of our exploration of the universe is financed by such federal agencies as the National Science Foundation and NASA. In this way, all of us, through a very small share of our tax dollars, get to be philanthropists for astronomy.

---

star is shifted to longer wavelengths. You can think of each spectral line that we observe as a sum or composite of spectral lines originating from different parts of the star's disk, all of which are moving at different speeds with respect to us. Each point has its own Doppler shift, so the absorption line we see is actually much wider in the spectrum than if the star were not rotating. If a star is rotating rapidly, all its spectral lines should be quite broad. In fact, astronomers call this effect line-broadening, and the amount of broadening can tell us the speed at which the star rotates (Figure 16.9).

Measurements of the widths of spectral lines show that many stars hotter than the Sun rotate in periods of only a day or two. The Sun, with its rotation period of about a month, rotates rather slowly. The rotation of most stars cooler than the Sun is slower still, and often cannot be measured with our present techniques.

Summary of Stellar Spectra

In 1835 the French philosopher Auguste Comte wrote a paper in which he asserted that it would never be possible by any means to study the chemical composition of stars. Yet within a few decades the development of astronomical spectroscopy, together with a rapidly maturing understanding of the ways atoms absorb and emit radiation, provided the tools to accomplish this "impossible" task.

In this chapter we have seen that spectrum analysis is an extremely powerful technique that allows the astronomer to learn all kinds of things about a star: its detailed chemical composition, as well as the temperature and pressure in its atmosphere. From the pressure, we get clues about its size. We can also measure its radial velocity and estimate its rotation. It is no wonder that astronomers spend much of their time obtaining and analyzing spectra.

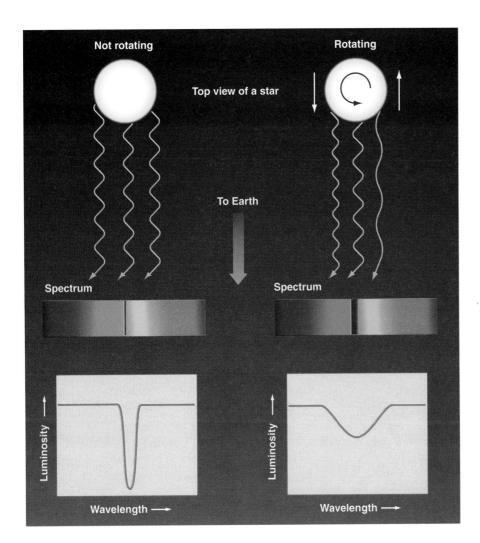

Figure 16.9
The rotation of a star broadens its spectral lines.

Summary

16.1 The total energy emitted per second by a star is called its **luminosity.** How bright a star looks to us is called its **apparent brightness.** For historical reasons, the apparent brightnesses of stars are often expressed in terms of **magnitudes.** If one star is five magnitudes brighter than another, it emits 100 times more energy. Since the apparent brightness of a star depends on its luminosity and distance, determination of apparent brightness and measurement of the distance to a star provide enough information to calculate its luminosity.

16.2 Stars have different colors, and these are indicators of temperature. The **color index** of a star is the difference in the magnitudes measured at any two different wavelengths. The difference between blue and visual magnitudes, B − V, is one frequently used color index; redder, cooler stars have more positive values of B − V.

16.3 The differences in the spectra of stars are principally due to differences in temperature, not composition. The spectra of stars are described in terms of seven **spectral classes.** In order of decreasing temperature, these spectral classes are O, B, A, F, G, K, and M. Spectra of stars of the same temperature but different atmospheric pressure have subtle differences, so spectra can be used to determine whether a star has a large radius and low atmospheric pressure (a **giant** star), or a small radius and high atmospheric pressure. Stellar spectra can also be used to determine the chemical composition of stars; hydrogen and helium make up most of the mass of all stars (just as they do in the Sun). Measurements of line shifts produced by the Doppler effect indicate the **radial velocity** of a star. Broadening of spectral lines by the Doppler effect is a measure of rotational velocity.

Review Questions

1. What two factors determine how bright a star appears to be in the sky?

2. Explain why color is a measure of a star's temperature.

3. What is the main reason that the spectra of stars are not all identical? Explain.

4. What elements are stars made of? How do we know this?

5. What two women astronomers made significant contributions to the understanding of stellar spectra? Discuss what each of them did.

6. Name at least three characteristics of a star that can be determined by measuring its spectrum. Explain how you would use a spectrum to determine these characteristics.

Thought Questions

7. If the star Sirius emits 23 times more energy than the Sun, why does the Sun appear brighter in the sky?

8. Draw a picture showing how two stars of equal intrinsic luminosity, one of which is blue and the other red, would appear on two images, one taken through a filter that passes mainly blue light, and the other through a filter that transmits mainly red light.

9. Table 16.1 lists the temperature ranges that correspond to various spectral types. What part of the star do these temperatures refer to?

10. Star A has lines of ionized helium in its spectrum, and star B has bands of titanium oxide. Which is hotter? Why? The spectrum of another star shows lines of ionized helium and also molecular bands of titanium oxide. What is strange about this spectrum? Can you suggest an explanation?

11. The spectrum of the Sun has hundreds of strong lines of non-ionized iron but only a few, very weak, lines of helium. A star of spectral type B has very strong lines of helium but very weak iron lines. Do these differences mean that the Sun contains more iron and less helium than the B star? Explain.

12. What are the approximate spectral classes of stars with the following characteristics?
  a. Balmer lines of hydrogen are very strong; some lines of ionized metals are present.
  b. Strongest lines are those of ionized helium.
  c. Lines of ionized calcium are the strongest in the spec-

trum; hydrogen lines show with only moderate strength; lines of neutral and ionized metals are present.
  d. Strongest lines are those of neutral metals and bands of titanium oxide.

13. Look at Appendix 13. Can you identify any relationship between the abundance of an element and its atomic weight? Are there any obvious exceptions to this relationship?

14. Appendix 10 lists the nearest stars. Are most of these stars hotter or cooler than the Sun? Do any of them emit more energy than the Sun? If so, which ones?

15. Look at Appendix 10. Which stars are the hottest? Which are the brightest? Are they the same?

16. What is the brightest star in the sky? The second brightest? What color is Betelgeuse? Use Appendix 11 to find the answers.

17. Why can only a lower limit to the rate of stellar rotation be determined from rotational broadening, rather than the actual rotation rate? (Refer to Figure 16.9.)

18. Most of the mass of every star (including the Sun) is made of hydrogen and helium. Why are these two elements the most common?

19. If you were so rich that you had $100 million to give away, what instrument or project in astronomy would you want to build or support? (This might be a good question to answer at the end of the semester.)

Problems

20. A fifth-magnitude star is about the faintest that can be seen without optical aid unless you have access to a very dark, unpolluted sky. How much fainter is it than a zero-magnitude star? A good pair of binoculars can reveal tenth-magnitude stars. How much fainter are these than stars of zero magnitude?

21. As seen from the Earth, the Sun has an apparent magnitude of about −26. What is the apparent magnitude of the Sun as seen from Saturn, about 10 AU away? (Remember that 1 AU is the distance from the Earth to the Sun.)

22. If a star has a color index of B − V = 2.5, how many times brighter does it appear in visual light than in blue light, relative to a standard star with B − V = 0?

23. What are the approximate spectral classes for stars whose wavelengths of maximum light have the following values? (See Section 4.2.)
  a. 290 nm
  b. 50 nm
  c. 600 nm
  d. 1200 nm
  e. 1500 nm

24. The following equation describes the quantitative relationship between the magnitudes and light flux received from two stars. If m_1 and m_2 are the magnitudes corresponding to stars from which we receive light flux in the amounts l_1 and l_2, the difference between m_1 and m_2 is defined by

$$m_1 - m_2 = 2.5 \log (l_2/l_1)$$

Use this equation to calculate the following:

a. The difference in magnitudes of two stars that differ in light flux received at the Earth by a factor of ten.

b. The difference in light flux received from two stars that differ by ten magnitudes.

c. The difference in magnitudes of two identical stars, one of which is ten times farther away than the other.

25. Appendix 11 lists the 20 brightest stars. How much more light flux do we receive from the brightest of these stars than from the faintest? Use the equation in Problem 24 to calculate the answer.

Suggestions for Further Reading

Fraknoi, A. and Freitag, R. "Women in Astronomy" in *Mercury*, Jan./Feb. 1992, p. 27. Part of an issue devoted to a discussion of women in astronomy.

Hearnshaw, J. "Origins of the Stellar Magnitude Scale" in *Sky & Telescope*, Nov. 1992, p. 494. A good history of how we have come to have this cumbersome system.

Hearnshaw, J. *The Analysis of Starlight*. 1986, Cambridge U. Press. A history of spectroscopy in astronomy.

Hirshfeld, A. "The Absolute Magnitude of Stars" in *Sky & Telescope*, Sep. 1994, p. 35.

Kaler, J. *Stars*. 1992, Scientific American Library/W. H. Freeman. Good modern review of our understanding of stars.

Kaler, J. *Stars and Their Spectra*. 1989, Cambridge U. Press. A detailed introduction to the field of spectroscopy and what it can tell us about the stars.

Kaler, J. "Origins of the Spectral Sequence" in *Sky & Telescope*, Feb. 1986, p. 129.

Kidwell, P. "Three Women of American Astronomy" in *American Scientists*, May/June 1990, p. 244. Focuses on Annie Cannon and Cecilia Payne.

Sneden, C. "Reading the Colors of the Stars" in *Astronomy*, Apr. 1989, p. 36. Discusses what we learn from spectroscopy.

Steffey, P. "The Truth about Star Colors" in *Sky & Telescope*, Sep. 1992, p. 266. About the color index, and how the eye and film "see" colors.

Welther, B. "Annie J. Cannon: Classifier of the Stars" in *Mercury*, Jan./Feb. 1984, p. 28.

Using **REDSHIFT** ™

1. Turn on only the stars and set the display to *Mercator projection* with a zoom factor to 0.2. Adjust the *Magnitude Limit* until only the brightest 10 stars appear.

Which star is brightest?

What is the brightest blue star?

What is the brightest red star?

Which star is most like the Sun?

How much brighter is the brightest star than the faintest?

2. Turn on the *Constellation Boundaries* (patterns off) and set the *Faint Magnitude* limit to 4.

Which star in Leo is closest?

Which of its stars is farthest?

Are the distances related to the stars' brightnesses?

Which star is hottest?

Which star is coolest?

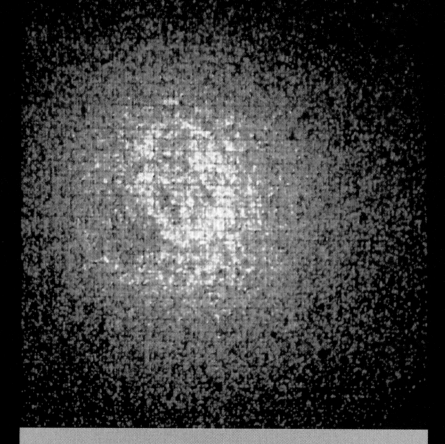

The determination of stellar masses depends on measuring the gravitational effects of one star on a companion star. With modern instruments and computer processing we can now measure binary star systems in which the stars are very close to each other. The top image shows a long time-exposure of the binary star Sigma Herculis. Atmospheric turbulence produces a blurred image of both stars that is about 2 arcsec in diameter. New techniques can remove this blurring and reveal two stars separated by just 0.07 arcsec. (Anthony Readhead/Palomar Observatory, Caltech)

CHAPTER 17

The Stars: A Celestial Census

Thinking Ahead

A typical person might live seven decades or so. A typical star lives for many billions of years. No one (and no culture on Earth) has ever watched a star go through all the stages of its life cycle. How then can we begin to understand the evolution of a star from its birth to its death?

> Raving politics, never at rest—as this poor earth's pale history runs—What is it all but a trouble of ants in the gleam of a million million of suns?
>
> Alfred Lord Tennyson in "Vastness" (1889).

Imagine for a moment that you are the captain of a well-equipped starship, sent from a civilization that orbits another star to study life-forms elsewhere. (You happen to resemble an intelligent cauliflower with appendages, but you have some excellent scientists among your crew.) You have arrived on Earth, and your schedule gives your crew exactly one Earth-day to learn all they can about the life cycle of the dominant life form on this new planet.[1] What would be your advice to the landing party?

The crew could pursue many strategies, after all. One might be for each cauliflower to watch a single member of the human species for a whole day, to see if any of them went through important life-cycle changes. For a random sample of humans, this would typically turn out to be a disappointingly unproductive strategy. A better approach might be for the crew to make a widespread survey, cataloging as many different humans as possible

---

[1] An interesting question is whether a new visitor to Earth would immediately infer that humans are the dominant life-form here. Some commentators have suggested that a more reasonable conclusion would be that cars rule the Earth. After all, they seem to swallow and disgorge humans (and other animals) at will, and humans are clearly their servants—washing, feeding, and polishing them with domestic regularity.

and then trying to determine which of the differences were significant for understanding how humans develop.

If you don't know anything about human beings to begin with (being an intelligent cauliflower, after all), those human characteristics that are important (and those that are simply random variations from one specimen to the next) would not be at all obvious. For example, you might note that humans come in different colors and hypothesize that they start life a rich, dark brown, getting lighter and lighter with age. The humans with the darkest skin would be the youngest, according to this theory. It's an interesting idea, but it would turn out to be completely wrong. Or you might note that humans come with a bit of fuzzy stuff on top of their heads, and make the very reasonable suggestion that the length of this fuzz is a good measure of age. This also would not get you very far down the road toward the cauliflower Nobel Prize.

On the other hand, there are characteristics that (properly understood) might help you pin down the human life cycle. For the first part of a human life, body mass and height are reasonably good indicators of age, for example. And the smoothness of our skin tends to decrease as we get older. To understand such subtle indicators would require a broad sampling of human characteristics and many years of careful study (even though your data would be collected in a single day).

Astronomers faced a similar problem with stars (Figure 17.1). Stars live such a long time that nothing much can be gained from staring at one for a human lifetime. It was necessary to measure the characteristics of many stars (to take a celestial census, in effect) and then determine which characteristics would help us understand the stars' life stories. Like the cauliflowers of our example, astronomers tried a variety of hypotheses about stars until they came up with the right approach to comprehending their development. But the key was making a thorough census of the stars around us.

17.1

A Stellar Census

Before we can make our own survey, we need to agree on a unit of distance appropriate to the objects we are studying. The stars are all so far away that kilometers (and even astronomical units) would be very cumbersome to use; so—as discussed in the Prologue—astronomers use a much larger "measuring stick" called the *light year (LY)*. A light year is the distance that light (the fastest signal we know) travels in one year. Since light covers an astounding 300,000 km per second, and since there are a lot of seconds in one year, a light year is a very large quantity: 9.5

(b)

(a)

Figure 17.1

(a) Stars, like people, come in a variety of colors. (Jeff Greenberg/Photo Researchers) (b) This image shows a technique invented by David Malin of the Anglo-Australian Observatory to demonstrate the colors of stars. We are looking at the constellation of Orion. The camera was held fixed, and the turning of the Earth caused the stars to drift across the image during the long exposure. As the picture was being taken, the photographer changed the focus of the lens to make each star into a wider and wider streak, thus bringing out the subtle colors. The three stars making the belt of Orion are in the center, the bright yellow-reddish star near the upper left is Betelgeuse, and the whitish-blue star at the lower right is Rigel. (© David Malin, Anglo-Australian Telescope Board)

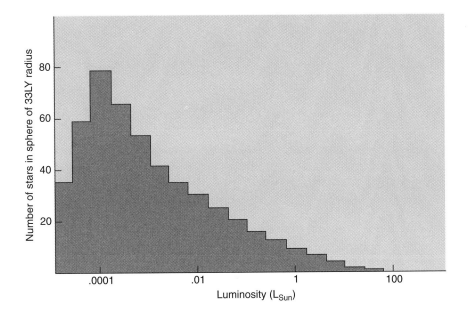

Figure 17.2
A luminosity function shows how many stars fall in each narrow range in luminosity. Shown here is the luminosity function of stars in our immediate neighborhood. The Sun's luminosity is 1 on this graph; note that most stars have a luminosity less than that.

trillion (9.5×10^{12}) km to be exact. (Bear in mind that the light year is a unit of *distance* even though the term *year* appears in it.) If you drove at the legal U.S. speed limit without stopping for food or rest, you would not arrive at the end of a light year in space until roughly 12 million years had passed. And the closest star is more than 4 LY away.

Notice that we have not yet said much about how such enormous distances can be measured. That is a complicated question, to which we will return in later chapters. For now, let us assume that distances have been measured for stars in our cosmic vicinity so that we can proceed with our census.

The Luminosity Function

As a first step, let's examine the stars in our immediate neighborhood. For example, Appendix 10 lists the stars found within 12 LY of the Sun, and also gives their luminosities, a characteristic defined in Chapter 16. (Remember that in order to understand how stars "work," we are interested in how much energy they put out, not how bright they happen to appear from Earth.) With surveys in hand, we can ask how many stars in a given volume of space are intrinsically very luminous, and how many are faint.

Figure 17.2 shows the results of counting stars in a volume of space slightly larger than the one for which data are given in Appendix 10. The stars in this region range from about 100 times the luminosity of the Sun, or 100 L_{Sun}, to less than one ten-thousandth L_{Sun}. If we divide these stars into groups according to luminosity, we can count how many fall into each narrow range. The relationship, shown in Figure 17.2, is called the *luminosity function*. (You are probably familiar with such graphs from surveys of people—for example, the number of residents in a city who fit into various income or age brackets.) As you can see, the Sun is more luminous than the vast majority of stars in our vicinity.

But does the local neighborhood contain samples of all types of stars? Surely the neighborhood in which you live does not contain all the types of people—distinguished according to age, education, income, race, etc.—that live in the entire country. For example, a few people do live to be over 100 years old, but there may be no such individual within several miles of where you live. In order to sample the full range of the human population, you would have to extend your census to a much larger area.

Similarly, there are some types of stars not found in our immediate neighborhood. A clue that we are missing something in our stellar census comes from the fact that only three first-magnitude stars—Sirius, Alpha Centauri, and Procyon—are found within 12 LY of the Sun (see Appendix 10). Why are we missing most of the stars that appear brightest in the sky when we take our census of the local neighborhood?

The answer, interestingly enough, is that the stars that appear brightest are *not* the closest ones to us. These stars are bright because they emit a very large amount of energy—so much, in fact, that they do not have to be nearby to look brilliant. You can confirm this by looking at Appendix 11, which gives distances for the 20 stars that appear brightest from Earth. In fact, it turns out that *most* of the stars visible without a telescope are hundreds of light years away and many times more luminous than the Sun. Among the 6000 stars visible to the unaided eye, at most 50 are intrinsically fainter than the Sun. Figure 17.3 is a graph showing the distribution of the luminosities of the 30 brightest-appearing stars. Notice that the most luminous of these stars emits nearly 100,000 times more energy than does the Sun.

These highly luminous stars are missing from the solar neighborhood because they are very rare. None of

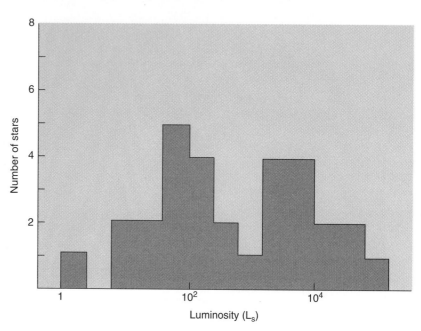

Figure 17.3

Distribution (bar graph) of the luminosity of the 30 brightest-appearing stars in our sky. We plot the number of stars in each range of luminosity.

them happens to be in the tiny volume of space immediately surrounding the Sun that was surveyed to get the data shown in Figure 17.2.

On the other hand, stars fainter than the Sun cannot be seen with the unaided eye unless they are *very* nearby. For example, stars with luminosities ranging from 10^{-2} to 10^{-4} L_{Sun} are very common, but a star with a luminosity of 10^{-2} L_{Sun} would have to be within 5 LY to be visible to the naked eye—and only three stars (all in one system) are this close to us. The nearest of these three stars, Proxima Centauri, cannot even be seen without a telescope because it has such a low luminosity. The common nearby stars typically do not send enough light across interstellar distances to be seen with the unaided eye, and just don't make it into a "top ten" list of stars that appear bright.

Let's again consider the highly luminous stars—those 100 or more times as luminous as the Sun. Although such stars are rare, they are visible to the unaided eye even when hundreds to thousands of light years away. For example, a star with a luminosity of 10^4 L_{Sun} can be seen without a telescope out to a distance of 5000 LY. Such stars are very rare, and we would not expect to find one nearby. The volume of space included within a distance of 5000 LY, however, is enormous; so even though they are intrinsically rare, many such highly luminous stars are readily visible to the unaided eye.

The contrast between these two samples of stars—those that are close to us and those that can be seen with the unaided eye—is an example of a *selection effect*. When a population of objects (stars in this example) includes a great variety of different types, we must be careful what conclusions we draw from an examination of any particular subgroup. Certainly we would be fooling ourselves if we assumed that the stars visible to the unaided eye are characteristic of the general stellar population; this subgroup is heavily weighted to the most luminous stars. It requires much more effort to assemble a complete data set for the nearest stars, since most are so faint that they can be observed only with a telescope. However, it is only by doing so that astronomers are able to work out the properties of the vast majority of the stars, which are actually much smaller and fainter than our own Sun.

The Density of Stars in Space

Now let's use our surveys to answer another question. What is the typical spacing between stars? There are at least (we may have missed a few) 59 stars within 16 LY of the Earth, counting the members of binary and multiple star systems, and the Sun itself. A sphere with a radius of 16 LY has a volume of about 17,000 cubic LY (LY^3). Since this volume of space contains at least 59 stars, the density of stars in space in the neighborhood of the Sun is about one star for every 300 LY^3. The average distance between stars is then about 7 LY (the cube root of 300). The Sun and Alpha Centauri, which are separated by 4.4 LY, are somewhat closer than the average. In any case, you can see that stars (and star systems) live in splendid isolation, with enormous gulfs of space among them.

Density is more typically expressed in g/cm^3, not stars per LY^3 as used in the preceding paragraph. To see how dense our neighborhood is in more common terms, assume the matter contained in stars could be spread out evenly over space. If a typical star has a mass 0.4 times that of the Sun, the average density of matter in the solar neighborhood is only about 3×10^{-24} g/cm^3—or about one hydrogen atom per cubic centimeter. (Notice that planets, if they exist elsewhere, have too little mass compared to stars for us to bother including them.) How does the density get to be so small? While stars have a lot of mass, the amount of space between the stars is so enormous that the average cosmic density of our vicinity is very low indeed.

Stellar Masses

We have seen that the Sun is more luminous than most stars. How does its mass compare with other stars? We will see as we study stars that mass is the prime indicator of what type and length of life a star will have. Yet it is not very easy to measure the mass of stars directly: somehow we need to put a star on the cosmic equivalent of a scale. Luckily, not all stars live like the Sun, in isolation from other stars. The best information about stellar masses comes from the study of **binary stars**—two stars that orbit around one another, bound together by gravity. We can use their pull on each other to measure their mass.

Many, but not all, stars are members of binary star systems. For example, among the 59 nearest stars, 27 (roughly one-half) are members of systems containing more than one star. Masses for stars can be calculated from measurements of their orbits, just as the mass of the Sun can be derived by measuring the orbits of the planets around it (see Chapter 2).

Binary Stars

Before we discuss in more detail how mass can be measured, we want to take a closer look at stars that come in pairs. The first binary star was discovered in 1650, less than half a century after Galileo began to observe the sky with a telescope. John Baptiste Riccioli, an Italian astronomer, noted that the star Mizar, in the middle of the Big Dipper's handle, appeared through his telescope as two stars. Since that discovery, thousands of binary stars have been cataloged. (Astronomers call any pair of stars close to each other in the sky *double stars*, but not all of these are two stars that are physically associated. Some are just chance alignments of stars that are actually at different distances from us.) Although stars most commonly come in pairs, there are also triple and quadruple systems.

One well-known binary star is Castor, located in the constellation of Gemini. By 1804 astronomer William Herschel, who also discovered the planet Uranus, had noted that the fainter component of Castor had slightly changed its position relative to the brighter component. (We use the term *component* to mean a member of a star system.) Here was evidence that one star was moving around another. It was the first evidence that gravitational influences exist outside the solar system. The orbital motion of the binary star Kruger 60 is shown in Figure 17.4. Binary star systems in which both of the stars can be seen with a telescope are called **visual binaries.**

A class of double stars in which both stars *cannot* be seen was discovered by Edward C. Pickering at Harvard in 1889. He was examining the spectrum of Mizar and found that the dark absorption lines in the brighter star's spectrum were usually double. Not only were there two lines where astronomers normally saw only one, but the spacing of the lines was constantly changing. At times the lines even became single. Pickering correctly deduced that the brighter component of Mizar, called Mizar A, is itself really two stars that revolve about each other in a period of 104 days. Stars like Mizar A, which appear as single stars when photographed or observed visually through the telescope, but which spectroscopy shows really to be double stars, are called **spectroscopic binaries.**

Mizar, by the way, is a good example of just how complex such star systems can be. Mizar has been known for centuries to have a faint companion called Alcor, which can be seen without a telescope. Mizar and Alcor form an *optical double*—a pair of stars that appear close together in the sky but do not orbit around each other. Through a telescope Mizar can be seen to have another, closer companion that does orbit around it; Mizar is thus a visual binary. The two components that make up this visual binary, known as Mizar A and Mizar B, are both spectroscopic binaries. So, Mizar is really a quadruple system of stars.

Strictly speaking, it is not correct to describe the motion of a binary star system by saying that one star orbits

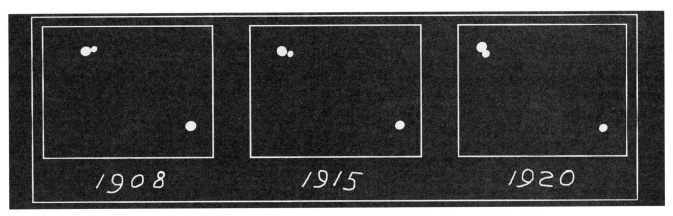

Figure 17.4
Revolution of a binary star. Three photographs covering a period of about 12 years show the mutual revolution of the two stars that make up the nearby star system Kruger 60. (Yerkes Observatory)

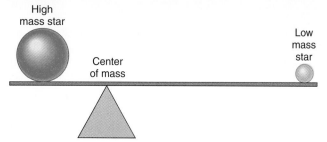

Figure 17.5
In a binary star system, both stars orbit their center of mass. A star with higher mass will be balanced closer to the center of mass, while a star with lower mass will be balanced farther from it.

the other. Gravitation is a *mutual* attraction. Each star exerts a gravitational force on the other, with the result that both stars orbit a point between them called the *center of mass*. Imagine that the two stars are seated, one at each end of a seesaw. The point at which the fulcrum would have to be located in order for the seesaw to balance is the center of mass, and it is always closer to the more massive star (Figure 17.5).

Figure 17.6 shows two stars moving around their center of mass, along with the spectrum we observe from the system at different times. When one star is approaching us relative to the center of mass, the other star is receding from us. In the top illustration, star A is moving toward us, so the lines in its spectrum are Doppler-shifted toward the blue end of the spectrum. Star B is moving away from us, so its spectrum shows a red shift. When we observe the composite spectrum of the two stars, each line appears double. When the two stars are both moving across our line of sight (neither away from nor toward us), however, they both have the same radial velocity (that of the pair's center of mass); hence the spectral lines of the two stars come together. This is shown in the second and fourth illustrations in Figure 17.6. A plot showing how the velocities of the stars change with time is called a *radial-velocity curve*; one for the binary system in Figure 17.6 is shown in Figure 17.7.

Masses from the Orbits of Binary Stars

We can estimate the masses of double star systems by using Newton's reformulation of Kepler's third law (discussed in Section 2.3). Kepler found that the time a planet takes to go around the Sun is related by a specific mathematical formula to its distance from the Sun. In our binary star situation, if two objects are in mutual revolution, then the period (P) with which they go around each other is related to the semimajor axis (D) of the orbit of one with respect to the other, according to the equation

$$D^3 = (M_1 + M_2)P^2$$

where D is in astronomical units, P is measured in years, and $M_1 + M_2$ is the sum of the masses of the two stars in units of the Sun's mass. Thus if we can observe the size of

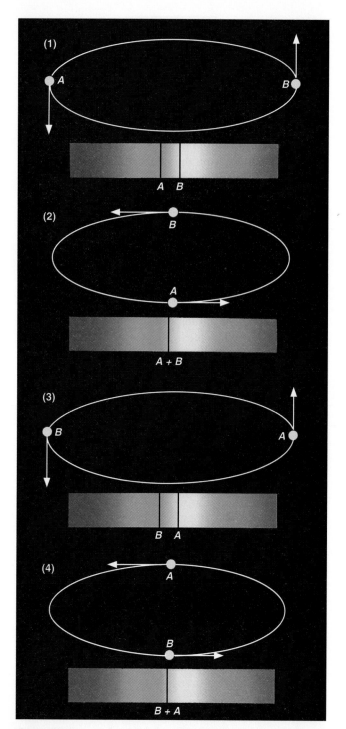

Figure 17.6
A schematic drawing of the motions of two stars orbiting each other. When one star is moving toward the Earth, the other is moving away; half a cycle later the situation is reversed. Doppler shifts cause the spectral lines to move back and forth. At times, lines from both stars can be seen well-separated from each other. When the two stars are moving perpendicular to our line of sight, the two lines are exactly superimposed, and so we see only a single spectral line.

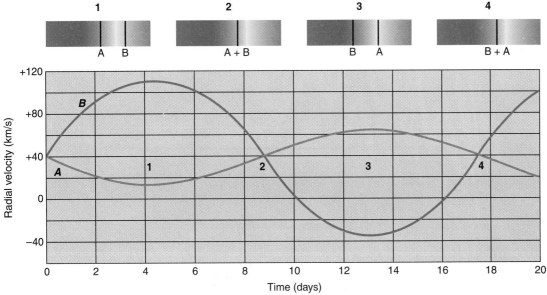

Figure 17.7
Radial-velocity curves for a spectroscopic binary system showing how the two components alternately approach and recede from the Earth. The positions on the curve corresponding to the illustrations in Figure 17.6 are marked.

the orbit and the period of mutual revolution of the stars in a binary system, we can calculate the sum of their masses.

Most spectroscopic binaries have periods ranging from a few days to a few months, with separations of usually less than 1 AU between their member stars. Recall that an AU is the distance from the Earth to the Sun, so this is a small separation and very hard to see at the distances of stars. This is why many of these systems are known to be double only through careful study of their spectra.

The mathematical analysis of a radial-velocity curve (such as the one in Figure 17.7) to determine the masses of the stars in a spectroscopic binary is complex in practice but not hard in principle. The speeds of the stars are measured from the Doppler effect. We then measure the period—how long the stars take to go through an orbital cycle—which can be determined from the velocity curve. Knowing how fast the stars are moving and how long they take to go around tells us the circumference of the orbit and hence the separation of the stars in kilometers or astronomical units. Using Kepler's law, the period and the separation allow us to calculate the sum of the stars' masses.

Of course, knowing the sum of the masses is not as useful as knowing the mass of each star separately. But the relative orbital speeds of the two stars can tell us how much of the total mass each star has. The more massive star has a smaller orbit and hence moves more slowly to get around in the same time. In practice we also need to know how the binary system is oriented in the sky to our line of sight; but if we do and the just-described steps are carried out carefully, the result is a measurement of the masses of each of the two stars in the system.

In short, a good understanding of the motion of two stars around a common center of mass, combined with the laws of gravity, allows us to measure the masses of stars in such systems. These measurements are absolutely crucial to developing a theory of how stars evolve. One of the best things about this method is that it is independent of the location of the binary system. It works as well for stars 1000 LY away as for those in our immediate neighborhood.

The Range of Stellar Masses

How large can the mass of a star be? We find that stars more massive than the Sun are rare. There are no stars within 30 LY of the Sun with masses greater than four times that of the Sun. Searches for massive stars at large distances from the Sun indicate that masses up to about 100 times that of the Sun do occur, but very few stars are this huge (Figure 17.8). There is no convincing evidence for the existence of any stars with masses significantly exceeding about 100 times the mass of the Sun.

According to theoretical calculations, the smallest mass that a true star can have is about 1/12 that of the Sun. By a true star, astronomers generally mean one that becomes hot enough to fuse hydrogen to form helium. Stars with masses smaller than the Sun's are very common, although they are difficult to detect if far away.

Objects with masses between 1/100 and 1/12 that of the Sun may produce energy for a brief time by means of nuclear reactions involving deuterium, but they do not become hot enough to force protons to combine to produce helium. Such objects are called **brown dwarfs,** and for

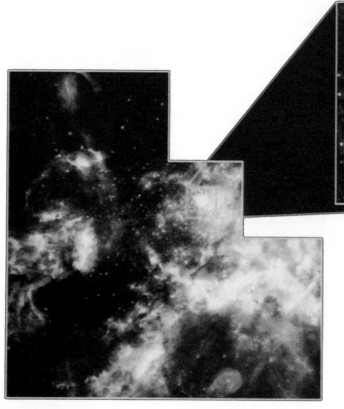

Figure 17.8

R 136, a cluster that contains many massive stars. Part of a giant region of cosmic raw material called the 30 Doradus nebula (larger image), this cluster of young hot stars (smaller image) contains some of the most massive stars known, including some with masses as large as 100 M$_{Sun}$. The nebula and cluster are 160,000 LY away in the Large Magellanic Cloud, a smaller galaxy that is a satellite of our Milky Way. The cluster is very difficult to make out clearly with telescopes on the ground; this image was one of the tests to which astronomers subjected the repaired Hubble Space Telescope. If it can distinguish stars in the R 136 cluster clearly (and see for yourself that it does), then the Space Telescope must be working pretty well. (NASA/STScI)

more than a decade astronomers have been trying to find examples of such "failed stars." This is not easy to do, since they are extremely dim and thus difficult to observe. Ingenious techniques are required to make sure that a very faint object is a true brown dwarf and not merely a very low-mass star.

In 1995 three groups of astronomers announced what appears to be convincing evidence for brown dwarfs in three different star systems. Two are located in a cluster of stars called the Pleiades, which is young enough so that any newly formed brown dwarfs have not had time to fade away to complete obscurity. One candidate, called PPL 15, appears to be just on the borderline between stars and brown dwarfs. The second, called Teide 1, seems less massive but still requires more investigation. However, the third candidate, in orbit around a nearby star called Gliese 229, appears to have somewhere between 20 and 40 times the mass of Jupiter, and is thus very likely a brown dwarf. Its spectrum, as observed with the giant Keck telescope, shows evidence of methane, a molecule that would be destroyed by heat in the outer layers of a star. Many groups around the world are continuing to hunt for other examples of these fascinating but elusive objects.

Still-smaller objects with masses less than 1/100 the mass of the Sun are called *planets.* They may radiate energy produced by the radioactive elements that they contain, and they may also radiate heat generated by slow gravitational contraction. However, their interiors will never reach temperatures high enough for nuclear reactions to take place. Jupiter, whose mass is about 1/1000 the mass of the Sun, is unquestionably a planet, for example. Until the 1990s we could only detect planets in our own solar system, but now we have begun to detect them elsewhere as well (see Chapter 20).

The Mass–Luminosity Relation

Now that we have measurements of the characteristics of many different types of stars, we can search for relationships among the characteristics. For example, we can ask whether the mass and luminosity of a star are related. It turns out that for most stars, they are: the more massive stars are generally also the more luminous. This relationship, known as the **mass–luminosity relation,** is shown graphically in Figure 17.9. Each point represents a star whose mass and luminosity are both known. Horizontal position on the graph shows the star's mass, given in units of the Sun's mass, and vertical position shows its luminosity in units of the Sun's luminosity.

Notice how good this relationship is: most stars fall along a line running from the lower left (low mass, low luminosity) corner of the diagram to the upper right (high mass, high luminosity) corner. About 90 percent of all stars obey the mass–luminosity relation illustrated in Figure 17.9. Later we will explore why such a relationship exists, and what we can learn from the roughly 10 percent of stars that "disobey."

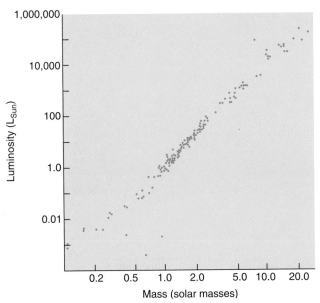

Figure 17.9
The mass–luminosity relation. The plotted points show the masses and luminosities of stars for which both of these quantities are known to an accuracy of 15 to 20 percent. The three points lying below the sequence of points are all white dwarf stars. (Adapted from data compiled by D. M. Popper)

17.3

Diameters of Stars

It is easy to measure the diameter of the Sun. Its angular diameter—that is, its apparent size on the sky—is about 1/2°. If we know the angle the Sun takes up in the sky and how far away it is, we can calculate its true (linear) diameter, which is 1.39 million km, or about 109 times the diameter of the Earth.

Unfortunately, the Sun is the only star whose angular diameter is easily resolved. All the other stars are so far away that they look like pinpoints of light through even the largest telescopes. (They often seem to be bigger, but that is merely distortion introduced by turbulence in the Earth's atmosphere.) Luckily, there are several techniques that astronomers can use to estimate the sizes of stars; we discuss two of them next.

Diameters of Stars Whose Light Is Blocked by the Moon

One method, which gives very precise diameters but can be used for only a few stars, is to observe the dimming of light that occurs when the Moon passes in front of a star. What astronomers measure (with great precision) is the time required for the star's brightness to drop to zero as the edge of the Moon moves across it. Since we know how rapidly the Moon moves in its orbit around the Earth, it is possible to calculate the angular diameter of the star. If the distance to the star is also known, this gives us its diameter in kilometers. This method works only for fairly bright stars that happen to lie along the zodiac, where the Moon (or, much more rarely, a planet) can pass in front of them as seen from the Earth.

Eclipsing Binary Stars

A second technique for measuring diameters works for those stars that are members of **eclipsing binaries,** and so we must make a brief detour from our main story to examine this type of star system. Some double stars are lined up in such a way that when viewed from the Earth, each star passes in front of the other during every revolution (Figure 17.10). When one star blocks the light of the other, preventing it from reaching the Earth, the luminos-

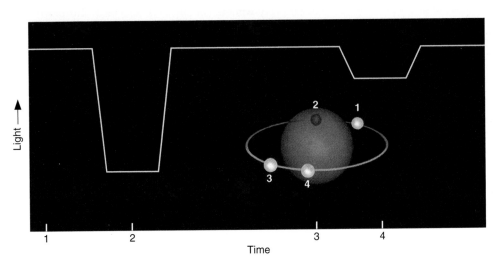

Figure 17.10
The light curve of a hypothetical eclipsing binary star with total eclipses (one star passes directly in front of and behind the other). The numbers indicate parts of the light curve corresponding to various positions of the smaller star in its orbit.

Making Connections

Astronomy and Mythology: Algol the Demon Star and Perseus the Hero

The name Algol comes from the Arabic *Ras al Ghul,* meaning "the demon's head." The word *ghoul* in English has the same derivation. As discussed in Chapter 1, many of the bright stars have Arabic names because during the long dark ages in medieval Europe, it was Arabic astronomers who preserved and expanded the Greek and Roman knowledge of the skies. The reference to the demon is part of the ancient Greek legend of the hero Perseus, who is commemorated by the constellation in which we find Algol, and whose adventures involve many of the characters associated with the northern constellations.

Perseus was one of the many half-god heroes fathered by Zeus (Jupiter in the Roman version), the king of the gods in Greek mythology. Zeus had, to put it delicately, a roving eye and was always fathering somebody or other with a human maiden who caught his fancy. (Perseus derives from *Per Zeus,* meaning "fathered by Zeus.") Set adrift with his mother by an (understandably) upset stepfather, Perseus grew up on an island in the Aegean Sea. The king there, taking an interest in Perseus' mother, tried to get rid of the young man by assigning him an extremely difficult task.

In a moment of overarching pride, a beautiful young woman named Medusa had compared her golden hair with that of the goddess Minerva. The Greek gods did not take kindly to being compared to mere mortals, and Minerva turned Medusa into a "Gorgon," a hideous, evil creature with writhing snakes for hair and a face that turned anyone who looked at it into stone. (In these mythological stories, there were no second or third chances; one strike and you were out.) Perseus was given the task of slaying the demon, which seemed like a pretty sure way to get him out of the way forever.

But because Perseus had a god for a father, some of the other gods gave him tools for the job, including Minerva's reflective shield and the winged sandals of Hermes (Mercury in the Roman story). By flying over her and looking only at her reflection, Perseus was able to cut off Medusa's head without ever looking at her directly. Taking her head (which could still turn onlookers to stone even without being attached to her body) with him, Perseus continued on to other adventures.

He next came to a rocky seashore, where boasting had gotten another family into serious trouble with the gods. Queen Cassiopeia had dared to compare her own beauty to that of the Nereids, sea nymphs who were daughters of Poseidon (Neptune in Roman mythology), the god of the sea. (Today one of the planet Neptune's moons is named after the Nereids.) Poseidon was so offended that he created a sea-monster named Cetus to devastate the kingdom. King Cepheus, Cassiopeia's beleaguered husband, consulted the oracle, who told him that he must sacrifice his beautiful daughter Andromeda to the monster.

When Perseus came along and found Andromeda chained to a rock near the sea, awaiting her fate, he rescued her by turning the monster to stone. (Scholars of mythology actually trace the essence of this story back to far-older legends from ancient Mesopotamia, in which the god-hero Marduk vanquishes a monster named Tiamat. Symbolically, a hero like Perseus or Marduk is usually associated with the Sun, the monster with the power of night, and the beautiful maiden with the fragile beauty of dawn, which the Sun releases after its nightly struggle with darkness.)

All the characters in these Greek legends can be found as constellations in the sky, not necessarily resembling their namesakes, but serving as reminders of the story. (See the star map for October, after the Appendices.) For example, vain Cassiopeia is sentenced to be very close to the celestial pole, rotating perpetually around the sky and hanging upside down every winter. The ancients imagined Andromeda still chained to her rock (it is much easier to see the chain of stars than to recognize the beautiful maiden in this star grouping). Perseus is next to her with the head of Medusa swinging from his belt. Algol represents this gorgon head, and has long been associated with evil and bad fortune in such tales. Some commentators have speculated that the star's change in brightness (which can be observed with the unaided eye) may have contributed to its unpleasant reputation, with the ancients regarding such a change as an evil sort of "wink."

---

ity of the system decreases, and astronomers say that an *eclipse* has occurred.

The first eclipsing binary was discovered very shortly after Pickering found that Mizar A is a spectroscopic binary, and the discovery helped solve a long-standing puzzle in astronomy. The star Algol, in the constellation of Perseus, changes its brightness in an odd but regular way. Normally Algol is a second-magnitude star, but at intervals of 2^d, 20^h, 49^m it fades to one-third of its regular brightness. After a few hours it brightens to normal again. This effect is easily seen even without a telescope if you know what to look for—try observing Algol if you have access to clear skies without too much light pollution. (See "Making Connections" above.)

In 1783, more than a century before the spectroscopic observations of stars became possible, a young

English astronomer named John Goodricke (profiled in Chapter 18) made a careful study of Algol. Even though Goodricke could neither hear nor speak, he made a number of major discoveries in the 21 years of his brief life. He suggested that Algol's unusual brightness variations might be due to an invisible companion that regularly passes in front of the brighter star and blocks its light. Unfortunately, Goodricke had no way to test this idea, since the equipment available at that time was not good enough to measure Algol's spectrum.

In 1889 the German astronomer Hermann Vogel demonstrated that, like Mizar, Algol is a spectroscopic binary. The spectral lines of Algol were not observed to be double because the fainter star of the pair gives off too little light compared with the brighter for its lines to be conspicuous in the composite spectrum. Nevertheless, the periodic shifting back and forth of the brighter star's lines gave evidence that it was revolving about an unseen companion. (The lines of both components need not be visible for a star to be recognized as a spectroscopic binary.)

The discovery that Algol is a spectroscopic binary verified Goodricke's hypothesis. The plane in which the stars revolve is turned nearly edgewise to our line of sight, and each star passes in front of the other during every revolution. (The eclipse of the fainter star in the Algol system is not very noticeable because the part of it that is covered contributes little to the total light of the system. This second eclipse can, however, be detected by careful measurements.) Any binary star produces eclipses if viewed from the proper direction, near the plane of its orbit, so that one star passes in front of the other (see Figure 17.10). But from our vantage point on Earth, only a few of these stars are oriented in this way.

Diameters of Eclipsing Binary Stars

We now turn back to the main thread of our story to discuss how all this can be used to measure the sizes of stars. The technique involves making a careful *light curve* of an eclipsing binary, a graph that plots how the brightness changes with time. Let us consider a hypothetical binary system in which the stars are very different in size, like those illustrated in Figure 17.11. To make life easy, we will assume that the orbit is viewed exactly edge-on. Even though we cannot see the two stars separately in such a system, the light curve can tell us what is happening. When the small star just starts to pass behind the large star (a point we call *first contact*), the brightness begins to drop. The eclipse becomes total (the small star is completely hidden) at the point called second contact. At the end of the total eclipse (third contact), the small star begins to emerge. When the small star has reached last contact, the eclipse is completely over.

To see how this allows us to measure diameters, look carefully at Figure 17.11. During the time interval between first and second contacts, the small star has moved a distance equal to its own diameter. During the time in-

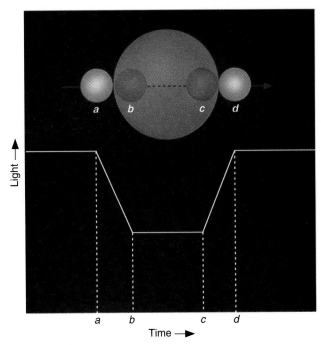

Figure 17.11
The light curve of a hypothetical eclipsing binary star, whose orbit we view exactly edge-on, in which the two stars fully eclipse each other. From the time intervals between contacts it is possible to estimate the diameters of the two stars.

terval from first to third contacts, the small star has moved a distance equal to the diameter of the large star. If the spectral lines of both stars are visible in the spectrum of the binary, the speed of the small star with respect to the large one can be measured from the Doppler shift. But knowing the speed with which the small star is moving and how long it took to cover some distance can tell us the span of that distance—in this case the diameters of the stars. The speed, multiplied by the time interval from first to second contact, gives the diameter of the small star. Multiply it by the time between first and third contact and you get the diameter of the large star.

In actuality, the orbits are generally not seen exactly edge-on, and the light from each star may be only partially blocked by the other. Furthermore, binary star orbits, just like the orbits of the planets, are ellipses, not circles. However, all these effects can be sorted out from very careful measurements of the light curve.

Stellar Diameters

The results of stellar size measurements confirm that most nearby stars are roughly the size of the Sun—with typical diameters of a million kilometers or so. Faint stars, as we might have expected, are generally smaller than more luminous stars. However, there are some dramatic exceptions to this simple generalization.

A few of the very luminous stars, those that are also red in color (indicating relatively low surface tempera-

TABLE 17.1
Measuring the Characteristics of Stars

Characteristic	Technique
Surface temperature	1. Determine the color (very rough).
	2. Measure the spectrum and get the spectral type.
Chemical composition	Determine which lines are present in spectrum.
Luminosity	Measure apparent brightness and compensate for distance.
Radial velocity	Measure the Doppler shift in the spectrum.
Rotation	Measure the width of spectral lines.
Mass	Measure the period and radial velocity curves of spectroscopic binary stars.
Diameter	1. Measure the way a star's light is blocked by the Moon.
	2. Measure the light curves and Doppler shifts for eclipsing binary stars.

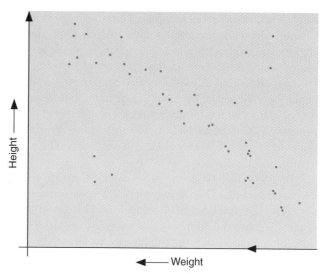

Figure 17.12
The plot of the height and weight of a representative group of human beings. Most lie along a "main sequence" representing normal people, but there are a few exceptions.

tures), turn out to be truly enormous. These stars are called, appropriately enough, *giants* or *supergiants*. An example is Betelgeuse, the second-brightest star in the constellation of Orion and one of the dozen brightest stars in our sky. Its diameter is greater than 10 AU, large enough to fill the entire inner solar system almost as far out as Jupiter. In Chapter 21 we will look in detail at the evolutionary process that leads to the formation of giant and supergiant stars.

17.4
The H–R Diagram

In this chapter and the previous one we have described some of the characteristics by which we might classify stars, and how those characteristics are measured. These ideas are summarized in Table 17.1. We have also given an example of a relationship between two of these characteristics in the mass–luminosity relation discussed earlier. When the characteristics of large numbers of stars were measured at the beginning of the 20th century, astronomers were able to begin a deeper search for patterns and relationships in these data.

To help understand what sorts of relationships might be found, let us return briefly to our intelligent cauliflowers who are trying to make sense of their data about human beings. Being good scientists, they might try plotting their data in different ways. Suppose they make a plot of the height of a good sampling of humans against their weight (which is a measure of their mass). Such a plot is shown in Figure 17.12 and has some interesting features. The way we have chosen to present our data, height in-

creases upward, while weight increases to the left. Notice that humans are not randomly distributed in that graph: most fall along a sequence that goes from the upper left to the lower right.

We can conclude from this graph that generally speaking, taller human beings weigh more, while shorter ones weigh less. This makes sense if you are familiar with the structure of human beings: typically, if we have bigger bones, we have more flesh to fill out our larger frame. It's not mathematically exact—there is a wide range of variation—but it's not a bad overall rule. And, of course, there are some dramatic exceptions. You occasionally see a short human who is very overweight and would thus be more to the bottom left of our diagram than the average sequence of people. Or you might have a very tall, skinny fashion model with great height but relatively small weight, who would be found near the upper right of the figure. A similar diagram has been found extremely useful for understanding the lives of stars.

In 1913 American astronomer Henry Norris Russell plotted the luminosities of stars against their spectral classes (a way of denoting their surface temperatures). This investigation, and a similar independent study in 1911 by Danish astronomer Ejnar Hertzsprung (Figure 17.13), led to an extremely important discovery concerning how these characteristics of stars are related.

Features of the H–R Diagram

Following Hertzsprung and Russell, let us plot the temperature (or spectral class) of a selected group of nearby stars against their luminosity and see what we find

(continued on page 354)

Henry Norris Russell

VOYAGERS IN ASTRONOMY

When Henry Norris Russell graduated from Princeton University, his work had been so brilliant that the faculty decided to create a new level of honors degree beyond "summa cum laude" for him. His students later remembered him as a man whose thinking was three times faster than just about anybody else's. His memory was so phenomenal, he could correctly quote an enormous number of poems and limericks, the entire Bible, tables of mathematical functions, and almost anything he had learned about astronomy. He was nervous, active, competitive, critical, and very articulate—he tended to dominate every meeting he attended. In outward appearance he was an old-fashioned product of the 19th century who wore high-top black shoes and high starched collars, and carried an umbrella every day of his life. His 264 papers were enormously influential in many areas of astronomy.

Born in 1877, the son of a Presbyterian minister, Russell showed early promise. When he was 12, his family sent him to live with an aunt in Princeton so he could attend a top preparatory school. He lived in the same house in that town until his death in 1957 (interrupted only by a brief stay in Europe for graduate work). He was fond of recounting that both his mother and his maternal grandmother had won prizes in mathematics, and that he probably inherited his talents in that field from their side of the family.

Before Russell, American astronomers devoted themselves mainly to surveying the stars and making impressive catalogs of their properties—especially their spectra (as described in Chapter 16). Russell began to see that interpreting the spectra of stars required a much more sophisticated understanding of the physics of the atom, a subject that was being developed by European physicists in the 1910s and 20s.

Russell embarked on a lifelong quest to ascertain the physical conditions inside stars from the clues in their spectra; his work inspired, and was continued by, a generation of astronomers, many trained by Russell and his collaborators.

Russell also made important contributions in the study of binary stars and the measurement of star masses, the origin of the solar system, the atmospheres of planets, and the measurement of distances in astronomy, among other fields. He was an influential teacher and popularizer of astronomy, writing a column on astronomical topics for *Scientific American* magazine for over 40 years. He and two colleagues wrote a textbook for college astronomy classes that helped train astronomers and astronomy enthusiasts over several decades. This book set the scene for the kind of textbook you are now reading, which not only lays out the facts of astronomy but explains how they fit together. Russell gave lectures around the country, often emphasizing the importance of understanding modern physics in order to grasp what was happening in astronomy.

Harlow Shapley, director of the Harvard College Observatory, called Russell "the dean of American astronomers." He was certainly regarded as the leader of the field for many years, and was consulted on many astronomical problems by colleagues from around the world. Today, one of the highest recognitions that an astronomer can receive is an award from the American Astronomical Society called the Henry Norris Russell Prize, set up in his memory.

Figure 17.13
Ejnar Hertzsprung (1873–1967) and Henry Norris Russell (1877–1957) independently discovered the relationship between the luminosity and surface temperature of stars that is summarized in what is now called the H–R diagram. (Sterrewacht Leiden and Princeton University Archives)

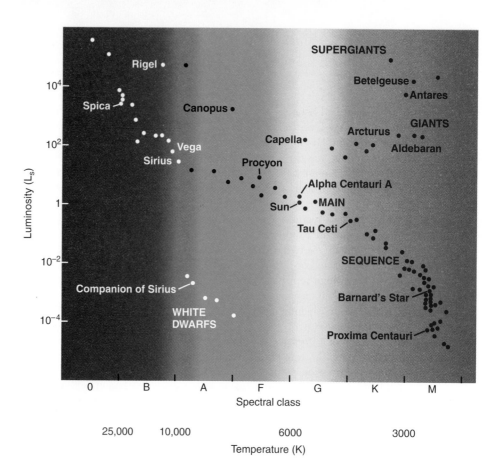

Figure 17.14

The H–R diagram for a selected sample of stars. Luminosity is plotted along the vertical axis. Along the horizontal axis we can plot either temperature or spectral type. Several of the brightest stars are identified by name.

(Figure 17.14). Such a plot is frequently called the **Hertzsprung–Russell diagram,** abbreviated **H–R diagram.** It is one of the most important and widely used diagrams in astronomy, with applications that extend far beyond the purposes for which it was originally developed nearly a century ago.

It is customary to plot H–R diagrams in such a way that temperature increases toward the left and luminosity toward the top. Notice the similarity to our plot of height and weight for people. Stars, like people, are not distributed over the diagram at random, as they would be if they exhibited all combinations of luminosity and temperature. Instead, we see that the stars cluster into certain parts of the H–R diagram. The great majority are aligned along a narrow sequence running from the upper left (hot, highly luminous) to the lower right (cool, less luminous). This band of points is called the **main sequence.** It represents a relationship between *temperature* and *luminosity* that is followed by most stars. Hotter stars are more luminous than cooler ones.

A number of stars, however, lie above the main sequence on the H–R diagram, in the upper right (cool, high luminosity) region. How can a star be at once cool, meaning each point on the star does not put out all that much energy, and yet very luminous? The only way is for the star to be enormous—to have so many points on its surface that the *total* energy output can still be large.

These stars must be giants or supergiants, the stars of huge diameter we discussed above.

The stars in the lower left (hot, low luminosity) corner of the diagram, on the other hand, have high surface temperatures, so that each point on a given star puts out a lot of energy. How then can the overall star be dim? It must be that it has a very small surface area; such stars are known as **white dwarfs** (white because the colors blend together to make them look bluish-white). We will say more about these puzzling objects in a moment. Figure 17.15 is a schematic H–R diagram for a large sample of stars, drawn to make the various types more apparent.

Now think back to our discussion of star surveys. It is difficult to plot an H–R diagram that is truly representative of all stars, because most stars are so faint that we cannot see those outside our immediate neighborhood. The stars plotted in Figure 17.14 were selected because their distances are known. This sample omits many intrinsically faint stars that are nearby but have not had their distances measured, so it shows fewer faint main-sequence stars than a "fair" diagram would. To be truly representative of the stellar population, an H–R diagram should be plotted for all stars within a certain distance. Unfortunately, our knowledge is reasonably complete only for stars within 10 to 20 LY of the Sun, among which there are no giants or supergiants. Still, from many surveys (and more can now be done with new, more powerful

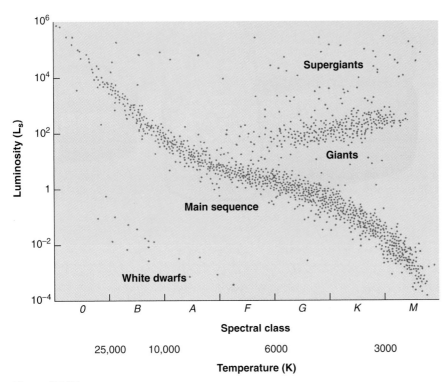

Figure 17.15
Schematic H–R diagram for many stars.

telescopes) we estimate that overall about 90 percent of the stars in our part of space are main-sequence stars, about 10 percent are white dwarfs, and fewer than 1 percent are giants or supergiants.

This result can be used directly to understand the lives of stars. Permit us another quick analogy with people. Suppose the intelligent cauliflowers return to your town and this time focus their attention on the location of young people, ages 6 to 18. Their survey teams fan out and take data about where such youngsters are found at all times during a 24-hour day. Some are found in the pizza parlor, others at home, others at the movies, many in school. After surveying a very large number of young people, the teams determine that, averaged over the course of the 24 hours, one-third of all youngsters are found in school.

How can we interpret this result? Does it mean that two-thirds of students are truants and the remaining one-third spend all their time in school? No, we must bear in mind that the survey teams counted youngsters throughout the full 24-hour day. Some survey teams worked at night, when most youngsters were at home asleep, and others in the late afternoon, when they were on their way home from school (and most likely to be enjoying a pizza). If the survey was truly representative, we *can* conclude, however, that if an average of one-third of all youngsters are found in school, then humans ages 6 to 18 must spend about one-third of their time in school.

We can do something similar for stars. We find that on average 90 percent of all stars are located on the main sequence of the H–R diagram. If we can identify some activity or life stage with the main sequence, then it follows that stars must spend 90 percent of their lives in that activity or life stage.

Understanding the Main Sequence

In Chapter 15 we discussed the Sun as a representative star. We saw that what stars such as the Sun "do for a living" is to convert hydrogen to helium deep in their interiors via the process of nuclear fusion, thus producing energy. We found that fusion is the means by which a star can maintain itself in equilibrium and continue to shine. The fusion of hydrogen to helium is an excellent source of energy for a star, because the bulk of every star consists of hydrogen atoms.

Our theoretical models of how stars evolve over time show us that a typical star will spend about 90 percent of its life fusing the abundant hydrogen in its core into helium. This then is a good explanation of why 90 percent of all stars are found on the main sequence in the H–R diagram. But if all the stars on the main sequence are doing the same thing (fusing hydrogen), why are they distributed along a sequence of points? That is, why do they differ in luminosity and surface temperature (which is what we are plotting on the H–R diagram)?

To help us understand how main-sequence stars differ, we can use one of the most important results from our studies of model stars (constructed on computers as described in Chapter 15). Astrophysicists have been able to

TABLE 17.2
Characteristics of Main-Sequence Stars

Spectral Type	Mass	Luminosity	Temperature	Radius
	(Sun = 1)	(Sun = 1)		(Sun = 1)
O5	40	7×10^5	40,000 K	18
B0	16	27×10^4	28,000 K	7
A0	3.3	55	10,000 K	2.5
F0	1.7	5	7,500 K	1.4
G0	1.1	1.4	6,000 K	1.1
K0	0.8	0.35	5,000 K	0.8
M0	0.4	0.05	3,500 K	0.6

show that the structure of stars that are in equilibrium and that derive all their energy from nuclear fusion is completely and uniquely determined by just two quantities—*total mass* and *composition.* This fact provides an interpretation of many features of the H–R diagram.

Imagine a cluster of stars forming from a cloud of interstellar "raw material" whose chemical composition is similar to the Sun's. All condensations that become stars will then begin with the same chemical composition and will differ from one another only in mass. Now suppose that we compute a model of each of these stars for the time at which it becomes stable and derives its energy from nuclear reactions, but before it has time to alter its composition appreciably as a result of these reactions.

The models calculated for these stars allow us to determine their luminosities, temperatures, and sizes. If we plot the results from the models—one point for each model star—on the H–R diagram, we get something that looks just like the main sequence we saw for real stars.

And here is what we find when we do this. The model stars with the largest masses are the hottest and most luminous, and they are located at the upper left of the diagram. The least-massive model stars are the coolest and least luminous, and they are placed at the lower right of the plot. The other model stars all lie along a line running diagonally across the diagram. The main sequence turns out to be a sequence of stellar masses.

This makes sense if you think about it. The most massive stars have the strongest gravitational pull, and can thus compress their centers to the greatest degree. This means they are the hottest inside and the best at generating energy from nuclear reactions deep within—hence they shine with the greatest luminosity and have the hottest surface temperatures. The stars with lowest mass, in turn, are coolest and thus the least luminous. Our Sun lies somewhere in the middle of these extremes (as you can see in Figure 17.14). The characteristics of representative main-sequence stars are listed in Table 17.2.

Note that this is exactly what we found earlier, in Section 17.2, when we examined the mass–luminosity rela-

tion for stars whose mass and luminosity we know (see Figure 17.9). We observed that 90 percent of all stars seem to follow the relationship; these are the 90 percent of all stars that lie on the main sequence in our H–R diagram. Our models and our observations agree.

What about the other stars on the real H–R diagram—the giants and supergiants and the white dwarfs? As we will see in the next few chapters, these are what main-sequence stars turn into as they age; they are the later stages in a star's life. As a star consumes its nuclear fuel, its source of energy changes, as do its chemical composition and interior structure. These changes cause the star to alter its luminosity and surface temperature so that it no longer lies on the main sequence.

Extremes of Stellar Luminosities, Diameters, and Densities

We can use the H–R diagram to explore the extremes in size, luminosity, and density found among the stars. Such extreme stars are not just interesting to fans of the *Guinness Book of World Records,* but can teach us a lot about how stars work. For example, we saw that the most massive main-sequence stars are the most luminous ones. We know of a few extreme stars that are a million times more luminous than the Sun, with masses up to 100 times the Sun's mass. These superluminous stars, most of which are at the upper left of the H–R diagram, are exceedingly hot, very blue stars of spectral type O. These are the stars that would be the most conspicuous at vast distances in space.

The cool supergiants in the upper right corner of the H–R diagram are as much as ten thousand times as luminous as the Sun. These stars also have diameters very much larger than that of the Sun. As discussed above, some supergiants are so large that if the solar system could be centered in one, the star's surface would lie beyond the orbit of Mars. We will have to ask, in coming chapters, what process can make a star swell up to such an enormous size, and how long these "swollen" stars can last in their distended state.

In contrast, the very common red, cool, low-luminosity stars at the lower end of the main sequence are much smaller and more compact than the Sun. An example of such a red dwarf is Ross 614B, with a surface temperature of 2700 K and only 1/2000 of the Sun's luminosity. We call such a star a dwarf because its diameter is only 1/10 that of the Sun. A star with such a low luminosity also has a low mass (about 1/12 that of the Sun). This combination of mass and diameter means that the star has an average density about 80 times that of the Sun. Its density must be higher, in fact, than that of any known solid found on the surface of the Earth. (Despite this, the star is made of gas throughout, because its center is so hot.)

The faint red main-sequence stars are not the stars of the most extreme densities, however. The white dwarfs, at the lower left corner of the H−R diagram, have densities many times greater still.

The White Dwarfs

The first white dwarf star was detected in 1862, but its spectrum was not obtained until 1914. Called Sirius B, it forms a binary system with Sirius, the brightest-appearing star in the sky. It eluded discovery and analysis for a long time because it is very, very faint. Although only 8 LY away, the white dwarf companion of Sirius is quite difficult to see without a rather large telescope (Figure 17.16). (Since Sirius is often called the Dog Star—being located in the constellation of Canis Major, the big dog—Sirius B is sometimes nicknamed the Pup.)

We have now found hundreds of white dwarfs. A good example of a typical white dwarf is the nearby star 40 Eridani B. Its surface temperature is a relatively hot 12,000 K, but its luminosity is only 1/275 L_{Sun}. Calculations show that its radius is only 1.4 percent of the Sun's, or about the same as that of the Earth, and its volume is 2.5×10^{-6} that of the Sun. Its mass, however, is 0.43 times the Sun's mass, just a little less than half. To fit such a substantial mass into so tiny a volume, the star's density must be about 170,000 times the density of the Sun, or more than 200,000 g/cm^3. A teaspoonful of this material would have a mass of some 50 tons! At such densities, matter cannot exist in its usual state; we will examine the peculiar behavior of this type of matter in Chapter 22. For

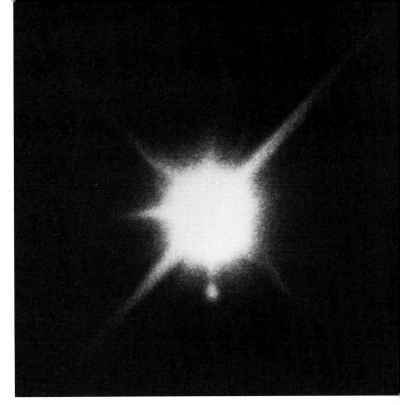

Figure 17.16
Sirius and its faint companion, the white dwarf star Sirius B. Sirius itself is overexposed on this image so that the fainter star can be seen. (Lick Observatory)

now, we should just note that white dwarfs are dying stars, reaching the end of their productive lives and ready for their stories to be over.

The British astrophysicist (and science popularizer) Arthur Eddington described the first known white dwarf this way: "The message of the companion of Sirius, when decoded, ran: 'I am composed of material three thousand times denser than anything you've ever come across. A ton of my material would be a little nugget you could put in a matchbox.' What reply could one make to something like that? Well, the reply most of us made in 1914 was, 'Shut up; don't talk nonsense.'" Today, however, astronomers not only accept that stars as dense as white dwarfs exist, but—as we will see—have found even denser objects in their quest to understand the evolution of different types of stars.

Summary

17.1 To understand the properties of stars, we must make wide-ranging surveys. We find the stars that appear brightest to our eyes are bright primarily because they are intrinsically very luminous, not because they are the closest to us. Most of the nearest stars are intrinsically so faint that they can be seen only with the aid of a telescope. The luminosity of stars ranges from about 10^{-4} L_{Sun} to more than 10^6 L_{Sun}. Stars with low luminosity are much more common than stars with high luminosity.

17.2 The masses of stars can be determined by analysis of the orbits of **binary stars.** The three types of double stars are **visual binaries, spectroscopic binaries,** and **eclipsing binaries.** Stellar masses range from about 1/12 to (rarely) 100 times the mass of the Sun. Objects having less mass than stars do are called **brown dwarfs** and *planets*. The most massive stars are, in most cases, also the most luminous, and this correlation is known as the **mass–luminosity relation.**

17.3 The diameters of stars can be determined by measuring the time it takes an object (the Moon, a planet, or a companion star) to pass in front of it and block its light. Diameters of members of **eclipsing binary** systems can be determined through analysis of their orbital motions.

17.4 The **Hertzsprung–Russell diagram,** or **H–R diagram,** is a plot of stellar luminosity as a function of surface temperature. Most stars lie on the **main sequence,** which extends diagonally across the H–R diagram from high temperature and high luminosity to low temperature and low luminosity. The position of a star along the main sequence is determined by its mass: high-mass stars emit more energy and are hotter than low-mass stars on the main sequence. Main-sequence stars derive their energy from the fusion of hydrogen to helium. About 90 percent of the stars in the solar neighborhood lie on the main sequence. Only 10 percent of the stars are **white dwarfs,** and fewer than 1 percent are *giants* or *supergiants*.

Review Questions

1. How does the intrinsic luminosity of the Sun compare with that of the 30 brightest stars? With that of the stars within 15 LY? Refer to Appendices 10 and 11.

2. Name and describe the three types of binary systems.

3. Describe two ways of determining the diameter of a star.

4. What are the largest and smallest known values of the mass, luminosity, surface temperature, and diameter of stars?

5. You are able to take spectra of both stars in an eclipsing binary system. List all properties of the stars that can be measured from their spectra and light curves.

6. Sketch an H–R diagram. Label the axes. Show where cool supergiants, white dwarfs, the Sun, and main-sequence stars are found.

Thought Questions

7. Is the Sun an average star? Why or why not?

8. Suppose you want to determine the average education level of people throughout the nation. Since it would be a great deal of work to survey every citizen, you decide to make your task easier by asking only the people on campus. Will you get the right answer? Will your survey be distorted by a selection effect? Explain.

9. Why do most known visual binaries have relatively long periods, and most spectroscopic binaries relatively short periods?

10. Figure 17.11 shows the light curve of a hypothetical eclipsing binary star in which the light of one star is completely blocked by another. What would the light curve look like for a system in which the light of the smaller star is only partially blocked by the larger one? Assume the smaller star is the hotter one. Sketch the relative positions of the two stars that correspond to various portions of the light curve.

11. There are fewer eclipsing binaries than spectroscopic binaries. Explain why. Within 50 LY of the Sun, visual binaries outnumber eclipsing binaries. Why? Which is easier to observe at large distances—a spectroscopic binary or a visual binary?

12. The eclipsing binary Algol drops from maximum to minimum brightness in about 4 hours, remains at minimum brightness for 20 min, and then takes another 4 hours to return to maximum brightness. Assume that we view this system exactly edge-on, so that one star crosses directly in front of the other. Is one star much larger than the other, or are they fairly similar in size?

13. Consider the following data on five stars:

Star	Apparent Magnitude	Spectrum
1	12	G, main sequence
2	8	K, giant
3	12	K, main sequence
4	15	O, main sequence
5	5	M, main sequence

  a. Which is the hottest?
  b. Coolest?
  c. Most luminous?
  d. Least luminous?
  e. Nearest?
  f. Most distant?

In each case, give your reasoning. (Recall that apparent magnitude is a measure of apparent brightness, where the *larger* the number, the *dimmer* the star appears to us.)

14. Which changes by the largest factor along the main sequence from spectral types O to M—mass, luminosity, or radius?

15. Suppose you want to search for main-sequence stars with very low mass using a space telescope. Will you design your telescope to detect light in the ultraviolet or the infrared part of the spectrum? Why?

16. An astronomer discovers a type-M star with a large luminosity. How is this possible? What kind of star is it?

17. Approximately 6000 stars are bright enough to be seen without a telescope. Are any of these white dwarfs? Use the information given in this chapter, and explain your reasoning.

Problems

18. Plot the luminosity functions of the nearest stars (see Appendix 10) and of the 20 brightest stars (Appendix 11). Explain how and why these two luminosity functions differ.

19. Find the combined mass of two stars in a binary system whose period of mutual revolution is two years, and for which the semimajor axis of the relative orbit is 2 AU.

20. We view a binary star exactly edge-on and observe eclipses. The star being eclipsed has an orbital velocity of 100 km/s. The time interval from first to second contact is $2^h 30^m$. The time from second to third contact is 10^h, and from third to fourth contact, $2^h 30^m$, again. What are the diameters of the two stars? How do these compare with the diameter of the Sun?

21. In Figure 17.7, is Star A or Star B more massive? Assume the orbit is viewed edge-on. What is the diameter of each star's orbit? If both stars are main-sequence stars, which is more luminous? Which is hotter?

22. Verify that a red dwarf with 1/12 the mass and 1/10 the radius of the Sun has a density 80 times that of the Sun. Calculate the density of Ross 614B, the red supergiant described in Section 17.4, which has 50 times the mass and 400 times the radius of the Sun. The outer parts of such a star would constitute an excellent laboratory vacuum.

23. Suppose you weigh 70 kg on the Earth. How much would you weigh on the surface of a white dwarf star the same size as the Earth but having a mass 300,000 times larger (nearly the mass of the Sun)?

Suggestions for Further Reading

Croswell, K. "The Grand Illusion: What We See Is Not Necessarily Representative of the Universe" in *Astronomy,* Nov. 1992, p. 44.

Croswell, K. "Visit the Nearest Stars" in *Astronomy,* Jan. 1987, p. 16. Explores the 19 nearest stars.

Davis, J. "Measuring the Stars" in *Sky & Telescope,* Oct. 1991, p. 361. Explains direct measurements of stellar diameters.

DeVorkin, D. "Henry Norris Russell" in *Scientific American,* May 1989.

Kaler, J. *Stars.* 1992, Scientific American Library/W. H. Freeman. Good modern introduction.

Kaler, J. "Journeys on the H–R Diagram" in *Sky & Telescope,* May 1988, p. 483.

Kopal, Z. "Eclipsing Binary Stars: Algol and Its Celestial Relations" in *Mercury,* May/June 1990, p. 88.

Moore, P. *Astronomers' Stars.* 1987, Norton. Focuses on 15 specific stars, including Mizar and Algol, that have been important in the development of astronomy.

Nielsen, A. "E. Hertzsprung—Measurer of Stars" in *Sky & Telescope,* Jan. 1968, p. 4.

Parker, B. "Those Amazing White Dwarfs" in *Astronomy,* July 1984, p. 15. Focuses on the history of their discovery.

Phillip, A. and Green, L. "Henry N. Russell and the H–R Diagram" in *Sky & Telescope,* April 1978, p. 306.

Using **REDSHIFT** ™

1. Turn on only the stars and set the *Faint Magnitude Limit* to magnitude 2. Turn off the undefined and composite spectral types and undefined luminosities. Select Earth under *Select Location.* Select a full sky view—.2 in zoom. Set Latitude on Earth as 0 degrees and viewing declination to 0. The sky will then be set to *Mercator projection,* full sky view. Count how many stars are in each spectral type and luminosity class.

Which types are the most common?

Explain why this would be true for the brightest stars in the sky.

2. Using the full-sky *Mercator projection,* compare the colors of the stars of magnitudes 0 to 3 and the colors of stars of magnitudes 5 to 6. Is there a difference?

What do the colors say about the stars we see with the naked eye?

3. Set the *Faint Magnitude Limit* so about 10 stars appear in the *Mercator* all-sky map. Click on each star to obtain its visual magnitude and spectral type. Plot each star in a spectral type versus magnitude diagram.

Does the plot reveal any useful information? Explain why.

This beautiful image shows a giant cluster of stars called 47 Tucanae, visible from the Earth's Southern Hemisphere. Such crowded groups, which astronomers call *globular clusters*, contain hundreds of thousands of stars, including some of the RR Lyrae variables discussed in this chapter.
(Photo by David Malin, courtesy of the Anglo-Australian Telescope Board)

CHAPTER 18

Celestial Distances

Thinking Ahead

When you are driving on a country road late at night, a point of light in the darkness can be a puzzling thing. Is it a nearby firefly, an oncoming motorcycle some distance away, or the porch light of a house much farther down the road? In the same way, astronomers are confronted with the question of how to tell the distances of stars when all that our eyes show us are faint points of light.

The determination of astronomical distances is central to understanding the nature of stars, but measuring such distances accurately is very difficult. After all, we cannot go out and lay a tape measure between the Sun and even the nearest star. Over the years, astronomers have developed a variety of clever techniques for estimating the vast distances that separate us from the stars. For nearby stars, we can use methods similar to the ones surveyors use here on Earth. For more distant objects, we have to apply some of the information about stellar luminosities and temperatures described in the previous two chapters, or use as a guidepost a special type of star that varies in brightness.

We now have a chain of methods for measuring cosmic distances, one that stretches from the Earth to the stars to the farthest reaches of the universe. One of the characteristics of that chain is that its links depend on one another: the measurement of distances to remote galaxies depends on the measurement of distances to the stars within our own Galaxy, which in turn depends on the accuracy of measurements within the solar system. The entire chain of cosmic distances is only as strong as its weakest

You have made the universe too large, says she. I protest, said I . . . when the Heavens were a little blue arch, stuck with stars, I thought the universe was too strait and close, I was almost stifled for want of air. But now [that] it is enlarged in height and breadth . . . I begin to breathe with more freedom, and think the universe to be incomparably more magnificent than it was before.

Bernard de Fontenelle in *Conversations on the Plurality of Worlds* (1686).

link, and so it is important that every link be as accurate as possible.

In this chapter we begin with the fundamental definitions of distances on Earth, and then extend our reach outward to the distant stars.

18.1

Fundamental Units of Distance

The first measures of distances were based on human dimensions—the inch as the distance between knuckles on the finger, or the yard as the span from the extended index finger to the nose of the British king. Later, the requirements of commerce led to some standardization of such units, but each nation tended to set up its own definitions. It was not until the middle of the 18th century that any real efforts were made to establish a uniform, international set of standards.

The Metric System

One of the enduring legacies of the Napoleonic era was the establishment of the *metric system* of units, officially adopted in France in 1799 and now used in most countries around the world. The fundamental metric unit of length is the *meter*, originally defined as one ten-millionth of the distance along the Earth's surface from the equator to the pole. French astronomers of the 17th and 18th centuries were pioneers in determining the dimensions of the Earth, so it was logical to use their information as the foundation of the new system.

Practical problems exist with a definition expressed in terms of the size of the Earth, since anyone wishing to determine the distance from one place to another can hardly be expected to go out and remeasure the planet. Therefore an intermediate standard meter consisting of a bar of platinum–iridium metal, was set up in Paris. In 1889, by international agreement, this bar was defined to be exactly 1 m in length, and precise copies of the original meter bar were made to serve as standards for other nations.

Other units of length are derived from the meter. Thus 1 kilometer (km) equals 1000 m, 1 centimeter (cm) equals 1/100 m, and so on. Even the old British and American units, such as the inch and the mile, are now defined in terms of the metric system.

Modern Redefinitions of the Meter

In 1960 the official definition of the meter was changed again. As a result of improved technology for generating spectral lines of precisely known wavelength, the meter was redefined to equal 1,650,763.73 wavelengths of a particular atomic transition in krypton-86. The advantage of this redefinition is that anyone with a suitably equipped laboratory can reproduce a standard meter, without reference to any particular metal bar.

In 1983 the meter was redefined once more, this time in terms of the velocity of light. At this point, the length of the standard unit of time, the *second,* had been fixed by international agreement as 9,192,631,770 times the frequency of a cesium-133 atomic clock. The meter was measured to be the distance light travels in a vacuum in a time interval of 1/299,792,458.6 s. This then defines the speed of light in a vacuum to be 299,792,458.6 m/s. Today, therefore, light travel-time provides us our basic unit of length. Putting it another way, a distance of one *light second* (LS) (the amount of space light covers in one second) is defined to be 299,792,458.6 m. We could just as well use the light second as the fundamental unit of length, but for practical reasons (and to respect tradition), we have defined the meter as a small fraction of the light second.

Distances Within the Solar System

The work of Copernicus and Kepler (see Chapters 1 and 2) established the *relative* distances of the planets—that is, how far from the Sun one planet is compared to another. But their work could not establish the *absolute* distances (in light seconds or meters or other standard units of length). This is like knowing the height of all the students in your class only as compared to the height of your astronomy instructor, but not in inches or centimeters. Somebody's height has to be measured directly.

Similarly, to establish absolute distances, astronomers had to measure one distance in the solar system directly. Estimates of the distance to Venus were made as Venus crossed the face of the Sun in 1761 and 1769, and an international campaign was organized to estimate the distance to the Earth-approaching asteroid Eros in the early 1930s. But not until the past three decades could planetary distances be measured with extremely high precision.

The key to our modern determination of solar system dimensions is *radar,* a type of radio wave that can bounce off solid objects (Figure 18.1). As discussed in several earlier chapters, by timing how long a radar beam (traveling at the speed of light) takes to reach another world and return, we can measure the distance involved very accurately. In 1961, radar signals were bounced off Venus for the first time, providing a direct measurement of the distance from Earth to Venus in terms of light seconds (the round-trip travel time of the radar signal). Subsequently, radar has also been used to determine the distances to Mercury, Mars, the satellites of Jupiter, the rings of Saturn, and several asteroids.

The "measuring stick" of distance astronomers use within the solar system is the *astronomical unit* (AU), the average distance from the Earth to the Sun. We then express all the other distances in the solar system in terms of the astronomical unit. Note, by the way, that it is not possible to use radar to measure the distance to the Sun directly because the Sun does not reflect radar very efficiently. But we can measure the distance to many other

Surveying the Stars

It is an enormous step to go from the planets to the stars. The nearest star is hundreds of thousands of astronomical units from the Earth. Yet in principle, we can survey distances to the stars using the same technique that a civil engineer employs to survey the distance to an inaccessible mountain or tree—the method of *triangulation.*

Triangulation

A practical example of triangulation is your own sense of depth perception. As you are pleased to discover every morning when you look in the mirror, your two eyes are located some distance apart and thus view the world from two different vantage points. This dual perspective allows you to get a general sense of how far away objects are.

To see what we mean, take a pen and hold it a few inches in front of your face. Look at it first with one eye (closing the other), and then switch eyes. Note how the pen seems to shift relative to objects across the room. Now hold the pen at arm's length: the shift is less. If you play with the pen for awhile, you will notice that the farther away you hold it, the less it seems to shift. Your brain automatically performs such comparisons and gives you a pretty good sense of how far away things in your immediate neighborhood are.

If your arms were made of rubber, you could stretch the pen far enough away from your eyes that the shift would become imperceptible. This is because our depth perception fails for objects more than a few tens of meters away. It would take a larger distance between viewing perspectives than the spacing between the eyes to see the shift of an object a city block or more from you.

Let's see how surveyors take advantage of this idea. Suppose you are trying to measure the distance to a tree across a deep river (Figure 18.2). You set up two observing stations some distance apart. That distance (AB in Figure 18.2) is called the *baseline.* Now the direction to the tree (C in the figure) in relation to the baseline is observed from each station. Note that C appears in different directions from the two stations. This apparent change in direction of the remote object due to a change in vantage point of the observer is called **parallax.** The parallax is also the angle that lines AC and BC make—in mathematical terms, the angle subtended by the baseline. A knowledge of the angles at A and B, and the length of the baseline, AB, allows the triangle ABC to be solved for any of its dimensions—say, the distance AC or BC. The solution could be accomplished by constructing a scale drawing, or by using the technique of trigonometry to make a numerical calculation. The greater the parallax, the nearer the object.

The farther away the tree (or astronomical object), the longer the baseline has to be to give us a reasonable measurement. Unfortunately, nearly all astronomical ob-

Figure 18.1
Radar telescope of the NASA Deep Space Network in California's Mojave Desert. This dish-shaped antenna can send and receive radar waves, and measure the distances to planets, satellites, and asteroids. (NASA/JPL)

solar system objects and use Kepler's laws to give us the length of the astronomical unit.

Years of painstaking analyses of radar measurements have led to a determination of the length of the astronomical unit to a precision of about one part in a billion. The length of 1 AU can be expressed in light travel time as 499.004854 LS, or about 8.3 light minutes (LM). If we use the definition of the meter given previously, this is equivalent to 1 AU = 149,597,892,000 m.

These distances are, of course, given here to a much higher level of precision than is normally needed. In this text, we are usually content to express numbers to a couple of significant places and leave it at that. For our purposes it will be sufficient to round off these numbers:

Speed of light: $c = 3.00 \times 10^8$ m/s $= 3.00 \times 10^5$ km/s

$$\text{Length of light second: LS} = 3.00 \times 10^8 \text{ m}$$
$$= 3.00 \times 10^5 \text{ km}$$

$$\text{Astronomical unit: AU} = 1.50 \times 10^{11} \text{ m} = 1.50 \times 10^8 \text{ km}$$
$$= 500 \text{ LS}$$

We now know the absolute distance scale within our own solar system with fantastic accuracy. This is the first link in the chain of cosmic distances.

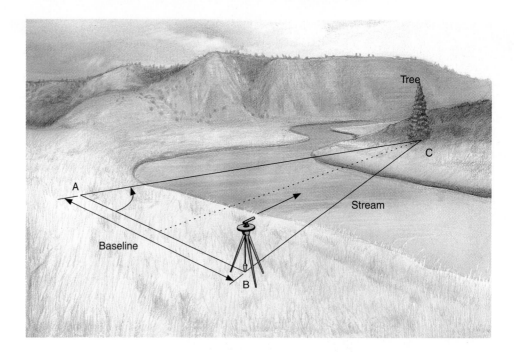

Figure 18.2
Triangulation allows us to measure distances to inaccessible objects. By getting the angle to a tree from two different vantage points, we can calculate the properties of the triangle they make, and thus the distance to the tree.

jects are very far away. To measure their distances requires either a very large baseline or highly precise angular measurements, or both. The Moon is the only object near enough that its distance can be found fairly accurately with measurements made without a telescope. Ptolemy determined the distance to the Moon correctly to within a few percent. He used the Earth itself as a baseline, measuring the position of the Moon relative to the stars at two different times of night.

With the aid of telescopes, later astronomers were able to measure the distances to the nearer planets or asteroids by using the Earth's diameter as a baseline. This is how the astronomical unit was first established. To reach for the stars, however, requires a much longer baseline for triangulation, and extremely sensitive measurements. Such a baseline is provided by the Earth's annual trip around the Sun.

Distances to Stars

As the Earth travels from one side of its orbit to the other, it graciously provides us with a baseline of 2 AU, or about 300 million km. Although this is a much bigger baseline than the diameter of the Earth, the stars are so far away that the resulting parallax shift is still not visible to the naked eye for even the closest stars.

In Chapter 1 we discussed how this perplexed the ancient Greeks, some of whom had suggested that the Sun might be the center of the solar system, with the Earth in motion around it. Aristotle and others argued, however, that the Earth could not be revolving about the Sun. If it were, they said, we would observe the parallax of the nearer stars against the background of more distant objects as we viewed the sky from different parts of the

Earth's orbit (Figure 18.3). Tycho Brahe advanced the same argument nearly 2000 years later, when his careful measurements of stellar positions with the unaided eye revealed no such shift. These early observers did not realize how truly distant the stars were, and did not have the tools to measure parallax shifts too small to be seen with the human eye.

By the 18th century, when there was no longer serious doubt about the Earth's revolution, it became clear that the stars must be extremely distant. Astronomers equipped with telescopes began to devise instruments capable of measuring the tiny shifts of nearby stars relative to the background of more distant (and thus unshifting) celestial objects. This was a significant technical challenge, since, even for the nearest stars, parallax angles are usually only a fraction of a second of arc. Recall that one second of arc is an angle of only 1/3600th of a degree (see Chapter 1). A coin the size of a quarter would appear to have a diameter of 1 arcsec if you were viewing it from a distance of about 3 miles, or 5 km. No wonder it took astronomers a while before they could measure such tiny shifts.

The first successful detections of stellar parallax were in the year 1838, when Friedrich Bessel (in Germany), Thomas Henderson (a Scottish astronomer working at the Cape of Good Hope), and Friedrich Struve (in Russia) independently measured the parallaxes of the stars 61 Cygni, Alpha Centauri, and Vega, respectively (Figure 18.4). Even the closest star, Alpha Centauri, showed a total displacement of only about 1.5 arcsec during the course of a year.

Figure 18.3 shows how such measurements work; seen from opposite sides of the Earth's orbit a star shifts positions when compared to a pattern of more distant

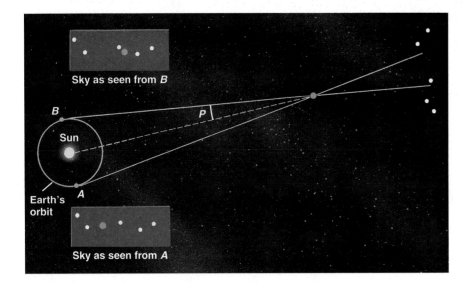

Figure 18.3
As the Earth revolves around the Sun, the direction in which we see a nearby star varies with respect to distant stars. We define the parallax of the nearby star to be one-half of the total change in direction, and usually measure it in arcseconds.

Sky as seen from *B*

B

Sun

Earth's orbit

A

Sky as seen from *A*

P

stars. Astronomers actually define parallax to be *one-half* the angle that a star shifts when seen from opposite sides of the Earth's orbit (the angle labeled *P* in Figure 18.3). The reason for this is just that they prefer to deal with a baseline of 1 AU instead of 2 AU.

Units of Stellar Distance

With a baseline of 1 AU, how far away would a star have to be to have a parallax of 1 arcsec? The answer turns out to be 206,265 AU, or 3.1×10^{13} km (in words, 31 million million kilometers). We give this unit a special name, the **parsec** (abbreviated **pc**)—derived from "the distance at which we have a **par**allax of one **sec**ond." The distance of

Figure 18.4
Friedrich Wilhelm Bessel (1784–1846) made the first authenticated measurement of the distance to a star (61 Cygni) in 1838, a feat that had eluded many dedicated astronomers for almost a century. (Yerkes Observatory)

a star in parsecs (*r*) is just the reciprocal of its parallax (*p*) in arcseconds. That is,

$$r = \frac{1}{p}$$

Thus a star with a parallax of 0.1 arcsec would be found at a distance of 10 pc, and one with a parallax of 0.05 arcsec would be 20 pc away.

Back in the days when most of our distances came from parallax measurements, this was a useful unit of distance, but it is not as intuitive as the **light year,** which we defined in Section 17.1. Light years have the advantage of being related directly to both the definition of the meter, which is expressed in terms of the speed of light, and the radar measurements that determine the length of the astronomical unit. In this text, we will use light years as our unit of distance, but many astronomers still use parsecs when they write technical papers or talk with each other at meetings. Conversion between the two distance units is simple: 1 pc = 3.26 LY, and 1 LY = 0.31 pc.

Another advantage of the light year as a unit is that it emphasizes the fact that as we look out into space, we are also looking back into time. The light that we see from a star 100 LY away left that star 100 years ago. What we study is not the star as it is now, but rather as it was in the past. The light from a distant galaxy that reaches our telescopes today most likely left its source before the Earth even existed.

ASTRONOMY BASICS

Naming Stars

If you've read this far, you have probably noticed that stars have a confusing assortment of names. Just look at the first three stars to have their parallax measured: 61 Cygni, Alpha Centauri, and Vega. Each of these names comes from a different tradition of designating stars.

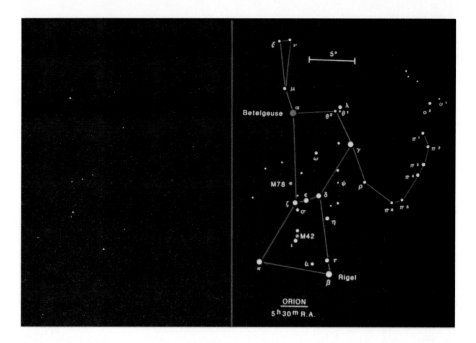

A photograph and diagram of the brightest objects in or near the star pattern of Orion the hunter (of Greek mythology) in the constellation of Orion. The Greek letters in Bayer's system are shown in yellow. The objects denoted M42 and M78 are not stars but *nebulae*—clouds of gas and dust; these numbers come from a list of "fuzzy objects" made by Charles Messier in 1781. (Richard Norton, Science Graphics)

The brightest stars have names that derive from the ancients. Some are from the Greek, such as Sirius, which means "the scorched one,"—a reference to its brilliance. A few are from Latin, but many of the best-known names are from Arabic because, as discussed in Chapter 1, much of Greek and Roman astronomy was "rediscovered" in Europe after the dark ages by means of Arabic translations. Vega, for example, means "swooping Eagle," and Betelgeuse (pronounced "Beetle-juice") means "right hand of the central one."

In 1603 the German astronomer Johann Bayer introduced a more systematic approach to naming stars. For each constellation, he assigned a Greek letter to the brightest stars, roughly in order of brightness. In the constellation of Orion, for example, Betelgeuse is the brightest star, so it got the first letter in the Greek alphabet—alpha—and is known as Alpha Orionis. (Orionis is the possessive form of Orion, so Alpha Orionis means "the first of Orion.") A star called Rigel, being the second brightest in that constellation, is called Beta Orionis. Since there are 24 letters in the Greek alphabet, this system allows the labeling of 24 stars in each constellation, hardly a complete sample.

In 1725 the English Astronomer Royal John Flamsteed introduced yet another system, in which the brighter stars eventually got a number in each constellation in order of their location or, more precisely, their right ascension. The system of sky coordinates that includes right ascension was discussed in Chapter 1. In this system Betelgeuse is called 58 Orionis and 61 Cygni is the 61st star in the constellation of Cygnus, the swan.

It gets worse. As astronomers began understanding more and more about stars, they drew up a series of specialized star catalogs, and fans of those catalogs began calling stars by their catalog numbers. If you look at Appendix 10—our list of the nearest stars (many of which are much too faint to get an ancient name, Bayer letter, or Flamsteed number)—you will see references to some of these catalogs. An

example is a set of stars labeled with a BD number, for "Bonner Durchmusterung." This was a mammoth catalog of over 324,000 stars in a series of zones in the sky, organized at the Bonn Observatory in the 1850s and 1860s. Keep in mind that this catalog was made before photography or computers came into use, so the position of each star had to be measured (at least twice) by eye, a daunting undertaking.

There is also a completely different system for keeping track of stars whose luminosity varies, and another for stars that brighten explosively at unpredictable times. Astronomers have gotten used to the many different star-naming systems, but students often find them bewildering and wish astronomers would settle down to one. Don't hold your breath: in astronomy, as in many fields of human thought, tradition holds a powerful attraction. Still, with high-speed computer databases to aid human memory, names may become less and less necessary. Today, astronomers often refer to stars by their precise locations on the sky rather than by their names or catalog numbers.

The Nearest Stars

No known star (other than the Sun) is within 1 LY or even 1 pc of the Earth. The stellar neighbors nearest to the Sun are three stars that make up a multiple system in the constellation of Centaurus. To the unaided eye the system appears as a single bright star, called Alpha Centauri, only 30° from the south celestial pole and hence not visible from the mainland United States. Alpha Centauri itself is a binary star—two stars in mutual revolution, too close together to be distinguished without a telescope. These two stars are 4.4 LY from us. Nearby is the third member of the system, a faint star known as Proxima Centauri. Proxima, with a distance of 4.3 LY, is slightly closer to us than the other two stars. (By the way, a few astronomers

Parallax and Space Astronomy

One of the most difficult things about precisely measuring the tiny angles of parallax shifts from Earth is that you have to observe the stars through our planet's atmosphere. As we saw in Chapter 5, the effect of the atmosphere is to spread out the points of starlight into fuzzy disks, making exact measurements of their positions more difficult. Astronomers have long dreamed of being able to measure parallax from space, and two instruments have recently turned this dream into reality.

The name of the Hipparcos satellite, launched in 1989 by the European Space Agency, is both an abbreviation for **Hi**gh **Pr**ecision **Par**allax **Col**lecting **S**atellite and a tribute to Hipparchus, the pioneering Greek astronomer whose work we discussed in Chapter 1. The satellite was designed to make the most accurate parallax measurements in history from 36,000 km above the Earth. However, the failure of its rocket motor meant it did not get the needed boost to reach the desired altitude, and it spent its four-year life in an elliptical orbit that varied from 500 to 36,000 km high. In this orbit, the satellite repeatedly passed through the Earth's radiation belts, which finally took its toll on the sensitive electronics.

Nevertheless, the satellite instruments were able to measure the positions of about 100,000 stars to an accuracy of 0.001 to 0.002 arcsec, as well as to make very accurate measurements of star motions. When these observations have been processed and published, they will form the most accurate catalog of stellar distances yet available.

The Hubble Space Telescope, whose on-board sensors must pinpoint with great accuracy the direction in which the telescope is pointing, can also perform precise parallax measurements for relatively nearby stars. In the next decade, astronomers will thus be able to know the distances of the stars within our own "corner" of the Galaxy with much greater precision. But no instruments, on Earth or in space, can overcome the natural limitations of this technique. Beyond a certain distance, one simply cannot triangulate; the more remote stars guard the secret of their distances too jealously.

Artist's conception of the Hipparcos satellite. (ESA)

---

have started to question whether Proxima is actually bound to the Alpha Centauri pair; some lines of evidence show that it may simply be passing close to Alpha temporarily.)

The nearest star visible without a telescope from most parts of the United States is the brightest-appearing of all the stars, Sirius, which has a distance of 8 LY. As we saw in Chapter 17, it too is a binary system, composed of a faint white dwarf orbiting a bluish-white main-sequence star. It is interesting to note that light reaches us from the Sun in about 8 min, and from Sirius in about 8 years.

The Extent of Parallax Measurements

Even though the parallaxes of stars are very small, today we can measure them to within a few thousandths of an arcsecond with ground-based telescopes. So far, parallaxes have been measured for more than 10,000 stars. Only for a fraction of them, however, are the parallaxes large enough (about 0.05 arcsec or more) to be measured with a precision of 10 percent or better, which is what astronomers need to make effective use of the distance measurement. A parallax of 0.05 arcsec corresponds to a distance of 20 pc, or a little more than 60 LY, a very small distance in the cosmic scheme of things. Measurements from space will soon expand the reach of accurate parallaxes to 1000 LY or more (see "Making Connections" box). However, even 1000 LY is merely 1 percent the size of our Galaxy's main disk. If we are going to reach very far away from our own neighborhood with our chain of methods for measuring cosmic distances, we need some completely new techniques.

Variable Stars: One Key to Cosmic Distances

Standard Bulbs Revisited

Let's briefly review why measuring distances to the stars is such a struggle. As discussed in Section 16.1, our problem is that stars come in a bewildering variety of intrinsic luminosities. (If stars were light bulbs, we'd say they come in a wide range of wattages.) Suppose, instead, that all stars had the same wattage or luminosity; in that case the more distant ones would always look dimmer, and we could tell how far away a star was simply by how dim it appeared. In the real universe, however, when we look at a star in our sky (with eye or telescope) and measure its apparent brightness, we cannot know whether it looks dim because it's a low-wattage bulb, because it is far away, or perhaps some of each.

Astronomers need to discover something else about the star that allows us to "read off" its intrinsic luminosity—in effect, to know what the star's true wattage is. With this information, we can then attribute how dim it looks from Earth to its distance. Recall that the apparent brightness of an object decreases with the square of the distance to that object. Therefore, if we know the luminosity of a star and its apparent brightness, we can calculate how far away it is.

Thus astronomers have searched for techniques that allow us to determine the luminosity of a star, and it is to these techniques that we turn next.

Variable Stars

The breakthrough in measuring distances to remote parts of our own Galaxy, and to other galaxies as well, came from the study of *variable stars*. Most stars are constant in their luminosity, at least to within a percent or two. Like the Sun, they generate a steady flow of energy from their interiors. However, some stars are seen to vary in brightness and for this reason are called variable stars. Many such stars vary on a regular cycle, like the flashing bulbs that decorate stores and homes during the winter holidays.

Let's define some tools to help us keep track of how a star varies. A graph that shows how the brightness of a variable star changes with time is called a **light curve** (Figure 18.5). The *maximum* is the point of the light curve where the star has its greatest brightness; the *minimum* is the point where it is faintest. If the light variations repeat themselves periodically, the interval between the two maxima is called the *period* of the star.

Pulsating Variables

There are two special types of variable stars for which—as we will see—measurements of the light curve give us accurate distances. These are the **cepheids** and the **RR Lyrae variables,** both of which are **pulsating variable stars.** A pulsating variable star actually changes its diameter with time—periodically expanding and contracting, as your chest does when you breathe. We now understand that these stars are going through a brief unstable stage in their lives. Their expansion and contraction can be measured by using the Doppler effect. The spectral lines shift toward the blue as the surface of the star moves toward us, and then to the red as it shrinks back. As the star pulsates, it changes overall color, indicating that its temperature is also varying. And, most important for our purposes, the luminosity of the pulsating variable also changes in a regular way as it expands and contracts.

Cepheid Variables

Cepheids are large, yellow, pulsating stars named for the first-known star of the group, Delta Cephei. The variability of Delta Cephei was discovered in 1784 by the young English astronomer John Goodricke (see "Voyagers in Astronomy" box). The star rises rather rapidly to maximum light and then falls more slowly to minimum light, taking a total of 5.4 days for one cycle. The curve in Figure 18.5 represents the light variation of Delta Cephei.

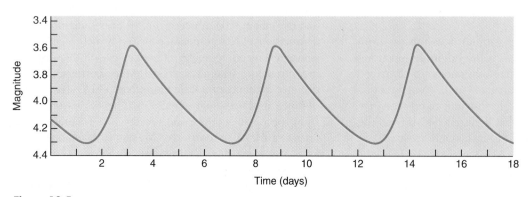

Figure 18.5
Light curve (plot of brightness as a function of time) of a typical cepheid variable.

John Goodricke (1764–1786)

The brief life of John Goodricke is a testament to the human spirit under adversity. Born deaf, and unable to speak, Goodricke nevertheless made a number of pioneering discoveries in astronomy through patient and careful observations of the heavens.

Born in Holland where his father was on a diplomatic mission, Goodricke was sent back to England at age eight to study at a special school for the deaf. He did sufficiently well to enter Warrington Academy, a secondary school that offered no special assistance for students with handicaps. His mathematics teacher there inspired an interest in astronomy, and in 1781, at age 17, Goodricke began observing the sky at his family home in York, England. Within a year he had discovered the brightness variations of the star Algol (discussed in Chapter 17), and suggested that an unseen companion star was causing the changes, a theory that waited over 100 years for proof. His paper on the subject was read before the Royal Society (the main British group of scientists) in 1783, and won him a medal from that distinguished group.

In the meantime, Goodricke had discovered two other stars that varied regularly, Beta Lyrae and Delta Cephei, both of which contin-

A portrait of John Goodricke by artist J. Scouler, which now hangs in the Royal Astronomical Society in London. There is some controversy about whether this is actually what Goodricke looked like, or whether the painting was much retouched to please his family. (Courtesy of the San Diego State University special collections library)

ued to interest astronomers for years to come. Goodricke shared his interest in observing with his older cousin, Edward Pigott, who went on to discover other variable stars during his much longer life. But Goodricke's time was quickly drawing to a close; at age 21, only two weeks after he was elected to the Royal Society, he caught a cold while making astronomical observations and never recovered.

Today the University of York has a building named Goodricke Hall, and a plaque that honors his contributions to science. Yet if you go to the churchyard cemetery where he is buried, an overgrown tombstone has only the initials "J.G." to show where he lies. Astronomer Zdenek Kopal, who has looked carefully into Goodricke's life, has speculated on why the marker is so modest: perhaps the rather staid Goodricke relatives were ashamed of having a "deaf-mute" in the family, and could not sufficiently appreciate how much a man who could not hear could nevertheless see.

Several hundred cepheid variables are known in our Galaxy. Most cepheids have periods in the range of 3 to 50 days, and luminosities that are about 1000 to 10,000 times greater than that of the Sun. Their variations in luminosity range from a few percent to a factor of 10. Polaris, the North Star, is a cepheid variable that for a long time varied by one-tenth of a magnitude, or by about 10 percent in visual luminosity, in a period of just under four days. Recent measurements indicate that the amount by which the brightness of Polaris changes is decreasing, and that sometime in the future this star will no longer be a pulsating variable. This is just one more piece of evidence that stars really do evolve and change in fundamental ways as they age.

The Period–Luminosity Relationship

The importance of cepheid variables lies in the fact that their periods and average luminosities are directly related. The longer the period (the longer the star takes to vary), the greater the luminosity. This **period–luminosity** relationship was a remarkable discovery, one for which astronomers still (pardon the expression) thank their lucky

stars. The period of such a star is easy to measure—a good telescope and a good clock are all you need. Once you have the period, the relationship (which can be put into precise mathematical terms) will give you the luminosity (the actual wattage) of the star. Astronomers can then compare this intrinsic brightness with the apparent brightness: as we saw, the difference between the two allows them to calculate the distance.

The relation between period and luminosity was discovered in 1908 by Henrietta Leavitt, a staff member at the Harvard College Observatory (one of a number of women working for low wages assisting Edward Pickering, the Observatory's director—see the "Voyagers in Astronomy" box on Annie Cannon in Chapter 16). She did not publish the results in a form that attracted the attention of astronomers until four years later. Some hundreds of cepheid variables had been discovered in the Large and Small Magellanic Clouds (Figure 18.6), two great star systems that are actually neighboring galaxies (although they were not known to be galaxies then).

These systems presented a wonderful opportunity to study the behavior of variable stars independent of their distance. For all practical purposes, the Magellanic

Figure 18.6

The Large Magellanic Cloud, a small irregular-shaped galaxy near our own Milky Way. It was in this galaxy that Henrietta Leavitt discovered the cepheid period–luminosity relationship. (National Optical Astronomy Observatories)

Clouds are so far away that astronomers can assume that all the stars in them are at roughly the same distance from us. (In the same way, all the suburbs of Los Angeles are roughly the same distance from New York City. Of course if you are *in* Los Angeles, you will notice distances between the suburbs; but compared to how far away New York City is, they seem small.) If all the variable stars in the Magellanic Clouds are at roughly the same distance, any difference in their apparent brightness must be a reflection of differences in their intrinsic luminosities.

Leavitt found that the brighter-appearing cepheids always have the longer periods of light variation. Thus, she reasoned, the period must be related to the luminosity of the stars. When Leavitt did this work, the distance to the Magellanic Clouds was not known, so she was only able to show differences in luminosities, not to calculate the actual cepheid luminosities themselves. To define the period–luminosity relationship with actual numbers (to *calibrate* it), astronomers first had to measure the actual distances to a few nearby cepheids in another way. (This was accomplished by finding cepheids associated in clusters with other stars whose distances could be estimated from their spectra, as discussed in the next section of this chapter.) Once the relationship was thus defined, it could give us the distance to any cepheid, wherever it might be located.

Here at last was the technique astronomers had been searching for to break the confines of distance that parallax imposed on them. Cepheids could be observed and monitored, it turned out, in many parts of our own Galaxy and in other galaxies as well. Astronomers, including Ejnar Hertzsprung and Harvard's Harlow Shapley, immedi-

ately saw the potential of the new technique; they and many others set to work exploring more-distant reaches of space using cepheids as signposts. As we will see, this work still continues, as the Hubble Space Telescope and other modern instruments try to identify and measure individual cepheids in galaxies farther and farther away (Figure 18.7).

RR Lyrae Stars

A related group of stars, whose nature was understood somewhat later than that of the cepheids, are called RR Lyrae variables, named for the star RR Lyrae, the best-known member of the group. More common than the cepheids but also less luminous, thousands of these pulsating variables are known in our Galaxy. The periods of RR Lyrae stars are always less than one day, and most of them change in brightness by less than one magnitude.

Astronomers have observed that the RR Lyrae stars occurring in any particular star cluster all have about the same apparent magnitude. Since stars in a cluster are all at approximately the same distance, it follows that RR Lyrae variables must all have nearly the same intrinsic luminosity—which turns out to be about 50 L_{Sun} (in this sense, RR Lyrae stars are a little bit like standard bulbs). Figure 18.8 displays the ranges of periods and absolute magnitudes for both the cepheids and the RR Lyrae stars.

RR Lyrae stars can be detected out to a distance of about 2 million LY, and cepheids to about 60 million LY. Compare these limits with parallaxes that even from space will probably not be measured for stars more distant than

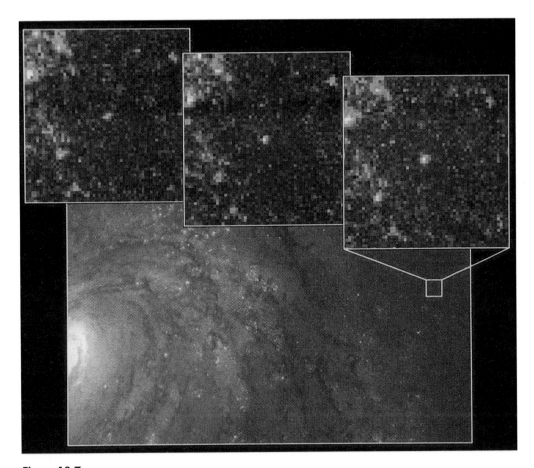

Figure 18.7
This image of part of the galaxy called M100 was taken with the Hubble Space Telescope in 1994. The insets show a single cepheid in the galaxy, going through its cycle of brightness variations. What makes this image remarkable is that M100 is 51 million LY away, the most distant galaxy in which individual cepheids had been identified and measured at that time. (Wendy Freedman, Carnegie Institution of Washington, and NASA)

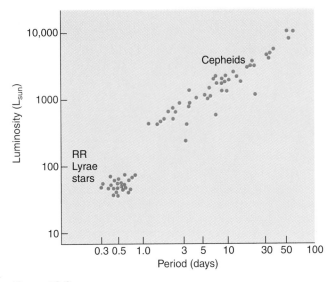

Figure 18.8
The period–luminosity relation for cepheid variables. Also shown are the period and luminosity for RR Lyrae stars.

about 1000 LY. You can see from this comparison just how important the discovery of the period–luminosity relationship was in enabling astronomers to extend their measurements of cosmic distances.

The H–R Diagram and Cosmic Distances

Distances from Spectral Types

As satisfying and productive as variable stars have been for distance measurement, their use still has many limitations. An obvious one is that you must first find a pulsating variable in a star group before this method will yield a distance. Suppose we need the distance to a star or stars where no variables are to be found nearby. In this case, the H–R diagram can come to our rescue.

If we can observe the spectrum of a star, we can estimate its distance from our understanding of the H–R diagram. As discussed in Chapter 16, a detailed examination of a stellar spectrum allows astronomers to classify the star into one of the *spectral types* indicating surface temperature. (The types are O, B, A, F, G, K, and M; each of these can be divided into numbered subgroups.) In general, however, the spectral type alone is not enough to al-

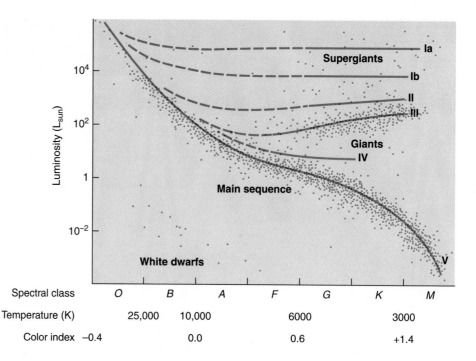

Figure 18.9
Luminosity classes for stars on the Hertzsprung–Russell diagram.

Spectral class	O	B	A	F	G	K	M
Temperature (K)		25,000	10,000		6000		3000
Color index	−0.4		0.0		0.6		+1.4

low us to estimate luminosity. Look again at Figure 17.14. A G2 star could be a main-sequence star with a luminosity of 1 L_{Sun}, or it could be a giant with a luminosity of 100 L_{Sun}, or even a supergiant with a still-higher luminosity.

But we can learn more from a star's spectrum than just its temperature. Remember, for example, that we can tell pressure differences in stars from the details of the spectrum (see Section 16.3). This is very useful, because giant stars are larger (and have lower pressures) than main-sequence stars, and supergiants are still larger than giants. If we look in detail at the spectrum of a star, we can determine whether it is a main-sequence star, a giant, or a supergiant.

Suppose, for example, that the spectrum, color, and other properties of a distant G2 star match those of the Sun exactly. It is then reasonable to conclude that this distant star is likely to be a main-sequence star just like the Sun, and to have the same luminosity as the Sun.

The most widely used system of star classification divides stars of a given spectral class into six categories called **luminosity classes.** These luminosity classes are denoted by Roman numerals as follows:

Ia: Brightest supergiants
Ib: Less-luminous supergiants
II: Bright giants
III: Giants
IV: Subgiants (intermediate between giants and main-sequence stars)
V: Main-sequence stars

The full spectral specification of a star includes its luminosity class. For example, a main-sequence star with spectral class F3 is written as F3 V. For an M2 giant, the specification would be M2 III. Figure 18.9 illustrates the approximate positions of stars of various luminosity classes on the H–R diagram. The dashed portions of the lines represent regions with very few or no stars.

With both its spectral and luminosity classes known, a star's position on the H–R diagram is uniquely determined. Since the diagram plots luminosity versus temperature, this means we can now read off the star's intrinsic luminosity. As before, if we know how luminous the star really is, and see how dim it looks, the difference allows us to calculate its distance. (For historical reasons, astronomers sometimes call this method of distance determination *spectroscopic parallax;* we find this name misleading, however, since the method has nothing to do with parallax.)

A Few Words About the Real World

Introductory textbooks such as ours work hard to present the material in a straightforward and simplified way. In doing so, we sometimes do our students a disservice by making scientific techniques seem *too* clean and painless. In the real world, the techniques we have just described turn out to be messy and difficult, and often give astronomers headaches that last long into the day!

For example, the relationships we have described—such as the period–luminosity relationship for certain variable stars—aren't exactly straight lines on a graph. The points representing many stars scatter widely when plotted, and thus the distances derived from them also have a certain built-in scatter or uncertainty.

The distances we measure with the methods we have discussed are therefore only accurate to within a certain percentage of error—sometimes 10 percent, sometimes 25 percent, sometimes as much as 50 percent or more. A 25 percent error for a star estimated to be 10,000 LY away means it could be anywhere from 7500 to 12,500 LY away. This would be an unacceptable uncertainty if you were loading fuel into a spaceship for a trip to the star, but is not a bad first figure to work with if you are an astronomer stuck on planet Earth.

Nor is the construction of H–R diagrams as easy as you might think at first. To make a good diagram, you need to measure the characteristics of many stars, which can be a time-consuming task. If the process is to advance our knowledge, the stars you measure must be far away; hence spectra are difficult to obtain. Astronomers and their graduate students may have to spend many nights at the telescope (and many days back home working with their data) before they get their distance measurement.

Nevertheless, with these tools—parallaxes for the nearest stars, RR Lyrae variable stars and the H–R diagram for clusters of stars in our own and nearby galaxies, and cepheids out to distances of 50 million LY—we can measure distances throughout our own Galaxy and beyond to neighboring stellar systems. Such distances, combined with measurements of composition, luminosity, and temperature made with the techniques described in Chapters 16 and 17, are the arsenal of information we need to trace the evolution of stars from birth to death, the subject to which we turn in the chapters that follow.

Summary

18.1 Early measurements of length were based on human dimensions, but today we use worldwide standards that specify lengths in units such as the meter. Distances within the solar system are now determined by timing how long it takes radar signals to travel from the Earth to the surface of a planet or other body, and then return.

18.2 Half the shift in a nearby star's position relative to very distant background stars, as viewed from opposite sides of the Earth's orbit, is called the **parallax** of that star and is a measure of its distance. The units used to measure stellar distance are the **light year (LY),** the distance light travels in one year, and the **parsec (pc),** the distance of a star with a parallax of 1 arcsec. One parsec = 3.26 LY. The first successful measurements of stellar parallaxes were reported in 1838. The star nearest the Sun is Alpha Centauri, which is actually a system of three stars at a distance of about 4.4 LY. Parallax measurements are a fundamental link in the chain of cosmic distances.

18.3 Cepheids and **RR Lyrae** stars are two types of **pulsating variable stars.** *Light curves* of these stars show that their luminosities vary with a regularly repeating period. Both types of variables obey a **period–luminosity** relationship, so measuring their periods can tell us their luminosities. Then we can calculate their distances by comparing their luminosities with their apparent brightnesses.

18.4 Stars with identical temperatures but different pressures (and diameters) have slightly different spectra. Spectral classification can therefore be used to estimate the **luminosity class** of a star, as well as its temperature. By seeing where the star is located on an H–R diagram, we can establish its luminosity. This, with the star's apparent brightness, again yields its distance.

Review Questions

1. Explain how parallax measurements can be used to determine distances to stars. Why can we not make accurate measurements of parallax beyond a certain distance?

2. Make up a table relating the following units of astronomical distance: kilometer, Earth radius, solar radius, astronomical unit, light year, and parsec.

3. Suppose you have discovered a new RR Lyrae variable star. What steps would you take to determine its distance?

4. Explain how you would use the spectrum of a star to estimate its distance.

5. Which method would you use to obtain the distance to each of the following?
   a. An asteroid crossing the Earth's orbit.
   b. A star astronomers believe to be no more than 50 LY from the Sun.
   c. A group of stars in the Milky Way Galaxy that includes a significant number of variable stars.
   d. A star that is not variable, but for which you can obtain a clearly defined spectrum.

6. What would be the advantage of making parallax measurements from Pluto rather than from Earth? Would there be a disadvantage?

7. Parallaxes are measured in fractions of an arcsecond. One arcsecond equals 1/60 arcmin; an arcminute is in turn 1/60°. To get some idea of how big 1° is, go outside at night and find the Big Dipper. The two pointer stars at the end of the bowl are 5.5° apart. The two stars across the top of the bowl are 10° apart. (Ten degrees is also about the width of your fist when held at arm's length and projected against the sky.) Mizar, the second star from the end of the Big Dipper's handle, appears double. The fainter star, Alcor, is about 12 arcmin from Mizar. For comparison, the diameter of the full Moon is about 30 arcmin. The belt of Orion is about 3° long. Why did it take until 1838 to make parallax measurements for even the nearest stars?

8. For centuries, astronomers wondered whether comets were true celestial objects, like the planets and stars, or a phenomenon that occurred in the atmosphere of the Earth. Describe an experiment to determine which of these two possibilities is correct.

9. The Sun is much closer to the Earth than are the nearest stars, yet it is not possible to measure accurately the parallax of the Sun relative to the stars by measuring its position directly. Explain why.

10. Parallaxes of stars are sometimes measured relative to the positions of galaxies or distant objects called quasars. Why is this a good technique?

11. Figure 18.5 is the light curve for the prototype cepheid variable Delta Cephei. How does the luminosity of this star compare with that of the Sun?

12. Look at Appendices 10 and 11. What percentage of the stars in each list are main-sequence stars (remember the luminosity classes)? Why is this percentage so different for the two lists of stars?

13. Suppose you measure the temperature of a star to be identical to that of the Sun. Is this enough information to determine its distance? Explain.

14. Which of the following can you determine about a star without knowing its distance: radial velocity, temperature, apparent magnitude, luminosity.

Problems

15. A radar astronomer who is new at the job claims she beamed radio waves to Jupiter and received an echo exactly 48 min later. Do you believe her? Why?

16. Demonstrate that a light year contains 9.46×10^{12} km. Demonstrate that 1 pc equals 3.086×10^{13} km and that it also equals 3.26 LY. Show your calculations.

17. Give the distances to stars having the following parallaxes:
   a. 0.1 arcsec
   b. 0.5 arcsec
   c. 0.005 arcsec
   d. 0.001 arcsec

18. Give parallaxes of stars having the following distances:
   a. 10 pc
   b. 3.26 LY
   c. 326 LY
   d. 10,000 pc

19. Consider a cepheid with a luminosity 10,000 times that of the Sun. If a given telescope and detector can detect a solar-type star out to a distance of only 1000 LY, to what distance can this same telescope observe the cepheid?

20. In the text we say that all of the cepheids in the Large Magellanic Cloud are nearly at the same distance away, and therefore their apparent brightnesses correlate with their periods. Look up the diameter and distance of the Large Magellanic Cloud in Appendix 12. What is the percentage difference in the distances of two stars, one of which is located on the edge of the galaxy closest to us, and the other on the far side? Suppose these stars have identical periods and identical intrinsic luminosities. How much will their apparent brightnesses differ? How will this difference affect the correlation between period and apparent brightness?

Suggestions for Further Reading

Croswell, K. "Visit the Nearest Stars" in *Astronomy*, Jan. 1987, p. 16. On the 19 nearest stars.

Ferris, T. *Coming of Age in the Milky Way.* 1988, Morrow. A history of how we established the scale of the cosmos.

Hirshfeld, A. "The Absolute Magnitude of Stars" in *Sky & Telescope*, Sep. 1994, p. 35. Good review, with charts.

Hodge, P. "How Far Away Are the Hyades?" in *Sky & Telescope*, Feb. 1988, p. 138. A history of how we measure distance to this important cluster.

Marschall, L. et al. "Parallax You Can See" in *Sky & Telescope*, Dec. 1992, p. 626. On measuring parallax for yourself.

Reddy, F. "How Far the Stars" in *Astronomy,* June 1983, p. 6. Nice summary of the entire chain of distances.

Rowan-Robinson, M. *The Cosmological Distance Ladder.* 1985, W. H. Freeman. Somewhat technical introduction to measuring distances in the universe.

Struve, O. "The First Determinations of Stellar Parallax" in *Sky & Telescope.* Vol. 16, 1956, pp. 9, 69.

Upgren, A. "New Parallaxes for Old" in *Mercury,* Nov./Dec. 1980, p. 143.

Using REDSHIFT ™

1. Zoom in on Corona Borealis and set the *Faint Magnitude Limit* to 5. Which star is brightest? Which star is intrinsically most luminous? Which star is the supergiant? Which star is on the main sequence?

Comparing the spectral type and visual magnitude to that of other stars in the constellation, can you deduce the luminosity class of Beta Coronae Borealis? Comparing the characteristics of Gamma Coronae Borealis to Kappa Coronae Borealis, can you predict the luminosity class for Kappa?

2. Set the *Faint Magnitude Limit* to 6 and center on Mensa in the Southern Hemisphere. Comparing its spectral type, luminosity class, and visual magnitude to other stars in the constellation, estimate the distance to Delta Mensae.

3. The distance given for Beta Orionis is incorrect. Compare spectral types and luminosity classes of the brightest 10 stars in Orion and use the inverse-square law of radiation to estimate the distance to Beta.

4. Turn the *Constellation Boundaries* on and set the *Faint Magnitude Limit* to 5.

Count the number of stars in Gemini in each spectral type (O, B, A, F, G, K, and M). Do the same for luminosity type (I, II, III, IV, and V).

Do Gemini's bright stars follow the same distributions of spectral type and luminosity class as for stars as a whole?

This region contains some of the most colorful interstellar clouds ever photographed. The blue area at the top is a cloud of dust surrounding the star Rho Ophiuchi; the blue color is starlight reflected from grains of dust. At the lower right, the bright star Antares, a cool red supergiant, is embedded in a thick dust cloud of its own making. (Immediately to the left of Antares is M4, a much more distant cluster of stars.) The reddish nebulae glow with light emitted from hydrogen atoms. (Photo by David Malin, © Royal Observatory, Edinburgh)

CHAPTER 19

Between the Stars: Gas and Dust in Space

Thinking Ahead

Astronomers examining the sky with telescopes come across an area that is generally crowded with stars except for a small dark section that has only a few. How can we determine if this dark region is simply a lack of stars in that particular direction, or a place where a cloud of some dark material blocks our view of stars behind it?

To begin our discussion of the life story of the stars, we must first examine where they come from. One of the most exciting discoveries of 20th-century astronomy was that our Galaxy contains not only stars, but vast quantities of what we might call "raw material" for stars—atoms or molecules of gas, and tiny solid particles we call dust. As we will see, the existence of this raw material helps us understand how new stars can continue to form, and gives us important clues about our own origins billions of years ago.

By earthly standards the space between stars is empty. No laboratory on Earth can produce so complete a vacuum. Yet this "emptiness" contains vast clouds of gas and solid particles. Some of these clouds are visible and are called **nebulae** (Latin for "clouds"). Others emit energy only at infrared or radio wavelengths. Still others make their existence known because they absorb some of the light passing through them. Astronomers refer

[Photographs] showed the entire group of stars [the Pleiades cluster] with an entangling system of nebulous matter, which seemed to bind together the different stars with misty wreaths and streams of filmy light . . . all of which is entirely beyond the keenest vision and the most powerful telescope.

E. E. Barnard, writing in *Popular Astronomy*, Vol. 6, p. 439 (1898)

to all the material between stars as **interstellar matter;** the collection of interstellar matter is called the *interstellar medium.*

Interstellar clouds do not last for the lifetime of the universe, but instead collide, coalesce, and grow. Some form stars within them that then inject energy into the cloud material and disperse it. When stars die they, in turn, eject some of their material into interstellar space. This material can then form new clouds and begin the cycle over again.

19.1

The Interstellar Medium

About 99 percent of the material between the stars is in the form of a *gas*—individual atoms or molecules. The most abundant elements in this gas are hydrogen and helium (which we saw are also the most abundant elements in the stars). The remaining 1 percent is solid—frozen particles sometimes called **interstellar grains** (Figure 19.1).

If all the interstellar gas were spread out smoothly, there would be about one atom of gas per cubic centimeter in interstellar space. (In contrast, the air in the room where you are reading this book has roughly 10^{19} atoms per cubic centimeter.) The dust grains are even scarcer. A cubic kilometer of space would contain only a few hundred to a few thousand tiny grains, each typically less than one ten-thousandth of a millimeter in diameter. These numbers are just averages, however, because the gas and dust are not distributed smoothly. Instead, as the pictures in this chapter show, the distribution is patchy and irregular, with the denser regions being referred to as "clouds."

In some clouds the density of gas and dust may exceed the average by as much as a thousand times or more, but even this density is more nearly a vacuum than any attainable on Earth. To show what we mean, let's imagine a vertical tube in the Earth's atmosphere with a cross section of 1 m². Such a tube of air would contain more atoms than we would find if we could extend that same tube from the top of the atmosphere all the way to the edge of the observable universe, 10 to 15 billion LY away.

While the density of interstellar matter is very low, the volume of space in which such matter is found is huge, and so its total mass is substantial. To see why, we must bear in mind that stars occupy only a tiny fraction of the volume of the Milky Way Galaxy. For example, it takes light only about 2 s to travel a distance equal to the radius of the Sun, but more than four *years* to travel from the Sun to the nearest star. Even if the spaces among the stars are sparsely populated, there's just a lot of space out there.

Astronomers estimate that the total mass of gas and dust in the Milky Way Galaxy is equal to about 5 percent of the mass contained in stars. Don't let that 5 percent fool you: this means that the mass of the interstellar mat-

Figure 19.1
Clouds of luminous gas and opaque dust, such as the nebulosity called NGC 3603, are found between the stars. A bright cluster of stars (bright white region near the bottom) energizes the surrounding gas, causing it to glow. The reddish glow is interrupted by lanes and clouds of gas mixed with enough dust to block the light from behind them. (Anglo-Australian Telescope Board)

ter in our Galaxy amounts to several billion times the mass of the Sun. There is plenty of raw material in the Galaxy to make many generations of new stars and planets (and perhaps even astronomy students).

ASTRONOMY BASICS

Naming the Nebulae

As you look at the captions for some of the spectacular photographs in this and the next chapter, you will notice the variety of names given to the nebulae. A few, which in small telescopes look like something recognizable, are sometimes named after the creatures or objects they resemble. Examples include the Crab, Tarantula, and Keyhole Nebulae. But most only have numbers that are entries in a catalog of astronomical objects.

Perhaps the best-known catalog of nebulae (as well as star clusters and galaxies) was compiled by the French astronomer Charles Messier (1730–1817). Messier's passion was discovering comets, and his devotion to this cause earned him the nickname "The Comet Ferret" from King Louis XV. When comets are first seen coming toward the Sun, they look like little fuzzy patches of light; in small telescopes, they are easy to confuse with nebulae or with groupings of many stars so far away that their light is all blended together. Time and again, Messier's heart leapt as he thought he had discovered one of his treasured comets, only to find that he had "merely" observed a nebula or cluster.

In frustration, Messier set out to catalog the position and appearance of over 100 objects that could be mistaken for comets. For him, this list was merely a tool in the far more important work of comet hunting. He would be very surprised if he returned today to discover that no one recalls his comets anymore, but that his catalog of "fuzzy things that are not comets" is still widely used. When the opening image in this chapter refers to M4, it denotes the fourth entry in Messier's list.

A far more extensive listing was compiled under the title of the New General Catalog (NGC) of Nebulae and Star Clusters in 1888 by John Dreyer, working at the observatory in Armagh, Ireland. He based his compilation on the work of William Herschel and his son John, plus many other observers who followed them. With the addition of two further listings (called the Index Catalogs), Dreyer's compilation eventually included 13,000 objects. Astronomers today still use his NGC numbers when referring to most nebulae and star groups.

19.2

Interstellar Gas

Some of the most spectacular astronomical photographs (Figure 19.2) show interstellar gas located near hot stars. This gas is heated to temperatures close to 10,000 K by the nearby stars; it then glows with the colors characteristic of the element hydrogen, which makes up the majority of the gas (about three-quarters by mass). The reddish color in the accompanying images is the telltale glow of hydrogen; ionized nitrogen also contributes. (The strongest of the lines in the visible spectrum of hydrogen is the red Balmer line; see Section 4.5.)

H II Regions—Gas Near Hot Stars

Hydrogen interstellar gas near very hot stars is ionized (the electron is stripped completely away) by the ultraviolet radiation from those stars. Since hydrogen is the main constituent of interstellar gas, we often characterize a region of space according to whether its hydrogen is neutral or ionized. A cloud of ionized hydrogen is called an **H II region.** (Spectroscopists use the Roman numeral I to indicate that an atom is neutral; successively higher Roman numerals are used for each higher stage of ionization. H II thus refers to hydrogen that has lost its one electron; Fe III is iron with two electrons missing.)

Figure 19.2
NGC 3576: a cloud of luminous gas surrounds a compact cluster of very hot stars. These hot stars ionize the hydrogen in the gas cloud, forming an H II region. When electrons then recombine with protons and move back down to the lowest energy orbit, emission lines are produced. The strongest hydrogen line that can be seen as visible light is in the red part of the spectrum, and is responsible for the color of the gas in the photograph.
(Anglo-Australian Telescope Board)

When the radiation from a hot star ionizes hydrogen in the surrounding gas, a neutral hydrogen atom is converted into a positive hydrogen ion (a proton) and a free electron. Such a detached proton won't remain alone forever when attractive electrons are around; it will capture a free electron, becoming neutral hydrogen once more. However, such a neutral atom can then absorb ultraviolet radiation again and start the cycle over. At a typical moment, most of the atoms are in the ionized state, which is why we call such areas H II regions.

The capture of the free electrons leads to the emission of light. The electrons cascade down through the various energy levels of the hydrogen atoms on their way to the lowest level, or ground state. During each transition downward, they give up energy in the form of light (see Chapter 4). This process of converting ultraviolet radiation into visible light is called *fluorescence.*

A fluorescent light works in basically the same way. Electrons within it are heated to high temperatures and then collide with atoms of mercury vapor in the tube. The mercury is excited to a high energy state because of these collisions. When the mercury atoms return to lower energy levels, they emit ultraviolet photons that in turn strike a phosphor-coated screen. The atoms in the screen absorb the ultraviolet photons and emit visible light by the process of fluorescence.

The interstellar gas contains other elements besides hydrogen. Many of them are also ionized in the vicinity of hot stars; they then capture electrons and emit light, just as hydrogen does, which allows them to be observed by astronomers. But generally, these atoms are outnumbered by the hydrogen atoms.

Neutral Hydrogen Clouds

Ionized hydrogen gas is the type of interstellar matter most often photographed, but observations show that most interstellar hydrogen is not ionized. The very hot stars required to produce H II regions are rare, and only a small fraction of interstellar matter is close enough to such hot stars to be ionized by them.

If interstellar gas located at large distances from stars is not ionized and cannot produce strong emission lines, how can we learn about such gas? One way is to search for the presence of atoms and molecules of interstellar matter in the spectra of stars. Cold clouds of gas can absorb some of the light from stars that lie behind them, producing dark spectral lines (as discussed in Chapter 4).

The first evidence for absorption by interstellar clouds came from the analysis of a spectroscopic binary star. While most of the lines in the spectrum of this binary shifted alternately from longer to shorter wavelengths and back again, as we would expect from the Doppler effect for one star in orbit around another, a few lines in the spectrum remained fixed in wavelength. Since both stars are moving in a binary system, lines that showed no motion puzzled astronomers. Subsequent work demonstrated that these lines were not formed in the star's atmosphere at all, but rather in a cold cloud of gas located between the Earth and the binary star.

The most conspicuous interstellar lines in the visible-light region of the spectrum are produced by atoms of sodium and calcium. We also see absorption in the spectrum produced by molecules (combinations of atoms) such as CN and CH. Ultraviolet observations made with orbiting telescopes have detected lines of carbon, hydrogen, oxygen, nitrogen, and other elements, as well as carbon monoxide and molecular hydrogen. We should note that although other elements may be more noticeable in such spectra, hydrogen is still the most abundant element in the colder regions (see next section).

The strengths of interstellar lines can be used to estimate how common each element is in the interstellar medium. For some elements, the interstellar abundances are about the same as their relative abundances in the Sun and other stars. For others, the relative abundance in interstellar space is noticeably lower. Interstellar gas has especially low abundances of elements that readily condense into solids (notably aluminum, calcium, and titanium, as well as iron, silicon, and magnesium). As we will see in Section 19.4, it is likely that many of the atoms of these elements have indeed combined to form tiny solid grains of interstellar dust. Thus they no longer show the spectral lines produced by atoms in the gaseous state.

Radio Observations of Cold Clouds: The 21-cm Line

By now you've probably gotten the message that hydrogen is the most common element out there. And most of the hydrogen in interstellar space is cold. Unfortunately, cold hydrogen—far from the radiation of hot stars—is not so easy to detect using visible light or ultraviolet radiation. Radio astronomers, however, discovered a way to learn about what turned out to be vast reservoirs of cold hydrogen gas in the Galaxy.

Cold clouds of interstellar gas emit spectral lines in the radio part of the electromagnetic spectrum. The most useful of these lines is produced by (no surprise) hydrogen at a wavelength of 21 cm. That's quite a long wavelength, implying that the wave has a low frequency and low energy. The energy is too low to come from electrons jumping between the energy levels discussed in Chapter 4. Where then does such a low-energy wave come from in hydrogen?

A hydrogen atom possesses a tiny amount of angular momentum because its electron is spinning on its axis and orbiting around the nucleus (proton). In addition, the proton has a spin of its own. If the proton and electron are spinning in opposite directions, the atom as a whole has a very slightly lower energy than if the two spins are aligned (Figure 19.3). If an atom in the lower energy state (spins opposed) acquires a small amount of energy, then the spins of the proton and electron can be aligned, leaving the atom in a slightly *excited state*. If the atom then loses that same amount of energy again, it returns to its ground state. The amount of energy involved corresponds to a wave of 21-cm wavelength.

Neutral hydrogen atoms can be excited in this way by collisions with electrons and other atoms. Such collisions are extremely rare in the sparse gases of interstellar space. An individual atom may wait many years before such an encounter aligns the spins of its proton and electron. Nevertheless, over many millions of years a significant fraction of the hydrogen atoms are excited by a collision. (Out there in cold space, that's about as much excitement as an atom typically experiences.) An excited atom can then lose its excess energy either by another collision or by giving off a radio wave with a wavelength of 21 cm. If there are no collisions, an excited hydrogen atom will wait an average of about 10 million years before emitting a photon and returning to its state of lowest energy.

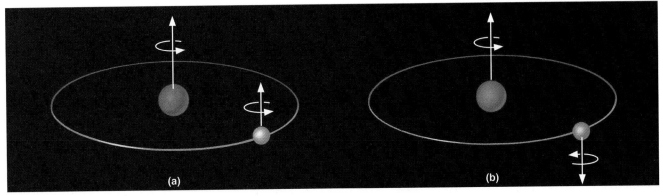

Figure 19.3
Formation of the 21-cm line. When the electron in a hydrogen atom is in the orbit closest to the nucleus, the proton and the electron may be spinning either in the same direction (left) or in opposite directions (right). When the electron flips over, the atom gains or loses a tiny bit of energy by either absorbing or emitting electromagnetic energy with a wavelength of 21 cm.

Equipment sensitive enough to detect the 21-cm line of neutral hydrogen became available in 1951. The mechanism for producing these waves was earlier described by Dutch astronomers, and they had actually prepared an experiment to detect 21-cm waves when a fire destroyed their equipment. As a result, two Harvard physicists, Harold Ewen and Edward Purcell, made the first detection (Figure 19.4), soon followed by confirmations from the Dutch and a group in Australia. Since

Figure 19.4
Harold Ewen in 1952, with the horn antenna that first detected 21-cm radiation, atop the Lyman Physics Laboratory at Harvard. The inset shows Edward Purcell, winner of the 1952 Nobel Prize in physics, a few years later. (Photos courtesy of E. M. Purcell and Harvard University)

that time many other radio lines produced by both atoms and molecules have been discovered (see following discussions).

Observations at 21 cm show that most of the neutral hydrogen in the Galaxy is confined to an extremely flat layer, less than 300 LY thick, that extends throughout the flat disk of the Milky Way. The 21-cm line is produced in cold hydrogen clouds that turn out to have temperatures of about 100 K. The diameters of these cold clouds range from about 3 to 30 LY; an imaginary spaceship traveling 1000 LY through space in the plane of the Galaxy would encounter, on average, only about two of these clouds. Overall, we can estimate that approximately 2 percent of interstellar space is filled with such cold clouds. The masses of the clouds are typically in the range 1 to 1000 times the mass of the Sun, with clouds of low mass being the most common.

Not all of interstellar hydrogen is found in these extremely cold clouds. Some hydrogen is warm, with temperatures of 3000 to 6000 K. At least 20 percent of the space between stars in the plane of the Milky Way is filled with such warm, neutral hydrogen. The warm hydrogen is not heated by nearby stars; what then warms it to such high temperatures? The answer came when even hotter interstellar gas was discovered by an orbiting observatory.

Ultra-Hot Interstellar Gas

The discovery of ultra-hot interstellar gas was a big surprise. Before the launch of astronomical observatories into space, models of the interstellar medium assumed that most of the region between stars was filled with cool hydrogen. But when astronomers were able to make ultraviolet observations from telescopes above the Earth's atmosphere, they discovered (for example) interstellar lines produced by oxygen atoms that had been ionized five times. To strip five electrons from their orbits around an oxygen nucleus requires a lot of energy. In fact, these observations imply that the temperature of the interstellar matter where these atoms occur must be approximately a *million* degrees.

It appears that the source of energy producing these remarkable temperatures is the explosion of massive stars at the ends of their lives. Such explosions, called *supernovae,* will be discussed in detail in Chapter 22. For now we'll just say that some stars, nearing the ends of their lives, become unstable and literally explode. These explosions send high-temperature gas, moving at velocities of thousands (and even tens of thousands) of kilometers per second, out into interstellar space, where it has a tremendous heating effect on the neighborhood gas.

Astronomers estimate that one supernova explodes roughly every 25 years somewhere in the Galaxy. On the average, the hot gas from a supernova will sweep through any given point in the Galaxy about once every 2 million years. At this rate, the sweeping action is continuous

enough to keep most of the space between clouds filled with gas at a temperature of a million degrees. When this ultra-hot gas comes into contact with a denser cloud, it heats the outer layers of the cloud to a temperature of a few thousand degrees.

Interstellar Molecules

As we just saw, visible-light and ultraviolet observations revealed the presence of a number of *simple* molecules among interstellar matter by absorbing some of the radiation of more distant stars. When more sophisticated equipment for obtaining spectra in radio and infrared wavelengths became available, astronomers found more complex molecules in interstellar clouds as well.

Just as atoms leave their "fingerprints" in the spectrum of visible light, so the vibration and rotation of atoms within molecules can leave spectral fingerprints in radio and infrared waves. If we spread out the radiation at such longer wavelengths, we can detect emission or absorption lines in the spectrum that are characteristic of specific molecules. Over the years, experiments in Earth laboratories have shown us the exact wavelengths associated with changes in the rotation and vibration of many molecules—giving us a template of possible lines against which we can now compare our observations of interstellar matter.

About a hundred kinds of molecules have been identified in interstellar space, most in giant clouds containing substantial amounts of gas and dust. These more complex molecules are mostly composed of combinations of hydrogen, oxygen, carbon, nitrogen, and sulfur atoms. Among them are molecules of H_2 (molecular hydrogen), water, ammonia, and hydrogen sulfide. Astronomers also identify a number of organic molecules (those associated with carbon chemistry on Earth), such as formaldehyde (used to preserve living tissues), hydrogen cyanide, and alcohol (see "Making Connections" box). Relatively heavy molecules such as HC_9N and $HC_{11}N$, which have long chains of carbon atoms, are found in some cold clouds. Other features in infrared spectra hint at complex rings of atoms; some of these are called *polycyclic aromatic hydrocarbons* (PAHs) by chemists because on Earth some of them have a strong smell or aroma. PAHs may have as many as 50 carbon atoms, and can be considered intermediate between molecules and solid grains.

The cold interstellar clouds also contain cyanoacetylene (HC_3N) and acetaldehyde (CH_3CHO), generally regarded as starting points for amino acid formation. These are the building blocks of proteins, and thus one of the fundamental chemicals from which living organisms on Earth are constructed. The presence of these organic molecules does not imply that life exists in space. But it does show that the chemical building blocks of life can form under a wide range of conditions in the universe. As we learn more about how complex molecules are pro-

Among the molecules astronomers have identified in interstellar clouds is alcohol, which comes in two varieties: methyl (or wood) alcohol, and ethyl alcohol (the kind you find in cocktails). Ethyl alcohol is a pretty complex molecule, written by chemists as C_2H_5OH. It is quite plentiful (relatively speaking)—in clouds where it has been identified we detect up to one molecule for every cubic meter. The largest of the cold clouds (which can be several hundred light years across) have enough ethyl alcohol to make 10^{28} fifths of liquor.

Spouses of future interstellar astronauts, however, need not fear that their wives or husbands will become interstellar alcoholics. Even if a spaceship were equipped with a giant funnel 1 km across, and could scoop it through such a cloud at the speed of light, it would take about a thousand years to gather up enough alcohol for one standard martini!

Furthermore, the very same clouds also contain water (H_2O) molecules. Your scoop would gather them up as well, and there are a lot more of them because they are simpler and thus easier to form. For the fun of it, one astronomical paper actually calculated the proof of a typical cloud. Proof is the ratio of alcohol to water in a drink, where 0 proof means all water, 100 proof means half alcohol and half water, and 200 proof means all alcohol. The proof of the interstellar cloud was only 0.2, hardly enough to blur your memory before a big astronomy exam.

duced in interstellar clouds, we gain an increased understanding of the kinds of processes that preceded the beginnings of life on Earth billions of years ago.

How do such fragile molecules survive in the harsh environment of interstellar space? Most of the more complex molecules would be dissociated (torn apart into individual atoms) by the short-wavelength radiation that comes from stars. Therefore these molecules can survive only where they are shielded from ultraviolet starlight. There are dense, dark, giant clouds that contain dust, which acts like a thick blanket of interstellar smog and keeps ultraviolet starlight from penetrating the interior of the cloud (although the radio waves that signal the presence of the molecules readily get out). It is in these clouds that we find molecules, in fact, these giant clumps of interstellar matter are called *molecular clouds.*

Some scientists also speculate that the surfaces of the dust grains (see Section 19.4)—which would seem very large if you were an atom—provide "nooks and crannies" where atoms can stick long enough to join with other atoms and form molecules. (Think of the dust grains as "interstellar social clubs" where lonely atoms can meet and form meaningful relationships.) Dusty clouds not only protect molecules but provide an environment that encourages their formation.

As we will see in the next chapter, the giant clouds of interstellar material where molecules are most common are among the most interesting structures in the Galaxy. It is here that we are most likely to see new stars forming from raw material, and beginning the life cycle that is our subject in the next several chapters.

19.3

A Model of the Interstellar Gas

Table 19.1 summarizes the characteristics of the various types of clouds that populate interstellar space. The contrasts are fascinating. There are cold, dense clouds in which hydrogen is not ionized. There are clouds so hot that atoms are mainly ionized and molecules cannot sur-

TABLE 19.1
Interstellar Gas

Type of Region	Temperature (K)	Density (number/cm³)	Description
H I: cold clouds	10^2	50	Hydrogen atoms in clouds typically 3–30 LY across.
H I: warm clouds	$3–6 \times 10^3$	0.3	Hydrogen is warmer but not ionized; found in clouds.
Hot gas	$10^5–10^6$	10^{-3}	Found throughout the Galaxy; hydrogen is ionized; heated by supernova explosions.
H II regions	10^4	$10^2–10^4$	Found near hot stars; hydrogen mostly ionized.
Giant molecular clouds	10	as high as 10^5	50–200 LY across; have dense clumps.

vive. The densities differ by a factor of a million, with the coolest clouds having the highest density. What do these data tell us about the interstellar medium? Where might we expect to find the various types of clouds? What is an individual cloud like? How do the clouds change as time passes?

Structure and Distribution of Interstellar Clouds

One important requirement for a model of the interstellar medium is that the clouds and the million-degree gas between them must be at approximately the same pressure. Suppose they were not. If the cloud pressure were higher, the cloud would expand until its pressure matched that of its environment. If, on the other hand, the pressure of the hot gas were greater than that of a cloud embedded in it, the hot gas would compress the cloud and force it to shrink until its pressure became high enough to resist further compression. Imagine a balloon as an analogy. If you add air to it, thereby increasing its pressure, the balloon will expand. If you squeeze it with your hands, you can make the balloon smaller because of the greater external pressure.

The pressure in a gas depends on both the temperature and the number of particles per cubic centimeter (its density). If the gas particles are hotter, their energy of motion is greater, and they collide harder with one another. If the density is greater, there are more collisions. For the gas pressures in two different regions next to each other to be the same, the region at higher temperature must have a lower density. A region with lower temperature, such as a cool cloud, must have a higher density. (In the same way, you can fit more people into a room when

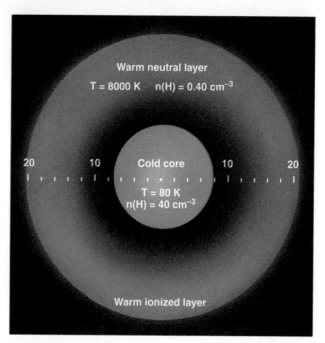

Figure 19.5
A typical interstellar cloud consists of a dense, cold core surrounded by a warm envelope. The horizontal scale shows the distance from the center in light years.

they are not moving around very much, say at a formal cocktail party. But if they want to do the waltz, fewer of them will fit.)

Figures 19.5 and 19.6 show in a schematic way what interstellar clouds look like based on the requirement that gas pressures must be nearly the same everywhere. Individual clouds are scattered at random throughout the Galaxy. The typical cloud may be a few tens of light years

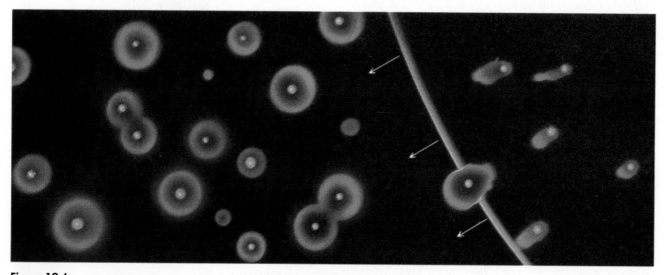

Figure 19.6
Interstellar clouds are embedded in hot, low-density gas heated to temperatures as high as a million degrees K by supernova explosions. In the upper right, a supernova remnant is shown sweeping through interstellar space. (Diagram adapted from work published in *The Astrophysical Journal* by C. McKee and J. Ostriker)

in diameter. The clouds are embedded in gas with a temperature of a million degrees or so, the legacy of exploding stars. The outer portions of the clouds are heated by this gas via conduction—the direct transfer of energy from the hot gas to the outer atoms of the clouds. The temperature of the outer portion of a typical cloud is thus raised to a few thousand degrees.

If a cloud is large enough, it can shield its innermost core from being heated, and the core may have a temperature 100 times lower and a density correspondingly 100 times higher. Typical values for a cloud core are a temperature of 80 K and a density of 40 hydrogen atoms per cubic centimeter.

Interstellar Matter Around the Sun

Let's take a brief aside to look at the distribution of interstellar matter in our own neighborhood. The Sun is located in a region where the density of interstellar matter is unusually low. The temperature of interstellar matter in the vicinity of the Sun is about 10^6 K, and its density is only 5×10^{-3} hydrogen atoms per cubic centimeter. This region of low-density gas, called the *Local Bubble*, extends to a distance of at least 300 LY from the Sun.

If interstellar space near the Sun contained the normal number of clouds, we would expect to have detected approximately 2000 of them within the Local Bubble. We have not: clouds of the type shown in Figure 19.5 are conspicuously absent. One possible explanation is that a supernova explosion in the past 100,000 to ten million years swept the region we now see as the Local Bubble nearly clean of interstellar clouds. Such an explosion would have heated the small amount of remaining gas to very high temperatures.

While typical cold hydrogen clouds are very rare within the Local Bubble, a few clouds do exist. The Sun itself seems to be inside a cloud with a density of 0.1 hydrogen atom per cubic centimeter and a temperature of 10,000 K. This cloud is so tenuous that it is referred to as Local Fluff. We do see one sizable warm cloud in the direction toward the center of our Galaxy but within 60 LY of the Sun. It may be that the Local Fluff is the warm, partially ionized edge of a denser, cooler cloud (see Figure 19.5) and that the Sun is just entering this cloud as it moves through the Galaxy.

Evolution of Interstellar Clouds

The model of interstellar gas described here presents a picture of the interstellar medium as it appears, on the average. In reality, interstellar gas pressures are not always balanced. Clouds collide and coalesce; they are torn apart by supernovae explosions and disrupted by stellar winds and radiation pressure from hot stars. The interstellar medium is constantly replenished with gas ejected by supernovae and hot stars. Initially, this gas forms relatively small clouds, then grow through collisions and mergers with other clouds. Ultimately the process leads to the formation of giant clouds with diameters as large as 200 LY, and masses that exceed 100,000 times that of the Sun. It is in these giant clouds that the most vigorous star formation occurs. The newly formed stars then go through their lives, and some (as we will see) become supernovae. As they explode they eject gaseous material, thus starting the cycle over again.

Cosmic Dust

Figure 19.7 shows a striking example of what is actually a common sight—a dark region on the sky that appears nearly empty of stars. For a long time astronomers debated whether these dark regions were empty "tunnels" through which we looked beyond the stars of the Milky Way Galaxy into intergalactic space, or clouds of some dark material that blocked the light of the stars beyond. American astronomer E. E. Barnard is generally credited with showing from his extensive series of nebulae photographs that the latter interpretation is the correct one (see "Voyagers in Astronomy" box).

Dark Nebulae

The dark cloud (called Barnard 86) seen in Figure 19.7 blocks the light of the many stars that lie behind it—note

Figure 19.7
A star cluster (NGC 6520) next to a dark cloud of interstellar matter (Barnard 86). Old stars in our Galaxy are yellowish in color and form the brightest part of the Milky Way. The dark cloud is seen in front of these background stars, visible only because it blocks out the light from the stars beyond. The small cluster of young blue stars may have begun, millions of years ago, as just such a dark cloud. (Anglo-Australian Observatory)

Edward Emerson Barnard

Born in 1857 in Nashville, Tennessee, two months after his father died, Edward Barnard grew up in such poor circumstances that he had to drop out of school at age nine to help support his ailing mother. He soon became assistant to a local photographer, where he learned to love both photography and astronomy, destined to become the dual passions of his life. He worked as a photographer's aide for 17 years, studying astronomy on his own. In 1883 he obtained a job as an assistant at the Vanderbilt University Observatory, which enabled him at last to take some astronomy courses.

Married in 1881, Barnard built a house for his family that he could ill afford. But as it happened, a patent medicine manufacturer offered a $200 prize (a lot of money in those days) for the discovery of any new comet. With the determination that became characteristic of him, Barnard spent every clear night searching for comets. He discovered seven of them between 1881 and 1887, earning enough money to make the payments on his home; this "Comet House" later became a local attraction. (By the end of his life, he had found 17 comets through diligent observation.)

In 1887 Barnard got a position at the newly founded Lick Observatory, where he soon locked horns with the director, Edward Holden, a blustering administrator who made Barnard's life miserable. (To be fair, Barnard soon tried to do the same for him.) Despite being denied the telescope time that he needed for his photographic work, Barnard in 1892 managed to discover the first new satellite found around Jupiter since Galileo's day, a stunning observational feat that earned him world renown. Now in a position to demand more telescope time, he perfected his photographic techniques and soon began to publish the best images of the Milky Way taken up to that time. It was during the course of this work that he began to examine the dark regions among the crowded star lanes of the Galaxy, and to realize that they must be vast clouds of obscuring material (rather than "holes" in the distribution of stars).

Astronomer/historian Donald Osterbrock has called Barnard an "observaholic": his daily mood seemed to depend entirely on how clear the sky promised to be for his night of observing. He was a driven, neurotic man, concerned about his lack of formal training, fearful of being scorned, and afraid that he might somehow slip back into the poverty of his younger days. He had difficulty taking vacations and lived for his work: only serious illness could deter him from making astronomical observations.

E. E. Barnard at the Lick Observatory 36-in refractor.
(Mary Lea Shane Archives of the Lick Observatory)

In 1895 Barnard, having had enough of the political battles at Lick, accepted a job at the Yerkes Observatory near Chicago, where he remained until his death in 1923. He continued his photographic work, publishing compilations of his images that became classic photographic atlases, and investigating the varieties of nebulae revealed in his photographs. He also made measurements of the sizes and features of planets, participated in observations of solar eclipses, and carefully cataloged dark nebulae (see Figure 19.7). In 1916 he discovered the star with the largest proper motion (see Chapter 16), the second-closest star system to our own, now called Barnard's Star in his honor.

---

how the regions in other parts of the photograph are crowded with stars. The only stars visible in the direction of Barnard 86 are those that happen to lie in front of it from our perspective. Barnard 86 is an example of a relatively dense cloud or *dark nebula* of tiny solid grains, which are often referred to as interstellar dust. Opaque clouds are conspicuous on any photograph of the Milky Way (see the figures in Chapter 24). The "dark rift," which runs lengthwise down a long part of the summer Milky Way and appears to split it in two, is produced by a collection of such obscuring clouds.

Dust clouds are invisible in the visible light region of the spectrum; we find them only when denser clouds block the light from stars behind them (Figure 19.8).

However, they glow brightly in the infrared. Small dust grains absorb visible light and ultraviolet radiation very efficiently. The grains are heated by the absorbed radiation, typically to temperatures from 20 to about 500 K, and reradiate this heat at infrared wavelengths. We can use Wien's law (see Section 4.2) to estimate where in the electromagnetic spectrum this radiation falls. For a temperature of 100 K, the maximum is at about 30 micrometers (μm; 1 μm = 10^{-4} cm), while grains as cold as 20 K will radiate most strongly near 150 μm. The Earth's atmosphere is opaque to radiation at these wavelengths, so emission by interstellar dust is best measured from space.

Observations from above the Earth's atmosphere by IRAS (the Infrared Astronomical Satellite) showed that

Figure 19.8
A dark dust cloud hides the light of stars behind it and is especially easy to see in front of a rich, starry background such as this one. A soft blue reflection nebula is visible in the center. (Photo by David Malin, © Anglo-Australian Telescope Board)

dust clouds are present throughout the plane of the Milky Way (Figure 19.9). The bright patches of emission have been given the name **infrared cirrus.** The closest infrared cirrus clouds are about 300 LY away.

Reflection Nebulae

The tiny interstellar dust grains absorb only a portion of the starlight they intercept. At least half of the starlight that interacts with a grain is merely scattered—that is, it is redirected helter skelter in all directions. Since neither the absorbed nor the scattered starlight reaches us directly, both absorption and scattering make stars look dimmer. The effects of both processes are called **interstellar extinction;** something astronomers must especially take into account when they observe stars in dusty or distant regions of our Galaxy.

Some dense clouds of dust are close to luminous stars, and scatter enough starlight to become visible. Such a cloud of dust, illuminated by starlight, is called a *reflection nebula* since the light we see is starlight reflected off the grains of dust. One of the best-known examples is the nebulosity around each of the brightest stars in the Pleiades cluster (Figure 19.10). The dust grains are small, and such small particles turn out to scatter light with blue wavelengths more efficiently than light at red wavelengths (Figure 19.11). A reflection nebula, therefore, usually appears bluer than its illuminating star.

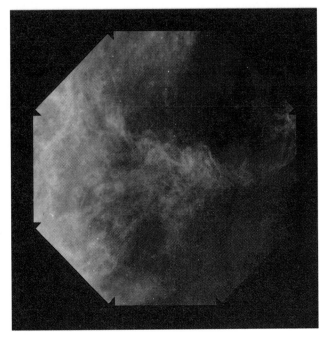

Figure 19.9
The Infrared Astronomical Satellite discovered patchy clouds of dust throughout the plane of the Milky Way Galaxy. These have been nicknamed "infrared cirrus" because of their resemblance to high clouds in the Earth's atmosphere. In this infrared image, the cirrus clouds are at the left, while three dusty nebulae can be seen to the right. (IPAC/JPL/NASA)

Figure 19.10
The Pleiades open star cluster. This cluster contains hundreds of stars (only a few are visible here) and is located about 400 LY from the Sun. The blue nebulosity is starlight reflected by interstellar dust in a cloud that the cluster happens to be passing through at the present time. (Royal Observatory Edinburgh/Anglo-Australian Telescope Board)

A similar scattering process explains why the Earth's atmosphere looks blue even though the gases that make up our air are transparent. As sunlight comes in, it scatters from the molecules of air. Again, the small size of the molecules means that the blue colors scatter much more efficiently than the greens, yellow, and reds. Thus the blue in sunlight is scattered out of the beam and all over the sky. The light from the Sun that comes to your eye, on the other hand, is missing some of its blue, so the Sun looks a bit yellower than it would from space.

Gas and dust are generally intermixed in space, although the proportions are not exactly the same everywhere. The presence of dust is apparent on many photographs of emission nebulae; Figure 19.12 shows a beautiful image of the Trifid Nebula in the constellation of Sagittarius, where we see an H II region surrounded by a blue reflection nebula. Which type of nebula appears brighter depends on the kind of stars that cause the gas

and dust to glow. Stars cooler than about 25,000 K have so little ultraviolet radiation of wavelengths shorter than 91.2 nm (required to ionize hydrogen) that the reflection nebulae around such stars outshine the emission nebulae. Stars hotter than 25,000 K emit enough ultraviolet energy that the emission nebulae produced around them generally outshine the reflection nebulae.

Interstellar Reddening

Seventy years ago, astronomers were puzzled by the existence of stars whose spectral lines indicated that they were intrinsically hot and blue, although their overall appearance showed the reddish colors of much cooler stars. Today we understand that the light from these stars is not only dimmed but also **reddened** by interstellar dust. Most of their violet, blue, and green light has been scattered or absorbed, but some of their orange and red light,

Figure 19.11
Interstellar dust scatters blue light more efficiently than red light, thereby making distant stars appear redder, and giving clouds of dust near stars a bluish hue. Here a red ray of light from a star comes straight through to the observer, while a blue ray is shown scattering. A similar scattering process makes the Earth's sky look blue.

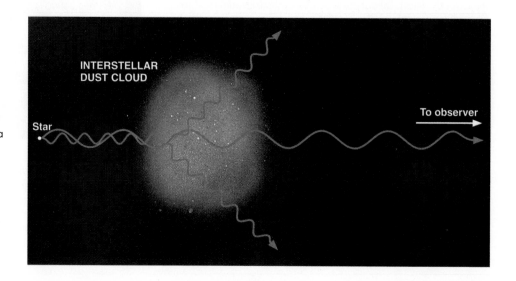

INTERSTELLAR DUST CLOUD

Star

To observer

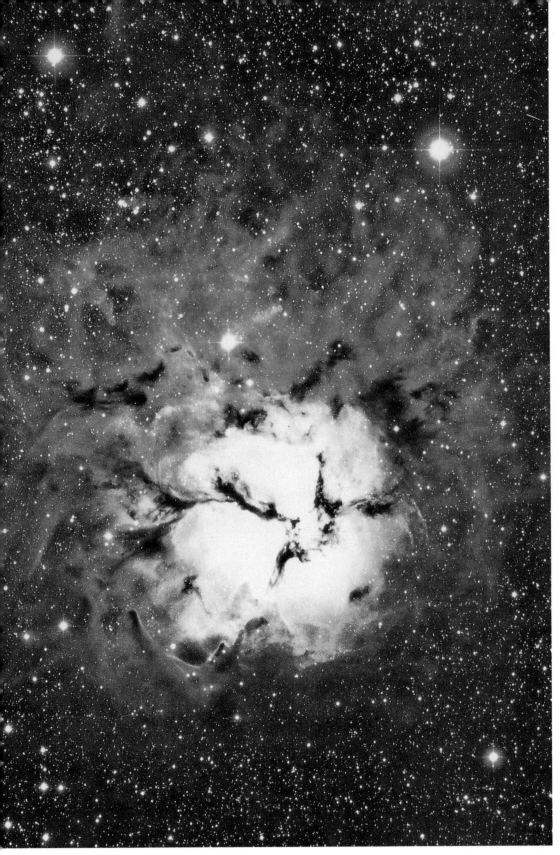

Figure 19.12
The Trifid Nebula (M20) in the constellation Sagittarius. In the reddish region, the hydrogen is ionized by nearby hot stars and glows through the process called fluorescence. The blue region is a reflection nebula. The Trifid Nebula is about 30 LY in diameter and about 3000 LY from the Sun. This beautiful image was processed by David Malin using special photographic techniques to bring out faint details. The bluish nebulosity can be seen surrounding the H II region. (Anglo-Australian Telescope Board)

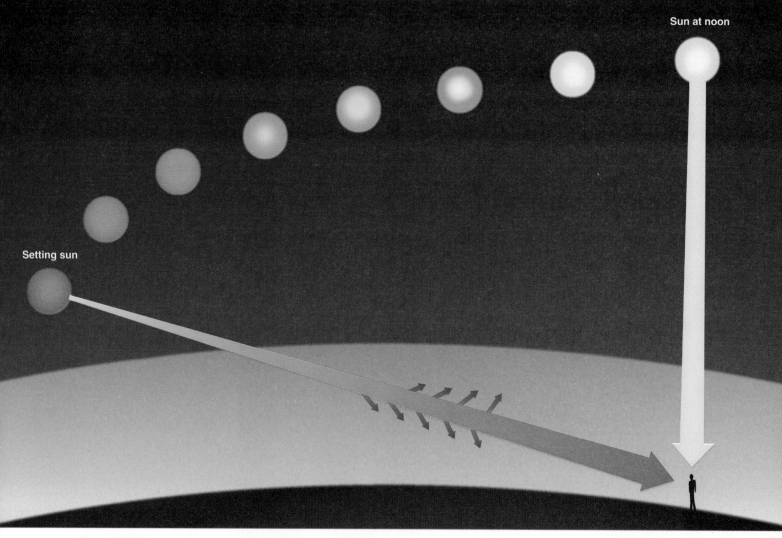

Figure 19.13
When light from the setting Sun traverses a long path through the Earth's atmosphere, blue light is scattered out of the direct light path, thereby making the Sun look redder.

with longer wavelengths, penetrates the obscuring dust and reaches Earth-based telescopes. (So, strictly speaking, *reddening* is not the most accurate term for this process, since no red color is added; instead, blues and related colors are subtracted, so it should more properly be called "deblueing.")

We have all seen an example of reddening. The Sun appears much redder at sunset than it does at noon. The lower the Sun is in the sky, the farther its light must travel through the atmosphere. Over this greater distance there is a higher probability that sunlight will be scattered. Since red light is less likely to be scattered than blue light, the Sun appears more and more red as it approaches the horizon (Figure 19.13).

The fact that light is reddened by interstellar dust means that long-wavelength radiation is transmitted more efficiently than short-wavelength radiation. Consequently, if we wish to see farther in a direction with considerable interstellar material, we should look at long wavelengths. This simple fact provides one of the motivations for the development of infrared astronomy. In the infrared region at 2 μm (2000 nm), for example, the obscuration is only one-sixth as great as in the visible region (500 nm), and we can therefore study stars that are more than twice as distant before their light is blocked by interstellar dust. This ability to see farther by observing in the infrared portion of the spectrum represents a major gain for astronomers trying to understand the structure of our Galaxy or probing its puzzling center (see Chapter 24).

Interstellar Grains

The preceding paragraphs have described the reddening and dimming of starlight by interstellar dust. But what exactly is this dust? Observations can give us some vital clues about the nature of the dust grains.

First of all, they tell us that the absorption of light is done by *solid particles* and not by interstellar gas. Except for specific spectral lines, atomic or molecular gas is almost transparent. Consider the Earth's atmosphere. Despite its incredibly high density compared with that of interstellar gas, it is so transparent as to be practically invisible (except when humans produce smog by filling it with solid particles of smoke and dirt; volcanoes, too, sometimes inject enough dust to make the sky look hazy). The quantity of *gas* required to produce the observed absorption of light in interstellar space would have to be enormous. The gravitational attraction of so great a mass of gas would produce effects upon the motions of stars that would be easily detected. Such effects are not observed, and thus the interstellar absorption cannot be the result of gases: the culprit must be something else.

Although gas does not absorb much light, we know from everyday experience that tiny solid or liquid particles can be very efficient absorbers. Water vapor in the air is quite invisible. When some of that vapor condenses into tiny water droplets, however, the resulting cloud is opaque. And dust storms, smoke, and smog offer familiar examples of the efficiency with which solid particles absorb light. On the basis of arguments like these, astronomers have concluded that widely scattered *solid* particles in interstellar space are responsible for the observed dimming of starlight. What are these particles made of? And how did they form?

Our observations show that a great deal of this dust exists; hence it must be primarily composed of elements that are abundant in the universe (and in interstellar matter). After hydrogen and helium, the most abundant elements are oxygen, carbon, and nitrogen. These three elements, along with magnesium, silicon, iron, and perhaps hydrogen itself, are thought to be the most important components of interstellar dust.

Observations support this hypothesis. Many heavy elements, including iron, magnesium, and silicon, are less abundant in interstellar gas than in the Sun and young stars. These heavy elements are assumed to be missing from the interstellar *gas* because they are condensed into solid particles of interstellar *dust.*

We also know from measurements of both total extinction and reddening that typical individual grains must be just slightly smaller than the wavelength of visual light. If the grains were a lot smaller, they would not block the light efficiently. Similarly, a wall of bowling balls, which are considerably smaller than the wavelength of radio waves, would not keep radio signals from reaching us inside a bowling alley.

On the other hand, if the dust grains were much larger than the wavelength of light, then starlight would not be reddened. Again, a bowling ball, which *is* much larger than the wavelength of light, would block both blue and red light with equal efficiency. In this way we can deduce that a characteristic interstellar dust grain contains 10^6 to 10^9 atoms and has a diameter of 10^{-7} to 10^{-8} m (10 to 100 nm). This is actually more like the specks of solid matter in cigarette smoke than the larger grains of dust you find under your desk when you are too busy studying astronomy to clean properly.

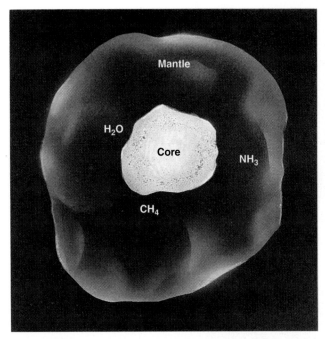

Figure 19.14
A typical interstellar grain is thought to consist of a core of rocky material (silicates), graphite, or possibly iron surrounded by a mantle of ices. Typical grain sizes are 10^{-7} to 10^{-8} m.

Observations of absorption features in spectra produced by interstellar material indicate that there must be many types of solid particles. Some interstellar grains apparently consist of a core of rock-like material (silicates); other grains appear to be nearly pure carbon (graphite). The nuclei of the grains are probably formed in shells of cooling gas ejected by red giants and other stars nearing the ends of their lives (these will be discussed in Chapter 22).

The grain nuclei may subsequently be incorporated into an interstellar cloud, where they can grow by gathering other atoms. The most widely accepted model pictures the grains with rocky cores and icy mantles (Figure 19.14). The most common ices in the grains are water (H_2O), methane (CH_4), and ammonia (NH_3).

19.5

Cosmic Rays

In addition to gas and dust, a third class of particles, noteworthy for the high speeds with which they travel, is found in interstellar space. They were first discovered in 1911 by the Austrian physicist Victor Hess, who flew simple instruments aboard balloons and demonstrated that high-speed particles are indeed coming to Earth from space (Figure 19.15). These particles, which became known as **cosmic rays** (an awkward name that is retained for historical reasons), resemble ordinary interstellar gas

Figure 19.15
Victor Hess returns from a 1912 balloon flight that reached an altitude of 5 1/3 km. It was on such balloon flights that Hess discovered cosmic rays. (Photo courtesy of Martin Pomerantz)

in terms of composition. In their behavior, however, cosmic rays differ radically from ordinary interstellar gas.

Composition of Cosmic Rays

Cosmic rays are high-speed atomic nuclei, electrons, and positrons. Velocities equal to 90 percent of the speed of light are typical. Most cosmic rays are hydrogen nuclei (protons) stripped of their accompanying electron. Helium and heavier nuclei constitute about 9 percent of the cosmic ray particles. The number of cosmic ray electrons is only about 2 percent the number of protons. Ten to 20 percent of the cosmic ray particles with masses equal to the mass of the electron carry positive charge rather than the negative charge that characterizes electrons. Such particles are called **positrons** and are a form of antimatter (see Section 15.2).

The abundances of various atomic nuclei in cosmic rays mirror the abundances in stars and interstellar gas, with one important exception. The light elements lithium, beryllium, and boron are far more abundant in cosmic rays than in the Sun and stars. These light elements are formed when high-speed cosmic ray nuclei of carbon, nitrogen, and oxygen collide with protons in interstellar space and break apart. (By the way, if you, like most readers, have not memorized all the elements and want to see how any of those we mention fit into the sequence of ele-

ments, you will find them all listed in Appendix 13 in order of the number of protons they contain.)

Cosmic rays reach the Earth in substantial numbers, and we can determine their properties either by capturing them directly, or by observing the reactions that occur when they collide with atoms in our atmosphere. The total energy deposited by cosmic rays in the Earth's atmosphere is only about one-billionth the energy received from the Sun, but it is comparable to the total energy received in the form of starlight. Some of the cosmic rays come to the Earth from the surface of the Sun. But most—as we will see—come from outside the solar system.

Origin of Cosmic Rays

There is a serious problem in identifying the source of cosmic rays. Since light travels in straight lines, we can tell where it comes from simply by looking. Cosmic rays are charged particles, and their direction of motion can be changed by magnetic fields. The paths of cosmic rays are curved both by magnetic fields in interstellar space, and by the Earth's own field. Calculations show that low-energy cosmic rays may spiral many times around the Earth before entering the atmosphere where we can detect them. If an airplane circles an airport many times before landing, it is impossible to determine the direction and city from which it came. So, too, after a cosmic ray circles the Earth several times, it is impossible to know where its journey began.

There are a few clues, however, about where cosmic rays might be generated. We know, for example, that magnetic fields in interstellar space are strong enough to keep all but the most energetic cosmic rays from escaping the Galaxy. It therefore seems likely that they are produced somewhere inside the Galaxy. The only likely exceptions are those with the very highest energy. Such cosmic rays move so rapidly that they are not significantly influenced by interstellar magnetic fields, and thus they could escape our Galaxy. By analogy, they could escape other galaxies as well, so some of the highest-energy cosmic rays that we detect may have been created in some distant galaxy. Still, most cosmic rays must have their source inside the Milky Way.

We can also estimate how far typical cosmic rays travel before striking the Earth. The light elements lithium, beryllium, and boron hold the key. Since they are formed when carbon, nitrogen, and oxygen strike interstellar protons, we can calculate how long, on the average, cosmic rays must travel through space in order to experience enough collisions to account for the relative abundances of lithium and the other light elements that they contain. It turns out that the required distance is about 30 times around the Galaxy. At speeds near the speed of light, it takes perhaps 3 to 10 million years for the average cosmic ray to travel this distance. This is only a small fraction of the age of the Galaxy or the universe, so cosmic rays must have been created fairly recently on a cosmic time scale.

The best candidates for a source of cosmic rays are the supernova explosions, which mark the deaths of massive stars. There are enough explosions, and the explosions generate enough energy, to account for the observed number of cosmic rays. What we do not yet know is what precise mechanism in these explosions accelerates pro-

tons and other atomic nuclei to the fast speeds we see. Some collapsed stars (including star remnants left over from supernova explosions) may, under the right circumstances, also serve as accelerators of particles. In any case, we again find that the raw material of the Galaxy is enriched by the life cycle of stars. In the next chapters we will see that as stars live and die, their material can be recycled in complex and intriguing ways.

Summary

19.1 About 5 percent of the visible matter in the Galaxy is in the form of gas and dust, which serves as the raw material for new stars. The material making up **interstellar matter** is distributed in a patchy way, sometimes becoming visible as glowing clouds of gas called **nebulae.** The most abundant elements in the interstellar gas are hydrogen and helium. About 1 percent of the interstellar matter is in the form of solid **interstellar grains.**

19.2 Interstellar gas near hot stars emits light by *fluorescence;* that is, light is emitted when an electron is captured by an ion and cascades down to lower energy levels. Glowing clouds of ionized hydrogen are called **H II regions** and have temperatures of about 10,000 K. Most hydrogen in interstellar space is not ionized and can best be studied by radio measurements of the 21-cm line. About 2 percent of interstellar space is filled with cold (100 K) clouds with densities of 50 atoms per cubic centimeter. Another 20 percent of the space between stars is filled with warm (3000 to 6000 K) neutral hydrogen clouds at a density of 0.3 atom per cubic centimeter. Some gas has been heated by supernova explosions to temperatures of 10^6 K.

19.3 Observations of interstellar gas are consistent with a model in which this gas is distributed in the form of clouds with cold, high-density cores surrounded by warm, lower-density envelopes, all embedded in a hot gas of very low density. The pressure in these three types of regions is approximately the same. The Sun is located at the edge of a low-density (0.1 atom per cubic centimeter; $T = 10^4$ K) cloud called the Local Fluff. The Sun and this cloud are located within the Local Bubble, a region extending to at least 300 LY from the Sun, within which the density of interstellar material is extremely low.

19.4 Interstellar dust grains absorb and scatter starlight. The effects of both processes are referred to as **interstellar extinction** and cause distant stars to appear fainter than they would if no dust grains lay along the path traversed by the starlight. Much of the dust is found in clouds called **infrared cirrus.** Since typical temperatures of the dust are 20 to 500 K, interstellar grains are best observed in the infrared. Because dust grains scatter blue light more efficiently than red light, stars seen through dust appear **reddened.** Interstellar grains typically have sizes comparable to the wavelength of light. They probably have a core of silicates or carbon surrounded by a mantle of such ices as water, ammonia, and methane.

19.5 Cosmic rays are particles that travel through interstellar space at a typical speed of 90 percent the speed of light. The most abundant elements in cosmic rays are the nuclei of hydrogen and helium, but **positrons** are also found. It is likely that many cosmic rays are produced in supernova explosions.

Review Questions

1. Identify several dark nebulae in photographs in this book (don't restrict yourself to this chapter alone). Give the figure numbers of the photographs, and specify where the dark nebulae are to be found on them.

2. Why do nebulae near hot stars look red? Why do dust clouds near stars look blue?

3. Describe the characteristics of the various types of interstellar gas clouds.

4. Prepare a table listing the different ways in which (a) dust and (b) gas can be detected in interstellar space.

5. Describe how the 21-cm line of hydrogen is formed. Why is this line an important tool for understanding the interstellar medium?

6. Describe the properties of the dust grains found in the space between stars.

7. Why is it difficult to determine where cosmic rays come from?

8. What causes reddening of starlight? Explain how the reddish color of the Sun's disk at sunset is caused by the same process.

9. Suppose a bright reflection nebula appears yellow. What kind of star is probably producing it?

10. Describe the spectrum of each of the following:
a. Starlight reflected by dust
b. A star behind invisible interstellar gas
c. An emission nebula

11. According to the text, a star must be hotter than about 25,000 K to produce an H II region. Both white dwarfs and main-sequence O stars have temperatures hotter than 25,000 K. Which type of star can ionize more hydrogen? Why?

12. From the comments in the text about which kinds of stars produce emission nebulae and which kinds are associated with reflection nebulae, what can you say about the temperatures of the stars in the Pleiades (see Figure 19.10)?

13. One way to calculate the size and shape of the Galaxy is to estimate the distances to faint stars from their observed apparent magnitudes, and to note the distance at which stars are no longer observable. The first astronomers to try this experiment did not know that starlight is dimmed by interstellar dust. Their estimates of the size of the Galaxy were much too small. Explain why.

14. Short-wavelength light is *absorbed* more efficiently by the Earth's atmosphere than is long-wavelength light. Sunburns are caused primarily by sunlight with wavelengths between 280 and 320 nm. The heat that we feel, however, is produced mainly by infrared radiation. Use this information to explain in general terms why it is easier to get sunburned at noon than in the late afternoon, even though the Sun feels nearly as hot at, say, 4:00 P.M. as it does at noon.

15. New stars form in regions where the density of gas and dust is high. Suppose you wanted to search for some recently formed stars. Would you more likely be successful if you observed at visible wavelengths or at infrared wavelengths? Why?

16. In big cities, you can see much farther on days without smog. Why?

17. The Sun is located in a region where the density of interstellar matter is low. Suppose that instead it were located in a dense cloud that dimmed visible light significantly. How would this have affected the development of civilization on Earth? For example, would it have presented a problem for early navigators?

Problems

18. Suppose that the average density of hydrogen gas in our Galaxy is one atom per cubic centimeter. If the Galaxy is a sphere with a diameter of 100,000 LY, how many hydrogen atoms are in the interstellar gas? What is the mass of this quantity of hydrogen?

19. H II regions can exist only if there is a nearby star hot enough to ionize hydrogen. Hydrogen is ionized only by radiation with wavelengths shorter than 91.2 nm. What is the temperature of a star that emits its maximum energy at 91.2 nm (use Wien's law)? Based on this result, what are the spectral types of those stars likely to provide enough energy to produce H II regions?

20. According to the text, there is about one atom of hydrogen in every cubic centimeter of interstellar space, and perhaps 10^3 dust grains per 10^9 m^3. If approximately 1 percent of the total mass of the interstellar medium is in the form of dust grains, what is the typical mass of a dust grain?

21. Suppose the density of a typical dust grain is 3 g/cm^3, and its mass is the value found in Problem 20; what is its radius? (Assume the grain is spherical, and remember that the volume of a sphere is given by the formula $V = (4/3)\pi R^3$.)

22. Suppose a star is behind a cloud of dust that dims its brightness by a factor of 100. Suppose you do not realize the dust is there. How much in error will your distance estimate be? Can you think of any measurement you might make to detect the dust?

Suggestions for Further Reading

Blitz, L. "Giant Molecular Cloud Complexes in the Galaxy" in *Scientific American,* Apr. 1982, p. 84.

Bowyer, S. et al. "Observing a Partly Cloudy Universe" in *Sky & Telescope,* Dec. 1994, p. 36. Review of the Extreme Ultraviolet Explorer mission and what it has shown us about the ISM.

Friedlander, M. *Cosmic Rays.* 1989, Harvard U. Press. The definitive introduction.

Helfand, D. "Fleet Messengers from the Cosmos" in *Sky & Telescope,* Mar. 1988, p. 265. Excellent summary of history and current understanding of cosmic rays.

Malin, D. *A View of the Universe.* 1993, Sky Publishing and Cambridge U. Press. Album by the skilled Australian astronomical photographer; includes some of the most dramatic color images of nebulae ever taken.

Marschall, L. "The Secrets of Interstellar Clouds" in *Astronomy,* Mar. 1982, p. 6.

Shore, L. and Shore, S. "The Chaotic Material Between the Stars" in *Astronomy,* June 1988, p. 6.

Smith, D. "Reflection Nebulae: Celestial Veils" in *Sky & Telescope,* Sep. 1985, p. 207.

Teske, R. "The Star That Blew a Hole in Space" in *Astronomy,* Dec. 1993, p. 31. On the Local Bubble.

Verschuur, G. "Interstellar Molecules" in *Sky & Telescope,* Apr. 1992, p. 379.

Verschuur, G. "Barnard's Dark Dilemma" in *Astronomy,* Feb. 1989, p. 30.

Verschuur, G. *Interstellar Matters.* 1989, Springer Verlag. Essays on dust and the interstellar medium; combines good historical material and reviews of modern topics.

Wynn-Williams, G. "Bubbles, Tunnels, Onions, Sheets: The Diffuse Interstellar Medium" in *Mercury,* Jan./Feb. 1993, p. 2.

Wynn-Williams, G. *The Fullness of Space: Nebulae, Stardust, and the Interstellar Medium.* 1992, Cambridge U. Press. A thorough introduction to the matter between stars.

Using **REDSHIFT** ™

1. Check the images in the *Photo Gallery* under The Galaxy, Interstellar Medium under Galactic, Gaseous Nebulae. Identify the following nebulae as principally bright, dark, or reflection. Also record secondary features of the nebulae that belong in a different class. Note: NGC 2024 is related to the Orion Nebula.

a. M8

b. M16

c. M42

d. NGC 2024

e. NGC 2237

f. NGC 6589-90

A portion of the Orion Nebula, about 1.6 LY across diagonally, recorded by the Hubble Space Telescope. It combines images taken in the red light of nitrogen gas, the green light of hydrogen, and the blue light of oxygen. The gases in the nebula are set to glow by the strong ionizing radiation from a brilliant O-type star that lies outside the region of this image, off to the bottom left. Visible throughout the region are a series of small, elongated "cocoons" of gas and dust, inside which stars are forming. Several of these can be seen in the inset, which shows a field only 0.14 LY across. Images such as this are good demonstrations of the capabilities of the repaired Hubble. (C. R. O'Dell/NASA)

CHAPTER 20

The Birth of Stars and the Search for Planets

Thinking Ahead

Although they had long suspected that other stars should have planets, astronomers were not able to discover a planet around a star like the Sun until 1995. Why did it take them so long?

Having examined the raw material from which stars form, we are ready to take a look at the process by which new stars emerge from interstellar matter. The mechanisms that make stars can sometimes also lead to one or more planets around those stars, so in this chapter we also examine the evidence for planets outside the solar system.

As we begin our study of the birth of stars, it may be useful to remember some of the key things we've learned so far. Table 20.1 on page 398 summarizes several important ideas about stars that were established in previous chapters.

The formation of stars is not merely a topic of historical interest. As we will see in this chapter, star formation is a continuous process that is going on *right now.* Stars of all masses, low as well as high, are being formed in the interiors of dust and gas clouds, throughout our Galaxy and others. In the last few years, new telescopes and instruments have enabled astronomers to look deep within nearby regions of star formation, and to obtain tantalizing glimpses of stars in the very earliest stages of their lives.

All these illustrious worlds,
and many more,
Which by the tube
astronomers explore . . .
Are suns, are centers,
whose superior sway,
Planets of various
magnitudes obey.

Sir Richard Blackmore in *The Creation*, Book II (quoted by A. Meadows in *The High Firmament*)

Figure 20.1

Stars form in clouds of gas and dust. M16 (the Eagle Nebula) is a large region of such cosmic raw material, visible because about 2 million years ago it produced a cluster of bright stars whose light ionizes the gas lying nearby. The cluster can be seen at the upper right of this image. The "elephant trunks" of dusty material visible near the center of the image are seen in much more detail on the cover of our text.
(Anglo–Australian Observatory Board)

20.1

Star Formation

If we want to find the youngest stars—those still in the process of formation—we must look in places with plenty of the raw material required to make stars. Since stars are made of gas, we focus our attention (and our telescopes) on the dense clouds of gas that dot the Milky Way (Figure 20.1).

Molecular Clouds: Stellar Nurseries

As we saw in Chapter 19, the most massive clouds—and some of the most massive objects in the Milky Way Galaxy—are the **giant molecular clouds.** These enormous reservoirs of gas and dust contain enough mass to make anywhere from a hundred to a million Suns, and the diameter of a typical cloud is 50 to 200 LY. Their name reflects the coldness of their interiors; with characteristic temperatures of about 10 K, most of their gas atoms are bound into molecules. Observations show that these clouds are the birthplaces of most stars in our Galaxy.

Molecular clouds are not smooth, but rather contain clumps or *dense cores* of material with very low temperatures and densities much higher (10^4 to 10^5 atoms per cubic centimeter) than is typical of most of the rest of the cloud. Both of these conditions—low temperature and high density—are just what is required to make new

stars. In order to form a star—that is, a dense, hot ball of matter capable of starting nuclear reactions deep within—we need a massive clump of atoms and molecules to shrink in radius and increase in density by a factor of nearly 10^{20}. Such a drastic collapse is brought about by the force of gravity.

TABLE 20.1
Basics About Stars from Earlier Chapters

- Stable (main-sequence) stars such as the Sun maintain equilibrium by producing energy through nuclear fusion in their cores. The ability to generate energy by fusion defines a star. (Sections 15.2, 15.3)
- Each second in the Sun approximately 600 million tons of hydrogen undergo fusion into helium, with about 4 million tons turning to energy in the process. This rate of hydrogen use means that eventually the Sun (and all other stars) will run out of central fuel. (Section 15.2)
- Stars come with many different masses, ranging from 1/12 M_{Sun} to roughly 100 M_{Sun}. There are far more low-mass than high-mass stars. (Section 17.2)
- The most massive main-sequence stars (spectral type O) are also the most luminous and have the highest surface temperature. The lowest-mass stars on the main sequence (spectral type M) are the least luminous and the coolest. (Section 17.4)
- A galaxy of stars such as the Milky Way contains enormous amounts of gas and dust—enough to make billions of stars like the Sun. (Section 19.1)

Figure 20.2
The star group named after Orion, the hunter of Greek mythology. Three stars close to each other in a line mark Orion's belt. The ancients imagined a sword hanging from the belt; one of the objects in this sword is the Orion Nebula. (Andrea Dupree, Harvard-Smithsonian CA, and Ronald Gilliland, NASA)

As we saw in Chapter 15 (and will continue to see in future chapters), the essence of the stars' life story is the ongoing competition between two forces: *gravity* and *pressure*. The force of gravity, pulling inward, tries to make a star collapse. Internal pressure produced by the motions of the gas atoms, pushing outward, tries to force the star to expand. When these two forces are in balance, the star is stable. Major changes in the structure of a star occur when one or the other of these two forces gains the upper hand. When a star is first forming, low temperature and high density both work to give gravity the advantage. Let's begin our examination of what happens by looking at a specific region of star formation.

The Orion Molecular Cloud

The best-studied of the stellar nurseries is in the constellation of Orion, the hunter, about 1500 LY away (Figure 20.2). A luminous cloud of dust and gas, called the Orion Nebula, can easily be seen with binoculars in the middle of the sword hanging from Orion's belt (Figure 20.3). This luminous material, however, is a mere "pimple" of light on the surface of a much larger molecular cloud. Most of the cloud does not glow with visible light, but betrays its presence by the radiation it gives off at infrared and radio wavelengths (Figure 20.4).

The Orion Nebula itself is a fascinating place, where a young cluster of hot stars is shining with great energy. Its radiation is opening up the cocoon of dust inside

Figure 20.3
A wide-angle view of some of the star-forming regions of Orion. The Orion Nebula is the fuzzy white object near the bottom. A cloud of dust that resembles a horsehead can be seen against the reddish glow of hydrogen in the upper left. The bright, over-exposed star above the horsehead and the one at the very top center of the image are two of the stars in Orion's belt. All the visible features of gas and dust are evidence of the much larger molecular cloud that lies behind them and that (at radio wavelengths) would fill most of the field shown here. (Anglo–Australian Observatory Board)

which it formed, exposing the stars to our view (Figure 20.5). Only a small number of stars in this Trapezium cluster can be seen with visible light, but infrared images—which penetrate the dust better—actually reveal over a thousand stars (Figure 20.6). The image that opens this chapter shows some of the wonderfully complex structure of gas, dust, and newly formed stars that characterizes this region.

(text cont. on page 401)

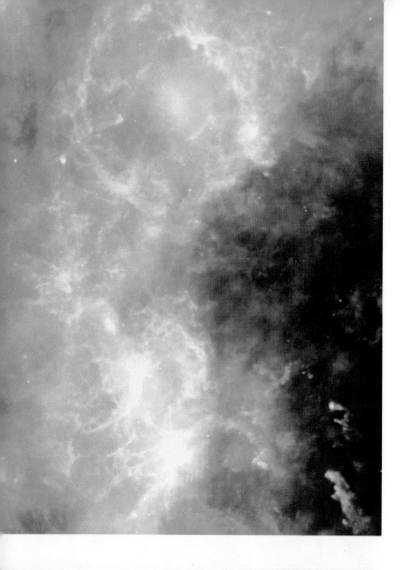

◄Figure 20.4
A wide-field view of Orion at three infrared wavelengths, taken with the Infrared Astronomical Satellite. The emission at 12 μm (typically coming from stars) is colored blue; emission at 60 μm is colored green; and emission at 100 μm (which comes from interstellar dust) is colored red. The two colorful regions seen in Figure 20.3 are the two bright yellow splotches at bottom right; the lower one is the Orion Nebula, and the higher one is the region of the horsehead, which can be seen as a small yellow finger poking out to the right. The horsehead area marks the region of Orion's belt. The red giant Betelgeuse is an intense point in the upper center of the image. To its right is a large ring that appears to be the remnant of an exploding star. Note how the infrared image reveals some of the vast quantity of dust in this region, and betrays the presence of the giant Orion Molecular Cloud. (Infrared Processing and Analysis Center/NASA)

Figure 20.5
In this wonderful image of the Orion Nebula, Australian photographer David Malin has used a special photographic technique called unsharp-masking to bring out subtle details. You can clearly see how the energy from the Trapezium cluster (the highly exposed white region) is opening up the gas and dust cloud within which it formed. The energy is ionizing the gas and reflecting from the dust to show the expanding bubble of the nebula. (Anglo–Australian Telescope Board)

◄Figure 20.6
The inner part of the Orion Nebula is seen in the infrared in this image taken with an electronic infrared camera on the 3.8-m United Kingdom Infrared Telescope on Mauna Kea. Blue colors represent radiation at 1.2 μm, green at 1.65 μm, and red at 2.2 μm. The hot O- and B-type stars of the Trapezium cluster are seen as blue in the center of the image. Dusty regions are seen as red and purple. The image covers a region of about 2 LY by 2 LY. About 500 stars can be found on this image, most belonging to the Trapezium cluster. This makes the region one of the densest groupings of stars astronomers have yet found. (© Mark McCaughrean 1989/NASA)

Interesting as the nebula may be, the full Orion molecular cloud is the truly impressive structure. In its long dimension it stretches over a distance of about 100 LY. The total quantity of molecular gas is about 200,000 times the mass of the Sun. A wave of star formation that began about 12 million years ago at one edge of this molecular cloud, near the western shoulder of the Orion star figure, has slowly moved through the cloud, leaving behind groups of newly formed stars. The stars in Orion's belt are about 8 million years old, and the stars near the Trapezium range in age from 300,000 to a million years.

Why is there so much material left around the newly formed stars in the Trapezium cluster? We have come to understand that star formation is not a very efficient process, using typically only a few percent—at most perhaps 25 percent—of the gas in a molecular cloud. The leftover material is heated either by the radiation of hot stars or by the explosions of the most massive stars that form. (We will see in later chapters that the most massive stars go through their lives very quickly, and end them by exploding.) Whether gently or explosively, the material in the neighborhood of the new stars is blown away into interstellar space. Older groups of stars can therefore be easily observed in visible light because they are no longer shrouded in dust and gas.

Because of the correlation between stellar ages and position in the Orion region, we know that star formation has been moving progressively through this molecular cloud. While we do not know what initially caused stars to begin forming in Orion, there is good evidence that the first generation of stars triggered the formation of additional stars, which in turn led to the formation of still more stars (Figure 20.7).

The basic idea is this: When a massive star is formed, it emits a large amount of ultraviolet radiation that heats the surrounding gas in the molecular cloud. This heating increases the pressure in the gas and causes it to expand.

When massive stars exhaust their supply of fuel, they explode, and the energy of the explosion also heats the gas. The hot gases burst into the surrounding cold cloud, compressing the material in it until the cold gas is at the same pressure as the expanding hot gas.

For the regions of cooler gas this can be a dramatic change. In the conditions typical of molecular clouds, the compression is enough to increase the gas density by a factor of 100. At densities this high, stars can begin to form in the compressed gas. Such a chain reaction—where the brightest and hottest stars of one area become the cause of star formation "next door"—seems to have occurred not only in Orion but also in many other molecular clouds.

The Birth of a Star

Although regions such as Orion can give us insight into how star formation might begin, its subsequent stages are still shrouded in mystery (and a lot of dust). There is almost a factor of 10^{20} difference between the density of a molecular cloud core and that of the youngest stars that can be detected. So far, we have been unable to get any *direct* evidence about what happens within a cloud as material comes together through the action of gravity and collapses through this range of densities to form a star.

Observations of this stage of stellar evolution are nearly impossible for several reasons. First, the dust-shrouded interiors of molecular clouds where stellar births take place cannot be observed with visible light. It is only with the new techniques of infrared and millimeter radio astronomy that we are able to make any measurements at all. In addition, the time scale for the initial collapse, measured in thousands of years, is very short, astronomically speaking. Since each star spends such a tiny fraction of its life in this stage, relatively few stars are going through the collapse process at any given time.

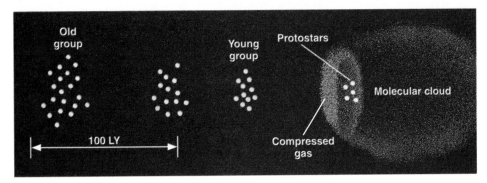

Figure 20.7
Schematic diagram showing how star formation can move progressively through a molecular cloud. The oldest group of stars lies to the left of the diagram and has expanded because of the motions of individual stars. Eventually the stars in the group will disperse and no longer be recognizable as a cluster. The youngest group of stars lies to the right, next to the molecular cloud. This group of stars is only 1 to 2 million years old. The pressure of the hot, ionized gas surrounding these stars compresses the material in the nearby edge of the molecular cloud and initiates the gravitational collapse that will lead to the formation of more stars.

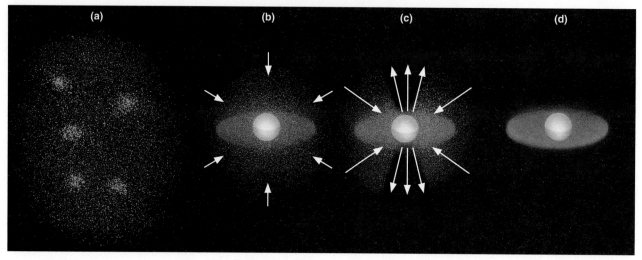

Figure 20.8

The formation of a star. (a) Dense cores form within a molecular cloud. (b) A protostar with a surrounding disk of material forms at the center of a dense core, accumulating additional material from the molecular cloud through gravitational attraction. (c) A stellar wind breaks out along the two poles of the star. (d) Eventually this wind sweeps away the cloud material and halts the accumulation of additional material, and a newly formed star surrounded by a disk becomes observable. (Based on drawings by F. Shu, F. Adams, and S. Lizano)

Furthermore, the collapse of a new star occurs in a region so small (0.3 LY) that in most cases we cannot resolve it with existing techniques. For all these reasons we have yet to catch a star in the act, so to speak, of its initial collapse. Nevertheless, through a combination of theoretical calculations and the limited observations available, astronomers have pieced together a picture of what the earliest stages of stellar evolution are likely to be.

During the period when a clump of matter in a molecular cloud is contracting to become a true star, we call it a **protostar.** Since a molecular cloud is more massive than a typical star by a factor of 100,000 or more, many protostars must form within each cloud.

The first step in the process of creating stars is the formation within the cloud—through a process we do not fully understand—of the dense cores of material discussed earlier (Figure 20.8a). These dense cores then attract additional matter because of the gravitational force they exert on the cloud material surrounding them. Eventually, the gravitational force of the infalling gas becomes strong enough to overwhelm the pressure exerted by the cold material that forms the dense cores. The material then undergoes a rapid collapse, the density of the core increases greatly as a result, and a protostar is formed.

According to the conservation of angular momentum (discussed in Chapter 2), a rotating body spins more rapidly as it decreases in size. In other words, if the object can turn its material around a smaller circle, it can move that material more quickly. This is exactly what happens when a protostar is forming: as it shrinks, its rate of spin goes up. Since the natural turbulence inside a molecular cloud tends to give any portion of it some initial spinning motion (even if it is very slow), all collapsing cores are expected to spin as they become protostars.

Rapid spin can begin to affect the shape of a ball of collapsing material. Rotation can keep material from falling inward at the equator of the spinning ball of gas (even while it continues falling in at the poles). You may have observed this same effect on the amusement park ride in which you stand with your back to a cylinder that is spun faster and faster. As you spin really fast, you are pushed against the wall so strongly that you could not possibly fall toward the center of the cylinder. So, too, the collapsing gas around a protostar can fall easily from directions away from the star's equator. But gas falling inward toward the equator is held back by the rotation, and forms an extended disk around the equator (as shown in Figure 20.8b).

The protostar and disk at this stage are embedded in an envelope of dust and gas, from which material is still falling onto the protostar. This dusty envelope blocks visible light, but infrared radiation can get through. As a result, a protostar itself in this phase of evolution is observable only in the infrared region of the spectrum. However, the dusty clouds of material around the star can sometimes be glimpsed when they are illuminated by the light of nearby stars that formed some time earlier, as in the Orion Nebula. The inset for the image that opens this chapter shows some beautiful examples of elongated dust clouds observed with the Hubble Space Telescope. Infrared observations reveal stars in the process of forming inside most of these clouds.

Winds and Jets

Observations show that a protostar eventually goes through a stage involving a powerful outflow of particles from its surface. This **stellar wind** consists mainly of protons (hydrogen nuclei) and electrons streaming away from the star at speeds of about 200 km/s (about 440,000 mi/h). Let us examine the effect of this wind on the material around the forming star.

When the wind first starts up, the disk of material around the star's equator blocks the wind in this direction. Where the wind particles *can* escape most effectively is in the direction of the star's poles (Figure 20.8c). If this model is correct, we should see beams of particles shooting out in opposite directions from the polar regions of new stars during this stage of their formation.

Such double beams (or *jets*) are precisely what astronomers observe around some protostars. Sometimes the material flowing out can be quite broad (Figure 20.9), but on other images, we see a remarkably narrow set of jets. On occasion, the jets collide with a somewhat denser lump of material nearby and excite its atoms to glow. These glowing regions, called *Herbig–Haro* (or *HH*) *objects* after the two astronomers who first identified them, allow us to trace the progress of the jets quite a distance from the star that produces them.

Figure 20.10 shows a series of recent images of Herbig–Haro objects, taken with the Hubble Space Telescope, that confirm our general picture of star formation but that show a considerable amount of complex structure in the jets and the clouds of material they energize. It may be that great clumps of material from the star's inner disk fall toward the star and then are caught up by the wind and blown outward, much as a clump of leaves can be

(text cont. on page 405)

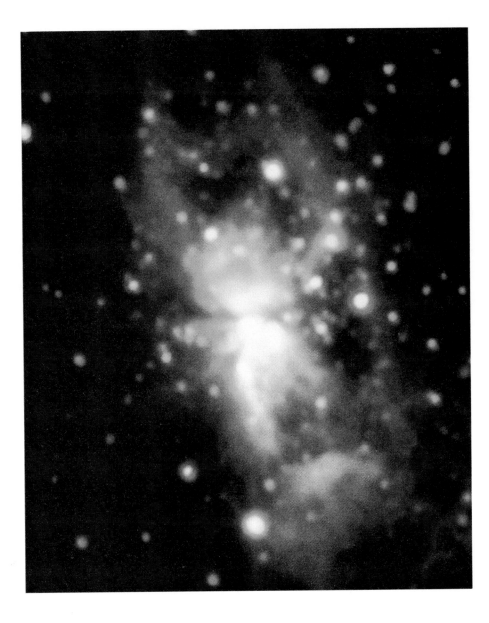

Figure 20.9
An infrared image of the protostar S106. Gas flows outward from this object in directions perpendicular to the dense disk that surrounds it. The disk is located where the outward flows appear to pinch together. (National Optical Astronomy Observatories)

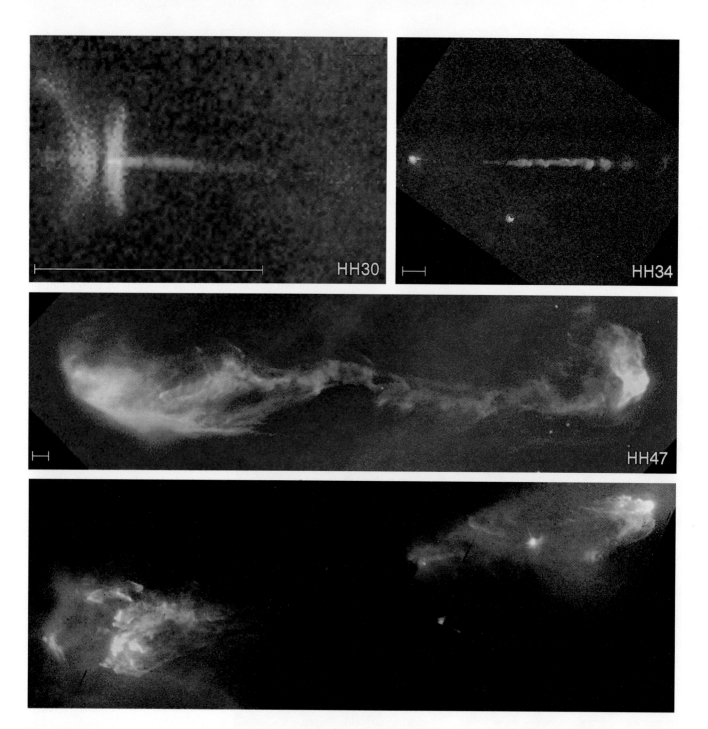

Figure 20.10

A collection of recent images taken with the Hubble Space Telescope, showing disks and jets from protostars. On the first three images the white bar at the bottom is the size of 1000 AU. (a) Upper left: HH 30, an edge-on disk around a protostar in the constellation of Taurus about 450 LY away. The reddish jet is amazingly narrow and shows knots of material that can be monitored to measure the rate at which material in the jet is moving. The speed turns out to be about a third of a million km/h. (C. Burrows and NASA) (b) Upper right: HH 34, a protostar in the vicinity of the Orion Nebula, displays a jet with beaded structure that gets wider as it gets farther from its star. (J. Hester and NASA) (c) Middle: HH 47, a protostar 1500 LY away (invisible inside a dusty disk at the left edge of the image), produces a very complicated jet. The star may actually be wobbling, perhaps because it has a companion. Light from the star illuminates the white region at the left because light can emerge perpendicular to the disk (just as the jet does). At right the jet is plowing into existing clumps of interstellar gas, producing a shock wave that resembles an arrowhead. (J. Morse and NASA) (d) Bottom: This view of HH 1 and 2 shows a classic double-beam jet emanating from a protostar (hidden in a dust disk in the center) in the constellation of Orion. Tip to tip, these jets are more than 1 LY wide. The bright regions (first identified by Herbig and Haro) are places where the jet is slamming into a clump of interstellar gas. (C. Burrows, J. Morse, J. Hester, and NASA)

blown away together in a windstorm on Earth. This might explain the clumpy structure we are beginning to see in many of these jets, but such ideas are still very speculative. Clearly, we have a lot more to learn about the details of these intriguing objects.

The wind from a forming star will ultimately sweep away its obscuring envelope of dust and gas, leaving behind the naked disk and protostar, which can now be seen with visible light (Figure 20.8c and 20.8d). We should note that at this point, the protostar itself is still contracting slowly and has not yet reached the main-sequence stage on the H–R diagram (a concept introduced in Chapter 17). But the disk can be more easily detected at this point, especially when observed at infrared wavelengths. In the last decade astronomers have accumulated a great deal of evidence for the presence of such disks around many newly forming stars.

This description of a protostar surrounded by a rotating disk of gas and dust sounds very much like what happened when the Sun and planets formed (see Section 13.3). Do the disks around protostars also form planets? We will return to this question at the end of the chapter.

20.2

The H–R Diagram and the Study of Stellar Evolution

One of the best ways to summarize all of these details about how a star or protostar changes with time is to use an H–R diagram. As a star consumes its nuclear fuel, its luminosity and temperature change. Thus its position on the H–R diagram, in which luminosity is plotted against temperature, also changes. As a star ages, we must replot it in different places on the diagram. Therefore astronomers often speak of a star *moving* on the H–R diagram, or of its evolution *tracing out a path* on the diagram. Of course, in this context, "tracing out a path" has nothing to do with the star's motion through space; this is just a shorthand way of saying that its temperature and luminosity change as it evolves.

To estimate just how much the luminosity and temperature of a star change as it ages, we must resort to calculations. In a theoretical study of stellar evolution, we compute a series of *models* for a star, each successive model representing a later point in time. Stars may change for a variety of reasons. Protostars, for example, change in size because they are contracting, and their temperature and luminosity change as they do so. After nuclear fusion begins in the star's core (see Chapter 15), main-sequence stars change because they are using up their nuclear fuel.

Given a model that represents a star at one stage of its evolution, we can calculate what it will be like at a slightly later time. At each step, the model predicts the lu-

minosity and size of the star, and from these we can figure out its surface temperature. A series of points on an H–R diagram, calculated in this way, allows us to follow the life changes of a star and hence is called its *evolutionary track*.

Evolutionary Tracks

Let's now use these ideas to follow the evolution of protostars that are on their way to becoming main-sequence stars. The evolutionary tracks of newly forming stars with a range of stellar masses are shown in Figure 20.11. These young stellar objects are not yet producing energy by nuclear reactions, but derive their energy from gravitational contraction—by the sort of process proposed for the Sun by Helmholtz and Kelvin in the last century (see Chapter 15).

Initially, a protostar remains fairly cool, with a very large radius and a very low density. It is transparent to infrared radiation, and the heat generated by gravitational contraction can be radiated away freely into space. Because heat builds up slowly inside the protostar, the gas pressure remains low, and the outer layers fall almost unhindered toward the center. Thus the protostar undergoes very rapid collapse, corresponding to the roughly vertical lines at the right of Figure 20.11. As the star shrinks its surface area gets smaller, and so its total luminosity decreases. The contraction stops only when the protostar becomes dense and opaque enough to trap the heat released by gravitational contraction.

Stars first become visible only after the stellar wind described earlier clears away the surrounding dust and gas, and this occurs near the point at which the evolutionary tracks in Figure 20.11 change from being nearly vertical to being nearly horizontal (although, as you can see, the details are not quite the same for stars of different mass). At this point, changes inside the contracting star keep the luminosity of stars like our Sun roughly constant, but start building up the surface temperature. Thus the star "moves" to the left in the H–R diagram.

To help you keep track of the various stages that stars undergo during their lives, it can be useful to compare the development of a star to that of a human being. (Clearly, you will not find an exact correspondence, but thinking through the stages in human terms may help you remember some of the ideas we are trying to emphasize.) Protostars might be compared with human embryos, as yet unable to sustain themselves, but drawing resources from their environment as they grow. Just as the birth of a child is the moment it is called upon to produce its own energy (through eating and breathing), so astronomers say that a star is born when it is able to sustain itself through nuclear reactions.

When the central temperature becomes high enough (about 10 million K) to fuse hydrogen into helium, we say that the star has *reached the main sequence* on the H–R diagram. (Astronomers call such a star a main-sequence

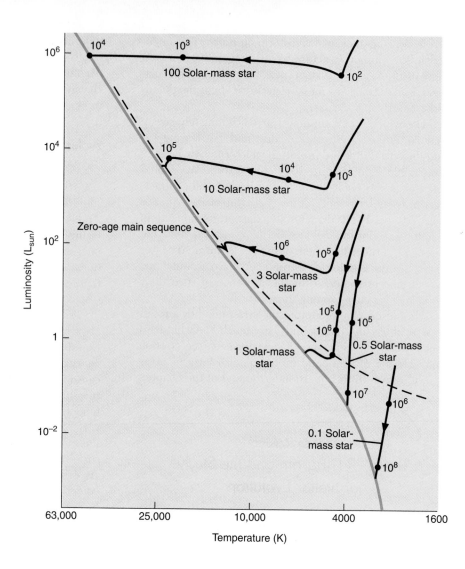

Figure 20.11

Evolutionary tracks for contracting protostars of different masses seen on an H–R diagram. The numbers next to each dark point on a track are the rough number of years it takes an embryo star to reach that stage in its life. You can see that the more mass a star has, the shorter the time it takes to go through each stage. Stars that lie above the dashed line would typically still be surrounded by infalling material and would be hidden by it.

star even when the discussion does not concern the H–R diagram.) It is now a full-fledged star, more or less in equilibrium, and its rate of change slows dramatically. Only the gradual depletion of hydrogen as it is transformed into helium in the core slowly changes the star's properties. The mass of a star determines exactly where it falls on the main sequence. As Figure 20.11 shows, massive stars on the main sequence have high temperatures and high luminosities. Low-mass stars have low temperatures and low luminosities.

Objects of extremely low mass never achieve high enough central temperatures to ignite nuclear reactions. The lower end of the main sequence stops where stars have a mass just barely great enough to sustain nuclear reactions at a sufficient rate to stop gravitational contraction. This critical mass is calculated to be about 1/12 the mass of the Sun. Objects below this critical mass are called either brown dwarfs or planets. Exactly where the dividing line should be drawn between brown dwarfs and planets is still a controversial subject among astronomers. Increasingly, "planet" is used to refer to an object formed in a disk surrounding a protostar, while a "brown dwarf," like a true star, forms at the center of the protostellar disk.

At the other extreme, the upper end of the main sequence terminates at the point where the energy radiated by the newly forming massive star becomes so great that it halts the accretion of additional matter. The upper limit of stellar mass is thought to be about 100 solar masses.

Evolutionary Time Scales

In general, a star's pre-main-sequence evolution slows down as it moves along its evolutionary track toward the main sequence. The numbers labeling the points on each track in Figure 20.11 are the times, in years, required for the embryo stars to reach those stages of contraction. The time for the whole evolutionary process, as the figure shows, depends on the mass of the star. Stars of mass much higher than the Sun's reach the main sequence in a few thousand to a million years. The Sun required millions of years before it was born. Tens of millions of years are required for stars of lower mass to evolve to the lower main sequence. (We will see that massive stars go through *all* stages of evolution faster than do low-mass stars.)

We will take up the subsequent stages in the life of a star in the next chapter, examining what happens after stars arrive on the main sequence and begin a "prolonged adolescence" of fusing hydrogen to form helium. But now we want to examine the connection between the formation of stars and planets.

Evidence That Planets Form Around Other Stars

Having developed on one, and finding it essential to our existence, we have a special interest in planets. Yet planets outside the solar system are extremely difficult to detect. The amount of light a planet reflects from its star is a depressingly tiny fraction of the light that star gives off. Furthermore, planets are lost in the glare of their much brighter parent stars. You might compare a planet orbiting a star to a mosquito flying around one of those giant spotlights at shopping center openings. Imagine viewing the scene from some distance away—say from an airplane. You could see the spotlight just fine, but what are your chances of catching the light reflected from the mosquito?

Despite such daunting difficulties, we would very much like to know whether planets are common around other stars. We can search for planets elsewhere *indirectly*—by observing their effects on the parent star, or by detecting a disk of material from which planets might be condensing.

Disks Around Protostars: Planetary Systems in Formation?

It is a lot easier to detect the raw material from which planets might be assembled than to detect planets after they are fully formed. The reasoning goes like this: As we saw in Chapter 13, planets form by the agglomeration of gas and dust particles in orbit around a newly created star. Each dust particle is heated by the young protostar and radiates in the infrared region of the spectrum. Before any planets form, we can detect such radiation from all of the individual dust particles that are destined to become part of the planets.

However, once those particles gather together and form a few planets, the overwhelming majority are hidden in the interiors of the planets where we cannot see them. All we can now detect is the radiation from the outside surfaces of the planets, which cover a drastically smaller area than the dusty disk from which they formed. The amount of infrared radiation is therefore greatest *before* the dust particles combine into planets. For this reason, our search for planets begins with a search for radiation from the material required to make them.

As we have seen in this chapter, the presence of a disk of gas and dust appears to be an essential part of star formation. Observations show that at least 50 percent of all protostars are surrounded by disks (Figure 20.12) that range in size from 10 to 1000 AU. For comparison, the average diameter of the orbit of Pluto—that is, the size of our own planetary system—is 80 AU. The mass contained in these disks is typically 1 to 10 percent of the mass of

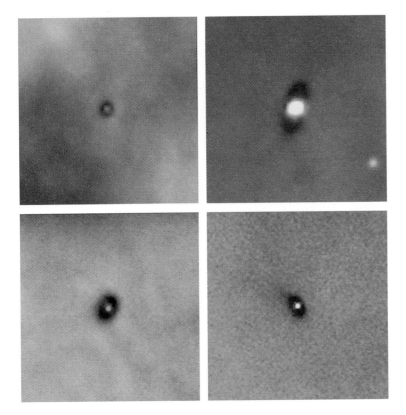

Figure 20.12
Hubble Space Telescope images of four disks around young stars in the Orion Nebula. The dark, dusty disks are seen silhouetted against the bright backdrop of the glowing gas in the nebula. The size of each image is about 30 times the diameter of our planetary system; this means the disks we see here range in size from two to eight times the orbit of Pluto. The red glow at the center of each disk is a young star, no more than a million years old. See this chapter's opening image for some disks in which stars are not easily visible.
(M. McCaughrean, C. R. O'Dell, and NASA)

our own Sun, which is more than the mass of all the planets in our solar system put together. Observations therefore demonstrate that a large fraction of stars begin their lives with enough material in the right place to form a planetary system.

Evolution of Proto-Planetary Disks

This result is encouraging, but the existence of rotating disks of raw material is still not enough. Do planets actually form? There is indirect evidence that they do. If we measure the temperature and luminosity of a protostar, we can place it in an H–R diagram like the one shown in Figure 20.11. By comparing the real star to our models of how protostars should evolve with time, we can estimate its age. What astronomers have discovered is that the properties of the disks change systematically with the ages of the encircled protostars.

In order to understand what the changes mean, we need to bear two things in mind. First, when we study the disks with infrared detectors, what we are actually able to measure is the heat radiation from dust. The dust absorbs energy radiated by the central protostar, and then the dust re-emits that energy as infrared radiation.

Second, because the protostar is the primary source of energy, the dust closest to the protostar is hottest. Temperatures in the disk at 0.1 AU from the star are about 1000 K; at 1 AU, the temperature is a few hundred degrees; and at their outer boundaries the disks are very, very cold—only a few tens of degrees. Therefore, if we observe some dust with temperatures of 1000 K or more in a given disk, we know that the dust extends nearly all the way to the surface of the star. If no hot matter is detectable, then the regions close to the star must be very nearly empty of dust.

What the observations show is that if a protostar is less than 3 million years old, its disk does contain hot dust. The disk, which is opaque to visible light, must therefore extend all the way from very close to the surface of the star out to tens or hundreds of astronomical units away. But by the time protostars reach the "advanced age" of 30 million years, their disks have thinned out and become transparent.

Most interestingly, around 3- to 30-million-year-old stars we find disks with outer parts that are still opaque but with so little hot material that we cannot detect the inner disk at all. That is, the inner regions close to the star are seriously depleted of dust. In these objects, the disk looks like a donut with the protostar centered in its hole.

How is it possible to have two rather sharply separated zones in the disk—one with a high density of dusty material and the other with almost none? This separation is especially puzzling since calculations show that matter from the outer, dense region should spiral inward quickly (in less than about 50,000 years) and fill the inner hole. This time span is pretty short compared to how much time the star spends in this general stage of its life. The fact that we see such holes around a large fraction of the protostars with ages between 3 and 30 million years indicates that there must be some mechanism to prevent the matter from flowing inward.

Calculations show that the formation of a Jupiter-sized planet would halt the inward flow. Suppose such a planet does form at a distance of a few astronomical units

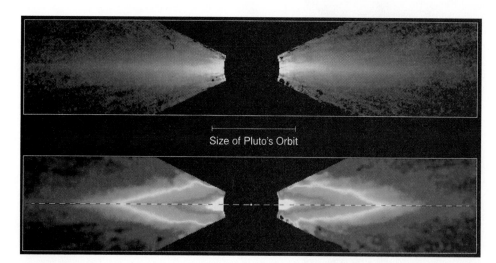

Size of Pluto's Orbit

Figure 20.13
The disk around the star Beta Pictoris, as seen with the Hubble Space Telescope. The star itself is blocked out so that the much fainter disk can be observed. The cleared-out area discussed in the text is also within the blocked-out area. (Top) A visible light image of the disk, shining with reflected starlight. Note the size of Pluto's orbit, given by the line below. (Bottom) Image-processing techniques have been used to assign different colors to each level of brightness in the disk. The pink-white inner part of the disk is slightly tilted relative to the plane of its outer parts (red-yellow-green). (C. Burrows, J. Krist, ESA/NASA)

from the protostar, presumably due to the progressive ag-glomeration of matter from the disk. This very agglomeration can contribute to the creation of dust-free regions. Calculations also show that any small dust particles and gas that were initially located in the region between the protostar and the planet, and that are not swept up by the planet, will accrete onto the star in about 50,000 years.

Matter lying outside the planet's orbit, on the other hand, is prevented from moving into the hole by the gravitational forces exerted by the planet. (We saw something similar in Saturn's rings, where the action of the shepherd moons keeps the material near the edge of the rings from spreading out.) The formation of giant planets can, according to the calculations, thus produce and sustain the holes in the disks surrounding very young stars.

This phenomenon can actually be seen in a star that is no longer a protostar, but has maintained a significant disk even as it has reached the main sequence. Called Beta Pictoris, it is located about 50 LY away in the southern constellation of Pictor. What distinguishes this star is that its disk (at least 2000 AU wide) can be detected with visible as well as with infrared light. At the heart of the disk is a clear zone about 50 AU wide—roughly the size of our own planetary system.

Figure 20.13 shows a recent Hubble Space Telescope image of the Beta Pictoris disk, where the star and the clear zone have been blocked out to reveal the structure of the disk in unprecedented detail. As the bottom image (in exaggerated color) shows, the inner part just outside the clear zone is slightly tilted relative to the outer parts (the dotted line shows the orientation of most of the disk). One possible interpretation of this inner tilt (which should long ago have been smoothed out in such a system) is that it is maintained by the presence of one or more planets in the clear area inside the disk. (If this interpretation is correct, then the Beta Pictoris disk may well be a dustier, thicker version of our solar system's Kuiper belt, discussed in Chapter 12.) Much more work will have to be done with this star before we can be sure about our conclusion, but it is intriguing to consider that more mature disks might also hold clues about the presence of planets around other stars.

20.4

Planets Beyond the Solar System: Search and Discovery

Once again, the fact that the formation of giant planets *could* explain such observations does not *prove* that such planets actually exist. A variety of techniques have been used to search for better evidence of planets, with some dramatic recent successes. Since planets are so difficult to image, the kinds of experiments possible with current technology look for the effects of surrounding planets on the central star—in particular, for changes in the star's motion through space caused by the orbital motion of surrounding planets.

Search for Orbital Motion

To understand how this approach works, consider a single Jupiter-like planet in orbit about a star. As described in Chapter 17, both the planet and the star in such systems actually revolve about their common center of mass. The sizes of their orbits are inversely proportional to their masses. Suppose the planet has a mass about one-thousandth that of the star; the size of the star's orbit is then one-thousandth that of the planet.

In other words, while the star will be much easier to see than the planet, its motion will be quite small. To get a sense of how difficult such an observation might be, let's first see how hard Jupiter (the most massive planet in our solar system) would be to detect from the distance of a nearby star. Consider an alien astronomer trying to observe our own system from Alpha Centauri, the closest star system to our own.

The diameter of Jupiter's apparent orbit viewed from this distance is 10 arcsec, and that of the Sun's orbit is 0.010 arcsec. Remember, even with the best telescope *we* presently have, these alien astronomers could not detect faint Jupiter directly. But if they could measure the apparent position of the Sun to this precision, they would see it describe an orbit of diameter 0.010 arcsec with a period equal to that of Jupiter, which is 12 years. In other words, if they watched it for 12 years, they would see the star "wiggle" back and forth in the sky by this minuscule fraction of a degree. From the observed motion and the period, they could deduce the mass of Jupiter and its distance using Kepler's laws.

It is just possible (but very difficult) to make astronomical measurements from the Earth's surface today with enough precision to detect Jupiter-mass companions around the nearer low-mass stars in this way. In the 1960s, after many decades of positional measurements with refracting telescopes, several astronomers thought they had discovered planets, but all of these claims have since been found to be in error. More recent applications of this technique with higher-precision, ground-based instruments have so far turned up nothing. With improved equipment, however, it would be possible to extend this search to several hundred nearby stars.

The technique just described relies on measuring the changes in the *position* of the star as its orbit moves it back and forth across the sky. But as the star and planet orbit each other, part of their motion will be in our line of sight. Such motion (as discussed in Chapter 4) can be measured using the Doppler effect and the star's spectrum.

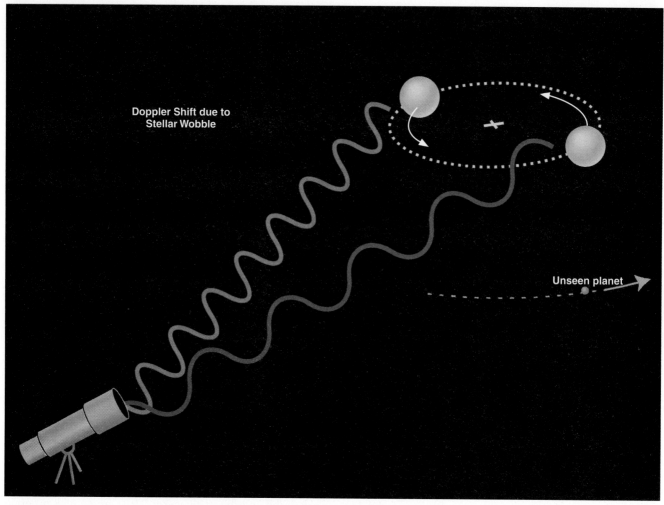

Figure 20.14
The motion of a star and a planetary companion around a common center of mass can be detected as a cyclical change in the Doppler shift of the star. When the star is moving away from us, the lines in its spectrum show a tiny redshift; when it is moving toward us they show a tiny blueshift. The change in color (wavelength) has been exaggerated here for educational purposes. In reality, the Doppler shifts we measure are extremely small and require sophisticated equipment to be detected. (Diagram courtesy of G. Marcy, San Francisco State University)

The Discovery of Planets

It was this method that finally allowed astronomers to establish the existence of planets around other stars like our own in 1995. With modern technology, astronomers can measure very tiny changes in the radial velocity of a star (its motion toward or away from us). As the star moves back and forth in orbit around the system's center of mass in response to the gravitational tug of an orbiting planet, the lines in its spectrum will shift back and forth (Figure 20.14).

Let's again consider the example of the Sun. Its radial velocity changes by about 13 m/s with a period of 12 years because of the gravitational force of Jupiter, and slightly more (15 m/s) if the effects of Saturn are also included. This corresponds to about 30 mi/h, roughly the speed at which many of us drive around town.

Detecting motion at this level in a star's spectrum presents an enormous technical challenge, but not an insurmountable one. Several groups of astronomers around the world have been actively engaged in searching for planets using specialized spectrographs designed for this purpose. Observers engaged in such programs must be patient as well as skillful, since the regular cycle of changes for a planet in an orbit like Jupiter's take a decade or so to become apparent. (Note that this means that the planets closest to their stars—and thus taking the shortest time to revolve around the center of mass—are the ones we are most likely to find at the beginning of such searches.)

The first planet discovered using this technique was found in 1995 by Michel Mayor and Didier Queloz of the Geneva Observatory, using an advanced-design spectrograph at the 1.9-m telescope of the Haute-Provence Ob-

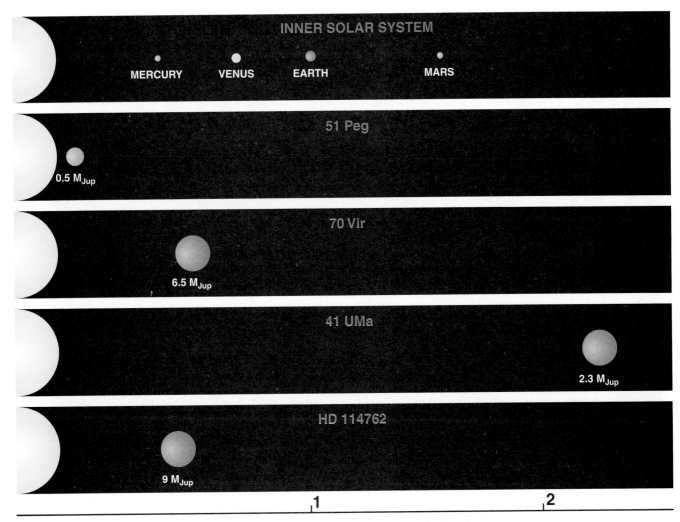

Figure 20.15
The relative positions of the planets around the first five normal stars known to have planetary companions. Other systems will probably have been found by the time you read this. The masses indicated for each planet are minimum masses. The sizes of objects are not to scale.

servatory in France. This planet orbits a main-sequence star resembling our Sun called 51 Pegasi, about 40 LY away. (The star can be found in the sky near the great square of Pegasus, the flying horse of Greek mythology, one of the best-known star patterns.) To everyone's surprise, the planet takes a mere 4.2 days to orbit around the star. Remember that Mercury, the innermost planet in our solar system, takes 88 days to go once around.

This means the planet must be very close to 51 Pegasi, circling it about 7 million km away (Figure 20.15). At that distance, the energy of the star should heat the planet's surface to a temperature close to 1000°C (a bit hot for future tourism). Since we do not know the angle at which we are seeing the planet's orbit, we can only calculate the minimum mass it could have, which is about half the mass of Jupiter (or more than 150 times that of the Earth). The actual mass may be somewhat larger than this; but, in any case, this is clearly a jovian and not a ter-

restrial-type planet. How a self-respecting jovian planet got so close to its star (it probably formed farther out in the system), whether or not the star's heat is continuously evaporating some of its outer material, and how frequently such close-in planets develop around stars are questions future observations and theoretical modeling will have to address.

With a "year" lasting only four Earth days, the effects of the new planet (for which some astronomers have already suggested the name Bellerophon, after the hero who rode Pegasus in Greek mythology) become evident in the spectrum of its star after only a few nights of observation. This enabled Geoffrey Marcy and Paul Butler of San Francisco State University to confirm its existence using the 3-m reflector at the Lick Observatory in California only one week after the Swiss team made their announcement. The two California astronomers and their students had been observing the spectra of about 120 nearby Sun-

TABLE 20.2
The First Planets Discovered Around Other Stars

Name of Star	Estimated Distance	Minimum Planet Mass	Period of Planet's Orbit	Semimajor Axis
	(LY)	(Jupiters)	(days)	(AU)
51 Pegasi	40–60	0.5	4.2	0.05
70 Virginis	30	6.5	116	0.5
47 Ursae Majoris	40	2.3	1100	2
HR 3522	44	0.8	14.76	0.1
HD 114762	100	9	84	0.38
HR 5185 (Tau Boötes)	62	3.9	3.3	0.05
PSR 1257 + 12	1600	A. 0.00005	25.3	0.19
		B. 0.011	66.5	0.36
		C. 0.009	98.2	0.47

like stars for over eight years with a spectrograph that can measure radial velocity changes to within 3 m/s. (Ironically, they had not included 51 Pegasi in their initial search program because it had been mistakenly listed as not quite resembling the Sun in a standard star catalog.)

In early 1996 Marcy and Butler were able to announce the discovery of planets around two other nearby stars, 70 Virginis and 47 Ursae Majoris. The two new planets are farther from their stars, taking 116 days and 1100 days, respectively, to go around. This puts them in orbits that are less shocking to our solar system prejudices—the first would be between the orbits of Mercury and Venus, the second between those of Mars and Jupiter. Their minimum masses are estimated around 6.5 and 2.3 Jupiters, making them quite substantial but still definitely planets and not brown dwarfs.

Two other stars, HR 3522 and HD 114762 (their names are merely entries in a lengthy star catalog), were soon added to the list of those with planetary companions. The planet around HD 3522 is similar to the one around 51 Pegasi in that its period is only a few days and its orbit is quite close to its star. The planet around HD 114762 takes 84 days to orbit its star, which means its orbit is just a bit smaller than Mercury's. It has a minimum mass of 9 Jupiters, the largest so far; it may well be a brown dwarf.

It is important to bear in mind that our current techniques can only detect planets with masses comparable to that of Jupiter. Planets with masses like Earth's might exist around these or other stars, but they would cause much smaller (and, at present, unobservable) Doppler shifts in their stars' spectra. Table 20.2 lists the planets we know about at the time this text is going to press, although others may have been discovered by the time you read this.

The discovery of planets around a number of Sun-like stars has ended centuries of conjecture and debate about the uniqueness of our solar system. It appears, as theorists have been suggesting for some time, that the formation of planets around the Sun was not a unique event, but a process that accompanies the formation of many stars. To the degree that planets are the only sort of habitats where we can imagine life as we know it, the discovery of these planets lends a new spirit of optimism to the search for life elsewhere, a subject to which we will return in the book's Epilogue.

Planets and Pulsars

To be absolutely fair, we should point out that the first planetary-mass objects discovered (in 1992) were not found around living stars like the ones discussed in the previous section, but around a kind of stellar "corpse" called a neutron star (to be described in Chapter 22). This particular neutron star (whose catalog designation is PSR 1257+12) has three and possibly four planets orbiting around it (also shown in Table 20.2). Many such dead stars give off sharp pulses of radio waves and are thus called *pulsars*. (If you are intrigued by how dead stars can emit sharp pulses of radiation, you now have another reason to read Chapter 22.)

In any case, astronomers can identify the motion of the neutron star around its center of mass by regular changes in the arrival of the pulses (in the same way that the motion of a living star can be seen in the Doppler shift of its spectrum). But there may be an important difference between the planets found around such stellar corpses and the planets discussed earlier. Because the formation of such a dead star generally involves the explosion of most of the original star, astronomers are convinced that the planet-mass companions of PSR 1257+12 are not "original" planets—in the sense that they existed during the star's life. Instead, they are probably "second generation planets" consisting of the remains of a nearby companion star that did not survive the explosion. Scoured by intense radiation from the spinning neutron star, such companions would be nothing like the planets in our own solar system. This is why more interest has greeted the discovery of planets around stars like the Sun.

Imaging Planets

What about looking for planets directly rather than just studying their parent stars? After all, "seeing is believing" is a very human prejudice. One future possibility might be to detect the (very slight) decrease in a star's light when one of its planets passes in front of it. But if *seeing* a planet means taking an image of it, then no extra-solar-system planet—even one as large as those we have discovered—could be seen with current equipment. Let's discuss the problems outlined at the start of Section 20.3 in a bit more detail.

Suppose, for example, that you were a great distance away and wished to detect reflected light from the planet Jupiter. Jupiter intercepts and reflects just about one-billionth of the Sun's radiation, so its apparent brightness in visible light is only a billionth that of the Sun. Smaller and less-reflective planets would be even more difficult to see. But the faintness of potential planets is not the biggest problem.

The real difficulty is that the faint light from a planet is swamped by the blaze of radiation from its parent star. If you are near-sighted, try looking at streetlights at night with your glasses off. You will see a halo of light surrounding every light. Bright stars seen through a telescope also appear to be surrounded by a bright halo of light (see the picture of Supernova 1987A, which does not appear as a sharp point of light, in Figure 22.10). In this case the problem is not that the telescope is near-sighted, but rather slight imperfections in its optics and atmospheric blurring prevent the light from coming to focus in a completely sharp point. Planets, if any, would lie within this halo, and their faint light could not be seen in the glare.

Astronomers are now trying to devise ways to suppress this glare. One of the best approaches appears to be observing in space in order to escape the effects of the Earth's atmosphere. The optimum wavelength range in which to observe is the infrared, since planets get brighter in the infrared while stars get fainter, thereby reducing the brightness contrast.

Even so, it may not be possible to obtain such images from the Earth's neighborhood. We are still deep within the Sun's *zodiacal cloud*—a thin cloud of dust left over from the formation of the planets—which tends to scatter and spread out starlight. It may be necessary to make the observations far from the Sun—say at the orbit of Jupiter—to minimize the scattered light from the star around which the planet orbits. It will also be necessary to use special techniques to artificially suppress the light from the central star. NASA is actively exploring the feasibility of such a mission sometime in the next century.

Even better than imaging would be a spectrum of a planet. For example, we have discussed that the oxygen in the Earth's atmosphere is unstable, combining easily with other elements. This means it has to be replenished through the action of the life-forms we call plants, which produce oxygen during photosynthesis. If we ever obtain a spectrum of a planet containing substantial amounts of oxygen in its atmosphere, we can conclude that the presence of life is very likely. It is not at all far-fetched to think that sometime in the next century such an experiment could be performed.

The search for planets around other stars seems to be a project whose time has come. If numerous large planets are out there, it is now within our capability to find them. The discovery of smaller planets like our Earth may still be some time away, but perhaps not impossible. Once we have found a representative number of planets, astronomers will for the first time be able to consider the formation of our own planetary system within a broader cosmic context.

Summary

20.1 Most stars form in **giant molecular clouds** that have masses as large as 10^6 times the mass of the Sun and typical diameters of 50 to 200 LY. The best-studied molecular cloud is Orion, where star formation began about 12 million years ago and is moving progressively through the cloud. Recently formed hot stars are exposing many stages of the process of star formation to our view in Orion. The formation of a star inside a molecular cloud begins with a dense core of material, which accretes matter and collapses due to gravity. The accumulation of material halts when the **protostar** develops a strong **stellar wind.** A turbulent cloud will form a rotating star with an equatorial disk of material. The wind tends to emerge more easily in the direction of a protostar's poles, leading to jets of material. These can collide with the material around the star and produce Herbig–Haro objects.

20.2 The evolution of a star can be described in terms of its temperature and luminosity changes, which can best be followed by plotting them on an H–R diagram. Protostars generate energy through gravitational contraction. The initial gravitational collapse takes several thousand years, after which a slow contraction typically continues for millions of years, until the star reaches the **main sequence** and nuclear reactions begin. The higher the mass of a star, the shorter the time it spends in each stage of evolution. Stars range from about 1/12 to 100 times the mass of the Sun.

20.3 There is observational evidence that most protostars are surrounded by a disk with a large enough diameter and enough mass (as much as 10 percent that of the Sun) to form a system of planets. Furthermore, some of these disks are observed to change with time. Initially, an opaque disk extends all the way to the surface of the protostar. After a few million years the inner part of the disk is cleared of dust, and the disk is then shaped like a donut with the protostar centered in the hole. The development of the hole can be explained if a large planet has formed at its outer boundary.

20.4 At present we can search for planets around nearby stars only by looking for the star's motion around the star–planet system's common center of mass. This can be done by looking either for changes in the star's position on the sky over time, or in its radial velocity (as seen in the Doppler shift of the lines in its spectrum). So far, planets have been detected around six Sun-like stars using this method (but more are being found regularly). Ambitious space experiments are now being planned that might make it possible to image extra-solar-system planets and even to obtain spectra of them. Oxygen would be an unmistakable indicator of biological activity. These experiments are not yet feasible but will probably be carried out sometime during the next century.

Review Questions

1. Give several reasons that the Orion Molecular Cloud is such a useful "laboratory" for studying the stages of star formation.

2. Why is star formation more likely to occur in cold molecular clouds than in regions where the temperature of the interstellar medium is several hundred thousand degrees?

3. Why have we learned a lot more about star formation since the invention of detectors sensitive to infrared radiation?

4. Describe what happens when a star forms. Begin with a dense core of material in a molecular cloud, and trace the evolution up to the point at which the newly formed star reaches the main sequence.

5. Why is it so hard to see planets around other stars, and so easy to see them around our own?

6. What techniques have been used to search for planets around other stars?

7. How were the first planets around stars like the Sun discovered?

8. Explain why taking an image of a planet around another star is so difficult, and how astronomers hope to overcome these difficulties in the future.

Thought Questions

9. A friend of yours, who did not do well in her astronomy class, tells you that she believes all stars are old and that none could possibly be born today. What arguments would you use to persuade her that stars are being born somewhere in the Galaxy during our lifetimes?

10. Look at the four stages in the birth of a star shown in Figure 20.8. In which stage(s) is the star visible in optical radiation? In infrared radiation? In which stage(s) is it generating energy by converting hydrogen to helium?

11. An enormously wealthy donor has just given a university in your state a large sum of money to "unlock the secrets of star birth." You are on a committee to decide how to spend the money. What kind of instruments would you buy or build? What kind of studies would you fund? Explain each answer.

12. Observations suggest that it takes more than 3 million years for the dust to begin clearing out of the inner regions of the disk surrounding protostars. Suppose this is the minimum time required to form a planet. Would you expect to find a planet around a ten-solar-mass star? (Refer to Figure 20.11.)

13. The evolutionary track for a star with one solar mass remains nearly vertical in the H–R diagram for awhile (see Figure 20.11). How is its luminosity changing during this time? Its temperature? Its radius? What is its source of energy?

14. Suppose you wanted to image a planet around another star. Would you try to observe in the optical or in the infrared? Why? Would the planet be easier to see if it were at 1 AU or 5 AU?

Problems

15. If a giant molecular cloud is 100 LY across and has a mass equivalent to 10,000 Suns, what is its average internal density? Suppose that half of the cloud's mass is made up of dust grains, each with a mass of 10^{-13} kg; how many such grains per cubic centimeter are in the cloud?

16. Use Figure 20.11 to determine how long, compared to the Sun, a star with ten times the Sun's mass would spend in the pre-main-sequence stages of its life. Express as a fraction.

17. An astronomer has just found a new planet around a Sun-like star. If its orbit has a period of 2.4 years, use Kepler's law to determine the size in astronomical units of its semimajor axis.

Suggestions for Further Reading

Caillault, J. et al. "The New Stars of M42" in *Astronomy,* Nov. 1994, p. 40. On studies of circumstellar disks in Orion.

Cohen, M. *In Darkness Born: The Story of Star Formation.* 1988, Cambridge U. Press. A fine introduction to the topic.

Fienberg, R. "Pulsars, Planets, Pathos" in *Sky & Telescope,* May 1992, p. 493. (Update May 1994, p. 10.)

Goldsmith, D. *The Astronomers.* 1991, St. Martin's Press. Chapter 7 includes a section on the birth of stars.

Lada, C. "Deciphering the Mysteries of Stellar Origins" in *Sky & Telescope,* May 1993, p. 18. Good review; fine beginning article.

Lemonick, M. "Searching for Other Worlds" in *Time,* Feb. 5, 1996, p. 52. On the discovery of more planets.

MacRobert, A. and Roth, J. "The Planet of 51 Pegasi" in *Sky & Telescope,* Jan. 1996, p. 38.

O'Dell, C. R. "Exploring the Orion Nebula" in *Sky & Telescope,* Dec. 1994, p. 20. Good review with recent Hubble results.

Stahler, S. "The Early Life of Stars" in *Scientific American,* July 1991, p. 48.

Appendix 1 lists some World Wide Web sites with information about searches for planets orbiting other stars, and for nebulae with protostars.

Using **REDSHIFT** ™

1. Display the 10 brightest stars in the constellation Boötes. Which is (are) most likely to have planets?

2. Use Figure 20.11 to estimate how long it took for each of these stars to reach the main sequence.

a. Beta Canes Venaticorum
b. Epsilon Eridani
c. Alpha Pictoris
d. Gamma Cassiopeiae
e. Procyon, Alpha Canis Minoris
f. Theta Ursae Majoris
g. Menkalinan, Beta Aurigae
h. Tau Herculis
i. Eta Cephei
j. Mu Herculis

One of the first images taken with the repaired Hubble Space Telescope shows R 136 (inset), a cluster of young stars at the center of the Tarantula Nebula (the larger image). The nebula is an enormous region of gas and dust located in one of our nearest neighbor galaxies, the Large Magellanic Cloud. If the Tarantula were at the distance of the Orion Nebula discussed in Chapter 20, it would fill over half the sky and produce enough light for you to read this book at night. Preliminary analysis indicates that the cluster has several thousand closely packed stars in it. (STScI/NASA)

CHAPTER 21

Stars: From Adolescence to Old Age

Thinking Ahead

While the Sun's output of heat and light seems steady and reliable now, the fact is that we humans have only been around for a tiny fraction of the Sun's life. We have good evidence that as billions of years go by, stars like the Sun change drastically. How will the evolution of our star affect the habitability of planet Earth?

W e now turn from the birth of stars and planets to the rest of their life stages. This is not an easy task, since, as we saw in Chapter 17, stars live much longer than astronomers. Thus we cannot hope to see the life story of any single star unfold before our eyes or telescopes. Like the rushed crew of our imaginary starship in Chapter 17 (with only one day to study the lives of the Earth's human inhabitants), we must survey as many of the stellar inhabitants in the Galaxy as possible. If we are lucky (and thorough), we can catch at least a few of them in each possible stage of their lives.

Surveys of stars in our own neighborhood and beyond do reveal a variety of stars with different characteristics (Figure 21.1). Some of the differences come about because stars have different masses and thus different temperatures and luminosities. But others are the result of changes that occur as a star ages. Through a combination of observation, theory, and clever detective work, we can use these differences to piece together the life story of a star.

The universe at large would suffer as little, in its splendor and variety, by the destruction of our planet, as the verdure and sublime magnitude of a forest would suffer by the fall of a single leaf.

Thomas Chalmers in *Discourses on the Christian Revelation Viewed in Connection with the Modern Astronomy* (1817)

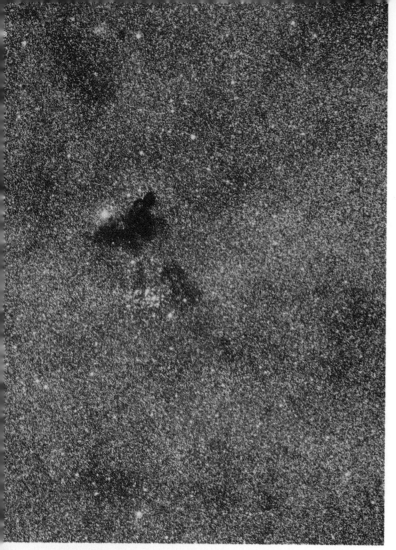

Figure 21.1
This snapshot of a crowded region of stars in the constellation of Sagittarius shows a yellowish population of older stars with a small cluster of younger bluish stars (NGC 6520), as well as a dark cloud (Barnard 86) inside which new protostars are forming. (© Anglo–Australian Telescope Board)

21.1

Evolution from the Main Sequence to Giants

One of the best ways to represent a "snapshot" of a group of stars is by plotting their properties on an H–R diagram. We have already used the H–R diagram to follow the evolution of protostars up to the time they reach the main sequence. Now let's see what happens next.

Once a star has reached the main-sequence stage of its life, it derives its energy almost entirely from the conversion of hydrogen to helium via the process of nuclear fusion (see Chapter 15). Since hydrogen is the most abundant element in stars, this process can help maintain the star's equilibrium for a long time. Thus all stars remain on the main sequence for most of their lives. Some as-

tronomers like to call this the star's "prolonged adolescence" or "adulthood" (continuing our analogy with the stages in a human life).

Since only 0.7 percent of the hydrogen used in fusion reactions is converted to energy, the star does not change its *total* mass appreciably during this long period. It does, however, change the chemical composition in its central regions, where the nuclear reactions occur: hydrogen is gradually depleted and helium accumulates. This change of composition forces the star to alter its structure, including its luminosity and size.

The original main sequence, corresponding to stars of different mass but roughly the same chemical composition, is sometimes called the **zero-age main sequence.** We use the term "zero-age" to note that each star reaches the main sequence when its hydrogen fusion reactions begin (the star's birth). Eventually, though, the point that represents any star on the H–R diagram will evolve away from the main sequence. That is, each star's luminosity and surface temperature (the characteristics recorded on the H–R diagram) will begin to change.

As helium accumulates in the center of a star, calculations show that the temperature and density in the inner region slowly grow, increasing the rate at which hydrogen fuses into helium. When it is hotter, each proton has more energy of motion on average; this means it will be more likely to interact with other protons and undergo fusion. (For the proton–proton cycle described in Chapter 15, the rate of fusion goes up roughly as the temperature to the fourth power. If the temperature were to double, the fusion would increase by a factor of 2^4, or 16 times.)

Consequently, the rate of nuclear energy generation increases with time, and the luminosity of the star gradually rises. Initially, however, these changes are small, and stars remain close to the zero-age main sequence for most of their lifetimes.

Lifetimes on the Main Sequence

How many *years* a star remains in the main-sequence stage depends on its mass. You might think that a more-massive star, having more fuel, would last longer, but it's not as simple as that. The lifetime of a star in a particular stage involving nuclear fusion depends on how much fuel it has, and on *how fast* it uses up that fuel. More-massive stars use up their fuel much more quickly than stars of low mass.

As we just saw, the rate of fusion is very strongly dependent on the star's inner temperature. And what determines how hot a star's central regions get? It is the mass of the star—the weight of the overlying layers—that controls how compressed and hot the star becomes at its core. The more massive the star, then, the faster it races through its storehouse of central hydrogen. Although massive stars *have* more fuel, they burn it so prodigiously that their lifetimes are much shorter than those of their low-mass counterparts.

TABLE 21.1
Lifetimes of Main-Sequence Stars

Spectral Type	Mass	Lifetime on Main Sequence
	(Mass of Sun = 1)	
O5	40	1 million years
B0	16	10 million years
A0	3.3	500 million years
F0	1.7	2.7 billion years
G0	1.1	9 billion years
K0	0.8	14 billion years
M0	0.4	200 billion years

The main-sequence lifetimes of stars of different masses are listed in Table 21.1. As this table shows, the most-massive stars spend only a few million years on the main sequence. A star of 1 solar mass remains there for about 10 billion years, while a star of about 0.4 solar mass has a main-sequence life of some 200 billion years, a value that is longer than the current age of the universe. (Bear in mind, however, that every star spends *most* of its total lifetime on the main sequence; stars spend an average of 90 percent of their lives peacefully fusing hydrogen into helium.)

Note that these results are not merely of academic interest. Human beings developed on a planet around a G-type star whose stable main-sequence lifetime was so long that it afforded life on Earth plenty of time to evolve. If we were to search for intelligent life like our own on planets around other stars, it would be a pretty big waste of time to search around O- or B-type stars. These stars remain stable for so short a time that the development of creatures complicated enough to take astronomy courses is very unlikely.

From Main-Sequence Star to Red Giant

Eventually all the hydrogen hot enough to undergo fusion is used up inside a star. The core (where nuclear reactions take place) now contains only helium, "contaminated" by whatever small percentage of heavier elements the star had to begin with. Energy can no longer be generated by hydrogen fusion in this helium core and, as we will see, helium requires far greater temperatures to undergo fusion. The helium in the core can be thought of as the accumulated "ash" from the nuclear "burning" of hydrogen during the main-sequence stage.

With nothing more to supply heat to the central region of the star, gravity again takes over: after a long period of stability, the core begins to contract. Once more the star's energy is partially supplied by gravitational energy, in the way described by Kelvin and Helmholtz (see

Section 15.1). As the star's core shrinks, the energy of the inward-falling material is converted to heat.

The heat generated in this way has an effect on the hydrogen that spent the whole long main-sequence time just outside the core. Like an understudy waiting in the wings of a hit play for a chance at fame and glory, this hydrogen was almost (but not quite) hot enough to undergo fusion and take part in the main action that sustains the star. Now, the additional heat produced by the shrinking core puts this hydrogen "over the limit," and a shell of hydrogen nuclei just outside the core becomes hot enough for hydrogen fusion to begin.

Energy now pours outward from this shell, flowing, as heat always does, to the cooler outer regions. This begins to heat up layers of the star farther out, causing them to expand a bit. Meanwhile the helium core continues to contract, producing more heat right around it, which leads to more fusion in the shell of fresh hydrogen. The fusion produces more energy, which also flows out into the upper layers of the star.

These changes result in a substantial and rather rapid readjustment of the star's entire structure, so that the star leaves the main sequence altogether. Most stars actually generate more energy in this stage than they did when hydrogen fusion was confined to the core; thus they increase in luminosity. The outer layers of the star begin to expand, and the star grows to enormous proportions—we say it becomes a *giant*.

Depending on their mass, these giant stars can become so large that if we were to replace the Sun with one of them, its outer atmosphere could extend to the orbit of Mars or even beyond. This is the next stage in the life of a star, as it moves (to continue our analogy with human lives) from its long period of youth into middle age. (After all, many human beings today also see their outer layers expand a bit during middle age.)

The expansion of the outer layers causes the temperature at the stellar surface to decrease. As it cools, the star's overall color becomes redder (we saw in Chapter 4 that in nature red corresponds to cooler temperature). So the star becomes simultaneously more luminous and cooler; on the H–R diagram this means it moves upward and to the right. The star becomes one of the *red giants* first discussed in Chapter 17 (Figure 21.2). You might say that these stars have "split personalities": their cores are contracting while their outer layers are expanding.

Models for Evolution to Red Giants

As discussed earlier, astronomers can construct computer models of stars with different masses and compositions to see how stars change throughout their lives. Figure 21.3, based on theoretical calculations by University of Illinois astronomer Icko Iben, shows an H–R diagram with several tracks of evolution from the main sequence to the red-giant stage. Tracks are shown for stars of several

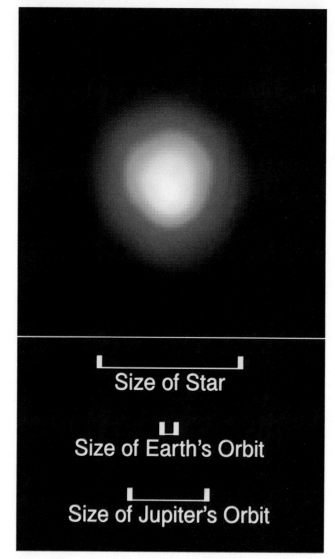

Figure 21.2
An easily observed example of a red giant star is Betelgeuse in the constellation of Orion (see Figure 20.2). Here we see an image taken in ultraviolet light with the Hubble Space Telescope—the first direct image ever made of the surface of another star. As shown by the bars at the bottom, Betelgeuse has an extended atmosphere so large that, if it were at the center of our solar system, it would stretch past the orbit of Jupiter. (A. Dupree, R. Gilliland, and NASA)

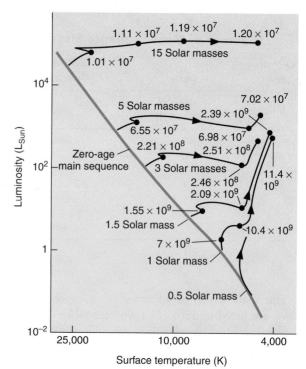

Figure 21.3
The predicted evolution of stars of different masses from the main sequence through the red-giant stage on the H–R diagram. The numbers show how many years the stars take to become red giants after leaving the main sequence. (Based on calculations by I. Iben)

21.2
Star Clusters

The description of stellar evolution just given is based on calculations. No star completes its main-sequence lifetime or its evolution to a red giant quickly enough for us to observe these structural changes as they happen. Fortunately, nature has provided us with a way to test our calculations.

Instead of observing the evolution of a single star, we can look at a group or *cluster* of stars. If a group of stars is very close together in space, held together by gravity, it is reasonable to assume that the individual stars in the group all formed at nearly the same time, from the same cloud, and with the same composition. We therefore expect that these stars will differ from one another only in mass and thus in how quickly they go through each stage of their lives.

Since stars with higher masses evolve more quickly, we can find clusters in which massive stars have already completed their main-sequence phase of evolution and become red giants, while stars of lower mass in the same cluster are still on the main sequence, or even undergoing

masses, with chemical compositions similar to that of the Sun. The red line is the initial or zero-age main sequence.

You can see how the changes inside the star we have been discussing force it to "move" on the H–R diagram. The numbers along the tracks in Figure 21.3 indicate the times, in years, required for the stars to reach those points in their evolution after leaving the main sequence. Once again, you can see that the more massive a star, the more quickly it goes through each stage in its life.

TABLE 21.2

Characteristics of Star Clusters

	Globular Clusters	Open Clusters	Associations
Number in Galaxy	150	Thousands	Thousands
Location in Galaxy	Halo and nuclear bulge	Disk (and spiral arms)	Spiral arms
Diameter (LY)	50–300	<30	100–500
Mass (solar masses)	10^4–10^6	10^2–10^3	10^2–10^3
Number of stars	10^4–10^6	50–1000	5–50
Color of brightest stars	Red	Red or blue	Blue
Luminosity of cluster (L_{Sun})	10^4–10^6	10^2–10^6	10^4–10^7

pre-main-sequence gravitational contraction. This way, we can see many stages of stellar evolution among the members of a single cluster.

There are three types of star clusters: globular clusters, open clusters, and stellar associations. Their properties are summarized in Table 21.2. As we will see, globular clusters contain only very old stars, while open clusters and associations contain young stars.

Globular Clusters

The **globular clusters** were given their name because of appearance. One of the most famous, M13 in the constellation of Hercules, passes nearly overhead on a summer evening at most places in the United States. As Figure 21.4 shows, M13 is a nearly symmetrical round system of many, many stars, with the highest concentration near the center. (The opening photo for Chapter 18, a wonderful image of another globular cluster, shows this type of concentration really well.) The brightest stars in globular clusters are red giants.

Most globular clusters contain hundreds of thousands of member stars. What would it be like to live inside such a cluster? In the densest part of M13, stars would be roughly a million times more concentrated than in our neighborhood. There would still be plenty of space between the stars, however. If the Earth orbited one of M13's inner stars, the nearest stars to our own would be *light months* away. They would still appear as points of light but would be brighter than any of the stars we see in our own sky. The Milky Way would probably be impossible to see through the bright haze of starlight produced by the cluster.

About 150 globular clusters are known in our Galaxy. Most of them are in a spherical halo (or cloud) surrounding the flat disk formed by the majority of the Galaxy's stars and interstellar matter. All the globular clusters are very far from the Sun, and some are found at distances of 60,000 LY or more from the galactic plane. The linear diameters of globular star clusters range from 50 LY to more than 300 LY. The most massive known globular cluster, called Omega Centauri (Figure 21.5), was recently found to contain enough mass to make 5 million stars like our Sun.

Open Clusters

Open clusters, on the other hand, are found in the disk of the Galaxy, often associated with interstellar matter. Open clusters contain far fewer stars than do globular clusters (Figure 21.6). The stars in open clusters usually appear well-separated from one another, even in the central regions, which explains the name we give this type of cluster. The Galaxy contains thousands of open clusters, but we can see only a small fraction of them. Interstellar dust (see Chapter 19) dims the light of more-distant clusters so much that they are undetectable.

Several open clusters are visible to the unaided eye. Most famous among them is the Pleiades (Figure 19.10),

Figure 21.4
The globular cluster M13. (U.S. Naval Observatory)

Figure 21.5
The most massive known globular cluster is Omega Centauri, whose central region is shown here. The field of view, about 90 LY across, contains hundreds of thousands of stars. The full cluster is actually significantly larger. (European Southern Observatory)

Figure 21.6
The Jewel Box (NGC 4755) is an open cluster of young, bright stars at a distance of about 8000 LY from the Sun. The name comes from its description by John Herschel as "a casket of variously colored precious stones." (Photo by David Malin; © Anglo–Australian Telescope Board)

which appears as a tiny group of six stars (some people can see even more) arranged like a dipper in the constellation of Taurus, the bull. A good pair of binoculars shows dozens of stars in the cluster, and a telescope reveals hundreds. The Hyades is another famous open cluster in Taurus. To the naked eye, it appears as a V-shaped group of faint stars marking the face of the bull. Telescopes show that the Hyades actually contains more than 200 stars.

Typical open clusters contain several dozen to several hundred member stars. Compared with globular clusters, open clusters are small, usually having diameters of less than 30 LY. A few brilliant stars in some open clusters, however, may cause them to outshine the far richer globular clusters.

Stellar Associations

An **association** appears as a group containing 5 to 50 hot, bright O and B stars scattered over a region of space some 100 to 500 LY in diameter. (Associations contain low-mass stars as well, but these are much fainter and less conspicuous.) Such a concentration of hot, luminous stars indicates that star formation in the association has occurred in the last million years or so. Since O and B stars go through their lives so quickly, they would not be around any more if they had formed much earlier. It is therefore not surprising that associations are typically found in regions rich in gas and dust.

Because associations, like ordinary open clusters, lie in regions occupied by interstellar matter, most are hidden from our view. There are probably several thousand undiscovered associations in our Galaxy.

21.3
Checking Out the Theory

Open clusters are younger than globular clusters, and associations are typically somewhat younger still. We know this because the stars in these different types of clusters are found in different places in the H−R diagram, and we can use their locations in combination with theoretical calculations to estimate their ages. The fact that stellar evolution theory can explain why the H−R diagrams of

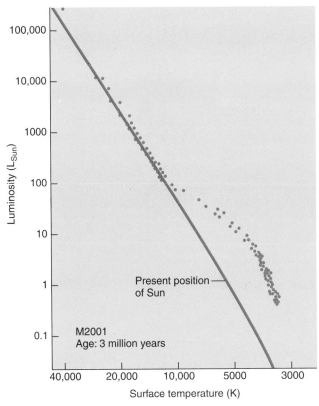

Figure 21.7
The H–R diagram of a hypothetical cluster at an age of 3 million years. Note that the high-mass (high-luminosity) stars have already arrived at the main-sequence stage of their lives, while the lower-mass (lower-luminosity) stars are still to the right of the zero-age main sequence, not yet hot enough to begin fusion of hydrogen.

Figure 21.8
The cluster NGC 2264, at a distance of 2500 LY. This region of newly formed stars is a complex mixture of red hydrogen gas ionized by hot embedded stars, dark obscuring dust lanes, and brilliant young stars. (Photo by David Malin; © Anglo–Australian Telescope Board)

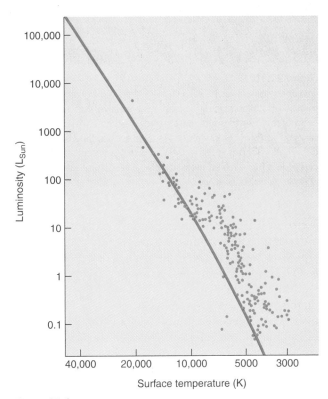

Figure 21.9
The H–R diagram for cluster NGC 2264. Compare to Figure 21.7; although the points scatter a bit more here, the theoretical and observational diagram are remarkably—and satisfyingly—similar. (Data by M. Walker)

globular clusters appear to be so different from those of open clusters and associations is one of the best arguments that the theory is right.

H–R Diagrams of Young Clusters

What does theory predict for the H–R diagram of a cluster whose stars have recently condensed from an interstellar cloud? After a few million years ("recently" for astronomers), the most-massive stars should have completed their contraction phase and be on the main sequence, while the less-massive ones should be off to the right, still on their way to the main sequence. These ideas are illustrated in Figure 21.7, which shows the H–R diagram calculated by R. Kippenhahn and his associates at Munich for a hypothetical cluster with an age of 3 million years.

There are real star clusters that fit this description. The first to be studied (in about 1950) was NGC 2264, which is still associated with the region of gas and dust from which it was born (Figure 21.8). Its H–R diagram is shown on Figure 21.9. Among the other star clusters in

Figure 21.10
The open star cluster NGC 3293. All the stars in such clusters form at about the same time. The most massive stars, however, exhaust their nuclear fuel more rapidly and hence evolve more quickly than stars of low mass. As stars evolve, they become redder. The bright orange star in NGC 3293 is the member of the cluster that has evolved most rapidly. (Photo by David Malin; © Anglo–Australian Telescope Board)

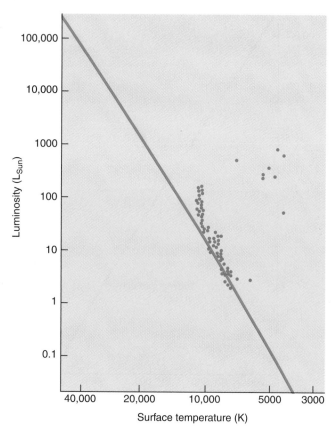

Figure 21.11
The H–R diagram for M41, a cluster older than NGC 2264. Note that M41 has several red giants and that some of its more-massive stars are no longer close to the zero-age main sequence (blue line).

such an early stage is the one in the middle of the Orion Nebula (shown in Figures 20.5 and 20.6).

As clusters get older, their H–R diagrams begin to change. After a short time—less than a million years after reaching the main sequence—the *most*-massive stars use up the hydrogen in their cores and evolve off the main sequence to become red giants. As more time goes on, stars of lower and lower mass begin to leave the main sequence and make their way to the upper right of the H–R diagram. Figure 21.10 is a photograph of NGC 3293, a somewhat older open cluster. One massive star has evolved to be a red giant and stands out as an especially bright orange member of the cluster.

Figure 21.11 shows the H–R diagram of the open cluster M41, which is roughly 100 million years old; a number of red giants have moved off to the right. Note

the gap that appears in this H–R diagram between the stars near the main sequence and the red giants. In the snapshot of stellar evolution represented by the H–R diagram, a gap does not necessarily represent a region of temperatures and luminosities that stars avoid. It simply represents a domain of temperature and luminosity through which a star moves very quickly as it evolves. We see a gap for M41 because at this particular moment we have not caught a star in the process of scurrying across this part of the diagram.

H–R Diagrams of Older Clusters

After 4 billion years have passed, the H–R diagram of a cluster will look quite a bit different. By this time, many more stars—including stars only a few times more massive than the Sun—have begun to leave the main sequence (Figure 21.12). This means that no stars are left near the top of the main sequence; only the low-mass stars near the bottom remain. The older the cluster, the lower the point on the main sequence where stars begin to move toward the red giant region. Astronomers actually use this "turn-off point" as a measure of the age of the cluster (compare Figures 21.9 and 21.12). To get actual

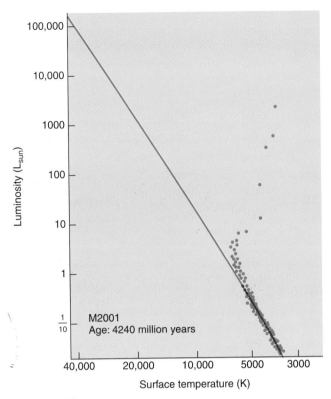

Figure 21.12

The H–R diagram of a hypothetical cluster at an age of 4.24 billion years. Note that most of the stars on the upper part of the main sequence have now turned off toward the red-giant region.

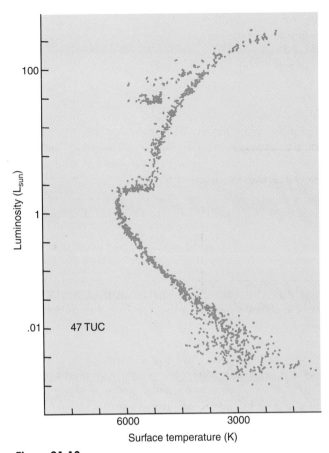

Figure 21.13

The H–R diagram of the globular cluster 47 Tucanae. Note the scale of luminosity is different from those of the other H–R diagrams in this chapter. We are focusing on the lower portion of the main sequence, the only part where stars still remain in this old cluster. (Data by J. Hesser and collaborators)

ages (in years) rather than relative ages, we must compare the appearance of our calculated H–R diagrams for different ages to actual H–R diagrams. Since the stars in a cluster formed at roughly the same time, this tells us the ages of the many individual stars that make up the cluster.

The oldest clusters are the globulars, typically about 15 billion years old. The H–R diagram of the globular cluster 47 Tucanae (pictured in the opening image for Chapter 18) is shown in Figure 21.13. Notice that its scale is different from the other H–R diagrams in this chapter. In Figure 21.12, for example, the luminosity scale on the left side of the diagram goes from 1/10 to 100,000 times the Sun's luminosity. But in Figure 21.13, the luminosity scale has been significantly reduced in extent. So many stars in this old cluster have had time to turn off the main sequence, only the very bottom of the main sequence remains.

Note that globular clusters have main sequences that turn off at a luminosity close to that of the Sun. Star formation in these systems evidently ceased billions of years ago, so no new stars are coming on to the main sequence to replace the ones that have turned off. Open clusters, on the other hand, are often found in regions of interstellar matter, where star formation can still take place. Apparently, new open clusters are still forming in our Galaxy. There are open clusters of all ages from less than 1 million

to several billion years, although no open cluster is as old as the globular clusters.

The globular clusters are the oldest structures in our Galaxy (and in other galaxies as well). Their study provides important clues about the age and early evolution of the Galaxy, and gives us a crucial reference point for understanding the age of the entire universe. We will return to their properties in future chapters.

21.4

Further Evolution of Stars

The stellar life story related so far applies to all stars: every one of them moves through the stages of contracting as a protostar, living most of its life as a stable main-sequence star, and eventually moving off the main sequence toward the red-giant region. (The pace at which the star goes through the stages depends, of course, on its mass, with more-massive stars evolving more quickly.) But after this point, the life stories of stars of different masses diverge, with a wider range of behavior possible according to mass, composition, and the presence of any nearby companion stars.

Because we have written this book for non-science students taking their first astronomy course, we will recount a somewhat simplified version of what happens to stars as they move toward the final stages in their lives. We will (perhaps to your heartfelt relief) not delve into all the possible ways stars can behave. Instead, we will focus on the key stages in the evolution of single stars, and show how the evolution of high-mass stars differs from that of low-mass stars (such as our Sun).

Helium Fusion

Let's begin by considering stars whose initial masses are comparatively low, no more than about two to three times the mass of our Sun. (That may not sound all that low, but we know that stars far more massive than this exist; we will see what happens to them in the next section.) Because there are many more of these low-mass stars than high-mass stars (see Chapter 17), the vast majority of stars—including our Sun—follow the scenario we are about to relate. By the way, we carefully used the term "initial masses" of stars because, as we will see, stars can lose quite a bit of mass in the process of aging and dying.

Remember that red giants have a helium core where no energy generation is taking place, surrounded by a shell where hydrogen is undergoing fusion. The core, however, is shrinking and growing hotter. Our studies of nuclear reactions in stars tell us that as the helium in the core heats up, no fusion reactions producing a stable end product occur until temperatures reach 100 million K. But once the dense core of the star reaches this extremely high temperature, three helium atoms can fuse to form a single carbon nucleus. (This process is called the **triple alpha process,** so named because nuclear physicists call the nucleus of the helium atom an alpha particle.) When the triple alpha process begins in the low-mass stars, our calculations show that the entire core is ignited in a quick burst of fusion called the **helium flash.**

ASTRONOMY BASICS

Stars in Your Little Finger

Stop reading for a moment and look at your little finger. It's full of carbon atoms because carbon is a fundamental chemical building block for life on Earth. Every one of those carbon atoms was once inside a red giant star, and was fused from helium nuclei in the triple alpha process. All the carbon on Earth—in you, in the charcoal you use for barbecuing, and in the diamonds you might exchange with your sweetheart—was "cooked" by previous generations of stars. How the carbon atoms (and other elements) made their way from

inside some of those stars to become part of the Earth is something we will discuss in the next chapter. For now, we want to emphasize that our description of stellar evolution is not merely of interest to specialists in astronomy, but is, in a very real sense, the story of our own cosmic "roots"—the history of how our atoms originated among the stars. We are made of star-stuff.

Becoming a Giant Again

After the helium flash occurs, the star continues to fuse the helium in its core for awhile, returning to the kind of equilibrium between pressure and gravity that characterized the main-sequence stage. Following the helium flash, the star's surface temperature increases and its overall luminosity decreases. The point representing the star on the H−R diagram thus moves to a new position to the left of and somewhat below its place as a red giant (Figure 21.14).

During this period, the core of the star is stable, fusing helium to form carbon. A newly formed carbon nucleus can sometimes be joined by another helium nucleus to produce a nucleus of oxygen.

At a temperature of 100 million degrees, the inner core is converting its helium fuel to carbon (and oxygen) at a rapid rate. Thus, the new period of stability cannot last very long: it is far shorter than the main-sequence stage. Soon all the helium hot enough for fusion will be used up, and again the inner core will not be able to generate energy via fusion. Once more, gravity will take over.

The situation is analogous to the end of the main-sequence stage when the star's central hydrogen was used up, but the star now has a somewhat more complicated structure. Again the star's core begins to collapse under its own weight. Heat released by the shrinking of the carbon and oxygen core flows into a shell of helium just above the core. This helium, which had not been hot enough for fusion into carbon earlier, is heated just enough for fusion to begin and to generate a flow of energy.

Farther out in the star is another shell where fresh hydrogen is fusing to form helium. Once again the outer regions of the star expand. Its brief period of stability over, the star moves back to the red giant domain on the H−R diagram for a brief time (Figure 21.14). In this stage, the star actually becomes somewhat brighter than it was during its first term as a giant, but it is merely a brief and final burst of glory.

Recall that the last time the star was in this predicament, helium fusion came to its rescue. The temperature at the star's center eventually became hot enough for the *product* of the previous step of fusion (helium) to become the *fuel* for the next step (helium fusing into carbon). But the step after the fusion of helium nuclei requires so hot a temperature that the kinds of lower-mass stars we are dis-

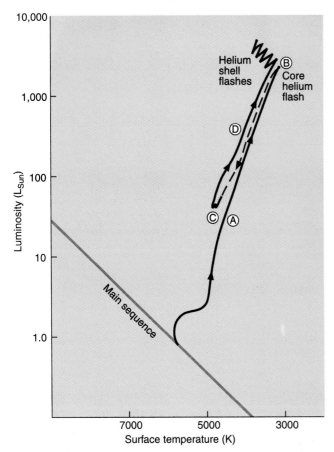

Figure 21.14
The evolution of a star like the Sun on an H–R diagram. Each stage is labeled with a letter: (A) the star evolves from the main sequence to be a red giant, decreasing in surface temperature and increasing in luminosity; (B) a helium flash occurs at this point, leading to a readjustment of the star's internal structure and to (C) a brief period of stability during which helium is fused to carbon and oxygen in the core (in the process the star becomes hotter and less luminous than it was as a red giant). (D) After the central helium is exhausted, the star becomes a giant again and moves to higher luminosity and lower temperature. By this time, however, the star has exhausted its inner resources and will soon begin to die. Where the evolutionary track becomes a dashed line, the changes are so rapid, they are difficult to model. (After calculations by Sackmann, Boothroyd, and Kraemer)

cussing simply cannot compress their cores to reach it. No further types of fusion are possible for such a star.

In stars with masses similar to that of the Sun, the formation of a carbon-oxygen core marks the end of the generation of nuclear energy at the center of the star. The star must now confront the fact that its death is near. We will discuss the death of stars in the next chapter, but in the meantime Table 21.3 summarizes the stages discussed so far in the life of a star with the mass of the Sun. One of the things that gives us confidence in our calculations of stellar evolution is that when we make H–R diagrams of older clusters, we see stars in each of the stages we have been discussing.

Mass Loss from Giant Stars

When stars become giants and their envelopes expand, they can begin to lose a substantial fraction of their mass into space. Astronomers have observed a strong wind of material flowing away from a number of giant stars, but the mechanism for generating the outflow is not very well understood. We estimate that by the time a star like the Sun reaches the point of the helium flash, for example, it can lose as much as 25 percent of its mass.

This is our first example of a kind of cosmic stellar recycling scheme. We saw that stars form from vast clouds of raw material (gas and dust). As they age, stars begin to return part of themselves to the galactic reservoirs of raw material. Eventually, some of the expelled material from aging stars may participate in the formation of new star systems.

However, the atoms returned to the Galaxy by an aging star are not always the same ones it received initially. The star has fused new elements over the course of its life, and during the red-giant stage material from the star's central regions is dredged up and mixed with its outer layers. As a result, the wind blows outward some particles that were "newly minted" inside the star. (As we will see, this mechanism is even more effective for high-mass stars, but it does work for stars like the Sun.)

21.5

The Evolution of More-Massive Stars

If what we have described so far were the whole story of the evolution of stars and elements, we would have a big problem on our hands. We will see in Chapter 28 that in our best models of the first few minutes of the universe, everything starts with the two simplest elements—hydrogen and helium (plus a tiny bit of lithium). All the predictions of the model imply that no heavier elements were produced at the beginning. This certainly fits with the observation—discussed in several earlier chapters—that most of the universe is made of hydrogen and helium even today.

Yet when we look around us on Earth, we see lots of other elements besides hydrogen and helium. These elements must have been made somewhere in the universe, *and the only place hot enough to make them is inside stars.* One of the fundamental discoveries of 20th-century astronomy is that the stars are the source of all the chemical richness that characterizes our world and our lives.

(text cont. on page 429)

The Red Giant Sun and the Fate of the Earth

How will the evolution of the Sun affect conditions on Earth? While the Sun has appeared reasonably steady in size and luminosity over recorded human history, that brief span means nothing compared to the kinds of time scales we have been discussing. Let's examine the long-term prospects for our planet.

The Sun took its place as a main-sequence star approximately 4.5 billion years ago. At that time it emitted only about 70 percent of the energy that it radiates today. As a result of the changes in its core caused by the buildup of helium, the Sun will continue to increase in luminosity as it ages. More and more radiation will reach each of the planets. Sooner or later—atmospheric models are not yet good enough to say exactly when, but estimates range from 500 million to 2 billion years—this increased heating of the Earth will cause a runaway greenhouse effect. The oceans will start to evaporate, resulting in more clouds and water vapor to hold in the heat. The Earth will start to resemble the Venus of today, and temperatures will become much too high for life as we know it.

The Sun will also expand as it ages, at first gradually as it completes the main-sequence stage, and then more quickly when it moves toward becoming a red giant. Our Sun will begin its climb on the H–R diagram to become a red giant about 6.5 billion years from now. In 7.8 billion years, it will have grown enough in size to engulf the planet Mercury. The surface of Venus will be molten, and temperatures on Earth will be greater than 750°C. Frozen water on the Galilean satellites of Jupiter will start to thaw out.

Then the Sun will experience a helium flash and shrink somewhat, but only temporarily. When it again becomes a giant, it will expand to a radius of 1 AU—the size of the Earth's orbit today. Mercury will again be inside the Sun, and friction with our star's outer atmosphere will make the planet spiral inward until it is completely vaporized.

The Earth and Venus will not be swallowed up, however, because they will no longer be in their present orbits. By this time, the Sun will have lost about 40 percent of its mass by means of its powerful red-giant winds. The gravitational force it exerts will therefore be smaller, and the size of the Earth's orbit (and those of the other planets) will increase accordingly. The Earth will be orbiting at about 1.7 AU. This is not enough to preserve our environment, however! The Sun will be about 5000 times brighter than it is now, and will heat the Earth to a temperature of 1300°C—hot enough to melt all the rocky material that makes up the continents.

From a personal perspective, such predictions are not especially worrisome. After all, no one alive today needs to take out an insurance policy against the Sun becoming very large and very bright billions of years from now. Some commentators feel that, given our inability to get along with each other, humans will have destroyed themselves long before the evolving Sun makes the Earth unlivable.

But if by some miracle our species manages to survive into that remote epoch, chances are that our technology and ability to manipulate the environment will have improved considerably. (Just think of how far we have come in ten thousand years. Imagine dropping off someone from ancient Egypt in a modern airport terminal!) If we survive a hundred million years, our species will probably leave the Earth, and it may become, at best, a museum world to which youngsters from many other worlds return to learn about their origins.

TABLE 21.3
The Evolution of a Star with the Sun's Mass

Stage	Time in This Stage (yrs)	Surface Temperature (K)	Luminosity (L$_s$)	Diameter (Diameter of Sun = 1)
Main Sequence	11 billion	6000	1	1
Becomes Red Giant	1.3 billion	down to 3100	up to 2300	165
Helium Fusion	100 million	4800	50	10
Giant Again	20 million	3100	5200	180

In coming to understand the life stages of lower-mass stars, we have already seen where carbon and oxygen are produced: they are the results of fusion inside stars that become red giants. But where do the heavier elements we know and love (such as the silicon and iron inside the Earth, and the gold and silver in our jewelry) come from? Such heavier elements are formed in the later evolution of more-massive stars.

Making New Elements in Massive Stars

Massive stars evolve in much the same way that the Sun does (but always more quickly)—up to the formation of a carbon-oxygen core. One difference is that for stars with more than about twice the mass of the Sun, helium begins fusion more gradually, rather than with a sudden flash. Also, when more-massive stars become red giants, they become so bright and large that we call them *supergiants*. Such stars can expand until their outer regions become as large as the orbit of Jupiter, which is precisely what the Hubble Space Telescope has shown for the star Betelgeuse (Figure 21.2). They also lose mass very effectively, producing more dramatic winds and outbursts as they age. Figure 21.15 shows a wonderful image of the massive star Eta Carinae, with a great deal of ejected material clearly visible.

But the crucial way that massive stars diverge from the story we have outlined is that they can start further kinds of fusion in their centers. The outer layers of a star whose mass is greater than about eight solar masses weigh enough to compress the carbon-oxygen core until it becomes hot enough to ignite carbon. Carbon can fuse into neon, still more oxygen, and finally silicon. After each of the possible sources of nuclear fuel is exhausted, the core contracts until it reaches a temperature high enough to lead to the fusion of still heavier nuclei.

Theorists have now found mechanisms whereby virtually all chemical elements of atomic weights up to that of iron can be built up by this **nucleosynthesis** (the making of new atomic nuclei) in the centers of the more-massive red giant stars. Moreover, our theories are able to predict the relative abundances with which they occur in nature; that is, the way stars build up elements during various fusion reactions can explain why some elements are common and others quite rare.

Our discussion has barely skimmed the surface of the many different nuclear pathways that can build up elements. Many of these reactions were worked out by nuclear physicists who became interested in applying their understanding of nuclear reactions to the world of astronomy. Primary among them was a group led by the late William Fowler of the California Institute of Technology, who received the 1983 Nobel Prize in physics for his work (Figure 21.16). While the details of these reactions are beyond the scope of this book, they account remarkably well for where the various chemical elements come from.

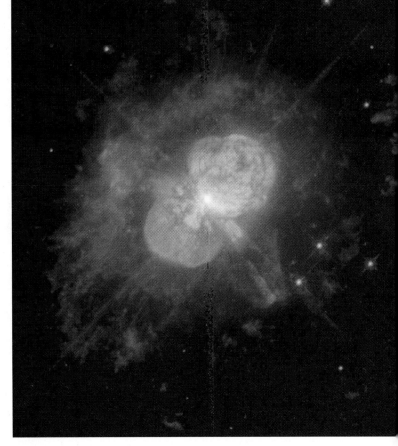

Figure 21.15
With a mass at least 100 times that of the Sun, the hot supergiant Eta Carinae is one of the most massive stars known. The Hubble Space Telescope took this image of some of the material it has ejected in the course of its evolution. The red outer region is material ejected in an outburst seen in 1841. Moving away from the star at a speed of about 1000 km/s, it is rich in nitrogen and other elements formed in the interiors of such massive stars. The inner blue-white region is material ejected at lower speeds and thus still closer to the star. It appears blue-white because it contains dust and reflects the light of Eta Carinae. (J. Hester and NASA)

Differences in Chemical Composition of Different Clusters

The fact that elements are synthesized in stars over time explains an otherwise puzzling fact about globular and open clusters. The two kinds of clusters do not have the same chemical composition. Hydrogen and helium, the most abundant elements in stars in the solar neighborhood, are also the most abundant constituents of stars in all kinds of clusters. The exact abundances of the elements heavier than helium, however, are not the same in different types of clusters.

In the Sun and most of its neighboring stars, the combined abundance (by mass) of the heavy elements is between 1 and 4 percent of the star's mass. The strengths of the heavy-element lines in the spectra of most open-

Figure 21.16
This 1981 photo shows the team of physicists and astronomers who in a famous 1957 paper first worked out the nuclear pathways by which the heavier elements formed. From left to right, they are Margaret Burbidge, William Fowler, Fred Hoyle, and Geoffrey Burbidge. (Photo courtesy of Caltech)

cluster stars show that they, too, have 1 to 4 percent of their matter in the form of heavy elements. Globular clusters, however, are a different story. The heavy-element abundance of stars in typical globular clusters is found to be only 1/10 to 1/100 that of the Sun.

Differences in chemical composition are related to when stars were formed. The very first generation of stars initially contained only hydrogen and helium. We have seen that these stars, to generate energy, created heavier elements in their interiors. In the last stages of their lives they ejected matter, now enriched in heavy elements, into the reservoirs of raw material between the stars. Such matter was then incorporated into a new generation of stars.

This means that the relative abundance of the heavy elements must be less and less as we look farther into the past. We saw that the globular clusters are much older than the open clusters and associations. Since globular-cluster stars formed much earlier than those in open clusters, they have only a relatively small abundance of elements heavier than hydrogen and helium.

As the Galaxy gets older, the proportion of heavier elements increases. This means that the first generation of stars that formed in our Galaxy was unlikely to have been accompanied by a planet like the Earth, full of silicon, iron, and many other heavy elements. Such a planet (and the astronomy students who live on it) were only possible after stars had a chance to make and recycle their heavier elements.

Approaching Death

Compared with the main-sequence lifetimes of stars, the events that characterize the last stages of stellar evolution pass very quickly. As the star's luminosity increases, its rate of nuclear fuel consumption goes up rapidly—just at that point in its life when its fuel supply is beginning to run down. It is as if a person suddenly did everything possible to hasten death, by overeating, overdrinking, smoking like a chimney, and so on.

After the prime fuel, hydrogen, is exhausted in a star's core, other sources of nuclear energy are, as we have seen, available to the star—in the fusion first of helium and then of other, more complex elements. But the energy yield of these reactions is much less than that of the fusion of hydrogen to helium. And to trigger these reactions, the central temperature must be higher than that required for the fusion of hydrogen to helium, leading to even more rapid consumption of fuel, and faster change. Clearly this is a losing game, and very quickly the star reaches its end. In doing so, however, some remarkable things can happen—as we will see in Chapter 22.

Summary

21.1 When stars are born, they lie on the **zero-age main sequence.** The amount of time a star spends in the main-sequence stage depends on its mass. The fusion of hydrogen to form helium changes the interior composition of a star, which in turn results in changes in temperature, luminosity, and radius. Eventually, as stars age, they evolve away from the main sequence to become red giants. The core of a red giant is contracting, but the outer layers are expanding as a result of fusion in a shell outside the core.

21.2 Calculations that show what happens as stars age can be checked by measuring the properties of stars in clusters. The members of a given cluster were formed at about the same time and have the same composition, so the comparison of theory and observations is fairly straightforward. There are three types of star clusters. **Globular clusters** have diameters of 50 to 300 LY, contain hundreds of thousands of old stars, and are distributed in a halo around the Galaxy. **Open clusters** typically contain hundreds of young to middle-aged stars, are located in the plane of the Galaxy, and have diameters of less than 30 LY. **Associations** are found in regions of gas and dust, and contain extremely young stars.

21.3 The H–R diagram of stars in a cluster changes systematically as the cluster grows older. The most-massive stars evolve the most rapidly. In the youngest clusters and associations, highly luminous blue stars are on the main sequence; the stars with the lowest masses lie to the right of the main sequence and are still contracting toward it. With passing time, stars of progressively lower mass evolve away from (turn off) the main sequence. In globular clusters, which have typical ages of about 15 billion years, there are no luminous blue stars at all. Astronomers can use the turn-off point of the main sequence to determine the age of a cluster.

21.4 After stars become red giants, their cores eventually become hot enough to produce energy by fusing helium to form carbon and oxygen. The fusion of three helium nuclei produces carbon through the **triple alpha process.** The rapid onset of helium fusion in the core of a star is called the **helium flash.** After this, the star becomes stable and reduces its luminosity and size briefly. In stars with masses similar to or less than that of the Sun, fusion stops after the helium in the core has been exhausted. Fusion of hydrogen and helium in shells around the contracting core makes the star a bright giant again, but only temporarily.

21.5 In stars with masses greater than about eight solar masses, nuclear reactions involving carbon, oxygen, and still-heavier elements can build up nuclei as heavy as iron. This creation of elements is called **nucleosynthesis.** These late stages of evolution occur very quickly. Ultimately all stars must use up all of their available energy supplies. In the process of dying, most stars eject some matter, enriched in heavy elements, into interstellar space where it can be used to form new stars. Each succeeding generation of stars contains a larger proportion of elements heavier than hydrogen and helium. This progressive enrichment explains why the stars in open clusters contain more heavy elements than do those in globular clusters, and tells us where most of the atoms on Earth and in our bodies come from.

Review Questions

1. Compare the following stages in the lives of a human being and a star: pre-natal, birth, prolonged adolescence, middle age, old age. What does a star with the mass of our Sun do in each of these stages?

2. What is the main factor that determines where a star falls along the main sequence?

3. What happens when a star exhausts hydrogen in its core, stopping the generation of energy by nuclear fusion of hydrogen to helium?

4. Describe the evolution of a star with a mass similar to that of the Sun, from the protostar stage to the time it becomes a red giant. First give the description in words and then sketch the evolution on an H–R diagram.

5. Describe the evolution of a star with a mass similar to that of the Sun, from just after it first becomes a red giant to the time it exhausts the last type of fuel its core is capable of fusing. After describing the stages in words, sketch them on an H–R diagram.

6. Suppose you have discovered a new star cluster. How would you go about determining whether it is an open or a globular cluster? List several characteristics that might help you decide.

7. Explain how an H–R diagram can be used to determine the age of a cluster of stars.

8. Where did the carbon atoms in the trunk of a tree on your college campus come from originally? Where did the neon in the fabled "neon lights of Broadway" come from originally?

9. Use star charts to identify at least one open cluster visible at this time of year. (Such charts can be found in *Sky & Telescope* and *Astronomy* magazines each month.) The Pleiades and Hyades are good autumn subjects, and Praesepe is good for springtime viewing. Go out and look at these clusters with binoculars and describe what you see.

10. Would you expect to find an Earth-like planet around a very low-mass star that formed right at the beginning of a globular cluster's life? Explain.

11. In the H–R diagrams for some young clusters, stars of *both* very low and very high luminosity are off to the right of the main sequence, whereas those of intermediate luminosity are on the main sequence. Can you offer an explanation? Sketch an H–R diagram for such a cluster.

12. If the Sun were a member of the cluster NGC 2264, would it be on the main sequence yet? Why?

13. Explain how you might decide whether stars in the red-giant region of the H–R diagram of a star cluster had evolved *away* from the main sequence, or were still evolving *toward* the main sequence.

14. If all the stars in a cluster have nearly the *same age*, why are clusters useful in studying evolutionary effects?

15. Suppose a star cluster were at such a large distance that it appeared as an unresolved spot of light through the telescope. What would you expect the overall color of the spot to be if it were the image of the cluster immediately after it was formed? How would the color differ after 10^{10} years? Why?

16. Suppose an astronomer told you she had found a type O main-sequence star that contained no elements heavier than helium. Would you believe her? Why?

Problems

17. Is the average mass of the stars in an association likely to be larger or smaller than the average mass in a globular cluster? Use the data in Table 21.2 to estimate the average mass of the stars in a globular cluster and in an association. Does this answer agree with your prediction?

18. What is the density in solar masses per cubic light year of the following clusters:

    a. A globular cluster 50 LY in diameter containing 10^5 stars

    b. A stellar association of 100 solar masses and 60 LY in radius

19. During the course of its main-sequence lifetime, a star converts about 10 percent of the hydrogen initially present into helium. How much does the mass of the star change as a result of fusion?

20. Stars exist that are as much as a million times more luminous than the Sun. Consider a star of mass 2×10^{32} kg and luminosity 4×10^{32} W. Assume that the star is 100 percent hydrogen, all of which can be converted to helium, and calculate how long it can shine at its present luminosity. (There are about 3×10^7 s in a year.)

21. Perform a similar computation for a typical star less massive than the Sun, such as one whose mass is 1×10^{30} kg and whose luminosity is 4×10^{25} W.

Suggestions for Further Reading

Darling, D. "Breezes, Bangs, and Blowouts: Stellar Evolution Through Mass Loss" in *Astronomy*, Sep. 1985, p. 78; Nov. 1985, p. 94.

Fortier, E. "Touring the Stellar Cycle" in *Astronomy*, Mar. 1987, p. 49. Observing objects that show the different stages of stellar evolution.

Kaler, J. "Giants in the Sky: The Fate of the Sun" in *Mercury*, Mar./Apr. 1993, p. 34.

Kaler, J. *Stars*. 1992, Scientific American Library/W. H. Freeman. Good modern introduction to stellar evolution.

Kaler, J. "The Largest Stars in the Galaxy" in *Astronomy*, Oct. 1990, p. 30. On red supergiants.

Mullan, D. "Caution! High Winds Beyond This Point" in *Astronomy*, Jan. 1982, p. 74. On stellar winds and mass loss.

Whitmire, D. and Reynolds, R. "The Fiery Fate of the Solar System" in *Astronomy*, Apr. 1990, p. 20. What will happen when the Sun becomes a red giant.

1. Use *RedShift* to determine the spectral type of each star listed below. Then use Table 21.1 to estimate how long each star will live on the main sequence; estimate the mass of each star; predict which stars will end their life after burning helium in their core and which will burn heavier elements in their core.

a. Beta Canes Venaticorum

b. Epsilon Eridani

c. Alpha Pictoris

d. Gamma Cassiopeiae

e. Procyon, Alpha Canis Minoris

f. Theta Ursae Majoris

g. Menkalinan, Beta Aurigae

h. Tau Herculis

i. Eta Cephei

j. Mu Herculis

2. To find which open and globular clusters are visible tonight to the naked eye or through binoculars, set the *Stellar Limiting Magnitude* to 5 and the *Deep-Sky Filter* to a faint magnitude of 5, with all objects except for star clusters turned off.

The Crab Nebula is the remnant of a stellar explosion first seen in 1054 A.D. Located in the constellation of Taurus about 6500 LY away, this object has been a key to understanding the death of massive stars and what happens to their material afterwards. (Max Planck Institute for Astronomy and Calar Alto Observatory; K. Meisenheimer and A. Quetz)

CHAPTER 22

The Death of Stars

Thinking Ahead

Several times in recorded history humanity awakened to find a new star in the sky, one so brilliant it was actually visible during the daytime. What do you think the reaction was to such a sight in ancient times? What would the public response be today if such a celestial event were to present itself with no warning?

He told me to look at my hand, for a part of it came from a star that exploded too long ago to imagine.

Paul Zindel in *The Effect of Gamma Rays on Man-in-the-Moon Marigolds.*

In the preceding two chapters we have followed the life story of stars, from the process of birth to the brink of death. Now we are ready to explore the ways that stars end their lives. Sooner or later every star exhausts its store of nuclear energy. Without an internal energy source, a star eventually gives way to the inexorable pull of gravity and collapses under its own weight. How that collapse takes place, and what kind of stellar "corpse" results from the process, depends on the star's mass at the time it collapses.

Bear in mind that a star's mass at the moment it is ready to die may be quite a bit smaller than it was at birth or during its long youth. As we noted in the last chapter, stars can lose a significant amount of mass in their middle and old ages. In this chapter we examine further ways that stars can lose mass in the last stages of their lives. Following the rough distinction made in the last chapter, we again discuss the stars of lower and higher mass separately.

Figure 22.1
Antares, found in the constellation of Scorpius, is one of the brightest stars in our sky. It is a supergiant with a large quantity of expelled material around it. Most of the energy from the star actually comes in the infrared region of the spectrum and cannot be seen on this visible-light image.
(© ROE/Anglo–Australian Telescope Board)

22.1

The Death of Low-Mass Stars

Let's begin with those stars whose final mass just before death is less than about 1.4 times the mass of the Sun. (We will explain why this is the crucial dividing line in a moment.) Note that most stars in the universe fall into this category. As we saw, the number of stars increases as mass decreases; as in the music business, only a few stars get to be superstars. Furthermore, many stars with an initial mass much greater than 1.4 solar masses (M_{Sun}) will be reduced to that level by the time they die. Our theoretical models indicate that stars starting with 5 M_{Sun} or less will surely be down to 1.4 M_{Sun} by the end; it may even be that stars with 10 M_{Sun} manage to lose enough mass to fit into this category.

Mass Loss

In Chapter 21 we left the life story of a star with a mass like the Sun's just as it was climbing upwards toward the giant region in the H–R diagram for the second time. Powered by fusion from both of its shells (one with helium fusion and one with hydrogen fusion), the star's outer layers expand and cool down again.

During this period, the star loses mass at an impressive rate. Over the tens of thousands of years that a star

like the Sun spends in this stage, it can lose a substantial fraction of its mass. As the ejected material expands and cools, its atoms can join together until they form molecules and, eventually, tiny particles of dust. We saw in Chapter 19 that this is one important source of the interstellar dust that is now prevalent throughout the star-forming regions of our Galaxy.

The shroud of dust makes the giant stars in this stage of their lives difficult to observe. In fact, only recently, through observations with infrared and microwave telescopes, has some of this expanding material been identified (Figure 22.1). Also, this period is a relatively brief one, and thus it is hard to catch stars "in the act." Still, we have now been able to confirm that they do pass through such a stage. And while the outside of the star is expanding and being expelled, the seeds for the star's death are being sown in its core.

The Contracting Core

Recall from Chapter 21 that during this time the core of the star was undergoing another crisis. Earlier in its life, during a brief stable period, helium in the core had gotten hot enough to fuse into carbon (and oxygen). But as the helium hot enough to do this had become exhausted, the star's core had once more found itself without a source of pressure to balance gravity, and so had begun to contract.

This collapse is the most devastating (and final) event in the life of the core. Because the star's mass is relatively low, it cannot ever get the temperature inside hot enough to begin another round of fusion (the way larger-mass stars can). The core thus continues to contract until it becomes so dense that a new and different way for matter to behave helps it achieve a final state of equilibrium. But this equilibrium is established only after the core has reached an enormous density, nearly a million times the density of water! In the process, what remains of the star (after all the mass loss has taken place) becomes one of the strange *white dwarfs* that we first met in Chapter 17.

Degenerate Stars

Since white dwarfs are far more dense than any substance on Earth, the behavior of matter inside them is different from anything we know from everyday experience. Under such compressed conditions, electrons actually resist any further compression and set up a powerful pressure inside the core. This pressure is the result of the fundamental rules that govern the behavior of electrons. According to these rules, which have been verified by studying how electrons behave under laboratory conditions, no two electrons can be in the same place at the same time doing the same thing. We specify the *place* of an electron by its precise position in space, and specify what it is doing by its motion and the way it is spinning.

The temperature in the interior of a star is always so high that the atoms are stripped of virtually all their electrons. In normal stars the density of matter is also relatively low, and the electrons are moving rapidly. It is very unlikely that any two of them will be in the same place moving in exactly the same way at the same time. But this all changes when a star exhausts its store of nuclear energy and begins its final collapse.

As the star's core contracts, electrons are squeezed closer and closer together. Eventually a star like the Sun becomes so dense that further contraction would require two or more electrons to violate the rule against occupying the same place and moving in the same way. Such a hot, dense gas is said to be **degenerate** (a term coined by physicists and not related to the electron's moral character). The electrons in a degenerate gas resist further crowding with overwhelming pressure. (It's as if the electrons said, "You can press inward all you want, but there is simply no room for any other electrons to squeeze in here without violating the rules of our existence.")

White Dwarfs

White dwarfs, then, are stars with degenerate electron cores that have stabilized the star against further collapse. Calculations showing that stars can do this, and what happens for stars of different masses, were first carried out by the Indian-American astrophysicist S. Chandrasekhar (see "Voyagers in Astronomy" box). He was able to show how

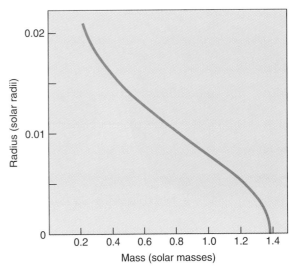

Figure 22.2
Theoretical relation between the masses and the radii of white dwarf stars. As you can see, the theory predicts that as the mass of the star increases (toward the right), its size gets smaller and smaller.

much a star will shrink before the degenerate electrons stop its contraction and hence what its final diameter will be (see Figure 22.2). Note that the larger the mass of the star, the *smaller* its radius.

A white dwarf with a mass like that of the Sun has a diameter about the same as that of the Earth. This is a remarkably small size for a star, and means that its material is very compressed. A teaspoonful of white dwarf material (if it could be brought to Earth in its compressed state) would weigh more than a full garbage truck. And, as we will see, such stars are quite hot on the outside—much hotter than the surface of the Sun.

Beyond the White Dwarf

When Chandrasekhar made his calculations about white dwarfs, he found something very surprising, which is clear on Figure 22.2 as well. According to the best theoretical models, a white dwarf with a mass of about 1.4 M_{Sun} or larger would have a radius of zero! What the calculations are telling us is that even the force of degenerate electrons cannot stop the collapse of a star with more mass than this. The maximum mass that a star can have and still become a white dwarf—1.4 M_{Sun}—is called the **Chandrasekhar limit.** Stars with masses exceeding this limit have a different kind of end in store—one that we will explore in the next section.

Planetary Nebulae

Before we describe the ultimate fate of white dwarfs, let's briefly return to the outer layers of our dying star. When the star was still in its giant state, some of its outer mate-

Subrahmanyan Chandrasekhar

Born in 1910 in Lahore, India, S. Chandrasekhar (known as Chandra to his friends and colleagues) grew up in a home that encouraged scholarship and an interest in science. His uncle, C. V. Raman, was a physicist who won the 1930 Nobel Prize. A precocious student, Chandra tried to read as much as he could about the latest ideas in physics and astronomy, although obtaining technical books was not easy in India at the time. He finished college at age 19 and won a scholarship to study in England. It was during the long boat voyage to get to graduate school that he first began doing calculations about the structure of white dwarf stars.

Chandra developed his ideas during and after his studies as a graduate student, showing—as we have discussed—that white dwarfs with masses greater than 1.4 times the mass of the Sun cannot exist, and that the theory predicts the existence of other kinds of stellar corpses as well. He wrote later that he felt very shy and lonely during this period, isolated from other students, afraid to assert himself, and sometimes waiting for hours to speak with some of the famous professors he had read about in India. His calculations soon brought him into conflict with certain distinguished astronomers, including Sir Arthur Eddington, who publicly ridiculed Chandra's ideas. At a number of meetings of astronomers, such leaders in the field as Henry Norris Russell refused to give Chandra the opportunity to defend his ideas, while allowing his more senior critics lots of time to criticize them.

Yet Chandra persevered, writing books and articles elucidating his theories, which turned out not only to be correct, but to lay the foundation for much of our modern understanding of the death of stars. (In 1983, he received the Nobel Prize in physics for this early work.)

In 1937 Chandra came to America and joined the faculty at the University of Chicago, where he remained for the rest of his life. There he devoted himself to research and teaching, making major contributions to many fields of astronomy—from our understanding of the motions of stars through the Galaxy to the behavior of the bizarre objects called black holes (see Chapter 23).

Perhaps because he remembered how alienated he had felt, Chandra spent a great deal of time with his graduate students, supervising the research of more than 50 PhD's during his life. He took his teaching responsibilities very seriously: during the 1940s, while based at the Yerkes Observatory, he willingly drove the more than 100-mile trip to the university each week to teach a class of only two students. As the university later reported, "Any concern about the cost-effectiveness of such a commitment was erased in 1957, when that entire class—consisting of T. D. Lee and C. N. Yang—won the Nobel Prize in physics."

Chandra also had a deep devotion to music, art, and philosophy, writing articles and books about the relationship between the humanities and science. He emphasized that "one can learn science the way one enjoys music or art. . . . Heisenberg had a marvelous phrase 'shuddering before the beautiful' . . . that is the kind of feeling I have."

S. Chandrasekhar (1910–1995)
(Courtesy of Emilio Segrè Visual Archives, Physics Today Collection)

rial was actually able to lift off. As a result, our star is now surrounded by one or more expanding shells of gas, each containing as much as 0.1 or 0.2 M_{Sun} of material (Figure 22.3).

This process strips our star of its outer layers, exposing the hotter regions underneath for the first time. As the star's core is heated by its compression, any remaining hydrogen is quickly fused into helium. As the core evolves to become a white dwarf, it becomes (as we saw) very hot—reaching surface temperatures of 100,000 K. Such stars are very strong sources of ultraviolet radiation, which pours out directly into the shell of recently expelled material.

As a result, this shell is heated, ionized, and set aglow (as are the H II regions around young hot stars described in Chapter 19). These glowing shells are among the most spectacular objects in the sky (Figure 22.4). They were given an extremely misleading name when first found: **planetary nebulae.** The name is derived from the fact that a few planetary nebulae, when viewed through a small telescope, bear a superficial resemblance to planets. Actually, they are thousands of times larger than the solar system and have nothing to do with planets, but once names are put into regular use in astronomy, it is extremely difficult to change them. There are tens of thousands of planetary nebulae in our own Galaxy, but many are hidden from view because their light is absorbed by interstellar dust.

Sometimes, as in Figure 22.5, a planetary nebula appears to be a ring. This is an illusion, caused by the fact that we are looking through more layers along the sides of the shell than in the middle. In the same way, the center of a soap bubble is often more transparent than its rim.

Figure 22.3
With the use of modern CCDs, the Ring Nebula (perhaps the best-known of all the planetary nebulae) reveals several shells of ejection. (George Jacoby and Bruce Balick, NOAO)

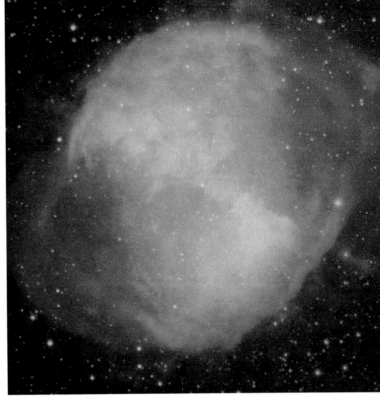

Figure 22.4
The Dumbbell Nebula is one of the most beautiful planetary nebulae in the sky. It is a shell of ejected material from a dying star, glowing because the hot star is giving off large amounts of ultraviolet radiation. (Max Planck Institute for Astronomy and Calar Alto Observatory; K. Meisenheimer and A. Quetz)

When the central star has had several episodes of ejection, or when the system contains more than one star, planetary nebulae can take on wonderfully complex structures, as shown in Figure 22.6.

Planetary nebula shells usually expand at speeds of 20 to 30 km/s, and a typical planetary nebula has a diameter of about 1 LY. If we assume that the gas shell has expanded at constant speed, we can calculate that the shells of all the planetary nebulae visible to us were ejected within the past 50,000 years or so. Still-older shells have expanded so much that they are simply too thin and tenuous to be seen. When we consider the relatively short time that each planetary nebula can be observed, and the number of such nebulae we see, we must conclude that a large fraction of all stars evolve through the planetary nebula phase. This confirms our view of planetary nebulae as a sort of "last gasp" of low-mass star evolution.

Figure 22.5
At a distance of about 400 LY, the Helix Nebula, NGC 7293, is the planetary nebula nearest the Sun. On photographs the Helix has a diameter about the same as that of the full Moon. The greenish color is produced by emission lines of ionized oxygen; the red is due to nitrogen and hydrogen. (Anglo-Australian Telescope Board)

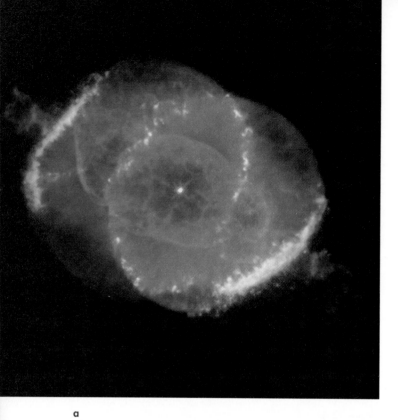

a

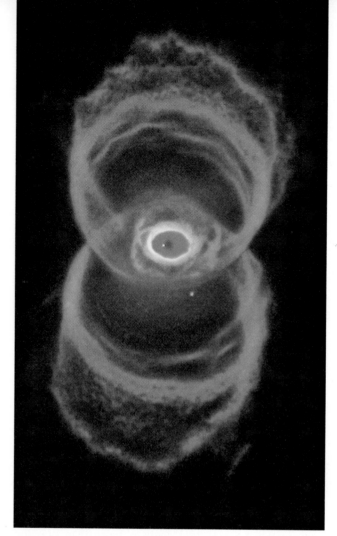

b

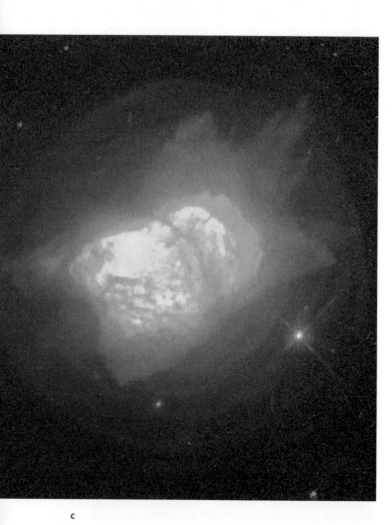

c

Figure 22.6

Three intriguing planetary nebulae as seen with the Hubble Space Telescope. (a) The Cat's Eye Nebula (NGC 6543) in the constellation of Draco shows several concentric shells of gas, jets of high-speed gas, and regions (in green) where the high-speed material is piling into previously ejected gas shells. The complexity of the features may mean that there are two stars at the center, both of which have ejected or are ejecting material in shells, rings around their equators, and/or jets from their polar regions. (P. Harrington, K. Borkowski, and NASA) (b) The Hourglass Nebula (MyCn 18), about 8000 LY away, also shows complicated structure. This picture is composed of three separate images taken in the light emitted by ionized nitrogen (red), hydrogen (green), and doubly ionized oxygen (blue). Notice that the star that gives rise to the material of the nebula is not in the center of the green region. Jets, rings, and companion stars may also be required to explain this complicated object. (R. Sahai, J. Trauger, NASA) (c) NGC 7027 is a planetary nebula located about 3000 LY from us in the constellation of Cygnus. This image combines both visible light and infrared information to bring out several episodes of ejection in the object. The initial (outer) blue shells were ejected much more symmetrically than the inner shells of material, which glow with infrared radiation from dust grains that have condensed within it. (H. Bond and NASA)

The Ultimate Fate of White Dwarfs

As a planetary nebula expands and fades away, the star inside—hot, dense, luminous, and dying—is now clearly revealed to our view. If the birth of a star is the onset of fusion reactions, then we must consider the end of all fusion reactions to be the time of a star's death. As the core is stabilized by degeneracy pressure, a last shudder of fusion passes through the outside of the star, consuming the little hydrogen still remaining. Now the star is a true white dwarf: it has no further source of energy, and so begins to cool. (Figure 22.7 shows the path of a star like the Sun on the H–R diagram during its final stages.)

The electrons in a degenerate gas move about, as do particles in any gas, but not with freedom. A particular electron cannot change position or momentum until another electron in an adjacent state gets out of the way. The situation is much like that in the parking lot after a big football game. Vehicles are closely packed, and a given car cannot move until the one in front of it moves, leaving an empty space to be filled. If one car leaves the lot, another can move ahead, and still others can follow behind it, producing a flow of cars toward the exit. It turns out

that heat can flow through an electron-degenerate gas, and the star can slowly lose heat into space.

Since the white dwarf can no longer contract (or produce energy through fusion), its only energy source is the heat represented by the motions of the atomic nuclei in its interior. The light it emits comes from this internal stored heat, which is substantial. Gradually, however, the white dwarf radiates away its heat and fades out. After many billions of years, the heat will be gone, the nuclei will cease their motion, and the white dwarf will no longer shine. It will then be a *black dwarf*—a cold stellar corpse with the mass of a star and the size of a planet. It will be composed mostly of carbon and oxygen, the product of the most advanced fusion reaction of which the star was capable.

We have one final surprise as we leave our low-mass star in the stellar graveyard. Calculations of what happens inside a degenerate star as it cools indicate that the atoms in essence "solidify" into a giant, highly compact lattice (organized rows of atoms just like in a crystal). When carbon is compressed and crystallized in this way, it becomes a diamond. A white dwarf star is the most impressive engagement present you could ever see, although any attempt to mine the diamond-like material inside would crush an ardent lover instantly!

Evidence for Significant Mass Loss

Our model of the evolution of low-mass stars into white dwarfs assumes there is significant mass loss in the red giant phase of stellar evolution; this is required to bring down the masses of many stars below the Chandrasekhar limit of 1.4 M_{Sun}. What evidence do we have that stars starting out significantly more massive than this limit actually do shed so much of their mass?

White dwarfs have been found in young, open clusters—clusters so young that only stars with masses greater than 5 M_{Sun} have had time to exhaust their supplies of nuclear energy and complete their evolution to the white dwarf stage. In the Pleiades, for example, stars with masses of 4 to 5 M_{Sun} are still on the main sequence. Yet this cluster also has at least one white dwarf. The star that turned into this dwarf must have had a main-sequence mass exceeding 5 M_{Sun}, since stars with lower masses have not yet had time to exhaust their stores of nuclear energy. It must have gotten rid of enough matter so that its mass at the time nuclear energy generation ceased was less than 1.4 M_{Sun}.

Another convincing piece of evidence can be seen in the presence of a white dwarf star around the bright star Sirius (see Chapter 17). Sirius and its white dwarf companion are part of a binary star system, and we presume both stars formed at the same time. The mass of Sirius is a little more than 2 M_{Sun} and it is still on the main sequence. The mass of the white dwarf is about the mass of the Sun (half of Sirius' mass), yet it has already died and become a white dwarf.

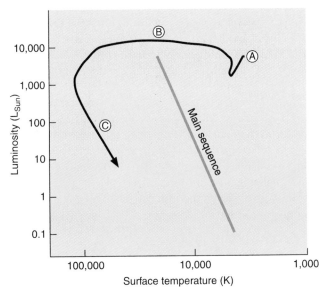

Figure 22.7
The evolutionary track on the H–R diagram for a star with a mass like the mass of the Sun which is nearing the end of its life. After the star becomes a giant again (point A on the diagram), it will lose more and more mass as its core begins to collapse. The mass loss will expose the hot inner core, which will appear at the center of a planetary nebula. In this stage the star moves across the diagram to the left as it becomes hotter and hotter during its collapse (point B on the diagram). At first the luminosity remains constant, but as the star begins to shrink significantly, it becomes less and less bright (point C). It is now a white dwarf and will slowly cool until all of its remaining store of energy is radiated away. (Assumes the Sun will lose about 46 percent of its mass during the giant stages; based on calculations by Sackmann, Boothroyd, and Kraemer)

Since more-massive stars evolve more quickly, the only way for Sirius' companion to be a white dwarf today is for it to have been *the more-massive star* when the pair formed. Thus it must have lost a substantial amount of its mass between the time it formed and now, when we see it as a white dwarf. Today astronomers believe that mass loss is an important part of the evolution of all stars.

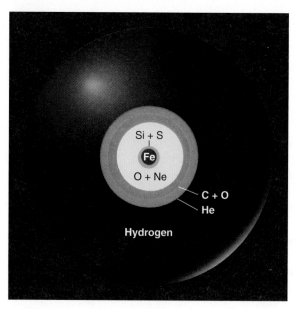

Figure 22.8
Just before its final gravitational collapse, a massive star resembles an onion. The iron core is surrounded by layers of silicon and sulfur, oxygen, neon, carbon mixed with some oxygen, helium, and finally hydrogen.

22.2

Evolution of Massive Stars: An Explosive Finish

Thanks to mass loss, stars with masses up to at least 5 (and perhaps even 10 M_{Sun}) probably end their lives as white dwarfs. But we know stars can have masses as large as 100 M_{Sun}. What is the ultimate fate of the higher-mass stars?

Nuclear Fusion of Heavy Elements

As discussed in Section 21.5, only after the helium in its core is exhausted does the evolution of a massive star take a significantly different course from that of lower-mass stars. In a massive star, the weight of the outer layers is sufficient to force the carbon and oxygen core to contract until it becomes hot enough to fuse carbon into neon. This cycle of contraction, heating, and the ignition of another nuclear fuel is repeated several more times. After each of the possible nuclear fuels is exhausted, the core contracts until it reaches a temperature high enough to fuse still-heavier nuclei. Massive stars go through these stages very, very quickly—far more quickly than the main-sequence and red giant stages. In really massive stars, some fusion stages toward the very end can take only months or even days!

How long can this process of peacefully building up elements by fusion go on? It turns out that there is a limit: the fusion of silicon into iron is the last step in the sequence of peaceful element production. The fusion of iron begins a rapid sequence of dramatic events that can actually force the star to explode.

Up to this point, a key result of each fusion reaction has been to *release* energy. This is because the nucleus of each fusion product has been a bit more stable than the nuclei that formed it. As discussed in Chapter 15, light nuclei give up some of their binding energy in the process of fusing into more tightly bound, heavier nuclei. It is this released energy that helps keep a star in balance. But of all the nuclei known, iron is the most tightly bound and thus the most stable.

You might think of the situation like this: all small nuclei want to "grow up" to be like iron, and are willing to pay (release energy) to move toward that goal. But iron is a mature nucleus with good self-esteem, perfectly content being iron; it *requires* payment (must absorb energy) to

change its stable nuclear structure. This is the exact opposite of what has happened in each nuclear reaction so far: instead of providing energy to balance the inward pull of gravity, any fusion of iron would *remove* some energy from the core of the star.

At this stage of its evolution, a massive star resembles an onion with an iron core. As we get farther from the center, we find shells of decreasing temperature in which nuclear reactions involving nuclei of progressively lower mass—silicon and sulfur, oxygen, neon, carbon, helium, and finally hydrogen—are taking place (Figure 22.8). What happens when the iron core is no longer a source of energy for the star?

A Ball of Neutrons

In effect, a massive star builds a white dwarf in its center where no nuclear reactions are taking place. For stars that begin their evolution with masses of at least 10 M_{Sun}, this white dwarf is made of iron. For stars with initial masses in the range 8 to 10 M_{Sun}, the white dwarf that forms the core is made of oxygen, neon, and magnesium because the star never gets hot enough to form elements as heavy as iron. Whatever its composition, the white dwarf embedded in the center of the star is supported against further gravitational collapse by degenerate electrons.

While no energy is being generated within the white-dwarf core of the star, fusion does still occur in shells surrounding the core. As the shells finish their fusion reactions and stop producing energy, the ashes of the last reaction fall onto the white-dwarf core, increasing its mass. Ultimately, the core is pushed over the Chandrasekhar limit of 1.4 M_{Sun}. That is, it becomes so massive that the force exerted by degenerate electrons is no longer great enough to resist gravity. What happens next is completely out of the realm of ordinary experience.

A stellar core containing degenerate electrons is still mostly empty space. Electrons and atomic nuclei are, after all, extremely small. The electrons and nuclei in a stellar core may be crowded compared to the air in your room, but there is still lots of space between them. As mass is added to the core, its density begins to increase. The electrons at first resist being crowded closer together, but if the mass grows larger than 1.4 M_{Sun}, they can no longer do so.

When the density reaches 4×10^{11} g/cm^3 (400 billion times the density of water), some electrons are actually *squeezed into the atomic nuclei*, where they combine with protons to form neutrons. (This can only happen at the outrageous densities found in collapsing stars.) Some of the electrons are now gone, and so the core's ability to resist the crushing mass of the star's overlying layers is reduced even more. The collapse becomes catastrophic. The atomic nuclei absorb more and more electrons and ultimately become so saturated with neutrons that they cannot hold onto them.

At this point the neutrons are squeezed out of the nuclei, now exerting a new force. As is true for electrons, it turns out that the neutrons strongly resist being in the same place and moving in the same way. The force that can be exerted by such *degenerate neutrons* is much greater than that produced by degenerate electrons, so they can ultimately resist the collapse. Calculations show that the upper limit on the mass of stars made only of neutrons is about 3 M_{Sun}.

In other words, if the collapsing core has less mass than this, it will reach a stable state as a crushed ball made mainly of neutrons, which astronomers call a **neutron star.** However, if the mass of the core is greater than this limit, then even neutron degeneracy cannot stop the core from collapsing, and the star becomes something unbelievably compressed called a *black hole*. Black holes are the subject of Chapter 23; for now, we restrict ourselves to those stars in which the collapse of the core is in fact halted by degenerate neutrons.

Collapse and Explosion

The collapse that takes place when electrons are absorbed into the nuclei is very rapid. In less than a second, the core, which originally was approximately the same diameter as the Earth, collapses to a diameter of less than 20 km. The speed with which material falls inward reaches one-fourth the speed of light. The collapse halts only when the density of the core exceeds the density of an atomic nucleus (which is the densest form of matter we know). A typical neutron star is so compressed that to duplicate its density we would have to squeeze all the people in the world into a single raindrop. That would give us one raindrop's worth of a neutron star.

Because of neutron degeneracy, this dense core strongly resists further compression, abruptly halting the collapse. The shock of the abrupt jolt generates waves throughout the outer layers of the star, causing those outer layers to blow off in a violent explosion called a **supernova.** Although the analogy is not exact, you might think of what happens as the parking lots at a local stadium fill up for a very popular rock concert. Suddenly, no further cars can squeeze into the parking structures. This causes a big chain reaction in the adjoining streets, where long lines of cars have formed.

Recently, both our models and observations of supernovae have led us to realize that the ghostly subatomic particles called *neutrinos*, introduced in Chapter 15, play a crucial role at this point. Each time an electron and a proton merge to make a neutron in the collapsing core, the merger releases a neutrino. Thus we expect vast numbers of neutrinos to be generated very rapidly as the neutron star forms.

While neutrinos ordinarily do not interact very much with ordinary matter (we accused them of being downright antisocial in Chapter 15), matter in the core of a collapsing star is far more compressed and thus more likely to interact with an enormous burst of neutrinos. We now believe that up to 90 percent of the energy of a supernova explosion may be carried by the neutrinos. They may also be responsible for getting the rest of the star to blow outward even while gravity is pulling everything inward.

Whatever the detailed mechanism responsible for the explosion of the star, we know such supernova explosions must occur because we can observe them happening in our Galaxy and in other galaxies as well. When they happen close-by, they can be among the most spectacular and important celestial events, as we will discuss in the next section. (Note that there are at least two different types of supernova explosions; the kind we have been describing are called, for historical reasons, Type II supernovae. We will describe how the types differ in Section 22.4.)

For now, Table 22.1 summarizes the discussion so far about what happens to stars of different initial masses at the ends of their lives. Like so much of our scientific understanding, this list represents a progress report: it is the best we can do with our present models and observations. The mass limits corresponding to various outcomes may change somewhat as models improve. There is much we do not yet understand about the details of what happens when stars die.

Supernova Observations

Supernovae were discovered long before astronomers realized that these spectacular cataclysms mark the death of stars (see "Making Connections" box). The word *nova* means "new" in Latin; before telescopes, when a star too dim to be seen with the unaided eye suddenly flared up in

(text cont. on page 445)

Supernovae in History

While many supernova explosions in our own Galaxy have gone unnoticed, a few were so spectacular that they were clearly seen and recorded by skywatchers and historians at the time. We can use these records, going back two millennia, to help us pinpoint where the exploding stars were and thus where to look for their remnants today.

The most dramatic supernova was observed in the year 1006 A.D. It appeared in May as a brilliant point of light visible during the daytime, perhaps 100 times brighter than the planet Venus. It was bright enough to cast shadows on the ground during the night, and was recorded with awe and fear by observers all over Europe and Asia. No one had seen anything like it before; Chinese astronomers, noting that it was a temporary spectacle, called it a "guest star."

Astronomers David Clark and Richard Stephenson have scoured records from around the world to find over 20 reports of the 1006 supernova. This has allowed them to determine with some accuracy where in the sky the explosion occurred. They place it in the modern constellation of Lupus; at roughly the position they have determined we do find a supernova remnant, now quite faint. From the way its filaments are expanding, it indeed appears to be about 1000 years old.

Another guest star was clearly recorded in Chinese records in July 1054 A.D. The remnant of that star, called the Crab Nebula, is shown in the opening image for this chapter. It is a marvelously complex object, whose study has been a key to understanding the death of massive stars. We estimate that it was about as bright as the planet Jupiter, nowhere near as dazzling as the 1006 event but still quite dramatic to anyone who kept track of objects in the sky. There is some evidence that Native Americans in New Mexico, for example, recorded the new star with the crescent moon nearby in a cave painting at a ceremonially important location. Another fainter supernova was seen in 1181 A.D.

The next supernova became visible in November 1572, and, being brighter than the planet Venus, was quickly spotted by a number of observers, including the young Tycho Brahe (see Chapter 2). His careful measurements of the star over a year and a half showed that it was not a comet or something in the Earth's atmosphere, since it did not move relative to the stars. He correctly deduced that it must be a phenomenon belonging to the realm of the stars, not of the solar system. The remnant of Tycho's Supernova (as it is now called) can still be detected in many different bands of the electromagnetic spectrum.

Not to be outdone, Johannes Kepler, Tycho Brahe's scientific heir, found his own supernova in 1604. Fainter than Tycho's, it nevertheless remained visible for about a year. Kepler wrote a book about his observations and conclusions that was read by many with an interest in the heavens, including Galileo.

No supernova has been spotted in our Galaxy for the past 300 years. Since the explosion of a visible supernova is a chance event, there is no way to say when the next one might occur. Around the world, dozens of professional and amateur astronomers keep a sharp lookout for "new" stars that appear overnight, hoping to be the first to spot the next guest star in our sky and make a little history themselves.

a

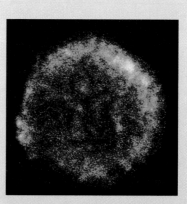

c

Several views of Tycho Brahe's 1572 supernova. (a) A broadsheet about the "new star" from 1573. (b) A modern map of the radio emission from the expanding shell of gas ejected by the supernova explosion. The different colors correspond to different intensities of radio emission, produced by extremely energetic electrons gyrating in a magnetic field. No central source has been found; apparently nothing remains of the star that exploded. (National Radio Astronomy Observatory/AUI) (c) As the debris from the explosion expands, it barrels into the surrounding interstellar gas, producing very high temperatures. The shocked material glows in x rays, as shown here in an image recorded with the Rosat x-ray satellite. (Max Planck Institute for Extraterrestrial Physics)

b

TABLE 22.1
The Ultimate Fate of Stars with Different Masses

Initial Mass (Mass of Sun = 1)	Final State at End of Its Life
<0.01	Planet
0.01 to 0.08	Brown dwarf
0.08 to 0.25	White dwarf made mostly of helium
0.25 to 8–10	White dwarf made mostly of carbon and oxygen
8–10 to 12	White dwarf made of oxygen–neon– magnesium*
12 to 40	Supernova explosion that leaves a neutron star
>40	Supernova explosion that leaves a black hole

* Stars in this mass range may produce a type of supernova different from the one we have discussed.

a brilliant explosion, observers concluded it must be a brand-new star. Twentieth-century astronomers reclassified the explosions with the greatest absolute luminosity as *super*novae.

From the historical records, from studies of the remnants of supernova explosions in our own Galaxy, and from analyses of supernovae in other galaxies, we estimate that, on average, one supernova explosion occurs somewhere in the Milky Way Galaxy every 25 to 100 years. Unfortunately, however, no supernova explosion has been detected in our Galaxy since the invention of the telescope. Either we have been exceptionally unlucky or, more likely, more-recent explosions have taken place in parts of the Galaxy where light is blocked from reaching us by interstellar dust.

At their maximum brightness, the most luminous supernovae have about 10 billion times the luminosity of the Sun. For a brief time, a supernova may outshine the entire galaxy in which it appears. After maximum brightness, the star fades in light and disappears from telescopic visibility within a few months or years. At the time of their outbursts, supernovae can eject material at typical velocities of 10,000 km/s (and speeds twice that high have been observed). A speed of 20,000 km/s corresponds to about 44 million mi/h, truly an indication of unimaginably great cosmic violence.

The Supernova Giveth and the Supernova Taketh Away

As most of the material that was once part of the star is blown into space by the supernova explosion, the life of a massive star comes to an end. But the death of each massive star is an important event in the history of its galaxy. The elements built up by fusion during the star's life are now "recycled" into space by the explosion, making them available (later) to form new stars and planets. Both because more of the star's material is lost into space, and because the ejection is so much more violent, the death of a massive star is a much more efficient recycler of newly minted starstuff than the more gentle death of a low-mass star. (To be fair, however, we should recall that there are a lot more low-mass stars than the kind of high-mass stars that make supernovae.)

The idea that supernova material becomes an important part of later generations of stars is not merely poetic speculation. We saw in Chapter 20 that the outward-moving shock of a supernova explosion compresses interstellar clouds and thus can help start the star-forming process. There is evidence in the detailed composition of ancient meteorites that the formation of our own solar system was triggered by a supernova in our neighborhood about 5 billion years ago.

But the supernova explosion has one further creative contribution to make, one we alluded to in the preceding chapter when we asked where the atoms in your jewelry came from. The supernova explosion produces a flood of energetic neutrons that barrel through the expanding material. These neutrons can be absorbed by iron and other nuclei, where they can turn into protons. Thus they build up the elements that are more massive than iron, including such terrestrial favorites as gold and silver. (This is the only place we know where such atoms as gold or uranium can be made.)

When supernovae explode, these elements (as well as the ones the star made during more stable times) are ejected into the existing gas between the stars, and mixed with it. Along with planetary nebulae, supernovae play a major role in building up the supply of chemical elements in the universe (Figure 22.9). Without them neither the authors nor the readers of this book would exist.

Furthermore, supernovae are thought to be the source of the high-energy *cosmic-ray* particles discussed in Section 19.5. Trapped by the magnetic field of the Galaxy, the particles from exploded stars continue to circulate around the vast spiral of the Milky Way. Scientists speculate that high-speed cosmic rays hitting the genetic material of Earth organisms over billions of years may have contributed to the steady *mutations*—subtle changes in the genetic code—that drive the evolution of life on our planet.

But, like so many creative events, supernovae also have a dark side. Suppose a life-form has the misfortune to develop around a star that happens to be close to a massive star destined to become a supernova. Such life-forms may find themselves snuffed out when the harsh radiation and high-energy particles from the explosion reach their world. If, as some astronomers speculate, life can develop on many planets around long-lived (lower-mass) stars, the suitability of that life's own star and planet may not be all that matters for its long-term evolution and survival. Life may well have formed around a number of

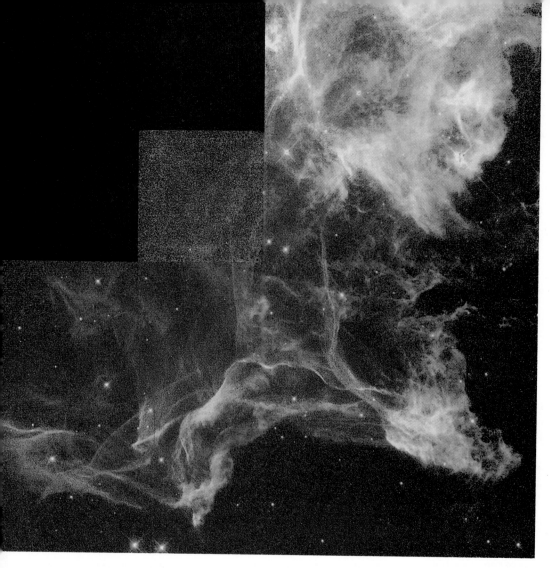

Figure 22.9

An image of a small portion of the old supernova remnant known as the Cygnus Loop, taken with the Hubble Space Telescope. The glowing gases mark the edge of a bubble-like blast wave produced by a stellar explosion that occurred about 15,000 years ago and is now approximately 130 LY wide. Here we see a region where the blast, moving from left to right, has hit a slightly denser cloud of interstellar gas and is causing it to glow. The image was taken through three different color filters: blue light shows doubly ionized oxygen atoms that glow with the energy of the blast-wave's passing; red light shows singly ionized sulfur atoms that have cooled off a bit since the shock passed them; and green shows light emitted by hydrogen atoms immediately behind the shock. In this way the supernova blast is mixing its atoms with those already in interstellar space, and compressing gas to help the formation of new generations of stars. (J. Hester/NASA)

pleasantly stable stars only to be wiped out because a nearby star went supernova.

What is a safe distance to be from a supernova explosion? A lot depends on the violence of the particular explosion, what type of supernova it is (see Section 22.4), and what level of destruction we are willing to accept. Calculations suggest that a supernova less than 50 LY away from us would certainly end all life on Earth, and that even one 100 LY away would have drastic consequences for the radiation levels here.

The good news is that there are at present no massive stars that promise to become supernovae within 50 LY of the Sun. (This is in part because the kinds of massive stars that become supernovae are overall quite rare.) The closest massive star to us, Spica (in the constellation of Virgo), is about 260 LY away, probably a safe distance.

Our ideas about supernovae are provocative, awe-inspiring, and occasionally even disturbing. What evidence do we have that supernova explosions really do have these consequences? Our best information about the vast energies involved comes from an event that was observed in 1987.

Supernova 1987A

Before dawn on February 24, 1987, Ian Shelton, a Canadian astronomer working at an observatory in Chile, pulled a photographic plate from the developer. Two nights earlier he had begun a survey of the Large Magellanic Cloud, a small galaxy that is one of the Milky Way's nearest neighbors in space. On his plate Shelton examined the Tarantula Nebula, a region of bright glowing gas where star formation is occurring. Nearby, where there should have been only faint stars, he saw a large bright spot. Concerned at first that his photograph was flawed, Shelton went outside to look at the Large Magellanic Cloud—and saw that a new object had indeed appeared in the sky (Figure 22.10). He soon realized he had discovered a supernova—one that could be seen with the unaided eye, despite the fact that it was 160,000 LY away.

Now known as SN 1987A, since it was the first supernova discovered in 1987, this brilliant newcomer to the southern sky gave astronomers their first opportunity to study the death of a relatively nearby star with modern instruments. Soon the world of astronomy was abuzz with

K, and the central density was about 5 g/cm³, or about five times the density of water.

By the time hydrogen was exhausted at the star's center, a helium core of about six times the mass of the Sun had developed, and hydrogen fusion was proceeding in a shell surrounding this core. The core contracted and grew hotter until it reached a temperature of 170 million K and a density of 900 g/cm³, at which time helium began to fuse to form carbon and oxygen. The surface of the star expanded to a radius of about 100 million km, a bit less than the distance from the Earth to the Sun. The star's luminosity nearly doubled, to 100,000 L_{Sun}, and it became a red supergiant. While in this stage the star lost some of its mass. This material has been detected by observations with the Hubble Space Telescope (Figure 22.11) because the light of the supernova flash is reflected from it.

Helium fusion lasted for only about 1 million years, forming a core of carbon and oxygen with a mass about four times that of the Sun. When helium was exhausted at the center of the star, the core contracted again, the surface also decreased in radius, and the star became a blue

Figure 22.10

Before-and-after pictures of the field around Supernova 1987A in the Large Magellanic Cloud. An arrow points to the star that exploded. It's easy to put such an arrow there *after* the explosion; astronomers can only wish such arrows appeared ahead of time so we could know which star was next. The difference in image quality between these pictures is an effect of the Earth's atmosphere, which was steadier when the plates used to make the pre-supernova picture were taken. (Anglo-Australian Telescope Board)

the news, and all available telescopes on Earth and in space were turned to the supernova. Because Shelton (and several other observers that same evening) had caught the star just as it was going off, theorists could at last test their detailed calculations of how massive stars die with data gathered at many different wavelengths.

By combining theory and observation, astronomers have reconstructed the life story of the star that became SN 1987A. Formed about 10 million years ago, it originally had a mass of about 20 M_{Sun}. For 90 percent of its life it lived quietly on the main sequence, converting hydrogen to helium. At this time its luminosity was about 60,000 times that of the Sun (L_{Sun}), and its spectral type was O. The temperature in its core was about 40 million

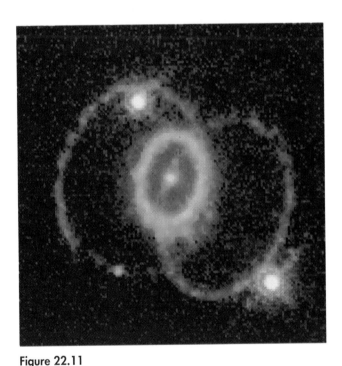

Figure 22.11

This image, taken with the repaired Hubble Space Telescope, shows (in yellow) a ring of stellar material ejected by the star that ultimately became SN 1987A. This material was ejected long before the supernova explosion and is now visible because the light of the supernova is reflected from it. Within 100 more years the expanding debris from the supernova, the elongated yellow blob in the center of the image, will plow into the ring and tear it apart. The outer rings (which are probably not in the same plane as the inner one) may or may not be earlier material released by the star; their origin is a source of great debate among astronomers. The blue stars are not associated with the supernova. (C. Burrows and NASA)

supergiant with a luminosity still about equal to 100,000 L_{Sun}. This is what it still looked like on the outside when it exploded. When the contracting core reached a temperature of 700 million K and a density of 150,000 g/cm^3, nuclear reactions began to convert carbon to neon, sodium, and magnesium. This phase lasted only about 1000 years.

The core, having exhausted carbon as a fuel, again contracted and heated—this time to a temperature of 1.5 billion K and a density of 10 million g/cm^3. The next stages in fusion, the conversion of neon to oxygen and magnesium, and then oxygen to silicon and sulfur, lasted for only a few years. At a temperature of 3.5 billion K and a density of 10^8 g/cm^3, iron began to form. Once iron was created, and the mass of the core exceeded 1.4 times that of the Sun, the collapse began. It was a catastrophic collapse, lasting only a few tenths of a second; the speed of infall in the outer portion of the iron core reached 70,000 km/s, about a fourth of the speed of light.

In the meantime, the outer shells of neon, helium, and hydrogen in the star did not yet know about the collapse. Information about the physical movement of different layers travels through a star at the speed of sound and cannot reach the surface in the few tenths of a second required for the core collapse to occur. Thus the surface layers of our star hung briefly suspended, much like a cartoon character that dashes off the edge of a cliff and hangs momentarily in space before realizing that he is no longer held up by anything.

The collapse of the core continued until the densities rose to several times that of an atomic nucleus. The resistance to further collapse then became very great, and the core rebounded. Infalling material ran into the "brick wall" of the rebounding core and was thrown outward with a great shock wave. Neutrinos poured out of the core, helping the shock wave blow the star apart. The shock reached the surface of the star a few hours later, and the star began to brighten into the supernova Ian Shelton observed.

Testing Our Theories

If this scenario (and all we have said about supernovae in general) is correct, we can make some predictions about the behavior of SN 1987A. If neutrinos play a crucial role in the explosion, vast quantities of them should come pouring out of the supernova even before the light emerges. We will examine the dramatic success of this prediction in the next section. In addition, the explosion should be rich in heavy elements just produced by the neutron-enrichment process described earlier. Even radioactive elements, ones that are unstable and decay pretty quickly into other elements, should be detectable in a supernova, since they will have *just formed*. (See Section 6.3 for a discussion of radioactivity.)

Figure 22.12 shows how the brightness of SN 1987A changed with time. In a single day the star soared in brightness by a factor of about 1000 and became just visi-

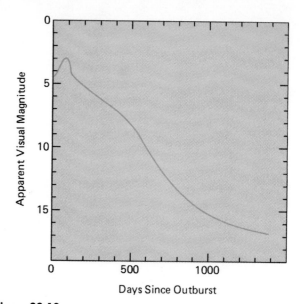

Figure 22.12
A graph that shows how the brightness of SN 1987A changed with time. (Courtesy N. Suntzeff/CTIO)

ble without a telescope. The star then continued to increase slowly in brightness until it was about the same apparent magnitude as the stars in the Little Dipper. Up until about day 40 after the outburst, the energy being radiated away was produced by the explosion itself. But what happened after that helped confirm our ideas about heavy element production. As the energy of the supernova explosion is radiated into space, we might expect the light from the explosion to fade away. But instead, SN 1987A remained bright, as a new source of energy from newly created radioactive elements came into play.

One of the elements formed in a supernova explosion is radioactive nickel, with an atomic mass of 56 (that is, the total number of protons plus neutrons in its nucleus is 56). Nickel-56 is unstable and changes spontaneously, with a half-life of about 6 days, to cobalt-56. It in turn decays with a half-life of about 77 days to iron-56, which is stable. Energetic gamma rays are emitted when these radioactive nuclei decay; they then serve as a new source of energy for the expanding layers. The gamma rays are absorbed in the overlying gas and re-emitted at visible wavelengths, keeping the remains of the star bright.

As you can see in Figure 22.12, astronomers did observe brightening due to radioactive nuclei in the first few months following the supernova's outburst, and then saw the extra light die away as more and more of the radioactive nuclei decayed to stable iron. The gamma-ray heating was responsible for virtually all of the radiation detected from SN 1987A after day 40. Some gamma rays also escaped directly without being absorbed. These were detected by Earth-orbiting telescopes at the wavelengths expected for the decay of radioactive nickel and cobalt, confirming our theory that new elements were formed in the crucible of the supernova.

Neutrinos from SN 1987A

If there had been any human observers in the Large Magellanic Cloud 160,000 years ago, the explosion we call SN 1987A would have been a brilliant spectacle in their skies. Yet we now know that less than 1/10 of 1 percent of the energy of the explosion appeared as visible light. About 1 percent of the energy was required to disrupt the star, and the rest was carried away by neutrinos. The overall energy in these neutrinos was truly astounding. In the first second, their total luminosity was 10^{46} W, which exceeds the luminosity of all the stars in all the galaxies in the part of the universe that we can observe. And the supernova generated this energy in a volume less than 50 km in diameter! Supernovae are by far the most violent events in the universe.

One of the most exciting results from observations of SN 1987A is that physicists actually detected these supernova neutrinos, thereby obtaining strong confirmation that the theoretical calculations of what happens when a star explodes are actually correct. The neutrinos were detected by two instruments, which might be called "neutrino telescopes," about 3 hours before the brightening of the star was first observed. (This is because the neutrinos get out of the exploding star more easily than light does.) Both neutrino telescopes, one in a deep mine in Japan and the other under Lake Erie, consist of several thousand tons of purified water surrounded by several hundred light-sensitive detectors. Incoming neutrinos interact with the water to produce positrons and electrons, which move rapidly through the water and emit deep blue light (Figure 22.13).

The Japanese system detected 11 neutrino events over an interval of 13 s, and the instrument beneath Lake Erie measured 8 events at the same time. Since the neutrino telescopes were in the Northern Hemisphere, and the supernova occurred in the Southern Hemisphere, the detected neutrinos had already passed through the Earth and were on their way back out into space when they were captured!

Only a few neutrinos were detected because the probability that they will interact with ordinary matter is very, very low. It is estimated that the supernova actually released 10^{58} neutrinos. About 50 billion of these passed through every square centimeter on the Earth, and about a million people experienced a neutrino interaction within their bodies. This interaction happened to only a single nucleus in each person, and thus had absolutely no biological effect; it went completely unnoticed by everyone concerned.

Since the neutrinos come directly from the heart of the supernova, their energies provide a measure of the temperature of the core as the star was exploding. The central temperature was about 200 billion K, a stunning figure to which no earthly analogy can bring much meaning. With neutrino telescopes we are peering into the final moment in the life stories of massive stars, and observing conditions beyond all human experience. Yet we are also seeing the unmistakable hints of our own origins.

22.3

Pulsars and the Discovery of Neutron Stars

After the supernova explosion fades away, all that is left behind is the neutron star. Neutron stars are the densest

 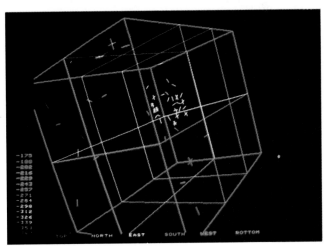

Figure 22.13
(Left) The neutrino detector under Lake Erie has 2048 light-sensitive tubes distributed around a tank that holds 8000 tons of pure water. (Right) A computer-generated display shows one of the neutrino detections on February 23, 1987. The yellow crosses and slashes near the center show which tubes were triggered by the passage of the neutrino. (F. Reines and J. C. van der Velde, IMB Collaboration)

TABLE 22.2 Properties of a Typical White Dwarf and Neutron Star		
Property	White Dwarf	Neutron Star
Mass (Sun = 1)	1.0 (always <1.4)	Always > 1.4 and < 3
Radius	5000 km	10 km
Density	5×10^5 g/cm^3	10^{14} g/cm^3

objects in the universe; the force of gravity at their surface is 10^{11} times greater than what we experience at the Earth's surface. The interior of a neutron star is composed of about 95 percent neutrons, with a small number of protons and electrons mixed in. In effect, a neutron star is a giant atomic nucleus, with a mass about 10^{57} times the mass of a proton. Its diameter is more like the size of a small town or an asteroid than a star. (Table 22.2 compares the properties of neutron stars and white dwarfs.) From this description, a neutron star probably strikes you as the object least likely to be observed from hundreds or thousands of light years away.

Yet, to the surprise of astronomers, neutron stars manage to signal their presence across vast gulfs of space. We have found hundreds of them around the Galaxy, and have thus been able to confirm some of the bizarre properties of neutron stars predicted by our theories.

The Discovery of Neutron Stars

In 1967 Jocelyn Bell, a research student at Cambridge University, was studying distant radio sources with a special detector that had been designed and built by her advisor Antony Hewish to find rapid variations in radio signals (Figure 22.14). The project computers spewed out reams of paper, showing where the telescope had surveyed the sky, and it was the job of Hewish's graduate stu-

dents to go through it all, searching for interesting phenomena. In September 1967 Bell discovered what she called "a bit of scruff"—a strange radio signal unlike anything seen before.

What Bell had found, in the constellation of Vulpecula, was a source of rapid, sharp, intense, and extremely regular pulses of radio radiation. Like the regular ticking of a clock, the pulses arrive precisely every 1.33728 s. Such exactness led the scientists to speculate that perhaps they had found signals from an intelligent civilization. Radio astronomers even half-jokingly dubbed the source "LGM" for "little green men." Soon, however, three similar sources were discovered in widely separated directions in the sky.

When it became apparent that this type of source was fairly common, astronomers concluded that they were highly unlikely to be signals from other civilizations. By today hundreds of such sources have been discovered; they are now called **pulsars,** short for pulsating radio sources. Antony Hewish won the 1974 Nobel Prize in physics for this work and other projects in radio astronomy.

The pulse periods of different pulsars range from a little longer than 1/1000 s to nearly 10 s. At first, the pulsars seemed particularly mysterious because nothing could be seen at their location on visible-light photographs. But then a pulsar was discovered right in the center of one of the best-known supernova remnants—the Crab Nebula. In addition to pulses of radio energy, we can observe pulses of visible light and x rays from the Crab as well (Figure 22.15).

The Crab Nebula is a fascinating object in many ways. Quite separate from the sharp pulses, the whole nebula glows with radiation at many wavelengths, and its overall energy output is over 100,000 times that of the Sun—not a bad trick for the remnant of a supernova that exploded in 1054 A.D., almost a thousand years ago. Can there be a connection between the pulsar and the large energy output of its host nebula?

Figure 22.14
Antony Hewish and Jocelyn Bell. (Courtesy, AIP Emilio Segrè Visual Archives, Weber Collection)

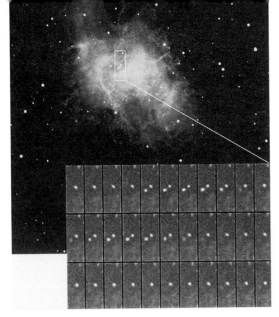

Figure 22.15
A series of photographs of the central part of the Crab Nebula taken by Nigel Sharp at Kitt Peak National Observatory. Note the star that seems to blink on and off: it is the pulsar, with a period of 0.033 s. (National Optical Astronomy Observatories)

A Spinning Lighthouse Model

The Crab pulsar itself emits considerably more energy than the Sun does. Yet that energy arrives in sharp bursts occuring 30 times each second. What type of object can emit such bursts of energy with a regularity that would be the envy of a Swiss watchmaker? As you might imagine, many ideas were proposed to account for the mysterious pulses. But most of them could not come up with a plausi-

ble physical mechanism to explain all the characteristics of the pulses, especially how rapid they are.

By applying a combination of theory and observation, astronomers have now determined that pulsars must be *spinning neutron stars.* Our model for how the pulses are generated involves something like a lighthouse on a rocky coast. To warn ships in all directions and yet not cost too much to operate, the light in a modern lighthouse turns, sweeping its beam across the dark sea. From the vantage point of a ship, you see a pulse of light each time the beam points in your direction. In the same way, something on a neutron star could sweep across the oceans of space, giving us a pulse of radiation each time the beam points at the Earth.

Neutron stars are ideal candidates for such a job because the collapse has made them so small that they can turn very rapidly. Recall the principle of the conservation of angular momentum. Even if the star was rotating very slowly when it was on the main sequence, its rotation had to speed up as it collapsed. Because the star's core shrinks until it is only 10 to 20 km across, it spins in only a fraction of a second. This is just the sort of time period we observe between pulsar pulses.

Any magnetic field that existed in the original star will be highly compressed when the core collapses to a neutron star. At the surface of the neutron star, protons and electrons are caught up in this spinning field and accelerated nearly to the speed of light. In only two places—at the north and south magnetic poles—can the trapped particles escape the strong hold of the magnetic field (Figure 22.16). Note that the magnetic north and south

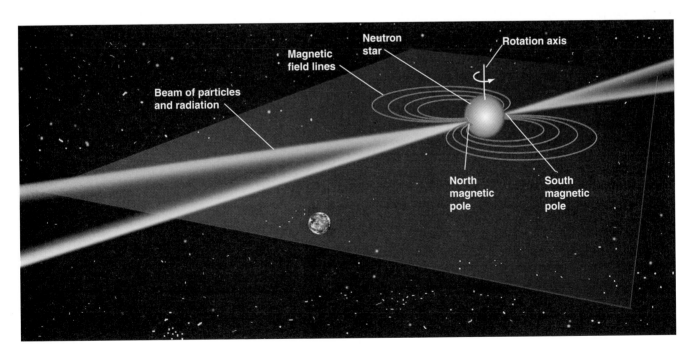

Figure 22.16
A diagram showing how emission at the magnetic poles of a neutron star can give rise to pulses of emission as the star rotates. This model requires that the magnetic poles be located in different places from the rotation poles.

poles do not have to be anywhere close to the north and south poles defined by the star's rotation. (We saw just such a difference between the two kinds of poles in the magnetic fields of Uranus and Neptune.)

At the two magnetic poles, the particles from the neutron star are focused into a narrow beam and come streaming out at enormous speeds, emitting energy over a broad range of the electromagnetic spectrum. The radiation itself is also confined to a narrow beam, explaining why the pulsar acts like a lighthouse. As the rotation carries first one and then the other magnetic pole of the star into our view, we see a pulse of radiation each time.

Proof of the Model

This model of a pulsar caused by beams of radiation from a highly magnetic and rapidly spinning neutron star is a very clever idea. But what evidence do we have that it is the correct model? A few pulsars are members of binary star systems for which enough information is available to calculate their masses using Kepler's laws. These pulsars have masses in the range of 1.4 to 1.8 times that of the Sun, just the sorts of masses that theorists predict neutron stars should have.

But there is an even better line of confirming argument, which brings us back to the Crab Nebula and its vast energy output. When the high-energy charged particles from the neutron star pulsar emerge along the poles of the magnetic field, they eventually hit the slower-moving material from the supernova. They energize this material and cause it to "glow" at many different wavelengths—just what we observe from the Crab Nebula. In a sense, the pulsar beams are a power source that renews the nebula long after the initial explosion of the star that made it.

Who pays the bills for all the energy we see coming out of a remnant like the Crab Nebula? After all, when energy emerges from one place, it must be depleted in another. The ultimate energy source in our model is the rotation of the neutron star, which propels charged particles outward and spins its magnetic field at enormous speeds. If our idea is correct, then as its energy is used to excite the Crab Nebula year after year, the pulsar inside the nebula will have to slow down because of lost energy. As it slows, the pulses eventually come a little less often; more time will elapse before the slower neutron star brings its beam back around.

While the Crab pulsar at first seemed rock-steady in the pacing of its pulses, several decades of careful observations have now clearly shown that they are in fact slowing down. Having measured how much the pulsar is slowing down, we can calculate how much rotation energy the neutron star is losing. It turns out to be the same as that emerging from the Crab Nebula in all its different forms of radiation. In other words, the rotating neutron star model can explain precisely why the Crab Nebula is glowing with the amount of energy we observe.

The Evolution of Pulsars

From observations of the several hundred pulsars discovered so far, astronomers have concluded that one new pulsar is born somewhere in the Galaxy every 25 to 100 years, the same rate at which supernovae are estimated to occur. Calculations suggest that the typical lifetime of a pulsar is about 10 million years; after that the neutron star no longer rotates fast enough to produce significant beams of particles and energy, and it is then no longer observable. We estimate that there are about 100 million neutron stars in our Galaxy.

According to present ideas, the Crab pulsar is rather young (only about 900 years old) and has a short period, while the other, older pulsars have already slowed to longer periods. Pulsars thousands of years old have lost too much energy to emit appreciably in the visible and x-ray wavelengths, and are observed only as radio pulsars; their periods are a second or more.

There is one other reason that we can see only a fraction of the pulsars in the Galaxy. Consider our lighthouse model again. On Earth, all ships approach on the same plane—the surface of the ocean—so the lighthouse can be built to sweep its beam over that surface. But in space, objects can be anywhere in three dimensions. As a given pulsar's beam sweeps over a circle in space, there is absolutely no guarantee that this circle will include the direction of Earth. In fact, if you think about it, many more circles in space will *not* include the Earth. Thus we estimate that we are unable to observe a large number of neutron stars whose beams miss us entirely.

At the same time, it turns out that only 3 of more than 400 pulsars discovered so far are embedded in the visible clouds of gas that mark the remnant of a supernova (Figure 22.17). This might at first seem mysterious, since we know that supernovae give rise to neutron stars and that we should expect each pulsar to have begun its life in a supernova explosion. But the lifetime of a pulsar turns out to be about 100 times longer than the length of time required for the expanding gas of a supernova remnant to disperse into interstellar space. Thus most pulsars are found with no other trace left of the explosion that produced them.

22.4

The Evolution of Binary Star Systems

The discussion of the life stories of stars presented so far has suffered from a bias—what we might call single-star chauvinism. Because the human race developed around a star that goes through life alone, we often tend to think of most stars in isolation. But as we saw in Chapter 17, it now appears that as many as half of all stars may develop in binary systems—those in which two stars are born enveloped in each other's gravitational embrace and go through life orbiting a common center of mass.

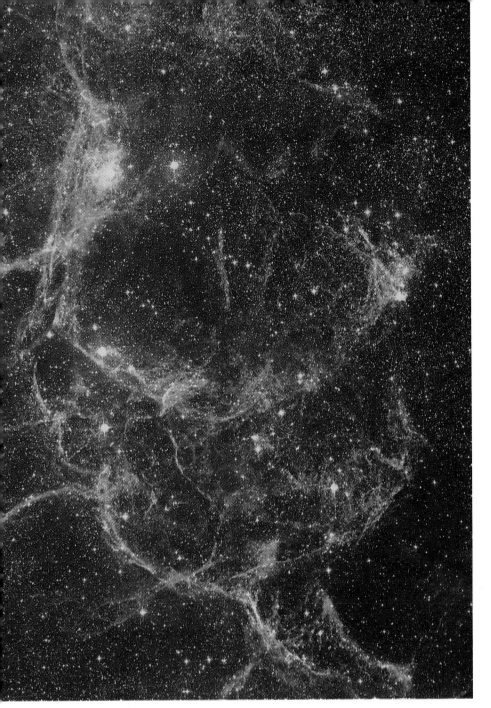

Figure 22.17
Here we see a small part of another (older) supernova remnant at whose center we can observe a pulsar. Located in the constellation of Vela, the supernova exploded over 10,000 years ago, and as a result the pulsar spins more slowly than the one in the Crab. The full remnant, most of which is not seen even on this wide-angle image, covers over 10° on the sky and is quite faint; special image-enhancement techniques were required to reveal the filaments we see. (© 1979 Royal Observatory, Anglo–Australian Telescope Board)

For these stars, the presence of a close-by companion can have a profound influence on their evolution. Stars can, under the right circumstances, exchange material, especially during the stages when one of them swells up into a giant or supergiant, or has a strong wind. When this happens and the companion stars are sufficiently close, material can flow from one star to another, decreasing the mass of the donor and increasing the mass of the recipient. Such *mass transfer* can be especially dramatic when the recipient is a stellar remnant such as a white dwarf or neutron star. While the detailed story of how such binary stars evolve is beyond the scope of our book, we do want to mention a few examples of how the stages described in this chapter may change when there are two stars in a system.

White Dwarf Explosions: The Mild Kind

Let's consider a system of two stars, in which one has become a white dwarf and the other is gradually transferring material onto it. As fresh hydrogen from the outer layers of its companion accumulates on the surface of the hot white dwarf, it begins to build up a layer of hydrogen. As more and more hydrogen accumulates and heats up on the surface of the degenerate star, the new layer eventually reaches a temperature that causes fusion to begin in a sudden, explosive way, blasting much of the new material away. The white dwarf quickly (but only briefly) becomes quite bright. To observers before the invention of the telescope, it seemed that a new star suddenly appeared, and they called it a **nova.** Novae fade away in a few months to a few years.

Hundreds of novae have been observed, each of them occurring in a binary star system, and each later showing a shell of expelled material. A number of stars have more than one nova episode, as more material accumulates on the white dwarf and the whole process repeats. As long as the episodes do not increase the mass of the white dwarf beyond the Chandrasekhar limit (by transferring too much mass), the white dwarf itself remains pretty much unaffected by the explosions on its surface.

White Dwarf Explosions: The Violent Kind

If a white dwarf accumulates matter from a companion star at a much faster rate, it can be pushed over the limit. With its mass now exceeding 1.4 times the mass of the Sun, such an object can no longer support itself as a white dwarf, and it begins to collapse. As it does so it heats up, and new nuclear reactions begin in the degenerate core. In less than one second an enormous amount of fusion takes place; the energy released is so great that it completely destroys the white dwarf. Gases are blown out into space at velocities of several thousand kilometers per second, and no central star remains behind.

Such an explosion is also called a supernova, since, like the destruction of a high-mass star, it can produce a huge amount of energy in a very short time, destroying the star in the process. We call these Type I supernovae (distinguished from the Type II's discussed earlier). Tycho's Supernova (see "Making Connections" box) appears to have been caused by such a white dwarf with an overly generous companion. (Very recent observations suggest that some mechanism allows white dwarfs with masses less than 1.4 M_{Sun} in binary systems to explode violently as well, although astronomers do not yet have a detailed model of how this happens.)

Neutron Stars with Companions

It is possible that a binary system can survive the explosion of one of its members. In that case, an ordinary star can share a system with a neutron star. If material is transferred from the "living" star to its "dead" (and highly compressed) companion, this material will be pulled in by the strong gravity of the neutron star. Such infalling gas will be compressed and heated to incredible temperatures. It will quickly become so hot that it will experience an explosive burst of fusion. The energies involved are so great that we would expect much of the radiation from such a burst to emerge as x rays and gamma rays.

And indeed, astronomers using high-energy observatories above the Earth's atmosphere (see Chapter 5) have recorded many objects that undergo just these types of x-ray and gamma-ray bursts. While there is much debate about the nature of the most energetic *bursters,* astronomers are reasonably confident that at least some of the x-ray bursts do come from material that falls onto the surface of a neutron star.

If the neutron star and its companion are positioned the right way, a significant amount of material can be transferred to the neutron star. If the neutron star is a pulsar, the new material can actually set it spinning faster (as spin energy is also transferred). Astronomers have found several pulsars spinning at a rate of almost *1000 times per second!* Such a rapid spin must have been externally caused. (Recall that the Crab Nebula pulsar, one of the youngest pulsars known, was spinning 30 times per second.) Indeed, some of the fast pulsars are clearly part of binary systems, while others may be alone only because they have consumed their former companions completely sometime in the past.

With this gruesome example, we have reached the end of our description of the final states of stars. Yet one piece of the story remains to be filled in. We saw that stars whose masses are less than 1.4 M_{Sun} at the time they run out of fuel end their lives as white dwarfs. Dying stars with masses between 1.4 and about 3 M_{Sun} become neutron stars. But, there are stars whose masses are greater than 3 M_{Sun} when they exhaust their fuel supplies; what becomes of them? The truly bizarre result of the death of such massive stellar cores is the subject of our next chapter.

Summary

22.1 During the course of their evolution, stars shed their outer layers and lose a significant fraction of their initial mass. Stars with masses up to about 10 M_{Sun} can lose enough mass to become *white dwarfs,* which have masses less than the **Chandrasekhar limit** (about 1.4 M_{Sun}). A typical white dwarf has a mass about the same as that of the Sun, and a diameter comparable to that of the Earth. The pressure exerted by **degenerate electrons** keeps white dwarfs from contracting to still smaller diameters. **Planetary nebulae** are shells of gas ejected by such stars, set to glowing by the ultraviolet radiation of the collapsing core. Eventually, white dwarfs cool off to become black dwarfs, stellar remnants made mainly of carbon and oxygen.

22.2 In a massive star, hydrogen fusion in the core is followed by several other fusion reactions involving heavier elements. Just before it exhausts all sources of energy, a massive star has an iron core surrounded by shells of silicon and sulfur, oxygen, neon, carbon, helium, and hydrogen. The fusion of iron requires energy (rather than releasing it). If the mass of a star's iron core exceeds the Chandrasekhar limit (but is less than 3 M_{Sun}), the core collapses until its density exceeds that of an atomic nucleus, forming a **neutron star** with a typical diameter of 20 km. The core rebounds and transfers energy outward, blowing off the outer layers of the star in a Type II **supernova** explosion. Studies of Supernova 1987A, including the detection of neutrinos, have confirmed theoretical calculations of what happens during such explosions.

22.3 At least some supernovae leave behind a highly magnetic, rapidly rotating neutron star, which can be observed as a **pulsar** if its beam of escaping particles and focused radiation is pointing toward us. Pulsars emit rapid pulses of radiation at regular intervals; their periods are in the range of 0.001 to 10 s. The rotating neutron star acts like a lighthouse, sweeping its beam in a circle. Pulsars lose energy as they age, the rotation slows, and their periods increase.

22.4 When a white dwarf or neutron star is a member of a close binary star system, its companion star can transfer mass to it. Material falling gradually onto a white dwarf can explode in a sudden burst of fusion and make a **nova.** If material falls rapidly onto a white dwarf, it can push it over the Chandrasekhar limit and cause it to explode completely as a Type I supernova. Material falling onto a neutron star can cause powerful bursts of x-ray and gamma-ray radiation.

Review Questions

1. How does a white dwarf differ from a neutron star? How does each one form? What keeps each from collapsing under its own weight?

2. Describe the evolution of a star with a mass like that of the Sun, from the main-sequence phase of its evolution until it becomes a white dwarf.

3. Describe the evolution of a massive star (say 20 times the mass of the Sun) up to the point at which it becomes a supernova. How does the evolution of a massive star differ from that of the Sun? Why?

4. How do the two types of supernovae discussed in this chapter differ? What kind of star gives rise to each type?

5. A star begins its life with a mass of 5 M_{Sun} but ends its life as a white dwarf with a mass of 0.8 M_{Sun}. List the stages in the star's life during which it most likely lost some of the mass it started with. How did mass loss occur in each stage?

6. If the formation of a neutron star leads to a supernova explosion, explain why only three out of the hundreds of known pulsars are found in supernova remnants.

7. How can the Crab Nebula shine with the energy of something like 100,000 Suns when the star that formed the nebula exploded almost 1000 years ago? Who pays the bills for much of the radiation we see coming from the nebula?

8. How is a nova different from a Type I supernova? How does it differ from a Type II supernova?

Thought Questions

9. You observe an expanding shell of gas through a telescope. What measurements would you make to determine whether you have discovered a planetary nebula or the remnant of a supernova explosion?

10. Arrange the following stars in order of age:
    a. A star with no nuclear reactions going on in the core, which is made primarily of carbon and oxygen.

    b. A star of uniform composition from center to surface; it contains hydrogen but has no nuclear reactions going on in the core.

    c. A star that is fusing hydrogen to form helium in its core.

    d. A star that is fusing helium to carbon in the core, and hydrogen to helium in a shell around the core.

e. A star that has no nuclear reactions going on in the core, but is fusing hydrogen to form helium in a shell around the core.

11. Would you expect to find any white dwarfs in the Orion Nebula? (See Chapter 20 to remind yourself of its characteristics.) Why or why not?

12. Suppose no stars more massive than about 2 M_{Sun} had ever formed. Would life as we know it have been able to develop?

13. Would you be more likely to observe a Type II supernova (the explosion of a massive star) in a globular cluster or in an open cluster? Why?

14. Astronomers believe there are something like 100 million neutron stars in the Galaxy. Yet we have only found

about 400 or so pulsars. Give several reasons that these numbers are so different. Explain each reason.

15. Would you expect to observe *every* supernova in our own Galaxy? Why or why not?

16. The Large Magellanic Cloud has about one-tenth the number of stars found in our own Galaxy. Suppose the mix of high- and low-mass stars is exactly the same in both galaxies. Approximately how often does a supernova occur in the Large Magellanic Cloud?

17. Look at the list of the nearest stars in Appendix 10. Would you expect any of these to become supernovae?

18. If most stars become white dwarfs at the ends of their lives, and the formation of white dwarfs is accompanied by the production of a planetary nebula, why don't we see many more planetary nebulae in the Galaxy than we do?

Problems

19. The gas shell of a particular planetary nebula is expanding at the rate of 20 km/s. Its diameter is 1 LY. Find its age. For this calculation, assume that there are 3×10^7 s/yr and 10^{13} km/LY.

20. Prepare a chart or diagram that exhibits the relative sizes of a typical red giant, the Sun, a typical white dwarf, and a neutron star of mass equal to the Sun's. You may have to be clever to devise such a diagram.

21. Suppose the central star of a planetary nebula is 16 times as luminous and 20 times as hot (about 110,000 K) as the

Sun. Find its radius in terms of the Sun's. Compare this radius with that of a typical white dwarf.

22. The luminosity of a supernova at maximum light is about 10^{10} L_{Sun}. The Sun would be just barely visible without a telescope at a distance of 10 parsecs (32.6 LY). How far away is it possible to see a supernova with the unaided eye?

23. Suppose the radius of the pulsar in the Crab Nebula is 10 km, and its rotation period is 0.033 s. What is the rotational velocity of the star at its equator?

Suggestions for Further Reading

See some of the references for the last chapter, plus:

Balick, B. et al. "The Shaping of Planetary Nebulae" in *Sky & Telescope*, Feb. 1987, p. 125.
Bethe, H. and Brown, G. "How a Supernova Explodes" in *Scientific American*, May 1985, p. 60.
Chapman, C. and Morrison, D. *Cosmic Catastrophes*. 1989, Plenum Press. Chapter 18 is on supernovae.
Filippenko, A. "A Supernova with an Identity Crisis" in *Sky & Telescope*, Dec. 1993, p. 30. Good review of supernovae in general, and of 1993J in M81.
Graham-Smith, F. "Pulsars Today" in *Sky & Telescope*, Sep. 1990, p. 240.
Greenstein, G. "Neutron Stars and the Discovery of Pulsars" in *Mercury*, Mar./Apr. 1985, p. 34; May/June 1985, p. 66.

Kaler, J. "The Smallest Stars in the Universe" in *Astronomy*, Nov. 1991, p. 50. On white dwarfs, neutron stars, and pulsars.
Kaler, J. "Realm of the Hottest Stars" in *Astronomy*, Feb. 1990, p. 22. On planetaries and their central stars.
Kawaler, S. and Winget, D. "White Dwarfs: Fossil Stars" in *Sky & Telescope*, Aug. 1987, p. 132.
Kirshner, R. "Supernova: The Death of a Star" in *National Geographic*, May 1988, p. 618. Excellent introduction for beginners on SN 1987A.
Marschall, L. *The Supernova Story*, 2nd ed. 1994, Princeton U. Press. The introduction of choice to supernovae and SN 1987A.
Soker, N. "Planetary Nebulae" in *Scientific American*, May 1992, p. 78.
Tierney, J. "The Quest for Order: Profile of S. Chandrasekhar" in *Science '82*, Sep. 1982, p. 68.
Wallerstein, G. and Wolff, S. "The Next Supernova" in *Mercury*, Mar./Apr. 1981, p. 44.

1. To see which planetary nebulae might be visible tonight through binoculars or a small telescope, set the *Stellar Limiting Magnitude* to 5 and the *Deep-Sky Filter* to a faint magnitude of 10, with all objects off except for planetary nebulae.

2. Under *Objects Filter* select *Deep Sky*. Turn off all objects except for Planetary Nebulae. Set the *Planetary Nebula* faint magnitude to 9 and count the number of nebulae. The number of stars visible to this limiting magnitude is roughly 250,000. Because most of these stars will go through the planetary nebula phase, the ratio of these two numbers gives an idea of the length of the planetary nebula phase.

Assuming the average star has a lifetime of 5 billion years, what is the length of the planetary nebula phase?

3. Using Table 22.1 and the masses for the stars found in *RedShift* exercise 21.1, determine which of these stars will end their lives as white dwarfs and which may end as neutron stars:

a. Beta Canes Venaticorum
b. Epsilon Eridani
c. Alpha Pictoris
d. Gamma Cassiopeiae
e. Procyon, Alpha Canis Minoris
f. Theta Ursae Majoris
g. Menkalinan, Beta Aurigae
h. Tau Herculis
i. Eta Cephei
j. Mu Herculis

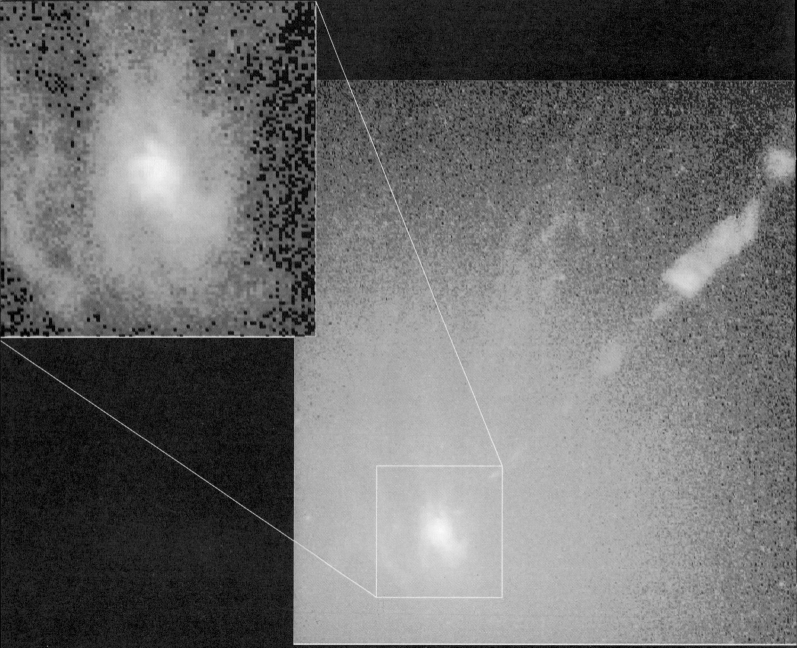

CHAPTER 23

Black Holes and Curved Spacetime

Thinking Ahead

Most of us think that time is like the flow of a river with strong current. Much like twigs in the water, we seem caught in the flow, moving together from the past into the future. Einstein showed, however, that the flow of time is not uniform for every observer in the universe; depending on the strength of gravity, the pace can vary. What is the flow of time like near the collapsed remnants of the most massive stars, where gravity is very strong?

> One thing is certain,
> and the rest debate—
> Light-rays, when near the Sun,
> do not go straight.
>
> Arthur S. Eddington (1920)

Most stars end their lives as white dwarfs and neutron stars. When a very massive star collapses at the end of its life, however, not even neutron degeneracy can support the core against its own weight. For a core whose mass is more than about three times that of the Sun (M_{Sun}), our theories predict that *no known force can stop it from collapsing forever!* Gravity simply overwhelms all other forces and crushes the core until it is infinitely small.

A star in which this occurs may become one of the strangest objects ever predicted by theory—a black hole. To understand what a black hole is, we need a theory that can describe the action of gravity under such extreme circumstances. Our best theory of gravity was put forward in 1916 by Albert Einstein and is called the theory of **general relativity.**

Figure 23.1
Albert Einstein has become a symbol for intellect in popular culture. The caption of this Italian ad translates to "Instinct says beer, reason says Carlsberg." (Photo courtesy of the archives, Caltech)

General relativity is one of the major intellectual achievements of the 20th century; if it were music, we would compare it to the great symphonies of Beethoven or Mahler. Until recently, however, scientists had little need for a better theory of gravity; Isaac Newton's ideas (see Chapter 2) are perfectly sufficient for most of the objects we deal with. In the past three decades, however, general relativity has become more than just a beautiful idea; it is now essential in understanding pulsars, quasars (which will be discussed in Chapter 26), and many other astronomical objects and events, including black holes.

We should perhaps mention that this is the point in an astronomy course when many students start to feel a little nervous (and perhaps wish they had taken botany or some other earthbound course to satisfy the science requirement). This is because in popular culture, Einstein has become a symbol for mathematical brilliance that is simply beyond the reach of most people (Figure 23.1). So when we mentioned that the theory of general relativity was Einstein's work, you, like many other students, may have shuddered just a bit, convinced that anything Einstein did was beyond your understanding. This popular view is unfortunate and mistaken. While the detailed calculations of general relativity do involve a good deal of higher mathematics, the basic ideas are not difficult to understand (and are, in fact, almost poetic in the way they give us a new perspective on the world).

23.1

Principle of Equivalence

The fundamental insight that led to the formulation of the theory of general relativity starts with a very simple thought: If you were able to jump off a high building and fall freely, you would not feel your own weight. Einstein built on this idea to reach sweeping conclusions about the very fabric of space and time itself. He called it the "happiest idea of my life."

Einstein himself pointed out an everyday example that illustrates this effect. Notice how your weight seems to be reduced in a high-speed elevator when it accelerates from a stop to a rapid descent. Similarly, your weight seems to increase in an elevator that starts to move quickly upward. This effect is not just a feeling you have: if you stood on a scale in such an elevator, you could measure your weight changing (you can actually perform this experiment in some science museums).

In a *freely falling* elevator, with no air friction, you would lose your weight altogether. Near weightlessness can be achieved by taking an airplane to high altitude and then dropping rapidly for a while. This is how NASA trains its astronauts for the experience of free-fall in space; the scenes of weightlessness in the movie *Apollo 13* were filmed in the same way.

A more formal way to state Einstein's idea is the following. Suppose we have a spaceship that contains a windowless laboratory equipped with all the tools needed to perform scientific experiments. One future day, imagine that an astronomer wakes up after a long night celebrating some scientific breakthrough and finds herself sealed into this laboratory. She has no idea how it happened, but notices that she is weightless. This could be because she is at rest, or moving at some steady speed through space, far away from any source of gravity (in which case she has plenty of time to wake up). But it could also be because she is falling freely toward a planet like the Earth (in which case she might first want to check her distance from the surface before making coffee).

What Einstein postulated is that there is *no* experiment she can perform inside the sealed laboratory to determine whether she is floating in space or falling freely in

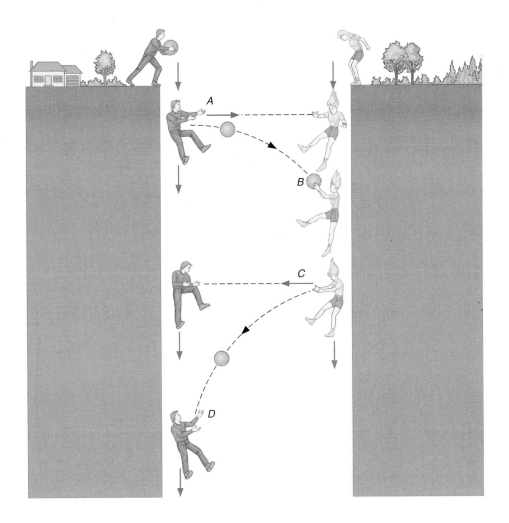

Figure 23.2
A brave couple playing catch as they descend into a bottomless abyss. Since the boy, girl, and ball all fall at the same speed, it appears to them that they can play catch by throwing the ball in a straight line between them. Within their world, there appears to be no gravity.

a gravitational field. As far as she is concerned, the two situations are completely *equivalent*. The idea that life in a freely falling laboratory is indistinguishable from, and hence equivalent to, life with no gravity is called the **equivalence principle.**

Gravity or Acceleration?

To explore the profound implications of this simple idea, let's consider as an example a foolhardy boy and girl who simultaneously jump from opposite banks into a bottomless chasm (Figure 23.2). If we ignore air friction, then we can say that while they fall, they both accelerate downward at the same rate and feel no external force acting on them. They can throw a ball back and forth, always aiming it straight at each other, as if there were no gravity. The ball falls at the same rate they do, so it can always remain in a line between them.

Such a game of catch is very different on the surface of the Earth. Everyone who grows up feeling gravity knows that a ball, once thrown, falls to the ground. Thus, in order to play catch with someone, you must aim the ball upward so that it follows an arc—rising and then falling as it moves forward—until it is caught at the other end.

Now suppose we isolate our freely falling boy, girl, and ball inside in a large box falling with them. No one inside the box is aware of any gravitational force. If the children let go of the ball, it doesn't fall to the bottom of the box or anywhere else, but merely sits there or moves in a straight line, depending on whether it is given any motion.

Astronauts in the Space Shuttle orbiting the Earth live in just such an accelerating environment (Figure 23.3). The Shuttle in orbit is falling freely around the Earth. While in free-fall, the astronauts live in a magical world where there seem to be no gravitational forces. One can give a wrench a shove, and it moves at constant speed across the orbiting laboratory. A pencil set in midair remains there, as if no force were acting on it.

Appearances are misleading however. There *is* a force in this situation. Neither the Shuttle nor the astronauts are *really* weightless, for they continually fall around the Earth, pulled by its gravity. But since all fall together—Shuttle, astronauts, wrench, and pencil—inside the Shuttle all gravitational forces appear to be absent.

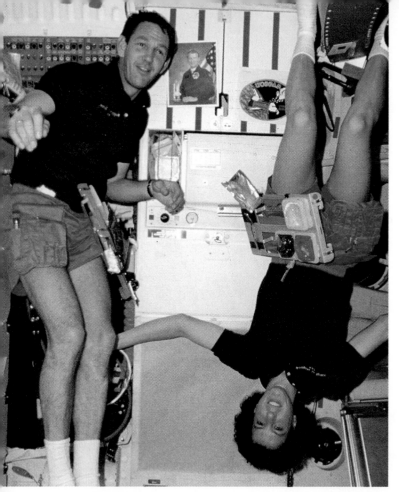

Figure 23.3
When the Space Shuttle is in free-fall in orbit, everything stays put or moves uniformly because there is no apparent gravitation acting inside the spacecraft. (NASA)

Thus the Shuttle provides an excellent example of the principle of equivalence—how local effects of gravity can be completely compensated by the right acceleration. To the astronauts, falling around the Earth creates the same effects as being far off in space, remote from all gravitational influences.

The Paths of Light and Matter

Einstein postulated that the equivalence principle is a fundamental fact of nature, and that there is *no* experiment inside the spacecraft by which an astronaut can ever distinguish between being weightless in remote space and being in free-fall near a planet like the Earth. This would apply to experiments done with beams of light as well. But the minute we use light in our experiments, we are led to some very disturbing conclusions.

It is a fundamental observation from everyday life that beams of light travel in straight lines. Imagine the Space Shuttle is moving through empty space far from any gravity. Send a laser beam from the back of the ship to the front and it will travel in a nice straight line and land on the front wall exactly opposite the point from which it left the rear wall. If the equivalence principle really applies universally, then this same experiment performed in free-fall around the Earth should give us the exact same result.

Imagine that the astronauts again shine a beam of light along the length of their ship. But, as shown in Figure 23.4, when the Shuttle is in free-fall, it falls a bit between the time the light leaves the back wall and the time it hits the front wall. (The amount of the fall is grossly exaggerated in Figure 23.4 for educational purposes.) Therefore, if the beam of light follows a straight line but the ship's path curves downward, then the light should strike the front wall at a point higher than the point from which it left.

However, this would violate the principle of equivalence—the two experiments would give different results. We are thus faced with giving up one of our two assumptions. Either the principle of equivalence is not correct, or light does not always travel in straight lines. As Einstein did, let's select the second one, always keeping in mind that such theoretical choices in science must be validated by testing their predictions against experimental results.

If we say that the principle of equivalence is right, the beam must arrive directly opposite the point from which it started in the ship. Then the light, like the ball the children were throwing back and forth, *must fall with the ship* if that ship is in orbit around the Earth (Figure 23.4). This would make its path curve downward, like the path of the ball, and thus the light would hit the front wall exactly opposite the spot from which it came.

Thinking this over, you might well conclude that it doesn't seem like such a big problem; why can't light fall the way balls do. But, as discussed in Chapter 4, light is profoundly different from balls. Balls have mass, while light does not.

Figure 23.4
In a spaceship moving to the left (in this figure) in its orbit about a planet, light is beamed from the rear, A, toward the front, B. Meanwhile the ship is falling out of its straight path (the amount it falls is exaggerated here). We might therefore expect the light to strike at B′, above the target in the ship. Instead, the light, bent by gravity, follows a curved path and strikes at C.

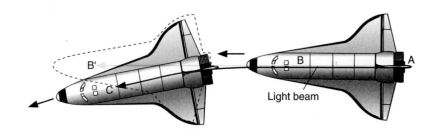

Light beam

Here is where Einstein's intuition and genius allowed him to make a profound leap. Einstein suggested that the light curves down to meet the front of the Shuttle because the Earth's gravity actually bends the fabric of space and time. This radical idea—which we will explain below—keeps the behavior of light the same in empty space and free-fall, but changes some of our most basic and cherished ideas about space and time. The reason we take Einstein's suggestion seriously is that, as we will see, experiments now clearly show his intuitive leap was correct.

Spacetime and Gravity

Is light actually bent from its straight-line path by Earth's gravity? Einstein preferred to think that it is space and time that are affected by gravity; light, and everything else that travels through space and time, then finds its path affected. As we will see, light always follows the shortest path—but that path may not always be straight. For example, on the curved surface of planet Earth, if you want to fly from Chicago to Rome, the shortest distance is not a straight line but the arc of a *great circle* (which we defined at the beginning of Chapter 3).

To show what Einstein's insight really means, let's first consider how we locate an event in space and time. For example, imagine you have to describe to worried school officials the fire that broke out in your room when your roommate tried making shish kebob in the fireplace. You explain that your dorm is at 6400 College Avenue, which locates the room in a left-right direction; you are on the 5th floor, which tells where you are in the up-down direction; and you are the 6th room back from the elevator, which tells where you are in the forward-backward direction. Then you explain that the fire broke out at 6:23 P.M. (but was soon brought under control). Any event in the universe can thus be pinpointed using the three dimensions of space and the one dimension of time.

Newton considered space and time completely independent, and that continued to be the accepted view until the beginning of the 20th century. But Einstein showed that there was an intimate connection between space and time, and that only by considering the two together—in what we call **spacetime**—can we build up a correct picture of the physical world. We examine spacetime a bit more in the next section.

The gist of Einstein's general theory is that the presence of mass (gravity) curves or warps the fabric of spacetime. When something else—a beam of light, an electron, or the Starship *Enterprise*—enters such a region of distorted spacetime, the path of that something else will be different from what it would have been in the absence of the mass. This idea is often summarized in the following way: matter tells spacetime how to curve, and the curvature of spacetime tells other matter how to move.

The amount of distortion in spacetime depends on the mass involved. Terrestrial objects, such as the book you are reading, have far too little mass to introduce any noticeable distortion. (If, like most people, you have never heard of distorted spacetime, that's the reason; it plays no role in life on Earth. Newton's view of gravity is just fine for building bridges, skyscrapers, or amusement-park rides.) It takes a mass like a star to produce measurable distortions, and a massive star is best. So you see, we *are* eventually going to talk about collapsing stars again, but not before discussing Einstein's ideas (and the evidence for them) in a bit more detail.

Spacetime Examples

How can we understand the distortion of spacetime by the presence of some (significant) amount of mass? Let's try the following analogy. You may have seen maps of New York City that squeeze the full three dimensions of this towering metropolis onto a flat sheet of paper, and still have enough information so tourists will not get lost. Let's do something similar with diagrams of spacetime.

Figure 23.5, for example, shows the progress of a motorist driving to the east on a stretch of road in Kansas where the countryside is absolutely flat. Since our motorist is traveling only in the east-west direction, and the terrain is flat, we can ignore the other two dimensions of space. The amount of time elapsed since he left home is shown on the vertical axis, and the distance traveled eastward is shown on the horizontal axis. From A to B he drove at a uniform speed; unfortunately, it was too fast a uniform speed and a police car spotted him. From B to C

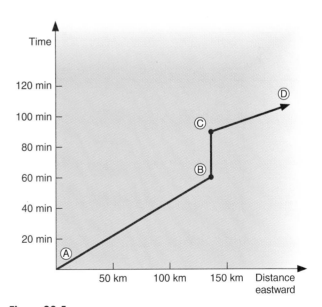

Figure 23.5
The progress of a motorist traveling across the Kansas landscape. Distance traveled is plotted along the horizontal axis. The time elapsed since the motorist left his starting point is plotted along the vertical axis.

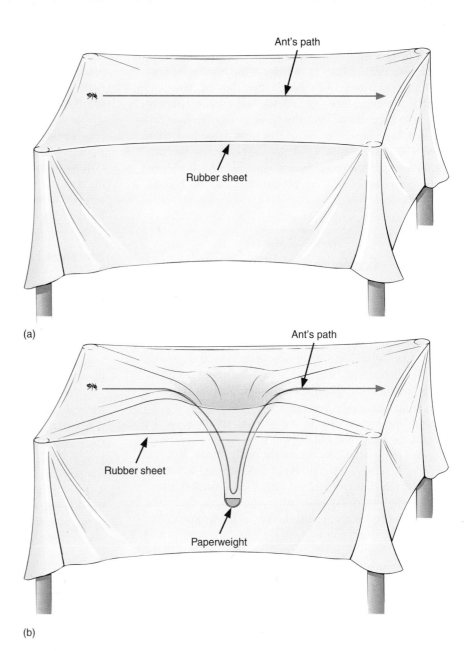

Ant's path

Rubber sheet

(a)

Ant's path

Rubber sheet

Paperweight

(b)

Figure 23.6
On a flat rubber sheet, trained ants have no trouble walking in straight lines. However, when a massive object puts a big depression into the sheet, then ants, who must walk where the sheet takes them, find their paths changed (warped) dramatically.

he stopped to receive his ticket and made no progress through space, only through time. From C to D he drove more slowly because the police car was behind him.

In the same way, we might (in our imaginations) squeeze the four dimensions of spacetime in the universe onto a flat sheet. To show the distortion, however, we will make it a rubber sheet that can stretch or warp if we put objects on it. Let's imagine we borrow one from a local hospital and stretch it taut on four posts.

To complete the analogy, we need something that travels in a straight line (as light does). Suppose we have an extremely intelligent ant that has been trained by some skilled psychologists to walk in a straight line. We reward the ant with a piece of sugar at the end of each run to reinforce the desired behavior.

We begin with just the rubber sheet and the ant, simulating empty space with no mass in it. We put the ant on one side of the sheet and it walks in a beautiful straight line over to the other side (Figure 23.6a). We next put a small grain of sand on the rubber sheet. While the sand does distort the sheet a tiny bit, this is not a distortion that we or the ant can measure. If we send the ant so it goes close to, but not on top of, the sand grain, it has little trouble walking in a straight line and earning its sugary reward.

Now we grab something with a little more mass—say a small pebble. It bends or distorts the sheet just a bit right around its position. If we send the ant into this region, it finds its path slightly altered by the distortion of the sheet. The distortion is not large, but if we follow the ant's path carefully, we notice it deviating slightly from a straight line.

The effect gets more noticeable as we increase the mass of the object we put on the sheet. Let's say we now use a massive paperweight. Such a heavy object distorts or warps the rubber sheet very effectively, putting good sag in it. We again send the ant so its path goes close to, but not on top of, the paperweight (Figure 23.6b). Far away from the paperweight, the ant has no trouble doing its walk.

But as the poor ant gets closer to the paperweight, it finds that the sheet is no longer straight! It has a big sag in it, down into which the ant is forced to go. Even worse, the ant must then climb up the other side of the sag, expending a lot of energy before it can return to walking on an undistorted part of the sheet. All this while, the ant is trying to move in a straight line, knowing that only straight-line walking will earn the reward. But through no fault of its own (after all, ants can't fly, so it has to stay on the sheet) the ant finds its path curved by the distortion of the sheet itself.

In the same way, according to Einstein's theory, we could say that light always "tries" to travel in a straight line through spacetime. But as the masses distorting spacetime get larger and larger, the shortest, most direct paths are no longer straight lines, but curves.

How large does a mass have to be before we can measure the distortion of spacetime? In 1916, when Einstein first proposed his theory, the grain of sand in our analogy might have been the entire Earth. There was, at that time, no way to measure the distortion of spacetime caused by a mass the size of our planet. To get the kind of small but measurable distortion caused by the pebble in our analogy, we needed the mass of the Sun. (We discuss how this effect was measured using the Sun in the next section.)

The paperweight in our analogy might be a white dwarf or neutron star. Around these compact, massive objects the distortion of spacetime is much more noticeable. And when, to return to the situation described at the beginning of the chapter, a star core with more than three times the mass of the Sun collapses forever, the distortions of spacetime can become truly mind-boggling!

23.3
Tests of General Relativity

What Einstein proposed was nothing less than a major revolution in our understanding of space and time. It was a new model of gravity, in which gravity was not seen as a force, but as something that changes the geometry of spacetime. Like all new ideas in science, no matter who advances them, Einstein's theory had to be tested by comparing its predictions against the experimental evidence. This was no easy matter, because the effects of the new view of gravity were apparent only when the mass was quite large. (For smaller masses, it required measuring techniques that would not become available until decades later.)

When the distorting mass is small, the predictions of general relativity must agree with those of Newton's the-

ory, which, after all, has served us admirably in our technology and in guiding space probes to the other planets. In familiar territory, therefore, the differences between predictions of the two theories are subtle and difficult to detect. Nevertheless, Einstein was able to demonstrate one proof of his theory that could be found in existing data, and to suggest another one that would be tested only a few years later.

The Motion of Mercury

Of the planets in our solar system, Mercury orbits closest to the Sun, and is thus located closest to where the Sun's gravity distorts spacetime. Einstein wondered if the distortion might produce a noticeable difference in the motion of Mercury that was not predicted by Newton's theory. It turned out that the difference was subtle but it was definitely there. Most important, it had already been measured.

Mercury has a highly elliptical orbit, so that it is only about two-thirds as far from the Sun at perihelion as it is at aphelion. It can be calculated that the gravitational effects (perturbations) of the other planets on Mercury should produce an advance of its perihelion. What this means is that each successive perihelion occurs in a slightly different direction as seen from the Sun (Figure 23.7).

According to Newtonian theory, the gravitational forces exerted by the planets will cause Mercury's perihelion to advance by about 531 arcsec per century. In the last century, however, it was observed that the actual advance is 574 arcsec per century. The discrepancy was first

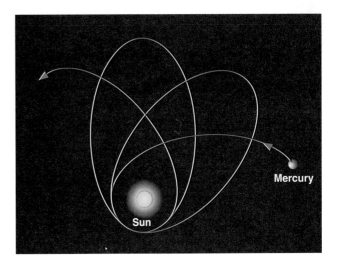

Figure 23.7
The major axis of the orbit of a planet, such as Mercury, rotates in space because of various perturbations. In Mercury's case the amount of rotation is larger than can be accounted for by the gravitational forces exerted by other planets; this difference is precisely explained by general relativity theory. The change from orbit to orbit has been significantly exaggerated on this diagram.

pointed out by Urbain Leverrier, the co-discoverer of Neptune. Just as discrepancies in the motion of Uranus allowed astronomers to discover the presence of Neptune, so it was thought the discrepancy in the motion of Mercury could mean the presence of an undiscovered inner planet. Astronomers searched for this planet near the Sun, even giving it a name—Vulcan, after the Roman god of fire. (The name would later be used for the home planet of a famous character on a popular television show about future space travel.)

But no planet has ever been found near the Sun, and the discrepancy was still bothering astronomers when Einstein was doing his calculations. General relativity, however, predicts that due to the curvature of spacetime there should be a tiny additional push on Mercury, over and above that predicted by Newtonian theory, at each perihelion. The result is to make the major axis of Mercury's orbit rotate slowly in space. The prediction of relativity is that the direction of perihelion should change by 43 arcsec per century. This is remarkably close to the observed discrepancy, and it gave Einstein a lot of confidence as he advanced his theory. The relativistic advance of perihelion was later also observed in the orbits of several asteroids that come close to the Sun.

Deflection of Starlight

Einstein's second test was something that had not been observed before and would thus provide an excellent proof of his theory. Since spacetime is more curved in regions where the gravitational field is strong, we would expect light passing very near the Sun to appear to follow a curved path (Figure 23.8), just like the ant in our analogy. Einstein calculated from general relativity theory that starlight just grazing the Sun's surface should be deflected by an angle of 1.75 arcsec.

We encounter a small "technical problem" when photographing starlight coming very close to the Sun: the Sun is an outrageously bright source of starlight itself! But during a total solar eclipse much of the Sun's light is blocked out, allowing the stars to be photographed. In a

paper published during World War I, Einstein (writing in a German journal) suggested that observation during an eclipse could detect the deflection of light passing near the Sun. A single copy of that paper, passed through neutral Holland, reached the British astronomer Arthur S. Eddington.

The technique involves taking a photograph six months earlier of the stars in front of which the eclipsed Sun will appear, and measuring the position of all the stars accurately. Then the same stars are photographed during the eclipse. This is when the starlight has to travel by the Sun through warped spacetime, and thus its path is no longer a straight line. As seen from Earth, the stars closest to the Sun will be "out of place"—slightly away from their regular positions.

The next suitable eclipse was on May 29, 1919. The British organized two expeditions to observe it, one on the island of Principe, off the coast of West Africa, and the other in Sobral, in northern Brazil. Despite some problems with the weather, both expeditions obtained successful photographs. The stars seen near the Sun were indeed displaced, and to the accuracy of the measurements, which was about 20 percent, the shifts were consistent with the predictions of relativity. It was a triumph that made Einstein a world celebrity.

The measurements made in 1919 were good enough to distinguish between Newton's theory of gravity, which predicts no deflection of starlight, and Einstein's theory of gravity, which does predict a deflection. Nevertheless, the accuracy of the measurements was still so low that they could not be considered a convincing demonstration of the theory's absolute correctness. Far higher accuracy has been obtained recently at radio wavelengths.

Radio waves traveling by the Sun should be deflected as well, and because the Sun is less bright in radio waves, such measurements can be carried out more easily. At first, as discussed in Chapter 5, radio astronomers could not pinpoint the exact direction of a radio wave source as accurately as could those working with visible light. But now, the technique of interferometry and the introduction of widely spaced arrays of radio antennas has allowed very

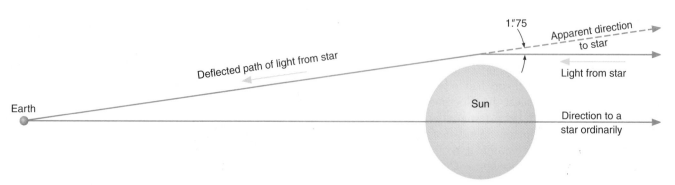

Figure 23.8
Starlight passing near the Sun is deflected slightly, so that it no longer travels in a straight line.

precise radio position measurements. As the Sun moves through the sky, measurements of the shifts in radio positions have allowed astronomers to confirm the predictions made by Einstein to within 1 percent.

Time in General Relativity

A Prediction

General relativity theory makes various predictions about the behavior of space and time. One of these predictions, put in everyday terms, is that *the stronger the gravity, the slower the pace of time.* Such a statement goes very much counter to our intuitive sense of time as a flow we all share. Time has always seemed the most democratic of concepts: all of us, regardless of wealth or status, appear to move together from the cradle to the grave in the great current of time.

But Einstein would argue that it only seems this way to us because all humans so far have lived and died in the gravitational environment of the Earth. We have had no chance to test the idea that the pace of time might depend on the strength of gravity because we have not experienced radically different gravities. And, the differences in the flow of time are extremely small until truly large masses are involved. Nevertheless, Einstein's prediction can now be tested, both on Earth and in space.

The Tests of Time

In an ingenious experiment in 1959, Robert Pound and Glenn Rebka used the most accurate atomic clock known to compare time measurements on the ground floor and the top floor of the physics building at Harvard University. For a clock, the experimenters used the frequency (the number of cycles per second) of gamma rays emitted by radioactive cobalt. Einstein's theory predicts that such a cobalt clock on the ground floor, being a bit closer to the Earth's center of gravity, should run very slightly slower than the same clock on the top floor. This is precisely what the experimenters observed. Later, atomic clocks were taken up in high-flying aircraft and even on one of the Gemini space flights. In each case, the clocks farther from the Earth ran a bit faster.

The effect is more pronounced if the gravity involved is the Sun's and not the Earth's. If stronger gravity slows the pace of time, then it will take longer for a light or radio wave that passes very near the edge of the Sun to reach the Earth than we would expect on the basis of Newton's law of gravity. (It also takes longer because spacetime is curved in the vicinity of the Sun.) The smaller the distance between the ray of light and the edge of the Sun at closest approach, the longer the delay in the arrival time.

In November 1976, when the two Viking spacecraft (see Chapter 9) were operating on the surface of Mars, Mars, as seen from Earth, went behind the Sun (Figure 23.9). Scientists had preprogrammed Viking to send a radio wave toward the Earth that would go extremely close to the outer regions of the Sun. According to general rela-

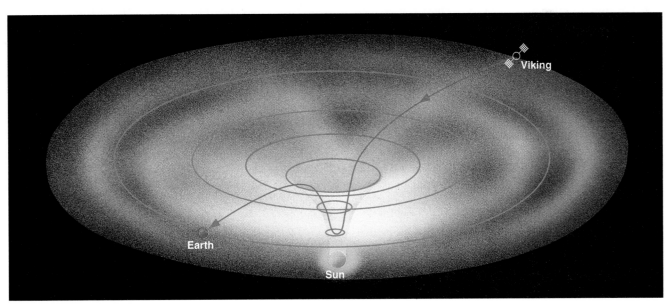

Figure 23.9
Radio signals from the Viking lander on Mars are delayed when they pass near the Sun, where spacetime is curved relatively strongly. In this picture, spacetime is pictured as a two-dimensional rubber sheet.

tivity, there would be a delay because the radio wave would be passing through a region where time ran more slowly. The experiment was able to confirm Einstein's theory to within 0.1 percent.

Gravitational Redshift

What does it mean to say that time runs more slowly? When light enters a region of strong gravity and time slows down, the light experiences a change in its frequency and wavelength; we call this a *gravitational redshift*.

To understand what happens in this situation, let's recall that a wave of light is a repeating phenomenon—crest follows trough follows crest with great regularity. In this sense, each light wave is a little clock, keeping time with its wave cycle. If stronger gravity slows down the pace of time, then the rate at which crest follows crest must slow down—the waves become less *frequent*. This means that there must be more space between the crests; in other words, the wavelength of the wave must increase. This kind of increase (when caused by the motion of the source) is just what we called a *redshift* in Chapter 4. Here, it is not motion but gravity that causes the redshift.

When Einstein's theory was first proposed, it was not possible to measure the gravitational redshift for the Earth or the Sun because the effects were too small for our instruments. But the much stronger gravity of white dwarf stars made them better candidates for observing such a redshift. For the average white dwarf, we do not know how to disentangle the redshift due to its gravity from the redshift caused by its motion through the Galaxy. But if a white dwarf is a member of a binary star system, the motions can be clarified. In such systems astronomers have successfully measured gravitational redshifts, starting in the 1920s.

Far-higher accuracy has been attained recently in the near-Earth environment with space-age technology. In the mid-1970s a hydrogen *maser*, a device (akin to a laser) that produces a microwave radio signal at a particular wavelength, was carried by a rocket to an altitude of 10,000 km. The rocket-borne maser was used to detect the radiation from a similar maser on the ground. The radiation showed a redshift due to the Earth's gravity that confirmed the relativity predictions to within a few parts in 10,000.

So far, then, each prediction of general relativity that can be tested has been confirmed within the accuracy of the experiments. Today, general relativity is accepted as our best description of gravity and is used by astronomers and physicists to understand the behavior of active galaxies, the beginning of the universe, and the subject with which we began the chapter—the death of truly massive stars.

Let's now apply what we have learned about gravity and spacetime curvature to the collapsing core in a very massive star. We saw that if the core's mass is greater than about 3 M_{Sun}, our theoretical understanding forces us to conclude that nothing can stop the core from collapsing forever. We will examine this situation from two perspectives—first from a pre-Einstein point of view, and then with the aid of general relativity.

Classical Collapse

Let's begin with a thought experiment. We want to know what speeds are required to escape from the gravitational pull of different objects. A rocket must be launched from the surface of the Earth at a very high speed if it is to escape the pull of the Earth's gravity. In fact, any object—rocket, ball, astronomy book—that is thrown into the air with a velocity less than 11 km/s will soon fall back to the Earth's surface. Only those objects launched with a velocity greater than this *escape velocity* can get away from the Earth.

The escape velocity from the surface of the Sun is higher yet—618 km/s. Now imagine that we begin to compress the Sun, forcing it to shrink in diameter. Recall that the pull of gravity depends on both the mass that is pulling you and your distance from the center of gravity of that mass. If the Sun is compressed, its *mass* will remain the same, but the *distance* between a point on the Sun's surface and the center will get smaller and smaller. Thus the pull of gravity for an object on the shrinking surface will get stronger and stronger (Figure 23.10).

When the shrinking Sun reaches the diameter of a neutron star (less than 100 km), the velocity required to escape its gravitational pull will be about half the speed of light. Suppose we continue to compress the Sun to a smaller and smaller diameter. (We saw this can't happen to our Sun in the real world because of electron degeneracy, but it will happen to the supermassive star core.) Ultimately, as the Sun shrinks the escape velocity will exceed the speed of light. If the speed you need to get away is faster than the fastest possible speed in the universe, then nothing, not even light, is able to escape. An object with such large escape velocity emits no light, and anything falling into it can never return.

In modern terminology, we call an object from which light cannot escape a **black hole,** a name suggested by the American scientist John Wheeler (Figure 23.11) in 1969. The idea that such objects might exist is, however, not a new one. Cambridge professor and amateur astronomer John Michell wrote a paper in 1783 about the possibility that stars with escape velocities exceeding that of light might exist. And in 1796 the French mathemati-

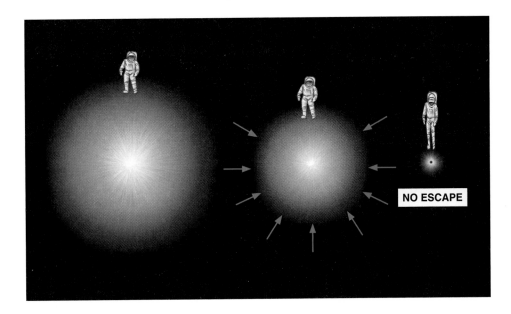

Figure 23.10
At left, an (imaginary) astronaut floats near the surface of a massive star-core about to collapse. As the same mass falls into a smaller sphere, the gravity at the surface goes up, making it harder to escape. Eventually the mass collapses into so small a sphere that the escape velocity exceeds the speed of light and nothing can get away. Note that the size of the astronaut has been exaggerated so you can see him. In the last picture the astronaut is also stretched and squeezed by the strong gravity.

cian Pierre Simon, Marquis de Laplace, made similar calculations using Newton's theory of gravity; he called the resulting objects "dark bodies."

While these early calculations provided a hint that something strange should be expected if very massive objects collapse under their own gravity, we really need general relativity theory to give an adequate description of what happens in such a situation.

Collapse with Relativity

General relativity tells us that gravity is really a curvature in spacetime. As gravity increases (as in the collapsing Sun of our thought experiment) the curvature gets larger and larger. Eventually, if the Sun could shrink down to a diameter of about 6 km, the curvature would become so great that light, trying to go in a straight line, would actually describe a circle and come back upon itself. Anything trying to escape the collapsing star would be unable to do so, because all paths "straight out" would curve back in.

Keep in mind that gravity is not pulling on the light. Gravity has curved spacetime, and light (like the trained ant of our earlier example), "doing its best" to go in a straight line, is now confronted with a world in which straight lines have become circular paths. The collapsing star is a *black hole*, in this view, because the very concept of "out" has no geometrical meaning. The star has become trapped in its own little pocket of spacetime, from which there is no escape.

The star's geometry cuts off communication with the rest of the universe at precisely the moment when,

in our earlier picture, the escape velocity becomes equal to the speed of light. The size of the star at this moment defines a surface that we call the **event horizon.** It's a wonderfully descriptive name: just as objects that sink below our horizon cannot be seen on Earth, so any event inside the event horizon can no longer affect the rest of the universe.

Imagine a future spacecraft foolish enough to sit close to the surface of a massive star that collapses in the way we have been describing. Perhaps the captain is asleep at the gravity meter, and before they can say Albert Einstein, they have collapsed with the star inside the event horizon. Frantically, they send an escape pod straight outward. But paths outward twist around to become paths inward, and the pod quickly returns. In desperation now, they send a radio message to their loved ones, bidding good-bye. But radio waves, like light, must travel through spacetime, and spacetime allows nothing to get out. Their woeful message returns to them unheard. Events inside the event horizon can *never again* affect events outside it.

Figure 23.11
John Wheeler, a brilliant physicist who did much pioneering work in general relativity theory, coined the term *black hole* in 1969. (Photo courtesy Roy Bishop)

The existence and behavior of an event horizon was first worked out by astronomer and mathematician Karl Schwarzschild (Figure 23.12). A member of the German army in World War I, he died in 1916 of an illness he contracted while doing artillery shell calculations on the Russian front. His paper on the theory of event horizons was among the last things he finished as he was dying; it was the first exact solution to Einstein's equations of general relativity. We call the radius of the event horizon the *Schwarzschild radius* in his memory.

The event horizon is the boundary of the black hole; calculations show that it does not get smaller once the star has collapsed inside it. It is the region that separates the things trapped inside it from the rest of the universe. Anything coming from the outside is also trapped once it comes inside the event horizon. The horizon's size turns out to depend only on the mass inside it. For a black hole of 1 M_{Sun}, the Schwarzschild radius is about 3 km; thus the entire black hole is about one-third the size of a neutron star of that same mass. Feed the black hole some mass, and the horizon will grow—but not very much. Doubling the mass will make a black hole 6 km in radius, still very tiny on the cosmic scale.

The event horizons of more-massive black holes have greater radii. For example, if a globular cluster of 100,000 stars could collapse to a black hole, it would be 300,000 km in radius, a little less than half the radius of the Sun. If the entire Galaxy could collapse to a black hole, it would

be only about 10^{12} km in radius—about a tenth of a LY. Smaller masses have correspondingly smaller horizons: for the Earth to become a black hole, it would have to be compressed to a radius of only 1 cm—about the size of a grape. A typical asteroid, if crushed to a small enough size to be a black hole, would have the dimensions of an atomic nucleus!

A Black Hole Myth

Much of the modern folklore about black holes is misleading. One idea you may have heard is that black holes are monsters that go about sucking things up with their gravity. Actually, it is only close to a black hole that the strange effects we have been discussing come into play. The gravitational attraction far away from a black hole is the same as that of the star that collapsed to form it.

Remember that the gravity of any star some distance away acts as if all of its mass were concentrated at a point in the center, which we call the center of gravity. For real stars, we merely *imagine* that all the mass is concentrated there; for black holes, we will see that all the mass *really is* concentrated at a point in the center. So if you are a star or distant planet, orbiting around a star that becomes a black hole, your orbit may not be significantly affected by the collapse of the star (although it may be affected by any mass loss that precedes the collapse). If, on the other hand, you venture close to the event horizon, it would be very hard for you to resist the pull of the warped spacetime near the black hole.

If another star or a spaceship were to pass one or two solar radii from a black hole, Newton's laws would be adequate to describe what would happen to it. Only very near the surface of a black hole is its gravitation so strong that Newton's laws break down. A massive star coming into our neighborhood would be far, far safer to us as a black hole than it would have been earlier in its life as a brilliant, hot star.

A Trip into a Black Hole

The fact that they cannot *see* inside black holes has not kept scientists from trying to calculate what they are like. One of the first things these calculations showed was that the formation of a black hole obliterates nearly all information about the star that collapsed to form it. Physicists like to say "black holes have no hair," meaning that nothing sticks out of a black hole to give us clues about what kind of star produced it. The only information a black hole can reveal about itself is its mass, its spin (rotation), and whether it has any electrical charge.

What happens to the collapsing star-core that made the black hole? Our best calculations predict that the material will continue to collapse under its own weight, forming an infinitely *squozen* point—a place of zero volume and infinite density—to which we give the name

(text cont. on page 472)

Figure 23.12
Karl Schwarzschild was the first to demonstrate mathematically that a black hole is possible. (Yerkes Observatory)

Gravity and Time Machines

Time machines are one of the favorite devices of science fiction. Such a device would allow you to move through time at a different pace or in a different direction from everyone else. General relativity suggests that it is possible, in theory, to construct a time machine using gravity that could take you into the future.

Let's imagine a place where gravity is terribly strong, such as a black hole. General relativity predicts that the stronger the gravity, the slower the pace of time. So imagine a future astronaut, with a fast spaceship, who volunteers to go on a mission to such a high-gravity environment. The astronaut leaves in the year 2200, just after graduating from college at age 22. She takes (let's say) exactly ten years to get to the black hole. Once there, she orbits some distance from it (taking care not to get pulled in).

She is now in a high-gravity realm where time passes much more slowly than it does on the Earth. This isn't just an effect on the mechanism of her clocks—*time itself* is running slowly. That means that every way she has of measuring time will give the same slowed-down reading when compared to time passing on Earth. Her heart will beat more slowly, her hair will grow more slowly, her antique wristwatch will tick more slowly, etc. She is not aware of this slowing down, since all her readings of time, whether made by her own bodily functions or with mechanical equipment, are measuring the same—slower—time. Meanwhile, back on Earth, time passes as it always does.

Our astronaut now emerges from the region of the black hole, her mission of exploration finished, and returns to Earth. Before leaving she carefully notes that (according to her timepieces) she spent about two weeks around the black hole. She then takes exactly ten years to return to Earth. Her calculations tell her that since she was 22 when she left the Earth, she will be 42 plus two weeks when she returns. So the year on Earth, she figures, should be 2242, and her classmates should now be approaching their midlife crises.

But our astronaut should have paid more attention in her astronomy class! Because time slowed down near the black hole, much less time passed for her than for the people on Earth. While her clocks measured two weeks spent near the black hole, more than 2000 weeks (depending on how close she got) could well have passed on Earth. That's equal to 40 years, meaning her classmates will be senior citizens in their 80s when she (a mere 42-year-old) returns. On Earth it will be not 2242, but 2282—and she will say that she has arrived in the future.

Is this scenario real? Well, it has practical flaws—we don't think there are any black holes close enough for us to reach in ten years, and we don't think any spaceship or human can survive around a black hole. But the key point, about the slowing down of time, is a natural consequence of Einstein's general theory of relativity, and we saw that its predictions have been confirmed by experiment after experiment.

Science-fiction writers who pay attention to developments in astronomy and physics have seized upon this idea for wonderful plot devices. In Larry Niven's novel *World Out of Time* (1976, Ballantine Books), the protagonist travels around an extremely massive black hole and arrives back on Earth some 3 million years in the future. And in Fred Pohl's *Gateway* (1977, Ballantine Books), the "hero" has to push his girlfriend into a black hole so he can escape its strong gravity. Not only can they never speak or touch again, he must then deal with the additional guilt of knowing that, now that he is far from the black hole, his time is passing much more quickly than hers. To him she seems almost suspended in time, and (if he had a good enough telescope) he could see her there for the rest of his life, looking back at him in disbelief that he could have done so horrible a thing. (Indeed, the "hero" spends much of the book talking to a robot psychiatrist named Albert about how guilty he feels.)

Gateway by F. Pohl
(Cover copyright Ballantine Books)

singularity. At the singularity, spacetime ceases to exist. The laws of physics as we know them break down. We do not yet have the physical understanding or the mathematical tools to describe the singularity itself. From the outside, however, the entire structure of a basic black hole (one that is not rotating) can be described as a singularity surrounded by an event horizon. In comparison with humans, black holes are really very simple objects!

Scientists have also calculated what would happen if a daring (perhaps we should say suicidal) astronaut were to fall into a black hole. Let's take up an observing position a long, safe distance away from the event horizon and watch this astronaut fall toward it. At first he falls away from us, moving ever faster, just as though he were approaching any massive star. However, as he nears the event horizon of the black hole, things change. As we saw, the strong gravitational field around the black hole will make his clocks run more slowly, when seen from our outside perspective.

If, as he approaches the event horizon, he sends out a signal once per second according to his clock, we will see the spacing between signals grow longer and longer until it becomes infinitely long when he reaches the event horizon. (If the infalling astronaut uses a blue light to send his signals every second, we will also see the light get redder and redder until its wavelength is nearly infinite.) As the spacing between clock ticks approaches infinity, it will appear to us that the astronaut is slowly coming to a stop, frozen in time at the event horizon.

In the same way, all matter falling into a black hole will also appear to an outside observer to stop at the event horizon, frozen in place and taking an infinite time to fall through it. For this reason, black holes are sometimes called *frozen stars.* But don't think that matter falling into a black hole will therefore be easily visible at the event horizon. The tremendous redshift will make it very difficult to observe any radiation from the "frozen" victims of the black hole.

This, however, is only how we, located far away from the black hole, see things. To the astronaut, his time goes at its normal rate and he falls right on through the event horizon into the black hole. (Remember, this horizon is not a physical barrier, but only a region in space where the curvature of spacetime makes escape impossible.)

You may have trouble with the idea that you and the astronaut have such very different ideas about what has happened. This is the reason Einstein's ideas about space and time are called theories of *relativity;* what each observer measures about the world depends on (is relative to) his frame of reference. The observer in strong gravity measures time and space differently from the one sitting in weaker gravity. When Einstein proposed these ideas, many scientists also had difficulty with the idea that two such very different views of the same event could be correct, each in its own "world," and tried to find a mistake in the calculations. There were no mistakes: we and the astronaut really would see the same fall into a black hole very differently.

For the astronaut, there is no turning back. Once inside the event horizon, the astronaut, any signals from his radio transmitter, and any cries of regret are doomed to remain hidden forever from the universe outside. He will, however, not have a long time (from his perspective) to feel sorry for himself as he approaches the black hole. Suppose he is falling feet first. The force of gravity exerted by the singularity on his feet will be slightly greater than on his head, and he will be stretched slightly. Since the singularity is a point, the left side of his body will be pulled slightly toward the right, and the right side slightly toward the left, bringing each side closer to the singularity. The astronaut will therefore be slightly squeezed in one direction and stretched in the other.

The Earth exerts similar *tidal forces* on an astronaut performing a spacewalk. In the case of the Earth, the tidal forces are so small that they pose no threat to the health and safety of the astronaut. Not so in the case of a black hole. Sooner or later, as the astronaut approaches the black hole, the tidal forces will become so great that the astronaut will be ripped apart. His legs will be ripped from his torso, his ankles from his legs, his toes from his feet, and so forth. Soon it will only be the individual atoms from his shredded body that will continue their inexorable fall into the singularity. (Jumping into a black hole is definitely a once-in-a-lifetime experience!)

One Way to Make a Black Hole

Theory tells us what a black hole must be like. But can black holes really exist? For example, we know of nothing in nature that could crush the Earth or the Sun into a small enough sphere to become a black hole. But, as we have seen, the collapse of a massive star-core may be exactly the kind of situation that could produce a black hole.

The critical question, and one that theory has not yet been able to answer, is this: Are the cores of some stars actually so massive that they exceed the upper limit on the mass of a neutron star (about three times the mass of the Sun)? If so, then nothing that we are presently aware of can halt their collapse, and they will form black holes.

If theory cannot yet tell us what conditions produce a black hole rather than a neutron star, then do observations provide any evidence that black holes actually do exist? At first, you might think it pretty hopeless to look for something many light years away that is on the order of 10 km across and completely black! But it *is* possible to detect the presence of a black hole by the unspeakable things it does to a companion star.

Evidence for Black Holes

As stars collapse into black holes, they leave behind their gravitational influence. We've already seen that a companion star some distance away may be unaffected by the collapse (although any accompanying supernova explosion may take its toll on the companion). If a member of a double-star system becomes a black hole, then we may be able to detect it by studying the motion and other properties of its companion.

Requirements for a Black Hole

So here is a prescription for finding a black hole: First look for a star whose motion (determined from the Doppler shift of its spectral lines) shows it to be a member of a binary star system. If both stars are visible, neither can be a black hole, so focus your attention just on those systems where only one star of the pair is visible even with our most sensitive telescopes.

Being invisible is not enough, however, because a relatively faint star might be unseen next to the glare of a brilliant companion, or in a shroud of dust. And even if the star really is invisible, it could be a neutron star, which does not emit any light. Therefore, we must also have evidence that the unseen star has a mass too high to be a neutron star, and that it is a collapsed object—one of extremely small size.

We can use Kepler's law (see Chapter 2) and our knowledge of the visible star to measure the mass of the invisible member of the pair. If the mass is greater than about 3 M_{Sun}, then we are likely seeing (or, more precisely, not seeing) a black hole—as long as we can make sure the object really is a collapsed star.

If matter falls toward a compact object of high gravity, the material is accelerated to high speed. Near the event horizon of a black hole, matter is moving at velocities that approach the speed of light. As the atoms rush chaotically toward the event horizon, they rub against each other; internal friction can heat them to temperatures of 100 million K or more. Such hot matter emits radiation in the form of flickering x rays. The last part of our prescription, then, is to look for a source of x rays associated with the binary system. Since x rays do not penetrate the Earth's atmosphere, such sources must be found using x-ray telescopes in space.

The infalling gas that produces the x-ray emission comes from the black hole's companion star. As we saw in Chapter 22, stars in close binary systems can exchange mass, especially as one of the members expands into a red giant. Suppose that one star in a double-star system has evolved to a black hole, and that the second star begins to expand in size. If the two stars are not too far apart, the outer layers of the expanding star may reach the point where the black hole exerts more gravitational force on them than do the inner layers of the red giant to which the atmosphere belongs. The outer atmosphere has then passed through the point of no return between the stars, and falls toward the black hole.

The mutual revolution of the giant star and the black hole causes the material falling toward the black hole to spiral around it rather than flowing directly into it. The infalling gas whirls around the black hole in a pancake of matter called an *accretion disk.* It is in the inner part of this disk that matter is revolving about the black hole so fast that internal friction heats it up to x-ray-emitting temperatures.

Another way to form an accretion disk in a binary star system is to have a powerful stellar wind come from the black hole's companion. Such winds are a characteristic of several stages in a star's life. Some of the ejected gas in the wind will then flow close enough to the black hole to be captured by it into the disk (Figure 23.13).

We should point out that, as often happens, these measurements are not quite as simple as they are described in introductory textbooks. In real life, Kepler's law only allows us to calculate the combined mass of the two stars in the binary system. We must learn more about the visible star of the pair and its history to disentangle its mass from the unseen companion. Also, the calculations are affected by how the orbit of the two stars is tilted toward the Earth, something we rarely know how to measure. And neutron stars can also have accretion disks that produce x rays, so astronomers must study the properties of these x rays carefully when trying to determine what kind of object is at the center of the disk. Nevertheless, the evidence for black holes in such systems is quite strong.

Candidate Objects

Because the x rays are such an important clue that we are dealing with a black hole that is having some material from its companion for lunch, the search for black holes had to await the launch of sophisticated x-ray telescopes into space. These instruments must have the resolution to distinguish the location of x-ray sources and allow us to match them to the positions of binary star systems. Such a case, we think, is the binary system that was our first black hole candidate—Cygnus X-1.

The visible star in Cygnus X-1 is spectral type B. Measurements of the Doppler shifts of the B star's spectral lines show that it has an unseen companion. The x rays flickering from it strongly suggest that it is a small collapsed object—but is it a neutron star or a black hole? (Remember, neutron stars are likely to be much more common, since they come from lower-mass stars, which are more numerous.) To find out, astronomers have tried to determine the mass of the invisible member of the pair: it appears to be nearly ten times that of the Sun. The companion is therefore too massive to be either a white dwarf or a neutron star.

There are several other binary systems that meet all the conditions to be prime black hole candidates. Table 23.1 lists the characteristics of some of our best cases.

Feeding a Black Hole

We saw that when an isolated star, even if it is in a binary star system, becomes a black hole, there is usually a limit to how much it can grow. After all, material must ap-

(text cont. on page 475)

Figure 23.13
Mass lost from a giant star through a stellar wind streams toward a black hole and swirls around it before finally falling in. In the inner portions of the accretion disk, the matter is revolving so fast that internal friction heats it to very high temperatures and x rays are emitted.

TABLE 23.1
Some Black Hole Candidates in Binary Star Systems

Name or Catalog Designation	Companion Star Spectral Type	Orbital Period	Black Hole Mass Estimates
		(days)	(M_{Sun})
Cygnus X-1	B supergiant	5.6	6–15
LMC X-3	B main sequence	1.7	4–11
A0620-00 (V616 Mon)*	K main sequence	7.8	4–9
GS2023+338 (V404 Cyg)	K main sequence	6.5	more than 6
GS2000+25 (QZ Vul)	K main sequence	0.35	5–14
GS1124-683 (Nova Mus 1991)	K main sequence	0.43	4–6
GRO J1655-40 (Nova Sco 1994)	F main sequence	2.4	4–5
H1705-250 (Nova Oph 1977)	K main sequence	0.52	more than 4

* As you can tell, there is no standard way of naming these candidates. The chain of numbers are the location of the source in right ascension and declination (the longitude and latitude system on the sky); the letter(s) preceding them designate the satellite that discovered the candidate—A for Ariel, G for Ginga, etc. The notations in parentheses are those used by astronomers who study binary star systems or novae.

proach close to the event horizon before the gravity is any different from that of the star that became the black hole. Out in the suburban regions of the Milky Way Galaxy where we live (see Chapter 24), stars and star systems are much too far apart for other stars to provide "food" to a hungry black hole.

But, as we will see, the central regions of galaxies are quite different from the outer parts. Here stars and raw material can be quite crowded together, and they can interact much more frequently with each other. Therefore, black holes in the centers of galaxies may have a much better opportunity to find mass close enough to their event horizons to pull it in. Black holes are not particular about what they "eat": they are happy to consume other stars, asteroids, gas, dust, and even other black holes. (If two black holes merge, you just get a black hole with more mass and a larger event horizon.)

As a result, black holes in crowded regions can grow, eventually swallowing thousands or even millions of times the mass of the Sun. Observations with the Hubble Space Telescope have recently shown dramatic evidence for the existence of a black hole containing about three billion M_{Sun} in a galaxy called M87 (see opening figure for this chapter). The feeding frenzy of such *supermassive black holes* may be responsible for some of the most energetic phenomena in the universe (see Chapters 24 and 26).

More and more evidence is accumulating that some stars end their lives as black holes. Furthermore, the observational tests of Einstein's general theory of relativity have convinced even the most skeptical scientists that his picture of warped or curved spacetime is indeed our best description of the effects of gravity. The ideas discussed in this chapter can seem strange and overwhelming, especially the first time you read them. But you have to admit that they make the universe even more interesting and bizarre than you might have thought before you took this class.

Summary

23.1 The **equivalence principle** is the foundation of **general relativity.** According to this principle, there is no way that anyone or any experiment in a sealed environment can distinguish between free-fall and the absence of gravity.

23.2 By considering the consequences of this principle, Einstein concluded that we live in a curved **spacetime.** The distribution of matter determines the curvature of spacetime; other objects (and even light) entering a region of spacetime must follow its curvature.

23.3 At low speeds and in weak gravitational fields, the predictions of general relativity agree with the predictions of Newton's theory of gravity. However, general relativity predicts, while Newtonian theory does not, that starlight (or radio waves) will be deflected when they pass near the Sun, and that the position where Mercury is at perihelion should change by 43 arcsec per century even if there are no other planets in the solar system to perturb its orbit. These predictions have been verified by experiment.

23.4 General relativity predicts that the stronger the gravity, the more slowly time must run. Experiments on Earth and with spacecraft have confirmed this prediction with remarkable accuracy.

23.5 Theory suggests that stars with stellar cores more massive than three times the mass of the Sun at the time they exhaust their nuclear fuel will collapse to become **black holes.** The surface surrounding a black hole, where the escape velocity equals the speed of light, is called the **event horizon,** and the radius of the surface is called the Schwarzschild radius. Nothing, not even light, can escape through the event horizon from the black hole. At its center each black hole has a **singularity,** a point of infinite density and zero volume. Matter falling into a black hole appears, as viewed by an outside observer, to freeze in position at the event horizon. However, if we were riding on the infalling matter, we would pass through the event horizon. In the process, tidal forces exerted by the singularity would tear our bodies apart.

23.6 Black holes that collapse from stars can only be detected in binary star systems having (a) one star of the pair that is invisible; (b) x-ray emission characteristic of an accretion disk around a compact object; and (c) an indication from the orbit and characteristics of the visible star that the mass of its invisible companion is greater than 3 M_{Sun}. Several systems with these characteristics have been found. More-massive black holes are also possible.

1. How does the principle of equivalence lead us to suspect that spacetime might be curved?

2. If general relativity offers the best description of what happens in the presence of gravity, why do physicists still make use of Newton's equations in describing gravitational forces on Earth (when building a bridge, for example)?

3. Einstein's general theory of relativity made predictions about the outcome of several experiments that had not yet been carried out at the time the theory was first published. Describe three experiments that verified the predictions of the theory.

4. If a black hole emits no radiation, why do astronomers and physicists today believe that the theory of black holes is correct?

5. What characteristics must a binary star have to be a good candidate for a black hole? Why is each of these characteristics important?

6. Suppose the Earth were to fall into a black hole. What would happen to it?

7. A student becomes so excited by the whole idea of black holes, he decides to jump into one.
   a. What is the trip like for him?
   b. What is it like for the rest of the class, watching from afar?

Thought Questions

8. Imagine that you have built a large room around the boy and girl in Figure 23.2, and that this room is falling at exactly the same rate as they are. Galileo showed that, if there is no air friction, light and heavy objects that are dropping due to gravity will fall at the same rate. Suppose this were not true, and that instead heavy objects fall faster; also suppose that the boy in Figure 23.2 is twice as massive as the girl. What would happen? Would this violate the equivalence principle?

9. A monkey hanging from a tree branch sees a hunter aiming a rifle directly at him. The monkey then sees a flash, telling him that the rifle has been fired. Reacting quickly, he lets go of the branch and drops so the bullet can pass harmlessly over his head. Does this act save the monkey's life? Why?

10. During the 1970s, the United States had in orbit a small space station called Skylab. Some of the astronauts exercised by running around the inside wall of their cylindrical vehicle.

How could they stay against the wall while running, rather than float aimlessly inside the Skylab? What physical principles were involved?

11. Why would we not expect to detect x rays from a disk of matter about an ordinary star?

12. Look elsewhere in this book for the necessary data, and indicate what the final stage of evolution—white dwarf, neutron star, or black hole—will be for each of the following kinds of stars.
   a. Spectral type-O main-sequence star
   b. B main-sequence star
   c. A main-sequence star
   d. G main-sequence star
   e. M main-sequence star

13. Which is likely to be more common in our Galaxy—white dwarfs or black holes? Why?

Problems

14. The formula for the radius of the event horizon is $R = 2GM/c^2$, where c is the speed of light. Show that this equation is equivalent to saying that the circumference of the event horizon is equal to 18.5 km times the mass of the black hole in units of the mass of the Sun.

15. What would be the circumference of a black hole with the mass of Jupiter?

16. Suppose that the Earth collapsed to the size of a golf ball.

   a. What would be the revolution period of the Moon around it, at a distance of 400,000 km?
   b. Of a spacecraft orbiting at a distance of 6000 km?
   c. Of a miniature spacecraft orbiting at a distance of 0.1 m?
   d. Calculate the orbital speed of this mini-spaceship, and compare it with the speed of light.

17. If the Sun could suddenly collapse to a black hole, how would the period of the Earth's revolution about it differ from what it is now?

Suggestions for Further Reading

Chaisson, E. *Relatively Speaking.* 1988, W. W. Norton. An astronomer introduces relativity and its astronomical applications.

Greenstein, G. *Frozen Star.* 1984, Freundlich Books. Wonderful, lucid introduction to the death of stars, with a good section on black holes.

Kaufmann, W. *Cosmic Frontiers of General Relativity.* 1977, Little, Brown. Good nontechnical primer on general relativity theory and its applications in astronomical contexts.

McClintock, J. "Do Black Holes Exist?" in *Sky & Telescope,* Jan. 1988, p. 28.

Overbye, D. "God's Turnstile: The Work of John Wheeler and Stephen Hawking" in *Mercury,* Jul./Aug. 1991, p. 98.

Parker, B. "Where Have All the Black Holes Gone?" in *Astronomy,* Oct. 1994, p. 36. On candidates within our Galaxy.

Thorne, K. *Black Holes and Time Warps.* 1994, W. W. Norton. The definitive introduction to black holes.

Wheeler, J. *A Journey into Gravity and Spacetime.* 1990, Scientific American Library/W. H. Freeman. A brilliant introduction by one of the foremost scientists of our time.

Will, C. *Was Einstein Right?—Putting General Relativity to the Test.* 1986, Basic Books. Superb guide to the many tests.

Zirker, J. "Testing Einstein's General Relativity During Eclipses" in *Mercury,* Jul./Aug. 1985, p. 98.

Using REDSHIFT ™

None of the companions to the black hole candidates in Table 23.1 are visible in *RedShift*.

1. If stars with initial masses greater than 40 solar masses may eventually become black holes, use Table 21.1 and *RedShift* to identify which main-sequence stars might become black holes.

2. If *RedShift* can display up to 250,000 stars, estimate the percentage of stars that end their lives as black holes.

Infrared radiation can penetrate the dust that lies between us and the center of the Milky Way. This infrared image shows the central 150 LY of our Galaxy. The bright region in the middle is the nucleus of the Galaxy, which harbors large amounts of gas, dense groups of stars, and perhaps a massive black hole. The image is a color-coded composite of views of the same scene at three infrared wavelengths: blue is 1.2 μm, green is 1.65 μm, and red is 2.2μm. (Ian Gatley, Michael Merrill, and Richard Joyce/National Optical Astronomy Observatories)

CHAPTER 24

The Milky Way Galaxy

Thinking Ahead

In the midst of a dark forest, a campfire illuminates the night. The light of the fire shows us the nearby trees and orients us to our immediate surroundings. But beyond the trees that catch the firelight lies a much larger darkness, where other trees and much else we don't know about can be found. In the same way, is it possible that as we examine our Galaxy of stars by the familiar light they cast, we are missing a much larger region whose nature we don't fully understand?

Having completed our survey of the life stories of the stars, we now turn our attention to the way they are grouped in the universe. We find stars arranged in vast islands scattered in the dark oceans of space—islands we call *galaxies*. Let us begin our examination of these great star systems by studying the one in which we find ourselves—the Milky Way Galaxy.

One of the most striking features in a truly dark sky is a band of faint white light that stretches from one horizon to the other (Figure 24.1). Because of its appearance, this band of light is called the Milky Way. In the Northern Hemisphere the band is brightest in the region of the constellation Cygnus, and is best viewed in the summer. In the Southern Hemisphere the Milky Way is even brighter—so bright, in fact, that native people in South America gave names to various portions of it just as northern astronomers gave constellation names to conspicuous groups of stars. Unfortunately, the faint light of the Milky Way is swamped by the artificial lighting in our urban areas (see "Making

Across the height of heaven there runs a road Clear when the night is bare, the Milky Way, Famed for its sheen of white . . .

Ovid, *Metamorphoses* (lines 169–175), quoted by E. C. Krupp in *Beyond the Blue Horizon* (1991)

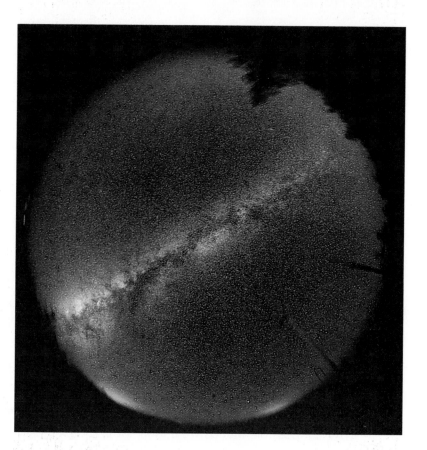

Figure 24.1
This full-sky view of the Milky Way, made with a special lens that can show a 360° panorama, was taken from the summit of Mount Graham, a 3200-m-high mountain in southeastern Arizona. Note the dark rift through the center of the Milky Way. Around the horizon in a counterclockwise direction, beginning from the bottom, you can see the lights of Willcox, Tucson, and Phoenix, as well as the silhouettes of test towers and fir trees. The brightest point-like object in the lower left of the picture is the planet Jupiter. (Roger Angel, Steward Observatory/University of Arizona)

Connections" box), so many city dwellers (perhaps you among them) have never seen the Milky Way.

Although the Milky Way was more easily visible to them on a cloudless night, the ancients did not really have a good idea what it was. The name comes from legends that compare its faint white splash of light to a stream of spilled milk. But folktales differ from culture to culture: one East African tribe thought of it as the smoke of ancient campfires; several Native American stories tell of a path across the sky traversed by sacred animals; and in Siberia it was known as the seam in the tent of the sky.

In 1610 Galileo made the first telescopic survey of the Milky Way and discovered that it is composed of a multitude of individual stars. From far away, the unaided eye sees their combined light as a splash of white. But why is the splash in the form of a strip or arc across the sky? All the stars we see with the unaided eye (and not just the arc of the Milky Way) are part of an enormous wheel or frisbee-shaped system of stars. The stars close to us lie all around us in every direction. But as we look at the more distant parts of the system, the sideways frisbee shape becomes more apparent. The frisbee is not oriented in the same direction as the plane of our solar system or the celestial equator, so we see the arc of the Milky Way flung diagonally across the sky.

We call this great stellar system, which we now know includes hundreds of billions of stars and much else, the **Milky Way Galaxy** or, more simply, just the **Galaxy.** The process of understanding the extent of the Galaxy, and our place in it, took many centuries.

24.1
The Architecture of the Galaxy

In 1785 William Herschel showed that the stellar system to which the Sun belongs has the shape of a wheel or disk (Figure 24.2). He used a telescope to count the numbers of stars that he saw in various directions. To understand how he did this, imagine that you are a member of the band standing in formation during halftime at a football game. If you count the band members that you see in different directions and get about the same number each time, you can conclude that the band has arranged itself in a circular pattern with you at the center. Since you see no band members above you or underground, you know that the circle made by the band is much flatter than it is wide.

Herschel saw much the same thing when he counted stars in different directions in the Galaxy. He found that most of the stars lay in a narrow band, the Milky Way, cir-

Figure 24.2
William Herschel (1738–1822), a German musician who emigrated to England and took up astronomy in his spare time, was the discoverer of infrared radiation and the planet Uranus. He built several large telescopes and made measurements of the Sun's place in the Galaxy, the Sun's motion through space, and the comparative brightnesses of stars. (NASA)

ter of the solar system, but at least our Sun appeared to be in the center of the distribution of stars.

We now know that Herschel was right about the shape of our system, but wrong about where the Sun lies within that disk. As we saw in Chapter 19, we live in a dusty Galaxy. Because interstellar dust absorbs the light from stars, Herschel could see only those stars within about 6000 LY of the Sun. This is a very small section of the much larger disk of stars that makes up the Galaxy: the full disk is actually about 100,000 LY across.

In the same way, if we stand on the streets of a large urban area on a smoggy day, we can see only a small section of the city; the dust and pollution in the air block the light from more distant neighborhoods. Herschel, confined to using visible light for his observations, had no way of penetrating the dust between the stars and discerning that the Sun is actually located fairly far from the center of the Galaxy.

Globular Clusters and the Center of the Galaxy

Until early in this century, astronomers generally accepted Herschel's conclusions. The discovery of the Galaxy's true size and our actual location came about largely through the efforts of Harlow Shapley (see "Voyagers in Astronomy" box). To understand what he did, let us return to the analogy of a smoggy day in a big city. Suppose that, despite your inability to see through the smog, you nevertheless need to get some sense of the extent of the city and where you are currently located.

Being a keen observer, you notice that the smog is actually confined to a relatively thin layer right at ground level (this is not true of real smog, but humor us for a minute). If you look upwards instead of through the smog, your view is somewhat clearer. As it happens, the local baseball team has just won the World Series, and all over

cling the sky, and that the numbers of stars were about the same in any direction around this band. He therefore concluded that the Sun must be near the center of a disk of stars (Figure 24.3). There was some philosophical comfort in this view: the Earth may not have been in the cen-

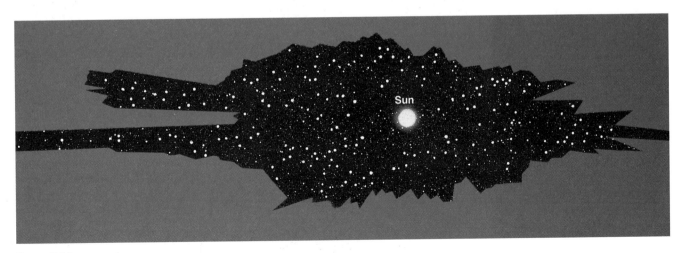

Figure 24.3
A copy of Herschel's diagram, showing a cross section of the Milky Way system based on counting stars in various directions. The large circle shows the location of the Sun.

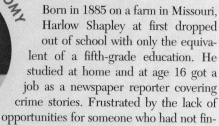

Harlow Shapley: Mapmaker to the Stars

Born in 1885 on a farm in Missouri, Harlow Shapley at first dropped out of school with only the equivalent of a fifth-grade education. He studied at home and at age 16 got a job as a newspaper reporter covering crime stories. Frustrated by the lack of opportunities for someone who had not finished high school, Shapley went back and completed a six-year high school program in only two years, graduating as class valedictorian.

In 1907, at age 22, he went to the University of Missouri, intent on studying journalism, but found that the School of Journalism would not open for a year. Leafing through the college catalog (or so he told the story later), he chanced to see "Astronomy" among the subjects beginning with "A." Recalling his boyhood interest in the stars, he decided to study astronomy for the next year (and the rest, as the saying goes, was history).

Upon graduation Shapley received a fellowship for graduate study at Princeton and began to work with the brilliant Henry Norris Russell (see "Voyagers in Astronomy" box in Chapter 17). For his PhD thesis, he made major contributions to the methods of analyzing the behavior of eclipsing binary stars. He was also able to show that cepheid variable stars are not binary systems, as some people thought at the time, but individual stars that pulsate with striking regularity.

Impressed with Shapley's work, George Ellery Hale offered him a position at the Mount Wilson Observatory, where the young man took advantage of the clear mountain air

Harlow Shapley (1885–1972)
(Harvard College Observatory Archives)

and the 60-in. reflector to do his pioneering study of variable stars in globular clusters.

Shapley subsequently accepted the directorship of the Harvard College Observatory, and over the next 30 years he and his collaborators made contributions to many fields of astronomy, including the study of neighboring galaxies, the discovery of dwarf galaxies, a survey of the distribution of galaxies in the universe, and much more. He wrote a series of nontechnical books and articles and became known as one of the most effective popularizers of astronomy. Shapley enjoyed giving lectures around the country, including at many smaller colleges where students and faculty rarely got to interact with scientists of his caliber.

During World War II Shapley helped rescue many scientists and their families from Eastern Europe; later he helped found UNESCO, the United Nations Educational, Scientific, and Cultural Organization. He wrote a little pamphlet called *Science from Shipboard* for men and women in the armed services who had to spend many weeks onboard transport ships to Europe. And during the difficult period of the 1950s when congressional committees began their "witch hunts" for communist sympathizers (including such liberal leaders as Shapley), he spoke out forcefully and fearlessly in defense of the freedom of thought. A man of many interests, he was fascinated by the behavior of ants, and wrote scientific papers about them as well as about galaxies.

By the time he died in 1972, Shapley was acknowledged as one of the pivotal figures of modern astronomy, a "20th century Copernicus" who had mapped the Milky Way and showed us our place in the Galaxy.

---

the city civic leaders have released giant balloons bearing the team name as part of the celebration. Because these balloons rise above the smog layer, they can be spotted much farther away. If the balloons have been released democratically all around the city, you can make a catalog of their locations and get a pretty good sense of the city's size and your location.

What Shapley realized is that the Galaxy has provided us with just such signposts above (and below) its dusty disk. They are the globular clusters (see Chapter 21)—collections of some 100,000 or more stars that can be seen (with telescopes) at much larger distances than stars within the Galaxy's disk.

Most globular clusters contain at least a few RR Lyrae variable stars that—along with the Cepheids—are excellent distance indicators (see Section 18.3). By comparing

the known intrinsic luminosity of these stars to how bright they appeared, he could derive their distance. (Recall that it is distance that makes the stars look dimmer than they would be "up close," and that we know the brightness fades as the distance squared.) Knowing the distance to any star in a cluster then tells us the distance to the cluster itself.

In 1917 Shapley used the distances and directions of 93 globular clusters to map out their positions in space (Figure 24.4). He found that the clusters are distributed in a spherical-shaped volume, which has its center not at the Sun but at a distant point along the Milky Way in the direction of Sagittarius. Shapley then made the bold assumption that the point on which the system of globular clusters is centered is also the center of the entire Galaxy. Today this assumption has been verified by many pieces of

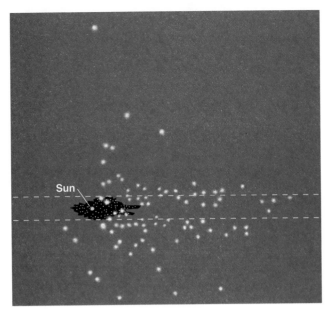

Figure 24.4
A copy of a diagram by Shapley, showing a map of the distribution of globular clusters and the positions of the Sun and the center of the Galaxy. The black area shows Herschel's old diagram (Figure 24.3), centered on the Sun, approximately to scale.

evidence, including the discovery that the Galaxy rotates around that same point, which is about 26,000 LY from the Sun.

Shapley's work showed once and for all that our star has no special place in the Galaxy. We are in a nondescript region of the Milky Way, one of at least 100 billion stars that circle the distant center of our Galaxy. Whatever importance humanity finds in the vastness of the cosmos is much more likely to come from our own actions than the location we have been assigned by nature.

Overview of the Galaxy

With modern instruments, astronomers can now penetrate the "smog" of the Milky Way. Just as you can listen to your favorite radio station whether or not it is smoggy in the city, astronomers can "tune in" on radio and infrared radiation from distant parts of the Galaxy. Measurements at these wavelengths (as well as observations of other galaxies like ours) have fleshed out Shapley's first picture of the Milky Way. We now know it is a thin, circular, rotating disk of luminous matter (Figure 24.5) distributed across a region about 100,000 LY in diameter, and about 1000 LY thick. (Given how thin the disk is, perhaps a French crepe is a more appropriate analogy than a frisbee.)

Bright stars, as well as the dust and gas from which stars form, are closely confined to the galactic disk. This disk, however, is embedded in a spherical **halo** of faint stars that extends to a distance of at least 50,000 LY from the galactic center. The globular clusters also trace out a large spherical halo around the galactic center.

It is not an easy task to pin down the characteristics of this halo. When we see a faint star above or below the plane of the Galaxy with a powerful telescope, it is rarely possible to determine its distance. Thus we do not know if it is an intrinsically faint star that is nearby, or a brighter star that is far away. Luckily, there are variable stars that are distance indicators in the halo: individual RR Lyrae stars have been found in significant numbers as far away as 50,000 LY above and below the galactic plane. This shows that the halo has an overall diameter of at least 100,000 LY.

Close in to the galactic center (within about 3000 LY), the stars are no longer confined to the disk, but form a **nuclear bulge** consisting of old stars. When we use ordinary visible radiation, we can only glimpse the stars in the bulge in directions where there happens to be rela-

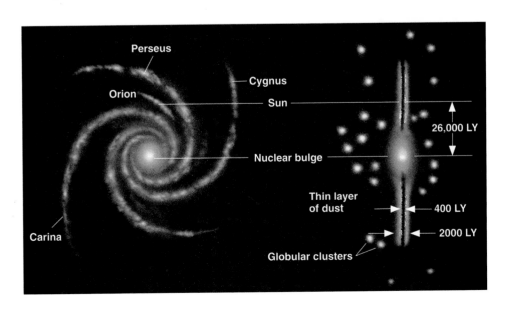

Figure 24.5
A schematic representation of the Galaxy. The Sun is located on the inside edge of the short Orion spur.

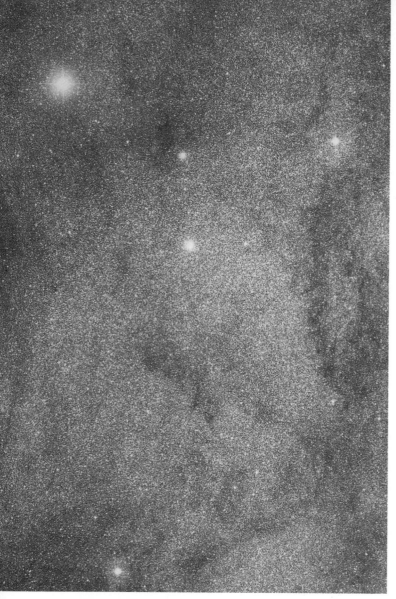

Figure 24.6
Although the nucleus of our Galaxy is hidden by dust at visible wavelengths, a few places where the obscuration is low allow us to see into the cloud of old stars concentrated toward the center and forming the nuclear bulge. This is a picture of such a region, called Baade's window after the astronomer who first pointed it out. The density of stars in this photograph peaks at a distance of about 26,000 LY. The brightest star in the picture is Gamma Sagittarii, a deep yellow star in the foreground. At a distance of about 100 LY, it can be seen with the unaided eye. (© Roe/AAT Board 1987)

Figure 24.7
This 1990 image presents a view of the inner part of the Milky Way Galaxy obtained by an experiment aboard the Cosmic Background Explorer Satellite (COBE). Taken in the near-infrared part of the spectrum, the image permits us to see, for the first time, the bulge of old stars that surround the center of our Galaxy. Redder colors correspond to regions with more dust, and this dust forms a thin disk of material that shows the plane of the Milky Way. The same dust, building up into a kind of interstellar smog, prevents us from seeing the inner parts of our Galaxy with visible light. (NASA).

tively little interstellar dust (Figure 24.6). The first picture that actually succeeded in showing the bulge as a whole was taken at infrared wavelengths (Figure 24.7).

The fact that much of the bulge is obscured by dust makes its shape difficult to determine. For a long time astronomers assumed it was spherical. However, infrared images and other data suggest that the bulge may be elongated—shaped like a very fat cigar. We know many other galaxies that also have bar-shaped concentrations of stars in their central regions; for that reason they are called barred spirals.

The stars and raw material in the disk of the Galaxy are not distributed smoothly. As the diagram in Figure 24.5 shows, the young stars in the disk are concentrated in a series of spiral arms. We will explore these features in more detail shortly. For now we want to note that establishing this overall picture of the Galaxy from our suburban, dust-shrouded vantage point has been one of the great achievements of 20th-century astronomy (and one that took decades of effort by astronomers working with a wide range of telescopes).

One thing that helped enormously was the discovery that our Galaxy is not unique in its characteristics. There are many other such spiral-shaped aggregations of stars in the universe. For example, the Milky Way resembles the Andromeda galaxy, which at a distance of about 2.3 million LY is one of our nearest neighbors (see Figure 24.11 later in the chapter). Just as you can get a much better picture of yourself if someone else takes the photo from some distance away, pictures (and other data) from nearby galaxies that resemble ours have been vital for understanding the properties of the Milky Way.

Interstellar Matter in the Galaxy

A major step in understanding the detailed structure of the Galaxy came with the discovery of 21-cm radio waves from cold hydrogen (see Section 19.2). Hydrogen is abundant throughout the Galaxy's disk, and radiation at this wavelength easily comes through dust. Furthermore, as we will see, different parts of the Galaxy move around the center at different rates. This means that various hydro-

Figure 24.9
The spiral galaxy NGC 2997 shines with the light of 100 billion or more stars. The two spiral arms, which appear to originate in the yellow nucleus, are peppered with bright red blobs of ionized hydrogen that are similar to regions of star formation in our own Milky Way Galaxy. The hot blue stars that generate most of the light in the arms of the galaxy form within these hydrogen clouds. (© 1980 Anglo-Australian Telescope Board)

gen clouds have different Doppler shifts as detected from Earth. Astronomers can use 21-cm radiation to sort out the different structures within the Galaxy, and also to get a much better sense of the full extent of its disk.

Radio observations of the 21-cm line show that the Galaxy's cold atomic hydrogen is confined to an extremely flat layer that is only about 400 LY thick (Figure 24.8). In the plane of the Galaxy, this hydrogen extends well beyond the Sun to a distance of about 80,000 LY from the galactic center. As discussed in Chapter 19, the Milky Way has plenty of raw material for making new stars in its disk, and it is in the disk that we see young, recently formed stars.

Dust, too, is confined to the disk, with the highest concentrations occurring in the spiral arms. The thickness of the dust layer is also about 400 LY. But it seems more concentrated toward the inner parts of the disk; very little dust is found outside the Sun's orbit around the galactic center.

The most-massive molecular clouds (discussed in Chapters 19 and 20) are found in the spiral arms of the Galaxy. In many cases, individual clouds have gathered into large complexes containing a dozen or more discrete clumps. Since the large molecular clouds and cloud complexes are the sites of star formation, most young stars are also found in spiral arms. Remember that the hot, brilliant stars are the easiest to see from a great distance. But these stars don't live very long, and are found during their brief lives quite close to the reservoirs of raw material that gave them birth. When we look at photographs of other spiral galaxies (Figure 24.9), it is these brilliant stars that we see clearly outlining the spiral arms.

Figure 24.8
A map showing the distribution of cold hydrogen in the Galaxy. Prepared from observations at 21 cm, the map is color-coded to mark different amounts of the gas: black and dark blue show regions with the least hydrogen, while red and white indicate the largest concentrations. Most of the hydrogen is in a thin disk that shows the plane of the Galaxy. The wisps and filaments of hydrogen above and below the disk are nearby structures, which we see all around us. (C. Jones and W. Forman/CfA, from data by many investigators)

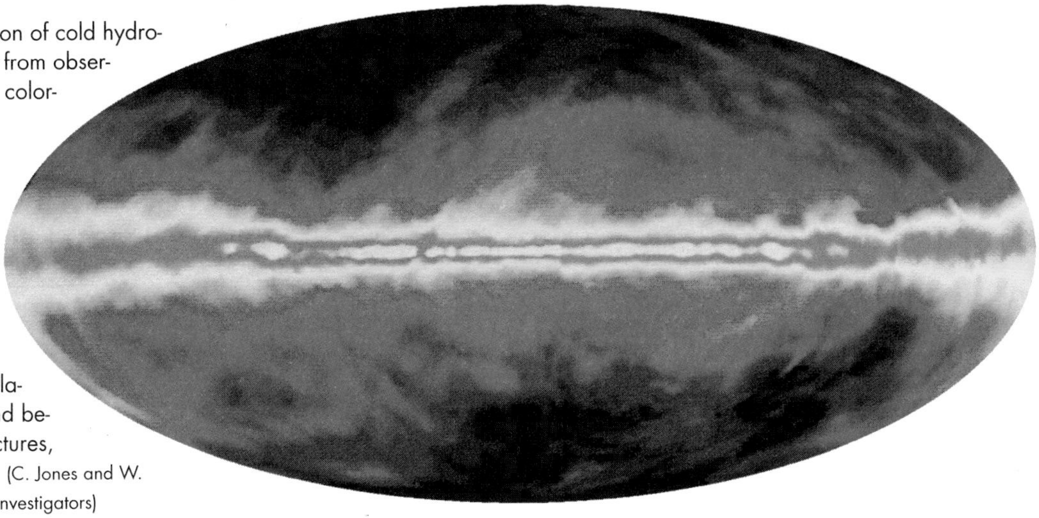

Spiral Structure of the Galaxy

The Arms of the Milky Way

Using the techniques just discussed, astronomers are beginning to pin down the precise layout of the Milky Way. It appears that the Galaxy has four major **spiral arms,** with some smaller spurs (shown in Figure 24.5). The Sun is near the inner edge of a short arm or spur called the *Orion arm,* which is about 15,000 LY long and contains such conspicuous features as the Cygnus Rift (the great dark nebula in the summer Milky Way) and the Orion Nebula.

More distant, and therefore less conspicuous, are the *Sagittarius–Carina* and *Perseus arms,* located, respectively, about 6000 LY inside and outside the Sun's position relative to the galactic center. Both of these arms, and the more distant *Cygnus arm,* are about 80,000 LY long. The fourth arm is unnamed and difficult to detect because it runs along the other side of the Galaxy. Radiation from it is mingled with radiation from the central regions of the Galaxy.

Formation of Spiral Structure

At the Sun's distance from its center, the Galaxy does not rotate like a solid wheel. Individual stars obey Kepler's third law, just as planets in our solar system do. Remember that Pluto takes longer than the Earth to complete one full circuit around the Sun. In just the same way, stars in larger orbits in the Galaxy trail behind those in smaller ones. This effect is called *differential galactic rotation.*

Differential rotation explains why so much of the interstellar material in our Galaxy is concentrated into elongated features that resemble spiral arms. No matter what the original distribution of the material might be, the differential rotation of the Galaxy can stretch it out into spiral features. Figure 24.10 shows the development of spiral arms from two irregular blobs of interstellar matter, as the portions of the blobs closest to the galactic center move fastest, while those farther out trail behind.

But this picture of spiral arms presents astronomers with an immediate problem. If that's all there were to the story, differential rotation would—over the 15-billion-year history of the Galaxy—have wound the Galaxy's arms tighter and tighter until all semblance of structure had disappeared. Somehow, the graceful spiral arms must maintain themselves over time, even as the material in them turns and turns around the center of the Milky Way.

We can think of the spiral arms as regions where matter is more densely concentrated. The arms are defined by the presence of a concentration, but individual stars and gas clouds may move into and out of the arms. To see what we mean, consider this analogy. Imagine that you are driving on a highway, and that cars with very nervous student drivers are moving unusually slowly in two of the lanes ahead of you. Traffic behind these cars will be forced to slow down, and the density of cars around them will increase. (So will the noise level, as people honk their horns and make rude remarks.)

Over time, individual cars like yours will manage to get past the slowly moving cars, but others will take their place in the traffic jam. Viewed from high above the highway, there will be a clump of cars moving along the freeway at the same speed as the two slowly moving cars. The clump will be moving more slowly than the traffic both in front of and behind the jam. The cars stuck in the clump will be different as time passes; each driver will only be caught in the traffic jam for a while.

If the same highway has other pairs of slow drivers farther back, there will be other clumps of traffic through which anyone on that road will have to move. In the same way, the "traffic" of stars and interstellar matter in the Galaxy slows down and clumps up when it meets the

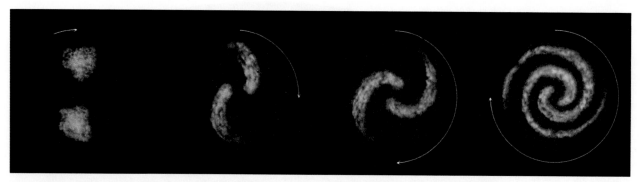

Figure 24.10
How spiral arms might form from irregular clouds of interstellar material stretched out by the different rotation rates throughout the Galaxy. The regions farthest from the galactic center rotate most slowly.

denser regions of the spiral arms. The concentrations of higher density rotate more slowly than does the actual material in the Galaxy, so that any group of stars, gas, and dust eventually passes through the spiral arms.

Astronomers have made models of how stars and gas clouds should move in response to the gravitational force exerted by matter in the Galaxy. The calculations show that objects moving around the Galaxy should indeed slow down slightly in the regions of the spiral arms, and linger there longer than elsewhere in their orbits. Thus a coiled wave of higher density builds up where the spiral arms are.

This theory for how spiral arms form is called the **spiral density wave** model. Spiral density waves have been observed directly in the rings of Saturn, another region containing a disk with many objects circling a common center. Astronomers are still debating how such density wave patterns begin, and how they can continue. But the model explains many of the observed properties of the Milky Way's disk (and those of other spiral galaxies).

As gas and dust clouds approach the inner boundaries of an arm and encounter the higher density of slowly moving matter, they collide with it. It is just where the shock of the collision occurs that the compression required for star formation is most likely to take place. This explains why the spiral arms are sites of vigorous star formation, and the home of the youngest stars. In some other galaxies, whose spiral arms we can view face-on, we see young stars, along with the densest dust clouds, occupying the inner boundaries of spiral arms—just as this theory predicts.

Not all the properties of each galaxy can be explained by the spiral density model. Even our galaxy has shorter spurs or segments between spiral arms that are not so easy to explain. Other spiral galaxies have even more chaotic or complicated patterns. Theorists have found other ways, unrelated to spiral density waves, to produce elongated structures that look like spiral arms. For example, great chain reactions of star formation may move progressively through molecular clouds and produce extended regions of young stars that mimic spiral arms (see Section 20.1). It is likely that more than one mechanism contributes to building up the specific design of galaxies such as the Milky Way.

24.3

Stellar Populations in the Galaxy

Two Kinds of Stars

The stars in our Galaxy can be roughly divided into two different groups, or populations, a discovery first made by Walter Baade of the Mount Wilson Observatory in southern California during World War II. As a German national, Baade was not allowed to do war research as many other scientists were, so he was able to make regular use of the Mount Wilson telescopes. Aided by the darker skies resulting from the wartime blackout of Los Angeles, he distinguished two distinct populations of stars in the nearby Andromeda galaxy (Figure 24.11).

Baade was impressed by the similarity of the stars in the Andromeda galaxy's nuclear bulge to those in our own Galaxy's globular clusters and the halo, and by the differences between all of these stars and those found in the spiral arms near the Sun. On this basis, he called all the stars in the halo, nuclear bulge, and globular clusters **population II,** and the bright blue stars in the spiral arms **population I.**

We now know that the populations differ not only in their locations in the Galaxy, but also in other characteristics such as chemical composition and age. Population I stars, found only in the disk, are especially concentrated in the spiral arms, and follow nearly circular orbits around the galactic center. Examples are bright supergiants, main-sequence stars of high luminosity (spectral classes O and B), and members of young open star clusters. Interstellar matter and molecular clouds are found in the same places as population I stars.

Population II stars show no correlation with the location of the spiral arms. These objects are found throughout the disk of the Galaxy, with the greatest concentration toward the nuclear bulge. Many follow elliptical orbits that carry them high above the galactic disk into the halo. Examples include planetary nebulae and RR Lyrae variables. The stars in globular clusters, found almost entirely in the halo and central bulge of the Galaxy, are also classified as population II.

Today we know much more about stellar evolution than did astronomers in the 1940s, and we can explain what causes the differences in the two stellar populations. One important difference has to do with the *ages* of the stars. Population I includes all the stars formed more recently: while some are as much as 10 billion years old, others are still forming today. The Sun, which is about 5 billion years old, is a good example of a population I star. Population II consists entirely of old stars that formed early in the history of the Galaxy: typical ages are 13 billion years and higher.

Clues from Chemical Composition

Measurements of stellar spectra show that a key difference between population I and population II stars is their chemical composition. Nearly all stars appear to be composed mostly of hydrogen and helium, but their abundances of the heavier elements differ. In the Sun and other population I stars, the heavy elements (those heavier than hydrogen and helium) account for about 1 to 4 percent of the total stellar mass. Population II stars in the outer galactic halo and in globular clusters have much lower abundances of the heavy elements—often

Figure 24.11
The spiral galaxy in Andromeda (M31) looks very much like our own Galaxy. Note the bulge of older, yellowish stars in the center, the bluer and younger stars in the outer regions, and the dust in the disk that blocks some of the light from the bulge. (Photo by Tony Hallas) *Inset:* Walter Baade (1893–1960). (Caltech Archives)

less than one-tenth or even one-hundredth the concentrations found in the Sun.

As discussed in earlier chapters, heavy elements are created in stars. They return to the Galaxy's reserves of raw material when stars die, only to be recycled into new generations of stars. Thus, as time goes on, stars are born with larger and larger supplies of heavier elements. The abundances of the elements demonstrate that the two stellar populations must have been born at different times. Population II stars seem to represent the early generations of the Galaxy. We might call them the senior citizens of the Milky Way; like many human seniors, they lived during simpler times (at least as far as the abundances of the elements are concerned).

The Real World

With rare exceptions, we should never trust any theory that divides the world into just two categories. Since Baade's pioneering work, we have learned that the notion that all stars can be characterized as either old and poor in heavy elements, or younger and rich in heavy elements, is an oversimplification. The idea of two populations helps organize our initial thoughts about the Galaxy, but we now know it cannot explain everything we observe.

For example, the stars in the nuclear bulge of the Galaxy are all fairly old. Their average age is in the range of 11 to 14 billion years, and none is younger than about 5 billion years old. Yet the abundance of heavy elements in these stars is about that of the Sun. Astronomers think that star formation in the crowded nuclear bulge occurred very rapidly just after the Milky Way Galaxy formed, so the stars there were soon enriched with heavy elements expelled in supernova explosions from the first generations of massive stars. Thus even stars that formed in the bulge 11 to 14 billion years ago started with a good supply of heavier elements.

Exactly the opposite situation occurs in the Small Magellanic Cloud, a small galaxy that orbits our own. Even the youngest stars in this galaxy are deficient in heavy elements, presumably because star formation has occurred so slowly that there have been, so far, relatively few supernova explosions. (Smaller galaxies also have more trouble holding onto the gas produced in supernova explosions long enough to recycle it; low-mass galaxies exert only a modest gravitational force, and the high-speed gas ejected by supernovae can easily escape from them.) The elements with which a star is endowed depend not only on when it forms in the history of its galaxy, but also on how many stars in its part of the galaxy have already completed their lives by the time the star is ready to form.

The Mass of the Galaxy

The Sun, like all the other stars in the Galaxy, orbits the center of the Milky Way. Our star's orbit is nearly circular and lies in the Galaxy's disk. The speed of the Sun in its orbit is about 220 km/s, which means it takes us approximately 225 million years to go once around. We call the period of the Sun's revolution about the nucleus the *galactic year;* it is a long time compared to human time scales. During the entire lifetime of the Earth, only about 20 galactic years have passed. And we have gone a mere fraction of the way around the Galaxy in all the time that humans have gazed into the sky.

Kepler Helps Weigh the Galaxy

We can use this information to estimate the mass of the Galaxy inside the Sun's orbit (just as we could "weigh" the Sun by monitoring the orbit of a planet around it—see Chapter 2). Let's assume that the Sun's orbit is circular and that the Galaxy is roughly spherical, both of which are fairly accurate assumptions. Long ago, Newton showed that if you have matter distributed in the shape of a sphere, then it's simple to calculate the pull of gravity on some object just outside that sphere: you can assume gravity acts as if all the matter were concentrated at a point in the center of the sphere. So, for our purposes we have a point in the center of the Galaxy, containing all the mass that lies inward of the Sun's position, and we have the Sun orbiting that point about 26,000 LY away.

This is the sort of situation to which Kepler's third law (as modified by Newton) can be directly applied. To use the formula in Section 2.3, we must put the distance in astronomical units and the period in years. There are 6.3×10^4 AU in 1 LY, so 26,000 LY becomes 1.6×10^9 AU. If it takes about 2.25×10^8 years for the Sun to orbit the center of the Milky Way, Kepler's law says:

$$M_{Galaxy} + M_{Sun} = a^3/P^2 = (1.6 \times 10^9)^3/(2.25 \times 10^8)^2$$
$$= \text{approximately } 10^{11} M_{Sun}$$

Note that Kepler's law gives you the sum of the masses, but since the mass of the Sun is completely trivial compared to the mass of the Galaxy, then for all practical purposes the result is the mass of the Milky Way. More sophisticated calculations based on complicated models give a similar result.

This estimate tells us only how much mass is contained in the volume inside the Sun's orbit. It is a good estimate for the *total* mass of the Galaxy if and only if there is hardly any mass outside the Sun's orbit. For many years astronomers thought this assumption was reasonable. The number of bright stars and the amount of *luminous matter* (meaning any material from which we can detect elec-

tromagnetic radiation) both drop off dramatically at distances of more than about 30,000 LY from the galactic center.

A Galaxy of Mostly Invisible Matter

In science, what seem to be reasonable assumptions can later turn out to be wrong (which is why we continue to do observations and experiments every chance we get). Observations now show that there is a lot more to the Milky Way than meets the eye (or the telescope)! While there is relatively little luminous matter beyond 30,000 LY, there turns out to be a lot of *invisible* material at large distances from the galactic center.

We can understand how astronomers reached this strange conclusion by remembering that, according to Kepler's third law, objects orbiting at large distances from a massive object move more slowly than do objects closer to that central mass. We have already seen an example of this idea in the case of the solar system. The outer planets move more slowly in their orbits than do those close to the Sun.

There are a few objects, including some gas clouds and globular clusters, that lie well outside the luminous boundary of the Milky Way. If most of the mass of our Galaxy were concentrated within the luminous region, then these very distant objects should travel around their galactic orbits at lower speeds than, for example, that of the Sun.

It turns out, however, that the few objects seen at large distances from the luminous boundary of the Milky Way Galaxy are *not* moving more slowly than the Sun. There are some globular clusters and RR Lyrae stars between 30,000 and 150,000 LY from the center of the Galaxy, and their orbital velocities are even greater than the Sun's (Figure 24.12).

What do these higher speeds mean? Kepler's third law tells us how fast objects must orbit a source of gravity if they are neither to fall in (because they move too slowly), or to escape (because they move too fast). If the Galaxy only had the mass calculated in the previous section, the high-speed outer objects should long ago have escaped the grip of the Milky Way. The fact that they have not done so means that our Galaxy must have more gravity than can be supplied by the luminous matter—in fact, a lot more gravity. The increasing speed of these outer objects tells us that the source of this gravity must extend outward from the center far beyond the Sun's orbit.

If the gravity were supplied by stars or something else that gives off radiation, we should have spotted this outer material long ago. We are therefore forced to the reluctant conclusion that this matter is *invisible* and has, except for its gravitational pull, gone entirely undetected!

Studies of the motions of the most remote globular clusters show that the total mass of the Galaxy out to a ra-

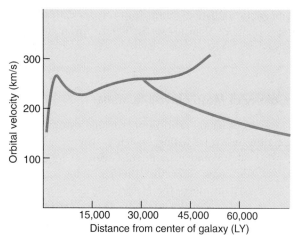

Figure 24.12
The orbital speed of carbon monoxide (CO) gas at different distances from the center of the Milky Way Galaxy is shown in red. The blue curve shows what the rotation curve would look like if all of the matter in the Galaxy were located inside a radius of 30,000 LY. Instead of going down, the speed of gas clouds farther out goes up, indicating a great deal of mass beyond the Sun's orbit.

dius of 150,000 LY is about 10^{12} M_{Sun}, ten times greater than the amount of matter inside the solar orbit. Theoretical arguments suggest that this **dark matter** (as astronomers have come to call it) is distributed in an enormous sphere around the Galaxy. Some astronomers call this the Galaxy's **corona,** a name taken from the very faint but extensive outer atmosphere that surrounds the Sun. But what could this corona be made of?

Let's look at a list of suspects taken from our study of astronomy so far. Since this matter is invisible, it clearly cannot be in the form of ordinary stars. And it cannot be gas in any form (remembering that there has to be a *lot* of it). If it were neutral hydrogen, its 21-cm radiation would have been detected. If it were ionized hydrogen, it should be hot enough to emit visible radiation. If a lot of hydrogen atoms out there had combined into hydrogen molecules, these should produce dark features in the ultraviolet spectra of objects lying beyond the Galaxy: such features have not been seen. Nor can the corona consist of interstellar dust, since dust in the required quantities would block the light from distant galaxies.

The dark matter cannot be black holes of stellar mass or old neutron stars, since the accretion of interstellar matter onto such objects would produce more x rays than are observed. Also, recall that the formation of black holes and neutron stars is preceded by supernova explosions, which scatter heavy elements into space to be incorporated into subsequent generations of stars. If the corona consisted of an enormous number of black holes and neutron stars, then the young stars we observe in our Galaxy

today would contain much larger abundances of heavy elements than they actually do.

The possibilities that remain to account for the dark matter in the Galaxy are low-mass objects such as planets or brown dwarfs, white dwarfs formed from an early generation of stars that have now cooled and ceased to shine, black holes with masses a million times the mass of the Sun, or exotic subatomic particles of a type not yet detected on Earth. Very sophisticated (and difficult) experiments are now under way to look for evidence in favor of one or another of these possibilities. Recent measurements suggest that white dwarfs make up at least some of the dark matter (see Chapter 27).

One important thing to note is that the problem of dark matter is not by any means confined to the Milky Way. Observations show that dark matter must also be present in other galaxies (whose outer regions also orbit too fast "for their own good") and, as we will see, even in great clusters of galaxies, whose members move around under the influence of far more gravity than can be accounted for by luminous matter alone.

Stop a moment and consider how astounding this conclusion is. Perhaps as much as 90 percent of the mass in our Galaxy (and many other galaxies) is invisible, and we do not even know what it is made of! The stars and raw material we *can* observe may be merely the tip of the cosmic iceberg (to use an old cliche); underlying it all may be other matter, perhaps familiar, perhaps startlingly new. Understanding the nature of this dark matter is one of the great challenges of astronomy today, and we will return to this problem in later chapters.

24.5

The Nucleus of the Galaxy

Another complex and puzzling region of the Galaxy is its center, which lies in the direction of the constellation Sagittarius. As discussed, we cannot see the nucleus in visible light because of absorption by the interstellar dust that lies between us and the galactic center (Figure 24.13). Light from the central region of the Galaxy is dimmed by a factor of a trillion (10^{12}). Infrared and radio radiation, however, whose wavelengths are long compared with the sizes of the interstellar grains, flows around the dust particles and can reach us (see the opening figure for this chapter). In fact, the very bright radio source in the nucleus, known as Sagittarius A, was the first cosmic radio source discovered.

Astronomers have used observations at all these wavelengths to try to determine what lies at the center of the Milky Way. It is a crowded, complicated region, full of gas, dust, stars, and stellar corpses; its exact geography is very difficult to sort out. Perhaps the question of most interest to astronomers is whether or not a massive black hole resides there. A great deal of effort has gone into trying to answer this question, and as we will see next, the jury is still out.

The Central Few Light Years

Let's take an imaginary journey to the heart of our Galaxy, starting from a distance of a few dozen light years and going as close to the center as observations permit. The first structure we encounter is a clumpy and rather irregular donut consisting of clouds of dust and gas, with much of the gas cool enough to form molecules. These clouds form a thick ring with an outer diameter of at least 25 LY and an inner diameter of about 10 LY. The individual molecular clouds typically have sizes in the range of 0.5 to 1.5 LY (Figure 24.14). The ring is rotating about the galactic center, but the motions of the individual clumps are turbulent, and the clouds sometimes collide with one another.

We can examine the gas in the ring with spectrographs, and use the Doppler effect to derive its orbital speed. Just as the Sun holds the planets in their orbits, so too there must be some mass inside the ring that keeps the fast-moving molecular clouds from flying off into space. We can use Kepler's third law as modified by Newton to estimate just how large this mass is. Given the observed rotational velocity of the ring, we calculate that the

Figure 24.13
This wide-angle picture covers over 50° of the sky in the direction of the center of the Milky Way. The center itself cannot be seen because of the vast quantities of dust in that direction.
(Anglo-Australian Telescope Board)

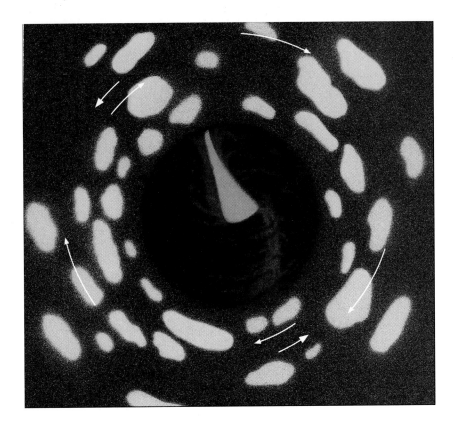

Figure 24.14
Schematic drawing of the central 20 LY of the Galaxy. Streamers of ionized gas are falling into the galactic center. Surrounding the central regions is a ring of dust, and beyond that lie individual clouds of gas. The reddish-brown region in this drawing corresponds to the region of radio emission shown in Figure 24.15.

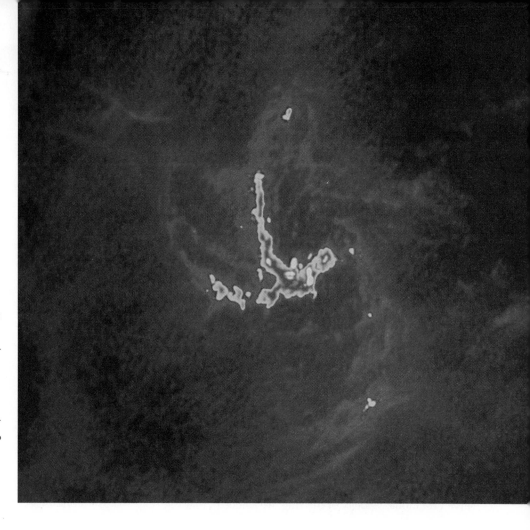

Figure 24.15
This radio image of a region approximately 10 LY across at the center of our Galaxy shows emission from hot gas. The colors indicate the amount of gas, ranging from the least (blue) to the most (red). The most likely interpretation of the image is that we see gas that is falling into the central region from the ring around it. The bright red region in the middle is the location of the compact radio source at the center of the Galaxy. (K. Lo et al., National Radio Astronomy Observatory/AUI)

total mass required is 2 to 5 million times the mass of the Sun. This is an impressive amount, but remember we are dealing with the crowded center of the Galaxy, so we should probably expect to find that it contains a lot of material. Let's move inside the ring to see whether we can figure out what and where this mass is.

To our surprise, just inside the ring we find hardly any interstellar matter at all. Measurements show that the density of gas and dust is 10 to 100 times higher in the ring than in the region just inside it. That is, the ring surrounds a cavity containing very little dust or cold gas. There are some streamers of hot, ionized gas (Figure 24.15), but the total mass of the ionized gas in the central few light years of the Galaxy is only about 100 M_{Sun}. This is not very much at all, considering how many stars we find in the inner region. Some remarkable event must have cleared out this cavity within the last 100,000 years; one possibility is an explosion of some kind at the center of the Galaxy, but we really don't know.

We do know that whatever cleared out the cavity must have happened recently (on a cosmic time scale), because material falling inward would soon blur the sharp edge that now separates the ring from its nearly empty interior. Collisions between the clouds also tend to eradicate individual clumps and produce a smooth ring of material, which is not what we observe.

Whatever accounts for the distribution of raw material in this area, the amount of gas we can measure in the cavity does not come close to accounting for the mass the ring motions require to be in the center of the Galaxy. What else is inside the ring? Infrared observations show a huge number of stars (perhaps as many as a million) packed into a region a few light years in diameter (Figure 24.16), where the average distance between stars is so small that light takes weeks, not years, to travel between neighboring stars. If there are planets around any of them, their nights must be bright with the brilliance of the star-studded sky.

But even the mass of all these stars is not enough to add up to the 2 to 5 million solar masses that we are looking for. To discover what else is inside the ring, we must continue our journey to the very core of this inner region (and think a bit about how we could detect a black hole if one is, in fact, there).

The Central Energy Source

As we saw in Chapter 23, showing the existence of a black hole is a challenge because the hole itself emits no radiation. We must examine the effect of the black hole on material outside the event horizon. One problem is that black hole event horizons are very small compared to the

Figure 24.16
This marvelous new infrared image shows a region at the center of the Galaxy that is about 6 LY across and is filled with closely packed stars. All the stars seen on this image are red giants; the differences in color are caused by the different amounts of reddening of their light by dust between us and the center. However, not all the reddish objects are stars. Some are regions of ionized hydrogen (HII), inside which there must be some hot stars whose energetic light does the ionizing. None of these objects can be seen with visible light since the dust completely obscures our view at visible wavelengths. (Ian Gatley and Michael Merrill/National Optical Astronomy Observatories)

kinds of measurements we can make from our vantage point 26,000 LY away. A black hole with a mass of about 3 million M_{Sun} would have a radius similar to that of our own Sun.

However, such a black hole might be surrounded by a whirling *accretion disk* of material not much bigger than the orbits of the planets in our own solar system. In theory, we could again measure the speed of the orbiting material to see how much mass would have to be squeezed inside to account for the motions. Unfortunately, we cannot see fine enough detail with existing telescopes to measure the velocities of gas and stars that are only a few tens of astronomical units from the center of the Galaxy.

Indirect evidence, however, suggests that the mass may be very highly concentrated at the core of the Milky Way. An intense radio source, called Sagittarius A°, is at or very near what we believe to be the exact center of the Galaxy. Measurements with the VLA radio telescope show that the diameter of this radio source is no larger than the diameter of Jupiter's orbit (10 AU), and it may even be somewhat smaller—close to the size we are looking for. The radio radiation would come from the disk around the black hole, from material being heated as it spirals inward toward the event horizon.

Another argument in favor of thinking that Sagittarius A° has a high mass is that it is stationary. Ordinary stars in

the galactic center change their positions by measurable amounts because of the gravitational tugs and pulls of other nearby stars, but Sagittarius A° does not. The most likely explanation is that Sagittarius A° is so much more massive than ordinary stars that it is not measurably moved by their gravitational influence.

Suppose there is a black hole of several million solar masses in the center of the Galaxy. Where did its mass come from? To start such a massive black hole, all you need is one very massive star early in the history of the Galaxy, going quickly through its life and dying by collapsing within its event horizon. Then gas, dust, and other stars that collide with this black hole could slowly be swallowed. Over billions of years, this could increase the mass of the original black hole enormously.

At the present time, matter is falling into the galactic center at the rate of about 1 M_{Sun} per 1000 years. If matter had been falling in at the same rate for about 5 billion years, then it would have easily been possible to accumulate the matter needed to form a black hole with a mass of several million solar masses. And it isn't merely gas and dust that could serve as a "meal" for such a black hole. The density of stars near the galactic center is such that we would expect a star to pass near the black hole and be swallowed by it every few thousand years.

So scientists find a great deal of circumstantial evi-

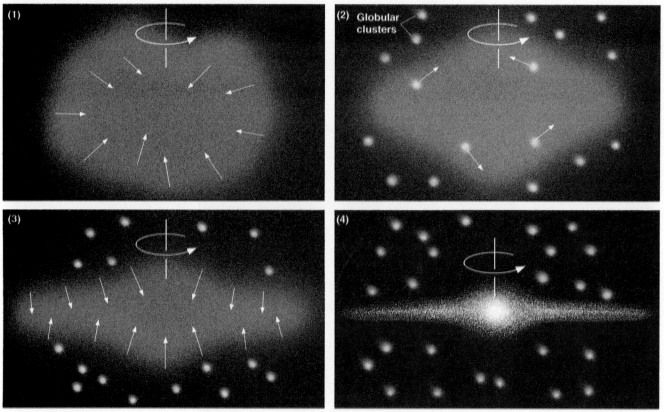

Figure 24.17
The Galaxy probably formed from an isolated, rotating cloud of gas that collapsed due to gravity. Halo stars and globular clusters either formed prior to the collapse or were formed elsewhere and attracted by gravity to the Galaxy early in its history. Stars in the disk formed late, and the gas from which they were made was contaminated with heavy elements produced in early generations of stars.

dence to support the idea that there is a black hole in the center of the Galaxy. But this may not be the only explanation. Over the next few years, astronomers will turn a host of new observational tools—gamma-ray, x-ray, infrared, and radio telescopes—toward the galactic center in an attempt to prove or disprove the existence of a black hole at the heart of the Milky Way. Whatever the final answer turns out to be, the possibility of such a supermassive black hole is not restricted to the center of our own Galaxy. As we will see in later chapters, there is also evidence that such black holes exist in other galaxies, and in the mysterious powerhouses called quasars as well.

24.6
The Formation of the Galaxy

Once astronomers understood what our Galaxy is like, they began to develop models for its formation and evolution. The first models were quite simple in that they assumed the Galaxy developed in isolation, uninfluenced by interactions with other galaxies. The flattened disk shape

of the Galaxy suggests that it formed through a process similar to the one that created the Sun and solar system (see Chapter 13). Building on this idea, early models assumed that the Galaxy, like the solar system, formed from a single rotating cloud.

The Protogalactic Cloud

Since the oldest stars—those in the halo and in globular clusters—are distributed in a sphere centered on the nucleus of the Galaxy, it makes sense to assume that such a *protogalactic cloud* was roughly spherical in shape. The oldest stars in the halo have ages of about 15 billion years, and so we estimate that the formation of the Galaxy began about that long ago. Then, just as in the case of the solar system, the protogalactic cloud collapsed and formed a thin rotating disk. Any stars born before the cloud collapsed did not participate in the collapse, but continue to orbit in the halo to the present day (Figure 24.17).

Gravitational forces caused the gas in the thin disk to fragment into clouds or clumps with masses like those of star clusters. These individual clouds then fragmented further to form individual stars. Since the oldest stars in

Light Pollution and the Milky Way

Bright lights have robbed most city dwellers of the pleasure of looking at stars. As we mentioned, many people in urban areas have never even seen the Milky Way. When light from street lamps and advertising signs shines into the sky, where it has no practical value (and wastes energy), we call it "light pollution." The amount of light that now flows upward has brightened the night sky so much that it is difficult or impossible to see the constellations, watch for meteor showers, or identify the occasional passing comet.

Artificial lights are also a problem for many major observatories, making it difficult for astronomers to detect faint stars and galaxies against the bright sky. More and more, astronomers must move their instruments to remote islands or mountains to escape the persistent glow of civilization.

But such lights, you may say, are the price we must pay for living in large cities. Aren't lights essential for safety and security? Fortunately, there is a solution that provides high-quality lighting while minimizing the amount of light pollution. The key first of all is to direct the light to where it is most useful. This means that outdoor lighting should be shielded so that light is emitted only downward to streets, sidewalks, and sports fields. Engineers have designed lighting fixtures that shield street lights in such a way that no light is projected upward. Some reflected light will brighten the sky no matter what we do, but with proper shielding and top reflectors, we could in theory direct much more of our light to ground level.

For astronomers, the concern is not just the amount of light shining upwards into the sky, but also the specific wavelengths in which that light is emitted. Much of modern astronomy depends on spectroscopy, where we must be able to detect light at the specific wavelengths corresponding to various elements. We therefore need a source of artificial light that will interfere with as few wavelengths as possible.

The accompanying diagram shows the spectra of three types of light sources. Two pollute at many wavelengths. But one, the low-pressure sodium lamp, emits light at only a few discrete wavelengths, while the rest of the spectrum is free of contaminating light. This is the kind of light that astronomers prefer cities to use. Fortunately, low-pressure sodium lights are extremely energy-efficient and thus are the most cost-effective form of street lighting. They are also the best light for visual acuity, since they emit most of their energy in the yellow part of the spectrum where the eye is most sensitive.

Check the lights in your own city. What type are they? (Your astronomy instructor may be able to lend you an inexpensive *diffraction grating* so you can observe the spectra of some street lights and compare them to the accompanying spectra.) Are the lights shielded? Notice the city lights next time you are in a plane at night. Do you see the lights themselves? If so, then some of the light is being emitted directly upward, where it does no good. Or do you see only the light reflected from the pavement? In this case, the lighting is properly shielded. If you live in or near a large city, check the sky. What part of it is contaminated with artificial light? In what directions is it brightest? What is the faintest star you can see? Now drive away from the city and notice how the sky changes. How far away do you have to go before the sky appears truly dark and you can see the Milky Way?

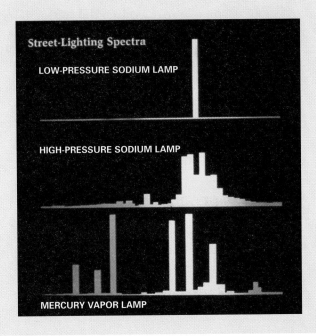

Comparison of the spectra of three forms of street lighting. Mercury vapor lights are the most common, but much of their energy is emitted in the red and blue regions of the spectrum, at wavelengths to which the eye is not very sensitive. In contrast, low-pressure sodium lamps emit all of their energy in a very narrow wavelength region where the eye is highly sensitive. This type of lighting leaves most of the spectrum free of radiation that would contaminate observations of faint stars and galaxies.

the disk are nearly as old as the stars in the halo, the collapse must have been rapid (astronomically speaking), requiring perhaps no more than a few hundred million years.

Recently, astronomers have begun to find hints that the evolution of the Galaxy has not been quite as peaceful as this traditional model suggests. Globular clusters differ in age by as much as 3 billion years; this means they could not all have formed during the few hundred million years before the interstellar matter collapsed to form the galactic disk, as the traditional model predicts. Studies also show that some globular clusters orbit the center of the Galaxy in a direction opposite that of the majority of the globulars and of the galactic disk. These "backward" globular clusters appear to be 2 to 3 billion years younger than globular clusters that rotate in the normal direction, and they tend to lie farther from the center of the Galaxy.

Collision Victims

Still more recently, in the direction of Sagittarius, astronomers have discovered a small new galaxy that is a satellite of our own Galaxy—just as the Moon is a satellite of the Earth. The Sagittarius dwarf galaxy is only 50,000 LY away from the nucleus of the Milky Way and is the closest galaxy known. It is very elongated, and its shape indicates that it is being torn apart by the Galaxy's gravitational force—just as Comet Shoemaker-Levy was torn apart when it passed too close to Jupiter (see Chapter 12).

Within another 100 million years, its stars will be captured and dispersed through the halo of the Milky Way.

The Milky Way may have more collisions in store. The Large and Small Magellanic Clouds (see Figure 14 in the Prologue), two other nearby satellite galaxies, are spiraling ever closer to our Galaxy and, according to calculations, will merge with it in the distant future. The Milky Way Galaxy has eight other small satellite galaxies, and at least some of these dwarfs appear to have been split off from the Magellanic Clouds when they passed close to the Milky Way.

All of this evidence suggests that the evolution of the Galaxy has been influenced by interactions and collisions with other galaxies. Astronomers now believe that the Milky Way formed in two stages. The first stage proceeded as shown in Figure 24.17. About 2 billion years after the formation of the majority of globular clusters, which rotate in the forward direction, the Galaxy acquired additional stars and globular clusters from one or more satellite galaxies that were captured when they ventured too close. (In the same way, you might recall, some of the satellites in our own solar system have retrograde orbits and are thought to be captured asteroids or comets.)

It appears likely that over the next few billion years, the Galaxy will gather still more stars from neighboring galaxies. We are coming to realize that environmental influences play an important role in determining the properties of our own Galaxy. In future chapters we will see that collisions and mergers are a major factor in the evolution of many other galaxies as well.

Summary

24.1 The Sun is located in the outskirts of the **Milky Way Galaxy.** The Galaxy consists of a disk containing dust, gas, and young stars; a **nuclear bulge** containing old stars; and a spherical **halo** containing very old stars, globular clusters, and RR Lyrae variables. Analysis by Shapley of the distribution of globular clusters gave the first indication that the Sun is not located at the center of the Galaxy. Radio observations at 21 cm show that cold hydrogen is confined to a flat disk with a thickness near the Sun of only 400 LY. Dust is found in the same locations; very little exists outside the Sun's orbit. The most-massive molecular clouds, where star formation is active, are concentrated in the spiral arms.

24.2 The Galaxy has four main **spiral arms** and several short spurs; the Sun is located on one of these spurs. Measurements show that the Galaxy does not rotate as a solid body, but instead its stars follow Kepler's laws; those closer to the galactic center complete their orbits more quickly. The **spiral density wave** theory is one way to account for the spiral arms. Calculations show that the gravitational forces within the Galaxy cause stars and gas clouds to slow down in the vicinity of the spiral arms, thereby leading to higher densities of material. When molecular clouds attempt to pass through these regions of higher density, star formation is triggered.

24.3 We can roughly divide the stars in the Galaxy into two categories. Old stars with few heavier elements are referred to as **population II** stars and are found in the halo, in globular clusters, and in the nuclear bulge. **Population I** stars, containing more heavier elements, are found in the disk and are especially concentrated in the spiral arms. The Sun is a member of population I. Population I stars formed after previous generations of stars produced heavy elements and ejected them into the interstellar medium.

24.4 The mass of the Galaxy can be determined by measuring the orbital velocities of stars or interstellar matter.

The Sun revolves completely around the galactic center in about 225 million years. The total mass of the Galaxy is about 10^{12} M_{Sun}, and about 90 percent of this mass consists of **dark matter** that emits no electromagnetic radiation and can be detected only because of the gravitational force it exerts on visible stars and interstellar matter. This dark matter is located mostly in the Galaxy's **corona;** its nature is not understood at present.

24.5 The central region of the Galaxy contains a ring of molecular clouds surrounding a cavity that is surprisingly empty of gas and dust, but that contains a very crowded cluster of stars. Measurements of the velocities of stars and gas show that the region within 2 LY of the galactic center contains a mass roughly 3 million times that of the Sun. A massive black hole can explain these observations,

but it is not yet certain that a black hole is the only possibility.

24.6 The Galaxy formed about 15 billion years ago. Models suggest that the stars in the halo and globular clusters formed first, while the Galaxy was spherical. The gas, somewhat enriched in heavy elements by the first generation of stars, then collapsed from a spherical distribution to a disk-shaped distribution. Stars are still forming today from the gas and dust that remain in the disk. Star formation occurs most rapidly in the spiral arms, where the density of interstellar matter is highest. There is evidence that the Galaxy captured additional stars and globular clusters from small galaxies that ventured too close to the Milky Way.

Review Questions

1. Explain why we see the Milky Way as a faint band of light stretching across the sky.

2. Explain where (and why) in a spiral galaxy you would expect to find globular clusters, molecular clouds, and atomic hydrogen.

3. Describe several characteristics that distinguish population I from population II stars.

4. Briefly describe the three parts of our Galaxy—the disk, the halo, and the corona.

5. Describe the evidence indicating that a black hole may be at the center of our Galaxy.

6. Explain why the abundances of heavy elements in stars correlate with their positions in the Galaxy.

Thought Questions

7. Suppose the Milky Way were a band of light extending only halfway around the sky (that is, in a semicircle). What, then, would you conclude about the Sun's location in the Galaxy? Give your reasoning.

8. The globular clusters revolve around the Galaxy in highly elliptical orbits. Where would you expect the clusters to spend most of their time? (Think of Kepler's laws.) At any given time, would you expect most globular clusters to be moving at high or low speeds with respect to the center of the Galaxy? Why?

9. Consider the following five kinds of objects: (1) open cluster, (2) giant molecular cloud, (3) globular cluster, (4) group of O and B stars, and (5) planetary nebulae.
   a. Which occur only in spiral arms?
   b. Which occur only in the parts of the Galaxy *other than* the spiral arms?
   c. Which are thought to be very young?
   d. Which are thought to be very old?
   e. Which have the hottest stars?

10. The dwarf galaxy in Sagittarius is the one closest to the Milky Way, yet it was discovered only in 1994. Can you think of a reason it was not discovered earlier? (*Hint:* Think about what else is in its constellation.)

11. Why does star formation occur primarily in the disk of the Galaxy?

12. Where in the Galaxy would you expect to find Type II supernovae, which are the explosions of massive stars that go through their lives very quickly? Where would you expect to find Type I supernovae, which involve the explosions of white dwarfs?

13. You are captured by space aliens who take you inside a complex cloud of gas and dust. To escape, you need to make a map of the cloud. Luckily, the aliens have a complete astronomical observatory with equipment for measuring all the bands of the electromagnetic spectrum. Use what you have learned in this chapter to explain what kinds of maps you would make of the cloud to plot your most effective escape route.

14. Suppose that stars evolved without losing mass—that once matter was incorporated into a star, it remained there forever. How would the appearance of the Galaxy be different from what it is now? Would there be population I and population II stars? What other differences would there be?

15. Use the data in Appendix 12 to identify the galaxies that are satellites of the Milky Way. How far away are they on average? How do their sizes compare with that of our own Galaxy?

16. Suppose the average mass of a star in the Galaxy were one-third of a solar mass. Use the value for the mass of the Galaxy given in the text to find how many stars the system contains.

17. Assume that the Sun orbits the center of the Galaxy at a speed of 220 km/s and a distance of 26,000 LY from the center.

    a. Calculate the circumference of the Sun's orbit, assuming it to be approximately circular.

    b. Calculate the Sun's period, the "galactic year."

    c. Use Newton's formulation of Kepler's third law to calculate the mass of the Galaxy inside the orbit of the Sun.

    d. It is estimated that the total mass of the Galaxy is ten times that within the Sun's orbit. If this mass were to collapse inside the orbit of the Sun, what would be the length of the galactic year?

    e. In this case, what would be the Sun's new orbital speed?

18. Construct a rotation curve for the solar system by using the orbital velocities of the planets. How does this curve differ from the rotation curve for the Galaxy? What does it tell you about where most of the mass in the solar system is concentrated?

19. Calculate the rotation speed of the gas ring at the galactic center if the mass inside the ring is $2 \times 10^6 \ M_{Sun}$ and the radius of the ring is 25 LY.

20. Calculate how much mass would fall into a black hole at the center of the Galaxy in a billion years at the current rate of infall of 1 M_{Sun} per 1000 years.

Suggestions for Further Reading

Bok, B. "Harlow Shapley and the Discovery of the Center of Our Galaxy" in Neyman, J., ed. *The Heritage of Copernicus.* 1974, MIT Press.

Croswell, K. *Alchemy of the Heavens.* 1995, Doubleday/Anchor. A popular-level review of current Milky Way research.

Dame, T. "The Molecular Milky Way" in *Sky & Telescope,* July 1988, p. 22.

Davis, J. *Journey to the Center of the Galaxy.* 1991, Contemporary Books. A journalist reviews our understanding of the Galaxy.

Ferris, T. *Coming of Age in the Milky Way.* 1988, Morrow. Historical survey.

Henbest, N. and Couper, H. *The Guide to the Galaxy.* 1994, Cambridge U. Press. Illustrated history of the growth of our understanding, plus a tour of the Milky Way.

Jayawardhana, R. "Destination: Galactic Center" in *Sky & Telescope,* June 1995, p. 26.

Mateo, M. "Searching for Dark Matter" in *Sky & Telescope,* Jan. 1994, p. 20.

Townes, C. and Genzel, R. "What Is Happening at the Center of Our Galaxy" in *Scientific American,* Apr. 1990, p. 46.

van den Bergh, S. and Hesser, J. "How the Milky Way Formed" in *Scientific American,* Jan. 1993, p. 72.

Verschuur, G. "In the Beginning" in *Astronomy,* Oct. 1993, p. 40. On globular clusters and the Galaxy.

Verschuur, G. "Journey into the Galaxy" in *Astronomy,* Jan. 1993, p. 32. Excellent tour.

1. Follow the directions for setting the *Mercator Projection* in Chapter 17 question #1. Choose the *Mercator Projection* with a *Zoom Factor* of 0.2. Compare the distributions of stars with spectral types O and B with that for types K and M.

Which plot shows the Milky Way and why?

Can you still see the Milky Way if you display all stars?

Select all the types of nebulae (stars and all other deep-sky objects off). Do they show the outline of the Milky Way?

Which object most clearly shows the direction of the Galaxy's center?

2. Display only galaxies. Why is there a band where galaxies do not appear?

This is part of the deepest image ever taken of the universe of galaxies. Two hundred seventy-six exposures, taken with the Hubble Space Telescope over 150 consecutive orbits, were combined to make an image that shows fainter galaxies than have ever been glimpsed before, including some that are a record-breaking 4 billion times fainter than the human eye can see. Hundreds of galaxies are visible in this section, which is about one-fourth of the full Hubble image. Some are so far away that we are seeing them as they were only a billion years or so after the universe began its expansion. The region photographed, a tiny patch of sky just above the Big Dipper, was selected because it is far from the disk of the Milky Way and contains very few nearby galaxies. (R. Williams, the Hubble Deep Field Team, & NASA)

CHAPTER 25

Galaxies

Thinking Ahead

In this chapter we reach vistas so great they completely stagger the human imagination. Galaxies dot the sky in every direction, as far as our telescopes permit us to peer. How can we measure the distances to galaxies so far away that their light takes millions and billions of years to reach us?

Our voyage now leaves the confines of the Milky Way as we begin our exploration of the realm of the other galaxies. Looking outward, we are awestruck at the vast number of "island universes"—each containing billions of stars. There are so many galaxies that a modern telescope allows us to see millions and millions of them just in the bowl of the Big Dipper. Like tourists from a small town, amazed by the extent and number of the great cities they are visiting, we are just coming to realize how much exists beyond the borders of our home.

We begin our exploration of galaxies with a guide to their properties, which is the thrust of this chapter. In the chapters that follow we will look more carefully at how such galaxies change with time, and at the role environment plays in their development.

The very idea that other galaxies exist was controversial for a long time. As late as 1920, many astronomers, including Harlow Shapley, thought the Milky Way encompassed all there was in the universe. The proof in 1924 that our Galaxy is not unique was one of the great advances in our understanding of our place in the universe.

There is nothing like astronomy to pull the stuff out of man.
His stupid dreams and red-rooster importance: let him count the star-swirls.

From the poem *Star-Swirls* by Robinson Jeffers (1924)

The Great Nebula Debate

Growing up at a time when the Hubble Space Telescope orbits above our heads, and giant telescopes are springing up on the great mountaintops of the world, you may be surprised to learn that we were not sure about other galaxies for such a long time. But bear in mind that today's giant telescopes and electronic detectors are recent additions to the astronomer's toolbox (see Chapter 5).

With the telescopes available in earlier centuries, galaxies looked like small fuzzy patches and were difficult to distinguish from the star clusters and gas and dust clouds that are part of our own Galaxy. All objects that were not sharp points of light were given the same name, "nebulae," the Latin word for clouds. Because even their precise shapes were often hard to make out, and no techniques had yet been devised for measuring their distances, the nebulae were the center of much debate and discussion among astronomers.

As early as the 18th century, the philosopher Immanuel Kant (1724–1804) suggested that some of the nebulae might be distant systems of stars (other Milky Ways). By 1908 nearly 15,000 nebulae had been cataloged and described. Some had been correctly identified as star clusters and others (such as the Orion Nebula) as gaseous nebulae. The nature of most of them, however, remained unexplained, particularly the ones that looked small and indistinct. (For more on how such nebulae are named, by the way, see the "Astronomy Basics" box in Chapter 19.)

If these nebulae were nearby, with distances comparable to those of observable stars, they were most likely clouds of gas within our Galaxy. If, on the other hand, they were remote, far beyond the edge of the Galaxy, they could be other systems containing billions of stars. To determine which idea was correct, astronomers had to find a way of measuring the distances to at least some of the nebulae. And for that, larger telescopes were necessary. When the 100-in. (2.5-m) telescope on Mount Wilson in southern California went into operation, astronomers finally had the tool to settle the controversy.

Working with the 100-in. telescope, Edwin Hubble (see "Voyagers in Astronomy" box) was able to resolve individual stars in several of the brighter spiral-shaped nebulae (including M31, the great spiral in Andromeda already mentioned in Chapter 24). Among these stars he discovered some faint variables that—when he analyzed their light curves—turned out to be cepheids. Here were reliable indicators that Hubble could use to measure the distances to the nebulae in the way that Henrietta Leavitt had pioneered (see Chapter 18). He estimated that the Andromeda galaxy was about 900,000 LY away from us. (Today we know it is actually slightly more than twice as distant as Hubble's estimate, but his conclusion about its true nature remains unchanged.)

No one in human history had ever measured a distance so great. The Milky Way could not possibly extend

Figure 25.1
The nearby spiral galaxy M83. This galaxy is about 10 million LY away and has a diameter of 30,000 LY. (© Anglo-Australian Telescope Board, 1977)

to cover so large a volume. The spiral nebulae had to be separate galaxies (Figure 25.1). When Hubble's paper on the distances to nebulae was read before a meeting of the American Astronomical Society on the first day of 1925, the entire room erupted in a standing ovation. A new era had begun in the study of the universe, and a new field—extragalactic astronomy—had just been born.

Types of Galaxies

Having established the existence of other galaxies, Hubble and others then began to observe them more closely—noting their shapes, their contents, and as many other properties as they could measure. This was a daunting task in the 1920s, when a single photograph or spectrum of a galaxy could take a full night of tireless observing. In recent decades, larger telescopes and electronic detectors have made this task less difficult, although the most distant galaxies (those that show us the universe in its earliest phases) still require enormous effort. (See this chapter's opening image, which required 10 days of observation with the Hubble Space Telescope.)

The first step in trying to understand a new type of object is often simply to describe its appearance. As it

Edwin Hubble: Expanding the Universe

The son of a Missouri insurance agent, Edwin Hubble graduated from high school at age 16. He excelled in sports, winning letters in track and basketball at the University of Chicago, where he studied both science and languages. Both his father and grandfather wanted him to study law, however, and he gave in to family pressure. He received a prestigious Rhodes scholarship to Oxford University in England, where he studied law with only middling enthusiasm. Returning to the United States, he spent a year teaching high school physics and Spanish, as well as coaching basketball, while trying to determine his life's direction.

The pull of astronomy eventually proved too strong to resist, and so he went back to the University of Chicago for graduate work. Just as he was about to get his degree and accept an offer to work at the soon-to-be-completed 100-in. telescope, the United States entered World War I and Hubble enlisted as an officer. Although the war ended by the time he had arrived in Europe, he received more officer's training abroad and enjoyed a brief time of further astronomical study at Cambridge before being sent home.

In 1919, at age 30, he joined the staff at Mount Wilson and began working with the world's largest telescope. Ripened by experience, energetic, disciplined, and a skillful observer, Hubble soon established some of the most important ideas in modern astronomy.

Edwin Hubble (1889–1953)
(Photo by J. Stokley/A.S.P. Archives)

He showed that other galaxies existed, classified them on the basis of their shapes, found a pattern to their motion (and thus put the notion of an expanding universe on a firm observational footing), and began a lifelong program to study the distribution of galaxies in the universe. Although a few others had glimpsed pieces of the puzzle, it was Hubble who put it all together and showed that an understanding of the large-scale structure of the universe was feasible.

His work brought Hubble much renown and many medals, awards, and honorary degrees. As he became better known (he was the first astronomer to appear on the cover of *Time* magazine), he and his wife enjoyed and cultivated friendships with movie stars and writers in southern California. Hubble was instrumental (if you'll pardon the pun) in the planning and building of the 200-in. telescope on Mount Palomar, and had begun to use it for studying galaxies when he was felled by a stroke in 1953. When astronomers built a space telescope that would allow them to extend Hubble's work to distances he could only dream about, it seemed natural to name it in his honor.

---

turns out, most of the bright nearby galaxies come in one of two shapes—they either have spiral arms, like our own Galaxy, or they appear to be elliptical (something like the Goodyear blimp). Many faint galaxies, on the other hand, have a more irregular shape.

Spiral Galaxies

Our own Galaxy and M31 (see Figure 24.11), which is believed to be much like it, are typical large **spiral galaxies.** They consist of a nucleus, a halo, a disk, and spiral arms. Interstellar material is usually spread throughout the disks of spiral galaxies. Bright emission nebulae and hot young stars are present, especially in the spiral arms (Figure 25.2), showing that new star formation is still going on.

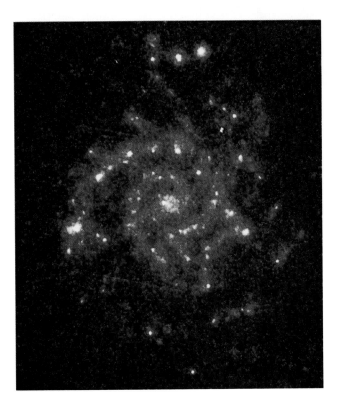

Figure 25.2
This ultraviolet image of the spiral galaxy M74 was taken in 1990 with NASA's Ultraviolet Imaging Telescope aboard the Space Shuttle. The colors have been added to show different intensities of ultraviolet light. The bright splotches outlining the spiral arms are regions of recent star formation where a cluster of hot stars is putting out a great deal of UV radiation. (NASA)

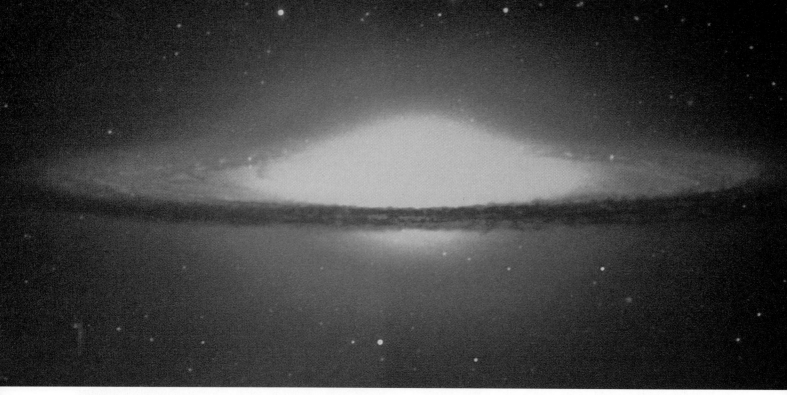

Figure 25.3
The Sombrero galaxy, M104. More than 24 million LY away, this galaxy is surrounded by a prominent dust ring almost 50,000 LY across. The central spheroidal region has the reddish appearance characteristic of an older population of stars. (European Southern Observatory)

Figure 25.4
NGC 1365, a barred spiral galaxy about 55 million LY away. This is a large galaxy, more luminous than either the Milky Way or M31. (European Southern Observatory)

The disks are often dusty, which is especially noticeable in those systems that we view almost edge-on (Figure 25.3). In galaxies we see face-on, the bright stars and emission nebulae make the arms of spirals stand out like those of a Fourth-of-July pinwheel (Figure 25.1). Open star clusters can be seen in the arms of nearer spirals, and globular clusters are often visible in their halos. Spiral galaxies contain a mixture of young and old stars, just as the Milky Way does. All spirals rotate, and the direction of their spin is such that the arms appear to trail, like the coattails of a brisk runner.

Perhaps a third or more of spiral galaxies have conspicuous bars of stars running through their nuclei. The spiral arms of such a system usually begin from the ends of the bar, rather than winding out directly from the nucleus. Such galaxies are called barred spirals (Figure 25.4). Some astronomers believe that almost all spirals, including the Milky Way, contain at least a weak bar.

In both normal and barred spirals, we observe a range of different shapes. At one extreme the nuclear bulge is large and luminous, the arms are faint and tightly coiled, and bright emission nebulae and supergiant stars are inconspicuous. At the other extreme the nuclear bulge is small—almost absent—and the arms are loosely wound. In these latter galaxies, luminous stars and emis-

Figure 25.5
Types of spiral galaxies, going from S0, which are mostly bulge and very little disk, to Sc, which are mostly disk and very little bulge. (Palomar Observatory, Caltech)

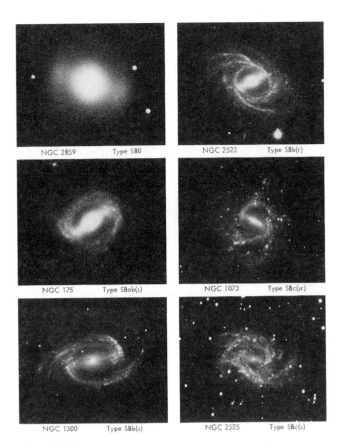

Figure 25.6
Types of barred spirals, paralleling the types of regular spirals. We suspect that most spirals may have some sort of bar-like structure in them, but it is much more pronounced in galaxies such as the ones shown here. (Palomar Observatory, Caltech)

sion nebulae are very prominent. Our Galaxy and M31 are both intermediate between the two extremes. Photographs of spiral galaxies, illustrating the different types, are shown in Figures 25.5 and 25.6.

The luminous parts of spiral galaxies range in diameter from about 20,000 to more than 100,000 LY, and the atomic hydrogen in the disks often extends to far greater diameters. There may also be considerable dark matter in them, in a much larger invisible corona, just as there is in the Milky Way. From the limited observational data available, their masses are estimated to range from 10^9 to 10^{12} M_{Sun}. The total luminosities of most spirals fall in the range of 10^8 to 10^{11} L_{Sun}. Our Galaxy and M31 are relatively large and massive, as spirals go.

Elliptical Galaxies

Elliptical galaxies consist almost entirely of old stars, and have shapes that are spheres or ellipsoids (somewhat squashed spheres). They contain no trace of spiral arms. Their light is dominated by older reddish stars (the population II stars discussed in Chapter 24), and in this respect ellipticals resemble the nuclear bulge and halo components of spiral galaxies. In the larger nearby ellipticals, many globular clusters can be identified (Figure 25.7).

Figure 25.7
This image, prepared with special photographic masking techniques, shows a swarm of globular clusters surrounding the giant elliptical galaxy M87. A jet of material is shooting out of the galaxy's core, which is not visible here. M87 is an active galaxy, with strong radio and x-ray emissions; such galaxies are discussed in Chapter 26. (Photo by David Malin; © Anglo-Australian Telescope Board)

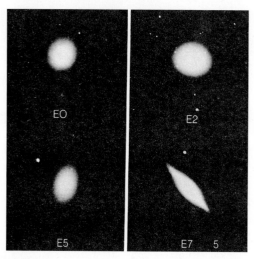

Figure 25.8
Types of elliptical galaxies. (Palomar Observatory, Caltech)

Figure 25.9
Leo I, a dwarf elliptical galaxy. (R. Schild, Center for Astrophysics)

Dust and emission nebulae are not conspicuous in elliptical galaxies, but many do contain a small amount of interstellar matter.

Elliptical galaxies show various degrees of flattening, ranging from systems that are approximately spherical to those that approach the flatness of spirals (Figure 25.8). The rare *giant ellipticals* (for example, M87 in Figure 25.7) reach luminosities of 10^{11} L_{Sun}. The mass in a giant elliptical can be as large as 10^{13} M_{Sun}. The diameters of these large galaxies extend over at least several hundred thousand light years and are considerably larger than the largest spirals. Although individual stars orbit around the center of an elliptical galaxy, the orbits are not all in the same direction as they are in spirals. Therefore, ellipticals don't rotate in a systematic way, and it is hard to estimate how much dark matter they contain.

This category of galaxies ranges all the way from the giants, just described, to *dwarf ellipticals,* which, astronomers have recently come to realize, may be the most common kind of galaxy. They escaped our notice for a long time because they are very faint and difficult to discern. An example of a dwarf elliptical is the Leo I system shown in Figure 25.9. There are so few bright stars in this galaxy that even its central regions are resolved. However, the total number of stars (most of which are too faint to show in our photo) is probably at least several million. The luminosity of this typical dwarf is approximately 10^6 L_{Sun}, about equal to that of the brightest globular clusters.

Intermediate between the giant and dwarf elliptical galaxies are systems such as M32 and NGC 205, two companions of M31. They can be seen in the photograph of M31 in Figure 24.11.

Irregular Galaxies

Hubble classified all star systems that did not have the regular shapes associated with the categories just described into the catchall bin of **irregular galaxies,** and we continue to use his term. These galaxies often appear chaotic, and many are undergoing relatively intense star formation activity; they contain both population I and population II stars. Nearby irregular galaxies, for which we can make good measurements, have lower masses and luminosities than typical spirals.

The two best-known irregular galaxies are the Large and Small Magellanic Clouds (Figures 25.10 and 25.11), which are among our nearest extragalactic neighbors. Their name reflects the fact that Ferdinand Magellan and his crew, making their round-the-world journey, were the first European travelers to notice them. Although not visible from the United States and Europe, these two systems are prominent from the Southern Hemisphere, where they look like wispy clouds detached from the Milky Way. They are only about one-tenth as distant as the Andromeda spiral. The Large Cloud contains the 30 Doradus complex (also known as the Tarantula Nebula), one of the largest and most luminous groupings of supergiant stars and associated gas known in any galaxy.

The Small Magellanic Cloud is greatly elongated and considerably less massive than the Large Cloud. The length of this narrow wisp of material is six times greater than its width, and the galaxy points directly toward our Galaxy like an arrow, stretching from about 150,000 LY out to 250,000 LY. The Small Cloud was probably the victim of a near-collision with the Large Cloud some 200 million years ago. It is now being pulled apart by the gravity of the Milky Way. Similar interactions may explain the strange shapes and active star-forming regions of other irregulars.

Do Galaxy Types Evolve?

Encouraged by the success of the H−R diagram for stars (see Chapter 17), astronomers studying galaxies hoped to find some sort of comparable scheme, where differences

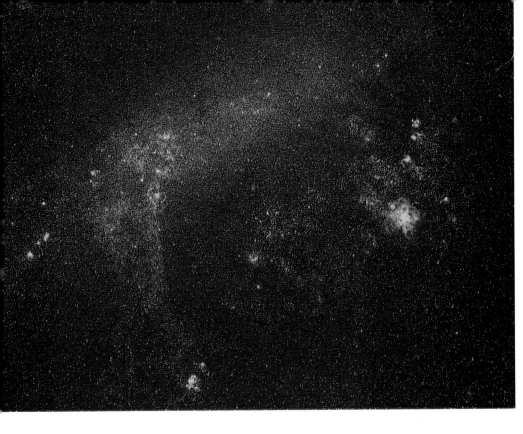

Figure 25.10
The Large Magellanic Cloud, a satellite of our own Galaxy, is visible to the naked eye from the Southern Hemisphere. The large red nebula (called the Tarantula) is the site of active star formation and contains many young supergiant stars. (National Optical Astronomy Observatories)

in appearance could be tied to different evolutionary stages. Wouldn't it be nice if every elliptical galaxy evolved into a spiral, for example, just as every main-sequence star evolves into a red giant? Several simple ideas of this kind were tried, some by Hubble himself, but none has stood the test of time (and observation).

While galaxies do change in appearance over billions of years, an isolated galaxy will not change from a spiral to an elliptical or vice versa. As spirals consume their gas, star formation stops, and the spiral arms gradually become less conspicuous. Over long periods, spirals therefore begin to look more like the galaxies at the top left of Figures 25.5 and 25.6 (which astronomers refer to as S0 types). We have also come to realize that collisions and mergers of smaller galaxies, including spirals, in the centers of dense galaxy clusters can build up the massive ellipticals such as M87. We will discuss the long-term evolution of galaxies in more detail in Chapter 27.

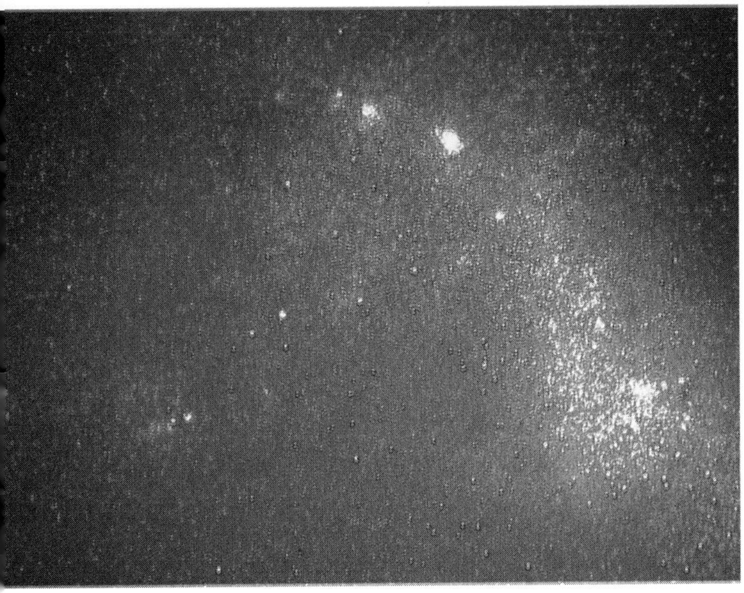

25.3

Properties of Galaxies

Masses of Galaxies

The technique for deriving the masses of galaxies is basically the same as that used to estimate the mass of the Sun, the stars, and our own Galaxy: we apply Kepler's third law as modified by Newton. When we measure how fast objects in the outer regions of the galaxy are orbiting the center, we can calculate how much mass is inside that orbit. For spiral galaxies, astronomers can measure the rotation speed by obtaining spectra of either stars or gas in the galaxy and looking for wavelength shifts produced by the Doppler effect.

Such observations of M31, for example, show it to have a mass (within the main visible part of the galaxy, out to a distance of 100,000 LY from the center) of about 4×10^{11} M_{Sun}, which is about the same as the mass of our own Galaxy. (The total mass of M31 is higher than 4×10^{11} M_{Sun} because we have not included the material that lies more than 100,000 LY from the center.) Like our own

Figure 25.11
The Small Magellanic Cloud. This dwarf irregular galaxy is another satellite of the Milky Way. (National Optical Astronomy Observatories)

TABLE 25.1
Characteristics of the Different Types of Galaxies

Characteristic	Spirals	Ellipticals	Irregulars
Mass (M_{Sun})	10^9 to 10^{12}	10^5 to 10^{13}	10^8 to 10^{11}
Diameter (thousands of LY)	15 to 150	3 to 600	3 to 30
Luminosity (L_{Sun})	10^8 to 10^{11}	10^6 to 10^{11}	10^7 to 2×10^9
Populations of stars	Old and young	Old	Old and young
Interstellar matter	Gas and dust	Almost no dust; little gas	Much gas; some have little dust, some much dust
Mass-to-light ratio in the visible part	2 to 10	10 to 20	1 to 10
Mass-to-light ratio for total galaxy	100	100	?

Galaxy, Andromeda appears to have a large amount of dark matter beyond its luminous boundary.

Elliptical galaxies do not rotate; for them we must use a slightly different technique to measure mass. Their stars are still moving in orbit around the galactic center, but not in the organized way that characterizes spirals. Since elliptical galaxies contain stars that are billions of years old, we can assume that the galaxies themselves are not flying apart. Therefore, if we can measure the various speeds with which the stars are moving in their orbits around the center of the galaxy, we can calculate how much mass the galaxy must contain in order to hold the stars within it.

In practice, the spectrum of a galaxy is a composite of the spectra of its many stars, whose different motions produce different Doppler shifts (some red, some blue). The result is that the lines we observe from the entire galaxy contain the combination of many Doppler shifts; they look much wider in a spectrum than would the same lines in a hypothetical galaxy in which the stars had no orbital motion. Astronomers call this phenomenon *line broadening*. The amount by which each line broadens indicates the range of speeds at which the stars are moving with respect to the center of the galaxy. The range of speeds depends, in turn, on the force of gravity that holds the stars within the galaxies. With information about the speeds, it is possible to calculate the mass of the elliptical galaxy.

Table 25.1 summarizes the range of masses (and other properties) of the various types of galaxies. The most-massive galaxies are the giant ellipticals, but the lowest-mass galaxies are ellipticals as well. On average, irregular galaxies have less mass than spirals.

Mass-to-Light Ratio

A useful way of characterizing a galaxy is by stating the ratio of its mass, in units of the Sun's mass, to its light output, in units of the Sun's luminosity. For the Sun, given this definition, the **mass-to-light ratio** is 1. Galaxies, however, are not just composed of stars that are like the Sun. The overwhelming majority of stars are less luminous than the Sun, and usually these stars contribute most to the mass of a system, without accounting for very much light. Since the average star has more mass than light, a galaxy's mass-to-light ratio is generally greater than 1. Galaxies in which star formation is still occurring tend to have mass-to-light ratios in the range of 1 to 10, while in galaxies consisting mostly of an older stellar population, such as ellipticals, the ratio is 10 to 20.

But these figures refer only to the inner, conspicuous parts of galaxies (Figure 25.12). In Chapter 24 we discussed the evidence for invisible matter in the outer regions of our own Galaxy, extending much farther from the galactic center than do the bright stars and gas. Recent measurements of the rotations of the outer parts of nearby galaxies, such as the Andromeda spiral, suggest that they, too, have extended distributions of dark matter around the visible disk of stars and dust. This largely invisible matter adds to the mass of the galaxy while contributing nothing to its luminosity, thus increasing the mass-to-light ratio. If dark invisible matter is present in a galaxy, its mass-to-light ratio can be as high as 100. The two different mass-to-light ratios measured for various types of galaxies are given in Table 25.1.

These measurements of other galaxies support the conclusion already reached from studies of the rotation of our own Galaxy—namely, that as much as 90 percent of all the material in the universe cannot at present be observed directly in any part of the electromagnetic spectrum. An understanding of the properties and distribution of this invisible matter is crucial to our understanding of galaxies. Through the gravitational force that it exerts, dark matter probably plays a dominant role in their formation and early evolution. As we will see in Chapter 28, it may also determine the ultimate fate of the universe.

There is an interesting parallel here between our time and the time during which Edwin Hubble was receiving his training in astronomy. By 1920 many scientists were aware that astronomy stood on the brink of important breakthroughs if only the nature and behavior of the nebulae could be settled with better observations. In the

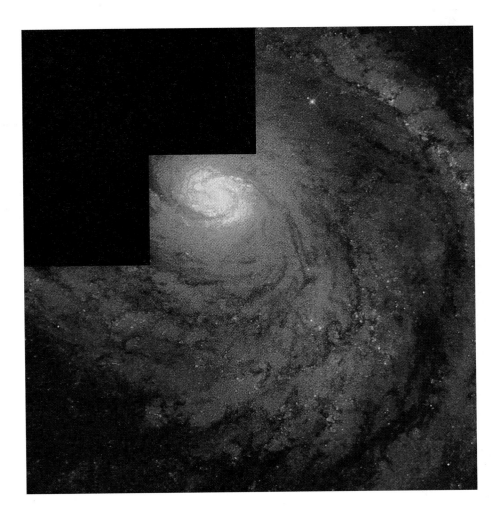

Figure 25.12
The Hubble Space Telescope took this dramatic image of part of the spiral galaxy M100, about 50 million LY away. With the Hubble, astronomers can make much better measurements of the galaxy's mass, structure, and distance. What cannot be seen on such images is any dark matter this galaxy contains. (J. Trauger and NASA)

same way, many astronomers today feel we may be closing in on a far more sophisticated understanding of the large-scale structure of the universe, if only we can learn more about the nature and properties of dark matter. If you follow astronomy in the newspapers or news magazines (as we hope you will), you should be hearing more about dark matter in the years to come.

25.4

The Extragalactic Distance Scale

To determine many of the key properties of a galaxy, such as its luminosity or its size, we must first know how far away it is. If we know the distance to a galaxy, we can convert how bright the galaxy appears to us in the sky into its true luminosity, because we know the precise way light is dimmed by distance. (The same galaxy ten times farther away, for example, would look a hundred times dimmer.) But the measurement of galaxy distances is one of the most difficult problems in modern astronomy: all the galaxies are far away, and most are so distant we cannot even make out individual stars in them.

For decades after Hubble's initial work, the techniques used to measure galaxy distances were relatively inaccurate, and different astronomers derived distances that differ by as much as a factor of two. Imagine if the distance between your home or dorm and your astronomy class were this uncertain; it would be difficult to make sure you got to class on time. In the past few years, however, astronomers have devised new techniques for measuring distances to galaxies; all of them give the same answer to within an accuracy of about 10 percent. As we will see, this means we may finally be able to make reliable estimates of the scale of the universe.

Variable Stars

Before we could measure distances to other galaxies, we first had to establish the scale of cosmic distances using objects in our own Galaxy. We described the chain of these distance methods in Chapter 18 (and we recommend that you review that chapter if it has been a while since you've read it). We saw that astronomers were especially delighted to learn how to measure distances using certain kinds of variable stars, including the bright cepheids. Such stars could be seen in star clusters within the Milky Way and even, as Hubble found, in nearby galaxies.

After the variables had been used to make distance measurements for a few decades, Walter Baade (whose work was discussed in Chapter 24) showed that there were actually two kinds of cepheids, and that astronomers had been unwittingly mixing them up. As a result, in the early 1950s all the distances to the galaxies had to be increased by about a factor of two. We mention this because we want you to bear in mind, as you read on, that science is always a progress report. Our first tentative steps in such difficult investigations are always subject to future revision as our techniques become more sophisticated.

The amount of work involved in finding cepheids and measuring their periods can be enormous. Hubble, for example, obtained 350 long-exposure photographs of the nearby spiral M31 over a period of 18 years, and identified only 40 cepheids. Even though cepheids are fairly luminous stars, they can be detected in only about 30 of the nearest galaxies with the world's largest ground-based telescopes. As mentioned in Chapter 18, one of the main projects being carried out with the Hubble Space Telescope is the measurement of cepheids in more-distant galaxies (out to at least 65 million LY), to improve the accuracy of the extragalactic distance scale (Figure 18.7).

Nevertheless, we can only use cepheids to measure distances within a small pocket of the universe of galaxies. After all, to use this method we must be able to resolve single stars and follow their subtle variations. Beyond a certain distance, even our finest space telescopes cannot help us do this. We needed to find other ways to measure the distances of galaxies.

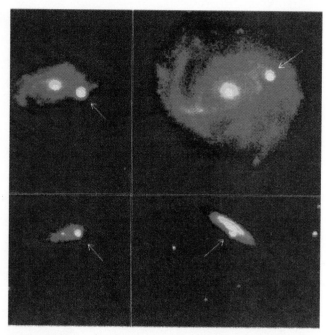

Figure 25.13
Four images of Type I supernovae in different galaxies, taken with the 1.2-m telescope at the Whipple Observatory. If Type I supernovae are all the same maximum luminosity, then those in more-distant galaxies will appear dimmer and can be used to estimate distances. (A. Riess et al./Harvard-Smithsonian Center for Astrophysics)

Standard Bulbs

We also discussed in Chapter 18 the great frustration that astronomers felt when they realized that the stars in general were not *standard bulbs*. If every light bulb in a huge auditorium is a standard 100-W bulb, then bulbs that look brighter to us must be closer, while those that look dimmer must be further away. If every star were a standard luminosity (or wattage), then we could similarly "read off" their distances based on how bright they appear to us. Alas, as we have seen, neither stars nor galaxies come in one standard-issue luminosity. Nonetheless, astronomers have been searching for objects out there that do act in some way like a standard bulb—that have the same intrinsic (built-in) brightness wherever they are.

A number of suggestions have been made for what sorts of objects might be standard bulbs, including the brightest supergiant stars, planetary nebulae (which give off a lot of ultraviolet radiation), and the average globular cluster in a galaxy. Particularly useful are Type I supernovae—those that involve the explosion of a white dwarf in a binary system (see Section 22.4). Observations show that supernovae of this type all reach nearly the same luminosity (about 10^{10} L_{Sun}) at maximum light. With such tremendous luminosities, these supernovae can be detected out to distances of at least 2 billion LY (Figure 25.13).

At very large distances we can use the total light emitted by an entire galaxy as a standard bulb. This technique, however, will not work for a single isolated galaxy because galaxies span an enormous range in intrinsic luminosity (see Table 25.1), and we cannot tell from appearance alone what the luminosity of any one galaxy really is. (For example, ellipticals with different luminosities often look similar when seen in isolation.) We can only tell which galaxies are highly luminous and which are not if we see a collection of them, with various brightnesses, side by side. Luckily, galaxies (as we will see in Chapter 27) are social creatures, and come in groups and clusters. Thus we can use the apparent brightness of the brightest elliptical members (say, the average of the brightest five) in a large cluster to estimate the distance to the cluster.

There is much debate among astronomers about how good these various standard bulbs are. It's as if, to continue our analogy, standard 100-W bulbs came from a slipshod manufacturer who made some of them 90 W and others 110 W. In this case you could only use the brightness of any given bulb to estimate its brightness plus or minus 10 percent. The universe is not in the business of making its objects into exact standard bulbs either, so there is still some variation in the luminosities of the best of these distance indicators. We must use such methods with a sense of their fallibility and calibrate them carefully every chance we get.

TABLE 25.2
Some Methods for Estimating Distances to Galaxies

Method	How Reliable	Galaxy Type	Approximate Distance Range
(SB = standard bulb)			(millions of LY)
Cepheid variables	Very	Spirals, irregulars	0–65
Brightest stars (SB)	Moderate	Spirals, irregulars	0–150
Planetary nebulae (SB)	Very	All	0–70
Globular clusters (SB)	Moderate	All	0–100
Surface brightness fluctuations	Very	Ellipticals	0–100
Type I supernovae (SB)	Very	All	0–2,000
Tully–Fisher method (21-cm line widths)	Very	Spirals, irregulars	0–300
Brightest galaxy in cluster (SB)	Very	Ellipticals in clusters	70–13,000
Redshifts (Hubble law)	Very	All	300–13,000

New Techniques

Some powerful new techniques for measuring distances have recently been developed. All start with observations of nearby galaxies whose distances have been measured through the use of cepheids and standard bulbs. The characteristics of these nearby galaxies are carefully calibrated and used to measure distances to galaxies so far away that individual stars and clusters can no longer be detected. Let's examine two of these techniques briefly.

The first method makes use of an interesting relationship noticed in the late 1970s by Brent Tully of the University of Hawaii and Richard Fisher of the National Radio Astronomy Observatory. They found that the luminosity of a spiral galaxy is related to its rotational velocity. The more mass a galaxy has, the faster the objects in its outer regions must orbit. A more massive galaxy has more stars in it, and is thus more luminous (ignoring dark matter for a moment). Using our terminology from the previous section, we can say that the mass-to-light ratios for various spiral galaxies are pretty similar.

Tully and Fisher used the 21-cm radiation from cold hydrogen to determine how rapidly material in a spiral galaxy was orbiting around its center. Since 21-cm radiation comes in a nice narrow line, the amount of broadening tells us the range of orbital velocities. The broader the line, the faster material is orbiting in the galaxy, and the more massive and luminous it turns out to be.

It is somewhat surprising that this technique works, since much of the mass associated with galaxies is dark matter, which does not contribute at all to the luminosity. There is also no obvious reason that the mass-to-light ratio should be similar for all spiral galaxies. Nevertheless, observations show that measuring the rotational velocity of a galaxy provides an accurate estimate of its intrinsic luminosity. Once we know how luminous the galaxy really is, we can compare the luminosity to the apparent brightness and use the difference to calculate its distance.

Another new technique involves measuring fluctuations in the apparent surface brightness of elliptical galaxies. This technique has been explored by John Tonry at MIT. Elliptical galaxies contain mostly very old stars and very little gas or dust. A perfectly sharp picture of an elliptical galaxy would look much like that of a globular cluster, with many individual stars appearing as discrete points of light. Even with the blurring caused by the Earth's atmosphere, the image of an elliptical galaxy does not appear perfectly smooth. Rather, it is mottled or bumpy due to the naturally lumpy distribution of light emitted by the individual stars belonging to that galaxy.

The amount of bumpiness depends on the distance to the galaxy. In a nearby galaxy we can see individual stars or clusters of stars, and the image has many bumps of varying brightness. In a very distant galaxy the stars cannot be resolved at all, and the image is smooth.

The distance to an elliptical galaxy can therefore be estimated by measuring the degree of bumpiness in the light distribution. This technique will not work for spiral galaxies, whose disks contain large, random amounts of dust that can also cause fluctuations in brightness distribution. But the method seems to give good results for elliptical galaxies, which are the very ones that cannot be effectively examined with the Tully–Fisher technique.

Table 25.2 lists the type of galaxy for which each of the techniques described here is useful, the range of distances over which the technique can be applied, and the reliability of the distance estimates derived with each technique.

25.5

The Expanding Universe

We now come to one of the most important discoveries ever made in astronomy—the pattern that underlies the motion of all galaxies. Before we describe how the discovery was made, we should point out that the first steps in the study of galaxies came at a time when the techniques of spectroscopy were also making great strides. Astronomers using larger telescopes could record the spectrum of a faint star or galaxy on more sensitive photographic plates, guiding their telescopes so they remained pointed to the same object for many hours. The resulting spectra of galaxies contained a wealth of information about the composition, mass, and motion of these great star systems.

Slipher's Moving Observations

Curiously, the discovery of the galaxies' motions began with the search for Martians and other solar systems. In 1894 the controversial (and wealthy) astronomer Percival Lowell established an observatory in Flagstaff, Arizona, to study the planets and to search for life in the universe (see Chapter 9). Lowell thought that the spiral nebulae might be solar systems in the process of formation—like the solar nebula described in Chapter 13. He therefore asked one of the observatory's young astronomers, Vesto M. Slipher (Figure 25.14), to photograph the spectra of some of the spiral nebulae to see if their spectral lines might show chemical compositions like those expected for newly forming planets.

The Lowell Observatory's major instrument was a 24-in. refracting telescope, which was not at all well suited to observations of faint spiral nebulae. With the technology available in those days, it took 20 to 40 hours to expose a good spectrum (in which Doppler shifts could reveal a galaxy's motion). This often meant continuing to expose the same photograph over several nights. Beginning in 1912, and making heroic efforts over a period of about 20

years, Slipher managed to photograph the spectra of more than 40 nebulae.

To his amazement, the Doppler shifts revealed huge speeds for the spiral nebulae. His first announcement came in 1914, years before Hubble showed that these objects were distant galaxies. A few spirals, such as M31, now known to be our close neighbors, turned out to be approaching us. But he found the overwhelming majority to be receding at speeds as high as 1800 km/s. Other observers soon confirmed his findings and made measurements of the motions of other spiral nebulae, all showing astounding **redshifts** (Doppler shifts in which the wavelength of the light gets longer) and thus rapid motion away from us.

The Hubble Law

The profound implications of Slipher's work became apparent only during the 1920s, when Hubble found ways of estimating the distances of the spiral nebulae. Hubble carried out the key observations in collaboration with a remarkable man, Milton Humason (Figure 25.15), who dropped out of school in the 8th grade and began his astronomical career by driving a mule train up the trail on Mount Wilson to the observatory. In those early days, supplies had to be brought up that way; even astronomers hiked up to the mountaintop for their turns at the telescope. Humason became interested in the work of the astronomers and, after marrying the daughter of the observatory's electrician, took a job as janitor there. After a time he became a night assistant, helping the astronomers run the telescope and take data. Eventually he made such a mark that he became a full astronomer at the observatory.

By the late 1920s Humason was collaborating with Hubble by photographing the spectra of faint galaxies

Figure 25.14
Vesto M. Slipher, 1875–1969. (Lowell Observatory)

Figure 25.15
Milton Humason worked with Hubble to establish the expansion of the universe. (Caltech Archives)

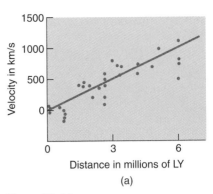

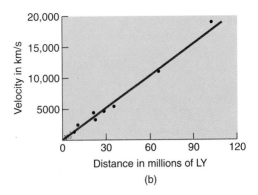

Figure 25.16

(a) Hubble's original velocity–distance relation, adapted from his 1929 paper in the *Proceedings of the National Academy of Sciences.* (b) Hubble and Humason's velocity–distance relation, adapted from their 1931 paper in *The Astrophysical Journal.* The red dots at the lower left are the points in the diagram in the 1929 paper. Comparison of the two graphs shows how rapidly the determination of galactic distances and redshifts progressed in the two years between these publications.

with the 100-in. telescope. (By then there was no question that the spiral nebulae were in fact galaxies.) When Hubble laid his own distance estimates next to Slipher's and Humason's measurements of the speed with which the galaxies were moving away, he found something stunning: there was a relationship between distance and velocity for galaxies. *The more distant the galaxy, the faster it was receding from us.*

In 1931 Hubble and Humason jointly published a classic paper in *The Astrophysical Journal,* the foremost technical publication read by U.S. astronomers (Figure 25.16). They compared distances and velocities of remote galaxies moving away from us at speeds as high as 20,000 km/s, and were able to show that the recession velocities of galaxies are directly proportional to their distances from us. We now know that this relationship holds for nearly every galaxy (except a few of the nearest ones) whose independent distance we have been able to measure! (The few galaxies approaching us turn out to be part of the Milky Way's own group, which has its own internal motions.)

Written as a formula, the relationship between velocity and distance is

$$v = H \times d$$

where v is the recession speed, d is the distance, and H is a number called the **Hubble constant.** This equation is now known as the **Hubble law.**

ASTRONOMY BASICS

Constants of Proportionality

Mathematical relationships such as the Hubble law are pretty common in life. To take a simple example, suppose you get hired by your college or university to call rich alumni and ask

for donations. You are paid $1.50 for each call; the more calls you squeeze in between studying astronomy and other courses, the more money you take home. We can set up a formula that connects p, your pay, and n, the number of calls:

$$p = A \times n$$

where A is the Alumni constant, with a value of 1.50 dollars. If you make 20 calls, you will earn 1.50 dollars times 20, or $30.

Suppose your boss forgets to tell you what you will get paid for each call. You can calculate the Alumni constant that governs your pay by keeping track of how many calls you make and noting your gross pay each week. If you make 100 calls the first week and are paid $150, you can deduce that the constant is 1.5 (in units of dollars per call). Hubble, of course, had no higher agent to tell him what his constant would be—he had to calculate its value from the measurements of distance and velocity.

Astronomers express the value of Hubble's constant in units that relate to how they measure speed and velocity for galaxies. In this book, we will use km/s per million LY as that unit. For many years, estimates of the value of the Hubble constant have been in the range of 15 to 30 km/s per million LY. The most recent work appears to be converging on a value near 25 km/s per million LY. If H is 25 km/s per million LY, a galaxy moves away from us at a speed of 25 km/s for every million light years of its distance. As an example, a galaxy 100 million LY away is moving away from us at a speed of 2500 km/s.

If every galaxy in the universe for which we can make distance measurements obeys this relationship, it cannot be a coincidence. (A few astronomers actually glimpsed this relationship in their data before Hubble's work, but dismissed it as a coincidence.) The Hubble law tells us something fundamental about the universe. Since all but the nearest galaxies appear to be in motion *away* from us, with the most distant ones moving the fastest, we must be living in an *expanding universe*. We will further explore the implications of this idea shortly, as well as in Chapter 27; for now we will just say that Hubble's observation underlies all our theories about the origin and evolution of the universe.

As we will see, Einstein's equations of general relativity (see Chapter 23), when solved for the behavior of the entire universe, implied that all the galaxies should show just this sort of expansion. But Einstein dismissed the idea as too disturbing, and introduced a kind of fudge factor into his equations to keep the universe at rest. Later he called his failure to predict what Hubble would soon discover "the greatest blunder of my life."

Hubble's Law and Distances

The regularity expressed in the Hubble law has a built-in bonus: it gives us a new way to determine the distances to remote galaxies. First we must reliably establish Hubble's constant by measuring *both* the distance and the velocity of many galaxies in many directions, to be sure the Hubble law is truly a universal property of galaxies. Then, a measurement of the speed with which a galaxy is moving away from us will tell us (by using the Hubble law) what its distance is. For example, if we find a galaxy moving away at 25,000 km/s, the Hubble law tells us it must be at a distance of a billion LY.

This is a very important technique, because, as we have seen, our best methods for determining galaxy distances don't take us much beyond 600 million LY (and they have many uncertainties). But to use the Hubble law as a distance indicator, all we need to do is get a spectrum of a galaxy and measure the Doppler shift, something astronomers are now very good at.

With large telescopes and modern spectrographs, such spectra can be taken of extremely faint galaxies. As we will see in the chapters to come, astronomers can now measure galaxies with redshifts that imply a velocity away from us over 90 percent the speed of light. At such high velocities, lines in the ultraviolet region of the spectrum—which normally cannot be observed from the ground—are shifted to yellow and red wavelengths and can be photographed as ordinary light.

Models for an Expanding Universe

At first, hearing about Hubble's law and being a fan of Copernicus and Shapley, you might be shocked. Are all the galaxies really moving away *from us?* Is there, after all, something special about our position in the universe? Worry not; the fact that galaxies obey the Hubble law only shows that the universe is expanding uniformly. A uniformly expanding universe is one that is expanding at the same rate everywhere. In such a universe, we and all other observers within it, no matter where they are located, *must* observe a proportionality between the velocities and distances of remote galaxies.

To see why, first imagine a ruler made of flexible rubber, with the usual lines marked off at each centimeter. Now suppose someone with strong arms grabs each end of the ruler and slowly stretches it so that, say, it doubles in length in 1 min (Figure 25.17). Consider an intelligent ant sitting on the mark at 2 cm—a point that is not at either end or in the middle of the ruler. He measures how fast other ants, sitting at the 4-, 7-, and 12-cm marks, move away from him as the ruler stretches.

The ant at 4 cm, originally 2 cm away from our ant, has doubled its distance in 1 min; it therefore moved away at a speed of 2 cm/min. The one at the 7-cm mark, which was originally 5 cm away from our ant, is now 10 cm away; it thus had to move at 5 cm/min. And the one that started at the 12-cm mark, which was 10 cm away from the ant doing the counting, is now 20 cm away, meaning it must have raced away at a speed of 10 cm/min. Ants at different distances move away at different speeds, and their speeds are proportional to their distances (just as the Hubble law states for galaxies). Yet all the ruler was doing was stretching uniformly.

Now let's repeat the analysis, but put the intelligent ant on some other mark—say, on 7 or 12. We discover that, as long as the ruler stretches uniformly, this ant also finds every other ant moving away at a speed proportional to its distance. In other words, the kind of relationship expressed by the Hubble law can be explained by a uniform stretching of the "world" of the ants. And *all* the ants on our simple diagram will see *the other ants* moving away from them as the ruler stretches.

Figure 25.17
Stretching a ruler. See text for explanation.

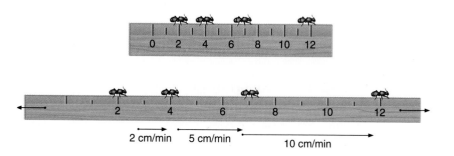

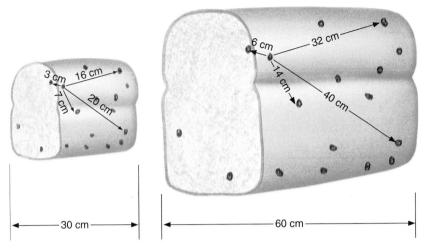

Figure 25.18
Expanding raisin bread.
See text for explanation.

For a three-dimensional analogy, let's look at the loaf of raisin bread in Figure 25.18. The chef has put too much yeast in the dough, and when she sets the bread out to rise, it doubles in size during the next hour, causing all the raisins to move farther apart. We again pick a representative raisin (that is not at the edge or the center of the loaf) and show the distances from it to several others in the figure (before and after the loaf expands).

Measure the increases in distance and calculate the speeds for yourself. Since each distance doubles during the hour, each raisin moves away from our selected raisin at a speed proportional to its distance. The same is true no matter which raisin you start with. Each raisin (if it could talk) would say, "Look! All the other raisins are expanding away from me. And their speeds are proportional to their distances."

Our two analogies are useful for clarifying our thinking, but you must not take them literally. On both the ruler and the raisin bread, there are points that are at the end or edge. You can use these to pinpoint the middle of the ruler and the loaf. While our models of the universe have some resemblance to the properties of the ruler and the loaf, we will see that they do not have a center or an edge.

What is useful to notice about both the ants and the raisins is that they themselves did not "cause" their motion. It isn't as if the raisins decided to take a trip away from each other, and then put on roller skates to get away. No, in both our analogies it was the stretching of the medium (the ruler or the bread) that moved the ants or the raisins farther apart. In the same way, we will see that the galaxies don't have rocket motors propelling them away from each other. Instead, they are passive participants in the expansion of space. As space stretches, the galaxies are carried farther and farther apart.

The expansion of the universe, by the way, does not imply that the individual galaxies and clusters of galaxies themselves are expanding. The raisins in our analogy do not grow in size as the loaf expands. Similarly, gravity holds galaxies and clusters together, and they merely get farther away from each other—without themselves changing in size—as the universe expands.

In the final chapter of our book, we will explore some of the implications of this expansion for the past history and ultimate fate of the universe. But first we turn to some of the most intriguing and energetic objects in the expanding universe.

Summary

25.1 Faint star clusters, clouds of glowing gas, dust clouds reflecting starlight, and galaxies all appeared as faint patches of light (or nebulae) in the telescopes available at the beginning of the 20th century. It was only when Hubble measured the distance to the Andromeda galaxy using cepheid variables in 1924 that the existence of other galaxies similar to the Milky Way in size and content was established.

25.2. The majority of bright galaxies are either **spirals** or **ellipticals.** Spiral galaxies contain both old and young stars, as well as interstellar matter, and have typical masses in the range of 10^9 to 10^{12} M$_{Sun}$. Our own Galaxy is a large spiral. Ellipticals are spheroidal or elliptical systems that consist almost entirely of old stars, with very little interstellar matter. Elliptical galaxies range in size from giants, more massive than any spiral, down to dwarfs, with masses of only about 10^6 M_{Sun}. A small percentage of galaxies with more chaotic shapes are classified as **irregulars.**

25.3 The masses of spiral galaxies are determined from measurements of their rates of rotation. The masses of elliptical galaxies are estimated from analyses of the motions of the stars within them. Galaxies can be characterized by their **mass-to-light ratios.** The luminous parts of galaxies with active star formation typically have mass-to-light ratios in the range of 1 to 10; the luminous parts of elliptical galaxies, which contain only old stars, typically have mass-to-light ratios of 10 to 20. The mass-to-light ratios of whole galaxies, including their outer regions, are as high as 100, indicating the presence of a great deal of dark matter.

25.4 Astronomers determine the distances to galaxies using a variety of methods, including the period–luminosity relationship for cepheid variables; objects such as Type I supernovae, which appear to be standard bulbs; and the Tully–Fisher method that connects the line-broadening of 21-cm radiation to the luminosity of spiral galaxies. Each method has limitations in terms of its precision, the kinds of galaxies it can be used with, and the range of distances over which it can be applied.

25.5 The universe is expanding. Observations show that the spectral lines of distant galaxies are **redshifted,** and that their recession velocities are proportional to their distances from us, a relationship known as the **Hubble law.** The rate of recession, called the **Hubble constant,** is approximately 25 km/s per million LY. We are not at the center of this expansion; an observer in any other galaxy would see the same expansion that we do.

Review Questions

1. Describe the main distinguishing features of spiral, elliptical, and irregular galaxies.

2. Why did it take so long for the existence of other galaxies to be established? What finally convinced astronomers that they do exist?

3. Explain what the mass-to-light ratio is and why it is smaller in regions of star formation in spiral galaxies than in the central regions of elliptical galaxies.

4. If we now realize dwarf ellipticals are the most common type of galaxy, why did they escape our notice for so long?

5. Describe the best ways to measure the distance to:
   a. a nearby spiral galaxy
   b. a nearby elliptical galaxy
   c. a distant isolated elliptical galaxy
   d. a distant isolated spiral galaxy
   e. a distant elliptical galaxy that is a member of a galaxy cluster

6. Why is the Hubble law considered one of the most important discoveries in the history of astronomy?

7. What does it mean to say that the universe is expanding? For example, is your astronomy classroom expanding?

Thought Questions

8. In 1920, Harlow Shapley and another astronomer, Heber Curtis, held a debate. Although it began as a discussion of one issue (the size of our Galaxy), it eventually came to involve the larger question of the existence of other galaxies. Use the resources in your library to look up information about this debate. Summarize the points that each participant made, and compare to what we know today. (*Hint:* You can start by looking up the works by Smith in "Suggestions for Further Reading.")

9. Where might the gas and dust (if any) in an elliptical galaxy come from?

10. Why can we not determine distances to galaxies by the same method used to measure the parallaxes of stars?

11. Starting with the determination of the size of the Earth, outline the steps necessary to obtain the distance to a remote cluster of galaxies.

12. Suppose that the Milky Way Galaxy were truly isolated, and that no other galaxies existed within 100 million LY. Suppose that galaxies were observed in larger numbers at distances greater than 100 million LY. Why would it be more difficult to determine accurate distances to those galaxies than if there were also galaxies relatively close-by?

13. Suppose you were Hubble and Humason, working on the distances and Doppler shifts of the galaxies. What sorts of things would you have to do to convince yourself (and others) that the relationship you were seeing between the two quantities was a real feature of the behavior of the universe? (For example, would data from two galaxies be enough to demonstrate the Hubble law?)

14. Measurements of the Andromeda galaxy show that it is rotating at a speed of 230 km/s, at a distance of about 100,000 LY or 6×10^9 AU from the center. Use Kepler's third law to calculate the mass of Andromeda. Is the true mass of the galaxy likely to be larger or smaller than this estimate? Why?

15. Calculate the mass-to-light ratio for a globular cluster with a luminosity of 10^6 L_{Sun} and 10^5 stars. (Assume that the average mass of a star in such a cluster is 1 M_{Sun}.) Do the same for a superluminous star of 100 M_{Sun} having the same luminosity of 10^6 L_{Sun}.

16. Plot the velocity–distance relation for the raisins in the bread analogy from the numbers given in Figure 25.18.

17. Repeat Problem 16, but use some other raisin for a reference and measure the distances with a ruler. Is your new plot the same as the last one?

18. Suppose a supernova explosion occurred in a galaxy at a distance of 10^8 LY. If we are only now detecting it, how long ago did the supernova actually occur? According to the Hubble law, what is the velocity with which this galaxy is moving away from us?

19. A cluster of galaxies is observed to have a radial velocity of 60,000 km/s. Find the distance to the cluster.

Suggestions for Further Reading

Christianson, G. *Edwin Hubble: Mariner of the Nebulae.* 1995, Farrar, Straus, Giroux. The definitive biography.

Dressler, A. *Voyage to the Great Attractor.* 1994, Knopf. Outstanding book by a noted astronomer on how modern extragalactic astronomy is done.

Eicher, D. "Candles to Light the Night" in *Astronomy,* Sep. 1994, p. 33. Introduction to standard bulbs and the cosmic distance scale.

Ferris, T. *The Red Limit.* 1983, Morrow. Well-written account of how we discovered the large-scale properties of the universe.

Freedman, W. "The Expansion Rate and Size of the Universe" in *Scientific American,* Nov. 1992, p. 76.

Hartley, K. "Elliptical Galaxies Forged by Collision" in *Astronomy,* May 1989, p. 42.

Hodge, P. "The Extragalactic Distance Scale: Agreement at Last?" in *Sky & Telescope,* Oct. 1993, p. 16.

Kinney, A. "Fourteen Billion Years Young" in *Mercury,* Mar./Apr. 1996, p. 29. On recent work with the Hubble Space Telescope.

Lake, G. "Understanding the Hubble Sequence [of Galaxies]" in *Sky & Telescope,* May 1992, p. 515.

Osterbrock, D. "Edwin Hubble and the Expanding Universe" in *Scientific American,* July 1993, p. 84.

Parker, B. "The Discovery of the Expanding Universe" in *Sky & Telescope,* Sep. 1986, p. 227.

Smith, R. "The Great Debate Revisited" in *Sky & Telescope,* Jan. 1983, p. 28.

Smith, R. *The Expanding Universe: Astronomy's Great Debate.* 1982, Cambridge U. Press.

Trefil, J. "Galaxies" in *Smithsonian,* Jan. 1989, p. 36. Nice long review article.

Using REDSHIFT ™

1. Click on *Deep Sky* in the *Objects Filter.* Turn on only the stars and galaxies. Set the *Stellar Limiting Magnitude* to 6 and the *Deep-Sky Limiting Magnitude* to 8 to see which galaxies you can view tonight through binoculars or a small telescope.

2. Display only galaxies with a limiting magnitude of 20; find a region with galaxies, center on it, and then set the *Zoom Level* to 15. Count the number of ellipticals, spirals, and irregulars.

Which type appears most common?

According to the chapter text, which is the most common type of galaxy? Can you explain any difference?

3. Set the *Limiting Magnitude* to select the 20 brightest galaxies. Which type is most common? Is it the same type as in Exercise 2? Explain any difference.

In this wonderful example of gravitational lensing, a cluster of galaxies (the yellow objects) is distorting the light of more-distant blue galaxies. The result is a complicated series of blue arcs that are—in a sense—ghost images of objects that would otherwise look like blue points of light. (Tony Tyson/AT&T Bell Laboratories)

CHAPTER 26

Quasars and Active Galaxies

Thinking Ahead

Suppose that at the center of a galaxy a massive black hole develops with access to a great deal of "fuel" (material relatively close to its event horizon). What would such a galaxy look like from far away?

During the first half of the 20th century, astronomers viewed the universe of galaxies as a mostly peaceful place. They assumed that galaxies had started billions of years ago and then evolved slowly as the populations of stars within them formed, aged, and died. That placid picture has completely changed in the last few decades of the 20th century.

Today astronomers can see that the universe is often shaped by violent events, including cataclysmic explosions of supernovae, collisions of whole galaxies, and the tremendous outpouring of energy as matter interacts in the environment surrounding massive black holes. The key event that led to our changed view of the nature of the universe was the discovery of the bizarre objects we now call *quasars*.

Quasar, quasar, burning bright

In the forests of the night

What immortal hand or eye

Can frame thy fearful luminosity.

With profound apologies to William Blake.

Quasars

The Sun does not radiate much energy at radio wavelengths. Astronomers expected that other stars would also be quiet in the radio region of the spectrum. They were, therefore, quite surprised when in 1960 two sources of radio waves were identified with what appeared to be faint blue stars (Figure 26.1). Naturally, astronomers rushed to take spectra of these radio "stars," but the spectra only deepened the mystery: they had emission lines that could not be identified with any known substance.

Huge Redshifts

Bear in mind that by the 1960s, astronomers had a century of experience in identifying elements and compounds in the spectra of stars. Elaborate tables had been published showing the lines that each element would produce under a wide range of conditions (see Chapter 16). A star with unidentifiable lines in the ordinary visible light spectrum was either an insult in the face of a hundred years of progress or, as turned out to be the case, a hint of something completely new.

The breakthrough came in 1963. At Caltech's Palomar Observatory, Maarten Schmidt (Figure 26.2) was puzzling over the spectrum of one of the radio stars, named 3C 273 because it was the 273rd entry in the third Cambridge catalog of radio sources. Schmidt recognized that the emission lines in its spectrum had the same spacing between them as the Balmer lines of hydrogen, but were shifted far to the red of the wavelengths at which the Balmer lines are normally located. Indeed, these lines were at such long wavelengths that if the redshifts were attributed to the Doppler effect, 3C 273 was receding from us at a speed of 45,000 km/s, or about 15 percent the speed of light!

After Schmidt's proposed identification of the lines in 3C 273, the puzzling emission lines in other star-like radio sources were re-examined to see if they, too, might be

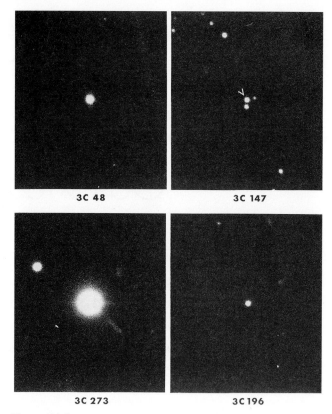

Figure 26.1
Quasi-stellar radio sources photographed with the 5-m Hale telescope. (Palomar Observatory, Caltech)

well-known lines with large redshifts. Such proved to be the case, but the other objects were found to be receding from us at even greater speeds. Since stars don't show Doppler shifts this large, no one had thought of considering high redshifts to be the cause of the strange spectra.

Their astounding speeds show that radio stars cannot possibly be nearby stars. Any true star moving at more than a few hundred kilometers per second would be able to overcome the gravitational force of the Galaxy and completely escape from it. (Eventually, astronomers also realized that there was more to some of these "stars" than just a point of light; see Figure 26.3.)

As we will see, radio stars only look like stars because they are compact and very far away. Since they superficially resemble stars, but do not share their properties, they were given the name *quasi-stellar radio sources*. Later, astronomers discovered objects with large redshifts that appear star-like but have no radio emission. Today, all these objects are referred to as *quasi-stellar objects* (*QSOs*) or, as they are more popularly known, **quasars.**

Figure 26.2
Maarten Schmidt (left), who solved the puzzle of the quasar spectra in 1963, shares a joke in this 1987 photo with Allan Sandage, who took the first spectrum of a quasar. Sandage has also been instrumental in measuring the value of the Hubble constant. Schmidt remains active in observing the properties and evolution of quasars. (Andrew Fraknoi)

(The name was soon appropriated by a manufacturer of home electronics.)

Today we know that about 90 percent of all known quasars are not radio sources. We're glad some do give off radio waves, however, because without that telltale oddity, it might have taken a lot longer to discover just how interesting these faint objects really are.

Thousands of QSOs have now been discovered, and spectra are available for a representative sample. All the spectra show large to very large redshifts (and none show blueshifts). The largest measured to date (March 1996) corresponds to a shift in wavelength of $\Delta\lambda/\lambda = 4.9$, where $\Delta\lambda$ is the difference between the wavelength that a spectral line would have if the emitting source were stationary, and the wavelength that it actually has. The stationary or laboratory wavelength is indicated by the Greek letter λ.

In the record-holding quasar, the first Lyman series line of hydrogen, with a laboratory wavelength of 121.5 nm in the ultraviolet portion of the spectrum, is shifted all the way through the visible region to 700 nm! At such high redshifts, the simple formula for converting a Doppler shift to speed (see Section 4.6) must be modified to take into account the effects of the theory of relativity.

Figure 26.3

A modern image of 3C 273 shows a brilliant (overexposed) point of light in the center, a fainter "host galaxy" around that point, and a wiggly jet that we now know measures some 150,000 LY in length. The four spikes coming from the central point are artifacts caused by the telescope mirror supports, but the jet is real. This image was taken with the High-Resolution Camera at the Canada-France-Hawaii Telescope on Mauna Kea. (R. McClure and J. Hutchings/DAO/National Research Council, Canada)

If we apply the relativistic form of the Doppler shift formula, we find that a redshift of 4.9 corresponds to a velocity of more than 94 percent the speed of light.

Explaining the Redshifts

What can these quasars be? At first, many theories were suggested to explain their properties. Could the large redshifts be caused by something else besides motion? Strong gravity causes a redshift (as we saw in Chapter 23), but such strong gravity would leave other evidence in the spectrum, and that evidence is simply not there. Could the quasars be high-speed projectiles shot out from galaxies by some unknown process? That might explain their high speeds, but then some galaxy somewhere would surely be shooting out a projectile that happened to come *toward us* and show a blueshift. Yet only redshifts have been observed.

Eventually, astronomers settled on the most straightforward explanation. If the quasars are like galaxies, and thus participating in the expansion of the universe, their redshifts make perfect sense. According to the Hubble law, their high redshifts simply mean that the quasars are far away from us. Could the QSOs be distant galaxies?

If they are, they certainly don't show the properties of normal galaxies. Galaxies contain stars, so the spectra of normal galaxies have absorption lines, as do the spectra of stars. Quasar spectra are dominated by emission lines. When they do have absorption lines, these frequently have redshifts that are less than the emission lines show. And we run into another difficulty when we try to account for the energy these objects emit.

The Luminosity Puzzle

If the quasars are as far away as their redshifts imply, then they must be incredibly luminous to be visible to us at all—far brighter than any normal galaxy could be. Even in visible light alone, most are far more energetic than the brightest elliptical galaxies. But quasars also emit energy at x-ray and ultraviolet wavelengths, and some are radio sources as well. When all their radiation is added together, some QSOs have total luminosities as large as 10^{14} L_{Sun}, or 10 to 100 times the brightness of the brighter elliptical galaxies.

Finding a mechanism to produce this much energy would be difficult under any circumstances. But there is an additional problem. When astronomers began looking for other quasars and then monitoring them carefully, they found that some vary in luminosity on time scales of months, weeks, or even, in some cases, days. This variation is irregular and can change the brightness of a quasar by a few tens of percent in both its light and radio output.

Think about what such a change in luminosity means. A quasar at its dimmest is still more brilliant than any normal galaxy. Now imagine that the brightness increases by 30 percent in a few weeks. Whatever mechanism is re-

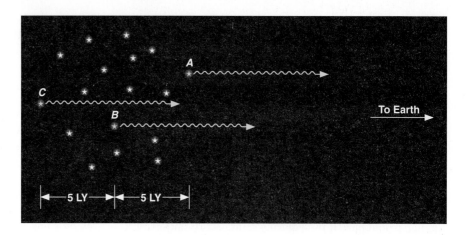

Figure 26.4

A diagram showing why light variations from a large region in space appear to last for an extended period of time as viewed from Earth. Suppose all the stars in this cluster, which is 10 LY across, brighten simultaneously and instantaneously. From Earth, star *A* will appear to brighten five years before star *B*, which in turn will appear to brighten five years earlier than star *C*. It will take ten years for an Earth observer to get the full effect of the brightening.

sponsible must be able to release new energy at rates that stagger our imaginations. The most dramatic changes in quasar brightness are equivalent to the energy released by 100,000 billion Suns. To produce this much energy we would have to convert the total mass of about ten Earths into energy every minute!

Moreover, because the fluctuations occur in such short times, the part of a QSO that is varying must be smaller than the distance light travels in the time it takes the variation to occur—typically a few months. To see why this must be so, let's consider a cluster of stars 10 LY in diameter (Figure 26.4) at a very large distance from the Earth (the Earth is off to the right in the figure). Suppose every star in this cluster somehow brightens simultaneously and remains bright. When the light from this event arrives at the Earth, we would first see the brighter light from stars on the near side; five years later we would see increased light from stars at the center. Ten years would pass before we detected more light from stars on the far side.

Even though all stars in the cluster brightened at the same time, the fact that the cluster is 10 LY wide means that ten years must elapse before the light from every part of the cluster reaches us. From Earth we would see the cluster get brighter and brighter, as light from more and more stars began to reach us, and not until ten years after the brightening began would we see the cluster reach maximum brightness. In other words, if an extended object suddenly flares up, it will seem to brighten over a period equal to the time it takes light to travel across the object from its far side.

We can apply this idea to brightness changes in quasars to estimate their diameters. Because QSOs typically vary (get brighter and dimmer) over periods of a few months, the region where the energy is generated can be no larger than a few light months. If it were larger, it would take longer than a few months for the light from the far side to reach us.

How large is a region of a few light months? Pluto, usually the outermost planet in our solar system, is about 5.5 light hours from us. The Oort comet cloud may extend out to about 10 light months, while the nearest star is 4 LY away. Clearly a region a few light months across is insignificantly small compared to a galaxy. And some quasars vary even more rapidly, which means their energy is generated in an even smaller region. Whatever mechanism powers the quasars must be able to generate more energy than that produced by an entire galaxy in a volume of space that, in some cases, is no larger than our solar system. When quasars were first discovered, no one could think of a way to explain so much energy in so small a space.

Solving the Quasar Paradox

In the 1960s, then, astronomers faced a dilemma. If the quasars were at the distances given by the Hubble law, some new mechanism was needed to explain their huge energy output. There *was* a way to keep the quasars from being such disturbingly strong cosmic powerhouses. If their redshifts were not caused by the expansion of the universe, then they would not be a measure of distance. The quasars could then be at any distance at all. If they were much closer to us, their energies would be correspondingly smaller and much easier to explain. But in that case, we had no explanation for what *did* cause their enormous redshifts. Either way, the quasars seemed to mean we were going to learn something new about the universe.

One way to decide between these two interesting alternatives was to get an independent estimate of how far away the quasars are. An enormous amount of effort, by astronomers all over the world, was directed to this task for about two decades. Some tried to show that there were quasars and galaxies that were physically connected but had *different* redshifts. Since the redshift of the *galaxy* is clearly caused by its expansion with the universe, this would show that at least some part of the quasar's redshift must have another cause. But none of the cases of connection was particularly convincing; most astronomers thought it was likely that the two objects just appeared to lie near each other on the dome of the sky. There seemed no reason to assume they were also at similar distances

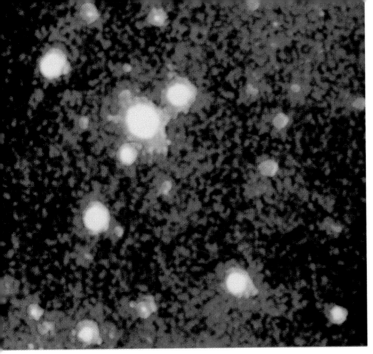

Figure 26.5
Quasar 3C 275.1, the first to be found at the center of a cluster of galaxies, appears as the brightest object near the center of this image. The quasar nucleus is surrounded by a gas cloud that is elliptical in shape. Its redshift indicates that this quasar and its cluster are both 7 billion LY away. (National Optical Astronomy Observatories)

from us. Another approach to the problem bore considerably more fruit.

As we will discuss in more detail in the next chapter, galaxies throughout the universe are found in groups and clusters. We know these groups are physically associated because they are not only close together in the sky, but their members show the same redshift. If quasars behave like galaxies (that is, if their redshifts are evidence of participation in the expansion of the universe), then we might find some quasars in distant clusters of galaxies. It is not enough for a quasar to be seen in the same area of the sky as a cluster; after all, that could be a coincidence, like seeing a bird and an airplane together, only to realize that the plane is really much farther away. To show that the quasar belongs to the cluster, the galaxies in the cluster and the quasar must share the same redshift.

This was not an easy idea to test. Quasars are typically far away; we see them because they are so luminous. Any ordinary galaxies with the same redshift (and thus the same large distance) as a quasar would be fainter and extremely difficult to detect. After much effort, astronomers have identified a number of faint galaxy clusters that contain quasars with the same redshifts as their other members (Figure 26.5). When only one such cluster-with-a-quasar was known, it was possible to suggest that the redshift similarity was a mere coincidence. Perhaps the quasar's redshift was caused by some other mechanism, and it just happened to lie in the same direction as the cluster.

Such coincidences do occasionally happen, but as gamblers trying to outsmart the house in Las Vegas can attest, they do not happen very often. As more and more clusters were found with quasars inside possessing the same redshift as all the cluster members, it became clear that the redshifts of quasars are caused by the same effect that causes the redshifts of ordinary galaxies—the expanding universe.

Other observations also support the hypothesis that quasars are located at the distances indicated by their redshifts. One key discovery was that many relatively nearby quasars are not true point sources but rather are embedded in a fainter, fuzzy-looking patch of light (Figure 26.6). The colors of this fuzz are like those of spiral galaxies. In a few cases, spectra have been obtained; they indicate that the light of the fuzz is derived from stars, clearly demonstrating that quasars are located in galaxies.

This still leaves us with the mystery of how the quasars can produce so much energy in such a tiny region. Clues to the answer come from the study of peculiar galaxies that provide a link between ordinary galaxies and the quasars.

Figure 26.6
The Hubble Space Telescope reveals the much fainter "host" galaxy around a quasar known only by its position in the sky (QSO 1229+204). The host galaxy is involved in a collision with a dwarf galaxy, which may be the reason the quasar at the center is bright. (J. Hutchings and NASA)

Figure 26.7
The Seyfert galaxy NGC 1566, at a distance of about 50 million LY, appears on this photograph to be a normal spiral. However, it has a very luminous nucleus with many of the characteristics of a quasar, though much less energetic. The active region at the center of NGC 1566 has been found to vary on a time scale of less than a month, indicating that it is extremely compact. Spectra show that hot gas near the tiny nucleus is moving at an abnormally high velocity, suggesting that it may be in orbit around a massive black hole. (Anglo-Australian Telescope Board)

26.2

Active Galaxies

When quasars were first discovered, they appeared to be much more luminous than any galaxies. And they are indeed more luminous than *normal* galaxies, but we have now found a class of galaxies that fills in the luminosity gap. These peculiar galaxies share many of the properties of the quasars, although to a less-spectacular degree. Members of this class are referred to as **active galaxies.** Since abnormal amounts of energy are produced in their centers, they are said to have **active galactic nuclei.** In effect, active galaxies have mini-quasars embedded in their nuclei.

Seyfert Galaxies

A good example of an active galaxy type is a group called *Seyfert galaxies*, which are spirals with point-like energetic nuclei (Figure 26.7). They are named after the astronomer who discovered the first examples. Like quasars, Seyferts have strong, broad emission lines in their spectra. Since stars typically produce absorption lines in the spectrum of a galaxy, emission lines indicate that there are hot gas clouds near the Seyfert nuclei. The width of the lines indicates that the gas is moving at speeds up to thousands of kilometers per second.

One of the most impressive images taken with the Hubble Space Telescope shows the nuclear region of the Seyfert galaxy NGC 1068 (Figure 26.8). The image, freed from the blurring of the Earth's atmosphere, shows the complex structure, including clouds of hot gas as small as 10 LY across, in the very center of the galaxy.

Some Seyferts, again like quasars, are radio or x-ray sources or both, and all emit strongly in the infrared. Some show brightness variations over a period of a few months, so, as we concluded for quasars, the region from which the radiation comes can be no more than a few light months across. The visible-light luminosities of Seyferts are about normal for spiral galaxies, but when their infrared emission is taken into account, their total luminosities are found to be about 100 times normal.

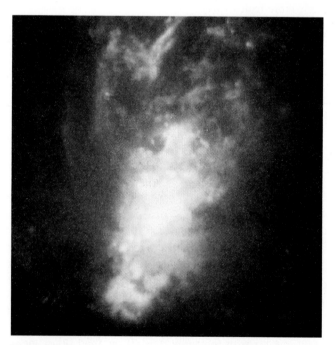

Figure 26.8
The central region of the Seyfert galaxy NGC 1068, about 150 LY across, can be seen in this image taken with the Hubble Space Telescope. Clouds of ionized gas as small as 10 LY in diameter are visible at the energetic center of this galaxy, which is about 60 million LY away in the constellation of Virgo. (D. Macchetto et al./NASA)

Studies of many such objects show that Seyfert and other peculiar galaxies tend to be more luminous than normal galaxies, but less luminous than quasars. Their bright but point-like nuclei indicate that enormous amounts of energy are being emitted from a small region at their centers. The crucial point about Seyferts and other active galaxies is that a significant fraction of their power comes from a source other than individual stars.

Active Elliptical Galaxies

Spirals are not alone in boasting peculiar activity in their centers. Some elliptical galaxies also have powerful central energy sources. The first clue came from radio observations. Many giant elliptical galaxies that appear comparatively normal when seen with visible light turn out to be powerful emitters of radio energy; hence we call them *radio galaxies*. The galaxy M87, discussed in Chapter 25, is a fine example. Figure 25.7 shows it looking deceptively

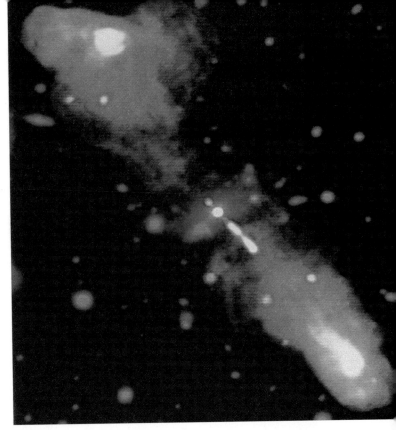

Figure 26.10

In this view of the elliptical radio galaxy 3C 219, a visible-light image (in blue) has been combined with a radio image (in red). The lobes of radio emission extend far beyond the visible parts of the galaxy; a short jet connects the lobes and the galaxy.

(D. Clark et al., National Radio Astronomy Observatory)

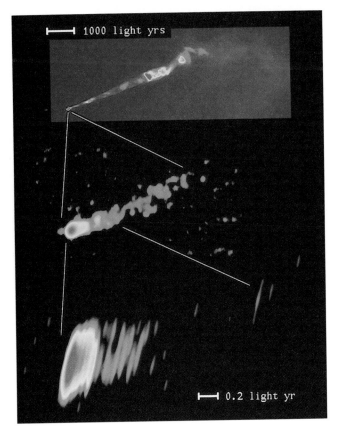

Figure 26.9

Three images of M87 made at radio wavelengths. The top view, made with the Very Large Array, shows the radio source at the galaxy's center (the red spot at left) and the full extent of the impressive jet. The center and bottom images show what can be done with the Very Long Baseline Array of radio telescopes, spanning the continent. They show the inner part of the jet with unprecedented detail; the smallest features we can discern are only light weeks in size. (J. Biretta, W. Junor, National Radio Astronomy Observatory)

peaceful; a radio image (Figure 26.9), however, allows us to "look" deeper inside the galaxy. It reveals an intense source of radio emission in the center, and a complicated jet of material extending from it for more than 6000 LY. (The jet is also visible in the opening image for Chapter 23.) Recent measurements show that knots of material in this jet are moving outward at speeds approaching two-thirds the speed of light. We will have more to say about what's happening at the center of M87 in the next section.

In some radio galaxies, most of the radio emission comes from small regions near their centers (reminding us of quasars). In others, we not only find bright sources in the nucleus, but also larger extended regions of radio emission surrounding the galaxy. In about three-quarters of the radio galaxies, the radio source is double, with most of the radiation coming from extended regions on opposite sides of each galaxy (Figure 26.10). Typically, the two emitting regions (called radio lobes) are far larger than the galaxy itself, and their centers are a few hundred thousand light years away from it. Often, observations also reveal two narrow jets of radio radiation pointing away from the galaxy toward the large extended sources. These jets are similar to the one in Figure 26.9, but can be more than a million LY long (Figure 26.11).

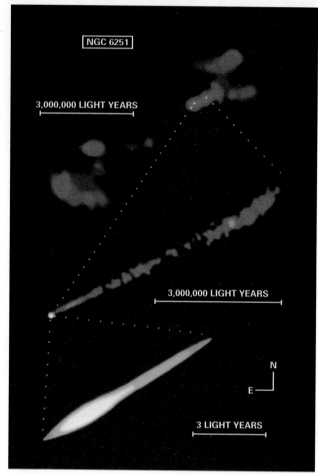

Figure 26.11
Several radio images of galaxy NGC 6251, whose unusually long jet is over 300,000 LY in length. The upper image shows two lobes of radio emission on either side of the galaxy, the middle image shows the long jet, and the bottom image shows one of the intense knots of emission in the jet. (National Radio Astronomy Observatory)

It appears that in all these objects hot ionized gases are "shot out" along the jets into the radio lobes by an extremely intense energy source in the nucleus of the galaxy. Some mechanism in the center is able to direct the flow of energy into narrowly focused beams that spread out and interact with colder neutral gas as they get beyond the confines of the galaxy. Interestingly, with modern instruments similar jets can now be seen in more than half of all quasars (Figures 26.3 and 26.12).

A Variety of Active Galactic Nuclei

Our examples barely scratch the surface of the many different types of active galaxies astronomers have uncovered in the last few decades. Although at first the variety in their appearance seemed bewildering, today we are coming to understand that the differences are often due to the angles at which we see the jets or lobes. (The dis-

tances of the galaxies and the environments in which their activity takes place also play a role.)

Remember that galaxies can be oriented in every possible way to our line of sight from Earth. Sometimes we look right down into the jet, as into the barrel of a gun; in these cases we see a brilliant spot. The jets of other galaxies shoot out to the side from our perspective, giving us the best view. Many times we see the central source, the jets, or the lobes at an angle, making our interpretations more difficult. And, of course, the farther away the object is, the more challenging is the task of observing it. Some of the quasars that looked like dim points of light with the technology of the 1960s now show jets coming from them, or the fuzzy hint of a surrounding galaxy.

Based on many such observations and decades of constructing theoretical models, most astronomers now view all of these objects as part of the same basic phenomenon. Some powerful source of energy at the centers

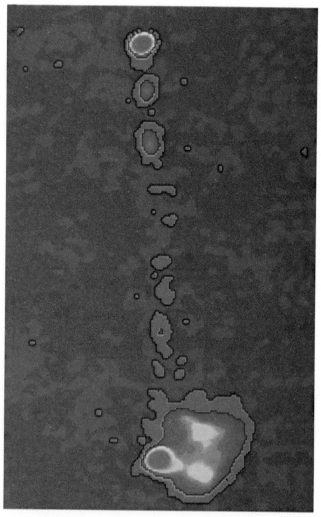

Figure 26.12
A radio image of the quasar 1007+417 taken with the VLA. The quasar is located at the bright point near the top of the picture; a jet leads to a lobe of radio emission at the bottom. (National Radio Astronomy Observatory)

of galaxies seems to be ultimately responsible for all the different forms of activity. In this view, the quasars are more-distant and extreme examples of this general behavior; often all we can see is the outrageously bright activity source in the center. (That quasars are extreme cases is born out by the fact that we have found only one quasar for every 10 million normal spiral galaxies.)

As we saw in Chapter 24, even our own Galaxy has a compact source of energy at its center. While the energy emitted by this compact source is tiny relative to that generated by a quasar, the center of the Milky Way may represent the low-energy extreme of the range of activity in compact galactic nuclei. Let us now turn to the key remaining question: what is the mechanism that can explain all this activity?

The Power Behind the Quasars

Once they realized that all these active objects have a compact central source of enormous energy in common, astronomers quickly began to come up with models to explain the energy source. These included collisions of stars in the crowded cores of galaxies; supermassive stars undergoing violent episodes in their regular life cycles; extraordinarily powerful supernovae; and other suggestions, some of which lay on the very frontiers of science.

Over the past few years, however, compelling evidence has been obtained to support the idea that quasars and other types of active galaxies derive their energy output from a massive black hole at the center. As we will discuss shortly, black holes are capable of producing large amounts of energy as long as material is falling toward them. (We saw this in Chapter 23 when we looked at binary star systems with black holes that were strong sources of x-ray emission.)

The ability of a black hole to produce energy depends on its mass and the amount of material that comes close enough to be "processed" by it. For a black hole with a billion Suns' worth of mass inside (10^9 M_{Sun}), even relatively modest amounts of additional material—only about 10 M_{Sun} per year—falling into the black hole would produce as much energy as a thousand normal galaxies. This could account for the total energy of a quasar.

Observational Evidence

Some of the strongest observational evidence for the existence of massive black holes in the centers of galaxies has been obtained with the Hubble Space Telescope. The opening figure for Chapter 23 shows that the stars at the center of the giant elliptical galaxy M87 become densely concentrated toward the center, forming a bright sharp core. The central density of stars in M87 is at least 300 times greater than expected for a normal giant elliptical

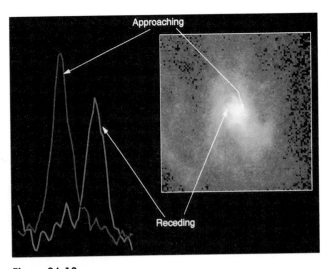

Figure 26.13
A disk of gas was discovered at the center of M87 with the Hubble Space Telescope. Observations made on opposite sides of the disk show that one side is approaching us (the spectral lines are blueshifted by the Doppler effect) while the other is receding (lines redshifted), a clear indication that the disk is rotating. The rotation speed is about 550 km/s or 1.2 million mi/h. Such a high rotation speed is clear evidence that there is a very massive black hole at the center of M87. (H. Ford et al. and NASA)

galaxy, and over 1000 times denser than the distribution of stars in the neighborhood of our own Sun. Detailed analysis of the distribution of light in M87 shows that a central black hole with a mass of close to 2.5 billion M_{Sun} would cause just this kind of concentration of stars.

With the Hubble, astronomers were able to identify and study an inner disk of hot (10,000 K) gas swirling around the center of M87 (Figure 26.13). It was surprising to find such hot gas in an elliptical galaxy, but the discovery was extremely useful for pinning down the existence of the black hole. Astronomers were able to measure the Doppler shift of spectral lines emitted by this gas, and to find its speed of rotation. They then used the speed to derive the amount of mass inside the disk—applying Kepler's third law. These measurements confirm that there is indeed a mass of about 2.5 billion M_{Sun} concentrated in a tiny region at the very center of M87. So much mass in a small volume of space must be a black hole.

We really don't know how such an enormous black hole was created. As discussed in earlier chapters, one possibility is that it formed from the merger of small black holes created by the explosion of massive stars when M87 was a young galaxy. However it got started, the black hole would have continued to grow by feeding on gas and stars that passed too close to it. As we will see in Chapter 27, whole galaxies sometimes collide and merge. Material from one or more galaxies swallowed by M87 may have fallen toward the black hole and added to its mass. (Such mergers may also explain why M87 is such a large elliptical galaxy.)

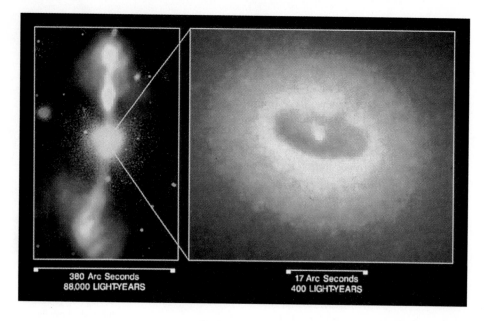

Figure 26.14
The picture on the left shows the active galaxy NGC 4261, located in the Virgo cluster. This elliptical galaxy—the white circular region in the center—is shown the way it looks in visible light, while the jets are seen at radio wavelengths. A Hubble Space Telescope image of the central portion of the galaxy is shown on the right. It contains a ring of dust and gas about 400 LY in diameter, surrounding what is probably a supermassive black hole. Note that the jets emerge from the galaxy in a direction perpendicular to the plane of the ring, consistent with the models shown in Figure 26.15. (W. Jaffe, H. Ford, and NASA)

380 Arc Seconds
88,000 LIGHT-YEARS

17 Arc Seconds
400 LIGHT-YEARS

The Hubble Space Telescope recently found similar evidence for a 2-billion-M_{Sun} black hole at the center of the galaxy NGC 3115 from the motions of the stars in its central region. A 1.2-billion-M_{Sun} black hole is required to explain observations of the elliptical galaxy NGC 4261 (Figure 26.14). Indeed, it is becoming clearer and clearer that many galaxies, including our neighbor the Andromeda galaxy, harbor black holes in their centers.

Energy Production Around a Black Hole

But how does a black hole become one of the most powerful sources of energy in the universe? As we saw in Chapter 23, a black hole itself can radiate no energy. The energy comes from material very close to the black hole, but not inside its event horizon. The black hole attracts matter—stars, dust, and gas—orbiting in the dense nuclear regions of the galaxy. This matter spirals in toward the spinning black hole and forms an accretion disk of material around it. As the material spirals ever closer to the black hole, it accelerates and becomes compressed, heating up to temperatures of millions of degrees. Such hot matter can radiate prodigious amounts of energy as it falls in toward the black hole.

To convince yourself that falling into a region with strong gravity can release a great deal of energy, imagine dropping your astronomy textbook out the window of the ground floor of the library. It will land with a thud, and maybe give a surprised pigeon a nasty bump, but the energy released by its fall will not be very great. Now take the same book up to the 15th floor of a tall building and drop it from there. For anyone below, astronomy could suddenly become a deadly subject; when the book hits, it does so with a great deal of energy. Dropping things from far away into the much stronger gravity of a black hole is much more effective. Just as the falling book can heat up the air, shake the ground, or produce sound energy that can be heard some distance away, so the energy of material falling toward a black hole can be converted to significant amounts of electromagnetic radiation.

The exact way in which the energy of infalling material is converted to radiation near a black hole is far more complicated than our simple example. To cite just one factor, the material does not go straight in, but whirls around in a chaotic accretion disk. In fact, to understand what happens in the "rough-and-tumble" region around a massive black hole, astronomers and physicists must resort to computer simulations (and they require supercomputers, fast machines capable of awesome numbers of calculations per second). The details of these models are beyond the scope of our book, but they look very promising.

Connecting the Model and the Observations

A number of the phenomena that we observe can be explained naturally in terms of this model. First, reasonable amounts of material falling into a very massive black hole *can* produce the vast quantities of energy that are emitted by quasars and active galactic nuclei. Detailed calculations show that about 10 percent of the mass of matter falling into a black hole can be converted to energy. This is a very efficient process: remember that during the entire course of the evolution of a star like the Sun, less than 1 percent of its mass is converted to energy by nuclear fusion.

The event horizon of a black hole is very small; recall from Chapter 23 that a black hole with a mass of a billion M_{Sun} would have a radius of approximately 3 billion km, about the distance between the Sun and the planet Uranus. The emission produced by infalling matter comes from a small volume of space closely surrounding the

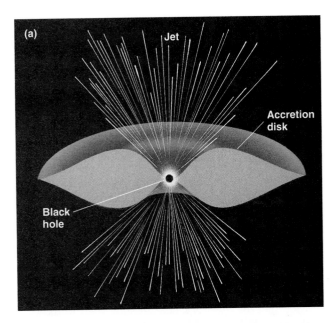

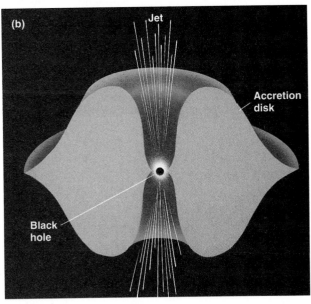

Figure 26.15

Schematic drawings of two accretion disks around large black holes. (a) A thin accretion disk. (b) A "fat" disk—the type needed to account for channeling outflow of hot material into narrow jets oriented perpendicular to the disk.

black hole. The small size of the radiating region is exactly what we need to explain the fact that quasars vary on time scales of weeks to months.

We also discussed that quasar spectra are dominated by strong emission lines. Such lines form when a hot glowing gas is present, which is just what we would expect to see around such an energetic black hole. Our models suggest that the broad emission lines are formed in relatively dense clouds within about half a light year of the black hole. These clouds are located farther out than the accretion disk and may provide additional fuel for the black hole in the future.

Radio Jets

So far our model seems to effectively explain the central energy source in quasars and active galaxies. But, as we have seen, quasars and other active galaxies also have long jets that glow with radio waves, light, and sometimes even x rays, and that extend far beyond the limits of the parent galaxy. Can we find a way for our black hole and its accretion disk to produce these jets as well?

A wide range of observations have traced the jets to within 3 to 30 LY of the parent quasar or galactic nucleus. While the black hole and accretion disk are typically smaller than 1 LY, we nevertheless presume that if the jets come this close, they probably originated in the vicinity of the black hole.

Why are energetic electrons and other particles ejected into jets, and often into two oppositely directed jets, rather than in all directions? Again, we must use theoretical models and supercomputer simulations of what happens when a lot of material whirls inward in a crowded black hole accretion disk. There is no agreement on exactly how jets form, but it is clear that any material escaping from the neighborhood of the black hole has an easier time doing so perpendicular to the disk.

In some ways the inner regions of black hole accretion disks resemble a baby that is just learning to eat by herself. As much food can sometimes wind up being spit out in various directions as going into the baby's mouth. In the same way, some of the material whirling inward toward a black hole finds itself under tremendous pressure, and orbiting with tremendous speed. Under such conditions, our simulations show that a significant amount of material can be flung outward—not back along the disk, where more material is crowding in, but above and below the disk. If the disk is thick (as it tends to be when a lot of material falls in quickly), it can channel the outrushing material into narrow beams (Figure 26.15).

Figure 26.16 is an artist's summary of some of the ideas we have been discussing. At the center of an active galaxy, a small bright spot marks the region of maximum radiation (and the greatest thickness for the accretion disk). On its perimeter the disk gets thinner. Two jets are seen emerging perpendicular to the disk. Not shown, of course, is the rest of the galaxy, which can sometimes obscure the inner regions and allow us only glimpses of the hungry monster within.

Evolution of Quasars

There is one other important fact about quasars that our black hole model can explain. Recall that when first introducing quasars, we mentioned that they generally tend to be far away—and are, in fact, the most distant objects we can detect in the universe. This result is very interesting, because when we see extremely distant objects, we are seeing them as they were long ago. Radiation from a quasar 8 billion LY away is telling us what that quasar was

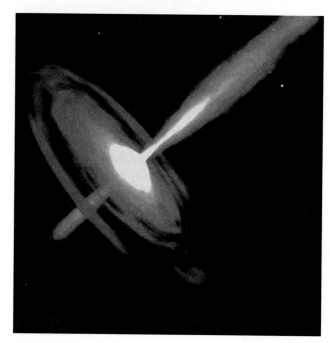

Figure 26.16
An artist's conception of a black hole accretion disk and perpendicular jets. (Painting by Dana Berry, STScI)

them. Why were there more quasars long ago (far away) than there are today (nearby)? Perhaps in the days of its youth, a black hole could find abundant fuel. But it may be that later in its life, much of the available fuel has been used up, and the hungry black hole has very little left with which to light up the galaxy's central regions.

In other words, if matter in the accretion disk is continually being depleted by falling into the black hole or being blown out from the galaxy in the form of jets, then a quasar can continue to radiate only as long as there is gas available to replenish the accretion disk. Where does this matter come from originally and how might it be replenished?

One possibility for the original fuel source is the very dense star clusters that form near the centers of galaxies. As these stars lose some of their mass during the normal course of stellar evolution—by means of stellar winds and supernova explosions—their combined material could provide fuel for the black hole. If stars venture too close to the black hole, it is also possible that its tidal forces are strong enough to tear the stars apart. Any raw material (gas and dust) near the center of the galaxy would also be a good source of fuel, and we would expect to find more gas in the central regions early in a galaxy's life than later on. (Recall that both ellipticals and the central bulges of spirals today have very little raw material left.)

As hinted above, an alternate source of fuel may come from collisions of galaxies. If two galaxies collide and merge, then gas and dust from one may come close enough to the black hole in the other to be devoured by it and so provide the necessary fuel. Astronomers are just beginning to understand how common collisions of galaxies are (see Chapter 27). But they present a possible way that a quasar or active galaxy might be restarted long after all the local fuel is used up.

like eight billion years ago, much closer to the time the galaxy formed. If there are more quasars far away, there must have been more quasars long ago.

Recently completed counts of quasars at different redshifts show us how dramatic this trend really is (Figure 26.17). The number of quasars was the greatest at the time when the universe was only 20 percent of its present age.

Our model says that quasars are black holes with enough fuel to make a brilliant accretion disk right around

Figure 26.17
A graph showing the number of quasars at earlier and earlier times (that is, farther and farther away). An age of 0 corresponds to the beginning of the universe; an age of 1 corresponds to the present time. Note that quasars were most abundant when the universe was about 20 percent of its current age.

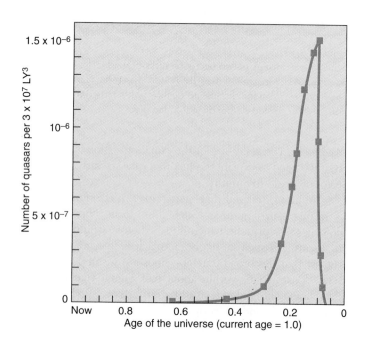

If this scenario is correct, we should generally see more quasars long ago (far away)—as we in fact do. And when we see a nearby quasar or active galaxy, we should check to see if it has been involved in any cosmic traffic accidents recently. And, indeed, many of the relatively nearby, still-active quasars appear to be embedded in galaxies that have experienced collisions with other galaxies. Gas and dust from these other galaxies have apparently been swept up by a dormant black hole and so provided the new source of fuel required to rekindle it.

This model also suggests that we should be able to detect low-level quasar-like activity in some nearby galaxies. That is, black holes might exist in the nuclei of nearby galaxies, but they would be producing only low levels of activity because they have already accreted nearly all of the matter available in their vicinity.

Emission lines like those seen in quasars and Seyferts but of much lower intensity are indeed seen in the nuclei of many otherwise normal nearby galaxies. Some astronomers speculate that a black hole may lurk in the center of every galaxy with a significantly crowded center, but that most of these black holes are now dark and quiet—mere shadows of their former selves. Whether or not this speculation is true remains to be seen.

26.4

Gravitational Lenses

Whatever the quasars are, their brilliance and large distance make them ideal probes of the far reaches of the universe. Like a distant searchlight that illuminates the hazards of the sea on a dark night, the radiation from distant quasars passes through or by whatever lies between them and our telescopes.

For example, while the spectra of quasars are dominated by emission lines, some quasars show absorption lines in their spectra as well. However, these absorption lines typically have smaller redshifts than the emission lines that characterize the quasar. Lines like this arise in clouds of gas within or between galaxies. These clouds are closer to us than the quasar, but difficult to make out in other ways. By studying them carefully, astronomers can learn about the distribution of raw material in the universe at different epochs.

But there is another—and very intriguing—way that the light from quasars can be affected on its way to us. Recall from Chapter 23 that the general theory of relativity predicts that light will be deflected in the vicinity of a strong gravitational field. This can happen when rays of light from a distant quasar pass near a compact massive galaxy on its way to Earth and are bent in a variety of ways. Calculations show that, like a reflection in a funhouse mirror, such bending can produce multiple or twisted images. These bizarre effects (called **gravitational lensing**) have now actually been observed.

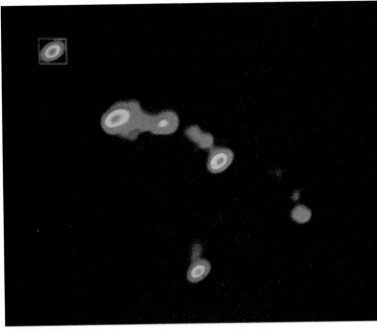

Figure 26.18
A radio image of the double quasar QSO 0957+561. A massive galaxy acting as a gravitational lens forms multiple images of a quasar lying far beyond it. The two images of the quasar are the bright point-like objects just above and below the center of the picture. The weak blue area just above the lower quasar image is the galaxy that acts as the gravitational lens. Several other radio sources are seen above and to the left of the double quasar. (National Radio Astronomy Observatory)

The First Gravitational Lens

In 1979 astronomers Dennis Walsh, Robert Carswell, and Ray Weymann of the University of Arizona noticed that a pair of quasars, very close to each other in the sky and known by the single name QSO 0957+561 (the numbers give their coordinates in the sky), were remarkably similar in appearance and spectra (Figure 26.18). Both had the same redshift. The astronomers suggested that the two quasars might actually be only one, and that the two images were produced by an intervening object—an effect first predicted by Einstein himself.

We now know that there is a dim galaxy lying in the same direction as one of the quasar images. In fact, the galaxy turns out to be a member of a galaxy cluster that is much closer to us than is the quasar. The geometry and estimated mass of the galaxy are just right to produce gravitational lensing. Figure 26.19 is a schematic diagram showing this effect.

Gravitational lenses can produce not only double images but also multiple images, arcs, or rings (Figure 26.20). A remarkable example is shown in this chapter's opening figure, in which we see the images of a blue background galaxy distorted into gracefully curving blue arcs by the foreground (yellow) cluster of galaxies.

(text cont. on page 533)

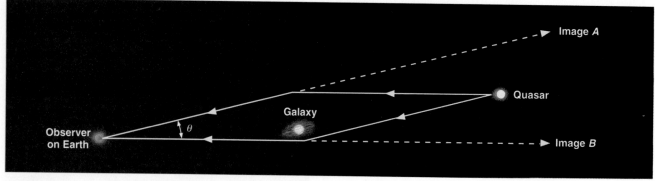

Figure 26.19
An explanation of how a gravitational lens could make two images, as in Figure 26.18. Two light rays from a distant quasar are shown being bent while passing a foreground galaxy; they then arrive together at Earth. Although the two beams of light contain the same information, they now appear to come from two different points on the sky. This sketch is oversimplified and not to scale, but it gives a rough idea of the lensing phenomenon.

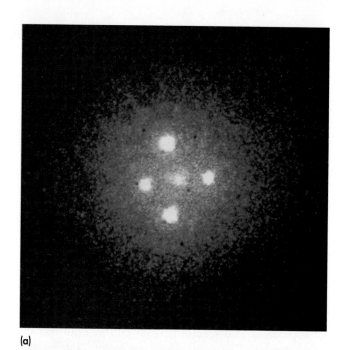

(a)

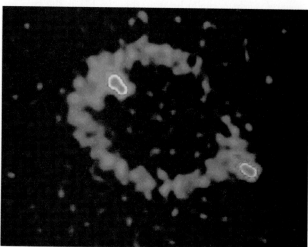

(b)

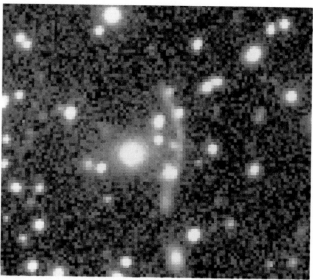

(c)

Figure 26.20
A gallery of gravitational lenses, with multiple or distorted images of distant sources. (a) The Hubble Space Telescope reveals four separate images of the same distant quasar, a configuration called an Einstein cross. The quasar is approximately 8 billion LY away. A galaxy (the slightly fuzzy object in the center) lying at a distance of 400 million LY serves as the gravitational lens. (NASA) (b) If the alignment is just right, a gravitational lens can turn a point source into what is affectionately called an Einstein ring. In this radio map, made with the Very Large Array, a dense cluster of galaxies creates two point images and a ring image of a distant quasar. (J. Hewitt, E. Turner, and NRAO) (c) The blue arc is a distant galaxy smeared out by the gravitational lensing effect of the foreground cluster of galaxies. (NOAO)

Quasars and the Attitudes of Astronomers

The discovery of quasars in the early 1960s was the first in a series of surprises astronomers had in store. Within another decade they would find neutron stars (in the form of pulsars), the first hints of black holes (in binary x-ray sources), and even the radio echo of the Big Bang itself. Many new possibilities seemed just around the horizon.

As Maarten Schmidt reminisced in 1988, "This had, I believe, a profound impact on the conduct of those practicing astronomy. Before the 1960s, there was much authoritarianism in the field. New ideas expressed at meetings would be instantly judged by senior astronomers and rejected if too far out." We saw a good example of this in the trouble Chandrasekhar had in finding acceptance for his ideas about the death of stars with cores greater than 1.4 M_{Sun}.

"The discoveries of the 1960s," Schmidt continued, "were an embarrassment, in the sense that they were totally unexpected and could not be evaluated immediately. In reaction to these developments, an attitude has evolved where even outlandish ideas in astronomy are taken seriously. Given our lack of solid knowledge in extragalactic astronomy, this is probably to be preferred over authoritarianism." [1]

This is not to say that astronomers (being human) don't continue to have prejudices and preferences. Those who had hoped to show that the redshifts of quasars were not connected with their distances (which was definitely a minority opinion), often felt excluded from meetings or from access to telescopes in the 1960s and 1970s. It's not so clear that they actually *were* excluded, as much as that they felt the very difficult pressure of knowing that most of their colleagues strongly disagreed with them. As it turned out, the evidence—which must ultimately decide all scientific questions—was not on their side either.

But today, as better instruments bring solutions to some problems and starkly illuminate our ignorance about others, the entire field of astronomy seems more open to discussing unusual ideas. Of course, before any hypotheses become accepted, they must be tested—again and again—against the evidence that nature itself reveals. Still, the many strange proposals published about what dark matter might be (we'll mention just a few in the next chapter) attest to the new openness that Schmidt described.

[1] From M. Schmidt: "The Discovery of Quasars" in *Modern Cosmology in Retrospect*, ed. B. Bertotti et al. (1990, Cambridge U. Press).

Gravitational Lenses and the Search for Dark Matter

Visible galaxies are not the only possible gravitational lenses. Concentrations of dark matter can also reveal their presence by producing this effect. Figure 26.21 shows a foreground cluster of yellowish galaxies; many blue arcs or small portions of arcs can also be seen in this image. Each arclet is the image of a very distant galaxy whose apparent shape was distorted when the light from it passed through the foreground cluster of yellowish galaxies.

These arclets can be used to map the total mass in the foreground cluster—including dark matter. The resulting maps show that the foreground galaxy cluster contains ten times more matter than is actually seen in visible light. If the dark matter were not there, we would see many fewer distorted images of the background blue galaxies. Astronomers hope to use such lens images to get a better sense of how much dark matter is contained in galaxies and clusters, but already the observations offer strong support for the idea that most of the mass in the universe is in the form of dark matter.

Figure 26.21
This image shows a cluster of galaxies (yellowish objects) that acts as a gravitational lens; it distorts the image of a very distant blue galaxy lying behind it, producing the blue arcs. From the shape and location of these arcs, astronomers can estimate how much dark matter is contained within the foreground cluster.
(Tony Tyson/AT&T Bell Labs)

Summary

26.1 The first **quasars** discovered looked like stars but had strong radio emission. The quasar spectra obtained so far show redshifts ranging from 15 to nearly 95 percent of the speed of light. The evidence shows that quasars are at the great distances implied by these redshifts, and that they therefore have 10 to 100 times the luminosity of the brighter normal galaxies. Their variations show that this tremendous energy output is generated in a small volume—in some cases, in a region no larger than our own solar system. Some quasars are members of small groups or clusters with the same redshift as the quasar. Other quasars have fuzz around them that has the spectrum of a normal galaxy, which shows that quasars are located inside galaxies.

26.2 Most astronomers now view quasars as the most extreme example of a class of peculiar galaxies that generate large amounts of energy in a small **active galactic nucleus.** Seyferts are an example of such **active galaxies.** Even some giant elliptical galaxies are strong radio sources. In many cases, there are jets of radio emission extending to large radio-emitting regions located on either side of the galaxy or quasar. Active galaxies have luminosities intermediate between those of normal galaxies and quasars.

26.3 Both active galactic nuclei and quasars derive their energy from material falling toward, and forming a hot accretion disk around, a massive black hole. Quasars were much more common billions of years ago than they are now, and astronomers speculate that they mark an early stage in the formation of galaxies, when fuel for the accretion disk and jets was more available. Quasar activity can apparently be retriggered by collisions between galaxies, which provide a new source of fuel to feed the black hole.

26.4 Quasars can be used as probes of distant reaches of the universe. For example, observations of quasars give evidence that the **gravitational lensing** effects predicted by general relativity actually do occur. Gravitational lenses can produce various types of quasar images, including double or multiple images and even arcs and rings. Analysis of lensed quasars supports the idea that most of the mass in the universe is in the form of dark matter.

Review Questions

1. Describe some differences between quasars and normal galaxies.

2. Describe the arguments supporting the idea that at least some quasars are at the distances indicated by their redshifts.

3. In what ways are active galaxies like quasars and different from normal galaxies?

4. Describe the process by which the action of a black hole can explain the energy radiated by quasars.

5. What is a gravitational lens? Show with a diagram how it is possible to see two quasars in the sky when in reality there is only one.

6. Describe the observations that convinced astronomers that M87 is an active galaxy.

Thought Questions

7. Suppose you observe a star-like object in the sky. How can you determine whether it is actually a star or a quasar?

8. Why don't any of the methods for establishing distances to galaxies, described in Chapter 25, work for quasars?

9. One of the early hypotheses to explain the high redshifts of quasars was that these objects had been ejected at very high velocities from galaxies. This idea was rejected, because no quasars with large blueshifts have been found. Explain why we would expect to see quasars with both blueshifted and redshifted lines if they were ejected from nearby galaxies.

10. If we see a double image of a quasar produced by a gravitational lens, and can obtain a spectrum of the galaxy that is acting as the gravitational lens, we can then put limits on the distance to the quasar. Explain how.

11. A friend of yours who has watched many *Star Trek* episodes says, "I thought that black holes pulled everything into them. Why then do astronomers think that black holes can explain the great *outpouring* of energy from quasars?" How would you respond?

Problems

12. Rapid variability in quasars indicates that the region in which the energy is generated must be small. Show why this is true. Specifically, suppose that the region in which the energy is generated is a transparent sphere 1 LY in diameter. Suppose that in 1 s this region brightens by a factor of 10 and remains bright for two years, after which it returns to its original luminosity. Draw its light curve (a graph of its brightness over time) as viewed from Earth.

13. A quasar has a redshift of $\Delta\lambda/\lambda = 4.1$. What is the observed wavelength of its Lyman α line of hydrogen, which has a laboratory or rest wavelength of 121.6 nm? Would this line be observable with a ground-based telescope in a quasar with zero redshift? Would it be observable from the ground in a quasar with a redshift of $\Delta\lambda/\lambda = 4.1$? Explain your answer.

Suggestions for Further Reading

Burns, J. "Chasing the Monster's Tail: New Views of Cosmic Jets" in *Astronomy*, Aug. 1990, p. 28.

Croswell, K. "Have Astronomers Solved the Quasar Enigma?" in *Astronomy*, Feb. 1993, p. 29.

Finkbeiner, A. "Active Galactic Nuclei: Sorting Out the Mess" in *Sky & Telescope*, Aug. 1992, p. 138. Good introduction by a science writer.

Osmer, P. "Quasars as Probes of the Distant and Early Universe" in *Scientific American*, Feb. 1982, p. 126.

Preston, R. "Beacons in Time: Maarten Schmidt and the Discovery of Quasars" in *Mercury*, Jan./Feb. 1988, p. 2.

Preston, R. *First Light*. 1987, Atlantic Monthly Books. On doing astronomy with the 200-in. telescope; contains a number of sections on quasar research.

Rees, M. "Black Holes in Galactic Centers" in *Scientific American*, Nov. 1990, p. 56.

Schild, R. "Gravity Is My Telescope" in *Sky & Telescope*, Apr. 1991, p. 375.

Smith, H. "Quasars and Active Galaxies" in *The Universe*, B. Preiss and A. Fraknoi, eds. 1987, Bantam.

Turner, E. "Gravitational Lenses" in *Scientific American*, July 1988, p. 54.

Weedman, D. "Quasars: A Progress Report" in *Mercury*, Jan./Feb. 1988, p. 12.

Wright, A. and Wright, H. *At the Edge of the Universe*. 1989, Ellis Horwood/John Wiley. Introductory book on extragalactic astronomy.

Using **REDSHIFT**™

1. Turn off the planets and stars, and set the deep-sky filter to show only quasars. Set the *Limiting Magnitude* to 20.

How many quasars do you see?

Is there a relation between quasar brightness and its redshift? (The redshift is given after "emission line" in the information window displayed for the object.)

2. Display only spiral galaxies and set the *Limiting Magnitude* to 20.

Estimate how many galaxies are displayed.

If quasars are related to spiral galaxies, how do you reconcile the number of spirals with the number of quasars counted in Exercise 1?

The inner parts of galaxies NGC 4038 and 4039, which are in the process of colliding. We can see many young star clusters, surrounded by reddish regions of ionized hydrogen, that formed as a result of the collision. The nuclei of the two galaxies can be seen as two diffuse yellowish "blobs" of light. Note the massive, dusty cloud of interstellar matter between the two galaxies. Such collisions alter the shape of each galaxy and result in bursts of star formation as the raw material in the galaxies interacts. (Photo by David Malin; copyright Anglo-Australian Telescope Board)

The inset is a negative image of the same galaxies with a larger field of view, showing the long streamers or tails of stars and gas pulled out of the galaxies by the interaction. (F. Schweizer, Carnegie Institution of Washington)

CHAPTER 27

The Organization of the Universe

Thinking Ahead

From the vantage point of our own island of stars, we look out and try to fathom how the other islands (galaxies) are organized in the dark ocean of space. How can we be sure that the islands we can see are representative of the whole universe?

"But I want to make sure of our whereabouts and whenabouts," said Van, "it is a philosophical need."

Vladimir Nabokov in *Ada* (1969)

Now that we are familiar with the different types of galaxies, we are ready to examine how they are organized in space and time. In this chapter we expand our perspective to the largest scales of all: we explore the distribution of galaxies over millions and billions of light years, and use that information to examine the way they change over the lifetime of the universe.

Only in the past decade have astronomers begun to make real progress in both these areas. Their work has been slow for several reasons. To understand the way galaxies are distributed in space, we must make a three-dimensional map of the positions of many galaxies. While it is easy to see where a galaxy lies on the dome of the sky, the third dimension—its distance away from us—is much more difficult to measure. (See the opening figure in the Prologue for a wonderful image of galaxies that lie at many different distances in the same part of the sky.) We must take a spectrum of each galaxy, collect enough light to measure a redshift, and then use the Hubble law to convert it to a distance. Only recently have we been able to do this for a sufficient number of remote galaxies to get a meaningful sample of the universe.

When it came to figuring out the evolution of galaxies, there were equally formidable difficulties. First, galaxies are made up of stars, and it was only after the evolution of individual stars was understood that astronomers could begin to explore how whole systems of stars change with time. Thirty years ago, any attempt to describe the evolution of galaxies would have been pointless: we simply did not know enough about the life histories of stars.

A second difficulty in the study of galaxies is the fact that they are very, very faint. Until recently it was extremely hard, even with large telescopes, to see individual stars in most galaxies, or to determine the shapes of the most remote galaxies. So measuring how they change with time became possible only when our instruments could show us the details of faint galaxies with greater resolution. Today, large telescopes on the ground and instruments like the Hubble in space are finally making such a task possible.

We do, however, find one built-in advantage in studying galactic evolution. As we have seen, the universe itself is a kind of time machine that permits us to observe remote galaxies as they were long ago. For the closest galaxies, the time the light takes to reach us is on the order of a few hundred thousand to a few million years. Typically not much changes over times that short. But when we observe a galaxy that is a billion LY distant, we are seeing it as it was when the light left it a billion years ago. By observing more and more distant objects, we look ever farther back toward a time when both galaxies and the universe were young (Figure 27.1).

Figure 27.1

An illustration of astronomical time travel. This true-color, long-exposure image, made during 48 orbits of the Earth with the Hubble Space Telescope, shows a small area in the direction of the constellation of Hercules. We can see galaxies an estimated 3 to 8 billion LY away, including a series of faint blue galaxies that were much more common in that earlier time than they are today. While the nature of these blue galaxies is still controversial, they appear blue because they are undergoing active star formation and making hot, bright blue stars. (R. Windhorst et al. and NASA)

27.1

The Distribution of Galaxies in Space

Celestial objects rarely journey through space alone. The Earth is one of nine planets orbiting the Sun. More than half of all stars are found in double or triple star systems, and many are members of star clusters. All of the stars and clusters in the Milky Way Galaxy are kept close together by the force of gravity, orbiting a common center. Astronomers wondered whether this cosmic togetherness persists on still-larger scales. Are galaxies found in groups as well, or do they all exist in isolation? And how, in general, are galaxies distributed in space? Are there as many in one direction of the sky as in any other, for example?

Edwin Hubble found answers to some of these questions only a few years after he first showed that other galaxies exist. As he examined galaxies all over the sky, Hubble made two discoveries that are crucial for studies of the evolution of the universe.

The Cosmological Principle

Hubble made his observations with what were then the world's largest telescopes—the 100-in. and 60-in. reflectors on Mount Wilson. Although these telescopes can probe to great depths, they can do so only in small fields of view. To photograph the entire sky with the 100-in. telescope would take thousands of years. So instead, Hubble sampled the sky in many regions, much as Herschel did with his star gauging (see Section 24.1). In the 1930s Hubble photographed 1283 sample areas, and on each print he carefully counted the numbers of galaxy images.

The first discovery Hubble made was that the number of galaxies visible in each area of the sky is about the same. (Strictly speaking this is true only if the light from distant galaxies is not absorbed by dust in our own Galaxy, but Hubble made corrections for this absorption.) He also found that the numbers of galaxies increase with faintness, as we would expect if the density of galaxies is about the same at all distances from us.

To understand what we mean, imagine you are taking snapshots in a crowded stadium during a sold-out concert. The people sitting near you look big, so only a few of them will fit into a photo. But if you focus on the people sitting in seats way on the other side of the stadium, they look so small that many more will fit into your picture. If all parts of the stadium have the same seat arrangements, then as you look farther and farther away, your photo will get more and more crowded with people. In the same way, as Hubble looked at fainter and fainter galaxies, he saw more and more of them.

Hubble's findings are enormously important, for they indicate that the universe is both **isotropic** and **homogeneous**—it looks the same in all directions, and a large volume of space at any given redshift is much like any other volume at that redshift. If that's so, it doesn't matter what section of it we observe (as long as it's a sizable portion)—any section will look the same as any other.

In other words, his results suggest not only that the universe is about the same everywhere, apart from changes with time, but also that the part we can see around us, aside from small-scale local differences, is representative of the whole. The idea that the universe is the same everywhere is called the **cosmological principle** and is the starting assumption for nearly all theories that describe the entire universe (see Chapter 28).

Without having established the cosmological principle, we could make no progress at all in studying the universe. Suppose our own local neighborhood were unusual in some way. Then we could no more understand what the universe is like than if we were marooned on a warm south seas island without outside communication and were trying to understand the geography of the Earth. From our limited vantage point, we could not know that some parts of the planet are covered with snow and ice, or that large continents exist with a much greater variety of terrain than that found on our island.

Hubble merely counted the numbers of galaxies in various directions without knowing how far away most of them were. But in the past decade astronomers have measured the velocities and distances of thousands of galaxies, and so built up a meaningful picture of the large-scale structure of the universe. In the rest of this section we describe what we know about the distribution of galaxies, beginning with those nearby.

The Local Group

The region of the universe for which we have the most detailed information is, as you would expect, our own local neighborhood. It turns out that the Milky Way Galaxy is a member of a small group of galaxies called, not too imaginatively, the **Local Group.** It is spread over about 3 million LY and contains at least 30 members. There are 3 large spiral galaxies (our own, the Andromeda galaxy, and M33), at least 12 dwarf irregulars, 2 intermediate ellipticals, and 14 known dwarf ellipticals (see Appendix 12).

New members of the Local Group are still being discovered. We mentioned in Chapter 24 that a dwarf galaxy only about 80,000 LY from the Earth and about 50,000 LY from the center of the Galaxy was recently found in the constellation of Sagittarius. (You may remember that this dwarf is actually venturing too close to the much larger Milky Way and will eventually be consumed by it.) Such newly discovered neighbors are typically very faint and often lie in directions where their light is dimmed by the dust in the disk of the Milky Way.

Figure 27.2 is a rough sketch of how the brighter members of the Local Group are distributed. The average of the motions of all the galaxies in the Group indicates that its total mass is about $5 \times 10^{12} \, M_{Sun}$. A substantial amount of this mass is in the form of dark matter.

Neighboring Groups and Clusters

Surveys of more-distant galaxies reveal that the clustering of our nearest neighbors into the Local Group is not exceptional. We find other galaxy groupings at every distance scale we have examined. To start locally, Figure 27.3 shows a galaxy called Dwingeloo 1 (don't you just love astronomical names!). Discovered in 1994, it is part of a small group about 10 million LY away.

However, small groups like ours are hard to notice once we turn our sight to much larger distances. Astronomers therefore focus their attention on more substantial groups called **galaxy clusters.** Such clusters are described as poor or rich depending on how many galaxies they contain. Note that as we examine clusters that are very far away, it gets more difficult to see any dwarf galaxies, and so these are typically not even counted when determining the richness of a cluster. Instead we must (by necessity) use the brighter galaxies in each cluster. (Some of this will now change as instruments such as the Keck and Hubble telescopes show us much fainter galaxies than we have been able to detect in the past.)

The nearest moderately rich galaxy cluster is called the Virgo cluster, after the constellation in which it is seen. It is about 50 million LY away and contains thousands of members, of which a few are shown in Figure 27.4. The giant elliptical (and very active) galaxy M87, which you came to know and love in Chapter 26, belongs to the Virgo cluster. Although M87 is not shown in Figure 27.4, two other giant ellipticals in the cluster are.

A good example of a cluster that is much larger than the Virgo complex is the Coma cluster, with a diameter of at least 10 million LY and thousands of observable galaxies (Figure 27.5). Lying about 250 to 300 million LY away, this cluster is centered on two giant ellipticals whose luminosities equal about 400 billion Suns. Dwarf galaxies are too faint to be seen at this distance, but if they are present, as we expect they are, then Coma likely contains tens of thousands of galaxies. The total mass of this cluster is about $4 \times 10^{15} \, M_{Sun}$ (enough mass to make 4 million billion stars like the Sun!).

Let's pause here for a moment of perspective. We are now discussing numbers by which even astronomers sometimes feel overwhelmed. The Coma cluster may have 10, 20, or 30 thousand galaxies, and each galaxy has billions and billions of stars. If you were traveling at the speed of light, it would still take you more than 10 million

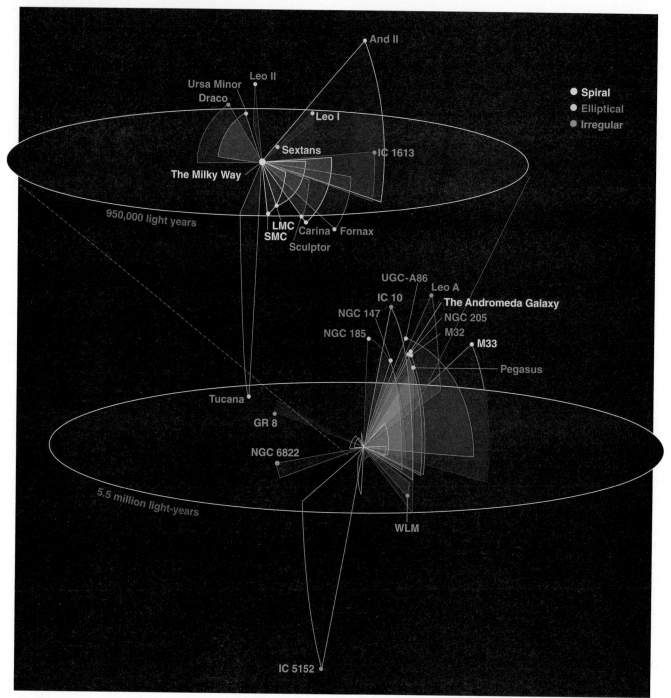

Figure 27.2
A three-dimensional view of some members of the Local Group of galaxies, with our Milky Way at the center. The exploded view at the top shows the region closest to the Milky Way. The three largest galaxies among the three dozen or so members of the Local Group are all spirals; most of the others are dwarf galaxies. (From an original drawing by Thomas L. Hunt, *Astronomy* Magazine; from data by Paul Hodge)

years (longer than the history of the human species) to cross this giant swarm of galaxies. And if you lived on a planet on the outskirts of one of these galaxies, many other members of the cluster would be close enough to be noteworthy sights in your nighttime sky.

Really rich clusters such as Coma are usually spherical in shape, with a high concentration of galaxies near the center. Giant elliptical galaxies are present in these central regions, but few if any spiral galaxies are found there. The spirals that do exist generally occur on the outskirts of clusters. We might say that ellipticals are highly "social": they are often found in groups, and very much enjoy "hanging out" with other ellipticals in crowded situations.

(text continued on page 542)

Figure 27.3
This member of a neighboring group of galaxies, called Dwingeloo 1 (after the location of the Dutch telescope that found it), was discovered in 1994. It is about 10 million LY away. The spiral galaxy is in the middle of the picture. The objects with colored spikes are bright stars in our own Galaxy that are overexposed. (S. Hughes and S. Maddox, Isaac Newton Telescope, Royal Greenwich Observatory)

Figure 27.4
The central region of the Virgo cluster of galaxies, the nearest rich cluster (50 million LY away). With its hundreds of bright galaxies, Virgo is the dominant feature of the Local Supercluster of galaxies. In this picture you can see two giant elliptical galaxies, M84 and M86, plus a number of spirals. (© Anglo-Australian Telescope Board)

Figure 27.5
The central part of the rich Coma cluster of galaxies, named because it can be seen in the constellation of Coma Berenices. The two dominant galaxies in this image are huge elliptical galaxies, and they have the yellowish color characteristic of old stars. (The bright blue object is a star in our own Galaxy.) More than 10,000 galaxies belong to this cluster. Astronomers see almost no spirals like the Milky Way in the central regions of such rich clusters. (National Optical Astronomy Observatories)

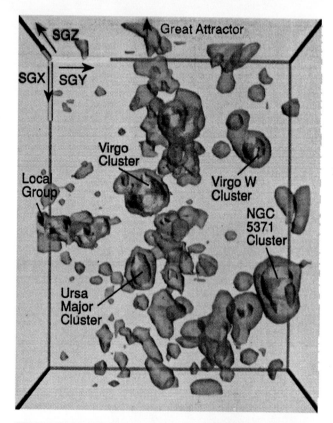

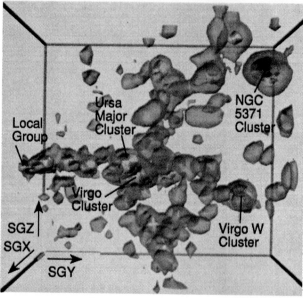

Figure 27.6
The distribution of galaxies in the Local Supercluster, a volume of space over 100 million LY across. Using a computer, astronomer Brent Tully has mapped the concentration of galaxies in an arbitrary box of space from two different perspectives. The Local Group is at the left wall in each case, and the Virgo cluster is in the center. The green areas show where the galaxies are found; the regions colored red and yellow have the highest density of galaxies. Galaxies are found in clumps and small groups, while much of space contains no galaxies at all. (Courtesy of Brent Tully, University of Hawaii)

Spirals, on the other hand, are more "shy": they are more likely to be found in poor clusters or on the edges of rich clusters. We will return to why these differences exist later in the chapter.

Superclusters and Voids

After astronomers discovered clusters of galaxies, they naturally wondered whether there were still-larger structures in the universe. Do the clusters of galaxies also gather together? It turns out they do, forming very large-scale structures that we call **superclusters.** And among the superclusters we have begun to glimpse great **voids,** emptier zones where few if any galaxies can be found.

The best-studied of the superclusters is the *Local Supercluster.* So-named because we are part of it, there is really nothing very "local" about it. With a diameter of at least 100 million LY and a mass of about 10^{15} M_{Sun}, it contains our Local Group of galaxies as well as the Virgo cluster, which lies near its center. The Milky Way lies on the outskirts of the Local Supercluster (Figure 27.6).

Perhaps the most surprising fact revealed by studies of the Local Supercluster is that space is mostly empty. Most of the galaxies are concentrated into clusters, and these clusters occupy only about 5 percent of the total volume of space contained within the boundaries of the Local Supercluster. We find similar results for other superclusters as well.

Slices of the Universe

Once we move outside the Local Supercluster, our view must take in enormous volumes of space. Surveying these distant regions is one of the most difficult and time-consuming projects in astronomy, although it is also one of our most important steps toward understanding the large-scale structure of the universe. How can we make such a task manageable?

The situation is similar to that faced by the first explorers in a huge, uncharted territory on Earth. Since there is only one band of explorers and an enormous amount of land, they have to make choices about where to go first. One strategy might be to strike out in a straight line in order to get a sense of the terrain. They might, for example, cross some mostly empty prairies and then hit a dense forest of trees. As they make their way through the forest, they learn how thick it is in the direction they are traveling, but not its width to their left or right. Then a river crosses their path; as they wade across, they can measure its width but learn nothing about its length. Still, as they go on in their straight line, they begin to get some sense of what the landscape is like and can make at least part of a map. Other explorers, striking out in other directions, will someday help fill in the remaining parts of that map.

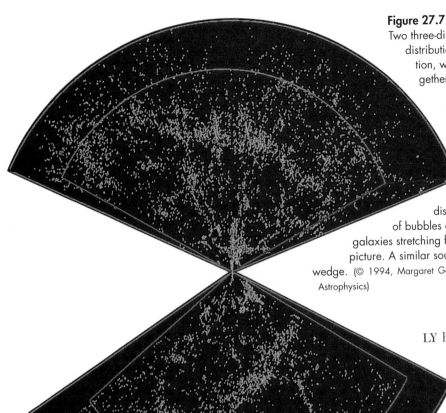

Figure 27.7

Two three-dimensional "slices of the universe" showing the distribution of galaxies in space as seen from our location, which is in the middle where the slices come together. The figure includes points for 9325 galaxies toward the north and south poles of the Milky Way for which redshifts have been measured. The upper and lower boundaries of the wedges are about 400 million LY from us. Note the concentration of galaxies in narrow bands or lanes, with large voids (some as wide as 150 million LY) between them. We would obtain a similar distribution if we took a slice through a collection of bubbles of various sizes. The Great Wall is the band of galaxies stretching from left to right across the middle of the upper picture. A similar southern wall runs diagonally across the southern wedge. (© 1994, Margaret Geller, J. Huchra, et al./Harvard-Smithsonian Center for Astrophysics)

Astronomers must make the same sort of choices. We cannot explore in every direction: there are far too many galaxies and far too few telescopes (and graduate students) to do the job. But we can pick a single direction or a small slice of the sky and start mapping the galaxies. Margaret Geller (see "Voyagers in Astronomy" box), John Huchra, and their students at the Harvard-Smithsonian Center for Astrophysics have examined such slices for a decade and a half; a computer plot of their results is shown in Figure 27.7. (In this figure, the radial velocities of the galaxies are plotted, but remember that according to the Hubble law, distance is proportional to radial velocity.) Other groups around the world are undertaking similar probes in other directions, and tens of thousands of galaxies have been mapped so far. Although this is still an insignificantly small section of what is out there, the first surveys have already yielded major surprises.

Galaxies are not arranged uniformly throughout the universe, but are found in huge filamentary superclusters that look like great arcs of inkblots splattered across a page. A dramatic example of such a filamentary structure, nicknamed the "Great Wall," is shown in Figure 27.7. A sheet of galaxies at least 500 million LY long, 200 million LY high, and about 15 million LY thick, the Great Wall lies an average of about 250 million LY from our own Galaxy. Its mass is estimated to be 2×10^{16} M_{Sun}, a factor of 10 greater than the mass of the Local Supercluster.

Separating the filaments and sheets are the voids, which look like huge empty bubbles walled in by the great arcs of galaxies. The existence of these voids came as a surprise when galaxy surveys first began showing them in the late 1970s and early 1980s. They have typical diameters of 150 million LY, with the clusters of galaxies concentrated along their walls. The whole arrangement of filaments and voids reminds us of a sink full of soap bubbles, the inside of a honeycomb, or a hunk of Swiss cheese with very large holes. If you take a good slice or cross-section through any of these, you will see something that looks like Figure 27.7.

Before the voids were discovered, most astronomers would probably have predicted that the regions between giant clusters of galaxies are filled with many small groups of galaxies, or even with isolated individual galaxies. Careful searches within these voids have so far confirmed that few galaxies of any kind are found there. Apparently, 90 percent of the galaxies occupy less than 10 percent of the volume of space.

We have said that the universe is expanding, but what exactly in Figure 27.7 is expanding? The galaxies and clusters of galaxies are held together by gravity and do not expand as the universe does. However, the voids do grow larger and the filaments move farther apart.

In this discussion we have not considered the presence of dark matter, which may make up a much larger part of many galaxies than the matter we can detect. Nor have we considered whether the voids might actually con-

Margaret Geller: Cosmic Surveyor

Born in 1947, Margaret Geller was the daughter of a chemist who encouraged her interest in science and helped her visualize the three-dimensional strucure of molecules as a child. (It was a skill that would come in very handy for visualizing the three-dimensional structure of the universe.) She remembers being bored in elementary school, but was encouraged to read on her own by her parents. Her recollections also include subtle messages from teachers that mathematics (her strong early interest) was not a field for girls, but she did not allow herself to be deterred.

Geller obtained a B.A. in physics from the University of California at Berkeley and became the second woman to receive a PhD in physics from Princeton. There, while working with James Peebles, one of the world's leading cosmologists (see Chapter 28), she became interested in problems relating to the large-scale structure of the universe. In 1980 she accepted a research position at the Harvard-Smithsonian Center for Astrophysics, one of the most dynamic institutions for astronomy research in the country. She saw that to make progress in understanding how galaxies and clusters are organized, a far more intensive series of surveys was required. Although it would not bear fruit for many years, Geller and her collaborators began the long, arduous task of mapping the galaxies.

Her team was fortunate to be given access to a telescope that could be dedicated to their project, the 60-in. reflector on Mount

Dr. Margaret Geller. (CfA)

Hopkins, near Tucson, Arizona, where they and their assistants continue to take spectra to determine galaxy distances. To get a slice of the universe, they point their telescope at a determined position in the sky, and then let the rotation of the Earth bring new galaxies into their field of view. In this way they have measured the positions and redshifts of over 15,000 galaxies, and made a wide range of interesting maps to display their data. Their surveys now include "slices" in both the Northern and Southern Hemispheres.

As news of her important work spread beyond the community of astronomers, Geller received a MacArthur Foundation Fellowship in 1990. These fellowships, popularly called "the MacArthur genius awards," are designed to recognize truly creative work in a wide range of fields. Geller continues to have a strong interest in visualization, and has (with filmmaker Boyd Estus) made several award-winning videos explaining her work to nonscientists (one is entitled *So Many Galaxies, So Little Time!*). She has appeared on a variety of national news and documentary programs, including the *MacNeil/Lehrer Newshour*, *The Astronomers*, and *The Infinite Voyage*. Energetic and outspoken, she has given talks on her work to many audiences around the country, and works hard to find ways to explain the significance of her pioneering surveys to the public.

> It's exciting to discover something that nobody's seen before. [To be] one of the first three people to ever see that slice of the universe . . . [was] sort of being like Columbus. . . . Nobody expected such a striking pattern!
> —Margaret Geller

---

tain dark matter. Many astronomers who make models of galaxy formation feel that dark matter plays a key role in determining the large-scale structure we observe. We will return to the role of dark matter in Section 27.4.

Implications of the Structure

Knowledge of exactly *how* galaxies are distributed may provide clues as to *why* they are distributed that way. The existence of such large filaments of galaxies and voids is a puzzle, because we have evidence (to be discussed in Chapter 28) that the universe was extremely smooth a few hundred thousand years after forming. The challenge for the theoretician is to understand how such a featureless universe changed into the complex and lumpy one that we see today.

For example, based on data such as that plotted in Figure 27.7, which shows that galaxies are distributed on the walls of bubble-like structures, some astronomers argued that matter used to be inside the bubbles and then was somehow cleaned out and pushed toward the walls. But how? One possibility is that there were giant explosions during the first few billion years of the universe that swept the gas into bubble-like shells, and that galaxies subsequently formed in those shells. Unfortunately, despite considerable ingenuity, astronomers have yet to devise a way to produce sufficiently energetic explosions to account for the largest voids.

Even if explosions do not seem to explain why the voids exist, this discussion should serve to show how measurements of galaxy distribution in space can help generate ideas about what kinds of events must have occurred when superclusters, clusters, and galaxies were just beginning to form. We will discuss more theories about the earliest stages of the universe in Chapter 28.

If all this structure exists, you may want to ask whether the universe can really be described as homogeneous and isotropic. The answer is probably yes, provided we consider regions of the universe large enough to include a number of superclusters and voids. This is similar to making a meaningful survey of small towns and wide-open spaces in the American midwest. If you live in a town and look only at the houses nearest to you, your perspective will be biased. Similarly, if you live on an isolated farm and look only at the nearest few acres, you will miss all the things that happen exclusively in urban areas. A representative survey must catch enough people or territory to include towns, villages, farms, and open country.

27.2

The Evolution of Galaxies: The Observations

Measuring the positions and distances of galaxies is not the only way to learn more about their origins. We can also examine their spectra, colors, and shapes at different periods of cosmic history to see how galaxies have evolved.

Spectra, Colors, and Shapes

The spectrum of a galaxy provides a great deal of information. We have already seen how the redshift can be used to estimate its distance. Studies of a galaxy's rotation can be used to calculate its mass. Detailed spectral line analysis can determine what types of stars inhabit a galaxy, what their compositions are, and whether a galaxy contains large amounts of interstellar matter.

Unfortunately, many galaxies are so faint that collecting enough light to produce a measurable spectrum is impossible. Astronomers thus have to use a much rougher guide to estimate what kinds of stars inhabit the faintest galaxies—their colors (Figure 27.8). To understand how this works, remember that hot luminous blue stars have lifetimes of only a few million years. If we see a very blue galaxy, we know that it must have many hot luminous blue stars, and that star formation must have occurred in the past few million years. A yellow or red galaxy, on the other hand, contains mostly old stars, formed billions of years before the light that we now see was emitted.

Another important clue to the nature of a galaxy is its shape. As we have seen, spiral galaxies contain young stars and large amounts of interstellar matter, while elliptical galaxies have mostly old stars and very little interstellar matter. Elliptical galaxies appear to have turned most of

Figure 27.8

Elliptical galaxy NGC 1199 and companions. NGC 1199 is the large elliptical galaxy at bottom left. It is accompanied by a face-on barred spiral galaxy (NGC 1189) seen toward the upper left; another spiral galaxy (NGC 1190) of spindly appearance is found near the right edge. We can see the yellow-orange color characteristic of older population II stars in the elliptical galaxy, and in the central bar and bulge of the barred spiral. Younger population I stars produce the bluish color of the barred spiral's disk. This picture shows how colors can be used to determine the ages of stars found in distant galaxies. (N. A. Sharp/NOAO)

their interstellar matter into stars many billions of years ago, while star formation has continued until the present day in spiral galaxies. If we can count the number of galaxies of each type that exist during each epoch of the universe, it will help us understand how the pace of star formation changes with time.

Unfortunately, very distant galaxies appear small on the sky. Often we cannot tell whether a distant galaxy is a spiral or an elliptical. One research program being carried out with the Hubble Space Telescope involves obtaining images of distant galaxies, unblurred by the Earth's atmosphere, in order to determine whether the same types of galaxies visible nearby in the present-day universe existed billions of years ago (see Figure 27.1 and the opening figure for Chapter 25).

The Ages and Composition of Galaxies

One of the most important conclusions from our observations of galaxies is that nearly all of them are very old. For example, our own Galaxy contains globular clusters with stars that are at least 13 billion years old, and some may be even older than that. The Milky Way must be at least as old as the oldest stars in it.

As we will discuss in Chapter 28, astronomers have traced the expansion of the universe backwards in time and discovered that the universe itself is not significantly older than the globular-cluster stars. Thus it appears that the globular-cluster stars in the Milky Way must have formed during the first billion years or so after the expansion began.

The light from the most-distant elliptical galaxies for which we have some composition information left them when the universe was only about half its present age. Yet they resemble nearby elliptical galaxies in their luminosity and colors, implying they have the same sorts of stars in them—even though the nearby galaxies are about twice as old. The similarity of ellipticals that span half the age of the universe suggests that star formation in this type of galaxy has either been absent or very slow for the last several billion years. Star formation in elliptical galaxies probably began less than a billion years or so after the universe began, and new stars continued to form for at most a few billion years.

The most-distant galaxy observed to date is so far away that the light we see now left it when the universe was about one-tenth its present age—or only about 1.5 billion years after the beginning. Yet the spectrum of this galaxy contains lines of heavy elements, including carbon, silicon, aluminum, and sulfur. These elements were not present when the universe began. This means that when the light from this galaxy was emitted, an entire generation of stars had already been born, lived out their lives, and died—spewing out the new elements made in the interiors of stars.

We can look even farther back in time by studying quasars (see Chapter 26); a few have been observed at even larger distances than the most-distant galaxies we can see. Remember that we have good evidence that quasars are found in the centers of galaxies. The gas in these distant quasars also contains some heavier elements such as carbon, nitrogen, and oxygen. Therefore, we again conclude that at least one generation of stars in the galaxy surrounding the quasar must already have completed its evolution before the light that we now see was emitted. Given the distance to quasars, this means that some stars must have formed when the universe was *even less* than a billion years old. Stars may already have been going through their life cycles before galaxies were fully developed. Any theory of galaxy formation must therefore include a way for star formation to begin close to the time that galaxies *start* their lives.

Star Formation in Different Galaxies

As we have discussed, the evolution of spiral and elliptical galaxies seems to differ in a fundamental way. In spirals, star formation is a continuous process that is still occurring today. In elliptical galaxies, even the youngest stars are older than the Sun. Since ellipticals contain very little dust or gas, star formation cannot take place in the present era.

Where did the gas and dust in ellipticals go? Much of it must have been consumed very rapidly during the formation of the first generations of stars. But star formation alone would not be efficient enough to consume all of the gas and dust originally present in elliptical galaxies. In any case, as stars evolve, they lose mass either via stellar winds, by forming planetary nebulae, or by exploding. Through all of these processes they replenish the interstellar material in their galaxy.

We must therefore conclude an efficient mechanism exists for removing gas and dust from elliptical galaxies. A clue to one such mechanism came from observations of galaxy clusters made with x-ray telescopes in orbit. In recent years, astronomers have discovered that galaxy clusters, particularly rich ones such as the Coma cluster, are usually sources of x rays (Figure 27.9). The x rays are produced by hot gas, with temperatures between 10 and 100 million degrees, located between the galaxies. This hot gas typically makes up a significant fraction of the total cluster mass. Where does it come from?

We noted earlier that spiral galaxies are absent from the central regions of rich galaxy clusters. The presence of

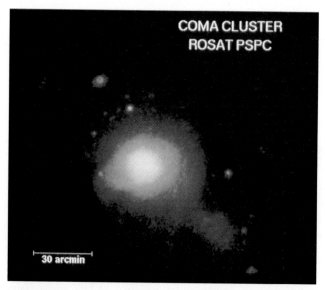

Figure 27.9
An x-ray image of the Coma cluster, taken with the Rosat x-ray satellite. You can see a large x-ray "cloud" of emission from hot gas in the cluster. Notice a smaller cloud near the bottom right, which appears to be a smaller cluster on a collision course with Coma. (Max Planck Institute for Extraterrestrial Physics)

hot gas and the absence of spirals are related. Lines in the x-ray spectrum indicate that the hot gas is not just hydrogen and helium, the composition we would expect if the gas were *primordial*—if it were the raw material from which the cluster galaxies originally formed. Instead, the hot gas contains heavy elements in abundances similar to that of the Sun. This means that the gas must have at some point passed through the interiors of stars, where heavy elements are built up. Since stars are found *in* galaxies and not between them, we conclude that the hot gas must have been processed inside the galaxies at one time.

As we saw in Chapter 22, massive stars not only make a variety of heavier elements during their lives, but they then have the courtesy to explode and recycle them into the gas among the stars. So over the billions of years, gas inside galaxies naturally builds up increasing concentrations of heavier elements. But how did this gas then get outside its parent galaxies in rich clusters such as Coma?

Galaxies in rich (and crowded) clusters are much more likely to collide with each other than in thin regions such as our Local Group. We look at collisions and mergers of galaxies in the next section, but for now we merely note that one effect of such collisions is to strip gas from both galaxies, allowing it to escape into the space around them. (See the "Astronomy Basics" box for more on why collisions among galaxies should not surprise you.)

In addition, the galaxies are in motion around the center of the cluster (just as stars and star clusters in our Galaxy are in motion around our center). As the galaxies move through the gas that is already out among them—traveling at speeds up to thousands of kilometers per second—the motion sweeps gas out of them. Figure 27.10 shows a beautiful example of a galaxy with windswept jets reminiscent of a long-haired bicyclist riding fast on a windy afternoon. Supernova explosions are also effective in driving gas out of galaxies.

Such removal of interstellar matter from galaxies not only increases the amount of gas between the galaxies, but *decreases* the amount of gas within each galaxy. As time passes, galaxies have less and less material with which to make new stars. Collisions in rich clusters also tend to destroy spiral structure. Ultimately, only rounded and featureless galaxies containing mostly old stars are left, which is just what we see in the centers of rich clusters. The swept-out gas remains hot because the galaxies continue to move through it at speeds up to thousands of kilometers per second, stirring and heating it.

In poor clusters, both the chances of collision and the speed with which galaxies move are significantly less. As a result, there is much less stripping of gas going on, and so more spirals can be seen throughout the cluster. Here again we meet the idea introduced in Chapter 25—that a galaxy's appearance depends not just on how it was born, but on the interactions with other galaxies in its environment.

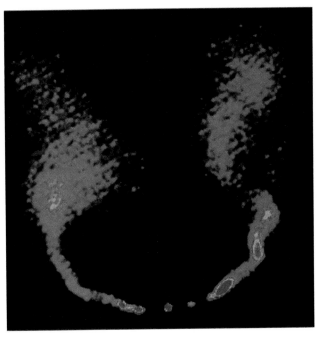

Figure 27.10
A radio image of galaxy NGC 1265, located in the Perseus cluster, which moves at about 2000 km/s through the hot gas that fills the cluster. Like a boat leaving a wake in a river, the jets (which are about 60,000 LY long) are swept back by the galaxy's motion through the gas. (National Radio Astronomy Observatory)

As a result, in the present era relatively few spirals are found near the centers of rich clusters. Conversely, isolated galaxies found in regions outside of clusters or groups of galaxies, where the density of material is low, are mostly spirals. Spiral galaxies, such as the Milky Way and the Andromeda galaxy, have managed to retain their gas and dust because they lie isolated in regions of space where the density of intergalactic gas is too low to sweep them clean.

ASTRONOMY BASICS

Why Galaxies Collide and Stars Rarely Do

Throughout this book we have emphasized the large distances between objects in space. You might therefore have been surprised to hear about frequent collisions between galaxies in a rich cluster. Yet (except for the very cores of galaxies) we have not worried at all about *stars* colliding. Let's see why there is a difference.

The reason is that stars are pathetically small compared to the distances between them. Let's use our Sun as an example. The Sun is about 1.4 million km wide, but is separated from the closest other star by about 4 LY, or about 38 trillion km. In other words, the Sun is 27 million of its own diameters from its nearest neighbor. This is typical of stars that are not in the nuclear bulge of a galaxy, or inside star clusters. Let's contrast this with the separation of galaxies.

The visible disk of the Milky Way is about 100,000 LY in diameter. We have three satellite galaxies that are just one or two Milky Way diameters away from us (and we have already seen that these will eventually collide with us). The closest major spiral is M31, about 2.4 million LY away. Therefore our nearest large galaxy neighbor is only 24 of our Galaxy's diameters from us. (And that does not even take into consideration that both galaxies probably have a much larger corona of dark matter!)

Galaxies in rich clusters are even closer together than the members of our poor Local Group. Thus the chances of galaxies colliding are far greater than the chances of stars in the disk of a galaxy colliding. And we should note that the difference between the separation of galaxies and stars also means that when galaxies do collide, their stars can often pass right by each other like ships passing unknowingly in the night.

Colliding Galaxies: Mergers and Cannibalism

It is becoming increasingly clear that collisions between galaxies play an important role in determining how galaxies evolve. During the past two decades, astronomers have begun to see mounting evidence of such collisions all around the universe. We have already discussed some of the effects of collisions. Let's take a closer look at what happens when galaxies collide.

Stars in colliding galaxies are not much affected. Since the stars are very far apart, a direct collision of two stars is very unlikely, although their orbits are altered. Interstellar matter, however, is much more affected by galaxy interactions. Interstellar gas clouds are large and likely to experience direct impacts with other clouds. These violent collisions compress the gas in the clouds, and the increased density can lead to star formation (Figure 27.11).

In some interacting galaxies, star formation is so intense that all the available gas is exhausted in only a few

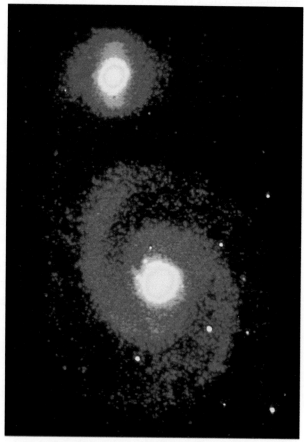

Figure 27.11
Visible light (left) and false-color infrared (right) images of M51, the Whirlpool galaxy. The outlying arm, which reaches to the companion galaxy, is much brighter in visible light than in the infrared. This difference in color suggests that most of the stars in this arm are hot young stars, which are blue in color, and that the formation of these stars was stimulated by the interaction of the two galaxies. (NOAO)

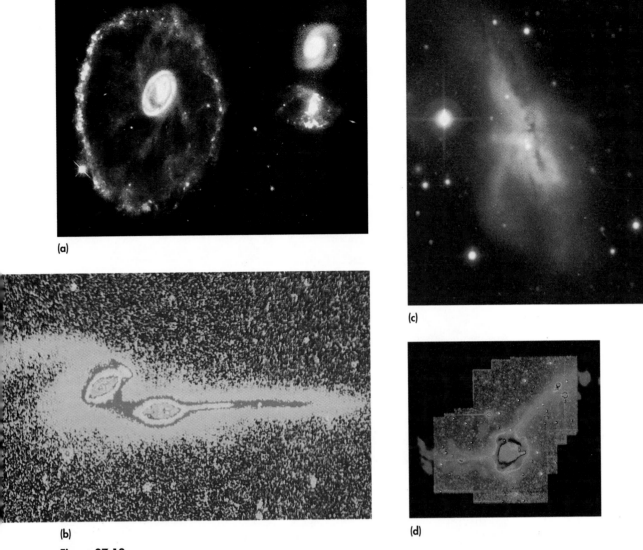

Figure 27.12

A gallery of interacting galaxies: (a) The Cartwheel galaxy, seen in this Hubble Space Telescope image, is the result of a head-on collision. The spiral galaxy at left collided with another galaxy (probably one of the two seen at right), which produced a ring of vigorous star formation (seen in blue). The ring is about 150,000 LY across and contains several billion new stars. (K. Borne and NASA) (b) NGC 4676 A and B are nicknamed "The Mice." This is a visible-light image that has been computer processed and color coded to show subtle details in the levels of light. You can see the long narrow tails of stars caused by the interactions of the two spirals. (NOAO) (c) NGC 6240 shows two galaxy nuclei quite close together in the center, with tails of material indicating that two spiral galaxies must have been involved in a collision. Observations with the IRAS satellite have shown that this galaxy puts out a tremendous amount of energy in the infrared. This is consistent with the idea that vigorous star formation in the center is warming vast quantities of dust (which hides much of this activity from our view in the visible region of the spectrum). (W. Keel and ESO) (d) This image of NGC 7252 combines visible light and radio views of the same pair of colliding galaxies. The red regions at the center of the collision are where we see the brightest visible light. The green areas are much fainter visible-light "tails" from stars and gas thrown out by the interaction. The blue color shows radio emission at 21 cm, the wavelength of hydrogen gas. Note that the hydrogen is found with the green tails and not in the red zone, confirming the idea that such collisions tend to strip gas from the galaxies involved. (J. Hibbard, J. van Gorkom, L. Schweizer, and National Radio Astronomy Observatory)

million years; the burst of star formation is clearly only a temporary phenomenon. Such bursts are very rare in isolated galaxies, but astronomers see them frequently in galaxies that are undergoing collisions. (A gallery of interesting colliding galaxies is shown in Figure 27.12.)

As you can see in these images, when galaxies interact, their shapes can change significantly. Great rings, huge tendrils of stars and gas, and other complex structures can form. Many of the larger irregular galaxies we observe may owe their chaotic shapes to past interactions.

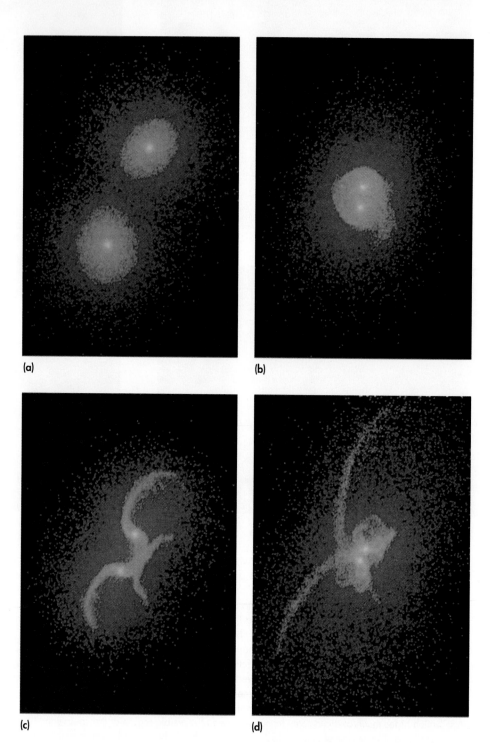

(a)

(b)

Figure 27.13
Given the right initial conditions, a computer simulation of the collision of two galaxies can produce a structure that strongly resembles NGC 4038/39, the pair of interacting galaxies shown in the opening figure of this chapter. The sequence shows the galaxies at (a) 60 million years, (b) 185 million years, (c) 310 million years, and (d) 435 million years after the interaction began. (Computer images courtesy of Josh Barnes, U. of Hawaii)

(c)

(d)

The details of galaxy collisions are complex and best simulated on a large computer (Figure 27.13). We lack much in our understanding of them, but it is nevertheless clear that collisions play a larger role in the development of galaxies than we thought even a decade ago.

If the collision is slow, the colliding galaxies may coalesce to form a single galaxy. When two galaxies of equal size are involved, we call such an interaction a **merger** (the term applied in the business world to two equal companies that join forces). But, small galaxies can also be swallowed by larger ones—**galactic cannibalism.**

The very large elliptical galaxies found in the centers of rich galaxy clusters probably form by cannibalizing a variety of smaller galaxies in their clusters. These "monster" galaxies frequently possess more than one nucleus, and have probably acquired their unusually high luminosities by swallowing nearby galaxies. The multiple nuclei are the remnants of their victims (Figure 27.14). Slow collisions and mergers can transform spiral galaxies into elliptical galaxies by stripping out their gas and causing great bursts of star formation that turns whatever gas remains into stars.

cally as interstellar matter has been used up, and the spirals are simply fainter in the current era.

The Evolution of Galaxies: The Theories

Having examined the observational evidence—much of it still fragmentary and inconclusive—we are now ready to take a look at some of the models that astronomers have proposed for how galaxies and groups of galaxies formed during the early history of the universe. Two possibilities have been explored in recent years.

Top-Down or Bottom-Up

Top-down theories assume that supercluster-sized concentrations of matter formed first and then fragmented to form galaxies. *Bottom-up theories* hypothesize that small structures formed first and then merged to build larger ones. If this bottom-up picture is correct, individual galaxies formed first and then gradually assembled to build clusters and then superclusters.

Both top-down and bottom-up theories assume that initially the universe was not absolutely smooth, but contained small fluctuations in density—regions where, by chance, more material had accumulated. (Some of this material, as we will see, may have been dark matter.) As the universe expanded, the regions of higher density acquired additional mass because they exerted a slightly larger than average gravitational force on surrounding material.

As in the case of star formation, the fate of these higher-density regions depended on the balance between pressure and gravity. Initially, each individual region expanded because the universe was expanding. However, if the inward pull of gravity exceeded the outward pressure, the individual region would ultimately have stopped expanding. It would have then begun to collapse to form a cluster of stars, a galaxy, a cluster of galaxies, or even a supercluster of galaxies.

Unfortunately, calculating what might have happened in the early universe is very difficult, primarily because we do not know what the dark matter is made of (see Section 27.4) and the most likely size of the first high-density regions to collapse is not clear. There are two possibilities that seem especially promising. In one, the typical initial condensations are very large and contain total masses equal to 10^{15} M$_{Sun}$, which is the mass of a supercluster. In the other, we start with clumps containing only 10^6 M$_{Sun}$, about the mass of a large globular cluster. Calculations indicate that condensations between these two sizes are not as likely to form.

Superclusters First

Top-down theories examine the consequences if very large-scale density fluctuations were the first to collapse.

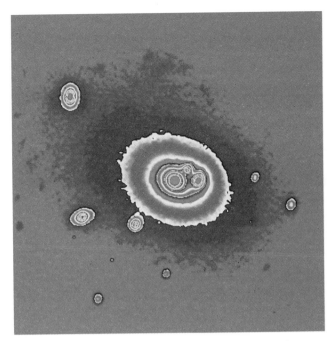

Figure 27.14
False-color image of light intensity in the multiple-nucleus galaxy NGC 6166 in the constellation of Hercules. This giant elliptical has apparently consumed several companion galaxies and lies at the center of a galaxy cluster. (R. Schild)

The Blue Galaxy Mystery

The galaxies we see today appear to differ from the types that were most common several billion years ago. Rich clusters at distances of about 5 billion years contain many more blue galaxies than do nearby rich clusters. These blue galaxies are mainly spirals (see Figure 27.1), and we have now observed them in a sufficient number of clusters to know that they were common when the universe was younger. Since a blue galaxy must contain young stars, this difference in color indicates that there were more spiral galaxies actively forming stars then than now—another indication that the rate of star formation has on average declined during the past 5 billion years or so.

Two processes have been suggested for producing a high rate of star formation in rich clusters at earlier times. First, as we saw earlier, collisions of galaxies can compress the gas within them and stimulate the formation of stars. Another agent could be the hot gas that exists in these clusters. A galaxy, as it moves on its orbit through the cluster, may run into a thicker clump of this high-temperature gas. In the ensuing collision, the cold molecular clouds in the galaxy may be compressed, again accelerating the rate of star formation.

Why have most of these luminous blue spirals vanished from the universe over the past 5 billion years? One possibility is that mergers have reduced the number of galaxies, thereby reducing the likelihood of collisions and the enhanced rates of star formation that they cause. Another is that the rate of star formation has slowed dramati-

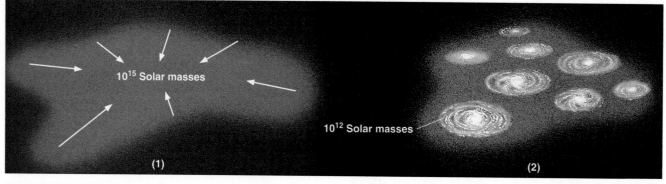

Figure 27.15
Schematic diagram showing how galaxies might have formed if large-scale supercluster structures formed first and then fragmented to form galaxies.

Calculations show that when gas clouds with masses of about 10^{15} M_{Sun} collapse, they form an irregular, pancake-shaped blob. Within the pancake many individual regions of high density are then formed, and these too will collapse under their own gravity. Stable structures can form only in regions with masses of less than 10^{12} M_{Sun} and diameters less than 300,000 LY—just the size of the galaxies we have now (Figure 27.15). Larger fragments are extremely diffuse and will be destroyed in collisions with other fragments before stars can form within them. Galaxy-sized fragments, however, can persist, and eventually the material within them can condense into stars and star clusters.

This top-down model has the advantage of being able to explain why galaxies are no larger than they are observed to be—but it has one fatal flaw. The original pancake takes a long time to collapse and fragment into still-smaller galaxy-sized structures. This model predicts that galaxies should still be forming in the present era. While astronomers have found a few nearby galaxies that may be young, it is clear that most of them formed billions of years ago. Another problem is that no way has yet been found to use top-down models to account for the existence of the large numbers of galaxies and quasars that were already present when the universe was only 10 percent of its present age.

Star Clusters First

The basic assumption of the bottom-up models is that the first higher-density regions to begin collapsing had masses of approximately a million M_{Sun}, about the same as a large globular cluster. Their collapse began when the universe was no more than 2 percent of its present age. As time passed, regions containing ever-larger mass, possibly as large as the mass of a giant elliptical, began to collapse as well. Galaxies could thus be formed either from the collapse of single galaxy-sized clouds or through the merger of several smaller structures.

Clusters of galaxies formed as individual galaxies congregated, drawn together by their mutual gravitational attraction (Figure 27.16). First a few

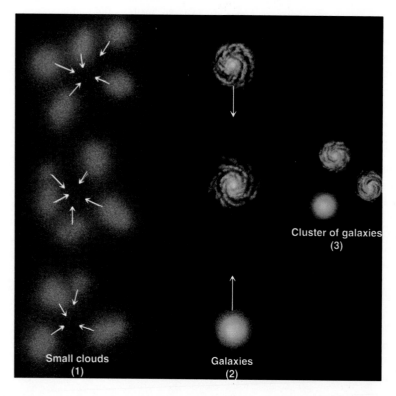

Figure 27.16
Schematic diagram showing how galaxies might have formed if small clouds formed first and then congregated to form galaxies and then clusters of galaxies.

galaxies came together to form groups, much like our own Local Group. Then the groups began combining to form clusters and eventually superclusters.

This model explains in a natural way why there are more small galaxies than large ones. The collapsing clouds are initially composed of gas, and they collide and merge to form galaxies. The more collisions and mergers that occur, the larger the galaxy that finally emerges. Since a given cloud is likely to experience only a few collisions and mergers, very large galaxies should be rare—just as 2-car accidents are more common than 50-car pileups. A detailed theory of galaxy building through this process, however, has not yet been worked out.

Giant elliptical galaxies are round, and occur in the regions of highest density. They were likely formed through the collision and merger of many smaller fragments. A collision of two systems of gas and stars tends to stir up the orbits of the individual stars within each system, and also to strip away from the system's outer regions any matter that is not held by gravity. The result of many collisions is thus to build round systems lacking extended disks.

According to this theory, spiral galaxies are formed in relatively isolated regions. A single cloud of gas collapses to a disk in which stars are then formed. A spiral might, through collisions with smaller systems, acquire some additional stars that populate its halo and nuclear bulge. Like the stars in ellipticals, which are also built through mergers, these stars are distributed in a spherical fashion. As the isolated cloud collapses to form a spiral, it may leave behind some fragments that become dwarf galaxies. Many spirals, including the Milky Way Galaxy, are surrounded by a swarm of small galaxies. As we have seen, the Milky Way is still capturing these small galaxies and adding them to its halo.

The bottom-up model predicts very few young galaxies in the present era, and indeed young nearby galaxies appear to be very rare. This model also predicts that clusters and superclusters should still be in the process of gathering together. Observations do in fact suggest that galaxy clusters are still accumulating their members.

Overall, the evidence appears to favor the idea that low-mass regions formed first, and that galaxies and clusters were formed as these regions interacted and merged under the influence of gravity. However, galaxy formation theory is still in its infancy, and cannot fully explain the individual differences we see in many galaxies and clusters. We are far from having a theory that can account for *all* of the observations, as we will see in Chapter 28.

27.4

A Universe of (Mostly) Dark Matter?

So far this chapter has focused almost entirely on matter that radiates electromagnetic energy. But, as we have pointed out in several earlier chapters, it is now clear that galaxies contain large amounts of *dark matter* as well. There may be much more dark matter, in fact, than matter we have detected—which means it could be foolish to ignore the effect of this material in our theories about the structure of the universe. (As many a ship captain in the polar seas found out too late, the part of the iceberg visible above the ocean's surface was not necessarily the only part he needed to pay attention to.) The dark matter could be extremely important in determining the evolution of galaxies and of the universe as a whole.

The idea that much of the universe is filled with dark matter may seem like a bizarre concept, but there is one historical example of dark matter that we have already described in this book. In the mid-19th century, measurements showed that the planet Uranus did not follow exactly the orbit predicted by adding up the gravitational forces of all of the known objects in the solar system. Its orbital deviations were attributed to the gravitational effects of an (at the time) invisible planet. Calculations showed where that planet had to be, and Neptune was discovered just about in the predicted location.

In the same way, astronomers are now trying to determine the location and amount of dark matter in galaxies by measuring its gravitational effects on objects we can see. And, by measuring the way that galaxies move in clusters, scientists are discovering that dark matter may play an important role in galaxy evolution as well. It appears that dark matter makes up at least 90 percent of all the matter in the universe. The following topics describe the search and evidence for the dark matter, and offer some speculations about what it might be made of.

Dark Matter in the Local Neighborhood

The first place we might look for dark matter is in our own solar system. Astronomers have examined the orbits of the known planets, and of spacecraft as they journey to the outer planets and beyond. No deviations have been found from the orbits predicted on the basis of the objects already discovered in our solar system, and so no evidence exists for large amounts of nearby dark matter.

Astronomers have also looked for evidence of dark matter in the region of the Milky Way Galaxy that lies within a few hundred light years of the Sun. In this vicinity most of the stars are restricted to a thin disk. It is possible to calculate how much mass the disk must contain in order to keep the stars from wandering far above or below it. The total matter that must be in the disk is less than twice the amount of luminous matter. This means that no more than half the mass near the Sun is dark.

Dark Matter Around Galaxies

In contrast with our local neighborhood, there is (as we saw in Chapter 24) evidence suggesting that 90 percent of the mass in the entire Galaxy may be in the form of a

Figure 27.17
The edge-on spiral galaxy NGC 5746 and a graph of the velocity with which it is rotating at each point. As is true of the Milky Way, its rotation does not decrease with distance from the center, as we would expect. This indicates the presence of a corona of dark matter, under whose influence the outer regions go around faster than the observed matter alone could explain.
(William Keel and ASP)

Dark Matter in Clusters of Galaxies

Galaxies in clusters orbit around the cluster's center of mass. It is not possible for us to follow a galaxy around its entire orbit. For example, it takes 10 billion years or more for the Andromeda and Milky Way galaxies to complete a single orbit around each other. It is possible, however, to measure the velocities with which galaxies in a cluster are moving, and then estimate what the total mass in the cluster must be to keep the individual galaxies from flying off into space. The observations indicate that the total amount of dark matter in clusters probably exceeds that contained within the galaxies themselves, indicating that dark matter exists between galaxies as well as inside them.

Dark Matter in Superclusters

The universe is expanding, but the expansion is not perfectly uniform. Some galaxies are moving away from us at a slightly faster than average rate. Others are moving away at a slower than average rate. Suppose, for example, that a galaxy lies outside but relatively close to a rich cluster of galaxies. The gravitational force of the cluster will tug on that neighboring galaxy and slow down the rate at which it moves away from the cluster due to the expansion of the universe.

As a specific example, consider the Local Group of galaxies, lying on the outskirts of the Virgo supercluster. The mass concentrated at the center of the Virgo cluster exerts a gravitational force on the Local Group. As a result, the Local Group is moving away from the center of the Virgo cluster at a velocity a few hundred kilometers per second slower than the Hubble law predicts. By measuring deviations from a smooth expansion, astronomers can estimate the total amount of mass contained in large clusters.

Astronomers have now measured accurate distances and velocities for thousands of galaxies within about 150 million LY of the Milky Way (Figure 27.18). Superim-

corona of dark matter. In other words, there could be nine times more dark matter than visible matter. The stars in the outer region are revolving very rapidly around the center of the Galaxy—so rapidly that the mass contained in all the stars and all the interstellar matter in the Galaxy cannot exert enough gravitational force to keep those distant stars in their orbits. The same result is found for other spiral galaxies as well (Figure 27.17).

Mathematical analyses of the rotation of spiral galaxies suggest that the dark matter is found in a large corona surrounding the luminous parts of each galaxy. The radius of these coronae may be as large as 300,000 LY.

Figure 27.18
A drawing showing the concentrations of matter within about 150 million LY of the Milky Way, shown at the center of the drawing. Our Galaxy is located in a region where the density is intermediate between that of the Great Attractor and a void. The mass concentration that corresponds to the Great Attractor is shown on the right-hand side of the drawing. All of the galaxies, including those in the Perseus–Pisces region, are flowing toward the Great Attractor at a velocity of about 425 km/s. You can think of it as a giant river in the sky, containing thousands of galaxies.

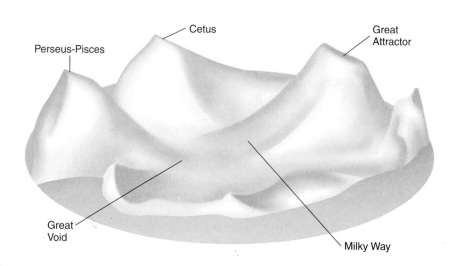

TABLE 27.1
Mass-to-Light Ratios

Type of Object	Mass-to-Light Ratio
Sun	1
Matter in vicinity of Sun	2
Total mass in Milky Way	10
Small groups of galaxies	50–150
Rich clusters of galaxies	250–300

posed on the more local motions we find a new and surprising trend. These galaxies tend to be flowing toward an enormous concentration of mass, dubbed the *Great Attractor*. The mass of the Great Attractor is estimated to be 3×10^{16} M$_{Sun}$, equivalent to tens of thousands of galaxies. This mass is much larger than the amount of luminous matter seen in this direction, and so most of the matter in the Great Attractor must be dark.

Mass-to-Light Ratio

Section 25.3 described the use of the mass-to-light ratio to characterize the matter in galaxies or clusters of galaxies. For systems containing mostly old stars, the mass-to-light ratio is typically 10 to 20, where mass and light are measured in units of the Sun's mass and luminosity. Mass-to-light ratios of 100 or more are a signal that a substantial amount of dark matter is present. Table 27.1 summarizes the results of measurements of mass-to-light ratios for various classes of objects. Very large mass-to-light ratios are found for all systems of galaxy size and larger, and this result indicates that dark matter is present in all of these types of objects. Dark matter apparently makes up most of the total mass of the universe. If observations of the motions of galaxies toward the Great Attractor prove to be correct, then as much as 99 percent of the mass in the universe may be dark.

The Search for Dark Matter

Suppose most of the universe really is made up of dark matter. How do we go about detecting it? The technique depends on its composition. For now, let's suppose that dark matter is made up of normal particles—protons and neutrons and electrons. We already know that these particles are not assembled into stars that shine, or we would see them. Neither can dark matter be in the form of dust and gas, or, again, we could detect it (see the discussion in Section 24.4). The protons and neutrons could be in the form of black holes, brown dwarfs, or white dwarfs. The latter two types of objects do emit some radiation but

have such low luminosities that they cannot be seen at distances greater than a few thousand light years. While we cannot *see* dark matter, astronomers can detect invisible assemblages of protons and neutrons because they act as gravitational lenses (see Chapter 26) and distort the radiation emitted by objects that lie behind them.

Two experiments have been designed to look for gravitational lensing by dark matter. The first involves the study of light from distant galaxies. As we saw in Chapter 26, foreground galaxies can act as a gravitational lens and distort the images of more-distant galaxies. These distorted images (such as Figure 26.21) can then be used to determine how much mass—including nonluminous dark matter—is present in the foreground cluster. The resulting maps prove that the foreground clusters of galaxies contain ten times more matter than is actually seen in visible light.

A second technique has been devised to look for black holes, brown dwarfs, and white dwarfs in the halo of our own Galaxy. These objects have been whimsically dubbed MACHOs (massive compact halo objects). If an invisible MACHO passes directly between a distant star and the Earth, it acts as a gravitational lens, focusing the light from the distant star and causing it to appear to brighten over a time interval of several days before returning to its normal brightness.

Research teams making observations of millions of stars in the Large Magellanic Cloud have recently reported several examples of the type of brightening expected if MACHOs are present in the halo of the Milky Way. From the number of such events, they can estimate how many MACHOs there are. It appears that MACHOs contribute no more than 50 percent of the dark matter in the Milky Way Galaxy.

Astronomers are eager to determine what the dark matter is, and this first result is a bit disappointing. Even if it is confirmed, we still don't know what at least 50 percent of the dark matter is made of. In the next chapter we will explore some possibilities. As we will see, it may turn out that most of the dark matter is composed of a new type of particle not yet detected on Earth.

If most of the universe is made of dark matter, we must consider what effect this material has on the clustering and evolution of galaxies. For example, while we do not presently detect anything inside the great voids, it is possible that they have significant amounts of dark matter within them. Dark matter may have played the role of "seed cores" in the formation of galaxies. And if many galaxies have large coronae made of dark matter, this will clearly influence how they will interact with one another and what shapes and types of galaxies their collisions will create. Astronomers armed with various theories are working hard to produce models of galaxy structure and evolution that take dark matter into account in just the right way. Unfortunately, it is too early to tell just what that "right way" is.

Summary

27.1 Counts of galaxies in various directions establish that the universe on the large scale is **homogeneous** and **isotropic** (the same everywhere and in all directions apart from evolutionary changes with time). The sameness of the universe everywhere is referred to as the **cosmological principle.** Galaxies tend to group together to form clusters. The Milky Way Galaxy is a member of the **Local Group,** which contains at least 30 member galaxies. Rich clusters (such as Virgo and Coma) contain thousands or tens of thousands of galaxies. **Galaxy clusters** often group together with other clusters to form large-scale structures called **superclusters,** which can extend over distances of several hundred million light years. Clusters and superclusters fill only a small fraction of space. Most of space consists of **voids** between superclusters, with nearly all galaxies confined to less than 10 percent of the total volume. Rich clusters of galaxies often contain hot (10^7 to 10^8 K) x-ray-emitting gas that has been stripped from member galaxies.

27.2 Observations place important constraints on models of galaxy formation. Galaxies were formed when the universe was no more than 1 or 2 billion years old. Star formation in spirals was much more active 5 billion years ago than it is today. There were probably more galaxies several billion years ago than there are today, with the number being reduced by collisions and mergers. Elliptical galaxies tend to be found in the centers of dense galaxy clusters, while spiral galaxies tend to be relatively isolated.

Galaxies in clusters are close enough together that collisions are likely. These collisions may trigger star formation through compression of interstellar clouds, cause the **merger** of galaxies of comparable size, or result in **galactic cannibalism,** in which a small galaxy is swallowed by a much larger one.

27.3 The challenge for galaxy formation theories is to show how an initially smooth distribution of matter can develop the structures—galaxies and galaxy clusters—that we see today. Calculations show that the first condensations of matter are likely to have contained either the mass of a galaxy supercluster or of a globular cluster. Observations seem to favor the initial condensation of globular-cluster-sized masses, which then congregate to form galaxies and clusters of galaxies.

27.4 The visible matter in the universe does not exert a large enough gravitational force to hold stars in their orbits within galaxies, or to hold galaxies in their orbits around other galaxies. There is at least 5 to 10 times, and perhaps as much as 100 times, more dark matter than luminous matter. Astronomers do not yet know whether the dark matter is made of ordinary matter—protons and neutrons, for example—or of some totally new type of particle not yet detected on Earth. Observations of gravitational lensing effects on distant objects have been used to detect some dark matter in the outer region of our Galaxy.

Review Questions

1. Explain what we mean when we call the universe homogeneous and isotropic. Would you say that the distribution of elephants on the Earth is homogeneous and isotropic? Why?

2. Describe the organization of galaxies into groupings, from the Local Group to superclusters.

3. Suppose you see a group of faint galaxies in a small part of the sky. How would you go about determining their distances?

4. What is the evidence that galaxies formed 1 or 2 billion years after the universe began?

5. Describe two possible ways in which galaxies might form. Which seems more likely? Why?

6. What is the evidence that a large fraction of the matter in the universe is invisible?

Thought Questions

7. Suppose you are standing in the center of a large, densely populated city that is exactly circular, surrounded by a ring of suburbs with lower-density population, surrounded in turn by a ring of farmland. Would you say the population distribution is isotropic? Homogeneous?

8. Use the data in Appendix 12 to determine which is more common in the Local Group—large luminous galaxies, or small faint galaxies. Which is more common—spirals or ellipticals?

9. Based on data in Appendix 12, would you describe the Milky Way Galaxy as a typical member of the Group? Why or why not?

10. Describe how you might use the color of a galaxy to determine something about what kinds of stars it contains.

11. Suppose a galaxy formed stars for a few million years and then stopped. What would be the most-massive stars on the main sequence after 500 million years? After 10 billion years? How would the color of the galaxy change over this time span? (Refer to Table 21.1.)

12. Suppose that the Milky Way Galaxy were truly isolated, with no other galaxies within 100 million LY. Then suppose that galaxies were observed in large numbers at distances greater than 100 million LY. Why would it be more difficult to determine accurate distances to those galaxies than if there were also galaxies relatively close by?

13. Given the ideas presented here about how galaxies form, would you expect to find a giant elliptical galaxy in the Local Group? Why or why not? *Is* there a giant elliptical in the Local Group?

14. Can an elliptical galaxy evolve into a spiral? Explain your answer.

15. Why do we know less about the formation of galaxies than about the formation of stars?

16. Suppose you developed a theory to account for the evolution of New York City. Would it most closely resemble a bottom-up or a top-down theory as we have applied those terms to galactic evolution?

17. Margaret Geller and John Huchra have been making maps by observing a slice of the universe and seeing where the galaxies lie within that slice. If the universe is isotropic and homogeneous, why do they need more than one slice? Suppose they now want to make each slice extend farther into the universe. What do they need to do?

Problems

18. Suppose that on one survey you count galaxies to a certain limiting faintness. On a second survey you count galaxies to a limit that is four times fainter.

    a. To how much greater distance does your second survey probe?

    b. How much greater is the volume of space you are reaching in your second survey?

    c. If galaxies are distributed homogeneously, how many times more of them would you expect to count on your second survey?

19. Calculate the mass-to-light ratios for the stars listed in Table 17.2. Can stars alone explain a mass-to-light ratio of 100, which is measured for some elliptical galaxies?

20. Show just how dominant the dark matter halo of our own Galaxy is by making a scale drawing. Use data given in the text for the diameter of the luminous disk and the sphere of dark matter that surrounds it.

Suggestions for Further Reading

Barnes, J. et al. "Colliding Galaxies" in *Scientific American,* Aug. 1991, p. 40.

Dressler, A. *Voyage to the Great Attractor.* 1994, A. Knopf. A noted astronomer describes how we find large-scale structure.

Geller, M. and Huchra, J. "Mapping the Universe" in *Sky & Telescope,* Aug. 1991, p. 134.

Hodge, P. "Our New Improved Cluster of Galaxies" in *Astronomy,* Feb. 1994, p. 26. On the Local Group.

Keel, W. "The Real Astrophysical Zoo: Colliding Galaxies" in *Mercury,* Mar./Apr. 1993, p. 44.

Lake, G. "Cosmology of the Local Group" in *Sky & Telescope,* Dec. 1992, p. 613.

Lemonick, M. *The Light at the Edge of the Universe.* 1993, Villard/Random House. A journalist's tour of extragalactic astronomy.

MacRobert, A. "Mastering the Virgo Cluster" in *Sky & Telescope,* May 1994, p. 42. How to observe the galaxies through small telescopes.

Parker, B. *Colliding Galaxies.* 1990, Plenum Press. Good nontechnical introduction.

Schramm, D. "Dark Matter and the Origin of Cosmic Structure" in *Sky & Telescope,* Oct. 1994, p. 28.

Using **REDSHIFT** ™

1. Spot the brightest galaxies in the Local Cluster by turning on only the stars and galaxies and setting the *Limiting Magnitude* to 6.5. All of these galaxies are visible with the unaided eye, but can also be viewed through binoculars or a small telescope.

2. Pick out the Virgo and Coma clusters by setting the *Limiting Magnitude* for galaxies to 14. You need at least a small telescope to see these galaxies.

From the map, what appears to be the most common type of galaxy in the Virgo cluster?

What would you expect the most common type to be? Can you explain any difference?

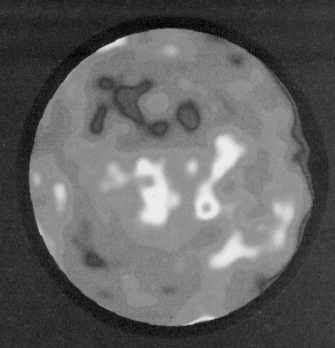

North Galactic Hemisphere

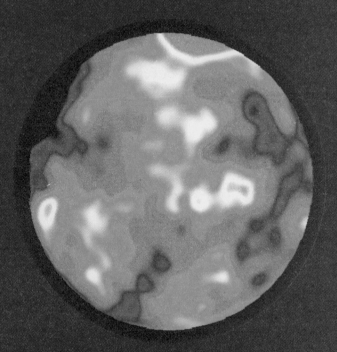

South Galactic Hemisphere

$-100\ \mu K$ $+100\ \mu K$

This false-color map shows tiny variations in the temperature of the cosmic microwave background, the radiation left over from the very hot, early phases of the universe. It is constructed from four years of observations with the Cosmic Background Explorer (COBE) satellite in Earth orbit. This radiation has been redshifted by the expansion of the universe until today it is mostly in the infrared and radio regions of the spectrum. The blue and red spots on the map correspond to regions of greater and lesser density in the early universe. It is out of such tiny fluctuations of density that all the structure we now observe in the universe grew. (C. Bennett, NASA)

CHAPTER 28

The Big Bang

Thinking Ahead

As we look farther and farther out in the universe, we look farther and farther back in time. What lies at the beginning of time? What is the earliest information about the universe that our instruments can detect?

We are now ready to complete our voyage through the universe by asking the largest questions that astronomers can ask. How did the entire universe come into being, how has it changed since the beginning, and what will its ultimate fate be? In past centuries such questions were considered the realm of religion and philosophy. The methods of science could not be applied to them because science depends on experiments and observations to decide among possible models or theories. Only recently have we been able to carry out the sorts of observations and experiments that can help us understand the past and future of the cosmos (Figure 28.1).

The branch of astronomy devoted to the study of the universe as a whole is called **cosmology.** An enormous field, it has been the subject of more popular books than any other astronomical subject. In this brief chapter we can give you only an overview of the highlights of current cosmological thinking. But at the end you will find a list of our favorite nontechnical books on the subject; we urge you to consult them if our discussion whets your appetite for more.

Let us begin by reviewing, in Table 28.1, some of the observational discoveries about the universe as a whole that have already been covered in this book. For example, as discussed in Chapter 25, we know from observations of redshifts and the Hubble law that the universe is expanding. Thus we do not need to consider

Making a model of the universe is like trying to pitch a tent on a moonless night in a howling Arctic wind. The tent is theory. The wind is experiment. Progress is made whenever a tent peg proves sturdy enough to hold.

Timothy Ferris in "Minds and Matter," a brief article on cosmology in *The New Yorker,* May 15, 1995

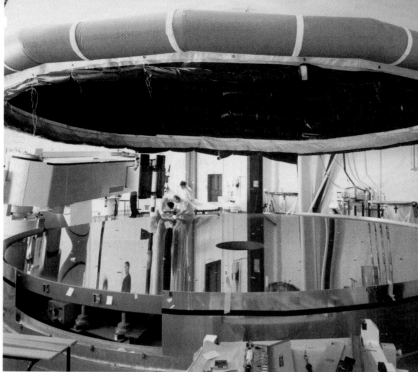

Figure 28.1
Two of the instruments that will be used to observe the most-distant objects in the universe to answer questions in cosmology: (Left) The Hubble Space Telescope, seen in the Shuttle during the December 1994 repair mission. (NASA) (Right) One of the 8.2-m-wide mirrors for the European Southern Observatory's Very Large Telescope being tested in a French optical workshop. It is the first of four such giant mirrors that will be coupled together on a mountaintop in Chile toward the end of this century to make the world's largest visible-light telescope. (ESO)

TABLE 28.1
Some Observed Characteristics of the Universe

1. All galaxies show a redshift proportional to distance, implying that the universe is expanding. (Chapter 25)
2. The distribution of galaxies on the largest scale is isotropic and homogeneous (the cosmological principle). (Chapter 27)
3. The contents of the universe evolve with time: hydrogen and helium are changed into heavier elements inside stars. (Chapters 15, 21, 22)
4. Gravity warps (curves) the fabric of spacetime. (Chapter 23)

---

any model that does not account in a natural way for this expansion. In the discussion that follows, we build on this fundamental observation (and others) to construct the best model we can currently make of the cosmos.

28.1

The Age of the Expanding Universe

With hindsight, it is surprising that scientists in the 1920s and 1930s were so shocked to discover that the universe is expanding. In fact, our theories of gravity demand that the universe *must* be either expanding or contracting. To

show what we mean, let's begin with a universe of finite size—say a giant ball of a thousand galaxies. All these galaxies attract each other because of their gravity. If they were initially stationary, they would inevitably begin to move closer together and eventually collide. They could avoid this collapse only if for some reason they happened to be moving away from each other at high speeds. In just the same way, a rocket launched at high enough speed can avoid falling back to Earth.

The problem of what happens in an infinite universe is harder to solve, but Einstein used his theory of general relativity to show that even infinite universes cannot be static. Since astronomers at that time did not yet know the universe was expanding, Einstein changed his equations with the introduction of an arbitrary new term called the **cosmological constant.** It represented a hypothetical force of repulsion that could balance gravitational attraction on the largest scales and permit galaxies to remain at fixed distances from one another. When Hubble and his co-workers reported that the universe was expanding, Einstein realized—to his chagrin—that the constant was not needed.

The Hubble Time

If we had a movie of the expanding universe, and ran the film backward, what would we see? The galaxies, instead of moving apart, would move *together*—getting closer and closer all the time. Eventually, we would find that *all* matter was once concentrated in an infinitesimally small volume. Astronomers identify this time with the *beginning of the universe*. The explosion of that concentrated universe at the beginning of time is called the **big bang** (not a bad term, since you can't have a bigger bang than one that creates the entire universe!). But when did this occur?

We can make a reasonable estimate of the time since the expansion began. To see how astronomers do this, let's begin with an analogy. Suppose your astronomy class decides to have a party (a kind of "big bang") at someone's home to celebrate the end of the semester. Unfortunately, everyone is celebrating with so much enthusiasm that the neighbors call the police, who arrive and send everyone away at the same moment. You get home at 2 A.M., still somewhat bitter about the way the party ended, and realize you forgot to look at your watch to see what time the police got there. But you use a map to measure that the distance between the party and your house is 40 km. And you also remember that you drove the whole trip at a steady speed of 80 km/h (since you were worried about the police cars following you). Therefore the trip must have taken:

$$\text{Time} = \frac{\text{Distance}}{\text{Velocity}} = \frac{40 \text{ km}}{80 \text{ km/h}} = 0.5 \text{ h}$$

So the party must have broken up at 1:30 A.M.

There were no humans around to look at their watches when the universe began, but we can use the same technique to estimate when the galaxies began moving away from each other. (Remember that, in reality, it is space that is expanding, not the galaxies that are somehow traveling on their own!) If we can measure how far apart the galaxies have gotten, and how fast they are moving, we can figure out how long a trip it's been.

Let's call the age of the universe measured in this way T_0. The time it has taken a galaxy to move a distance, d, away from the Milky Way (remember that at the beginning the galaxies were all together in a very tiny volume) is (as in our example)

$$T_0 = d/v,$$

where v is the velocity of the galaxy. Since individual galaxies have their own local motions, we want to make measurements not for just one galaxy, but for a good sample of them. If we can measure the speed with which many different galaxies are moving away, and also the distances between them, we can establish how long ago the expansion began.

Making such measurements should sound very familiar. This is just what Hubble and many astronomers after him needed to do in order to establish the Hubble law and the Hubble constant. We learned in Chapter 25 that a

Figure 28.2
A key target for measuring distances to galaxies (and thus both the Hubble constant and the age of the universe) with the Hubble Space Telescope is this cluster of galaxies in the southern constellation of Fornax. You can see a number of bright elliptical galaxies on this wide-angle photograph, and a beautiful barred spiral in the lower right. (© Anglo-Australian Telescope Board)

galaxy's distance and its velocity in the expanding universe are related by

$$v = H \cdot d$$

where H is the Hubble constant. Combining these two expressions gives us

$$T_0 = d/v = d/(H \cdot d) = 1/H$$

We see, then, that the work of calculating this time was already done for us when astronomers measured the Hubble constant. The age of the universe estimated in this way turns out to be just the *reciprocal of the Hubble constant* (that is, $1/H$). This age estimate is sometimes called the *Hubble time*. For a Hubble constant of 25 km/s per million LY, the Hubble time is about 13 billion years.

Remember that to determine the Hubble constant (and thus our rough gauge of the age of the universe), astronomers have to be able to measure the velocities and distances of many galaxies. We can measure velocity using Doppler shift of the lines in a galaxy's spectrum, but—as we saw in Chapter 25—distances are much more difficult to obtain (Figure 28.2). Controversy still exists over the accuracy of our various methods. If distance estimates were in error by a factor of two, the Hubble constant would be wrong by a factor of two, and the estimate of the age of the universe would also be wrong by a factor of two. Thus the age of the universe and the extragalactic distance scale are inextricably connected.

Over the past 20 years, estimates of the Hubble constant have ranged from about 15 to 35 km/s per million LY. Since the Hubble constant is the reciprocal of the age, the bigger the constant, the faster the universe expands, and hence the younger it must be. The Hubble time implied by these values ranges from 20 billion years on the old end to about 10 billion years on the young end. In just the past five years, several new techniques for estimating distances have all converged on a Hubble constant of about 25 km/s per millon LY, which makes the Hubble time about 13 billion years. (In units used by professional astronomers and often quoted in the press, 25 km/s per million LY corresponds to 80 km/s per million parsecs; the current most likely value of the Hubble constant appears to be 70–75 km/s per million parsecs or 21–23 km/s per million LY, which we have rounded to 25 km/s per million LY for this text.)

The Role of Deceleration

The Hubble time turns out to be the maximum possible age of the universe, but its actual age may be less, because in figuring out the Hubble time we have assumed that the universe has always been expanding at the same rate. Continuing with our analogy, this is equivalent to assuming that you traveled home from the party at a constant rate, when in fact this may not have been the case. At first, angry about having to leave, you may have driven fast, but then as you calmed down—and thought about police cars on the highway—you may have begun to slow down until you were driving at a more socially acceptable speed (such as 80 km/h). In this case, given that you were driving faster at the beginning, the trip home would have taken less than a half-hour.

In the same way, in calculating the Hubble time, we have assumed that H has been constant throughout all of time. This may not be a good assumption. Matter creates gravity, whereby all objects pull on all other objects. This mutual attraction will slow the expansion as time goes on, which means that in the past, the rate of expansion must have been faster than it is today. How much faster depends on the importance of gravity in slowing the expansion. We say the universe has been *decelerating* since the beginning.

If the universe were nearly empty, the role of gravity would be minor. Then the deceleration would be close to zero and the universe would have been expanding at a constant rate. But in a universe with any significant density of matter, the deceleration means that the expansion is slower now than it used to be. In this case, the age of the universe is smaller than our estimate assuming a con-

stant expansion rate. If we use the current rate of expansion to estimate how long it took the galaxies to reach their current separations, we will overestimate the age of the universe—just as we may have overestimated the time it took for you to get home from the party.

We will see next that the age for a universe that is decelerating just enough to eventually stop its expansion is two-thirds of the maximum age. Therefore, astronomers believe that the range of likely ages for a Hubble constant of 25 km/s per million LY is between 9 and 13 billion years.

Comparing Ages

How else can we estimate the age of the universe besides measuring the rate at which it expands? One way is to find the oldest objects whose ages we can measure. After all, the universe has to be at least as old as the oldest objects in it. In our Galaxy and others, the oldest stars are found in the globular clusters (Figure 28.3), which can be dated using the methods described in Section 21.3.

Our best models indicate that some globular clusters are at least 15 billion years old. If it took a billion years or so after the expansion began for the first stars to form, which theorists say is likely, then globular cluster ages suggest that the universe is 16 billion years old. This result is inconsistent with the age estimated from the rate of expansion for a Hubble constant of 25!

Astronomers are working hard to determine whether there is a real inconsistency in the different approaches to estimating the age of the universe. Many things can go

Figure 28.3
Globular clusters, such as 47 Tucanae, are among the oldest objects in our Galaxy and can be used to estimate its age. This image was taken with the 1-m Schmidt telescope at La Silla in Chile. (European Southern Observatory)

wrong in arriving at the numbers we have cited. The observational determinations of the Hubble constant still have some uncertainties. If the constant were 20 km/s, for example, the range of ages measured from the expansion would be 10.5 to 16 billion years.

The models of stellar evolution that we use to estimate the ages of the oldest stars might be inaccurate. In that case, the globular clusters may not be as old as we currently think. And it may turn out that our models of the universe are too simple—that some of our assumptions about its various properties are not quite right. (For example, Einstein's cosmological constant may not be equal to zero after all.) Given the uncertainties, we have in many places in this book referred to the age of the universe in round numbers as 10 to 15 billion years.

One of the major challenges of modern astrophysics is to reconcile the various estimates of the age of the universe. Good agreement would support our current models of its birth and evolution. Proof of disagreement could lead to a revolution in our understanding comparable to what occurred when Hubble discovered the redshift–distance relation. It is much too early to declare a major crisis in astronomy (as some popular accounts of the controversy have described it). On the other hand, astronomers would sleep better if the different ages could be brought closer together.

The Geometry of Spacetime

We now turn to another of the basic observed features of the universe, which we must include as a prime ingredient in our modeling recipe. As we saw in Chapter 23, the effect of gravity is to bend or curve the fabric of space and time. We observed that this effect is difficult to measure on Earth, but it becomes overwhelming as we approach the event horizon of a black hole.

Einstein's equations of general relativity describe the relationship between gravity and spacetime for any system, including the entire universe. The predictions of general relativity are wonderfully mind-boggling. We describe them briefly in this section, but warn you that we are unable to do them full justice.

Playing with Dimensions

To appreciate what happens when the gravity of all the matter in the universe affects the curvature of spacetime for the whole universe, we must first discuss the concept of *dimensions*. A dimension can be defined as an independent option to move. For example, if you are standing at the foot of a wall with a ladder on it, you have three independent options to move (Figure 28.4). You can go to the left or the right; you can step forward or backward; and you can climb up or down. Each of these motions is independent, in the sense that you can do one without doing any of the others. In most life situations, however, you combine the three: for example, you might climb a grand curving staircase by moving up, forward, and to the right simultaneously.

We live in a world of three dimensions and take them pretty much for granted. With minimal effort we can

  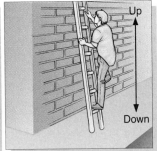

The 3 dimensions (options to move)

Figure 28.4

In a world of three dimensions, there are three independent options to move: forward–backward, left–right, and up–down.

move in any of them, and everything from our bodies to the most distant galaxies extends in all three of the dimensions we have discussed. What general relativity says—translated into everyday language—is that gravity can curve or warp the universe in a *fourth* dimension. (This is *not* time, but a fourth *spatial* dimension.)

Such a dimension is impossible to picture in your mind (although there are now very nice computer programs that can show three-dimensional slices of four-dimensional solids—much as an architect's drawings show two-dimensional slices through a three-dimensional house). Since no one (including your authors) can think in four dimensions, let's try to understand the properties of such a universe by reducing our dimensions by one. We will examine one possible two-dimensional universe that is curved in the third dimension.

A Balloon Analogy

Imagine that the "universe" of interest to us is the outer skin of a large spherical balloon. We can represent the galaxies by pasting grains of sand onto this skin. If we are restricted to moving on the *surface* of the balloon—and not inside or above it—then we are dealing with a two-dimensional world. (Check this by counting the number of options to move: you can move forward–backward or left–right, but you cannot move up–down and remain in the world of the skin.) But the two-dimensional universe of the balloon's surface is curved in the third dimension—it is wrapped around the space inside the balloon.

Let's explore the properties of our two-dimensional balloon-skin world. We imagine that we ourselves are tiny two-dimensional specks that live in this world. If we travel around, we can determine some of its characteristics (Figure 28.5):

- If you keep going in a straight line (say always forward), you eventually come back to where you started. The world of the balloon skin has no *edge*.
- Since you can go around in any direction, there is no point in this world that could be considered a *center*. All points on the skin could equally well call themselves the center; no point is any more special than any other.
- If someone starts blowing up the balloon, the grains of sand pasted to the skin move away from each other. Each piece of the balloon skin stretches equally. All the grains move away from each other, but as they look out at other grains, they see something interesting. The more balloon there is between two sand grains, the faster they appear to move away from each other. This is the Hubble law! (We saw the same thing with our raisin bread in Chapter 25.)
- As the balloon is blown up, we find its surface gets bigger and bigger. Where does this new surface come from? It is being stretched out of the existing balloon skin. And the motion of the dust grains is caused by this stretching (not by anything the grains themselves are doing).

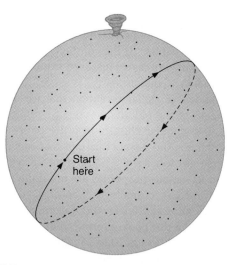

Figure 28.5
A "universe" that consists of only the surface of a balloon (with grains of sand pasted to the skin) has neither center nor edge. If you move straight ahead in one direction, you eventually return to where you started. If you blow up such a balloon, all the grains of sand move away from all the other grains.

If we were one of the two-dimensional specks living on the balloon skin, we would find our world mighty strange! A world with no center, no edge, and straight lines that come back upon themselves—all of its properties would go against our two-dimensional intuition. Yet, as a three-dimensional astronomy student, you probably don't find the situation all that odd. After all, the surface of the Earth is just this kind of "world." Our planet's *surface* has no center and no edge, and if you keep going straight ahead on it, you come back to where you started.

A speck might also ask where in the world of the balloon surface the expansion began. The only proper answer is that the world of the balloon skin began expanding everywhere at once. There is no grain of sand, no spot of balloon skin that was uniquely the site of the first expansion. As air filled the balloon, all points on its skin moved away from all other points at the same time.

The difference between you and a two-dimensional speck is that you can picture what is happening in one more dimension than the specks can. The surface of the balloon (or the Earth) has the "strange properties" we have outlined because it is curved in another (third) dimension—a dimension that the poor speck simply cannot imagine.

The kind of curved "universe" we have been describing in our analogy is called a **closed universe** because as you keep going, it closes in on itself until you eventually get back to where you started. The event horizon of a black hole encloses this sort of a "pocket universe." We hasten to point out that this is not the only kind of curvature that can exist. In an **open universe** the opposite occurs: as you move outward, more space than you expect opens up in front of you. An open universe curves away from where you started in all directions. It is much more difficult to picture, but it too has neither center nor edge.

The Curvature of Space

Now let's apply to the real universe what we learned in our analogy. Suppose the gravity of all the matter in the universe curves space in some (to us unimaginable) fourth dimension. Then we—like the specks on the balloon skin—might discover some intuitively odd things. For example, there would be no center or edge to the universe. Every galaxy would have an equal right to call itself the center or to demand that no galaxy be considered the center.

The expansion of the universe means that space stretches, with more space emerging from the existing space. The galaxies move apart because space is stretching and carrying them apart. The more distant galaxies, which have more space between us and them, stretch away the fastest, giving us the Hubble law.

And if an enterprising travel guide wanted to start tours to the place where the expansion began, this effort would be doomed to failure. The expansion of the universe began *everywhere at once* throughout the universe we can see.

Just as a two-dimensional speck could not fathom the idea that there might be a third dimension in which its world is bent, so we three-dimensional creatures resist the idea that there might be a fourth dimension of which we cannot become aware through our senses. But remember that in science, it is not whether a model or theory is likeable or simple that counts; rather, the ultimate judge must be whether experiments support the theory.

What evidence do we have that gravity can warp or bend three-dimensional space into a fourth dimension? We have already discussed a number of experiments confirming this view in Chapters 23 and 26, including the deflection of light and the advance in the perihelion of Mercury, as well as the existence of gravitational lenses in the realm of the galaxies. Today, scientists have learned to accept both the curvature of space and their inability to picture it.

The Redshift Re-Examined

If it is space that is stretching, and not the galaxies that are moving through space, why do the galaxies show redshifts in their spectra? When you were young and naive—three chapters ago—it was fine to discuss the redshifts of distant galaxies as if they resulted from their motion away from us. But now that you are an older and wiser student of cosmology, this view will simply not do!

A more accurate view of the redshifts of galaxies is that the waves are stretched by the stretching of the space they travel through. Think about the light from a remote galaxy. As it moves away from its source, the light has to travel through space. If space is stretching during all the time the light is traveling, the light waves will be stretched as well. As we saw in Chapter 4, a redshift is a stretching of waves—the wavelength of each wave increases. (Continuing with our balloon analogy, if we paint thin wavy

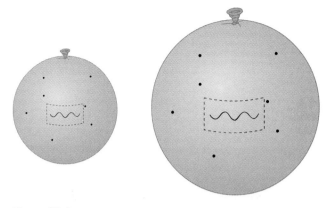

Figure 28.6
As the surface of a balloon expands, a wave on its surface stretches. For light waves, the increase in wavelength would be seen as a redshift.

lines on our expanding balloon's skin to represent light waves, the crests of those waves will get farther and farther apart as the balloon is stretched; see Figure 28.6.) Light from more distant galaxies is stretched more than light from closer ones—and thus shows a greater redshift.

28.3

Models of the Universe

Having examined a few of the key ideas in cosmology in more detail, we are now ready to look at some of the specific models astronomers have made to describe the large-scale behavior of the universe. These models begin with the facts outlined in Table 28.1. Then they make predictions about how the universe has evolved so far and what will happen to it in the future.

The Expanding Universe (with Deceleration)

Every model of the universe must include the expansion we observe. In addition, the cosmological principle shows that the universe is homogeneous. As a result, the expansion rate must be uniform (the same everywhere apart from any possible deceleration over time), causing the universe to undergo a uniform change in *scale* over time. By scale we mean, for example, the distance between two clusters of galaxies. It is customary to represent the scale by a scale factor R; if R doubles, then the distance between the clusters has doubled. Since the universe is expanding at the same rate everywhere, the change in R tells us how much it has expanded (or contracted) at any given time. For a static universe, R would be constant. In an expanding universe, R increases with time.

The simplest scenario of an expanding universe would be one in which R increases with time uniformly. But, as we have seen, the effect of gravity makes this case

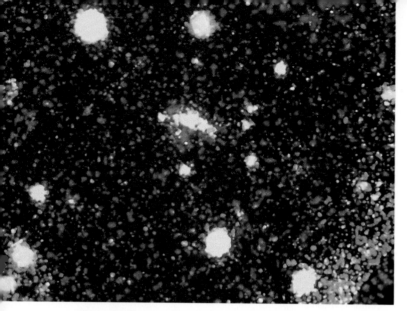

Figure 28.7

This image of one of the most-distant known galaxies, 4C 41.17, combines two views taken with the fully operational Keck telescope in the infrared with one visible-light view taken with the 4-m telescope on Kitt Peak. The galaxy has a redshift of 3.8, which means that the light we see in the infrared (colored red and green on the image) was emitted as visible light. The light we see as visible light (blue) was actually emitted as ultraviolet. The shape of the galaxy (the blob in the center) suggests it is interacting with a companion that is not visible on the image. The only way we have of estimating the distance of remote galaxies such as this is the Hubble law. This means the actual distance to 4C 41.17 cannot be determined until we know the deceleration of the universe.

unlikely. Gravity decelerates the expanding universe. We can make different models depending on the amount of gravity and thus the amount of deceleration. In some models—as we will see—the universe expands forever. In others it stops expanding and starts to contract. If we could measure the precise amount by which the universe is decelerating, we could select the correct model and collect our Nobel Prize.

Unfortunately, it is very difficult to estimate the rate of deceleration. One way might be to measure the distances and speeds of very distant galaxies. We are seeing these as they were long ago, and so we could see how much faster they were moving when the universe was young. However, think about how we actually measure the distance to very remote groups of galaxies (Figure 28.7). We use the Hubble law, which allows for only one rate of expansion, not for a rate that changes with time. So when we use this method to estimate how far the galaxies are from us, we are *assuming* that we live in a uniformly expanding universe.

To check on how much the universe is slowing down, we need an *independent* way of measuring the distance. For example, if a certain type of galaxy were a standard bulb (see Chapter 27), we could use its apparent brightness to tell us how far away each galaxy is. Alas, not only are galaxies not standard bulbs at any *given* time in his-

tory, but we also know that their brightness can change over time. As we have seen, galaxies evolve, experience collisions and mergers, and sometimes undergo bursts of star formation that make them unusually bright for several million years. Although we can't use whole galaxies as standard bulbs, we saw that Type I supernovae may play such a role. They can be seen out to distances of a few billion light years, and astronomers are now trying to estimate the deceleration by finding such supernovae in distant galaxies.

The Cosmic Tug-of-War

Since the deceleration is produced by gravity, another approach to estimating its rate is to measure how much material the universe contains. Here is where the cosmological principle really comes in handy. Since the universe is the same all over, we only need to measure how much material exists in a (large) representative sample of it. (Such sampling is how pollsters can describe how we feel about political issues without asking each and every person in the country.) What astronomers look at is the *average density* of the universe.

By average density we mean the mass of matter (including the equivalent mass of energy)[1] that would be contained in each unit of volume (say, 1 cm³) if all the stars, galaxies, and other objects were taken apart, atom by atom, and if all those particles, along with the light and other energy, were distributed throughout all of space with absolute uniformity. If the average density is low, the universe will not decelerate very much and can expand forever. High average density, on the other hand, means that a lot of gravity is pulling the galaxies together, and so the expansion will eventually stop.

In a sense, we can say that there is a "tug-of-war" going on between the expansion of the universe, which pushes everything apart, and gravity, which pulls everything together. One of the great questions of modern astronomy is who will win this tug-of-war; the answer can tell us what the ultimate fate of the universe will be.

Since we know how fast the galaxies are moving away from each other, we can calculate the **critical density** for the universe—the mass per unit volume that will be just enough to slow the expansion to zero at some time infinitely far in the future. If the actual density is higher than this critical density, then the expansion will ultimately reverse and the universe will begin to contract. If the actual density is lower, then the universe will expand forever.

These various possibilities are illustrated in Figure 28.8. Time increases to the right, and the scale, R, increases upward in the figure. Today, marked "present" along the time axis, R is increasing in each model. We are

---

[1] By equivalent mass we mean that which would result if the energy were turned into mass using Einstein's formula, $E = mc^2$.

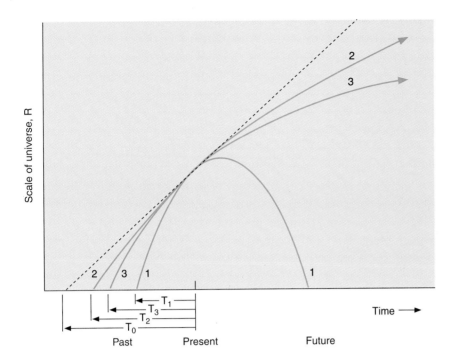

Figure 28.8
A plot of R, the scale of the universe, against time for various cosmological models. Curve 1 represents a closed universe (where the density is greater than the critical value), curve 2 represents an open universe, and curve 3 is a universe with a critical density. The dashed line is for an empty universe, one in which the expansion is not slowed by gravity. Note that the age of the universe equals the Hubble time only for this (unlikely) possibility. In all the other cases, the universe is less old than the Hubble time.

still "early" in the history of the universe, so the galaxies are expanding no matter which model is right. (The same situation holds for a baseball thrown high into the air. While it may eventually fall back down, at the beginning of the throw it moves upward.) The straight dashed line corresponds to the empty universe with no deceleration; it intercepts the time axis at a time, T_0 (the Hubble time), in the past. The other curves represent models with varying amounts of deceleration, starting from the Big Bang at shorter times in the past. Let's take a look at the future according to the different models.

Facing the Closed Future

One of the things we learn from general relativity is that the future of the expansion and the geometry of space-time are intimately related. The differently curved universes we explored briefly in Section 28.2 correspond to different futures. For example, let's take curve 1 in Figure 28.8. In this case the actual density of the universe is higher than the critical density. This universe will stop expanding some time in the future and begin contracting. Eventually the scale drops to zero, which means that space will have shrunk down to infinitely small size. The noted physicist John A. Wheeler (who gave black holes their name) calls this the "big crunch," because matter, energy, space, and time would all be crushed out of existence.

This scenario corresponds to the *closed* geometry discussed earlier, and is therefore called the closed universe. It is closed in two senses: at any time, space curves so that if you could keep going, you would eventually come back to where you started; and at the end of time, space closes down on itself. (See "Making Connections" box for more on what happens toward the end.)

It is tempting to speculate that another big bang might follow the "crunch," giving rise to a new expansion phase, and then another contraction—perhaps oscillating between successive big bangs indefinitely in the past and future. Such speculation is sometimes referred to as the *oscillating theory* of the universe, but it is not really a theory because we know of no mechanism that can produce another big bang. General relativity predicts, instead, that at the crunch the universe would collapse into a universal black hole. The oscillating model is more a philosophical idea than a scientific one; scientific hypotheses, after all, must be tested with experiments. There is no way to test whether another cycle of the universe could arise from the big crunch because nothing—not matter, not energy, not space, not time—could survive the big crunch.

Facing the Open Future

If the density of the universe is less than the critical density, (curve 2 in Figure 28.8), gravity is never important enough to stop the expansion, and so the universe expands forever. This corresponds to the (more difficult to imagine) open geometry discussed earlier, and is thus called an open universe. Such a universe is infinite and always has even more room in it than naive three-dimensional observers would expect. In this case, time and space begin with the Big Bang, but they have no end; the universe simply continues expanding, always a bit more slowly as time goes on. Groups of galaxies eventually get so far apart that it would be difficult for observers in any of them to see the others (see "Making Connections" box).

At the critical density (curve 3) the universe can just barely expand forever. The critical-density universe has an age of exactly two-thirds T_0, where T_0 is the age of the

What Might It Be Like in the Distant Future?

Some say the world will end in fire,
Some say in ice.
From what I've tasted of desire
I hold with those who favor fire.

—From the poem "Fire and Ice"
by Robert Frost (1923)

Given the destructive power of impacting asteroids, expanding red giants, and nearby supernovae, our species may not be around in the remote future. Nevertheless, you might enjoy speculating about what it would be like to live in a much, much older universe.

The far future in a *closed* universe would be exciting but not very healthy for living things. As space contracted (R would get smaller and smaller), the galaxies would see each other's light blueshifted and not redshifted; that is, the compression of space would compress the waves. But shorter waves have more energy. Thus the temperature of the background radiation (see Section 28.4) would increase with time.

As the universe got smaller, it would become easier to circumnavigate. Rays of light could eventually make it around the universe and come back to where they started. In theory, you could see a star die in one direction, and then see the light of its birth come around from the other direction. "Ghost galaxies"—images of galaxies whose light had gone completely around the universe—would become visible, adding to the number of galaxies we observed.

Ultimately, as space shrank and the temperature of the background radiation increased, the temperature of space would become higher than that of any planets. Heat would then flow from space to a planet, instead of the other way around (which is what we now take for granted). As a planet heated up, any life-forms on its surface would be "cooked" by the heat of space. As the radiation of space got even hotter, heat would flow from space into the stars, breaking them apart. Their end could be mourned only briefly, however, because soon all matter and energy would be crushed out of existence by the closing down of spacetime.

The *open* universe scenario is unsettling in a completely different way. In this case the universe would expand forever (R would just increase) and the clusters of galaxies would spread ever farther apart with time. As the eons passed, the universe would get thinner, colder, and darker.

Within each galaxy, stars would continue to go through their lives, eventually becoming white dwarfs, neutron stars, and black holes. Low-mass stars might take a long time to finish their evolution, but in this model we would literally have all the time in the world. Ultimately, even the white dwarfs would cool down to be black dwarfs, any neutron stars that revealed themselves as pulsars would stop spinning, and black holes with accretion disks would complete their "meals." The final states of stars would be dark and difficult to observe.

Therefore the light that now reveals galaxies to us would eventually go out. Even if a small pocket of raw material were left in one unsung corner of a galaxy, ready to be turned into a fresh cluster of stars, we would only have to wait until the time that their evolution was also done. And time is one thing this model of the universe has plenty of. There would surely come a time when all the stars were out, galaxies were as dark as space, and no source of heat remained to help living things survive. Then the lifeless galaxies would just continue to move apart in their lightless realm.

If these views of the future seem discouraging (from a human perspective), you might take heart in the knowledge that science is always a progress report. The most-advanced ideas about the universe from a hundred years ago now strike us as rather primitive. It may well be that our best models of today will in a hundred or a thousand years also seem rather childish, and that there are other factors determining the ultimate fate of the universe of which we are still completely unaware.

empty universe. Interestingly, this model (called a **flat universe**) has zero curvature, and resembles the three-dimensional universe you probably expected before you took this class. Universes that will expand forever have ages between two-thirds T_0 and T_0. Universes that will someday begin to contract have ages less than two-thirds T_0. The various models we have described are summarized in Table 28.2.

Who's Winning the Tug-of-War?

What kind of universe do we live in? Is its density larger or smaller than the critical density? The critical density depends on H_0. If the Hubble constant is 25 km/s per million LY, the critical density is about 10^{-29} g/cm^3. Since we cannot find a way at present to measure the deceleration directly, astronomers have been focusing their efforts on finding the actual density of the universe.

There are several observational tests by which we can try to determine the average density of matter in space. One way is to count all the galaxies out to a given distance and use estimates of their masses, including dark matter, to calculate the average density. Such estimates indicate a density of about 1 to 2×10^{-30} g/cm^3 (10 to 20 percent of critical) and suggest that the universe is open. However,

TABLE 28.2
Some Models of the Universe

Kind of Universe	Age	Average Density	Ultimate Fate of Universe
(Model)	(billions of years)*	(g/cm^3)	
Closed	Less than 9	More than 10^{-29}	Stop and contract
Flat	9	About 10^{-29}	Barely expand forever
Open	9–13	Less than 10^{-29}	Expand forever

* The numbers here assume a Hubble constant of 25 km/s per million LY. If the Hubble constant is smaller than this, the ages increase correspondingly.

we may have underestimated the amount of dark matter in galaxy clusters, and there may also be a lot of dark matter between galaxy clusters (including in the great voids) where we cannot detect it. Therefore, we cannot really be certain that we live in an open universe based on this measurement.

The ages of stars also suggest that we live in an open universe. The best estimate of H_0 is 25 km/s per million LY, which corresponds to a maximum age of 13 billion years. If we live in a critical-density universe, the actual age would be only two-thirds of 13 billion years, which is 9 billion years. It seems highly unlikely that our models of how stars evolve, which give ages of 15 billion years for the oldest stars, are bad enough to give stellar ages that are wrong by a factor of nearly two. As you can see in Table 28.2, an open universe provides the oldest possible age.

All such conclusions, however, depend on the models of the universe we choose to test. There are more-complex cosmological models—such as those where Einstein's cosmological constant (the repulsion force) is not zero—in which the density comes closer to the critical value. In this introductory text we cannot explore these alternative models; suffice it to say that astronomers are now trying to design observational tests to determine whether more-complex models are required.

TABLE 28.3
Lookback Times for Different Redshifts

Redshift	Percent of Current Age of Universe When the Light Was Emitted*
0	100% (now)
0.5	55%
1.0	35%
2.0	19%
3.0	12%
4.0	9%
4.5	8%
4.9	7%
Infinite	0%

* Assumes a universe with a density equal to the critical density.

Lookback Time

In Chapter 25 we discussed how we can use the Hubble law to measure the distance to a galaxy. The Hubble law applies quite simply to galaxies that are not moving too fast (that is, are not too far away). Once we get to large distances, we are looking so far into the past that we must take the deceleration of the universe into account. Since we do not know how much the universe is decelerating, we must *assume* one of the models of the universe to be able to convert large redshifts into distances.

This is why astronomers squirm when reporters and students ask them *exactly* how far away some newly discovered distant quasar or galaxy is. We really can't give an answer without first explaining the model of the universe we are assuming in calculating it (by which time a reporter or student is long gone!).

Once we assume a model, we can use it to calculate the *lookback time* for an object—the measure of how long ago the light we see was emitted from it. As an example, Table 28.3 lists the lookback times as fractions of the current age of the universe (assuming a model in which the universe has critical density). The numbers are not so important, but notice that as we find objects with higher and higher redshifts, we are looking back to very small fractions of the age of the universe.

We have already learned some important things by observing objects at large redshifts. For example, we saw that quasars were most abundant when the universe was roughly 20 percent of its current age (Figure 26.17). At a still-earlier time, quasars were exceedingly rare. Perhaps galaxies had not yet formed, or had formed only recently and not yet had time to produce massive black holes. Such observations provide clear evidence that our universe is evolving with time.

28.4

The Beginning of the Universe

As we look farther and farther back, we see the galaxies and quasars thin out, and we approach the era when matter had still not settled down into the structures we observe today. What were things like when the universe was young, and space had not yet stretched very significantly? In other words, what was it like just after the Big Bang?

The History of the Idea

It is one thing to say the universe had a beginning (as the equations of general relativity imply), and quite another to describe that beginning. The Belgian priest and cosmologist Georges Lemaître (1894–1966) was probably the first to propose a specific model for the Big Bang itself (Figure 28.9). He envisioned all the matter of the universe starting in one great bulk he called the *primeval atom*, which then broke into tremendous numbers of pieces. Each of these pieces continued to fragment further until they became the present atoms of the universe, created in a vast nuclear fission. In a popular account of his theory, Lemaître wrote, "The evolution of the world could be compared to a display of fireworks just ended—some few red wisps, ashes and smoke. Standing on a well-cooled cinder we see the slow fading of the suns and we try to recall the vanished brilliance of the origin of the worlds."

Physicists today know much more about nuclear physics than was known in the 1920s, and have shown that the primeval fission model cannot be correct. Yet Lemaître's vision was in some respects quite prophetic. We still believe that everything was together at the beginning; it was just not in the form of matter as we now know it.

In the 1940s the American physicist George Gamow (Figure 28.10) suggested a universe with a different kind of beginning—involving nuclear fusion instead of fission.

Figure 28.10
This montage of images shows George Gamow emerging like a genie from a bottle of YLEM, a Greek term for the original substance from which the world formed. Gamow revived the term to describe the material of the hot Big Bang. Flanking him are Robert Herman (left) and Ralph Alpher, with whom he collaborated in working out the physics of the Big Bang. The modern composer Karlheinz Stockhausen was inspired by Gamow's ideas to write a piece of music called *Ylem,* in which the players actually move away from the stage as they perform, simulating the expansion of the universe. (Courtesy of Ralph Alpher)

He worked out the details with Ralph Alpher, and they published the results in 1948. (Gamow, who had a quirky sense of humor, decided at the last minute to add the name of physicist Hans Bethe to their paper, so that the coauthors would be Alpher, Bethe, and Gamow, a pun on the first three letters of the Greek alphabet: alpha, beta, and gamma.) Gamow's universe started with fundamental particles that built up the heavy elements by fusion in the Big Bang.

His ideas were close to our modern view, except we now know that only the three lightest elements—hydrogen, helium, and a small amount of lithium—were formed in appreciable abundances at the beginning. The heavier elements formed later in stars. Since the 1940s many astronomers and physicists have worked on a detailed theory of what happened in the early stages of the universe. The result of their efforts is now called the *standard model* of the Big Bang.

Figure 28.9
Abbé Georges Lemaître (1894–1966), Belgian cosmologist, studied theology at Mechelen, and mathematics and physics at the University of Leuven. It was there that he began to explore the expansion of the universe, and postulated its explosive beginning. He actually predicted the Hubble law two years before its verification, and he was the first to seriously consider the physical processes by which the universe began. (Yerkes Observatory)

Standard Model

Three basic ideas hold the key to tracing the changes that occurred during the first few minutes after the universe began. The first is that the universe cools as it expands, much as gas cools when sprayed from an aerosol can. Figure 28.11 shows how the temperature changes with the passage of time. In the first fraction of a second, the uni-

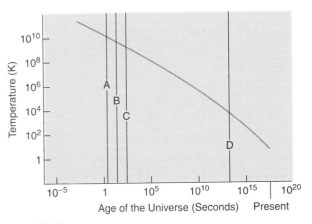

Figure 28.11
This graph shows how the temperature of the universe varies with time as predicted by the standard model of the Big Bang. Note that both the temperature (vertical axis) and the time in seconds (horizontal axis) change over vast scales on this compressed diagram. The vertical line labeled A designates approximately the time at which neutrinos stop interacting with matter. B denotes the time when the temperature becomes too cool for new positrons and electrons to form from radiation. Helium synthesis occurs at time C, and the universe becomes transparent to radiation at time D. The line showing the drop in temperature ends at the point in time labeled "the present."

verse was unimaginably hot. By the time 0.01 s had elapsed, the temperature had dropped to 100 billion (10^{11}) K. After about 3 min, it had fallen to about 1 billion (10^9) K, still some 70 times hotter than the interior of the Sun. After a few hundred thousand years, the temperature was down to a mere 3000 K, and the universe has continued to cool since that time.

All of these temperatures but the last are derived from theoretical calculations, since (obviously) no one was there to measure them directly. As we will see, however, we have actually detected the feeble glow of radiation emitted at a time when the universe was a few hundred thousand years old. Indeed, the fact that we have done so is one of the strongest arguments in favor of the Big Bang model.

The second step in understanding the evolution of the universe is to realize that at very early times it was so hot that it contained mostly radiation (and not the matter that we find dominating today). The photons—the packets of pure electromagnetic energy described in Chapter 4—that filled the universe could collide and produce material particles. That is, under the conditions just after the Big Bang, energy could turn into matter (and matter could turn into energy). We can calculate how much mass is produced from a given amount of energy by using Einstein's formula $E = mc^2$ (see Chapter 15).

The idea that energy can turn into matter is a new one for many students since it is not part of our everyday experience. (That's because when we compare the universe today to what it was like right after the Big Bang, we live in cold, hard times! The photons in the universe today

typically have far less energy than the amount required to make new matter.) In Chapter 15 we briefly mentioned that when subatomic particles of matter and *antimatter* collide, they turn into pure energy. But is it really true that energy can turn into matter (as our theories predict)? This process has been observed in particle accelerators around the world. If we have enough energy, under the right circumstances, new particles of matter (and antimatter) are indeed created.

Our third key point is that the hotter the universe (see Figure 28.11), the more energetic the photons available to make matter (and antimatter). To take a specific example, at a temperature of 6 billion (6×10^9) K, the collision of two typical photons can create an electron and its antimatter counterpart, a positron. The much more massive proton can be created only in an environment with a temperature in excess of 10^{14} K.

The Evolution of the Early Universe

Keeping these three ideas in mind, we can trace the evolution of the universe from the time it was about 0.01 s old and had a temperature of about 100 billion K. Why not begin at the very beginning? There are as yet no theories that allow us to penetrate to a time before about 10^{-43} s. When the universe was that young, its density was so high that the theory of general relativity is not adequate to describe it; at present, we have no theory that can deal with such extreme conditions.

Scientists, by the way, have been somewhat more successful in describing the universe when it was older than 10^{-43} s but still less than about 0.01 s old. During that time the universe was filled with energy and strongly interacting subatomic particles. Although the theory of these particles is difficult to deal with, very recently theoretical physicists have begun to speculate about what things may have been like during this very, very early time. We will look at some of their speculations later in this chapter.

The universe was a somewhat more familiar-sounding place by 0.01 s after the beginning. At that time it consisted of a soup of matter and radiation; the matter included protons and neutrons, leftovers from an even younger and hotter universe. Each particle collided rapidly with other particles. The temperature was no longer high enough to allow colliding photons to produce neutrons and protons, but it was sufficient for the production of electrons and positrons (Figure 28.12a). There was probably also a sea of exotic subatomic particles that would later play a role as dark matter. All the particles jiggled about on their own; it was still much too hot for protons and neutrons to combine to form the nuclei of atoms.

Our picture is of a seething cauldron of a universe, with photons colliding and interchanging energy, sometimes being destroyed to create a pair of particles. The particles in the universe also collide with one another. Frequently, a matter particle and an antimatter particle

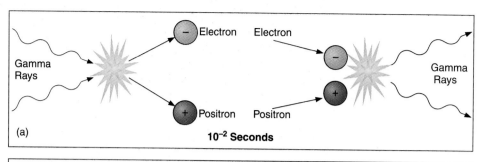

(a)

10⁻² Seconds

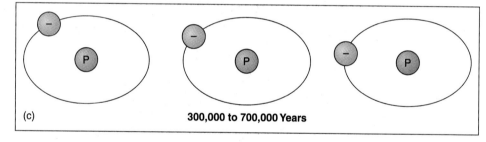

Figure 28.12
A summary of some of the interactions involving particles in the early universe. (a) In the first fractions of a second, when the universe was very hot, energy was converted to particles and antiparticles. The reverse reaction also happened: a particle and antiparticle could collide and produce energy. (b) As the temperature of the universe decreased, the energy of typical photons became too low to create matter. Instead, existing particles fused to create such nuclei as deuterium and helium. (c) Later it became cool enough for electrons to settle down with nuclei and make neutral atoms. Most of the universe was still hydrogen.

(b) **3 Minutes**

(c) **300,000 to 700,000 Years**

meet and turn each other into a burst of gamma-ray radiation.

Among the particles created in the early phases of the universe was the ghostly neutrino (see Chapter 15), which today interacts only very rarely with ordinary matter. In the crowded conditions of the very early universe, however, neutrinos ran into so many electrons and positrons that they experienced frequent interactions despite their "antisocial" natures.

By the time the universe was a little more than 1 s old, the density had dropped to the point where neutrinos no longer interacted with matter, but simply traveled freely through space. In fact, these neutrinos should now be all around us. Since they have been traveling through space unimpeded (and hence unchanged) since the universe was 1 s old, measurement of their properties would offer one of the best tests of the Big Bang model. Unfortunately, the very characteristic that makes them so useful—the fact that they interact so weakly with matter that they have survived unaltered for all but the first second of time—also renders them undetectable, at least with present techniques. Perhaps someday someone will devise a way to capture these elusive messengers from the past.

Atoms Form

When the universe was about 3 min old and its temperature was down to about 900 million K, protons and neutrons could combine without being immediately disrupted by high-energy photons. They began to form the simplest stable nuclei—deuterium (heavy hydrogen), helium, and lithium. At higher temperatures these atomic nuclei had immediately been blasted apart by interactions with photons, and so could not survive. But at the temperatures and densities reached between 3 and 4 min after the beginning, deuterium lasted long enough that collisions could convert some of it to helium, which consists of two protons and two neutrons (Figure 28.12b). In essence, the entire universe was acting the way centers of stars do today—fusing new elements from simpler components.

This burst of cosmic fusion was only a brief interlude, however. The universe was expanding and cooling down. This meant that no elements beyond lithium could form (it got too cool too fast), and even the light elements stopped forming after a few minutes. In the cool universe we know and love today, the fusion of new elements is limited to the centers of stars and the explosions of supernovae.

Still, the fact that the Big Bang model allows the creation of a good deal of helium is the answer to a long-standing mystery in astronomy. Put simply, there is just too much helium in the universe to be explained by what happens inside stars. All the generations of stars that have produced helium in their centers just cannot account for the quantity of helium we observe. Furthermore, even the oldest stars and the most-distant galaxies show significant amounts of helium. These observations find a natural explanation in the synthesis of helium by the Big Bang itself during the first few minutes of time. We estimate that ten times more helium was manufactured in the first 3 min of the universe than in all the generations of stars during the succeeding 10 to 15 billion years.

There is even more we can learn from the way the early universe made atomic nuclei. It turns out that *all* of the deuterium in the universe was formed during the first 3 min. In stars, any region hot enough to fuse two protons to form a deuterium nucleus is also hot enough to change it further—either by destroying it through a collision with an energetic photon, or by converting it to helium through nuclear reactions.

The amount of deuterium produced in the first 3 min of creation depends on the density of the universe at the time deuterium was formed. If the density was high, nearly all the deuterium would have been converted to helium through interactions with protons, just as it is in stars. If the density was low, then the universe expanded and thinned out rapidly enough that some deuterium survived. The amount of deuterium we see today thus gives us a clue to the density of the universe when it was about 3 min old. Theoretical models can relate the density then to the density now; thus measurements of the abundance of deuterium today can give us an estimate of the current density of the universe.

The deuterium measurements indicate that the present-day density is about 10^{-31} g/cm^3, suggesting that the universe will expand forever. There is a possible loophole, however. The deuterium abundance is determined by the density of protons and neutrons, since these are the particles that interact to form it. From the deuterium abundance we know that not enough protons and neutrons are present, by a factor of about 100, to produce a critical-density universe. If, however, there are dark matter parti-

cles of some other kind that are not involved in nuclear reactions, then it is still possible that we live in a critical-density universe. This is an additional reason for thinking that dark matter may be made of some exotic, unknown kind of particle, and not combinations of protons and neutrons like the readers of this book. We discuss this possibility in the last section of the chapter.

The Universe Becomes Transparent

Although fusion stopped after a few minutes, the universe continued to resemble the interior of a star for a few hundred thousand years: it remained hot and opaque, with radiation being scattered from one particle to another. It was still too hot for electrons to "settle down" and become associated with a particular nucleus. And electrons are especially effective at scattering photons, thus ensuring that no radiation ever got very far in the early universe without having its path changed. In a way, the universe was like an enormous crowd right after a popular concert; if you get separated from a friend, even if he is wearing a flashing button, it is impossible to see through the dense crowd to spot him. Only after the crowd clears is there a path for the light from his button to reach you.

Not until a few hundred thousand years after the Big Bang, when the temperature had dropped to about 3000 K and the density of atomic nuclei to about 1000 per cubic centimeter, did the electrons and nuclei combine to form stable atoms of hydrogen and helium (Figure 28.12c). With no free electrons to scatter photons, the universe became transparent for the first time in cosmic history. From this point on, matter and radiation interacted much less frequently; we say that they *decoupled* from each other and evolved separately. If we are to detect the light of the early universe, it will come from this decoupling time when radiation was first allowed to move over significant distances.

One billion years after the Big Bang, stars and galaxies had begun to form. Deep in the interiors of stars matter was reheated, nuclear reactions were ignited, and the more gradual synthesis of the heavier elements began. In the meantime, the radiation from the decoupling time continued to cool as space stretched and all radiation became redshifted (thus carrying less and less energy). As billions of years passed, the afterglow of the Big Bang faded away.

We conclude this quick tour of our model of the early universe with a reminder. You must not think of the Big Bang as a *localized* explosion *in space*—like an exploding superstar. There were no boundaries and no site of the explosion. It was an explosion *of space* (and matter and energy) that happened everywhere in the universe. All matter and energy that exist today, including the particles of which you are made, came from the Big Bang. We were, and still are, in the midst of the Big Bang; it is all around us.

Discovery of the Cosmic Background Radiation

In the late 1940s Ralph Alpher and Robert Herman (see Figure 28.10) realized that just before the universe became transparent, it must have been radiating like a blackbody at a temperature of 3000 K (see Chapter 4). If we could have seen that radiation just after neutral atoms formed, it would have resembled radiation from a reddish star. It was as if a giant fireball filled all of the universe.

But that was at least 10 billion years ago, and in the meantime the scale of the universe has increased a thousandfold. This expansion has increased the wavelength of the radiation by a factor of 1000, and, according to Wien's law, has correspondingly lowered the temperature by a factor of 1000 (see Section 4.2). Alpher and Herman predicted that the glow from the fireball should now be at radio wavelengths, and should resemble the radiation from a blackbody at a temperature only a few degrees above absolute zero. Since the fireball was everywhere throughout the universe, the radiation left over from it should also be everywhere. There was no way at the time they published their conclusion to observe such radiation from space, so the prediction was forgotten.

In the mid-1960s, in Holmdel, New Jersey, Arno Penzias and Robert Wilson of AT&T's Bell Laboratories were using a delicate microwave antenna (Figure 28.13) to measure the intensity of radio radiation all around the sky (to check on possible sources of radiation that might interfere with communications satellites). They were plagued with some unexpected background noise, just like static on a radio, that they could not get rid of. The puzzling thing about this radiation was that it seemed to be coming from all directions at once. This is very unusual in astronomy; after all, most radiation has a specific direction where it is strongest—the direction of the Sun, or a supernova remnant, or the disk of the Milky Way, for example.

Penzias and Wilson at first thought that any radiation appearing to come from all directions must be coming from inside their telescope, so they took everything apart to look for the source of the noise. They even found that some pigeons had roosted inside the big horn and had left (as Penzias delicately put it) "a layer of white, sticky, dielectric substance coating the inside of the antenna." However, nothing they did could reduce the background radiation to zero, and they reluctantly came to accept that it must be real, and coming from space.

Penzias and Wilson were not cosmologists, but as they began to discuss their puzzling discovery with other scientists, they were quickly put in touch with a group of astronomers and physicists at Princeton University (a short drive away) who had—as it happened—been redoing the calculations of Gamow's group from the 1940s, and realized that the radiation from the decoupling time should be detectable as a faint afterglow of radio waves. Their work predicted that the temperature corresponding to this **cosmic background radiation (CBR)** should be about 3 K; Penzias and Wilson found the intensity of the radiation they had discovered to match a blackbody with just that temperature.

Many other experiments on Earth and in space soon confirmed the discovery: the radiation was indeed coming from all directions (it was isotropic) and matched the predictions of the Big Bang theory with remarkable precision. Penzias and Wilson had inadvertently observed the glow from the primeval fireball. They received the Nobel Prize for their work in 1978. And just before his death in 1966, Lemaître learned that his "vanished brilliance" had been discovered and confirmed.

Figure 28.13
Robert Wilson (right) and Arno Penzias pose in front of the horn-shaped antenna with which they discovered the cosmic background radiation. The photo was taken in 1978, just after they received the Nobel Prize in physics. (AT&T Bell Laboratories)

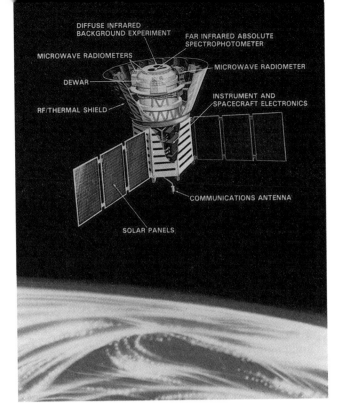

Figure 28.14
An artist's conception of COBE, the satellite designed to explore the cosmic background radiation at infrared and microwave wavelengths. Various instruments aboard the probe are marked. (NASA)

Properties of the Background Radiation

Accurate measurements of the CBR have now been made with a satellite orbiting the Earth. Named the Cosmic Background Explorer (COBE), it was launched by NASA on November 18, 1989 (Figure 28.14). The data it received quickly showed that the CBR closely matches that expected from a blackbody with a temperature of 2.73 K

(Figure 28.15). This is exactly the result expected if the CBR is indeed redshifted radiation emitted by a hot gas that filled all of space shortly after the universe began.

The first important conclusion from measurements of the CBR, therefore, is that the universe we have today has evolved from a hot, uniform state. This observation provides direct support for the idea that we live in an evolving universe.

A second result (seen with other instruments and now confirmed by COBE) is that the CBR appears to be slightly hotter in one direction than in the exact opposite direction in the sky. This difference comes about because of our own motion through space. If you approach a blackbody, its radiation is slightly Doppler-shifted to shorter wavelengths and therefore resembles radiation from a slightly hotter blackbody. If you move away from a blackbody, its radiation is redshifted and resembles that from a slightly cooler object. Since the CBR fills the universe, we can determine how we are moving by noting the directions in which it is redshifted and blueshifted, and the amount of the shift.

The small temperature differences of the CBR indicate that the Sun, the Milky Way, and the whole Local Group of galaxies are moving at a speed of about 500 km/s in the general direction of the constellation Hydra. Note that this motion is in addition to the motion of these galaxies resulting from the overall expansion of the universe. As we saw in Chapter 27, the extra motion is probably caused by the gravitational attraction of an unusually dense concentration of luminous dark matter (the Great Attractor) that pulls the Local Group toward Hydra.

A third conclusion from the COBE observations is that the early universe was a quiet place. The intensity of the CBR follows a blackbody curve with no excess radiation at any wavelength. If there had been violent events when the universe was very young, those events would

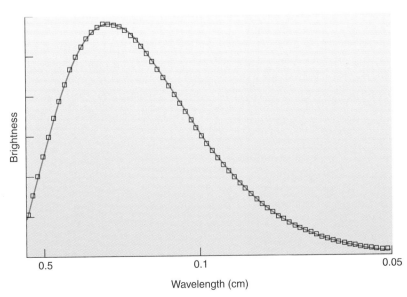

Figure 28.15
The solid line shows how the intensity of radiation should change with wavelength for a blackbody with a temperature of 2.73 K. The boxes show the intensity of the cosmic background radiation as measured at various wavelengths by COBE's instruments. The fit is perfect. When this graph was first shown at a meeting of astronomers, they gave it a standing ovation.

have distorted the smooth curve shown in Figure 28.15. Astronomers had, for example, speculated that there might have been a generation of massive stars that formed and completed their evolution through the supernova phase prior to the formation of galaxies.

It was known even before the launch of COBE that the CBR is extremely *isotropic*. In fact, its uniformity is one of the best confirmations of the cosmological principle. Ground-based measurements showed that if we look at places in the sky that differ in direction by less than a degree, any fluctuations in the intensity of the CBR are less than a few parts in ten thousand.

But according to our theories, the temperature could not have been *perfectly* uniform when the CBR was emitted. After all, the CBR is radiation that was scattering from the particles in the universe at the time of decoupling. If the radiation were completely smooth, then all those particles must have been distributed through space absolutely evenly. Yet it is those particles that have become all the galaxies and stars (and astronomy students) that now inhabit the cosmos. Had they been completely smoothly distributed, they could not have formed all the large-scale structure now present in the universe—the clusters and superclusters of galaxies discussed in the last few chapters.

The early universe must have had tiny density fluctuations from which such structure could evolve. Regions of higher-than-average density would have attracted additional matter and eventually grown into the galaxies and clusters that we see today. These regions would appear to us to have lower-than-average temperatures. (This is because a denser region has more gravity and thus redshifts the radiation, causing it to look a tiny bit cooler.) Therefore, if the seeds of present-day galaxies existed at the time the CBR was emitted, we should see some slight changes in the CBR temperature as we look in different directions in the sky.

Scientists working with the data from the COBE satellite have indeed detected very subtle temperature differences present in the CBR (see the opening figure for this chapter). The temperature variations are typically only 16 millionths of a degree K. The regions of lower-than-average temperature come in a variety of sizes, but even the smallest of the cool areas detected by COBE is far too large to be the precursor of an individual galaxy, or even a supercluster of galaxies. Thus, while the COBE results are a dramatic confirmation that fluctuations did exist in the density of the primeval fireball, astronomers still need to find a way to turn these fluctuations into the structure we presently observe.

New ground-based experiments have found CBR temperature fluctuations on scales of about 20 arcmin, which correspond approximately to the size of a supercluster. Instruments on Earth and in space will be trying to find even smaller variations in years to come. However these observations turn out, it is remarkable that we are now able to see the earliest stages of the birth of structure in the universe.

28.5

The Inflationary Universe

Problems with the Standard Big Bang Model

The hot Big Bang model that we have been describing is remarkably successful. It accounts for the expansion of the universe, explains the observations of the CBR, and correctly predicts the abundances of the light elements. As it turns out, this model also predicts that there should be exactly three types of neutrinos in nature, and this prediction has been confirmed by experiments with high-energy accelerators.

The theory is not complete, however. The standard Big Bang model does not explain why there is more matter than antimatter in the universe, nor does it account for the origin of the density fluctuations that ultimately grew into galaxies. It also does not explain the remarkable *uniformity* of the universe. The CBR is the same, no matter in which direction we look, to an accuracy of about 1 part in 100,000. This sameness might be expected if all the parts of the visible universe were in contact at some point, and had time to come to the same temperature. In the same way, if we put some ice into a glass of water and wait a while, the ice will melt and the water will cool down until they are the same temperature.

However, if we accept the standard Big Bang model, all parts of the visible universe were *not* in contact at any time. The fastest that information can go from one point to another is the speed of light. There is a maximum distance that light can have traveled from any point since the time the universe began. This distance is called that point's *horizon distance*, because anything farther away is "below its horizon"—unable to make contact with it. (We saw a similar use of the word *horizon* when we discussed black holes in Chapter 23.) One region of space separated by more than the horizon distance from another has been completely isolated from it through the entire history of the universe.

If we measure the CBR in two opposite directions in the sky, we are observing regions that were significantly beyond each other's horizon distance at the time the CBR was emitted. *We* can see both regions, but *they* can never have seen each other! Why, then, are their temperatures so precisely the same? According to the standard Big Bang model, they have never been able to exchange information, and there is no reason they should have identical temperatures. (It's a little like seeing the clothes students wear at two schools in different parts of the world become identical, without the students ever having been in contact.) The only explanation is simply that the universe somehow started out being absolutely uniform (which is like saying all students were born liking the same clothes). Scientists are always uncomfortable when they must appeal to a special set of initial conditions to account for what they see.

Another problem with the standard Big Bang model is that it does not explain why the density of matter in the universe is so close to the critical density. As we have seen, current observations are unable to tell us whether the expansion of the universe will continue forever or come to a halt and reverse itself. The interesting point, however, is that we find ourselves in a universe that is so nearly balanced between these two possibilities that we cannot yet determine which is correct. The density could have been, after all, so low that it would be obvious that the expansion of the universe will continue forever. Alternatively, there could have been so much matter that the universe was already beginning to contract. Instead, the amount of matter present is within a factor of ten or so of the value that corresponds to precise balance between these two situations. The standard Big Bang model offers no explanation of why this should be the case.

To understand the new ideas that seek to explain these characteristics of the universe, we must first digress and talk about the forces acting on subatomic particles. Then we will return to discussing the grand picture of how the universe might have evolved.

Grand Unified Theories

In physical science, the term *force* (see Chapter 2) is used to describe anything that can change the motion of a particle or body. One of the remarkable discoveries of modern science is that all known physical processes can be described through the action of four forces—gravity, electromagnetism, the strong nuclear force, and the weak nuclear force (Table 28.4).

Although gravity is perhaps the most familiar to you, and certainly appears strong if you jump off a tall building, the force of gravity between two elementary particles—say two protons—is by far the weakest of the four forces. Electromagnetism, which includes both magnetic and electrical forces, holds atoms together and produces the electromagnetic radiation that we use to study the universe. The weak nuclear force is only weak in comparison to its strong "cousin," but is in fact much stronger than gravity.

Both the weak and strong nuclear forces differ from the first two forces in that they act only over very small distances—those comparable to the size of an atomic nucleus or less. The weak force is involved in radioactive de-

cay and in reactions that result in the production of neutrinos. The strong force holds protons and neutrons together in an atomic nucleus (as we saw in Chapter 15).

Physicists have wondered why there are four forces in the universe—why not 300 or, preferably, just one? An important hint comes from the name *electromagnetic* force. For a long time scientists thought that the forces of electricity and magnetism were separate, but James Clerk Maxwell (see Chapter 4) was able to *unify* these forces—to show that they are aspects of the same phenomenon. In the same way, many scientists (including Einstein) have wondered if the four forces we now know could also be unified. Physicists have developed models, called **grand unified theories (GUTs),** to unify three of the four forces.

In these theories, the strong, weak, and electromagnetic forces, are not three independent forces, but instead are different manifestations or aspects of what is, in fact, a single force. The theories predict that at high enough temperatures there would be only one force. At lower temperatures (like the ones in the universe today), however, this single force has changed into three different forces (Figure 28.16). Just as different gases freeze at different temperatures, we can say that the different forces "froze out" of the unified force at different temperatures. Unfortunately, the temperatures at which the three forces were one are so high that they cannot be reached in any terrestrial laboratory. Only the early universe, at times prior to 10^{-35} s, was hot enough to unify these forces. (Many physicists think that gravity is also unified with the other forces at still-higher temperatures.)

The Inflationary Hypothesis

Some forms of the GUTs predict that a remarkable event occurred when the universe was about 10^{-35} s old and the forces were starting to separate. The equations of general relativity, combined with the special state of matter at that time, predict that gravity could briefly have been a repulsive force. In our own time, gravity is an attractive force that slows the expansion of the universe, but for a brief instant near 10^{-35} s after the expansion began, gravity could actually have accelerated the expansion. It is as if the cosmological constant had, for a brief instant, not been equal to zero, and hence there was a tremendous repulsion in the universe.

TABLE 28.4
The Forces of Nature

Force	Relative Strength Today	Range	Important Applications
Gravity	1	Whole universe	Motions of planets, stars, galaxies
Electromagnetism	10^{36}	Whole universe	Atoms, molecules, electricity, magnetic fields
Weak nuclear	10^{33}	10^{-17} m	Radioactive decay
Strong nuclear	10^{38}	10^{-15} m	The existence of atomic nuclei

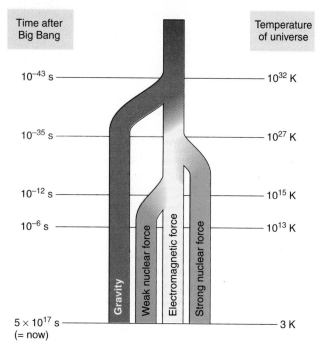

Time after Big Bang

10⁻⁴³ s

10⁻³⁵ s

10⁻¹² s

10⁻⁶ s

5 × 10¹⁷ s
(= now)

Gravity

Weak nuclear force

Electromagnetic force

Strong nuclear force

Temperature of universe

10^{32} K

10^{27} K

10^{15} K

10^{13} K

3 K

Figure 28.16
The strength of the four forces depends on the temperature of the universe. This diagram shows that at very early times when the temperature of the universe was very high, all four forces resembled one another and were indistinguishable. As the universe cooled, the forces took on separate and distinctive characteristics.

A model universe in which this rapid, early expansion occurs is called an **inflationary universe.** The inflationary universe is identical to the Big Bang universe for all time after the first 10^{-30} s. Prior to that, there was a brief period of extraordinarily rapid expansion or inflation during which the scale of the universe increased by a factor of about 10^{50} times more than predicted by standard Big Bang models (Figure 28.17). As the universe expanded, its temperature dropped below the critical value at which all three forces behave in a symmetrical fashion. In the cooler, asymmetrical universe the nuclear forces dominated the electromagnetic force; they continue to do so in our world today.

Prior to the inflation, all the parts of the universe that we can now see were so small and close to each other that they could exchange information. That is, the horizon distance included all of the universe that we can now observe. Before inflation occurred, there was adequate time for the observable universe to homogenize itself and come to the same temperature. Then inflation expanded those regions tremendously, so that many parts of the universe are now beyond each other's horizon.

Another appeal of the inflationary model is its prediction that the density of the universe should be exactly equal to the critical density. To see why this is so, remember that the density of the universe is intimately connected with the curvature of space. A high-density uni-

verse, for example, would have a closed geometry (like the surface of the balloon we discussed). But the period of inflation was equivalent to blowing up the balloon to a tremendous size. The universe became so big that from our vantage point, no curvature should be visible. (In the same way, the Earth's surface is so big that it looks flat to us in any location.)

If the universe inflated to look completely flat locally, what is the density that goes with a flat geometry? Recall that it is the critical density. Thus a universe that looks flat should have just the right average density to give us the critical value. This prediction is a two-edged sword. It explains why the universe is so close to critical density. But if it turns out that the universe is *not* at critical density, it would be a serious blow to the inflationary model.

So what is the density of the universe? Table 28.5 summarizes our earlier discussion of the amount of matter estimated to be present in various types of astronomical objects. Luminous matter in galaxies contributes less than 1 percent of the mass required to reach critical density. Even if we add the invisible dark matter that is detected through its gravitational influence on luminous objects, we are up to only 10 to 20 percent of the critical density.

Furthermore, the abundance of deuterium indicates that protons and neutrons amount to no more than about 5 percent, and perhaps as little as 1 percent, of the critical density. The only observational evidence that immediately supports the idea that we live in a critical-density universe is the motion of nearby galaxies toward the Great

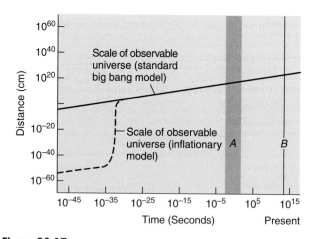

Figure 28.17
How the scale factor of the observable universe changes with time for the standard Big Bang model (solid line) and for the inflationary model (dashed line). (Note that the time scale at the bottom is extremely compressed.) During inflation, regions that were very small and in contact with each other are suddenly blown up to be much larger and outside each other's horizon distance. The two models are the same for all times after 10^{-30} s. Electrons, positrons, and the lightest atomic nuclei are formed during the time interval labeled A. The universe becomes transparent to radiation at the time designated B. "Now" is at the right edge of the figure.

TABLE 28.5
What Different Kinds of Objects Contribute to the Density of the Universe (Dark Matter Is Included)

Object	Density as a Percent of Critical Density
Stars	Less than 1%
Individual galaxies	1–3%
Rich clusters of galaxies	10–20%
Large unseen masses such as the Great Attractor	50–100%

Attractor—implying that some large amount of mass is pulling us in that direction. These observations are so new that some independent confirmation will be required before we can really be confident about the conclusion. But if they are right, the additional mass in the universe must not be made of protons or neutrons.

Dark Matter Possibilities

What we have concluded, then, is that in order for the universe to be at critical density, as inflationary models require, most of its matter must be invisible, *and* cannot be made of protons and neutrons. How should astronomers go about looking for such dark matter? The techniques depend on what we think it might be made of.

One possibility is that the dark matter is made of neutrinos with sufficient mass to produce a critical-density universe. The mass of the neutrino is surely very small, but some tentative experimental evidence suggests that it is not actually zero (see Chapter 15). Attempts are now being made with high-energy accelerators to measure the mass of the neutrino by detecting the change in momentum or energy of another particle that collides with it. Only after the mass is determined will it be possible to calculate the total mass of neutrinos in the universe.

Another experiment involves the study of neutrinos emitted by supernovae. If a supernova were to go off in our own Galaxy, then we could measure the arrival times of the three different types of neutrinos. If they arrive at different times, then at least some of them must have mass. If neutrinos have zero mass, then all of the types will travel at the speed of light and arrive at the same time.

Another possible form dark matter can take is some type of elementary particle that we have not yet detected here on Earth—a particle that has mass and exists in sufficient abundance to reach critical density. GUTs predict the existence of such particles. One class of particles has been given the name WIMPs, which stands for *weakly interacting massive particles*. Since these particles do not participate in nuclear reactions leading to the production of deuterium, the deuterium abundance puts no limits on how many WIMPs might be in the universe. (A number

of other exotic particles have also been suggested as prime constituents of dark matter, by the way.)

If WIMPs do exist, then some of them should be passing through our physics laboratories right now. The trick is to catch them. Since by definition they interact only weakly (infrequently) with other matter, the chances that they will have a measurable effect are small. Nevertheless, physicists are now devising experiments to try to detect them. The basic idea is that a WIMP moving through such a detector might collide with an atomic nucleus and cause it to move. This motion would then be transferred to other particles in the detector, causing a very(!) small change in temperature. Experiments based on this idea will be conducted in the next decade, but are extremely difficult to carry out.

Dark Matter and the Formation of Galaxies

Inflationary models require dark matter for another reason, one related to the formation of galaxies. As we have seen, galaxies must have grown from density fluctuations in the early universe. The observations with COBE give us information on the size of those fluctuations. It turns out that the density variations are too small, at least according to our current theories, to have formed galaxies in the first billion years or so after the Big Bang. Yet observations of quasars suggest that galaxies were indeed formed that early.

The COBE data, however, give us information about density fluctuations only for the type of matter that interacts with radiation. Suppose there is a type of matter that does not interact with light at all—namely dark matter. This matter could have much greater variations in density, which we would not be able to detect because dark matter does not affect the intensity of the radiation measured by COBE. This dark matter might form a kind of gravitational trap that could have begun to attract ordinary matter immediately after the universe became transparent. As ordinary matter became increasingly concentrated, it could have turned into galaxies more quickly thanks to these traps.

"I CAN'T TELL YOU WHAT'S IN THE DARK MATTER SANDWICH. NO ONE KNOWS WHAT'S IN THE DARK MATTER SANDWICH."

(© 1996 Ted Goff)

For an analogy, imagine a boulevard with traffic lights every half-mile or so. Suppose you are part of a motorcade of cars accompanied by police who lead you past each light, even if it is red. So, too, when the early universe was opaque, radiation carried ordinary matter with it, sweeping past the concentrations of dark matter. Now suppose the police leave the motorcade, and the lights all turn red at the same time. The red lights act as traffic traps; approaching cars now have to stop, and so they bunch up. Likewise, after the early universe became transparent, ordinary matter interacted with radiation only occasionally and so could fall into the dark-matter traps.

The size of the gravitational traps depends on the nature of the dark matter. Suppose it is moving near the speed of light—astronomers call this *hot dark matter*—as neutrinos would. Then small-scale density fluctuations are smoothed out by the rapidly streaming particles as they move from high- to low-density regions. In this case, large-scale structure would form first. If, on the other hand, the dark matter moves slowly—we call this *cold dark matter*—then the particles do not have time to move far enough to smooth out small-scale density fluctuations. In this case, relatively small structures, the size of globular clusters or individual galaxies, are likely to form first.

Neither hot nor cold dark matter is entirely successful in explaining the distribution of galaxies discussed in Chapter 27. Hot dark matter models predict that all galaxies should be found in large sheet-like structures, which is not seen. Cold dark matter cannot produce voids, walls, and long structures such as the Great Wall. Now theories are being developed that contain both hot and cold dark matter. Even though current models are not adequate to explain how galaxies form, it is important to note that galaxies are difficult to form at all unless a substantial amount of dark matter of some kind is present.

To sum up, attempts to understand what the standard Big Bang model does not explain—isotropy, the density of the universe, the formation and distribution of galaxies—have suggested radical new ideas about the universe, which must be tested by observations. If it is really true that most of the matter in the universe is made of some type of particle that we have not yet discovered, then we must accept the challenge of trying to detect it. The search is on—in huge accelerators, in university laboratories around the world, and in deep underground mines, where scientists are trying to trap elusive dark matter particles just as they once succeeded in capturing neutrinos. There are exciting years ahead in cosmology.

Conclusion

Thus the explorations of space end on a note of uncertainty. . . . With increasing distance our knowledge fades . . . and we search among ghostly errors of measurement for landmarks that are scarcely substantial.

—E. Hubble in *The Realm of the Nebulae* (1936)

You may have found this brief discussion of dark matter and cosmology a bit frustrating. We have offered glimpses of theories and observations, but have been unable to provide satisfying answers to some of the problems we have raised. These ideas are at the forefront of modern science, where questions are more numerous than answers, and much more work is needed before we can see clearly. Bear in mind that less than a century has passed since Hubble demonstrated the existence of other galaxies. The quest to understand how the universe of galaxies came to be will keep astronomers busy for a long time to come.

Summary

28.1 Cosmology is the study of the organization and evolution of the universe. The universe is expanding, and this is one of the starting observational points for modern cosmological theories. Before Hubble showed that the universe was expanding, Einstein introduced a **cosmological constant** into his equations to counterbalance gravity and make the universe static. From the expansion rate of the universe we can estimate that all of the matter within it was concentrated in an infinitesimally small volume roughly 9 to 15 billion years ago, a time we call the **Big Bang.** The age of the universe is uncertain because we do not know how much gravity *decelerates* the universe.

28.2 The mass in the universe curves the fabric of space-time. A curved universe has neither center nor edge. In a **closed universe,** which is finite, if you keep going straight ahead you eventually return to where you started. In an **open universe,** which is infinite, not only do you not return to your starting point, but more space than you expect opens up. The Big Bang happened throughout space, everywhere at once. Galaxy redshifts are a result of the stretching of space.

28.3 The factor that controls the ultimate fate of the universe is the density of matter (and energy), which is related to the curvature. The **critical density** is that required to stop the expansion at a time infinitely far in the future. If the density is higher than this, then the rate of expansion will slow and reverse direction so that the galaxies all come together again (a closed universe). Observations, however, suggest that the density is actually so low that the expansion will continue forever (an open universe). A universe with critical density would have zero curvature; we call this model a **flat universe.**

28.4 The universe cools as it expands. The energy of photons is determined by their temperature, and calculations show that in the hot early universe photons had so much energy that when they collided with one another they could produce material particles. As the universe expanded and cooled, protons and neutrons formed first; then came electrons and positrons. Next, fusion reactions produced deuterium, helium, and lithium nuclei. Finally, the universe became cool enough to form neutral hydrogen atoms. At this *decoupling* time the universe became transparent to radiation. Scientists have detected the **cosmic background radiation (CBR)** from the hot early universe. Measurements with the COBE satellite show that the CBR is a blackbody with a temperature of 2.735 K. Tiny fluctuations in the CBR may be showing us the seeds of large-scale structure in the universe.

28.5 The Big Bang model does not explain why the CBR has the same temperature in all directions. Neither does it explain why there was originally more matter than antimatter, nor why the density of the universe is so close to critical density. New **grand unified theories (GUTs),** which predict a period of very rapid expansion, or **inflation,** when the universe was 10^{-35} s old, are being developed to try to explain these observations. One prediction of these new theories is that the density of the universe should be exactly equal to the critical density. Most observations are inconsistent with such a high density. Determination of the quantity and composition of dark matter is crucial to understanding the early history of the universe.

Review Questions

1. What are the basic observations about the universe that any theory of cosmology must explain?

2. Describe three possible futures for the universe. What property of the universe determines which of these possibilities is the correct one?

3. Which formed first in the early universe—protons and neutrons, or electrons and positrons? Why?

4. Which formed first—hydrogen nuclei or hydrogen atoms? Explain the sequence of events that led to each.

5. Describe at least two characteristics of the universe that are explained by the standard Big Bang model.

6. Describe two properties of the universe that are not explained by the standard Big Bang model (without inflation). How does inflation explain these two properties?

7. Why do astronomers believe there must be dark matter that is not in the form of atoms with protons and neutrons?

8. What is the most useful probe of the early evolution of the universe—a giant elliptical galaxy, or an irregular galaxy such as the Large Magellanic Cloud? Why?

9. What are the advantages and disadvantages of using quasars to probe the early history of the universe?

10. Suppose someone proposed a model with the Great Attractor as the center of the universe, and with all the galaxies and galaxy clusters falling in toward it. Give some arguments against this idea.

11. Suppose the universe expands forever. Describe what will become of the radiation from the primeval fireball. What will the future evolution of galaxies be like? Could life as we know it survive forever in such a universe? Why?

12. Some theorists argue that the universe is at just the critical density. Do the current observations support this hypothesis?

13. Summarize the evidence for the existence of dark matter in the universe.

14. In this text we have discussed numerous motions of the Earth as it travels through space with the Sun. Describe as many of these as you can.

15. There are a variety of ways of estimating the ages of various objects in the universe. Describe some of these ways, and indicate how well they agree with one another and with the age of the universe itself as estimated by its expansion.

16. Since the time of Copernicus, each revolution in astronomy has moved humans farther from the center of the universe. Now it appears that we may not even be made of the most common form of matter. Trace the changes in scientific thought about the central nature of the Earth, the Sun, and our Galaxy on a cosmic scale. Explain how the notion that most of the universe is made of dark matter continues this "Copernican tradition."

17. Construct a time line for the universe, and indicate when various significant events, from the beginning of the expansion to the formation of the Sun to the appearance of humans on Earth, occurred.

18. The Andromeda galaxy is approaching the Sun at a velocity of about 300 km/s. Does this indicate that the universe is not expanding? Compare this velocity with that of the Sun in its orbit around the center of the Galaxy. Suppose Andromeda is orbiting the Milky Way with a period of 10 billion years. What velocity would it have?

19. Show that if $H = 25$ km/s per million LY, then the maximum age of the universe is approximately 13 billion years.

20. Suppose $H = 15$ km/s per million LY. What is the maximum age of the universe?

21. It is possible to derive the age of the universe, given the value of the Hubble constant and the distance to a galaxy. Consider a galaxy at a distance of 400 million LY, receding from us at a velocity, v. If the Hubble constant is 25 km/s per million LY, what is its velocity? How long ago was that galaxy right next door to our own Galaxy if it has always been receding at its present rate? Express your answer in years. Since the universe began when all galaxies were very close together, this number is an estimate for the age of the universe.

Boslaugh, J. *Masters of Time: Cosmology at the End of Innocence.* 1992, Addison-Wesley. A journalist reports from the frontiers of cosmological research.

Brush, S. "How Cosmology Became a Science" in *Scientific American,* Aug. 1992, p. 62.

Croswell, K. "A Milestone in Fornax" in *Astronomy,* Oct. 1995, p. 42. On efforts to determine the age of the universe.

Davies, P. *The Last Three Minutes.* 1994, Basic Books. Introduction to the ultimate fate of the universe.

Davies, P. "Everyone's Guide to Cosmology" in *Sky & Telescope,* March 1991, p. 250. Good introductory article.

Ferris, T. *The Red Limit,* 2nd ed. 1983, Morrow. An eloquent history of modern cosmology.

Fienberg, R. "COBE Confronts the Big Bang" in *Sky & Telescope,* July 1992, p. 34. Good summary of the temperature fluctuations discovery.

Gribbin, J. *In Search of the Big Bang.* 1986, Bantam. Excellent beginner's introduction to ideas in cosmology.

Guth, A. and Steinhardt, P. "The Inflationary Universe" in *Scientific American,* May 1984, p. 116.

Halliwell, J. "Quantum Cosmology and the Creation of the Universe" in *Scientific American,* Dec. 1991, p. 76.

Harrison, E. *Cosmology.* 1981, Cambridge U. Press. A fine, erudite textbook, full of good examples.

Overbye, D. *Lonely Hearts of the Cosmos.* 1991, Harper Collins. Wonderful introduction to cosmology today, with a focus on the people involved.

Roth, J. and Primack, J. "Cosmology: All Sewn Up or Coming Apart at the Seams?" in *Sky & Telescope,* Jan. 1996, p. 20.

Smoot, G. and Davidson, K. *Wrinkles in Time.* 1993, Morrow. The full story of the COBE discoveries, by one of the team leaders and a science journalist.

1. Display only spiral galaxies and set the *Limiting Magnitude* to 10. Under *Find Objects*, choose Messier Objects in *Deep Sky*. Center the *Display* on galaxy M106 and set the *Zoom Factor* to .9. Print the map. Reset the *Zoom Factor* to 0.925 and print a second map. Imagine the two maps represent different ages of the universe; the zoom factors then correspond to different amounts of expansion by the universe.

Overlay the two maps and notice the "expansion" of the galaxies. (You may need to darken the center of each galaxy marker and hold the sheets toward a light.) Compare the expansion of a galaxy near and one far from M106.

Can you explain the expansions in terms of Hubble's redshift law?

Pick a galaxy other than M106 and line up the two markers for the galaxy.

Does the "expansion" still occur?

Can you use the expansion to locate the center of the universe? Explain your answer.

2. Display only spiral galaxies and set the *Limiting Magnitude* to 10. Center the display on the galaxy M106 and set the *Zoom Factor* to 0.9. Print the map. (You can use the map from Exercise 1.) Reset the *Zoom Factor* to 1.2 and print a second map.

Pick five widely separated galaxies contained in both maps and number them. Select one galaxy to represent the Galaxy; use a millimeter ruler to measure the distances of the other galaxies from the "Galaxy," first on the 0.9-zoom map and then on the other map. Because the distances represent galaxy positions at different ages of the universe, the difference between the distances for each galaxy is a velocity of expansion. To find the rates of expansion, take the difference between the distances for each galaxy and divide it by the distance for the 1.2-zoom map.

How is this related to the Hubble constant?

Choose a different galaxy to represent the Galaxy. Repeat the measurements and find the expansion rate. Did it change?

Light echoes from Supernova 1987A. The expanding shell of light from the exploding star reflects from two dusty regions around the supernova, making two circles of light. To bring out the faint light echoes, David Malin subtracted from the echo image a photo of the area taken before the explosion. Since the stars and unrelated tendrils of gas have not changed, they subtract out, leaving black areas, while the yellow light of the supernova echo stands out clearly in contrast. Most of the atoms in our bodies were recycled through the action of such supernovae. (© Anglo-Australian Telescope Board)

Epilogue: Cosmic Evolution and Life Elsewhere

Our voyages have taken us through billions of years of time and billions of light years of space. It is time to sum up where we have been and what we have learned, and to ask one remaining question. Are we the only creatures in the universe thinking about its origin and evolution, or could there possibly be other intelligent life-forms among the stars who also enjoy considering such questions?

We shall not cease from exploration,
And the end of all our exploring,
Will be to arrive where we started,
And know the place for the first time.

T. S. Eliot, *Little Gidding* (from *The Four Quartets* in *The Collected Poems of T. S. Eliot,* 1934, 1936)

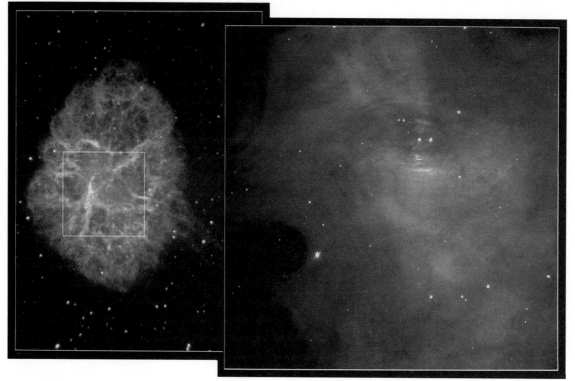

Figure E.1

The Crab Nebula is the remnant of a supernova first seen in July 1054. Now almost 11 LY across, this remnant still glows with tremendous energy in many bands of the electromagnetic spectrum. It is powered inside by a pulsar, a spinning neutron star whose beams sweep over the Earth like a lighthouse 30 times a second. The left-hand image, taken with a ground-based telescope, shows the full nebula. The blue glow in the center is powered by the pulsar. The right-hand image, taken with the Hubble Space Telescope, is a close-up of the Crab's central region. The pulsar itself can be seen as the left member of a pair of stars near the center of the frame. The Hubble was able to observe wisps of material streaming away from the pulsar at half the speed of light! (J. Hester, P. Scowen, and NASA)

What Were the Atoms in Your Body Doing Billions of Years Ago?

Like all good voyages, this one—we hope—has taught you as much about yourself as about the places we have visited. Let's review what we have learned about the history of the cosmos by examining the history of the atoms in your body before they became a part of you.

After the universe cooled sufficiently for atoms to exist, all matter consisted of hydrogen and helium (with a very small amount of lithium). The hydrogen atoms in the water and other hydrogen-rich molecules in your body formed at this early time; they are the oldest atoms that are part of you. But it is not possible to make an organism as complex and interesting as you with only the first three elements. Before something like you could evolve, several generations of stars had to go through their cycles of birth, life, and death.

As we have seen throughout this book, the only place new elements can be synthesized (now that the universe has cooled) is inside stars. We know from observations of distant quasars that stars must have formed within the first billion years or so, because quasar spectra already show the presence of some of the heavier elements. As the years went on, generations of stars produced more and more of the heavier elements. The most-massive stars not only produced the greatest variety of new nuclei, but then had the courtesy to explode (Figure E.1), scattering the newly minted atoms into space.

Over the years, thanks to supernovae, the gas between the stars became increasingly enriched with heavier materials. In the cooler outer layers of old stars, atoms frequently combined into solid particles that we call interstellar dust. The next generations of stars and planets, containing atoms of carbon, nitrogen, silicon, iron, and the rest of the familiar elements, then formed from reservoirs of enriched gas and dust. One of the most remarkable discoveries of modern astronomy is that life on Earth is mostly composed of just those elements that stars find easiest to make.

About 5 billion years ago, possibly prodded by the shock wave of a nearby supernova, a cloud of gas and dust in this cosmic neighborhood began to collapse under its own weight. Out of this cloud formed the Sun and its

Figure E.2
Comet Hyakutake captured by amateur astrophotographer Robert Provin of California State University, Northridge. This 12-min exposure was taken on a clear night in the Mojave Desert with a camera and telephoto lens mounted on a telescope tripod. As the moving comet was held steady in the camera, the stars appeared to streak. (R. Provin)

planets, together with all the smaller bodies, such as comets, that also orbit the Sun (Figure E.2). The third planet from the Sun, as it cooled, developed an atmosphere that served to moderate temperature extremes and allow the formation of large quantities of liquid water on its surface. The chemicals available on the cooling Earth may have been further enriched by the addition of molecules frozen in the nuclei of comets that eventually collided with our planet.

The chemical variety and moderate conditions on Earth eventually led to the formation of self-reproducing molecules and the beginnings of life. Over the billions of years of Earth history, that life slowly evolved and became more complex. The course of evolution was punctuated by occasional planet-wide changes caused by collisions with those planetesimals (or their fragments) that had not been incorporated into the Sun or one of its accompanying worlds. Mammals may owe their domination of the Earth's surface to just such a collision 65 million years ago.

Through many twisting turns, the course of evolution on Earth produced a creature with self-consciousness, able to ask questions about its own origins and place in the cosmos (Figure E.3). Like most of the Earth, this creature is composed of atoms that were forged in the centers or explosions of stars—in this case assembled rather cleverly into brains, kidneys, fingers, and faces. We might say that through the thoughts of human beings, the matter in the universe can become aware of itself.

The atoms in your body are merely on loan to you from the lending library of atoms that make up our local corner of the universe. Atoms of many kinds circulate through your body, and then leave it with the breath you exhale and the food you eat and excrete. Even the atoms that take up more permanent residence in your tissues will not be part of you much longer than you are alive. Ultimately, you will return your atoms to the vast reservoir of the Earth, where they will be incorporated into other structures and maybe even other living things in the millennia to come.

This picture of *cosmic evolution,* of our descent from the stars, has been obtained through the efforts of scientists in many fields over many decades. Some of its details are, as we saw throughout the book, still tentative and incomplete, but we feel reasonably confident in its broad outlines. While we do not claim to know the full story of how the universe and we ourselves evolved over the course of cosmic history, it is remarkable how much we *have* been able to learn in the short time we have had instruments to probe the physical nature of the stars.

The Copernican Principle

Our study of astronomy has also shown that we have always been wrong in the past whenever we have claimed that the Earth is somehow unique. Copernicus and

Figure E.3
Human beings have the intellect to wonder about their planet and what lies beyond it. Through them, the universe becomes aware of itself. (Photo by A. Fraknoi)

Galileo showed that the Earth was not the center of the solar system, but merely one of a number of bodies orbiting the Sun. Our study of the stars has demonstrated that the Sun itself is a rather undistinguished star, living peacefully through its main-sequence stage like so many billions of others. There seems nothing special about our position in the Milky Way, and nothing surprising about our Galaxy's position in either its own group or its supercluster.

The recent discovery of planets around other stars confirms our ideas that our planetary system is not the only one, and that the formation of planets is probably a natural consequence of the formation of many kinds of stars. Now that our technology is sufficiently advanced to detect planets elsewhere, they seem to be turning up with impressive regularity in the nearby star systems where we can most easily search. While our current techniques of finding planets allow us only to identify Jupiter-mass bodies, there is no reason to believe that other planetary systems could not contain planets like the Earth as well.

Philosophers of science sometimes call this idea — that there is nothing all that special about our place in the universe — the *Copernican principle*. Although it may be tempting to consider ourselves the central focus of all creation, no evidence for such a belief is found in any of the observations discussed in this book.

Most scientists, therefore, would be surprised if the beginning or evolution of life were absolutely limited to the surface of our planet, and had happened nowhere else. There are, after all, billions of stars in our Galaxy with main-sequence lifetimes long enough for life to have developed on a planet around them. Astronomers and biologists have thus long conjectured that a series of events similar to those on the early Earth has probably led to living organisms around other stars. And, where conditions are right, such life may well have evolved to become what we would call intelligent — that is, aware of and interested in its own cosmic history. (In this sense, we must conclude — with tongue firmly in cheek — that taking an astronomy class is the supreme example of intelligent behavior in the universe!)

Such arguments from the Copernican principle — however interesting they may be for philosophers — are nonetheless insufficient for scientists. We would like to find actual evidence for the existence of intelligent life elsewhere. Despite the sensationalistic claims in the tabloid media, no such evidence has yet been found. But because many scientists feel that such a discovery would be a defining moment in the history of the human species, a number of searches for extraterrestrial life have already been carried out, and others are under way.

The Building Blocks of Life

While no unambiguous evidence has yet been found for life beyond Earth, its chemical building blocks have been detected in a wide range of extraterrestrial environments.

Meteorites, such as the one found near Murchison, Australia, have yielded a variety of amino acids (the building blocks of proteins) whose chemical structures clearly denote extraterrestrial origins (see Chapter 13). When we examine the evaporated material around comets such as Comet Halley, we find a number of *organic* molecules — those that on Earth are associated with the chemistry of life.

One of the most surprising results of modern radio astronomy, as we saw in Chapter 19, was the discovery of organic molecules in the giant gas and dust clouds between stars. We find such molecules most readily in regions where the interstellar dust is most abundant — precisely those regions in which star formation (and probably planet formation) happens most easily.

Starting in the early 1950s, scientists have tried to duplicate in their laboratories the chemical pathways that led to life on our planet. In a series of experiments pioneered by Stanley Miller and Harold Urey at the University of Chicago, biochemists have simulated conditions on the early Earth and been able to produce many of the fundamental building blocks of life — including those that go into forming proteins and nucleic acids (Figure E.4). While their accomplishment is still far from making life in the laboratory, it does show that the first steps on the long road to life may not be as difficult as once thought.

Already there are suggestions that these steps once led to life on our neighbor planet Mars. As we saw in Chapter 9, the Viking spacecraft did not find any indication of life existing there today, although its observations suggest that conditions on the red planet were more conducive to the formation of life billions of years ago. In 1996, detailed laboratory analysis of the one ancient martian rock sample available on earth — the 4.5-billion-year-old meteorite called ALH 84001 — showed that this bit of martian crust experienced wet conditions about 3.5 billion years ago. At that time, water flowed on Mars (see Chapter 9) and left traces of carbonate minerals and some organic compounds embedded in the meteorite. Some scientists even suspect that tiny structures seen in this rock at high magnification might be the fossilized remains of ancient microbial life. If there are indeed fossils in ancient martian rock, then life would have begun on Mars independently of the Earth and it would be one of the most spectacular discoveries in the history of science. But even if these specific features turn out to be of inorganic origin, the work on ALH 84001 has already verified that wet, apparently Earth-like conditions once existed on Mars. The atmospheres of the jovian planets and of Saturn's satellite Titan do show evidence of organic molecules, but none that have developed into life-forms. It may be that conditions more similar to the early Earth (or early Mars) are required for life to develop, or that we simply do not understand the many difficulties involved in the formation of molecules that can make copies of themselves. In laboratories around the world, biochemists are trying to find answers to these questions.

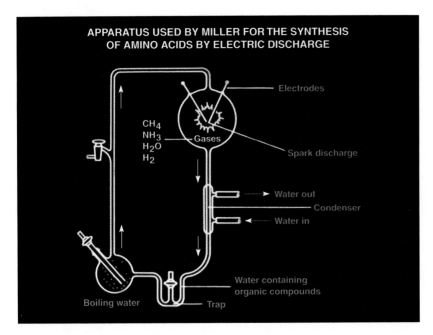

Figure E.4
The Miller–Urey experiment, performed in 1953, simulated conditions on the early Earth. An "atmosphere" consisting of methane, ammonia, water vapor, and hydrogen was subjected to electrical sparks to simulate lightning. Water at the bottom of the apparatus provided an "ocean" into which materials synthesized in the atmosphere could fall. When its contents were analyzed, they were found to contain a variety of amino acids, the building blocks of proteins. (F. Drake and the Astronomical Society of the Pacific)

Searching for Life Elsewhere

Suppose, for the sake of argument, that we take an optimistic view and assume that life has developed around many other stars, perhaps on planets that resemble the Earth in chemically significant ways. Suppose further that in some of these cases, life has evolved to be intelligent (self-aware and interested in what is happening elsewhere in the universe). How can we find out about, and perhaps even make contact with, such life-forms?

This problem is similar to making contact with people who live in a remote part of the Earth. If students in the United States want to converse with students in Australia, for example, they have two choices. Either one group gets on an airplane and travels to meet the other, or they communicate via some message medium (today, probably by telephone, fax, e-mail, or short-wave radio). Given how expensive airline tickets are, most students would probably select the message route.

In the same way, if we want to get in touch with intelligent life around other stars, we can travel, or we can try to exchange messages. Because of the great distances involved, interstellar space travel is either very slow or very expensive. The fastest spacecraft the human species has built so far would take almost 80,000 years to get to the nearest star. While we could certainly design a faster craft, the more quickly we require it to travel, the greater the energy cost involved. To reach neighboring stars in less than a human life span, we would have to travel close to the speed of light. In that case, however, the expense would become truly astronomical.

The late Bernard Oliver, vice president of the Hewlett–Packard Corporation and an engineer with an abiding interest in life elsewhere, made a revealing calculation about the costs of rapid space travel. Since we do not know what sort of technology we (or other civilizations) might someday develop, Oliver considered a trip to the nearest star in a spaceship with a "perfect engine"— one that would convert its fuel into energy with 100 percent efficiency. (No future technology can possibly do better than this. In reality, nature is unlikely to yield an efficiency even close to the perfect value; just think how much of the energy released by the fuel in a car is wasted.) Even with a perfect engine, the energy cost of a single round-trip journey at 70 percent the speed of light would be equivalent to about *500,000 years worth of total U.S. electrical energy consumption!* Congress is unlikely to fund such a program in the near future.

In case you are wondering why this figure is so high, you must remember that the voyagers could not depend on finding "gas stations" open at their destination. Therefore they would have to carry the fuel for the return legs of the journey with them, and getting all that fuel up to 70 percent the speed of light would be very expensive. The important thing about Oliver's calculation is that it does not depend on present-day technology (since it assumes a perfect engine), but only on the known laws of science. What it shows is that no matter who does the traveling, it is expensive to go fast enough to get to the stars within the course of a single human life.

Figure E.5
The image engraved on the plaques aboard the Pioneer 10 and 11 spacecraft. The human figures are drawn in proportion to the spacecraft, which is shown behind them. The Sun and planets in the solar system can be seen at the bottom, with the trajectory that the spacecraft followed. The lines and markings in the left center show the positions and pulse periods for a number of pulsars, to locate the spacecraft's launch in space and time. (Recall that the periods of pulsars lengthen as time goes on, so the exact timing of the pulses might tell a sophisticated group of alien astronomers when the craft began its journey.) (NASA)

This is one reason astronomers are so skeptical about claims that unidentified flying objects (UFOs) are spaceships from extraterrestrial civilizations. Given the distance and expense involved, it seems unlikely that the dozens of UFOs (and, recently, even UFO abductions) reported each year could all be visitors from other stars so fascinated by Earth civilization that they are willing to expend fantastically large amounts of energy or time to reach us.

In fact, a sober evaluation of UFO reports often converts them to IFOs (identified flying objects), or NFOs (not-at-all flying objects). While some are hoaxes, others are natural phenomena such as ball lightning, fireballs, bright planets, or even flocks of birds with reflective bellies. Still others are human craft, such as private planes with some lights missing, or classified military airplanes. In fact, not a single UFO has ever left behind any physical evidence that can be tested in an Earth laboratory and thus shown to be of nonterrestrial origin.[1]

Some visionaries have suggested that we might someday overcome the long-time/large-energy demands of space travel by hollowing out asteroids and sending them to the stars with large colonies of people on board (equipped with all the necessary provisions for a really long trip). While the original settlers would never see the stars, their far-future descendants might arrive and start a settlement on an Earth-like planet somewhere. This is, for now, only the stuff of science fiction.

Messages on Spacecraft

In the real world we do have four spacecraft—two Pioneers and two Voyagers—which, having finished their program of planetary exploration, are now leaving the solar system. At their coasting speeds, they may well take hundreds of thousands or millions of years to get anywhere close to another star. On the other hand, they were the first products of human technology to leave our home system, and so we wanted to put messages on board to show where they came from.

Each Pioneer carries a plaque with a pictorial message engraved on a gold-anodized aluminum plate (Figure E.5). The Voyagers, launched in 1977, have audio and video records attached (Figure E.6), which allowed the inclusion of over 100 photographs and a selection of music from around the world. (Included among the excerpts from Bach, Beethoven, folk music, tribal chants, etc., is one piece of rock and roll—"Johnny Be Goode" by Chuck Berry.) Given the enormity of the space between stars in our section of the Galaxy, it is very unlikely that these messages will ever be received by anyone. They are more like a note in a bottle thrown into the sea by a shipwrecked sailor, with no realistic expectation of its being found, but a slim hope that perhaps someday, somehow, someone will know of his fate.

---

[1] If you are interested in pursuing the topic of what UFOs are and aren't, we recommend the following books: Klass, P. *UFO Abductions: A Dangerous Game* (1988, Prometheus Books); Peebles, C. *Watch the Skies: A Chronicle of the Flying Saucer Myth* (1994, Smithsonian Institution Press); and Shaeffer, R. *The UFO Verdict: Examining the Evidence* (1981, Prometheus Books).

The Voyager Message

An excerpt from the Voyager record:

> We cast this message into the cosmos. It is likely to survive a billion years into our future, when our civilization is profoundly altered. . . . If [another] civilization intercepts Voyager and can understand these recorded contents, here is our message:
>
> This is a present from a small, distant world, a token of our sounds, our science, our images, our music, our thoughts, and our feelings. We are attempting to survive our time so we may live into yours. We hope, someday, having solved the problems we face, to join a community of galactic civilizations. This record represents our hope and our determination, and our good will in a vast and awesome universe.

—Jimmy Carter, President of the United States of America, June 16, 1977

Communicating with the Stars

If direct visits to stars are unlikely, we must turn to the other alternative for making contact—exchanging messages. Here the news is a lot better. As we have seen throughout this book, we already know (and have learned to use) a messenger—electromagnetic radiation—that moves through space at the fastest speed in the universe. Traveling at the speed of light, radiation reaches the nearest star in only four years, and does so at a fraction of the cost of sending material objects. These advantages are so clear and obvious that we assume they will occur to any other species of intelligent beings who develop technology.

However, we have access to a wide spectrum of electromagnetic radiation, ranging from the longest-wavelength radio waves to the shortest-wavelength gamma rays. Which would be the best for interstellar communication? It would not be smart to select a wavelength that is easily absorbed by interstellar gas and dust, or one that is unlikely to penetrate the atmosphere of a planet like ours. Nor would we want to pick a wave that has lots of compe-

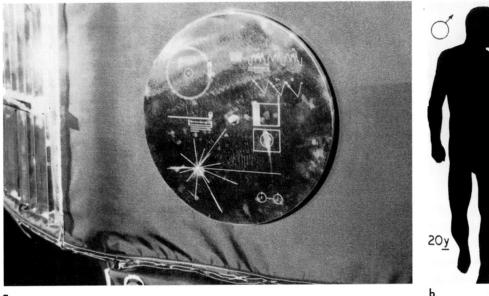

a

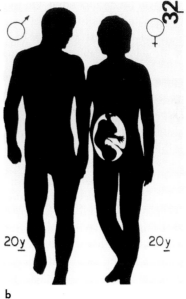

b

Figure E.6
(a) Encoded onto a gold-coated copper disk, the Voyager record contains 118 photographs, 90 min of music from around the world, greetings in almost 60 languages, and other audio material. It is a summary of the sights and sounds of Earth. (b) One of the images encoded onto the record. Originally, the team devising the record wanted to send a photograph, from a medical book, of a nude man and a pregnant woman. However, NASA was concerned about offending some people on Earth, and so artist Jon Lomberg drew a silhouette version, which allowed the fetus inside the woman to be shown as well. (NASA)

tition for attention in our neighborhood. For example, it would be difficult (and pretty dumb) for us to put together a signal from our civilization in the visible light region of the spectrum. How could it compete with our extremely strong local source of light, the Sun?

One final criterion makes the selection an easy one: we want the radiation to be inexpensive to produce in large quantities. When we consider all these requirements, radio waves win hands down. Being the lowest-frequency (and lowest-energy) band of the spectrum, they are not very expensive to produce (permitting us to use them extensively for communications on the Earth). They are not significantly absorbed by interstellar dust and gas (which is what allows us to use them to map the structure of our Galaxy). With some exceptions, they easily pass through the Earth's atmosphere, and through the atmospheres of the other planets we are acquainted with. And the Sun does not put out a large quantity of radio waves, which means that a radio message has a realistic chance of being "heard" above the local noise.

The Cosmic Haystack

For these reasons, many astronomers have decided that the radio band is probably the best place in the spectrum for communication among intelligent civilizations. Having made such a decision, however, we still have many questions and a daunting task ahead of us. Shall we *send* a message, or try to *receive* one? Obviously, if every civilization decides to receive only, then no one will be sending, and everyone will be disappointed. On the other hand, it may be appropriate for us to *begin* by listening, since we are likely among the most primitive civilizations in the Galaxy who are interested in exchanging messages.

We do not make this statement to insult the human species (which—with certain exceptions—we are rather fond of). Instead, we base it on the fact that we have had the ability to receive (or send) a radio message across interstellar distances for only a few decades. Compared to the ages of the stars and the Galaxy, this is a mere instant. If there are civilizations out there who are ahead of us in development by even a short time (in the cosmic sense), they are likely to have a head start of many, many years. If there are civilizations behind us, chances are they are sufficiently far behind that they have not yet developed radio communications. In other words, we, who have just started, may well be the "youngest" species in the Galaxy with this capability. Just as the youngest members of a community are often told to be quiet and listen to their elders for a while before they say something foolish, so we may want to begin our exercise in extraterrestrial communication by listening.

Even restricting our activities to listening, however, leaves us with an array of challenging questions. For example, you know from your own experience with radio transmissions that a typical signal only comes in on one channel (which means it is carried by one small frequency

| TABLE E.1 |
| **The Cosmic Haystack Problem: Some Questions About an Extraterrestrial Message** |

- From what direction (which star) is the message coming?
- On what channels (or frequencies) is the message being broadcast?
- How wide in frequency is the channel?
- How strong is the signal (can our radio telescopes detect it)?
- Is the signal continuous, or does it shut off at times (as, for example, a lighthouse beam does when it turns away from us)?
- Does the signal drift (change) in frequency because of the changing relative motion of the source and the receiver?
- How is the message encoded in the signal (how do we decipher it)?
- Can we even recognize a message from a completely alien species? Might it take a form we don't at all expect?

band of radio waves). The owners of your favorite station are confident—because there aren't that many channels on the radio dial—that you will find them despite this (although a few actually transmit over two different frequencies, one on the AM band and one on the FM band). Many of us, when we first arrive in a new city, scan up and down the radio band until we find the stations we like. But if we have only an AM radio in our car, and the stations playing our favorite music are all FM, then we are out of luck.

In the same way, it would be very expensive (and perhaps even ill-mannered) for an extraterrestrial civilization to broadcast on a huge number of channels. Most likely, they select one or a few channels for their particular message. But the radio band of the electromagnetic spectrum contains an astronomically large number of possible channels. How can we know in advance which one they have selected, and how they have coded their message into the radio signal?

If your radio has a poor antenna, then it may not pick up the signal from a weak station some distance away. You may not learn about the existence of that station (and others like it) until you buy better equipment. The same will be true for interstellar transmissions. If an extraterrestrial civilization's signal is just too weak for our present-day radio telescopes, they may be broadcasting their little alien hearts out, but we will miss it completely.

Table E.1 summarizes these and other factors that scientists must grapple with when trying to tune into radio messages from distant civilizations. Because their success depends on either guessing right about so many factors, or else searching through all the possibilities for each factor, some scientists have compared their quest to looking for a needle in a haystack. Thus they like to say that the list of factors in Table E.1 defines the *cosmic haystack problem.*

Figure E.7
A 25th anniversary photo of some members of the Project Ozma team standing in front of the 85-ft radio telescope with which the 1960 search for extraterrestrial messages was performed. Frank Drake is in the back row, second from the right. (National Radio Astronomy Observatory)

Radio Searches

Although the cosmic haystack problem seems daunting, many other research problems in astronomy also require a large investment of time, equipment, and patient effort. And, as several astronomers have pointed out, if we don't search, we're sure not to find anything. Thus several groups of radio astronomers have undertaken searches for extraterrestrial messages during the last three decades.

The very first search for such radio signals was conducted by astronomer Frank Drake in 1960, using the 85-ft antenna at the National Radio Astronomy Observatory (Figure E.7). Called Project Ozma, after the queen of the exotic land of Oz in the children's stories of Frank L. Baum, his experiment involved looking at about 7200 channels and two nearby stars over a period of 200 hours. Although he found nothing, he demonstrated the feasibility of such a search, and set the stage for the more sophisticated projects that followed. (It is interesting to note

that what took 200 hours in 1960 could be done with today's automated systems in about a thousandth of a second.)

Since 1960 almost 60 radio searches have been carried out by scientists around the world, each exploring a minuscule region of the cosmic haystack. Although a number of interesting signals have been found, none has met the crucial test of being detected more than once, so it could be checked. Scientists are continuing several of the searches, always trying to improve their equipment and beat the odds against finding that elusive needle of a message.

In 1992 NASA began the most comprehensive search for radio messages ever undertaken, only to have Congress cut the funding for their project after less than a year. Using private donations, the nonprofit SETI (Search for Extra-Terrestrial Intelligence) Institute has undertaken to continue the search. They now call it Project Phoenix, since the program has risen from the ashes of its funding crisis.

Using modern electronics and computers, the Phoenix system can "listen in" on 28 million channels simultaneously. Its software checks promising signals and alerts the experimenters if an interesting signal persists on any one channel, or moves between channels because of Doppler shift. The intent of the project (which resumed in 1995) is to search 2 billion channels for each of about 1000 nearby stars. To make it onto their list, a star must be roughly similar to the Sun and at least 3 billion years old. That means it could have had enough time for intelligent life to develop on any Earth-like planets around it. The first 200 stars, visible from the Southern Hemisphere, were searched using the 64-m radio antenna near Parkes, Australia (Figure E.8). So far, no signal has been found; the search will continue using other large radio telescopes.

What If We Succeed?

No one can predict when or whether such searches will be successful. It may well be that civilizations technologically far in advance of our own use other forms of communication that we are not yet aware of. After all, 150 years ago we did not have an inkling of the possibilities of radio communication, while today it is difficult to imagine our civilization without it. On the other hand, we would never dream of giving a preschooler a book like the one you are reading. Young children learning to read are given very simple books until they have mastered the basics. We hope that advanced civilizations remember their own youth and send out messages that even youngsters like us can find and interpret.

What will happen if we do find a radio signal that is unambiguously the product of an extraterrestrial intelligence? The existence of the signal itself will be of tremendous philosophical importance, demonstrating that we are

(text cont. on page 595)

a

Figure E.8
(a) The 64-m radio telescope at Parkes, Australia, was used in 1995 to search for radio signals from possible extraterrestrial civilizations around 200 stars. (b) Phoenix project scientists Jill Tarter and Peter Backus are shown at the telescope controls during their observing run. (Photos by Seth Shostak, SETI Institute)

b

not alone in the cosmos. But unless we are able to interpret the message, there may not be much practical value in the discovery. If we can eventually work out a method of mutual communication, however, an interesting question will arise: who will speak for planet Earth?

Suppose we know that a star 35 LY away has a technological civilization around it, and that they have given us a kind of code by which we can make ourselves understood. (An easy way to begin interacting might be to send pictures.) Who decides what to send? Does the whole planet try to agree on one set of messages, or can any individual or group send a separate communication? There may be many countries, religious groups, cultural organizations, corporations, and individuals who can afford a radio antenna and are interested in getting "their message" out. Among ourselves, we rarely speak with one voice; should we try to do so in addressing the universe? Confronting such questions may be a good test of whether there is intelligent life on Earth.

Conclusion

Whether or not we ultimately turn out to be the only intelligent species in our part of the Galaxy, our exploration of the cosmos will surely continue. A humble acknowledgment of how much we have left to learn is one of the fundamental hallmarks of science. This should not, however, prevent us from feeling exhilarated about how much we have already managed to discover, and curious about what else we might find out.

Our progress report on the ideas of astronomy ends here, but we hope that your interest in the universe does not. We hope you will keep up with developments in astronomy through the media, or by going to an occasional public lecture by a local scientist. Who, after all, can even guess all the amazing things that future research projects will reveal about both the universe and our connection with it?

Appendix 1
Astronomy on the World Wide Web

by David Bruning (*Astronomy* Magazine)
and Andrew Fraknoi (Foothill College)

With its almost unlimited capability of storing information and images, the World Wide Web has captured the imaginations of professional and amateur astronomers. New Web sites are springing up faster than any mere mortal can keep track of, and you can spend many hours surfing the Web without ever returning to Earth (or completing all the homework you have due in your other courses!).

In this Appendix we have listed some of our favorite Web sites, with particular attention to those that will be most useful for students in introductory astronomy classes. Our brief list is a tiny fraction of all the astronomy sites out there. Luckily, one of the best aspects of the Web is the ability to connect sites that are related and to switch between them with the click of a button. Many of these sites will point you to others, which will point you to yet others, which will . . . well, you get the general idea.

If you are not familiar with the Web, check with the library or computer center at your college or university. Most schools now offer access and training for the Web, and several on-line services such as *CompuServe* and *America Online* offer access to the Web. But because surfing the Web can be so hypnotic, be sure you understand in advance how much you are being charged for each hour of connection.

One small warning: unlike a textbook, which is checked for accuracy and reviewed by experts in the field, the Web is unedited. Anyone can post anything they wish, and "search engines" (see below) can turn up postings that could be written by an expert, a beginner, or an out-and-out crackpot. As you gather information from a search on the Web, keep track of the *source* of each piece of information. Notice the institution or person it comes from; if something sounds strange, check with your instructor before accepting everything at face value. Despite what some Web sites claim, for example, there is no face built by aliens on Mars, no record of UFO landings hidden by the U.S. military, and no evidence that ancient astronauts built the pyramids because humans were too stupid to do it themselves!

You can access the Web sites listed below by typing the Uniform Resource Locator (URL) in the Open URL or Open Location window of your Web browser. Each code begins with the name of the protocol by which information is transferred; in most cases, this is the "hyper-text transfer protocol" or http. (In some browsers, you do not even need to type "http://" but only what comes afterwards.) A much longer list of astronomy sites with easy links is kept by one of us (DB) at:

http://www.kalmbach.com/astro/HotLinks/HotLinks.html

Table of Contents

1. Some Astronomical Institutions of Special Interest

- Anglo-Australian Observatory (includes wonderful color astronomical photographs of nebulae and galaxies taken by David Malin, some of which appear in this book):
  http://www.aao.gov.au
- Astronomical Society of the Pacific (an organization of professional and amateur astronomers dedicated to public education; has interesting links to other sites):
  http://www.physics.sfsu.edu/asp/asp.html
- *Astronomy* magazine (a magazine for amateur astronomers and students; its site has good information on astronomy as a hobby, astronomy clubs, and other astronomy sites):
  http://www.kalmbach.com/astro/astronomy.html
- Explorations in Education (Space Telescope Science Institute) (features electronic picture books on astronomy and space topics, organized by subject):
  http://www.stsci.edu/exined-html/exined-home.html
- Jet Propulsion Laboratory (a NASA center run by the California Institute of Technology; has good information on current projects in planetary exploration):
  http://www.jpl.nasa.gov/
- The Planetary Society (active space interest group founded by Carl Sagan; encourages the exploration of the planets and the search for life elsewhere):
  http://wea.mankato.mn.us/tps/
- *Sky & Telescope* magazine (monthly publication for amateur

astronomers; a site with up-to-date information on
sky events):

    http://www.skypub.com/

- Star°s Family of Astronomy Resources (includes a mammoth
database of astronomical organizations and people which can
be searched):

    http://cdsweb.u-strasbg.fr/starsfamily.html

- Students for Exploration and Development of Space (SEDS)
(an active site with many images and lots of information):

    http://www.seds.org

2. Selected Telescopes and Observatories

- Compton Gamma-Ray Observatory:

    http://cossc.gsfc.nasa.gov/cossc/descriptions/cgro.html

- Dominion Astrophysical Observatory (largest Canadian observatory):

    http://www.dao.nrc.ca/DAO/DAO-homepage.html

- European Southern Observatory (operates a number of large
telescopes):

    http://http.hq.eso.org/eso-homepage.html

- Global Oscillation Network Group (for helioseismology):

    http://helios.tuc.noao.edu/gonghome.html

- Harvard College Observatory:

    http://cfa-www.harvard.edu/hco-home.html

- Infrared Space Observatory (ISO):

    http://isowww.estec.esa.nl

- Keck Observatory (10-m telescopes):

    http://astro.caltech.edu/keck.html

- Mauna Kea Tour (site of many large telescopes in Hawaii):

    http://www.ifa.hawaii.edu/images/aerial-tour/

- Mount Wilson Observatory:

    http://www.mtwilson.edu

- National Optical Astronomy Observatories (includes Kitt
Peak):

    http://www.noao.edu/noao.html

- National Radio Astronomy Observatory:

    http://info.aoc.nrao.edu/

- Royal Greenwich Observatory (England):

    http://cast0.ast.cam.ac.uk/RGO/RGO.html

- Space Telescope Science Institute (where to find all the wonderful Hubble Space Telescope images, including those so recent they didn't get into this book):

    http://www.stsci.edu/public.html

- X-Ray Timing Explorer (XTE):

    http://heasarc.gsfc.nasa.gov/docs/xte/xte_1st.html

- Guide to ground-based solar and astrophysical observatories:

    http://ranier.oact.hq.nasa.gov/Sensors_page/
      Ground Observ.html

3. The Solar System in General

- Tours of the solar system and general information:

    http://seds.lpl.arizona.edu/nineplanets/nineplanets/
      nineplanets.html
    http://www.jpl.nasa.gov/tours
    http://www.c3.lanl.gov/~cjhamil/SolarSystem/
      homepage.html
    http://www.fourmilab.ch/solar/solar.html
    http://www.nosc.mil/planet-earth/planets.html

- Images of the planets:

    http://www-pdsimage.jpl.nasa.gov/PIA/

    http://astrosun.tn.cornell.edu/
    http://cdwings.jpl.nasa.gov/pds/

4. Specific Worlds in the Solar System

- See section 3, above, for sites that also include information on
each world; for example:

    http://www.c3.lanl.gov/~cjhamil/SolarSystem/Venus.html

- **Venus**

    Surface of the planet:

      http://stoner.eps.mcgill.ca/bud/first.html

    Browse the Magellan images:

      http://delcano.mit.edu/cgi-bin/midr-query

- **Earth**

    Create your own views:

      http://www.fourmilab.ch/earthview/vplanet.html

    Images from the Spaceborne Imaging Radar project:

      http://www.jpl.nasa.gov/sircxsar/

    Aurora predictions:

      http://www.pfrr.alaska.edu/~pfrr/AURORA/PREDICT/
        CURRENT.HTML

- **Moon**

    Phases:

      http://dragon.aoc.nrao.edu/casey-cgi/moon.cgi/today

    Clementine mission:

      http://www.nrl.navy.gov/Clementine

- **Mars**

    Interactive Mars map:

      http://www.c3.lanl.gov/~cjhamil/Browse/mars.html

    Mars atlas:

      http://ic-www.arc.nasa.gov/ic/projects/bayes-group/
        Atlas/Mars

    Center for Mars Exploration:

      http://cmex-www.arc.nasa.gov/

- **Jupiter**

    Galileo mission:

      http://www.noao.edu/galileo
      http://www.jpl.nasa.gov:80/galileo/index.html
      http://ccf.arc.nasa.gov/galileo_probe/

- **Saturn**

    Appearance of rings:

      http://ringside.arc.nasa.gov/www/rpx/viewer/
        rpx_viewer.html

    Cassini mission:

      htpp://www.jpl.nasa.gov/cassini/

5. Comets, Asteroids, Meteorites

- **Comets**

    A wealth of information about comets (has comet news, lists
    which comets are currently visible, images of many comets,
    elements of comet orbits, etc.):

      http://encke.jpl.nasa.gov/

    Shoemaker-Levy 9 impact at Jupiter:

      http://newproducts.jpl.nasa.gov/s19/

    Comet Hale-Bopp:

      http://www.halebopp.com/index.html
      http://encke.jpl.nasa.gov/hale_bopp_info.html

    Comets of the past:

      http://medicine.wustl.edu/~kronkg/past_comets.html

    Rosetta comet lander mission:

      http://champwww.jpl.nasa.gov/champollion/index.html

Comet observing tips (British Astronomical Association):
http://www.ast.cam.ac.uk/~jds/
Comets orbiting beyond Neptune (Kuiper belt objects):
http://cfa-www.harvard.edu/cfa/ps/lists/TNOs.html

- **Asteroids**
  Minor Planet Center:
  http://cfa-www.harvard.edu/cfa/ps/mpc.html
  Near-Earth Asteroid Rendezvous Mission:
  http://nssdc.gsfc.nasa.gov/planetary/near.html
  http://hurlbut.jhuapl.edu/NEAR/
  Centaurs:
  http://cfa-www.harvard.edu/cfa/ps/lists/
  Centaurs.html
- **Meteorites**
  General guide:
  http://www.c3.lanl.gov/~cjhamil/
  SolarSystem/meteorite.html
  Meteorites in Antarctica:
  http://exploration.jsc.nasa.gov/curator/antmet/
  antmet.html

6. The Sun

- Tour of sun (National Solar Observatory):
  http://blazing.sunspot.noao.edu/Exhibit/Exhibit.html
- Daily Sun images and space weather:
  http://www.sel.noaa.gov/
  http://www.sel.bldrdoc.gov/today.html
- Movie of Sun's surface:
  http://www.erim.org/algs/PD/pd_home.html
- Satellite missions and tutorials on the Sun:
  http://wwwssl.msfc.nasa.gov/ssl/pad/solar
- Solar and Heliospheric Observatory (SOHO) satellite:
  http://sohowww.nascom.nasa.gov/
- Ulysses satellite:
  http://ulysses.jpl.nasa.gov/

7. Stars and Stellar Evolution

- Nearest stars (list):
  http://proxima.astro.virginia.edu/~pai/
  Recons/nearest25.html
- Planets around other stars:
  http://cannon.sfsu.edu/~gmarcy/
- Stellar properties:
  http://www.astro.washington.edu/strobel/
  star-props/star-props.html
- Tutorials:
  http://altair.syr.edu:2024/SETI/TUTORIAL/
- Variable star observing:
  http://www.aavso.org/

8. Nebulae (Clouds of Raw Material, Planetary Nebulae, Supernova Remnants, etc.)

- Images of star-forming regions:
  http://donald.phast.umass.edu/gs/wizimlib.html
- Images of nebulae:
  Anglo-Australian Observatory (David Malin):

http://www.aao.gov.au/images.html
Bill Arnett's Nebula Page:
http://www.seds.org/billa/twn/
Planetary Nebula Sampler:
http://www.noao.edu/jacoby/pn_gallery.html

9. Galaxies

- Images of Galaxies:
  Anglo-Australian Observatory (David Malin):
  http://www.aao.gov.au/galaxies.html
  Messier catalog (University of Arizona):
  http://www.seds.org/messier/index.html
  http://zebu.uoregon.edu/messier.html
  William Keel (University of Alabama):
  http://www.astr.ua.edu/choosepic.html
- Interacting galaxies: http://crux.astr.ua.edu/

10. Cosmology

- Cosmic Background Explorer satellite:
  http://www.gsfc.nasa.gov/astro/cobe/cobe_home.html
- Gamma-ray bursts, debate about:
  http://antwrp.gsfc.nasa.gov/diamond_jubilee/debate.html
- Images and animations:
  http://www.ncsa.uiuc.edu/General/NCSAExhibits.html
- Lab exercise on the Hubble Law:
  http://www.gettysburg.edu/project/physics/clea/
  CLEAhome.html
- MACHO project (search for dark matter):
  http://wwwmacho.anu.edu.au/
- OGLE project (dark matter search):
  http://www.astrow.edu.pl
- On-line text on cosmology:
  http://uu-gna.mit.edu:8001/uu-gna/text/astro/universe/
  index.html
- Simulation of early universe:
  http://zeus.ncsa.uiuc.edu:8080/GC3_Home_Page.html
- Tutorial on cosmology:
  http://altair.syr.edu:2024/SETI/TUTORIAL/bigbang.html

11. The Search for Life Elsewhere

- Planets around other stars:
  http://cannon.sfsu.edu/~gmarcy/
- SETI Institute (rich source of material on radio searches):
  http://www.seti-inst.edu/
- Tutorial on SETI (Syracuse University):
  http://altair.syr.edu:2024/SETI/seti.html

12. Observing Sky Events (Eclipses, Conjunctions, Meteor Showers, etc.)

- General:
  Abrams Planetarium Sky Calendar:
  http://www.pa.msu.edu/abrams/
  Mt. Wilson On-line Stargazer Map:
  http://www.mtwilson.edu/services/starmap.html
  Links to backyard astronomy:
  http://www.kalmbach.com/astro/astronomy.html

Another monthly calendar of sky events:
http://www.nscee.edu/~drdale/onOrbit_05_95/
SkyCalendar.html

Sky almanac from *Astronomy* Magazine:
http://www.kalmbach.com/astro/SkyEvents/SkyEvents.html

- Eclipses:
    Animation:
    http://ageninfo.tamu.edu/eclipse/
    Guide to eclipses:
    http://www.c3.lanl/gov/~cvjhamil/SolarSystem/education/
    eclipses.html
    Solar eclipse bulletins:
    http://umbra.nascom.nasa.gov/sdac.html
- Meteor showers:
    http://medicine.wustl.edu/~kronkg/meteor_shower.html
- Radio astronomy for amateurs:
    http://www.rmplc.co.uk/eduweb/sites/trao/index.html

13. Miscellaneous Sites Related to Astronomy

- Ancient astronomy:
    http://kira.pomona.claremont.edu/
- Astronomical League (umbrella organization of amateur astronomy clubs in the U.S.):
    http://www.mcs.net/~bstevens/al
- Astronomy software reviews (by John Mosley for *Sky & Telescope):*
    http://www.skypub.com/software/mosley.html
- Black holes and general relativity:
    The National Center for Supercomputing Applications (has images relating to black holes and current research in general relativity theory):
    http://jean-luc.ncsa.uiuc.edu/
    Visual Trip to a black hole:
    http://cossc.gsfc.nasa.gov/htmltest/rjn_bht.html
- Elements, table of:
    http://www.cchem.berkeley.edu/Table/index.html
- History of astronomy links:
    http://aibn55.astro.uni-bonn.de:8000/
    ~pbrosche/astoria.html
- Image processing (obtain the superb free program *NIH Image* from):
    http://rsb.info.nih.gov/nih-image
- Lab exercises (CLEA Project):
    http://www.gettysburg.edu/project/physics.clea/
    CLEAhome.html
- Light pollution (and what we can do about it):
    http://www.ida.org/
- NASA information on:
    NASA itself:
    http://www.nasa.gov
    Space Shuttle missions:
    http://shuttle.nasa.gov
    Overview of NASA information on Web:
    http://www.gsfc.nasa.gov/NASA_homepage.html

Astronauts and human spaceflight:
http://images.jsc.nasa.gov

History of spaceflight:
http://www.ksc.nasa.gov/history/history.html

Space movie archive:
http://www.univ-rennes1.fr/ASTRO/anim-e.html

- Planetaria around the world:
    http://www.lochness.com
    http://www.kalmbach.com/astro/SpacePlaces/
    SpacePlaces.html
- Skyview (show maps of any part of the sky at different wavelengths):
    http://skyview.gsfc.nasa.gov/skyview.html
- This month in space and astronomy history (calendar of significant events):
    http://newproducts.jpl.nasa.gov/calendar/history/html

14. Sites with Good Listings of Other Astronomy Sites

- Astro Web (sorted by category):
    http://fits.cv.nrao.edu/www/astronomy.html
- Astronomical information on the internet:
    http://ecf.hq.eso.org/astro-resources.html
- *Astronomy* Magazine:
    http://www.kalmbach.com/astro//HotLinks.html
- Guide to on-line data in astronomy:
    http://www.hq.eso.org/online-resources-paper/rrn.html
- Kennedy Observatory list of astronomy WWW resources:
    http://www.cs.dal.ca/~andromed/
- Sonoma State University:
    http://yorty.sonoma.edu/people/faculty/tenn/
    Astronomy Links.html
- Space mission acronyms and links:
    http://ranier.oact.hq.nasa.gov/Sensors_page/
    MissionLinks.html
- U.S. Geological Survey:
    http://info.er.usgs.gov/network/science/astronomy/
- WebStars (astrophysics sites):
    http://guinan.gsfc.nasa.gov/WebStars.html

15. Search Engines (How to Find More Information)

- SavvySearch (searches via 24 search engines for you):
    http://savvy.cs.colostate.edu:2000/
- Metacrawler (searches via 8 engines):
    http://metacrawler.cs.washington.edu:8080/home.html
- Alta Vista (powerful search engine):
    http://www.altavista.digital.com/
- Lycos:
    http://lycos.cs.cmu.edu
- Yahoo:
    http://www.yahoo.com

Appendix 2
Sources of Astronomical Information

1. Popular-level Astronomy Magazines

Astronomy Magazine (Kalmbach Publishing, P. O. Box 1612, Waukesha, WI 53187) The largest circulation astronomy publication in the world; it is very colorful and basic.

Griffith Observer (Griffith Observatory, 2800 E. Observatory Rd., Los Angeles, CA 90027) A small magazine specializing in historical topics.

Mercury Magazine (Astronomical Society of the Pacific, 390 Ashton Ave., San Francisco, CA 94112) The magazine of the largest public education society in astronomy with many interesting features.

Planetary Report (The Planetary Society, 65 N. Catalina Ave., Pasadena, CA 91106) The magazine of a large membership organization that promotes the exploration of the solar system.

Stardate Magazine (McDonald Observatory, RLM 15.308, University of Texas, Austin, TX 78712) Accompanies the popular public radio program on astronomy.

Sky & Telescope (P.O. Box 9111, Belmont, MA 02178) Found in most libraries, this is the popular astronomy magazine "of record" with many excellent articles on astronomy and amateur astronomy.

The following magazines (found in most libraries) contain frequent reports on astronomical developments: *Discover*, *National Geographic*, *Science News*, and *Scientific American*.

2. Organizations for Astronomy Enthusiasts

American Association of Variable Star Observers, 25 Birch St., Cambridge, MA 02138 (617-354-0484). An organization of amateur astronomers devoted to making serious observations of stars whose brightness changes.

Astronomical League, c/o Berton Stevens, 2112 Kingfisher Lane E., Rolling Meadows, IL 60008 (708-398-0562). Umbrella organization of amateur astronomy clubs in the U.S. Write to them to find out where the club closest to you is located. (First check with your instructor or the September issue of *Sky & Telescope* each year.) If you write to this all-volunteer organization, be sure to enclose a stamped self-addressed envelope.

Astronomical Society of the Pacific, 390 Ashton Ave., San Francisco, CA 94112 (415-337-1100). This international organization brings together scientists, teachers, and people with a general interest in astronomy; its name is merely a reminder of its origins on the West Coast of the U.S. in 1889. Write for their catalog of interesting astronomy materials or information about their national meetings and astronomy expos.

Committee for the Scientific Investigation of Claims of the Paranormal (CSICOP), P. O. Box 703, Buffalo, NY 14226 (716-636-1425). An organization of scientists, educators, magicians, and other skeptics that seeks to inform the public about the rational perspective on such pseudosciences as astrology, UFOs, psychic power, etc. Publishes *The Skeptical Inquirer* magazine, full of great debunking articles, and holds meetings and workshops around the world.

International Dark Sky Association, c/o David Crawford, 3545 N. Stewart, Tucson, AZ 85716. Small non-profit organization devoted to fighting light pollution and educating politicians, lighting engineers, and the public about the importance of not spilling light where it will interfere with astronomical observations.

The Planetary Society, 65 N. Catalina Ave., Pasadena, CA 91106 (818-793-1675). Large national membership organization founded by Carl Sagan and others; lobbies for more planetary exploration and SETI; publishes a colorful magazine and has a catalog of reasonably priced slides, videos, and gift items.

The Royal Astronomical Society of Canada, 136 Dupont St., Toronto, Ontario M5R 1V2 (416-924-7973). The main organization of amateur astronomers in Canada, with local centers in each province. Write for information on contacting the center near you.

3. Selected Institutions with Services and Materials for the Public

Jet Propulsion Laboratory, Public Information Office, 4800 Oak Grove Dr., Pasadena, CA 91109. This NASA center will usually respond to public inquiries about planetary missions with information pamphlets and lithographs.

NASA Headquarters, Public Information Branch, Washington, DC 20546. You can obtain information about NASA missions and projects. Leave plenty of time to get a response.

Space Telescope Science Institute, Public Information Office, 3700 San Martin Dr., Baltimore, MD 21218. Will often respond to intelligently worded requests for information or pictures from the Hubble Space Telescope.

Appendix 3
Glossary

NOTE: The number in parentheses is the chapter where the term is first defined or explained (P means the prologue; E means the epilogue).

absolute brightness (magnitude) *See* luminosity.

absolute zero A temperature of $-273°C$ (or 0 K), where all molecular motion stops. (4)

absorption spectrum Dark lines superimposed on a continuous spectrum. (4)

accelerate To change velocity; to speed up, slow down, or change direction. (1)

accretion Gradual accumulation of mass, as by a planet forming by the building up of colliding particles in the solar nebula or gas falling into a black hole. (13)

accretion disk A disk of matter spiraling in toward a massive object; the disk shape is the result of conservation of angular momentum. (23)

active galactic nucleus A galaxy is said to have an active nucleus if unusually violent events are taking place in its center, emitting large quantities of electromagnetic radiation. Seyfert galaxies and quasars are examples of galaxies with active nuclei. (26)

active region Areas on the Sun where magnetic fields are concentrated; sunspots, prominences, and flares all tend to occur in active regions. (14)

albedo The fraction of incident sunlight that a planet or minor planet reflects.

alpha particle The nucleus of a helium atom, consisting of two protons and two neutrons.

amplitude The range in variability, as in the light from a variable star.

angular diameter Angle subtended by the diameter of an object.

angular momentum A measure of the momentum associated with motion about an axis or fixed point. (2)

antapex (solar) Direction away from which the Sun is moving with respect to the local standard of rest.

Antarctic Circle Parallel of latitude 66°30′S; at this latitude the noon altitude of the Sun is 0° on the date of the summer solstice.

antimatter Matter consisting of antiparticles: antiprotons (protons with negative rather than positive charge), positrons (positively charged electrons), and antineutrons. (15)

aperture The diameter of an opening, or of the primary lens or mirror of a telescope. (5)

apex (solar) The direction toward which the Sun is moving with respect to the local standard of rest.

aphelion Point in its orbit where a planet (or other body) is farthest from the Sun. (2)

apogee Point in its orbit where an Earth satellite is farthest from the Earth. (2)

apparent brightness A measure of the observed light received from a star or other object at the Earth; i.e., how bright an object appears in the sky, as contrasted with its luminosity. (4)

Arctic Circle Parallel of latitude 66°30′N; at this latitude the noon altitude of the Sun is 0° on the date of the winter solstice.

array (interferometer) A group of several telescopes that is used to make observations at high angular resolution. (5)

association A loose group of young stars whose spectral types, motions, or positions in the sky indicate that they have probably had a common origin. (21)

asterism An especially noticeable star pattern in the sky, such as the Big Dipper. (1)

asteroid An object orbiting the Sun that is smaller than a major planet, but that shows no evidence of an atmosphere or of other types of activity associated with comets. Also called a minor planet. (6)

asteroid belt The region of the solar system between the orbits of Mars and Jupiter in which most asteroids are located. The main belt, where the orbits are generally the most stable, extends from 2.2 to 3.3 AU from the Sun. (11)

astrology The pseudoscience that deals with the supposed influences on human destiny of the configurations and locations in the sky of the Sun, Moon, and planets; a primitive belief system that had its origin in ancient Babylonia. (1)

astronomical unit (AU) Originally meant to be the semimajor axis of the orbit of the Earth; now defined as the semimajor axis of the orbit of a hypothetical body with the mass and period that Gauss assumed for the Earth. The semimajor axis of the orbit of the Earth is 1.000000230 AU. (2)

atom The smallest particle of an element that retains the properties that characterize that element.

atomic mass unit *Chemical:* one-sixteenth of the mean mass of an oxygen atom. *Physical:* one-twelfth of the mass of an atom of the most common isotope of carbon. The atomic mass unit is approximately the mass of a hydrogen atom, 1.67×10^{-27} kg.

atomic number The number of protons in each atom of a particular element.

atomic weight The mean mass of an atom of a particular element in atomic mass units.

aurora Light radiated by atoms and ions in the ionosphere, mostly in the magnetic polar regions. (7)

autumnal equinox The intersection of the ecliptic and celestial equator where the Sun crosses the equator from north to south. A time when every place on Earth has 12 hours of daylight and 12 hours of night. (3)

axis An imaginary line about which a body rotates.

azimuth The angle along the celestial horizon, measured eastward from the north point to the intersection of the horizon with the vertical circle passing through an object.

Balmer lines Emission or absorption lines in the spectrum of hydrogen that arise from transitions between the second (or first excited) and higher energy states of the hydrogen atom. (4)

bands (in spectra) Emission or absorption lines, usually in the spectra of chemical compounds, so numerous and closely spaced that they coalesce into broad emission or absorption bands.

bar A force of 100,000 newtons acting on a surface area of 1 square meter is equal to 1 bar. The average pressure of the Earth's atmosphere at sea level is equal to 1.013 bars. (7)

barred spiral galaxy Spiral galaxy in which the spiral arms begin from the ends of a "bar" running through the nucleus rather than from the nucleus itself. (25)

barycenter The center of mass of two mutually revolving bodies.

basalt Igneous rock, composed primarily of silicon, oxygen, iron, aluminum, and magnesium produced by the cooling of lava. Basalts make up most of Earth's oceanic crust and are also found on other planets that have experienced extensive volcanic activity. (7)

Big Bang theory A theory of cosmology in which the expansion of the universe is presumed to have begun with a primeval explosion. (28)

binary star Two stars revolving about each other. (17)

binding energy The energy required to separate completely the constituent parts of an atomic nucleus. (15)

blackbody A hypothetical perfect radiator, which absorbs and re-emits all radiation incident upon it. (4)

black dwarf A presumed final state of evolution for a low-mass star, in which all of its energy sources are exhausted and it no longer emits significant radiation. (22)

black hole A collapsed massive star (or other collapsed body) whose velocity of escape is equal to or greater than the speed of light; thus no radiation can escape from it. (23)

Bohr atom A particular model of an atom, invented by Niels Bohr, in which the electrons are described as revolving about the nucleus in circular orbits. (4)

brown dwarf An object intermediate in size between a planet and a star. The approximate mass range is from about 1/100 of the mass of the Sun up to the lower mass limit for self-sustaining nuclear reactions, which is 0.08 solar mass. (17)

carbonaceous meteorite A primitive meteorite made primarily of silicates but often including chemically bound water, free carbon, and complex organic compounds. Also called carbonaceous chondrites. (13)

carbon-nitrogen-oxygen (CNO) cycle A series of nuclear reactions in the interiors of stars involving carbon as a catalyst, by which hydrogen is transformed to helium. (15)

Cassegrain focus An optical arrangement in a reflecting telescope in which light is reflected by a second mirror to a point behind the primary mirror.

CBR *See* cosmic background radiation.

CCD *See* charge-coupled device.

cD galaxy A supergiant elliptical galaxy frequently found at the center of a cluster of galaxies.

celestial equator A great circle on the celestial sphere 90° from the celestial poles; where the celestial sphere intersects the plane of the Earth's equator. (1)

celestial meridian An imaginary line on the celestial sphere passing through the north and south points on the horizon and through the zenith.

celestial poles Points about which the celestial sphere appears to rotate; intersections of the celestial sphere with the Earth's polar axis. (1)

celestial sphere Apparent sphere of the sky; a sphere of large radius centered on the observer. Directions of objects in the sky can be denoted by their position on the celestial sphere.

center of gravity Center of mass.

center of mass The average position of the various mass elements of a body or system, weighted according to their distances from that center of mass; that point in an isolated system that moves with constant velocity, according to Newton's first law of motion. (17)

cepheid variable A star that belongs to a class of yellow supergiant pulsating stars. These stars vary periodically in brightness, and the relationship between their periods and luminosities is useful in deriving distances to them. (18)

Chandrasekhar limit The upper limit to the mass of a white dwarf (equals 1.4 times the mass of the Sun). (22)

charge-coupled device (CCD) An array of electronic detectors of electromagnetic radiation, used at the focus of a telescope (or camera lens). A CCD acts like a photographic plate of very high sensitivity. (5)

chemical condensation sequence The calculated chemical compounds and minerals that would form at different temperatures in a cooling gas of cosmic composition; used to infer the composition of grains that formed in the solar nebula at different distances from the protosun. (13)

chromosphere That part of the solar atmosphere that lies immediately above the photospheric layers. (14)

circular satellite velocity The critical speed that a revolving body must have in order to follow a circular orbit. (2)

circumpolar zone Those portions of the celestial sphere near the celestial poles that are either always above or always below the horizon.

closed universe A model of the universe in which the curvature of space is such that straight lines eventually curve back upon themselves; in this model, the universe expands from a big bang, stops, and then contracts to a big crunch. (28)

cluster of galaxies A system of galaxies containing several to thousands of member galaxies. (27)

color index Difference between the magnitudes of a star or other object measured in light of two different spectral regions, for example, blue minus visual ($B - V$) magnitudes. (16)

coma (of comet) The diffuse gaseous component of the head of a comet; i.e, the cloud of evaporated gas around a comet nucleus. (12)

comet A small body of icy and dusty matter that revolves about the Sun. When a comet comes near the Sun, some of its material vaporizes, forming a large head of tenuous gas, and often a tail. (6)

compound A substance composed of two or more chemical elements.

conduction The transfer of energy by the direct passing of energy or electrons from atom to atom.

conservation of angular momentum The law that the total amount of angular momentum in a system remains the same (in the absence of any force not directed toward or away from the point or axis about which the angular momentum is referred).

constellation One of 88 sectors into which astronomers divide the celestial sphere; many constellations are named after a prominent group of stars within them that represents a person, animal, or legendary creature from ancient mythology. (P)

continental drift A gradual movement of the continents over the surface of the Earth due to plate tectonics. (7)

continuous spectrum A spectrum of light composed of radiation of a continuous range of wavelengths or colors rather than only certain discrete wavelengths. (4)

convection The transfer of energy by moving currents in a fluid. (7)

core (of a planet) The central part of a planet, consisting of higher density material. (7)

corona (of Galaxy) A region lying above and below the plane of the Galaxy out to much greater distances than the material that gives off electromagnetic radiation. (24)

corona (of Sun) Outer atmosphere of the Sun. (14)

coronal hole A region in the Sun's outer atmosphere where visible coronal radiation is absent. (14)

cosmic background radiation (CBR) The microwave radiation coming from all directions that is believed to be the redshifted glow of the Big Bang. (28)

cosmic rays Atomic nuclei (mostly protons) that are observed to strike the Earth's atmosphere with exceedingly high energies. (19)

cosmological constant A term in the equations of general relativity that represents a repulsive force in the universe. The cosmological constant is usually assumed to be zero. (28)

cosmological principle The assumption that, on the large scale, the universe at any given time is the same everywhere— isotropic and homogeneous. (27)

cosmology The study of the organization and evolution of the universe. (28)

crater A circular depression (from the Greek word for cup), generally of impact origin.

crescent moon One of the phases of the Moon when it appears less than half full.

critical density In cosmology, the density that provides enough gravity to bring the expansion of the universe just to a stop after infinite time. (28)

crust The outer layer of a terrestrial planet. (6)

dark matter Nonluminous mass, whose presence can be inferred only because of its gravitational influence on luminous matter. Dark matter may constitute as much as 99 percent of all the mass in the universe. The composition of the dark matter is not known. (24)

dark nebula A cloud of interstellar dust that obscures the light of more distant stars and appears as an opaque curtain.

declination Angular distance north or south of the celestial equator. (3)

degenerate gas A gas in which the allowable states for the electrons have been filled; it behaves according to different laws from those that apply to "perfect" gases and resists further compression. (22)

density The ratio of the mass of an object to its volume. (2)

deuterium A "heavy" form of hydrogen, in which the nucleus of each atom consists of one proton and one neutron. (15)

differential galactic rotation The rotation of the Galaxy, not as a solid wheel, but so that parts adjacent to one another do not always stay close together. (24)

differentiation (geological) A separation or segregation of different kinds of material in different layers in the interior of a planet. (6)

disk (of Galaxy) The central plane or "wheel" of our Galaxy, where most of the luminous mass is concentrated. (24)

dispersion Separation, from white light, of different wavelengths being refracted by different amounts. (4)

Doppler effect Apparent change in wavelength of the radiation from a source due to its relative motion away from or towards the observer. (4)

Earth-approaching asteroid An asteroid with an orbit that crosses the Earth's orbit or that will at some time cross the Earth's orbit as it evolves under the influence of the planets' gravity. *See also* near-Earth object.

eccentricity (of ellipse) Ratio of the distance between the foci to the major axis. (2)

eclipse The cutting off of all or part of the light of one body by another; in planetary science, the passing of one body into the shadow of another. (3)

eclipsing binary star A binary star in which the plane of revolution of the two stars is nearly edge-on to our line of sight, so that the light of one star is periodically diminished by the other passing in front of it. (17)

ecliptic The apparent annual path of the Sun on the celestial sphere. (1)

effective temperature *See* temperature (effective).

ejecta Material excavated from an impact crater, such as the blanket of material surrounding lunar craters and crater rays.

electromagnetic force One of the four fundamental forces or interactions of nature; the force that acts between charges and binds atoms and molecules together.

electromagnetic radiation Radiation consisting of waves propagated through the building up and breaking down of electric and magnetic fields; these include radio, infrared, light, ultraviolet, x rays, and gamma rays. (4)

electromagnetic spectrum The whole array or family of electromagnetic waves, from radio to gamma rays. (4)

electron A negatively charged subatomic particle that normally moves about the nucleus of an atom.

element A substance that cannot be decomposed, by chemical means, into simpler substances.

elementary particle One of the basic particles of matter. The most familiar of the elementary particles are the proton, neutron, and electron. (15)

ellipse A curve for which the sum of the distances from any point on the ellipse to two points inside (called the foci) is always the same. (2)

elliptical galaxy A galaxy whose appearance resembles a solid made of a series of ellipses and that contains no conspicuous interstellar material.

ellipticity The ratio (in an ellipse) of the major axis minus the minor axis to the major axis.

emission line A discrete bright line in the spectrum. (4)

emission nebula A gaseous nebula that derives its visible light from the fluorescence of ultraviolet light from a star in or near the nebula.

emission spectrum A spectrum consisting of emission lines. (4)

energy level A particular level, or amount, of energy possessed by an atom or ion above the energy it possesses in its least energetic state; also used to refer to the states of energy an electron can have in an atom. (4)

epicycle A circular orbit of a body in the Ptolemaic system, the center of which revolves about another circle (the deferent). (1)

equator A great circle on the Earth, 90° from (or equidistant from) each pole. (1)

equinox One of the intersections of the ecliptic and celestial equator; one of the two times during the year when the length of the day and night are the same. (3)

equivalence principle Principle that a gravitational force and a suitable acceleration are indistinguishable within a sufficiently local environment. (23)

escape velocity The velocity a body must achieve to break away from the gravity of another body and never return to it. (2)

eucrite meteorite One of a class of basaltic meteorites believed to have originated on the asteroid Vesta.

event horizon The surface through which a collapsing star passes when its velocity of escape is equal to the speed of light; that is, when the star becomes a black hole. (23)

excitation The process of imparting to an atom or an ion an amount of energy greater than it has in its normal or least-energy state. (4)

exclusion principle *See* Pauli exclusion principle.

extinction Reduction of the light from a celestial body produced by the Earth's atmosphere, or by interstellar absorption.

extragalactic Beyond our own Milky Way Galaxy.

eyepiece A magnifying lens used to view the image produced by the objective of a telescope.

fall (of meteorites) Meteorites seen in the sky and recovered on the ground. (13)

fault In geology, a crack or break in the crust of a planet along which slippage or movement can take place, accompanied by seismic activity. (7)

field A mathematical description of the effect of forces, such as gravity, that act on distant objects. For example, a given mass produces a gravitational field in the space surrounding it, which produces a gravitational force on objects within that space.

find (of meteorites) A meteorite that has been recovered but was not seen to fall. (13)

fireball A spectacular meteor, seen for more than an instant in the sky. (13)

fission The breakup of a heavy atomic nucleus into two or more lighter ones. (15)

flare A sudden and temporary outburst of light from an extended region of the Sun's surface. (14)

fluorescence The absorption of light of one wavelength and re-emission of it at another wavelength; especially the conversion of ultraviolet into visible light.

flux The rate at which energy or matter crosses a unit area of a surface.

focal length The distance from a lens or mirror to the point where light converged by it comes to a focus. (5)

focus (of ellipse) One of two fixed points inside an ellipse from which the sum of the distances to any point on the ellipse is a constant. (2)

focus (of telescope) Point where the rays of light converged by a mirror or lens meet.

forbidden lines Spectral lines that are not usually observed under laboratory conditions because they result from atomic transitions that are highly improbable.

force That which can change the momentum of a body; numerically, the rate at which the body's momentum changes. (2)

Fraunhofer line An absorption line in the spectrum of the Sun or of a star.

Fraunhofer spectrum The array of absorption lines in the spectrum of the Sun or a star.

frequency Number of vibrations per unit time; number of waves that cross a given point per unit time (in radiation). (4)

fusion The building up of heavier atomic nuclei from lighter ones. (15)

galactic cannibalism The process by which a larger galaxy strips material from a smaller one. (27)

galactic cluster An "open" cluster of stars located in the spiral arms or disk of the Galaxy. (21)

galaxy A large assemblage of stars; a typical galaxy contains millions to hundreds of billions of stars.

Galaxy The galaxy to which the Sun and our neighboring stars belong; the Milky Way is light from remote stars in the disk of the Galaxy. (24)

gamma rays Photons (of electromagnetic radiation) of energy higher than those of x rays; the most energetic form of electromagnetic radiation. (4)

general relativity theory Einstein's theory relating acceleration, gravity, and the structure (geometry) of space and time. (23)

geocentric Centered on the Earth.

giant (star) A star of large luminosity and radius. (16)

giant molecular cloud Large, cold interstellar clouds, with diameters of dozens of light years and typical masses of 10^5 solar masses; found in the spiral arms of galaxies, these clouds are where massive stars form. (20)

globular cluster One of about 120 large spherical star clusters that form a system of clusters centered on the center of the Galaxy. (21)

grand unified theories (GUTs) Physical theories that attempt to describe the four interactions (forces) of nature as different manifestations of a single force. (28)

granite The type of igneous silicate rock that makes up most of the continental crust of the Earth. (7)

granulation The rice-grain-like structure of the solar photosphere; granulation is produced by upwelling currents of gas that are slightly hotter, and therefore brighter, than the surrounding regions, which are flowing downward into the Sun. (14)

gravity The mutual attraction of material bodies or particles.

gravitational energy Energy that can be released by the gravitational collapse, or partial collapse, of a system; i.e., by particles that fall in toward the center of gravity.

gravitational lens A configuration of celestial objects, one of which provides one or more images of the other by gravitationally deflecting its light. (27)

gravitational redshift The redshift of electromagnetic radiation caused by a gravitational field. The slowing of clocks in a gravitational field. (23)

great circle Circle on the surface of a sphere that is the curve of intersection of the sphere with a plane passing through its center. (3)

greenhouse effect The blanketing (absorption) of infrared radiation near the surface of a planet by, for example, carbon dioxide in its atmosphere. (7)

ground state The lowest energy state of an atom.

H or H_0 *See* Hubble constant.

H I region Region of neutral hydrogen in interstellar space. (19)

H II region Region of ionized hydrogen in interstellar space. (19)

half-life The time required for half of the radioactive atoms in a sample to disintegrate. (6)

halo (of galaxy) The outermost extent of our Galaxy or another, containing a sparse distribution of stars and globular clusters in a more or less spherical distribution. (21)

heavy elements In astronomy, usually those elements of greater atomic number than helium.

helio- Prefix referring to the Sun.

heliocentric Centered on the Sun. (1)

helium flash The nearly explosive ignition of helium in the triple-alpha process in the dense core of a red giant star. (21)

Herbig–Haro (HH) object Luminous knots of gas in an area of star formation, which are set to glow by jets of material from a protostar. (20)

hertz A unit of frequency: one cycle per second. Named for Heinrich Hertz, who first produced radio radiation.

Hertzsprung–Russell (H–R) diagram A plot of luminosity against surface temperature (or spectral type) for a group of stars. (17)

highlands (lunar) The older, heavily cratered crust of the Moon, covering 83 percent of its surface and composed in large part of anorthositic breccias. (8)

homogeneous Having a consistent and even distribution of matter that is the same everywhere. (27)

horizon (astronomical) A great circle on the celestial sphere 90° from the zenith; more popularly, the circle around us where the dome of the sky meets the Earth. (1)

horoscope A chart used by astrologers, showing the positions along the zodiac and in the sky of the Sun, Moon, and planets at some given instant and as seen from a particular place on Earth—usually corresponding to the time and place of a person's birth. (1)

Hubble constant Constant of proportionality between the velocities of remote galaxies and their distances. The Hubble constant is thought to lie in the range of 15 to 30 km/s per million LY. (25)

Hubble law (or the law of the redshifts) The radial velocities of remote galaxies are proportional to their distances from us. (25)

hydrostatic equilibrium A balance between the weights of various layers, as in a star or the Earth's atmosphere, and the pressures that support them. (15)

hypothesis A tentative theory or supposition, advanced to explain certain facts or phenomena, which is subject to further tests and verification. (P)

igneous rock Any rock produced by cooling from a molten state. (7)

inclination (of an orbit) The angle between the orbital plane of a revolving body and some fundamental plane—usually the plane of the celestial equator or of the ecliptic.

inertia The property of matter that requires a force to act on it to change its state of motion; the tendency of objects to continue doing what they are doing in the absence of outside forces. (2)

inertial system A system of coordinates that is not itself accelerated, but that either is at rest or is moving with constant velocity.

inflationary universe A theory of cosmology in which the universe is assumed to have undergone a phase of very rapid expansion during the first 10^{-30} s. After this period of rapid expansion, the Big Bang and inflationary models are identical. (28)

infrared cirrus Patches of interstellar dust, which emit infrared radiation and look like cirrus clouds on the images of the sky produced by the Infrared Astronomy Satellite. (19)

infrared radiation Electromagnetic radiation of wavelength longer than the longest (red) wavelengths that can be perceived by the eye, but shorter than radio wavelengths. (4)

interference A phenomenon of waves that mix together such that their crests and troughs can alternately reinforce and cancel one another. (5)

international date line An arbitrary line on the surface of the Earth near longitude 180° across which the date changes by one day. (3)

interstellar dust Tiny solid grains in interstellar space, thought to consist of a core of rock-like material (silicates) or graphite surrounded by a mantle of ices. Water, methane, and ammonia are probably the most abundant ices. (19)

interstellar extinction The attenuation or absorption of light by dust in the interstellar medium. (19)

interstellar medium *or* interstellar matter Interstellar gas and dust. (19)

inverse-square law (for light) The amount of energy (light) flowing through a given area in a given time (flux) decreases in proportion to the square of the distance from the source of energy or light. (4)

ion An atom that has become electrically charged by the addition or loss of one or more electrons. (4)

ionization The process by which an atom gains or loses electrons.

ionosphere The upper region of the Earth's atmosphere in which many of the atoms are ionized.

ion tail (of comet) *See* plasma.

irregular galaxy A galaxy without rotational symmetry; neither a spiral nor an elliptical galaxy. (25)

irregular satellite A planetary satellite with an orbit that is retrograde, or of high inclination or eccentricity.

isotope Any of two or more forms of the same element, whose atoms all have the same number of protons but different numbers of neutrons. (4)

isotropic The same in all directions. (27)

joule The metric unit of energy; the work done by a force of 1 newton (N) acting through a distance of 1 m.

jovian planet *or* giant planet Any of the planets Jupiter, Saturn, Uranus, and Neptune in our solar system, or planets of roughly that mass and composition in other planetary systems. (6)

kinetic energy Energy associated with motion.

Kuiper belt A reservoir of cometary material just beyond the orbit of Pluto. (12)

laser An acronym for *light amplification by stimulated emission of radiation*; a device for amplifying a light signal at a particular wavelength into a coherent beam.

latitude A north-south coordinate on the surface of the Earth; the angular distance north or south of the equator measured along a meridian passing through a place. (3)

law of areas Kepler's second law: the radius vector from the Sun to any planet sweeps out equal areas in the planet's orbital plane in equal intervals of time. (2)

law of the redshifts *See* Hubble law.

leap year A calendar year with 366 days, inserted approximately every 4 years to make the average length of the calendar year as nearly equal as possible to the tropical year. (3)

light *or* **visible light** Electromagnetic radiation that is visible to the eye.

light curve A graph that displays the time variation of the light from a variable or eclipsing binary star. (17)

light year The distance light travels in a vacuum in one year; 1 LY = 9.46×10^{12} km, or about 6×10^{12} mi. (P)

line broadening The phenomenon by which spectral lines are not precisely sharp but have finite widths.

line profile A plot of the intensity of light versus wavelength across a spectral line.

Local Group The cluster of galaxies to which our Galaxy belongs. (28)

local standard of rest A coordinate system that shares the average motion of the Sun and its neighboring stars about the galactic center.

Local Supercluster The supercluster of galaxies to which the Local Group belongs. (28)

longitude An east-west coordinate on the Earth's surface; the angular distance, measured east or west along the equator, from the Greenwich meridian to the meridian passing through a place. (3)

luminosity The rate at which a star or other object emits electromagnetic energy into space. (16)

luminosity class A classification of a star according to its luminosity within a given spectral class. Our Sun, a G2V star, has luminosity class V. (18)

luminosity function The relative numbers of stars (or other objects) of various luminosities. (17)

lunar eclipse An eclipse of the Moon. (3)

Lyman lines A series of absorption or emission lines in the spectrum of hydrogen that arise from transitions to and from the lowest energy states of the hydrogen atoms.

Magellanic Clouds Two neighboring galaxies visible to the naked eye from southern latitudes.

magma Mobile, high-temperature molten state of rock, usually of silicate mineral composition and with dissolved gases and other volatiles. (7)

magnetic field The region of space near a magnetized body within which magnetic forces can be detected.

magnetic pole One of two points on a magnet (or the Earth) at which the greatest density of lines of force emerge. A compass needle aligns itself along the local lines of force on the Earth and points more or less toward the magnetic poles of the Earth.

magnetosphere The region around a planet in which its intrinsic magnetic field dominates the interplanetary field carried by the solar wind; hence, the region within which charged particles can be trapped by the planetary magnetic field. (7)

magnitude A measure of the amount of light flux received from a star or other luminous object. (1)

main sequence A sequence of stars on the Hertzsprung–Russell diagram, containing the majority of stars, that runs diagonally from the upper left to the lower right. (17)

major axis (of ellipse) The maximum diameter of an ellipse. (2)

mantle (of Earth) The greatest part of the Earth's interior, lying between the crust and the core. (7)

mare (pl. maria) Latin for "sea"; name applied to the dark, relatively smooth features that cover 17 percent of the Moon. (8)

mass A measure of the total amount of material in a body; defined either by the inertial properties of the body or by its gravitational influence on other bodies.

mass extinction The sudden disappearance in the fossil record of a large number of species of life, to be replaced by new species in subsequent layers. Mass extinctions are indications of catastrophic changes in the environment, such as might be produced by a large impact on the Earth. (7)

mass-light ratio The ratio of the total mass of a galaxy to its total luminosity, usually expressed in units of solar mass and solar luminosity. The mass-light ratio gives a rough indication of the types of stars contained within a galaxy and whether or not substantial quantities of dark matter are present. (26)

mass-luminosity relation An empirical relation between the masses and luminosities of many (principally main-sequence) stars. (17)

Maunder Minimum The interval from 1645 to 1715 when solar activity was very low. (14)

mean solar day Average length of the apparent solar day.

merger (of galaxies) When galaxies (of roughly comparable size) collide and form one combined structure. (27)

meridian (celestial) The great circle on the celestial sphere that passes through an observer's zenith and the north (or south) celestial pole. (3)

meridian (terrestrial) The great circle on the surface of the Earth that passes through a particular place and the North and South Poles of the Earth.

Messier catalog A catalog of nonstellar objects compiled by Charles Messier in 1787 (includes nebulae, star clusters, and galaxies).

metals In general, any element or compound whose electron structure makes it a good conductor of electricity. In astronomy, all elements beyond hydrogen and helium. (16)

metamorphic rock Any rock produced by the physical and chemical alteration (without melting) of another rock that has been subjected to high temperature and pressure. (7)

metastable level An energy level in an atom from which there is a low probability of an atomic transition accompanied by the radiation of a photon.

meteor The luminous phenomenon observed when a small piece of solid matter enters the Earth's atmosphere and burns up; popularly called a "shooting star." (13)

meteorite A portion of a meteoroid that survives passage through the atmosphere and strikes the ground. (6)

meteoroid A particle or chunk of typically rocky or metallic material in space before any encounter with the Earth.

meteor shower Many meteors appearing to radiate from a common point in the sky caused by the collision of the Earth with a swarm of solid particles, typically from a comet. (13)

micron Old term for micrometer (10^{-6} meter).

microwave Shortwave radio wavelengths.

Milky Way The band of light encircling the sky, which is due to the many stars and diffuse nebulae lying near the plane of the Galaxy. (24)

minerals The solid compounds (often primarily silicon and oxygen) that form rocks.

minor planet *See* asteroid.

model atmosphere *or* **photosphere** The result of a theoretical calculation of the run of temperature, pressure, density, and so on, through the outer layers of the Sun or a star.

molecule A combination of two or more atoms bound together; the smallest particle of a chemical compound or substance that exhibits the chemical properties of that substance.

momentum A measure of the inertia or state of motion of a body; the momentum of a body is the product of its mass and velocity. In the absence of a force, momentum is conserved. (2)

near-Earth object (NEO) A comet or asteroid whose path intersects the orbit of the Earth. (11)

nebula Cloud of interstellar gas or dust. (19)

neutrino A fundamental particle that has little or no rest mass and no charge but that does have spin and energy. Neutrinos rarely interact with ordinary matter. (15)

neutron A subatomic particle with no charge and with mass approximately equal to that of the proton.

neutron star A star of extremely high density composed almost entirely of neutrons. (12)

nonthermal radiation *See* synchrotron radiation.

nova A star that experiences a sudden outburst of radiant energy, temporarily increasing its luminosity by hundreds to thousands of times. (22)

nuclear Referring to the nucleus of the atom.

nuclear bulge Central part of our Galaxy.

nuclear transformation The change of one atomic nucleus into another, as in nuclear fusion. (15)

nucleosynthesis The building up of heavy elements from lighter ones by nuclear fusion. (21)

nucleus (of atom) The heavy part of an atom, composed mostly of protons and neutrons, and about which the electrons revolve. (4)

nucleus (of comet) The solid chunk of ice and dust in the head of a comet. (11)

nucleus (of galaxy) Central concentration of matter at the center of a galaxy. (24)

occultation The passage of an object of large angular size in front of a smaller object, such as the Moon in front of a distant star or the rings of Saturn in front of the Voyager spacecraft. (11)

Oort comet cloud The large spherical region around the Sun from which most "new" comets come; a reservoir of objects with aphelia at about 50,000 AU, or extending about a third of the way to the nearest other stars. (11)

opacity Absorbing power; capacity to impede the passage of light. (15)

open cluster A comparatively loose or "open" cluster of stars, containing from a few dozen to a few thousand members, located in the spiral arms or disk of the Galaxy; sometimes referred to as a galactic cluster. (21)

open universe A model of the universe in which gravity is not strong enough to bring the universe to a halt; it expands forever. In this model the geometry of spacetime is such that if you go in a straight line, you not only can never return to where you started, but even more space opens up than you would expect from Euclidean geometry. (28)

optical In astronomy: relating to the visible-light band of the electromagnetic spectrum. Optical observations are those made with visible light.

optical double star Two stars at different distances that are seen nearly lined up in projection so that they appear close together, but that are not really gravitationally associated.

orbit The path of a body that is in revolution about another body or point.

oscillation A periodic motion; in the case of the Sun, a periodic or quasi-periodic expansion and contraction of the whole Sun or some portion of it. (14)

ozone A heavy molecule of oxygen that contains three atoms rather than the more normal two. Designated O_3.

parabola A conic section of eccentricity 1.0; the curve of the intersection between a circular cone and a plane parallel to a straight line in the surface of the cone.

parallax An apparent displacement of a nearby star that results from the motion of the Earth around the Sun; numerically, the angle subtended by 1 AU at the distance of a particular star. (18)

parsec A unit of distance in astronomy, equal to 3.26 light years. At a distance of 1 parsec, a star has a parallax of one arcsecond.

Pauli exclusion principle Quantum mechanical principle by which no two particles of the same kind can have the same position and momentum.

peculiar velocity The velocity of a star with respect to the local standard of rest; that is, its space motion, corrected for the motion of the Sun with respect to our neighboring stars.

penumbra The outer, not completely dark part of a shadow; the region from which the source of light is not completely hidden. (3)

perfect radiator or blackbody A body that absorbs and subsequently re-emits all radiation incident upon it.

periastron The place in the orbit of a star in a binary-star system where it is closest to its companion star.

perigee The place in the orbit of an Earth satellite where it is closest to the center of the Earth. (2)

perihelion The place in the orbit of an object revolving about the Sun where it is closest to the Sun's center. (2)

period-luminosity relation An empirical relation between the periods and luminosities of certain variable stars. (18)

perturbation The disturbing effect, when small, on the motion of a body as predicted by a simple theory, produced by a third body or other external agent. (2)

photochemistry Chemical changes caused by electromagnetic radiation. (10)

photometry The measurement of light intensities.

photon A discrete unit of electromagnetic energy. (4)

photosphere The region of the solar (or a stellar) atmosphere from which continuous radiation escapes into space. (14)

pixel An individual picture element in a detector; for example, a particular silicon diode in a CCD.

plage A bright region of the solar surface observed in the monochromatic light of some spectral line. (14)

Planck's constant The constant of proportionality relating the energy of a photon to its frequency.

planet Any of the nine largest bodies revolving about the Sun, or any similar bodies that may orbit other stars. Unlike stars, planets do not (for the most part) give off their own light, but only reflect the light of their parent star. (1)

planetarium An optical device for projecting on a screen or domed ceiling the stars and planets and their apparent motions in the sky.

planetary nebula A shell of gas ejected from, and enlarging about, a certain kind of extremely hot star that is nearing the end of its life. (22)

planetesimals The hypothetical objects, from tens to hundreds of kilometers in diameter, that formed in the solar nebula as an intermediate step between tiny grains and the larger planetary objects we see today. The comets and some asteroids may be leftover planetesimals. (6)

plasma A hot ionized gas.

plate tectonics The motion of segments or plates of the outer layer of the Earth over the underlying mantle. (7)

polar axis The axis of rotation of the Earth; also, an axis in the mounting of a telescope that is parallel to the Earth's axis.

Population I and II Two classes of stars (and systems of stars), classified according to their spectral characteristics, chemical compositions, radial velocities, ages, and locations in the Galaxy. (24)

positron An electron with a positive rather than negative charge; an antielectron. (15)

potential energy Stored energy that can be converted into other forms; especially gravitational energy.

precession (of Earth) A slow, conical motion of the Earth's axis of rotation, caused principally by the gravitational pull of the Moon and Sun on the Earth's equatorial bulge. (1)

precession of the equinoxes Slow westward motion of the equinoxes along the ecliptic that results from precession.

pressure Force per unit area; expressed in units of atmospheres or pascals.

prime focus The point in a telescope where the objective focuses the light.

prime meridian The terrestrial meridian passing through the site of the old Royal Greenwich Observatory; longitude 0°. (3)

primitive In planetary science and meteoritics, an object or rock that is little changed, chemically, since its formation, and hence representative of the conditions in the solar nebula at the time of formation of the solar system. Also used to refer to the chemical composition of an atmosphere that has not undergone extensive chemical evolution. (7)

primitive meteorite A meteorite that has not been greatly altered chemically since its condensation from the solar nebula; called in meteoritics a chondrite (either ordinary chondrite or carbonaceous chondrite). (13)

primitive rock Any rock that has not experienced great heat or pressure and therefore remains representative of the original condensates from the solar nebula—never found on any object large enough to have undergone melting and differentiation. (7)

principle of equivalence Principle that a gravitational force and a suitable acceleration are indistinguishable within a sufficiently local environment. (23)

prism A wedge-shaped piece of glass that is used to disperse white light into a spectrum.

prominence A phenomenon in the solar corona that commonly appears like a flame above the limb of the Sun. (14)

proper motion The angular change per year in the direction of a star as seen from the Sun. (16)

proton A heavy subatomic particle that carries a positive charge; one of the two principal constituents of the atomic nucleus.

proton–proton cycle A series of thermonuclear reactions by which nuclei of hydrogen are built up into nuclei of helium. (15)

protoplanet *or* -star *or* -galaxy The original material from which a planet (or a star or galaxy) condensed. (20)

pulsar A variable radio source of small angular size that emits very rapid radio pulses in very regular periods that range from fractions of a second to several seconds. (22)

pulsating variable A variable star that pulsates in size and luminosity. (18)

quantum efficiency The ratio of the number of photons incident on a detector to the number actually detected.

quantum mechanics The branch of physics that deals with the structure of atoms and their interactions with one another and with radiation.

quasar An object of very high redshift that looks like a star, presumed to be extragalactic and highly luminous; an active galactic nucleus. (26)

radar The technique of transmitting radio waves to an object and then detecting the radiation that the object reflects back to the transmitter; used to measure the distance to, and motion of, a target object. (5)

radial velocity The component of relative velocity that lies in the line of sight; motion toward or away from the observer. (4)

radial velocity curve A plot of the variation of radial velocity with time for a binary or variable star.

radiant (of meteor shower) The point in the sky from which the meteors belonging to a shower seem to radiate. (13)

radiation A mode of energy transport whereby energy is transmitted through a vacuum; also the transmitted energy itself. (4, 15)

radiation pressure The transfer of momentum carried by electromagnetic radiation to a body that the radiation impinges upon.

radioactive dating The technique of determining the ages of rocks or other specimens by the amount of radioactive decay of certain radioactive elements contained therein; something you do when you are desperate on Saturday night. (6)

radioactivity (radioactive decay) The process by which certain kinds of atomic nuclei naturally decompose, with the spontaneous emission of subatomic particles and gamma rays. (6)

radio galaxy A galaxy that emits greater amounts of radio radiation than average. (26)

radio telescope A telescope designed to make observations in radio wavelengths.

reddening (interstellar) The reddening of starlight passing through interstellar dust, caused because dust scatters blue light more effectively than red. (19)

red giant A large, cool star of high luminosity; a star occupying the upper right portion of the Hertzsprung–Russell diagram.

redshift A shift to longer wavelengths of light, typically a Doppler shift caused by the motion of the source away from the observer.

reducing In chemistry, referring to conditions in which hydrogen dominates over oxygen, so that most other elements form compounds with hydrogen. In very reducing conditions free hydrogen (H_2) is present and free oxygen (O_2) cannot exist.

reflecting telescope A telescope in which the principal optical component (objective) is a concave mirror. (5)

reflection nebula A relatively dense dust cloud in interstellar space that is illuminated by reflected starlight. (19)

refracting telescope A telescope in which the principal optical component (objective) is a lens or system of lenses. (5)

refraction The bending of light rays passing from one transparent medium (or a vacuum) to another.

relativistic particle (or electron) A particle (electron) moving at nearly the speed of light.

relativity A theory formulated by Einstein that describes the relations between measurements of physical phenomena by two different observers who are in relative motion at constant velocity (the special theory of relativity) or that describes how a gravitational field can be replaced by a curvature of space-time (the general theory of relativity).

resolution The degree to which fine details in an image are separated, or the smallest detail that can be discerned in an image. (5)

resonance An orbital condition in which one object is subject to periodic gravitational perturbations by another, most commonly arising when two objects orbiting a third have periods of revolution that are simple multiples or fractions of each other. (11)

retrograde (rotation or revolution) Backward with respect to the common direction of motion in the solar system; counter-clockwise as viewed from the north, and going from east to west rather than from west to east. (1)

retrograde motion An apparent westward motion of a planet on the celestial sphere or with respect to the stars.

revolution The motion of one body around another.

rift zone In geology, a place where the crust is being torn apart by internal forces, generally associated with the injection of new material from the mantle and with the slow separation of tectonic plates. (7)

right ascension A coordinate for measuring the east-west positions of celestial bodies; the angle measured eastward along the celestial equator from the vernal equinox to the hour circle passing through a body. (3)

rotation Turning of a body about an axis running through it.

RR Lyrae variable One of a class of giant pulsating stars with periods less than one day. (18)

runaway greenhouse effect A process whereby the heating of a planet leads to an increase in its atmospheric greenhouse effect and thus to further heating, thereby quickly altering the composition of its atmosphere and the temperature of its surface. (9)

satellite A body that revolves about a planet.

Schwarzschild radius *See* event horizon.

scientific method The procedure scientists follow to understand the natural world: (1) the observation of phenomena or the results of experiments; (2) the formulation of hypotheses that describe these phenomena and that are consistent with the body of knowledge available; (3) the testing of these hypotheses by noting whether or not they adequately predict and describe new phenomena or the results of new experiments; (4) the modification or rejection of hypotheses that are not confirmed by observations or experiment. (P)

sedimentary rock Any rock formed by the deposition and cementing of fine grains of material. (7)

seeing The unsteadiness of the Earth's atmosphere, which blurs telescopic images. Good seeing means the atmosphere is steady. (5)

seismic waves Vibrations traveling through the Earth's interior that result from earthquakes. (7)

seismology (solar) The study of small changes in the radial velocity of the Sun as a whole or of small regions on the surface of the Sun. Analyses of these velocity changes can be used to infer the internal structure of the Sun. (15)

seismology (terrestrial) The study of earthquakes, the conditions that produce them, and the internal structure of the Earth as deduced from analyses of seismic waves. (7)

semimajor axis Half the major axis of a conic section, such as an ellipse.

SETI The search for extraterrestrial intelligence, usually applied to searches for radio signals from other civilizations. (E)

Seyfert galaxy A galaxy belonging to the class of those with active galactic nuclei; one whose nucleus shows bright emission lines; one of a class of galaxies first described by C. Seyfert. (26)

shepherd satellite Informal term for a satellite that is thought to maintain the structure of a planetary ring through its close gravitational influence. (11)

sidereal period The period of revolution of one body about another measured with respect to the stars.

sidereal time Time on Earth measured with respect to the stars, rather than the Sun; the local hour angle of the vernal equinox. (3)

sidereal year Period of the Earth's revolution about the Sun with respect to the stars. (3)

sign (of zodiac) Astrological term for any of 12 equal sections along the ecliptic, each of length 30°. Because of precession, these signs today are no longer lined up with the constellations from which they received their names. (1)

singularity A theoretical point of zero volume and infinite density to which any object that becomes a black hole must collapse, according to the general theory of relativity. (23)

SNC meteorite One of a class of basaltic meteorites now believed by many planetary scientists to be impact-ejected fragments from Mars.

solar activity Phenomena of the solar atmosphere: sunspots, plages, and related phenomena. (14)

solar antapex Direction away from which the Sun is moving with respect to the local standard of rest.

solar apex The direction toward which the Sun is moving with respect to the local standard of rest.

solar eclipse An eclipse of the Sun by the Moon, caused by the passage of the Moon in front of the Sun. Solar eclipses can occur only at the time of new moon. (3)

solar motion Motion of the Sun, or the velocity of the Sun, with respect to the local standard of rest.

solar nebula The cloud of gas and dust from which the solar system formed. (6)

solar seismology The study of pulsations or oscillations of the Sun in order to determine the characteristics of the solar interior. (14)

solar system The system of the Sun and the planets, their satellites, the minor planets, comets, meteoroids, and other objects revolving around the Sun. (6)

solar time A time based on the Sun; usually the hour angle of the Sun plus 12 h. (3)

solar wind A flow of hot charged particles leaving the Sun. (14)

solstice Either of two points on the celestial sphere where the Sun reaches its maximum distances north and south of the celestial equator; time of the year when the daylight is the longest or the shortest. (3)

spacetime A system of one time and three spatial coordinates, with respect to which the time and place of an event can be specified. (23)

space velocity *or* **space motion** The velocity of a star with respect to the Sun.

spectral class (or type) The classification of stars according to their temperatures using the characteristics of their spectra; the types are O B A F G K M. (16)

spectral line Radiation at a particular wavelength of light produced by the emission or absorption of energy by an atom.

spectral sequence The sequence of spectral classes of stars arranged in order of decreasing temperatures of stars of those classes. (16)

spectrometer An instrument for obtaining a spectrum; in astronomy, usually attached to a telescope to record the spectrum of a star, galaxy, or other astronomical object. (4)

spectroscopic binary star A binary star in which the components are not resolved optically, but whose binary nature is indicated by periodic variations in radial velocity, indicating orbital motion. (17)

spectroscopic parallax A parallax (or distance) of a star that is derived by comparing the apparent magnitude of the star with its absolute magnitude as deduced from its spectral characteristics.

spectroscopy The study of spectra.

spectrum The array of colors or wavelengths obtained when light (or other radiation) from a source is dispersed, as in passing it through a prism or grating. (4)

speed The rate at which an object moves without regard to its direction of motion; the numerical or absolute value of velocity.

spicule A jet of rising material in the solar chromosphere. (14)

spiral arms Arms (or long denser regions) of interstellar material and young stars that wind out in a plane from the central nucleus of a spiral galaxy. (24)

spiral density wave A mechanism for the generation of spiral structure in galaxies; a density wave interacts with interstellar matter and triggers the formation of stars. Spiral density waves are also seen in the rings of Saturn.

spiral galaxy A flattened, rotating galaxy with pinwheel-like arms of interstellar material and young stars winding out from its nucleus. (25)

spring tide The highest tidal range of the month, produced when the Moon is near either the full or the new phase. (3)

standard bulb An astronomical object of known luminosity; such an object can be used to determine distances.

star A sphere of gas shining under its own power.

star cluster An assemblage of stars held together by their mutual gravity.

Stefan-Boltzmann law A formula from which the rate at which a blackbody radiates energy can be computed; the total rate of energy emission from a unit area of a blackbody is proportional to the fourth power of its absolute temperature.

stellar evolution The changes that take place in the characteristics of stars as they age.

stellar model The result of a theoretical calculation of the physical conditions in the different layers of a star's interior.

stellar parallax *See* parallax.

stellar wind The outflow of gas, sometimes at speeds as high as hundreds of kilometers per second, from a star.

stony-iron meteorite A type of meteorite that is a blend of nickel-iron and silicate materials. (13)

stony meteorite A meteorite composed mostly of stony material. (13)

stratosphere The layer of the Earth's atmosphere above the troposphere (where most weather takes place) and below the ionosphere. (7)

strong nuclear force *or* **strong interaction** The force that binds together the parts of the atomic nucleus. (15)

subduction zone In terrestrial geology, a region where one crustal plate is forced under another, generally associated with earthquakes, volcanic activity, and the formation of deep ocean trenches. (7)

summer solstice The point on the celestial sphere where the Sun reaches its greatest distance north of the celestial equator; the day with the longest amount of daylight. (3)

Sun The star about which the Earth and other planets revolve.

sunspot A temporary cool region in the solar photosphere that appears dark by contrast against the surrounding hotter photosphere. (14)

sunspot cycle The semiregular 11-year period with which the frequency of sunspots fluctuates. (14)

supercluster A large region of space (more than 100 million LY across) where groups and clusters of galaxies are more concentrated; a cluster of clusters of galaxies. (27)

supergiant A star of very high luminosity and relatively low temperature. (17)

supernova An explosion that marks the final stage of evolution of a star. A Type I supernova occurs when a white dwarf accretes enough matter to exceed the Chandrasekhar limit, collapses, and explodes. A Type II supernova marks the final collapse of a massive star. (22)

surface gravity The weight of a unit mass at the surface of a body.

synchrotron radiation The radiation emitted by charged particles being accelerated in magnetic fields and moving at speeds near that of light.

tail (of a comet) *See* dust tail of a comet *and* plasma.

tangential (transverse) velocity The component of a star's space velocity that lies in the plane of the sky.

tectonic Activity and motion that result from expansion and contraction of the crust of a planet.

temperature A measure of how fast the particles in a body are moving or vibrating in place; a measure of the average heat energy in a body.

temperature (Celsius; formerly centigrade) Temperature measured on scale where water freezes at 0° and boils at 100°.

temperature (color) The temperature of a star as estimated from the intensity of the stellar radiation at two or more colors or wavelengths.

temperature (effective) The temperature of a blackbody that would radiate the same total amount of energy that a particular object, such as a star, does.

temperature (excitation) The temperature of a star as estimated from the relative strengths of lines in its spectrum that originate from atoms in different stages of excitation.

temperature (Fahrenheit) Temperature measured on a scale where water freezes at 32° and boils at 212°.

temperature (ionization) The temperature of a star as estimated from the relative strengths of lines in its spectrum that originate from atoms in different stages of ionization.

temperature (Kelvin) Absolute temperature measured in Celsius degrees, with the zero point at absolute zero.

temperature (radiation) The temperature of a blackbody that radiates the same amount of energy in a given spectral region as does a particular body.

terrestrial planet Any of the planets Mercury, Venus, Earth, or Mars; sometimes the Moon or Pluto are included in the list. (6)

theory A set of hypotheses and laws that have been well demonstrated to apply to a wide range of phenomena associated with a particular subject.

thermal energy Energy associated with the motions of the molecules or atoms in a substance.

thermal equilibrium A balance between the input and outflow of heat in a system.

thermal radiation The radiation emitted by any body or gas that is not at absolute zero.

thermonuclear energy Energy associated with thermonuclear reactions or that can be released through thermonuclear reactions.

thermonuclear reaction A nuclear reaction or transformation that results from encounters between particles that are given high velocities (by heating them). (15)

tidal force A differential gravitational force that tends to deform a body. (3)

tidal stability limit The distance—approximately 2.5 planetary radii from the center—within which differential gravitational forces (or tides) are stronger than the mutual gravitational attraction between two adjacent orbiting objects. Within this limit, fragments are not likely to accrete or assemble themselves into a larger object. Also called the Roche limit.

tide Deformation of a body by the differential gravitational force exerted on it by another body; in the Earth, the deformation of the ocean surface by the differential gravitational forces exerted by the Moon and Sun. (3)

transition region The region in the Sun's atmosphere where the temperature rises very rapidly from the relatively low temperatures that characterize the chromosphere to the high temperatures of the corona. (14)

triple-alpha process A series of two nuclear reactions by which three helium nuclei are built up into one carbon nucleus. (21)

tropical year Period of revolution of the Earth about the Sun with respect to the vernal equinox.

Tropic of Cancer The parallel (circle) of latitude 23.5° N.

Tropic of Capricorn The parallel (circle) of latitude 23.5° S.

troposphere Lowest level of the Earth's atmosphere, where most weather takes place. (7)

turbulence Random motions of gas masses, as in the atmosphere of a star.

21-cm line A line in the spectrum of neutral hydrogen at the radio wavelength of 21 cm. (19)

ultraviolet radiation Electromagnetic radiation of wavelengths shorter than the shortest visible wavelengths; radiation of wavelengths in the approximate range 10 to 400 nm. (4)

umbra The central, completely dark part of a shadow. (3)

uncertainty principle Heisenberg uncertainty principle. It is fundamentally impossible to make simultaneous measurements of a particle's position and velocity with infinite accuracy.

universe The totality of all matter, radiation and space; everything accessible to our observations.

variable star A star that varies in luminosity. (18)

velocity The speed and the direction a body is moving; e.g., 44 km/s toward the north galactic pole.

velocity of escape The speed with which an object must move in order to enter a parabolic orbit about another body (such as the Earth), and hence move permanently away from the vicinity of that body. (2)

vernal equinox The point on the celestial sphere where the Sun crosses the celestial equator passing from south to north; a time in the course of the year when the day and night are roughly equal. (3)

very-long-baseline interferometry (VLBI) A technique of radio astronomy whereby signals from telescopes thousands of kilometers apart combined to obtain very high resolution by letting waves from different sites interfere with each other. (5)

visual binary star A binary star in which the two components are telescopically resolved. (17)

void A region between clusters and superclusters of galaxies that appears relatively empty of galaxies. (27)

volatile materials Materials that are gaseous at fairly low temperatures. This is a relative term, usually applied to the gases in planetary atmospheres and to common ices (H_2O, CO_2, and so on), but it is also sometimes used for elements such as cadmium, zinc, lead, and rubidium that form gases at temperatures up to 1000 K. (These are called volatile elements, as opposed to refractory elements.)

volume A measure of the total space occupied by a body. (2)

watt A unit of power (energy per unit time).

wavelength The spacing of the crests or troughs in a wave. (4)

weak nuclear force *or* **weak interaction** The nuclear force involved in radioactive decay. The weak force is characterized by the slow rate of certain nuclear reactions—such as the decay of the neutron, which occurs with a half-life of 11 min.

weight A measure of the force due to gravitational attraction.

white dwarf A star that has exhausted most or all of its nuclear fuel and has collapsed to a very small size; such a star is near its final stage of life. (17)

Wien's law Formula that relates the temperature of a blackbody to the wavelength at which it emits the greatest intensity of radiation. (4)

winter solstice Point on the celestial sphere where the Sun reaches its greatest distance south of the celestial equator; the time of the year with the shortest amount of daylight. (3)

Wolf-Rayet star One of a class of very hot stars that eject shells of gas at very high velocity.

x rays Photons of wavelengths intermediate between those of ultraviolet radiation and gamma rays.

x-ray stars Stars (other than the Sun) that emit observable amounts of radiation at x-ray frequencies.

year The period of revolution of the Earth around the Sun. (1)

Zeeman effect A splitting or broadening of spectral lines due to magnetic fields. (14)

zenith The point on the celestial sphere opposite to the direction of gravity; or the direction opposite to that indicated by a plumb bob; the point directly above the observer. (1)

zero-age main sequence Main sequence on the H–R diagram for a system of stars that have completed their contraction from interstellar matter and are now deriving all their energy from nuclear reactions, but whose chemical composition has not yet been altered by nuclear reactions. (21)

zodiac A belt around the sky 18° wide centered on the ecliptic. (1)

zone of avoidance A region near the Milky Way where obscuration by interstellar dust is so heavy that few or no exterior galaxies can be seen.

Appendix 4
Powers-of-Ten Notation

In astronomy (and other sciences), it is often necessary to deal with very large or very small numbers. In fact, when numbers get truly large in everyday life, such as the national debt in the U.S., we call them astronomical. Among the ideas astronomers must routinely deal with is that the Earth is 150,000,000,000 m from the Sun, and the mass of the hydrogen atom is 0.00000000000000000000000000167 kg. No one in his or her right mind would want to continue writing so many zeros!

Instead, scientists have agreed on a kind of shorthand notation, which is not only easier to write, but (as we shall see) makes multiplication and division of large and small numbers much easier. If you have never used this *powers-of-ten notation* or *scientific notation*, it may take a bit of time to get used to it, but you will soon find it much easier than keeping all those zeros.

Writing Large Numbers

The convention in this notation is that we generally have *only one number to the left of the decimal point*. If a number is not in this format, it must be changed. The number 6 is already in the right format, because for integers, we understand there to be a decimal point to the right of them. So 6 is really 6., and there is indeed only one number to the left of the decimal point. But the number 165 (which is 165.) has three numbers to the left of the decimal point, and is thus ripe for conversion.

To change 165 to proper form, we must make it 1.65 and then keep track of the change we have made. (Think of the number as a weekly salary, and suddenly it makes a lot of difference whether we have $165 or $1.65.) We keep track of the number of places we moved the decimal point by expressing it as a power of ten. So 165 becomes 1.65×10^2 or 1.65 multiplied by ten to the second power. The small raised 2 is called an *exponent*, and it tells us how many times we moved the decimal point to the left.

Note that 10^2 also designates 10 squared or 10×10, which equals 100. And 1.65×100 is just 165, the number we started with. Another way to look at scientific notation is that we separate out the messy numbers out front, and leave the smooth units of ten for the exponent to denote. So a number like 1,372,568 becomes 1.372568 times a million (10^6) or 1.372568 times 10 multiplied by itself 6 times. We had to move the decimal point six places to the left (from its place after the 8) to get the number into the form where there is only one digit to the left of the decimal point.

The reason we call this powers-of-ten notation is that our counting system is based on increases of ten; each place in our numbering system is ten times greater than the place to the right of it. As you have probably learned, this got started because human beings have ten fingers and we started counting with them. It is interesting to speculate that if we ever meet intelligent life-forms with only eight fingers, their counting system would probably be a powers-of-eight notation!

So, in the example we started with, the number of meters from the Earth to the Sun is 1.5×10^{11}. Elsewhere in the book, we mention that a string a light year long would fit around the Earth's equator 236 million or 236,000,000 times. In scientific notation, this would become 2.36×10^8. Now if you like expressing things in millions, as the annual reports of successful companies do, you might like to write this number as 236×10^6. However, the usual convention is to have only one number to the left of the decimal point.

Writing Small Numbers

Now take a number like 0.00347, which is also not in the standard (agreed-to) form for scientific notation. To get it into that format, we must make the first part of it 3.47 by moving the decimal point three places to the right. Note that this motion to the right is the opposite of the motion to the left that we discussed above. To keep track, we call this change negative and put a minus sign in the exponent. Thus 0.00347 becomes 3.47×10^{-3}.

In the example we gave at the beginning, the mass of the hydrogen atom would then be written as 1.67×10^{-27} kg. In this system, one is written as 10^0, a tenth as 10^{-1}, a hundredth as 10^{-2}, and so forth. Note that any number, no matter how large or how small, can be expressed in scientific notation.

Multiplication and Division

The powers-of-ten notation is not only compact and convenient, it also simplifies arithmetic. To multiply two numbers expressed as powers of ten, you need only multiply the numbers out front and then add the exponents. If there are no numbers out front, as in $100 \times 100,000$, then you just add the exponents ($10^2 \times 10^5 = 10^7$). When there are numbers out of front, you do have to multiply them, but they are much easier to deal with than numbers with many zeros in them.

Some examples:

$$3 \times 10^5 \times 2 \times 10^9 = 6 \times 10^{14}$$

$$0.04 \times 6,000,000 = 4 \times 10^{-2} \times 6 \times 10^6$$
$$= 24 \times 10^4 = 2.4 \times 10^5$$

Note in the second example that when we added the exponents, we treated negative exponents as we do in regular arithmetic (-2 plus 6 equals 4). Also, notice that the first result we got had a 24 in it, which was not in the usual form, having two places to the left of the decimal point, and we therefore changed it.

To divide, you divide the numbers out front and subtract the exponents. Here are several examples:

$$1,000,000 \div 1000 = 10^6 \div 10^3 = 10^{6-3} = 10^3$$

$$9 \times 10^{12} \div 2 \times 10^3 = 4.5 \times 10^9$$

$$2.8 \times 10^2 \div 6.2 \times 10^5 = 0.452 \times 10^{-3} = 4.52 \times 10^{-4}$$

If this is the first time that you have met scientific notation, we urge you to practice using it (you might start by solving the exercises below). Like any new language, the notation looks complicated at first, but gets easier as you practice it.

Exercises

1. On April 8, 1996, the Galileo spacecraft was 775 million kilometers from Earth. Convert this number to scientific notation. How many astronomical units is this? (An astronomical unit is the distance from the Earth to the Sun; see above, but remember to keep your units consistent!)

2. During the first six years of its operation, the Hubble Space Telescope circled the Earth 37,000 times, for a total of 1,280,000 km. Use scientific notation to find the number of kilometers in one orbit.

3. In a college cafeteria, a soybean-vegetable burger is offered as an alternative to regular hamburgers. If 489,875 burgers were eaten during the course of a school year, and 997 of them were veggie-burgers, what fraction of the burgers does this represent?

4. In a June 1990 Gallup poll, 27 percent of adult Americans thought that alien beings have actually landed on Earth. The number of adults in the U.S. in 1990 was (according to the census) about 186,000,000. Use scientific notation to determine how many adults believe aliens have visited the Earth.

5. In 1995, 1.7 million degrees were awarded by colleges and universities in the U.S. Among these were 41,000 PhD degrees. What fraction of the degrees were PhD's? Express this number as a percent.

Appendix 5
Units Used in Science

In the American system of measurement (originally developed in England), the fundamental units of length, weight, and time are the foot, pound, and second, respectively. There are also larger and smaller units, which include the ton (2240 lb), the mile (5280 ft), the rod (16½ ft), the yard (3 ft), the inch (1/12 ft), the ounce (1/16 lb), and so on. Such units, whose origins in decisions by British royalty have been forgotten by most people, are quite inconvenient for conversion and arithmetic computation.

In science, therefore, it is more usual to use the metric system, which has been adopted in virtually all countries except the United States. Its great advantage is that every unit increases by a factor of ten, instead of the strange factors in the American system. The fundamental units of the metric system are

length: 1 meter (m)

mass: 1 kilogram (kg)

time: 1 second (s)

A meter was originally intended to be 1 ten-millionth of the distance from the equator to the North Pole along the surface of the Earth. It is about 1.1 yd. A kilogram is the mass that on Earth results in a weight of about 2.2 lb. The second is the same in metric and American units.

The most commonly used quantities of length and mass of the metric system are the following:

Length

1 kilometer (km) = 1000 meters = 0.6214 mile
1 meter (m) = 0.001 km = 1.094 yards = 39.37 inches
1 centimeter (cm) = 0.01 meter = 0.3937 inch
1 millimeter (mm) = 0.001 meter = 0.1 cm
1 micrometer (μm) = 0.000001 meter = 0.0001 cm
1 nanometer (nm) = 10^{-9} meter = 10^{-7} cm

To convert from the American system, here are a few helpful factors:

1 mile = 1.6093 km
1 inch = 2.5400 cm

Mass

Although we don't make the distinction very carefully in everyday life on Earth, strictly speaking the kilogram is a unit of mass (measuring how many atoms a body has) and the pound is a unit of weight (measuring how strongly the Earth's gravity pulls on a body).

1 metric ton = 10^6 grams = 1000 kg (and it produces a weight of 2.2046×10^3 lbs on Earth)

1 kg = 1000 grams (and it produces a weight of 2.2046 lbs on Earth)

1 gram (g) = 0.0353 oz (and the equivalent weight is 0.0022046 lb)

1 milligram (mg) = 0.001 g

And a weight of 1 lb is equivalent on Earth to a mass of 0.4536 kg, while a weight of 1 oz is produced by a mass of 28.3495 g.

Temperature

Three temperature scales are in general use:

1. Fahrenheit (F); water freezes at 32°F and boils at 212°F.
2. Celsius or centigrade° (C); water freezes at 0°C and boils at 100°C.
3. Kelvin or absolute (K); water freezes at 273 K and boils at 373 K.

All molecular motion ceases at −459°F = −273°C = 0 K, a temperature called *absolute zero*. Kelvin temperature is measured from this lowest possible temperature. It is the temperature scale most often used in astronomy. Kelvins are degrees that have the same value as centigrade or Celsius degrees, since the difference between the freezing and boiling points of water is 100 degrees in each.

On the Fahrenheit scale, the difference between the freezing and boiling points of water is 180 degrees. Thus, to convert Celsius degrees or Kelvins to Fahrenheit, it is necessary to multiply by 180/100 = 9/5. To convert from Fahrenheit to Celsius degrees or Kelvins, it is necessary to multiply by 100/180 = 5/9. The full conversion formulas are:

K = °C + 273
°C = 0.555 × (°F − 32)
°F = (1.8 × °C) + 32

°Celsius is now the name used for centigrade temperature; it has a more modern standardization but differs from the old centigrade scale by less than 0.1°.

Appendix 6
Some Useful Constants for Astronomy

Physical Constants

speed of light $(c) = 2.9979 \times 10^8$ m/s

gravitational constant $(G) = 6.672 \times 10^{-11}$ N m^2/kg^2

Planck's constant $(h) = 6.626 \times 10^{-34}$ joule·s

mass of a hydrogen atom $(m_H) = 1.673 \times 10^{-27}$ kg

mass of an electron $(m_e) = 9.109 \times 10^{-31}$ kg

Rydberg constant $(R) = 1.0974 \times 10^7$ per m

Stefan-Boltzmann constant $(\sigma) =$
  5.670×10^{-8} joule/(s·m^2·deg^4)

constant in Wien's law $(\lambda_{max} T) = 2.898 \times 10^{-3}$ m·deg

electron volt (energy) $(eV) = 1.602 \times 10^{-19}$ joules

energy equivalent of 1 ton TNT $= 4.3 \times 10^9$ joules

Astronomical Constants

astronomical unit $(AU) = 1.496 \times 10^{11}$ m

light year $(LY) = 9.461 \times 10^{15}$ m

parsec $(pc) = 3.086 \times 10^{16}$ m $= 3.262$ LY

sidereal year $(yr) = 3.158 \times 10^7$ s

mass of Earth $(M_E) = 5.977 \times 10^{24}$ kg

equatorial radius of Earth $(R_E) = 6.378 \times 10^6$ m

obliquity of ecliptic $(\epsilon) = 23° 27'$

surface gravity of Earth $(g) = 9.807$ m/s^2

escape velocity of Earth $(v_E) = 1.119 \times 10^4$ m/s

mass of Sun $(M_{Sun}) = 1.989 \times 10^{30}$ kg

equatorial radius of Sun $(R_{Sun}) = 6.960 \times 10^8$ m

luminosity of Sun $(L_{Sun}) = 3.83 \times 10^{26}$ watts

solar constant (flux of energy received at Earth) $S =$
  1.37×10^3 watts/m^2

Hubble constant $(H_0) =$
  approximately 25 km/sec per million LY

Appendix 7
Physical Data for the Planets

Planet	Diameter (km)	Diameter (Earth = 1)	Mass (Earth = 1)	Mean Density (g/cm³)	Rotation Period (days)	Inclination of Equator to Orbit (°)	Surface Gravity (Earth = 1)	Velocity of Escape (km/s)
Mercury	4,878	0.38	0.055	5.43	58.6	0.0	0.38	4.3
Venus	12,104	0.95	0.82	5.24	−243.0	177.4	0.91	10.4
Earth	12,756	1.00	1.00	5.52	0.997	23.4	1.00	11.2
Mars	6,794	0.53	0.107	3.9	1.026	25.2	0.38	5.0
Jupiter	142,800	11.2	317.8	1.3	0.41	3.1	2.53	60
Saturn	120,540	9.41	94.3	0.7	0.43	26.7	1.07	36
Uranus	51,200	4.01	14.6	1.2	−0.72	97.9	0.92	21
Neptune	49,500	3.88	17.2	1.6	0.67	29	1.18	24
Pluto	2,200	0.17	0.0025	2.0	−6.387	118	0.09	1

Orbital Data for the Planets

Planet	Semimajor Axis AU	Semimajor Axis 10⁶ km	Sidereal Period Tropical Years	Sidereal Period Days	Mean Orbital Speed (km/s)	Orbital Eccentricity	Inclination of Orbit to Ecliptic (°)
Mercury	0.3871	57.9	0.24085	87.97	47.9	0.206	7.004
Venus	0.7233	108.2	0.61521	224.70	35.0	0.007	3.394
Earth	1.0000	149.6	1.000039	365.26	29.8	0.017	0.0
Mars	1.5237	227.9	1.88089	686.98	24.1	0.093	1.850
(Ceres)	2.7671	414	4.603		17.9	0.077	10.6
Jupiter	5.2028	778	11.86		13.1	0.048	1.308
Saturn	9.538	1427	29.46		9.6	0.056	2.488
Uranus	19.191	2871	84.07		6.8	0.046	0.774
Neptune	30.061	4497	164.82		5.4	0.010	1.774
Pluto	39.529	5913	248.6		4.7	0.248	17.15

Adapted from *The Astronomical Almanac* (U.S. Naval Observatory), 1981.

Appendix 8
Satellites of the Planets

Planet	Satellite Name	Discovery	Semimajor Axis (km × 1000)	Period (days)	Diameter (km)	Mass (10²⁰ kg)	Density (g/cm³)
Earth	Moon	—	384	27.32	3476	735	3.3
Mars	Phobos	Hall (1877)	9.4	0.32	23	1×10^{-4}	2.0
	Deimos	Hall (1877)	23.5	1.26	13	2×10^{-5}	1.7
Jupiter	Metis	Voyager (1979)	128	0.29	20	—	—
	Adrastea	Voyager (1979)	129	0.30	40	—	—
	Amalthea	Barnard (1892)	181	0.50	200	—	—
	Thebe	Voyager (1979)	222	0.67	90	—	—
	Io	Galileo (1610)	422	1.77	3630	894	3.6
	Europa	Galileo (1610)	671	3.55	3138	480	3.0
	Ganymede	Galileo (1610)	1,070	7.16	5262	1482	1.9
	Callisto	Galileo (1610)	1,883	16.69	4800	1077	1.9
	Leda	Kowal (1974)	11,090	239	15	—	—
	Himalia	Perrine (1904)	11,480	251	180	—	—
	Lysithea	Nicholson (1938)	11,720	259	40	—	—
	Elara	Perrine (1905)	11,740	260	80	—	—
	Ananke	Nicholson (1951)	21,200	631 (R)	30	—	—
	Carme	Nicholson (1938)	22,600	692 (R)	40	—	—
	Pasiphae	Melotte (1908)	23,500	735 (R)	40	—	—
	Sinope	Nicholson (1914)	23,700	758 (R)	40	—	—
Saturn	Unnamed	Voyager (1985)	118.2	0.48	15?	3×10^{-5}	—
	Pan	Voyager (1985)	133.6	0.58	20	3×10^{-5}	—
	Atlas	Voyager (1980)	137.7	0.60	40	—	—
	Prometheus	Voyager (1980)	139.4	0.61	80	—	—
	Pandora	Voyager (1980)	141.7	0.63	100	—	—
	Janus	Dollfus (1966)	151.4	0.69	190	—	—
	Epimetheus	Fountain, Larson (1980)	151.4	0.69	120	—	—
	Mimas	Herschel (1789)	186	0.94	394	0.4	1.2
	Enceladus	Herschel (1789)	238	1.37	502	0.8	1.2
	Tethys	Cassini (1684)	295	1.89	1048	7.5	1.3
	Telesto	Reitsema et al. (1980)	295	1.89	25	—	—
	Calypso	Pascu et al. (1980)	295	1.89	25	—	—
	Dione	Cassini (1684)	377	2.74	1120	11	1.4
	Helene	Lecacheux, Laques (1980)	377	2.74	30	—	—
	Rhea	Cassini (1672)	527	4.52	1530	25	1.3
	Titan	Huygens (1655)	1,222	15.95	5150	1346	1.9
	Hyperion	Bond, Lassell (1848)	1,481	21.3	270	—	—
	Iapetus	Cassini (1671)	3,561	79.3	1435	19	1.2
	Phoebe	Pickering (1898)	12,950	550 (R)	220	—	—
Uranus	Cordelia	Voyager (1986)	49.8	0.34	40?	—	—
	Ophelia	Voyager (1986)	53.8	0.38	50?	—	—
	Bianca	Voyager (1986)	59.2	0.44	50?	—	—
	Cressida	Voyager (1986)	61.8	0.46	60?	—	—
	Desdemona	Voyager (1986)	62.7	0.48	60?	—	—
	Juliet	Voyager (1986)	64.4	0.50	80?	—	—
	Portia	Voyager (1986)	66.1	0.51	80?	—	—
	Rosalind	Voyager (1986)	69.9	0.56	60?	—	—

(Table continued on page A-24)

Satellites of the Planets (Continued)

Planet	Satellite Name	Discovery	Semimajor Axis (km × 1000)	Period (days)	Diameter (km)	Mass (10²⁰ kg)	Density (g/cm³)
	Belinda	Voyager (1986)	75.3	0.63	60?	—	—
	Puck	Voyager (1985)	86.0	0.76	170	—	—
	Miranda	Kuiper (1948)	130	1.41	485	0.8	1.3
	Ariel	Lassell (1851)	191	2.52	1160	13	1.6
	Umbriel	Lassell (1851)	266	4.14	1190	13	1.4
	Titania	Herschel (1787)	436	8.71	1610	35	1.6
	Oberon	Herschel (1787)	583	13.5	1550	29	1.5
Neptune	Naiad	Voyager (1989)	48	0.30	50	—	—
	Thalassa	Voyager (1989)	50	0.31	90	—	—
	Despina	Voyager (1989)	53	0.33	150	—	—
	Galatea	Voyager (1989)	62	0.40	150	—	—
	Larissa	Voyager (1989)	74	0.55	200	—	—
	Proteus	Voyager (1989)	118	1.12	400	—	—
	Triton	Lassell (1846)	355	5.88 (R)	2720	220	2.1
	Nereid	Kuiper (1949)	5,511	360	340	—	—
Pluto	Charon	Christy (1978)	19.7	6.39	1200	—	—

Appendix 9
Upcoming (Total) Eclipses

1. Total Eclipses of the Sun

Date	Duration of Totality (min)	Where Visible
1997 March 9	2.8	Siberia, Arctic
1998 Feb. 26	4.4	Central America
1999 Aug. 11	2.6	Central Europe, Central Asia
2001 June 21	4.9	Southern Africa
2002 Dec. 4	2.1	South Africa, Australia
2003 Nov. 23	2.0	Antarctica
2005 April 8	0.7	South Pacific Ocean
2006 March 29	4.1	Africa, Asia Minor, U.S.S.R.
2008 Aug. 1	2.4	Arctic Ocean, Siberia, China
2009 July 22	6.6	India, China, South Pacific
2010 July 11	5.3	South Pacific Ocean
2012 Nov. 13	4.0	Northern Australia, South Pacific
2013 Nov. 3	1.7	Atlantic Ocean, Central Africa
2015 March 20	4.1	North Atlantic, Arctic Ocean
2016 March 9	4.5	Indonesia, Pacific Ocean

Date	Duration of Totality (min)	Where Visible
2017 Aug. 21	2.7	Pacific Ocean, U.S.A., Atlantic Ocean
2019 July 2	4.5	South Pacific, South America
2020 Dec. 14	2.2	South Pacific, South America, South Atlantic Ocean
2021 Dec. 4	1.9	Antarctica
2023 April 20	1.3	Indian Ocean, Indonesia
2024 April 8	4.5	South Pacific, Mexico, East U.S.A.
2026 Aug. 12	2.3	Arctic, Greenland, North Atlantic, Spain
2027 Aug. 2	6.4	North Africa, Arabia, Indian Ocean
2028 July 22	5.1	Indian Ocean, Australia, New Zealand
2030 Nov. 25	3.7	South Africa, Indian Ocean, Australia

2. Some Upcoming Total Lunar Eclipses

1997 Sep. 16	2003 May 16	2007 March 3
2000 Jan. 21	2003 Nov. 9	2007 Aug. 28
2000 July 16	2004 May 4	2008 Feb. 21
2001 Jan. 9	2004 Oct. 28	

Appendix 10

The Nearest Stars

Name (Catalog number)	Distance (LY)	Spectral Type	Location[1] RA	Location[1] Dec	Luminosity (Sun = 1)
Sun	—	G2V	—	—	1.0
Proxima Centauri	4.2	M5V	14 30	−62 41	6×10^{-6}
Alpha Centauri A	4.3	G2V	14 33	−60 50	1.5
Alpha Centauri B	4.3	K0V	14 33	−60 50	0.5
Barnard's Star	6.0	M4V	17 57	+04 33	4×10^{-4}
Wolf 359 (Gliese 406)	7.8	M6V	10 56	+07 03	2×10^{-5}
Lalande 21185 (HD 95735)	8.2	M2V	11 04	+36 02	5×10^{-3}
Luyten 726-8 A	8.6	M5V	01 38	−17 58	6×10^{-5}
Luyten 726-8 B (UV Ceti)	8.6	M6V	01 38	−17 58	4×10^{-5}
Sirius A	8.6	A1V	06 45	−16 43	24
Sirius B	8.6	w.d.[2]	06 45	−16 43	3×10^{-3}
Ross 154 (Gliese 729)	9.6	M4V	18 50	−23 49	5×10^{-4}
Ross 248 (Gliese 905)	10.3	M6V	23 42	+44 12	1×10^{-4}
Epsilon Eridani	10.7	K2V	03 33	−09 27	0.3
Ross 128 (Gliese 447)	10.8	M4V	11 48	+00 49	3×10^{-4}
Luyten 789-6 A	11.1	M5V[3]	22 39	−15 20	1×10^{-4}
Luyten 789-6 B	11.1	—	22 39	−15 20	—
Luyten 789-6 C	11.1	—	22 39	−15 20	—
BD +43°44 A (Gliese 15 A)	11.3	M1V	00 18	+44 61	6×10^{-3}
BD +43°44 B (Gliese 15 B)	11.3	M3V	00 18	+44 61	4×10^{-4}
Epsilon Indi	11.3	K5V	22 03	−56 47	0.14
61 Cygni A	11.3	K5V	21 07	+38 45	8×10^{-2}
61 Cygni B	11.3	K7V	21 07	+38 45	4×10^{-2}
BD +59°1915 A (Gliese 725 A)	11.4	M3V	18 43	+59 37	3×10^{-3}
BD +59°1915 B (Gliese 725 B)	11.4	M4V	18 43	+59 37	2×10^{-3}
Tau Ceti	11.4	G8V	01 44	−15 56	0.45
Procyon A	11.4	F5IV	07 39	+05 13	7.7
Procyon B	11.4	w.d.[2]	07 39	+05 13	6×10^{-4}
CD −36°15693 (Lacaille 9352)	11.5	M2V	23 06	−35 52	1×10^{-2}
GJ 1111 (G51-15)	11.8	M7V	08 29	+26 47	1×10^{-5}
GJ 1061	12.0	M5V	03 36	−44 30	8×10^{-5}
Luyten 725-32 (YZ Ceti)	12.2	M5V	01 12	−18 04	3×10^{-4}
BD +5°1668 (Gliese 273)	12.3	M4V	07 28	+05 17	1×10^{-3}
CD −39°14192 (Gliese 825)	12.6	M0V	21 17	−38 52	3×10^{-2}
Kapteyn's Star	12.6	M0V	05 11	−44 56	4×10^{-3}

[1] Location (right ascension and declination) given for Epoch 2000.0
[2] White dwarf star
[3] The stars in this system are so close to each other, it is not possible to measure their spectral types and luminosities separately.

With many thanks to Dr. Todd Henry, Space Telescope Science Institute, and the RECONS team for updated information. For the latest data on nearby star systems, see their web page at http://proxima.astro.virginia.edu/~pai/Recons

Appendix 11
The Brightest Stars

NOTE: These are the stars that *appear* the brightest visually, as seen from our vantage point on Earth. They are not necessarily the stars that are intrinsically the brightest.

Name[1]	Luminosity (Sun = 1)	Distance[2] (LY)	Spectral Type	ProperMotion (arcsec/yr)	Right Ascension (Epoch 2000.0) (h)	(m)	Declination (Epoch 2000.0) (deg)	(min)
Sirius (α CMa)	40	9	A1V	1.33	06	45.1	−16	43
Canopus (α Car)	1500	98	F01	0.02	06	24.0	−52	42
Alpha Centauri	2	4	G2V	3.68	14	39.6	−60	50
Arcturus (α Boo)	100	36	K2III	2.28	14	15.7	+19	11
Vega (α Lyr)	50	26	A0V	0.34	18	36.9	+38	47
Capella (α Aur)	200	46	G5III	0.44	05	16.7	+46	00
Rigel (β Ori)	8×10^4	815	B8Ia	0.00	05	12.1	−08	12
Procyon (α Cmi)	9	11	F5IV-V	1.25	07	39.3	+05	13
Betelgeuse (α Ori)	10^5	500	M2Iab	0.03	05	55.2	+07	24
Achernar (α Eri)	500	65	B3V	0.10	01	37.7	−57	14
Beta Centauri	9300	300	B1III	0.04	14	03.8	−60	22
Altair (α Aql)	10	17	A7IV-V	0.66	19	50.8	+08	52
Aldebaran (α Tau)	200	20	K5III	0.20	04	35.9	+16	31
Spica (α Vir)	6000	260	B1V	0.05	13	25.2	−11	10
Antares (α Sco)	10^4	390	M1Ib	0.03	16	29.4	−26	26
Pollux (β Gem)	60	39	K0III	0.62	07	45.3	+28	02
Fomalhaut (α PsA)	50	23	A3V	0.37	22	57.6	−29	37
Deneb (α Cyg)	8×10^4	1400	A2Ia	0.00	20	41.4	+45	17
Beta Crucis	10^4	490	B0.5IV	0.05	12	47.7	−59	41
Regulus (α Leo)	150	85	B7V	0.25	10	08.3	+11	58

[1] The brightest stars typically have names from antiquity (although most fainter stars are not given names, but merely catalog designations). Next to each star's ancient name, we have put its name in the system originated by Bayer (see the "Astronomy Basics" box in Chapter 18.) The abbreviations of the constellations are given in Appendix 14.

[2] The distances of the more remote stars are estimated from their spectral types and apparent brightnesses and are only approximate. The luminosities for those stars are approximate to the same degree.

Appendix 12
The Brightest Members of the Local Group

Galaxy	Type[1]	Right Ascension[2] (h)	(m)	Declination (Degrees)	Distance[3] (1000 LY)	Absolute Magnitude	Apparent Magnitude	Diameter (1000 LY)
Milky Way	Sbc	17	46	−29		−20.6		130
Andromeda; M31; NGC224	Sb	00	43	+41	2200	−21.6	4.4	200
M33; NGC598	Sc	01	34	+31	2500	−19.1	6.3	45
Large Magellanic Cloud	Irr	05	24	−70	160	−18.4	0.6	20
Small Magellanic Cloud	Irr	00	53	−73	300	−17.0	2.8	15
IC10	Irr	00	20	+59	4000	−16.2	11.7	6
NGC205	E5pec	00	40	+42	2200	−15.7	8.6	10
M32; NGC221	E2	00	43	+41	2200	−15.5	9.0	5
NGC6822	Irr	19	45	−15	1700	−15.1	9.3	8
WLM	Irr	00	02	−15	2000	−15.0	11.3	7
IC5152	Sd	22	03	−51	2000	−14.6	11.7	5
NGC185	E3pec	00	39	+48	2200	−14.6	10.1	6
IC1613	Irr	01	05	+02	2500	−14.5	10.0	12
NGC147	E5	00	33	+48	2200	−14.4	10.4	10
Leo A	Irr	09	59	+31	5000	−13.5	12.7	7
Pegasus	Irr	23	29	+15	5000	−13.4	12.4	8
Fornax	E3	02	40	−34	500	−12.9	8.5	3
GR8	Irr	12	59	+14	4000	−11.0	14.6	0.2
DDO210	Irr	20	47	−13	3000	−11.0	15.3	4
Sagittarius Dwarf[3]	Dwarf E	18	50	−32	80	?	?	25
Sagittarius	Irr	19	30	−18	4000	−10.6	15.6	5
Sculptor	E3	00	60	−34	300	−10.6	9.1	1
Andromeda I	E3	00	46	+38	2200	−10.6	14.0	2
Andromeda III	E5	00	35	+36	2200	−10.6	14.0	3
Andromeda II	E2	01	16	+33	2200	−10.6	14.0	2.3
Pisces; LGS3	Irr	01	03	+22	3000	−9.7	15.5	0.5
Leo I	E3	10	09	+12	600	−9.6	11.8	1
Leo II	E0	11	14	+22	600	−9.2	12.3	0.5
Ursa Minor	E5	15	09	+67	300	−8.2	11.6	1
Draco	E3	17	20	+58	300	−8.0	12.0	0.5
Carina	E4	06	42	−51	300	>−5.5	>13.0	0.5
Sgr(Anon)	Epec				50			

[1] S mean spiral, Sb means barred spiral, E means elliptical, Irr means irregular; the numbers represent subgroups into which Hubble and others divided these broad categories.
[2] Coordinates are given for Epoch 2000.0
[3] Many of the distances (and therefore the diameters calculated from them) are only approximate.
[4] This close neighbor galaxy is so extended on the sky that giving a magnitude would not make sense.

Data courtesy of Paul Hodge, University of Washington.

Appendix 13
The Chemical Elements

Element	Symbol	Atomic Number	Atomic Weight* (Chemical Scale)	Number of Atoms per 10^{12} Hydrogen Atoms
Hydrogen	H	1	1.0080	1×10^{12}
Helium	He	2	4.003	8×10^{10}
Lithium	Li	3	6.940	2×10^3
Beryllium	Be	4	9.013	3×10^1
Boron	B	5	10.82	9×10^2
Carbon	C	6	12.011	4.5×10^8
Nitrogen	N	7	14.008	9.2×10^7
Oxygen	O	8	16.00	7.4×10^8
Fluorine	F	9	19.00	3.1×10^4
Neon	Ne	10	20.183	1.3×10^8
Sodium	Na	11	22.991	2.1×10^6
Magnesium	Mg	12	24.32	4.0×10^7
Aluminum	Al	13	26.98	3.1×10^6
Silicon	Si	14	28.09	3.7×10^7
Phosphorus	P	15	30.975	3.8×10^5
Sulfur	S	16	32.066	1.9×10^7
Chlorine	Cl	17	35.457	1.9×10^5
Argon	Ar(A)	18	39.944	3.8×10^6
Potassium	K	19	39.100	1.4×10^5
Calcium	Ca	20	40.08	2.2×10^6
Scandium	Sc	21	44.96	1.3×10^3
Titanium	Ti	22	47.90	8.9×10^4
Vanadium	V	23	50.95	1.0×10^4
Chromium	Cr	24	52.01	5.1×10^5
Manganese	Mn	25	54.94	3.5×10^5
Iron	Fe	26	55.85	3.2×10^7
Cobalt	Co	27	58.94	8.3×10^4
Nickel	Ni	28	58.71	1.9×10^6
Copper	Cu	29	63.54	1.9×10^4
Zinc	Zn	30	65.38	4.7×10^4
Gallium	Ga	31	69.72	1.4×10^3
Germanium	Ge	32	72.60	4.4×10^3
Arsenic	As	33	74.91	2.5×10^2
Selenium	Se	34	78.96	2.3×10^3
Bromine	Br	35	79.916	4.4×10^2
Krypton	Kr	36	83.80	1.7×10^3
Rubidium	Rb	37	85.48	2.6×10^2
Strontium	Sr	38	87.63	8.8×10^2
Yttrium	Y	39	88.92	2.5×10^2
Zirconium	Zr	40	91.22	4.0×10^2
Niobium (Columbium)	Nb(Cb)	41	92.91	2.6×10^1
Molybdenum	Mo	42	95.95	9.3×10^1
Technetium	Tc(Ma)	43	(99)	—
Ruthenium	Ru	44	101.1	68
Rhodium	Rh	45	102.91	13
Palladium	Pd	46	106.4	51
Silver	Ag	47	107.880	20
Cadmium	Cd	48	112.41	63
Indium	In	49	114.82	7

* Where mean atomic weights have not been well determined, the atomic mass numbers of the most stable isotopes are given in parentheses.

(Table continued on page A-30)

Element	Symbol	Atomic Number	Atomic Weight* (Chemical Scale)	Number of Atoms per 10^{12} Hydrogen Atoms
Tin	Sn	50	118.70	1.4×10^2
Antimony	Sb	51	121.76	13
Tellurium	Te	52	127.61	1.8×10^2
Iodine	I(J)	53	126.91	33
Xenon	Xe(X)	54	131.30	1.6×10^2
Cesium	Cs	55	132.91	14
Barium	Ba	56	137.36	1.6×10^2
Lanthanum	La	57	138.92	17
Cerium	Ce	58	140.13	43
Praseodymium	Pr	59	140.92	6
Neodymium	Nd	60	144.27	31
Promethium	Pm	61	(147)	—
Samarium	Sm(Sa)	62	150.35	10
Europium	Eu	63	152.00	4
Gadolinium	Gd	64	157.26	13
Terbium	Tb	65	158.93	2
Dysprosium	Dy(Ds)	66	162.51	15
Holmium	Ho	67	164.94	3
Erbium	Er	68	167.27	9
Thulium	Tm(Tu)	69	168.94	2
Ytterbium	Yb	70	173.04	8
Lutecium	Lu(Cp)	71	174.99	2
Hafnium	Hf	72	178.50	6
Tantalum	Ta	73	180.95	1
Tungsten	W	74	183.86	5
Rhenium	Re	75	186.22	2
Osmium	Os	76	190.2	27
Iridium	Ir	77	192.2	24
Platinum	Pt	78	195.09	56
Gold	Au	79	197.00	6
Mercury	Hg	80	200.61	19
Thallium	Tl	81	204.39	8
Lead	Pb	82	207.21	1.2×10^2
Bismuth	Bi	83	209.00	5
Polonium	Po	84	(209)	—
Astatine	At	85	(210)	—
Radon	Rn	86	(222)	—
Francium	Fr(Fa)	87	(223)	—
Radium	Ra	88	226.05	—
Actinium	Ac	89	(227)	—
Thorium	Th	90	232.12	1
Protactinium	Pa	91	(231)	—
Uranium	U(Ur)	92	238.07	1
Neptunium	Np	93	(237)	—
Plutonium	Pu	94	(244)	—
Americium	Am	95	(243)	—
Curium	Cm	96	(248)	—
Berkelium	Bk	97	(247)	—
Californium	Cf	98	(251)	—
Einsteinium	E	99	(254)	—
Fermium	Fm	100	(253)	—
Mendeleevium	Mv	101	(256)	—
Nobelium	No	102	(253)	—
Lawrencium	Lr	103	(262)	—
Rutherfordium	Rf	104	(261)	—
Hahnium	Ha	105	(262)	—
Seaborgium	Sg	106	(263)	—
Nielsbohrium	Ns	107	(262)	—
Hassium	Hs	108	(264)	—
Meitnerium	Mt	109	(266)	—
Ununnilium	Uun	110	(269)	—
Unununium	Uuu	111	(272)	—

* Where mean atomic weights have not been well determined, the atomic mass numbers of the most stable isotopes are given in parentheses.

Appendix 14

The Constellations

Constellation (Latin Name)	Genitive Case Ending	English Name or Description	Abbreviation	Approximate Position	
				α h	δ °
Andromeda	Andromedae	Princess of Ethiopia	And	1	+40
Antila	Antilae	Air pump	Ant	10	−35
Apus	Apodis	Bird of Paradise	Aps	16	−75
Aquarius	Aquarii	Water bearer	Aqr	23	−15
Aquila	Aquilae	Eagle	Aql	20	+5
Ara	Arae	Altar	Ara	17	−55
Aries	Arietis	Ram	Ari	3	+20
Auriga	Aurigae	Charioteer	Aur	6	+40
Boötes	Boötis	Herdsman	Boo	15	+30
Caelum	Caeli	Graving tool	Cae	5	−40
Camelopardus	Camelopardis	Giraffe	Cam	6	+70
Cancer	Cancri	Crab	Cnc	9	+20
Canes Venatici	Canum Venaticorum	Hunting dogs	CVn	13	+40
Canis Major	Canis Majoris	Big dog	CMa	7	−20
Canis Minor	Canis Minoris	Little dog	CMi	8	+5
Capricornus	Capricorni	Sea goat	Cap	21	−20
Carina*	Carinae	Keel of Argonauts' ship	Car	9	−60
Cassiopeia	Cassiopeiae	Queen of Ethiopia	Cas	1	+60
Centaurus	Centauri	Centaur	Cen	13	−50
Cepheus	Cephei	King of Ethiopia	Cep	22	+70
Cetus	Ceti	Sea monster (whale)	Cet	2	−10
Chamaeleon	Chamaeleontis	Chameleon	Cha	11	−80
Circinus	Circini	Compasses	Cir	15	−60
Columba	Columbae	Dove	Col	6	−35
Coma Berenices	Comae Berenices	Berenice's hair	Com	13	+20
Corona Australis	Coronae Australis	Southern crown	CrA	19	−40
Corona Borealis	Coronae Borealis	Northern crown	CrB	16	+30
Corvus	Corvi	Crow	Crv	12	−20
Crater	Crateris	Cup	Crt	11	−15
Crux	Crucis	Cross (southern)	Cru	12	−60
Cygnus	Cygni	Swan	Cyg	21	+40
Delphinus	Delphini	Porpoise	Del	21	+10
Dorado	Doradus	Swordfish	Dor	5	−65
Draco	Draconis	Dragon	Dra	17	+65
Equuleus	Equulei	Little horse	Equ	21	+10
Eridanus	Eridani	River	Eri	3	−20
Fornax	Fornacis	Furnace	For	3	−30
Gemini	Geminorum	Twins	Gem	7	+20
Grus	Gruis	Crane	Gru	22	−45
Hercules	Herculis	Hercules, son of Zeus	Her	17	+30
Horologium	Horologii	Clock	Hor	3	−60
Hydra	Hydrae	Sea serpent	Hya	10	−20
Hydrus	Hydri	Water snake	Hyi	2	−75
Indus	Indi	Indian	Ind	21	−55
Lacerta	Lacertae	Lizard	Lac	22	+45

(Table continued on A–32)

The Constellations (Continued)

Constellation (Latin Name)	Genitive Case Ending	English Name or Description	Abbreviation	α h	δ °
Leo	Leonis	Lion	Leo	11	+15
Leo Minor	Leonis Minoris	Little lion	LMi	10	+35
Lepus	Leporis	Hare	Lep	6	−20
Libra	Librae	Balance	Lib	15	−15
Lupus	Lupi	Wolf	Lup	15	−45
Lynx	Lyncis	Lynx	Lyn	8	+45
Lyra	Lyrae	Lyre or harp	Lyr	19	+40
Mensa	Mensae	Table Mountain	Men	5	−80
Microscopium	Microscopii	Microscope	Mic	21	−35
Monoceros	Monocerotis	Unicorn	Mon	7	−5
Musca	Muscae	Fly	Mus	12	−70
Norma	Normae	Carpenter's level	Nor	16	−50
Octans	Octantis	Octant	Oct	22	−85
Ophiuchus	Ophiuchi	Holder of serpent	Oph	17	0
Orion	Orionis	Orion, the hunter	Ori	5	+5
Pavo	Pavonis	Peacock	Pav	20	−65
Pegasus	Pegasi	Pegasus, the winged horse	Peg	22	+20
Perseus	Persei	Perseus, hero who saved Andromeda	Per	3	+45
Phoenix	Phoenicis	Phoenix	Phe	1	−50
Pictor	Pictoris	Easel	Pic	6	−55
Pisces	Piscium	Fishes	Psc	1	+15
Piscis Austrinus	Piscis Austrini	Southern fish	PsA	22	−30
Puppis*	Puppis	Stern of the Argonauts' ship	Pup	8	−40
Pyxis* (= Malus)	Pyxidus	Compass of the Argonauts' ship	Pyx	9	−30
Reticulum	Reticuli	Net	Ret	4	−60
Sagitta	Sagittae	Arrow	Sge	20	+10
Sagittarius	Sagittarii	Archer	Sgr	19	−25
Scorpius	Scorpii	Scorpion	Sco	17	−40
Sculptor	Sculptoris	Sculptor's tools	Scl	0	−30
Scutum	Scuti	Shield	Sct	19	−10
Serpens	Serpentis	Serpent	Ser	17	0
Sextans	Sextantis	Sextant	Sex	10	0
Taurus	Tauri	Bull	Tau	4	+15
Telescopium	Telescopii	Telescope	Tel	19	−50
Triangulum	Trianguli	Triangle	Tri	2	+30
Triangulum Australe	Trianguli Australis	Southern triangle	TrA	16	−65
Tucana	Tucanae	Toucan	Tuc	0	−65
Ursa Major	Ursae Majoris	Big bear	UMa	11	+50
Ursa Minor	Ursae Minoris	Little bear	VMi	15	+70
Vela*	Velorum	Sail of the Argonauts' ship	Vel	9	−50
Virgo	Virginis	Virgin	Vir	13	0
Volans	Volantis	Flying fish	Vol	8	−70
Vulpecula	Vulpeculae	Fox	Vul	20	+25

* The four constellations Carina, Puppis, Pyxis, and Vela originally formed the single constellation, Argo Navis.

Appendix 15

The Messier Catalog of Nebulae and Star Clusters

M	NGC or (IC)	Right Ascension (1980) h	m	Decli- nation (1980) °	'	Apparent Visual Magnitude	Description
1	1952	5	33.3	+22	01	8.4	"Crab" nebula in Taurus; remains of SN 1054
2	7089	21	32.4	−0	54	6.4	Globular cluster in Aquarius
3	5272	13	41.2	+28	29	6.3	Globular cluster in Canes Venatici
4	6121	16	22.4	−26	28	6.5	Globular cluster in Scorpius
5	5904	15	17.5	+2	10	6.1	Globular cluster in Serpens
6	6405	17	38.8	−32	11	5.5	Open cluster in Scorpius
7	6475	17	52.7	−34	48	3.3	Open cluster in Scorpius
8	6523	18	02.4	−24	23	5.1	"Lagoon" nebula in Sagittarius
9	6333	17	18.1	−18	30	8.0	Globular cluster in Ophiuchus
10	6254	16	56.1	−4	05	6.7	Globular cluster in Ophiuchus
11	6705	18	50.0	−6	18	6.8	Open cluster in Scutum Sobieskii
12	6218	16	46.3	−1	55	6.6	Globular cluster in Ophiuchus
13	6205	16	41.0	+36	30	5.9	Globular cluster in Hercules
14	6402	17	36.6	−3	14	8.0	Globular cluster in Ophiuchus
15	7078	21	28.9	+12	05	6.4	Globular cluster in Pegasus
16	6611	18	17.8	−13	47	6.6	Open cluster with nebulosity in Serpens
17	6618	18	19.6	−16	11	7.5	"Swan" or "Omega" nebula in Sagittarius
18	6613	18	18.7	−17	08	7.2	Open cluster in Sagittarius
19	6273	17	01.4	−26	14	6.9	Globular cluster in Ophiuchus
20	6514	18	01.2	−23	02	8.5	"Trifid" nebula in Sagittarius
21	6531	18	03.4	−22	30	6.5	Open cluster in Sagittarius
22	6656	18	35.2	−23	56	5.6	Globular cluster in Sagittarius
23	6494	17	55.8	−19	00	5.9	Open cluster in Sagittarius
24	6603	18	17.3	−18	26	4.6	Open cluster in Sagittarius
25	(4725)	18	30.5	−19	16	6.2	Open cluster in Sagittarius
26	6694	18	44.1	−9	25	9.3	Open cluster in Scutum Sobieskii
27	6853	19	58.8	+22	40	8.2	"Dumbbell" planetary nebula in Vulpecula
28	6626	18	23.2	−24	52	7.6	Globular cluster in Sagittarius
29	6913	20	23.3	+38	27	8.0	Open cluster in Cygnus
30	7099	21	39.2	−23	16	7.7	Globular cluster in Capricornus
31	224	0	41.6	+41	10	3.5	Andromeda galaxy
32	221	0	41.6	+40	46	8.2	Elliptical galaxy; companion to M31
33	598	1	32.7	+30	33	5.8	Spiral galaxy in Triangulum
34	1039	2	40.7	+42	43	5.8	Open cluster in Perseus
35	2168	6	07.5	+24	21	5.6	Open cluster in Gemini
36	1960	5	35.0	+34	05	6.5	Open cluster in Auriga
37	2099	5	51.1	+32	33	6.2	Open cluster in Auriga
38	1912	5	27.3	+35	48	7.0	Open cluster in Auriga
39	7092	21	31.5	+48	21	5.3	Open cluster in Cygnus
40	—	12	21	+59	—	—	Close double star in Ursa Major
41	2287	6	46.2	−20	43	5.0	Loose open cluster in Canis Major
42	1976	5	34.4	−5	24	4	Orion nebula
43	1982	5	34.6	−5	18	9	Northeast portion of Orion nebula

(Table continued on A-34)

The Messier Catalog of Nebulae and Star Clusters (Continued)

M	NGC or (IC)	Right Ascension (1980) h	m	Declination (1980) °	′	Apparent Visual Magnitude	Description
44	2632	8	39	+20	04	3.9	Praesepe; open cluster in Cancer
45	—	3	46.3	+24	03	1.6	The Pleiades; open cluster in Taurus
46	2437	7	40.9	−14	46	6.6	Open cluster in Puppis
47	2422	7	35.7	−14	26	5.0	Loose group of stars in Puppis
48	2548	8	12.8	−5	44	6.0	"Cluster of very small stars"
49	4472	12	28.8	+8	06	8.5	Elliptical galaxy in Virgo
50	2323	7	02.0	−8	19	6.3	Loose open cluster in Monoceros
51	5194	13	29.1	+47	18	8.4	"Whirlpool" spiral galaxy in Canes Venatici
52	7654	23	23.3	+61	30	8.2	Loose open cluster in Cassiopeia
53	5024	13	12.0	+18	16	7.8	Globular cluster in Coma Berenices
54	6715	18	53.8	−30	30	7.8	Globular cluster in Sagittarius
55	6809	19	38.7	−30	59	6.2	Globular cluster in Sagittarius
56	6779	19	15.8	+30	08	8.7	Globular cluster in Lyra
57	6720	18	52.8	+33	00	9.0	"Ring" nebula; planetary nebula in Lyra
58	4579	12	36.7	+11	55	9.9	Spiral galaxy in Virgo
59	4621	12	41.0	+11	46	10.0	Spiral galaxy in Virgo
60	4649	12	42.6	+11	40	9.0	Elliptical galaxy in Virgo
61	4303	12	20.8	+4	35	9.6	Spiral galaxy in Virgo
62	6266	16	59.9	−30	05	6.6	Globular cluster in Scorpius
63	5055	13	14.8	+42	07	8.9	Spiral galaxy in Canes Venatici
64	4826	12	55.7	+21	39	8.5	Spiral galaxy in Coma Berenices
65	3623	11	17.9	+13	12	9.4	Spiral galaxy in Leo
66	3627	11	19.2	+13	06	9.0	Spiral galaxy in Leo; companion to M65
67	2682	8	50.0	+11	53	6.1	Open cluster in Cancer
68	4590	12	38.4	−26	39	8.2	Globular cluster in Hydra
69	6637	18	30.1	−32	23	8.0	Globular cluster in Sagittarius
70	6681	18	42.0	−32	18	8.1	Globular cluster in Sagittarius
71	6838	19	52.8	+18	44	7.6	Globular cluster in Sagittarius
72	6981	20	52.3	−12	38	9.3	Globular cluster in Aquarius
73	6994	20	57.8	−12	43	9.1	Open cluster in Aquarius
74	628	1	35.6	+15	41	9.3	Spiral galaxy in Pisces
75	6864	20	04.9	−21	59	8.6	Globular cluster in Sagittarius
76	650	1	41.0	+51	28	11.4	Planetary nebula in Perseus
77	1068	2	41.6	−0	04	8.9	Spiral galaxy in Cetus
78	2068	5	45.7	0	03	8.3	Small emission nebula in Orion
79	1904	5	23.3	−24	32	7.5	Globular cluster in Lepus
80	6093	16	15.8	−22	56	7.5	Globular cluster in Scorpius
81	3031	9	54.2	+69	09	7.0	Spiral galaxy in Ursa Major
82	3034	9	54.4	+69	47	8.4	Irregular galaxy in Ursa Major
83	5236	13	35.4	−29	31	7.6	Spiral galaxy in Hydra
84	4374	12	24.1	+13	00	9.4	Elliptical galaxy in Virgo
85	4382	12	24.3	+18	18	9.3	Elliptical galaxy in Coma Berenices
86	4406	12	25.1	+13	03	9.2	Elliptical galaxy in Virgo
87	4486	12	29.7	+12	30	8.7	Elliptical galaxy in Virgo
88	4501	12	30.9	+14	32	9.5	Spiral galaxy in Coma Berenices
89	4552	12	34.6	+12	40	10.3	Elliptical galaxy in Virgo
90	4569	12	35.8	+13	16	9.6	Spiral galaxy in Virgo

M	NGC or (IC)	Right Ascension (1980)		Decli- nation (1980)		Apparent Visual Magnitude	Description
		h	m	°	'		
91	omitted	—	—	—	—	—	
92	6341	17	16.5	+43	10	6.4	Globular cluster in Hercules
93	2447	7	43.7	−23	49	6.5	Open cluster in Puppis
94	4736	12	50.0	+41	14	8.3	Spiral galaxy in Canes Venatici
95	3351	10	42.9	+11	49	9.8	Barred spiral galaxy in Leo
96	3368	10	45.7	+11	56	9.3	Spiral galaxy in Leo
97	3587	11	13.7	+55	07	11.1	"Owl" nebula; planetary nebula in Ursa Major
98	4192	12	12.7	+15	01	10.2	Spiral galaxy in Coma Berenices
99	4254	12	17.8	+14	32	9.9	Spiral galaxy in Coma Berenices
100	4321	12	21.9	+15	56	9.4	Spiral galaxy in Coma Berenices
101	5457	14	02.5	+54	27	7.9	Spiral galaxy in Ursa Major
102	5866(?)	15	05.9	+55	50	10.5	Spiral galaxy (identification in doubt)
103	581	1	31.9	+60	35	6.9	Open cluster in Cassiopeia
104*	4594	12	39.0	−11	31	8.3	Spiral galaxy in Virgo
105*	3379	10	46.8	+12	51	9.7	Elliptical galaxy in Leo
106*	4258	12	18.0	+47	25	8.4	Spiral galaxy in Canes Venatici
107*	6171	16	31.4	−13	01	9.2	Globular cluster in Ophiuchus
108*	3556	11	10.5	+55	47	10.5	Spiral galaxy in Ursa Major
109*	3992	11	56.6	+53	29	10.0	Spiral galaxy in Ursa Major
110*	205	0	39.2	+41	35	9.4	Elliptical galaxy; companion to M31

* Not in Messier's original (1781) list; added later by others.

Index

Note: Figures, tables, and footnotes are indicated by *italics*, "t", and "n" respectively.

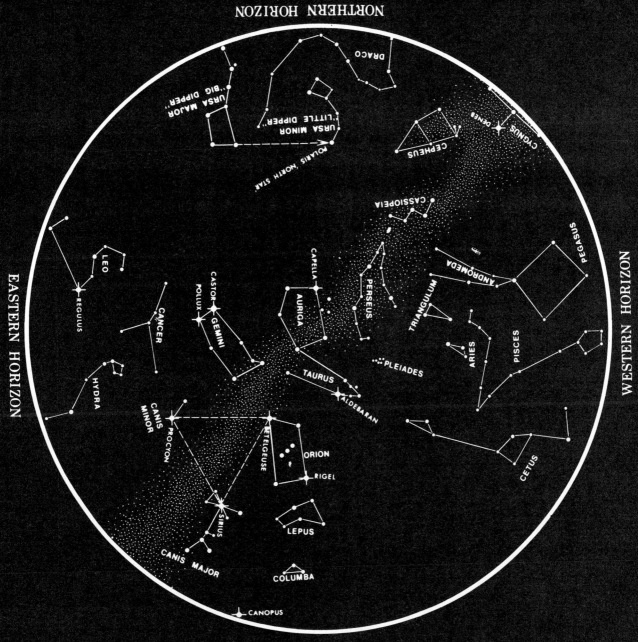

THE NIGHT SKY IN JANUARY

SOUTHERN HORIZON

Latitude of chart is 34°N, but it is practical throughout the continental United States.

To use: Hold chart vertically and turn it so the direction you are facing shows at the bottom.

Chart time (Local Standard):
10 p.m. First of month
9 p.m. Middle of month
8 p.m. Last of month

Star Chart from *GRIFFITH OBSERVER*, Griffith Observatory, Los Angeles

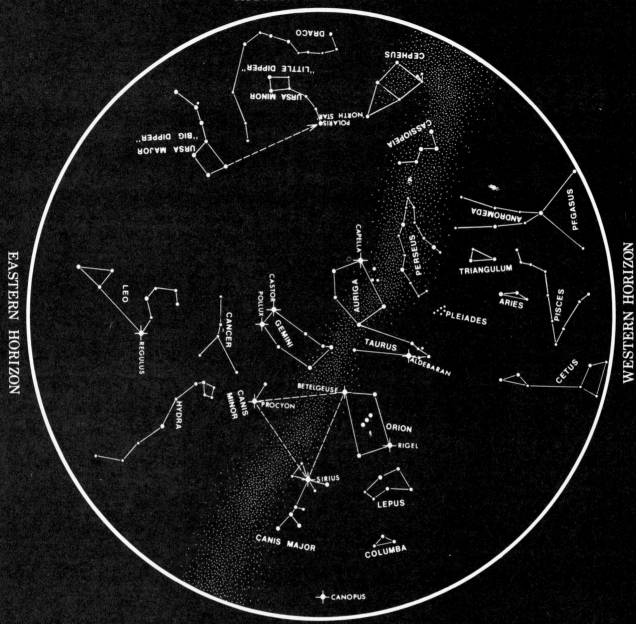

THE NIGHT SKY IN FEBRUARY

Latitude of chart is 34°N, but it is
practical throughout the continental
United States.

To use: Hold chart vertically and turn
it so the direction you are facing
shows at the bottom.

Chart time (Local Standard):

10 p.m. First of month

9 p.m. Middle of month

8 p.m. Last of month

Star Chart from *GRIFFITH OBSERVER*, Griffith Observatory, Los Angeles

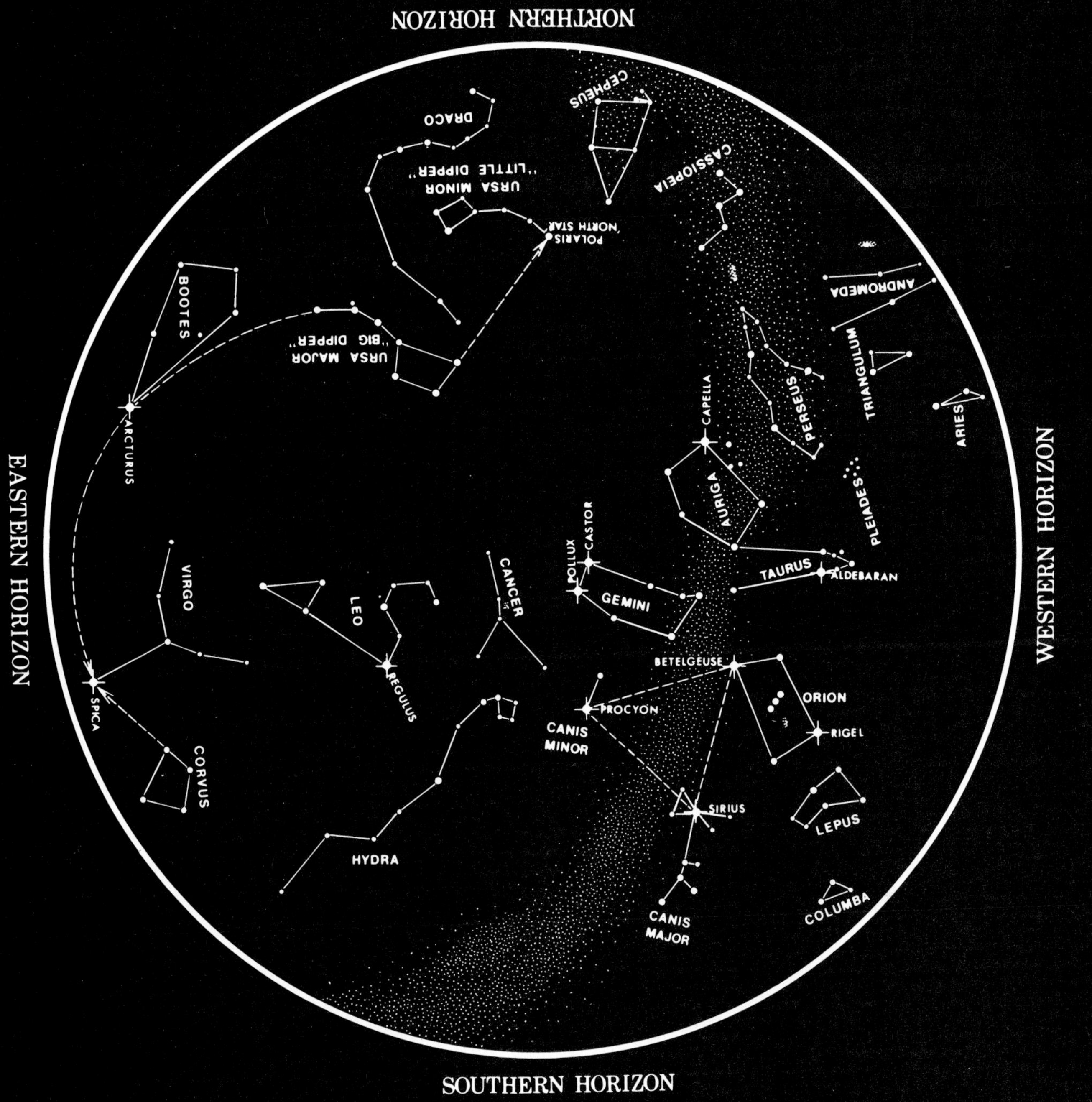

THE NIGHT SKY IN MARCH

Latitude of chart is 34° N, but it is practical throughout the continental United States.

To use: Hold chart vertically and turn it so the direction you are facing shows at the bottom.

Chart time (Local Standard):

10 p.m. First of month

9 p.m. Middle of month

8 p.m. Last of month

Star Chart from *GRIFFITH OBSERVER*, Griffith Observatory, Los Angeles

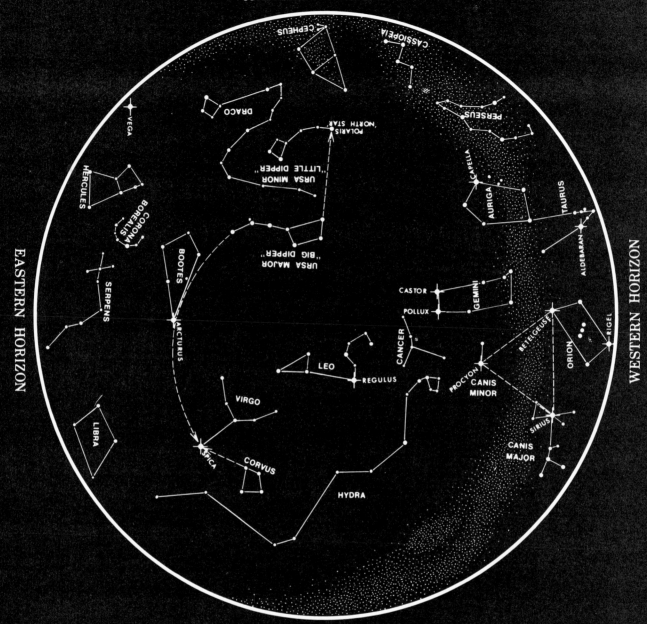

THE NIGHT SKY IN APRIL

Latitude of chart is 34°N, but it is
practical throughout the continental
United States.

To use: Hold chart vertically and turn
it so the direction you are facing
shows at the bottom.

Chart time (Local Standard):

10 p.m. First of month

9 p.m. Middle of month

8 p.m. Last of month

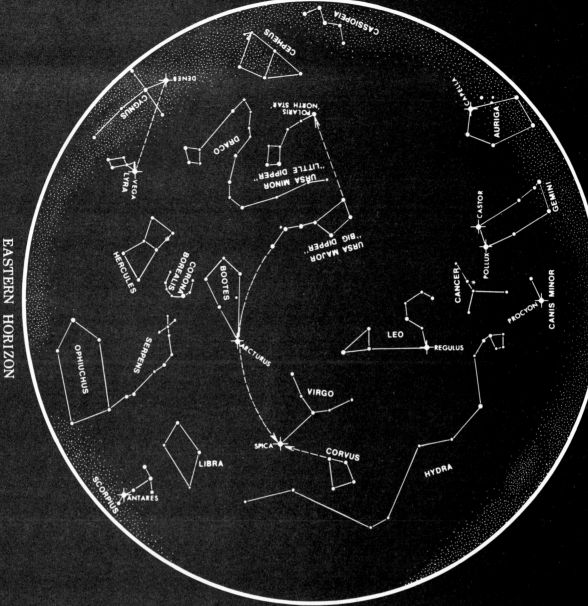

EASTERN HORIZON

WESTERN HORIZON

SOUTHERN HORIZON

THE NIGHT SKY IN MAY

Latitude of chart is 34°N, but it is practical throughout the continental United States.

To use: Hold chart vertically and turn it so the direction you are facing shows at the bottom.

Chart time (Local Standard):

10 p.m. First of month

9 p.m. Middle of month

8 p.m. Last of month

Star Chart from *GRIFFITH OBSERVER*, Griffith Observatory, Los Angeles

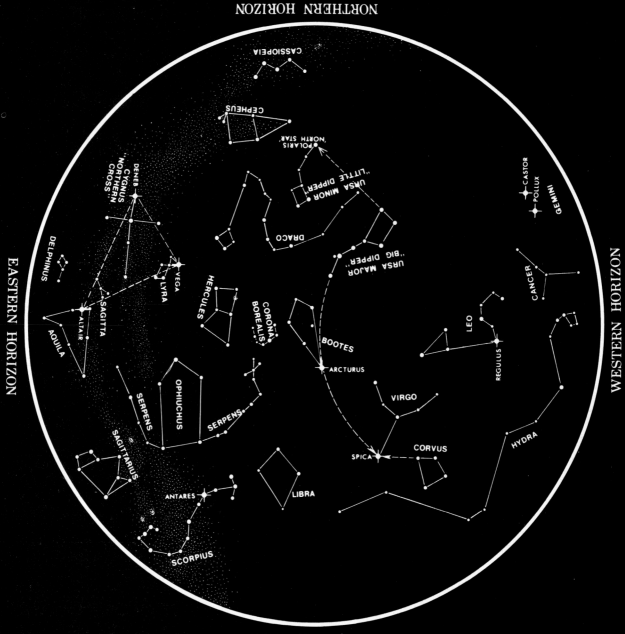

NORTHERN HORIZON

CASSIOPEIA

CEPHEUS

POLARIS "NORTH STAR"

"LITTLE DIPPER"

URSA MINOR

DRACO

CASTOR
POLLUX
GEMINI

DENEB
CYGNUS
"NORTHERN
CROSS"

"BIG DIPPER"
URSA MAJOR

CANCER

DELPHINUS

VEGA
LYRA

HERCULES

CORONA
BOREALIS

LEO

REGULUS

EASTERN HORIZON

AQUILA

ALTAIR

SAGITTA

BOOTES

ARCTURUS

VIRGO

WESTERN HORIZON

OPHIUCHUS

SERPENS

SERPENS

SPICA

CORVUS

HYDRA

SAGITTARIUS

ANTARES

LIBRA

SCORPIUS

SOUTHERN HORIZON

THE NIGHT SKY IN JUNE

Latitude of chart is 34°N, but it is practical throughout the continental United States.

To use: Hold chart vertically and turn it so the direction you are facing shows at the bottom.

Chart time (Local Standard):

10 p.m. First of month

9 p.m. Middle of month

8 p.m. Last of month

Star Chart from *GRIFFITH OBSERVER*, Griffith Observatory, Los Angeles

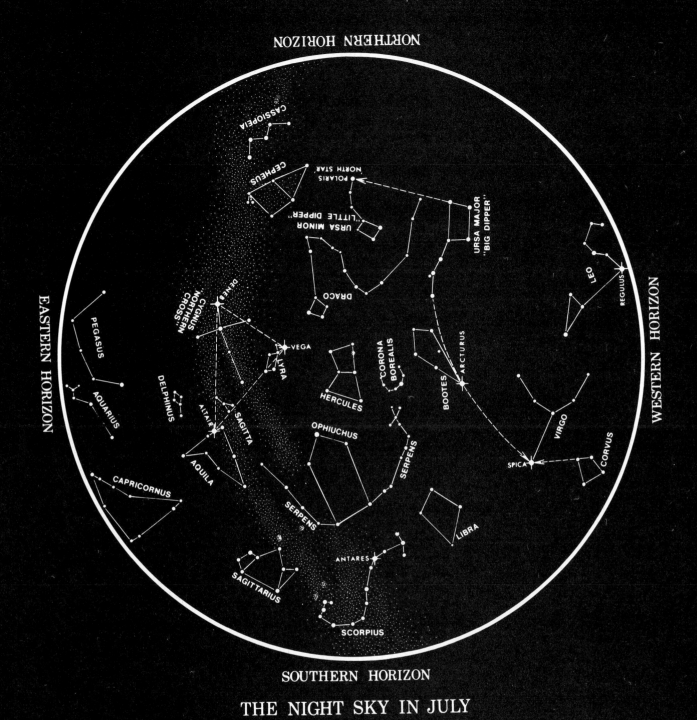

THE NIGHT SKY IN JULY

Latitude of chart is 34°N, but it is
practical throughout the continental
United States.

To use: Hold chart vertically and turn
it so the direction you are facing
shows at the bottom.

Chart time (Local Standard):
10 p.m. First of month
9 p.m. Middle of month
8 p.m. Last of month

Star Chart from *GRIFFITH OBSERVER*, Griffith Observatory, Los Angeles

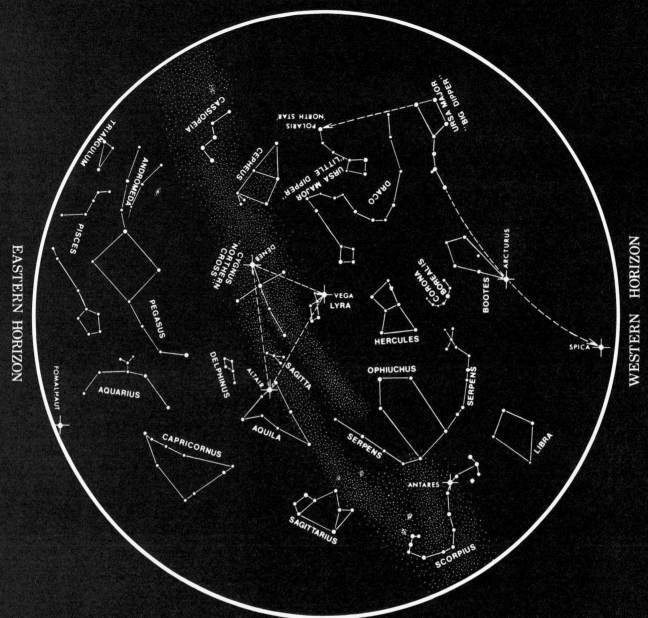

THE NIGHT SKY IN AUGUST

Latitude of chart is 34°N, but it is practical throughout the continental United States.

To use: Hold chart vertically and turn it so the direction you are facing shows at the bottom.

Chart time (Local Standard):

10 p.m. First of month

9 p.m. Middle of month

8 p.m. Last of month

Star Chart from *GRIFFITH OBSERVER*, Griffith Observatory, Los Angeles

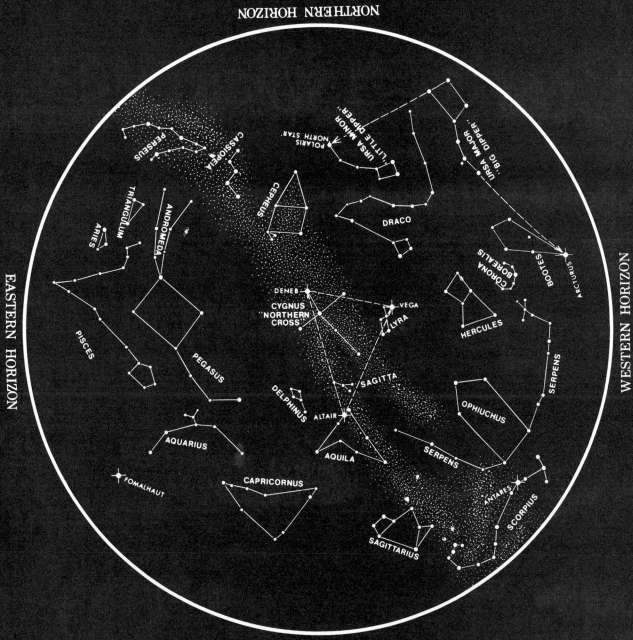

THE NIGHT SKY IN SEPTEMBER

Latitude of chart is 34°N, but it is practical throughout the continental United States.

To use: Hold chart vertically and turn it so the direction you are facing shows at the bottom.

Chart time (Local Standard):
10 p.m. First of month
9 p.m. Middle of month
8 p.m. Last of month

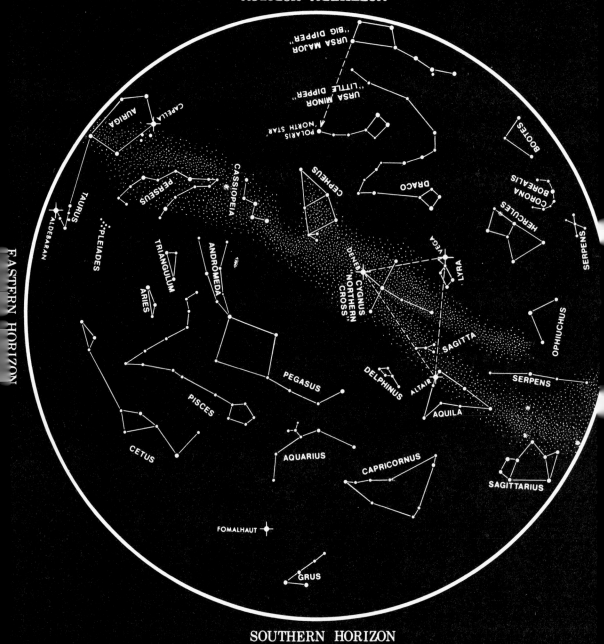

EASTERN HORIZON

SOUTHERN HORIZON

THE NIGHT SKY IN OCTOBER

Latitude of chart is 34°N, but it is practical throughout the continental United States.

To use: Hold chart vertically and turn it so the direction you are facing shows at the bottom.

Chart time (Local Standard):

10 p.m. First of month

9 p.m. Middle of month

8 p.m. Last of month

; Angeles

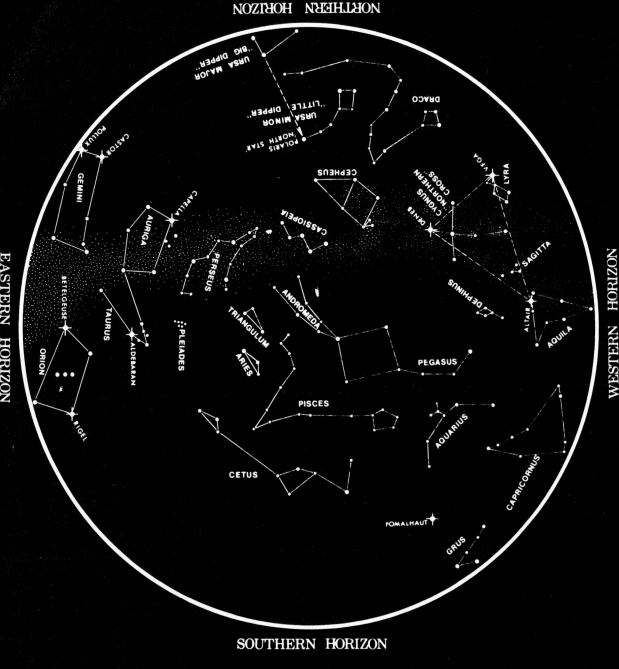

THE NIGHT SKY IN NOVEMBER

Latitude of chart is 34°N, but it is practical throughout the continental United States.

To use: Hold chart vertically and turn it so the direction you are facing shows at the bottom.

Star Chart from *GRIFFITH OBSERVER*, Griffith Observatory, Los Angeles

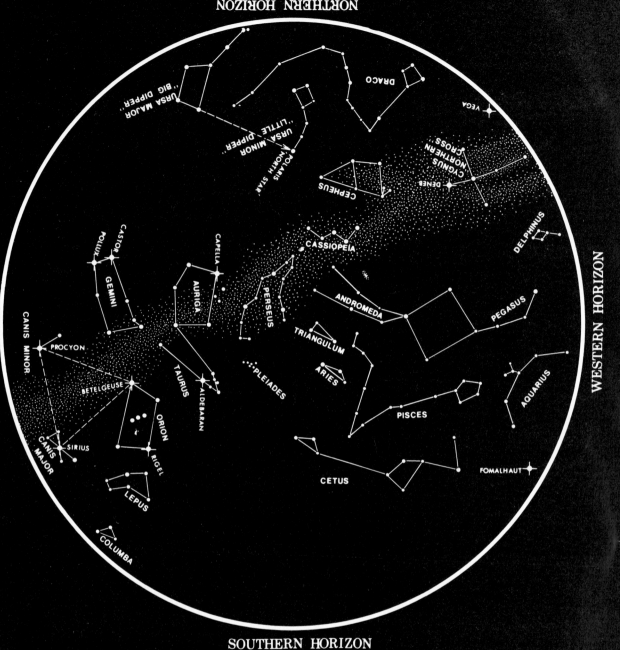

THE NIGHT SKY IN DECEMBER

Latitude of chart is 34°N, but it is practical throughout the continental United States.

To use: Hold chart vertically and turn it so the direction you are facing shows at the bottom.

Chart time (Local Standard):

10 p.m. First of month
9 p.m. Middle of month
8 p.m. Last of month

Star Chart from *GRIFFITH OBSERVER*, Griffith Observatory, Los Angeles